T0074019

Data Mining: The Textbook

Charu C. Aggarwal

Data Mining

The Textbook

Charu C. Aggarwal
IBM T.J. Watson Research Center
Yorktown Heights
New York
USA

A solution manual for this book is available on Springer.com.

ISBN 978-3-319-14141-1 ISBN 978-3-319-14142-8 (eBook)
DOI 10.1007/978-3-319-14142-8

Library of Congress Control Number: 2015930833

Springer Cham Heidelberg New York Dordrecht London

Printed on acid-free paper

Springer is part of Springer Science+Business Media (www.springer.com)

To my wife Lata,
and my daughter Sayani

Contents

Preface

"*Data is the new oil.*"– Clive Humby

The field of data mining has seen rapid strides over the past two decades, especially from the perspective of the computer science community. While data analysis has been studied extensively in the conventional field of probability and statistics, *data mining* is a term coined by the computer science-oriented community. For computer scientists, issues such as scalability, usability, and computational implementation are extremely important.

The emergence of data science as a discipline requires the development of a book that goes beyond the traditional focus of books on only the fundamental data mining courses. Recent years have seen the emergence of the job description of "data scientists," who try to glean knowledge from vast amounts of data. In typical applications, the data types are so heterogeneous and diverse that the fundamental methods discussed for a multidimensional data type may not be effective. Therefore, more emphasis needs to be placed on the different data types and the applications that arise in the context of these different data types. A comprehensive data mining book must explore the different aspects of data mining, starting from the fundamentals, and then explore the complex data types, and their relationships with the fundamental techniques. While fundamental techniques form an excellent basis for the further study of data mining, they do not provide a complete picture of the true complexity of data analysis. This book studies these advanced topics without compromising the presentation of fundamental methods. Therefore, this book may be used for both introductory and advanced data mining courses. Until now, no single book has addressed all these topics in a comprehensive and integrated way.

The textbook assumes a basic knowledge of probability, statistics, and linear algebra, which is taught in most undergraduate curricula of science and engineering disciplines. Therefore, the book can also be used by industrial practitioners, who have a working knowledge of these basic skills. While stronger mathematical background is helpful for the more advanced chapters, it is not a prerequisite. Special chapters are also devoted to different aspects of data mining, such as text data, time-series data, discrete sequences, and graphs. This kind of specialized treatment is intended to capture the wide diversity of problem domains in which a data mining problem might arise.

The chapters of this book fall into one of three categories:

- **The fundamental chapters:** Data mining has four main "super problems," which correspond to clustering, classification, association pattern mining, and outlier analysis. These problems are so important because they are used repeatedly as building blocks in the context of a wide variety of data mining applications. As a result, a large amount of emphasis has been placed by data mining researchers and practitioners to design effective and efficient methods for these problems. These chapters comprehensively discuss the vast diversity of methods used by the data mining community in the context of these super problems.

- **Domain chapters:** These chapters discuss the specific methods used for different *domains* of data such as text data, time-series data, sequence data, graph data, and spatial data. Many of these chapters can also be considered application chapters, because they explore the specific characteristics of the problem in a particular domain.

- **Application chapters:** Advancements in hardware technology and software platforms have lead to a number of data-intensive applications such as streaming systems, Web mining, social networks, and privacy preservation. These topics are studied in detail in these chapters. The domain chapters are also focused on many different kinds of applications that arise in the context of those data types.

Suggestions for the Instructor

The book was specifically written to enable the teaching of both the basic data mining and advanced data mining courses from a single book. It can be used to offer various types of data mining courses with different emphases. Specifically, the courses that could be offered with various chapters are as follows:

- **Basic data mining course and fundamentals:** The basic data mining course should focus on the fundamentals of data mining. Chapters 1, 2, 3, 4, 6, 8, and 10 can be covered. In fact, the material in these chapters is more than what is possible to teach in a single course. Therefore, instructors may need to select topics of their interest from these chapters. Some portions of Chaps. 5, 7, 9, and 11 can also be covered, although these chapters are really meant for an advanced course.

- **Advanced course (fundamentals):** Such a course would cover advanced topics on the fundamentals of data mining and assume that the student is already familiar with Chaps. 1–3, and parts of Chaps. 4, 6, 8, and 10. The course can then focus on Chaps. 5, 7, 9, and 11. Topics such as ensemble analysis are useful for the advanced course. Furthermore, some topics from Chaps. 4, 6, 8, and 10, which were not covered in the basic course, can be used. In addition, Chap. 20 on privacy can be offered.

- **Advanced course (data types):** Advanced topics such as text mining, time series, sequences, graphs, and spatial data may be covered. The material should focus on Chaps. 13, 14, 15, 16, and 17. Some parts of Chap. 19 (e.g., graph clustering) and Chap. 12 (data streaming) can also be used.

- **Advanced course (applications):** An application course overlaps with a data type course but has a different focus. For example, the focus in an application-centered course would be more on the modeling aspect than the algorithmic aspect. Therefore, the same materials in Chaps. 13, 14, 15, 16, and 17 can be used while skipping specific details of algorithms. With less focus on specific algorithms, these chapters can be covered fairly quickly. The remaining time should be allocated to three very important chapters on data streams (Chap. 12), Web mining (Chap. 18), and social network analysis (Chap. 19).

The book is written in a simple style to make it accessible to undergraduate students and industrial practitioners with a limited mathematical background. Thus, the book will serve both as an introductory text and as an advanced text for students, industrial practitioners, and researchers.

Throughout this book, a vector or a multidimensional data point (including categorical attributes), is annotated with a bar, such as $\overline{X}$ or $\overline{y}$. A vector or multidimensional point may be denoted by either small letters or capital letters, as long as it has a bar. Vector dot products are denoted by centered dots, such as $\overline{X} \cdot \overline{Y}$. A matrix is denoted in capital letters without a bar, such as R. Throughout the book, the $n \times d$ data matrix is denoted by D, with n points and d dimensions. The individual data points in D are therefore d-dimensional row vectors. On the other hand, vectors with one component for each data point are usually n-dimensional column vectors. An example is the n-dimensional column vector $\overline{y}$ of class variables of n data points.

Acknowledgments

I would like to thank my wife and daughter for their love and support during the writing of this book. The writing of a book requires significant time, which is taken away from family members. This book is the result of their patience with me during this time.

I would also like to thank my manager Nagui Halim for providing the tremendous support necessary for the writing of this book. His professional support has been instrumental for my many book efforts in the past and present.

During the writing of this book, I received feedback from many colleagues. In particular, I received feedback from Kanishka Bhaduri, Alain Biem, Graham Cormode, Hongbo Deng, Amit Dhurandhar, Bart Goethals, Alexander Hinneburg, Ramakrishnan Kannan, George Karypis, Dominique LaSalle, Abdullah Mueen, Guojun Qi, Pierangela Samarati, Saket Sathe, Karthik Subbian, Jiliang Tang, Deepak Turaga, Jilles Vreeken, Jieping Ye, and Peixiang Zhao. I would like to thank them for their constructive feedback and suggestions. Over the years, I have benefited from the insights of numerous collaborators. These insights have influenced this book directly or indirectly. I would first like to thank my long-term collaborator Philip S. Yu for my years of collaboration with him. Other researchers with whom I have had significant collaborations include Tarek F. Abdelzaher, Jing Gao, Quanquan Gu, Manish Gupta, Jiawei Han, Alexander Hinneburg, Thomas Huang, Nan Li, Huan Liu, Ruoming Jin, Daniel Keim, Arijit Khan, Latifur Khan, Mohammad M. Masud, Jian Pei, Magda Procopiuc, Guojun Qi, Chandan Reddy, Jaideep Srivastava, Karthik Subbian, Yizhou Sun, Jiliang Tang, Min-Hsuan Tsai, Haixun Wang, Jianyong Wang, Min Wang, Joel Wolf, Xifeng Yan, Mohammed Zaki, ChengXiang Zhai, and Peixiang Zhao.

I would also like to thank my advisor James B. Orlin for his guidance during my early years as a researcher. While I no longer work in the same area, the legacy of what I learned from him is a crucial part of my approach to research. In particular, he taught me the importance of intuition and simplicity of thought in the research process. These are more important aspects of research than is generally recognized. This book is written in a simple and intuitive style, and is meant to improve accessibility of this area to both researchers and practitioners.

I would also like to thank Lata Aggarwal for helping me with some of the figures drawn using Microsoft Powerpoint.

Author Biography

Charu C. Aggarwal is a Distinguished Research Staff Member (DRSM) at the IBM T. J. Watson Research Center in Yorktown Heights, New York. He completed his B.S. from IIT Kanpur in 1993 and his Ph.D. from the Massachusetts Institute of Technology in 1996.

He has worked extensively in the field of data mining. He has published more than 250 papers in refereed conferences and journals and authored over 80 patents. He is author or editor of 14 books, including the first comprehensive book on outlier analysis, which is written from a computer science point of view. Because of the commercial value of his patents, he has thrice been designated a Master Inventor at IBM. He is a recipient of an IBM Corporate Award (2003) for his work on bio-terrorist threat detection in data streams, a recipient of the IBM Outstanding Innovation Award (2008) for his scientific contributions to privacy technology, a recipient of the IBM Outstanding Technical Achievement Award (2009) for his work on data streams, and a recipient of an IBM Research Division Award (2008) for his contributions to System S. He also received the EDBT 2014 Test of Time Award for his work on condensation-based privacy-preserving data mining.

He has served as the general co-chair of the IEEE Big Data Conference, 2014, and as an associate editor of the IEEE Transactions on Knowledge and Data Engineering from 2004 to 2008. He is an associate editor of the ACM Transactions on Knowledge Discovery from Data, an action editor of the Data Mining and Knowledge Discovery Journal, editor-in-chief of the ACM SIGKDD Explorations, and an associate editor of the Knowledge and Information Systems Journal. He serves on the advisory board of the Lecture Notes on Social Networks, a publication by Springer. He has served as the vice-president of the SIAM Activity Group on Data Mining. He is a fellow of the ACM and the IEEE, for "contributions to knowledge discovery and data mining algorithms."

Chapter 1

An Introduction to Data Mining

"Education is not the piling on of learning, information, data, facts, skills, or abilities – that's training or instruction – but is rather making visible what is hidden as a seed."—Thomas More

1.1 Introduction

Data mining is the study of collecting, cleaning, processing, analyzing, and gaining useful insights from data. A wide variation exists in terms of the problem domains, applications, formulations, and data representations that are encountered in real applications. Therefore, "data mining" is a broad umbrella term that is used to describe these different aspects of data processing.

In the modern age, virtually all automated systems generate some form of data either for diagnostic or analysis purposes. This has resulted in a deluge of data, which has been reaching the order of petabytes or exabytes. Some examples of different kinds of data are as follows:

- *World Wide Web:* The number of documents on the indexed Web is now on the order of billions, and the invisible Web is much larger. User accesses to such documents create Web access logs at servers and customer behavior profiles at commercial sites. Furthermore, the linked structure of the Web is referred to as the *Web graph*, which is itself a kind of data. These different types of data are useful in various applications. For example, the Web documents and link structure can be mined to determine associations between different topics on the Web. On the other hand, user access logs can be mined to determine frequent patterns of accesses or unusual patterns of possibly unwarranted behavior.

- *Financial interactions:* Most common transactions of everyday life, such as using an automated teller machine (ATM) card or a credit card, can create data in an automated way. Such transactions can be mined for many useful insights such as fraud or other unusual activity.

C. C. Aggarwal, *Data Mining: The Textbook*, DOI 10.1007/978-3-319-14142-8_1

- *User interactions:* Many forms of user interactions create large volumes of data. For example, the use of a telephone typically creates a record at the telecommunication company with details about the duration and destination of the call. Many phone companies routinely analyze such data to determine relevant patterns of behavior that can be used to make decisions about network capacity, promotions, pricing, or customer targeting.

- *Sensor technologies and the Internet of Things:* A recent trend is the development of low-cost wearable sensors, smartphones, and other smart devices that can communicate with one another. By one estimate, the number of such devices exceeded the number of people on the planet in 2008 [30]. The implications of such massive data collection are significant for mining algorithms.

The deluge of data is a direct result of advances in technology and the computerization of every aspect of modern life. It is, therefore, natural to examine whether one can extract *concise* and possibly *actionable* insights from the available data for application-specific goals. This is where the task of data mining comes in. The raw data may be arbitrary, unstructured, or even in a format that is not immediately suitable for automated processing. For example, manually collected data may be drawn from heterogeneous sources in different formats and yet somehow needs to be processed by an automated computer program to gain insights.

To address this issue, data mining analysts use a pipeline of processing, where the raw data are collected, cleaned, and transformed into a standardized format. The data may be stored in a commercial database system and finally processed for insights with the use of analytical methods. In fact, while data mining often conjures up the notion of analytical algorithms, the reality is that the vast majority of work is related to the data preparation portion of the process. This pipeline of processing is conceptually similar to that of an actual mining process from a mineral ore to the refined end product. The term "mining" derives its roots from this analogy.

From an analytical perspective, data mining is challenging because of the wide disparity in the problems and data types that are encountered. For example, a commercial product recommendation problem is very different from an intrusion-detection application, even at the level of the input data format or the problem definition. Even within related classes of problems, the differences are quite significant. For example, a product recommendation problem in a multidimensional database is very different from a social recommendation problem due to the differences in the underlying data type. Nevertheless, in spite of these differences, data mining applications are often closely connected to one of four "super-problems" in data mining: association pattern mining, clustering, classification, and outlier detection. These problems are so important because they are used as building blocks in a majority of the applications in some indirect form or the other. This is a useful abstraction because it helps us conceptualize and structure the field of data mining more effectively.

The data may have different formats or *types.* The type may be quantitative (e.g., age), categorical (e.g., ethnicity), text, spatial, temporal, or graph-oriented. Although the most common form of data is multidimensional, an increasing proportion belongs to more complex data types. While there is a conceptual portability of algorithms between many data types at a very high level, this is not the case from a practical perspective. The reality is that the precise data type may affect the behavior of a particular algorithm significantly. As a result, one may need to design refined variations of the basic approach for multidimensional data, so that it can be used effectively for a different data type. Therefore, this book will dedicate different chapters to the various data types to provide a better understanding of how the processing methods are affected by the underlying data type.

A major challenge has been created in recent years due to increasing data volumes. The prevalence of continuously collected data has led to an increasing interest in the field of *data streams*. For example, Internet traffic generates large streams that cannot even be stored effectively unless significant resources are spent on storage. This leads to unique challenges from the perspective of processing and analysis. In cases where it is not possible to explicitly store the data, all the processing needs to be performed in real time.

This chapter will provide a broad overview of the different technologies involved in preprocessing and analyzing different types of data. The goal is to study data mining from the perspective of different problem abstractions and data types that are frequently encountered. Many important applications can be converted into these abstractions.

This chapter is organized as follows. Section 1.2 discusses the data mining process with particular attention paid to the data preprocessing phase in this section. Different data types and their formal definition are discussed in Sect. 1.3. The major problems in data mining are discussed in Sect. 1.4 at a very high level. The impact of data type on problem definitions is also addressed in this section. Scalability issues are addressed in Sect. 1.5. In Sect. 1.6, a few examples of applications are provided. Section 1.7 gives a summary.

1.2 The Data Mining Process

As discussed earlier, the data mining process is a pipeline containing many phases such as data cleaning, feature extraction, and algorithmic design. In this section, we will study these different phases. The workflow of a typical data mining application contains the following phases:

1. *Data collection:* Data collection may require the use of specialized hardware such as a sensor network, manual labor such as the collection of user surveys, or software tools such as a Web document crawling engine to collect documents. While this stage is highly application-specific and often outside the realm of the data mining analyst, it is critically important because good choices at this stage may significantly impact the data mining process. After the collection phase, the data are often stored in a database, or, more generally, a *data warehouse* for processing.

2. *Feature extraction and data cleaning:* When the data are collected, they are often not in a form that is suitable for processing. For example, the data may be encoded in complex logs or free-form documents. In many cases, different types of data may be arbitrarily mixed together in a free-form document. To make the data suitable for processing, it is essential to transform them into a format that is friendly to data mining algorithms, such as multidimensional, time series, or semistructured format. The multidimensional format is the most common one, in which different *fields* of the data correspond to the different measured properties that are referred to as *features*, *attributes*, or *dimensions*. It is crucial to extract relevant features for the mining process. The feature extraction phase is often performed in parallel with data cleaning, where missing and erroneous parts of the data are either estimated or corrected. In many cases, the data may be extracted from multiple sources and need to be *integrated* into a unified format for processing. The final result of this procedure is a nicely structured data set, which can be effectively used by a computer program. After the feature extraction phase, the data may again be stored in a database for processing.

3. *Analytical processing and algorithms:* The final part of the mining process is to design effective analytical methods from the processed data. In many cases, it may not be

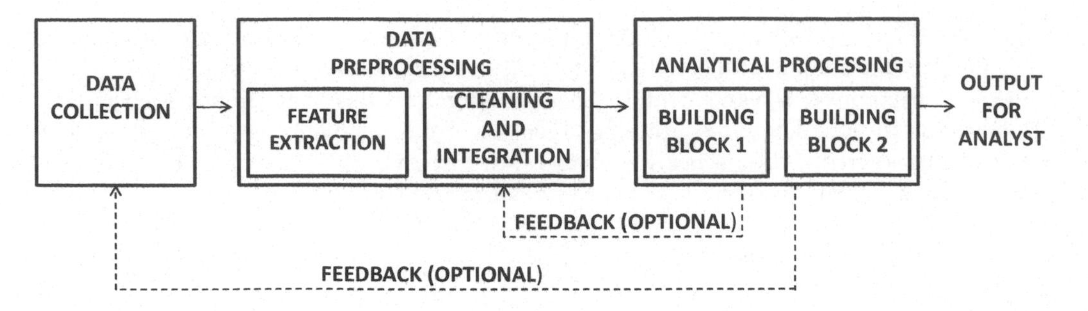

Figure 1.1: The data processing pipeline

possible to directly use a standard data mining problem, such as the four "superproblems" discussed earlier, for the application at hand. However, these four problems have such wide coverage that *many* applications can be broken up into components that use these different building blocks. This book will provide examples of this process.

The overall data mining process is illustrated in Fig. 1.1. Note that the analytical block in Fig. 1.1 shows multiple building blocks representing the design of the solution to a particular application. This part of the algorithmic design is dependent on the skill of the analyst and often uses one or more of the four major problems as a building block. This is, of course, not always the case, but it is frequent enough to merit special treatment of these four problems within this book. To explain the data mining process, we will use an example from a recommendation scenario.

Example 1.2.1 *Consider a scenario in which a retailer has Web logs corresponding to customer accesses to Web pages at his or her site. Each of these Web pages corresponds to a product, and therefore a customer access to a page may often be indicative of interest in that particular product. The retailer also stores demographic profiles for the different customers. The retailer wants to make targeted product recommendations to customers using the customer demographics and buying behavior.*

Sample Solution Pipeline In this case, the first step for the analyst is to collect the relevant data from two different sources. The first source is the set of Web logs at the site. The second is the demographic information within the retailer database that were collected during Web registration of the customer. Unfortunately, these data sets are in a very different format and cannot easily be used together for processing. For example, consider a sample log entry of the following form:

```
98.206.207.157 - - [31/Jul/2013:18:09:38 -0700] "GET /productA.htm
HTTP/1.1" 200 328177 "-" "Mozilla/5.0 (Mac OS X) AppleWebKit/536.26
(KHTML, like Gecko) Version/6.0 Mobile/10B329 Safari/8536.25"
"retailer.net"
```

The log may contain hundreds of thousands of such entries. Here, a customer at IP address 98.206.207.157 has accessed productA.htm. The customer from the IP address can be identified using the previous login information, by using cookies, or by the IP address itself, but this may be a noisy process and may not always yield accurate results. The analyst would need to design algorithms for deciding how to filter the different log entries and use only those which provide accurate results as a part of the *cleaning and extraction* process. Furthermore, the raw log contains a lot of additional information that is not necessarily of any use to the retailer. In the *feature extraction* process, the retailer decides to create one record for each customer, with a specific choice of features extracted from the Web page accesses. For each record, an attribute corresponds to the number of accesses to each product description. Therefore, the raw logs need to be processed, and the accesses need to be aggregated during this *feature extraction* phase. Attributes are added to these records for the retailer's database containing demographic information in a *data integration phase.* Missing entries from the demographic records need to be estimated for further *data cleaning.* This results in a single data set containing attributes for the customer demographics and customer accesses.

At this point, the analyst has to decide how to use this cleaned data set for making recommendations. He or she decides to determine similar groups of customers, and make recommendations on the basis of the buying behavior of these similar groups. In particular, the *building block* of clustering is used to determine similar groups. For a given customer, the most frequent items accessed by the customers in that group are recommended. This provides an example of the entire data mining pipeline. As you will learn in Chap. 18, there are many elegant ways of performing the recommendations, some of which are more effective than the others depending on the specific definition of the problem. Therefore, the entire data mining process is an art form, which is based on the skill of the analyst, and cannot be fully captured by a single technique or building block. In practice, this skill can be learned only by working with a diversity of applications over different scenarios and data types.

1.2.1 The Data Preprocessing Phase

The data preprocessing phase is perhaps the most crucial one in the data mining process. Yet, it is rarely explored to the extent that it deserves because most of the focus is on the analytical aspects of data mining. This phase begins after the collection of the data, and it consists of the following steps:

1. *Feature extraction:* An analyst may be confronted with vast volumes of raw documents, system logs, or commercial transactions with little guidance on how these raw data should be transformed into meaningful database features for processing. This phase is highly dependent on the analyst to be able to abstract out the features that are most relevant to a particular application. For example, in a credit-card fraud detection application, the amount of a charge, the repeat frequency, and the location are often good indicators of fraud. However, many other features may be poorer indicators of fraud. Therefore, extracting the right features is often a skill that requires an understanding of the specific application domain at hand.

2. *Data cleaning:* The extracted data may have erroneous or missing entries. Therefore, some records may need to be dropped, or missing entries may need to be estimated. Inconsistencies may need to be removed.

3. *Feature selection and transformation:* When the data are very high dimensional, many data mining algorithms do not work effectively. Furthermore, many of the high-dimensional features are noisy and may add errors to the data mining process. Therefore, a variety of methods are used to either remove irrelevant features or transform the current set of features to a new data space that is more amenable for analysis. Another related aspect is data *transformation*, where a data set with a particular set of attributes may be transformed into a data set with another set of attributes of the same or a different type. For example, an attribute, such as age, may be partitioned into ranges to create discrete values for analytical convenience.

The data cleaning process requires statistical methods that are commonly used for missing data estimation. In addition, erroneous data entries are often removed to ensure more accurate mining results. The topics of data cleaning is addressed in Chap. 2 on data preprocessing.

Feature selection and transformation should not be considered a part of data preprocessing because the feature selection phase is often highly dependent on the specific analytical problem being solved. In some cases, the feature selection process can even be tightly integrated with the specific algorithm or methodology being used, in the form of a *wrapper model* or *embedded model*. Nevertheless, the feature selection phase is usually performed before applying the specific algorithm at hand.

1.2.2 The Analytical Phase

The vast majority of this book will be devoted to the analytical phase of the mining process. A major challenge is that each data mining application is unique, and it is, therefore, difficult to create general and reusable techniques across different applications. Nevertheless, many data mining formulations are repeatedly used in the context of different applications. These correspond to the major "superproblems" or building blocks of the data mining process. It is dependent on the skill and experience of the analyst to determine how these different formulations may be used in the context of a particular data mining application. Although this book can provide a good overview of the fundamental data mining models, the ability to apply them to real-world applications can only be learned with practical experience.

1.3 The Basic Data Types

One of the interesting aspects of the data mining process is the wide variety of data types that are available for analysis. There are two broad types of data, of varying complexity, for the data mining process:

1. *Nondependency-oriented data:* This typically refers to simple data types such as multidimensional data or text data. These data types are the simplest and most commonly encountered. In these cases, the data records do not have any specified dependencies between either the data items or the attributes. An example is a set of demographic records about individuals containing their age, gender, and ZIP code.

2. *Dependency-oriented data:* In these cases, implicit or explicit relationships may exist between data items. For example, a social network data set contains a set of *vertices* (data items) that are connected together by a set of *edges* (relationships). On the other hand, time series contains implicit dependencies. For example, two successive values collected from a sensor are likely to be related to one another. Therefore, the time attribute implicitly specifies a dependency between successive readings.

In general, dependency-oriented data are more challenging because of the complexities created by preexisting relationships between data items. Such dependencies between data items need to be incorporated directly into the analytical process to obtain contextually meaningful results.

Table 1.1: An example of a multidimensional data set

Name	Age	Gender	Race	ZIP code
John S.	45	M	African American	05139
Manyona L.	31	F	Native American	10598
Sayani A.	11	F	East Indian	10547
Jack M.	56	M	Caucasian	10562
Wei L.	63	M	Asian	90210

1.3.1 Nondependency-Oriented Data

This is the simplest form of data and typically refers to *multidimensional data.* This data typically contains a set of *records.* A record is also referred to as a *data point*, *instance*, *example*, *transaction*, *entity*, *tuple*, *object*, or *feature-vector*, depending on the application at hand. Each record contains a set of *fields*, which are also referred to as *attributes*, *dimensions*, and *features.* These terms will be used interchangeably throughout this book. These fields describe the different properties of that record. Relational database systems were traditionally designed to handle this kind of data, even in their earliest forms. For example, consider the demographic data set illustrated in Table 1.1. Here, the demographic properties of an individual, such as age, gender, and ZIP code, are illustrated. A multidimensional data set is defined as follows:

Definition 1.3.1 (Multidimensional Data) *A multidimensional data set $\mathcal{D}$ is a set of n records, $\overline{X_1} \ldots \overline{X_n}$, such that each record $\overline{X_i}$ contains a set of d features denoted by $(x_i^1 \ldots x_i^d)$.*

Throughout the early chapters of this book, we will work with multidimensional data because it is the simplest form of data and establishes the broader principles on which the more complex data types can be processed. More complex data types will be addressed in later chapters of the book, and the impact of the dependencies on the mining process will be explicitly discussed.

1.3.1.1 Quantitative Multidimensional Data

The attributes in Table 1.1 are of two different types. The age field has values that are numerical in the sense that they have a natural ordering. Such attributes are referred to as *continuous*, *numeric*, or *quantitative.* Data in which all fields are quantitative is also referred to as *quantitative data* or *numeric data.* Thus, when each value of x_i^j in Definition 1.3.1 is quantitative, the corresponding data set is referred to as quantitative multidimensional data. In the data mining literature, this particular subtype of data is considered the most common, and many algorithms discussed in this book work with this subtype of data. This subtype is particularly convenient for analytical processing because it is much easier to work with quantitative data from a statistical perspective. For example, the mean of a set of quantitative records can be expressed as a simple average of these values, whereas such computations become more complex in other data types. Where possible and effective, many data mining algorithms therefore try to convert different kinds of data to quantitative values before processing. This is also the reason that many algorithms discussed in this (or virtually any other) data mining textbook assume a quantitative multidimensional representation. Nevertheless, in real applications, the data are likely to be more complex and may contain a mixture of different data types.

1.3.1.2 Categorical and Mixed Attribute Data

Many data sets in real applications may contain categorical attributes that take on *discrete unordered* values. For example, in Table 1.1, the attributes such as gender, race, and ZIP code, have discrete values without a natural ordering among them. If each value of x_i^j in Definition 1.3.1 is categorical, then such data are referred to as *unordered discrete-valued* or *categorical.* In the case of *mixed attribute* data, there is a combination of categorical and numeric attributes. The full data in Table 1.1 are considered mixed-attribute data because they contain both numeric and categorical attributes.

The attribute corresponding to gender is special because it is categorical, but with only two possible values. In such cases, it is possible to impose an artificial ordering between these values and use algorithms designed for numeric data for this type. This is referred to as *binary* data, and it can be considered a special case of either numeric or categorical data. Chap. 2 will explain how binary data form the "bridge" to transform numeric or categorical attributes into a common format that is suitable for processing in many scenarios.

1.3.1.3 Binary and Set Data

Binary data can be considered a special case of either multidimensional categorical data or multidimensional quantitative data. It is a special case of multidimensional categorical data, in which each categorical attribute may take on one of at most two discrete values. It is also a special case of multidimensional quantitative data because an ordering exists between the two values. Furthermore, binary data is also a representation of setwise data, in which each attribute is treated as a set element indicator. A value of 1 indicates that the element should be included in the set. Such data is common in market basket applications. This topic will be studied in detail in Chaps. 4 and 5.

1.3.1.4 Text Data

Text data can be viewed either as a string, or as multidimensional data, depending on how they are represented. In its raw form, a text document corresponds to a *string.* This is a dependency-oriented data type, which will be described later in this chapter. Each string is a sequence of characters (or words) corresponding to the document. However, text documents are rarely represented as strings. This is because it is difficult to directly use the ordering between words in an efficient way for large-scale applications, and the additional advantages of leveraging the ordering are often limited in the text domain.

In practice, a *vector-space representation* is used, where the frequencies of the words in the document are used for analysis. Words are also sometimes referred to as *terms.* Thus, the precise ordering of the words is lost in this representation. These frequencies are typically normalized with statistics such as the length of the document, or the frequencies of the individual words in the collection. These issues will be discussed in detail in Chap. 13 on text data. The corresponding $n \times d$ data matrix for a text collection with n documents and d terms is referred to as a *document-term matrix.*

When represented in vector-space form, text data can be considered multidimensional quantitative data, where the attributes correspond to the words, and the values correspond to the frequencies of these attributes. However, this kind of quantitative data is special because most attributes take on zero values, and only a few attributes have nonzero values. This is because a single document may contain only a relatively small number of words out of a dictionary of size 10^5. This phenomenon is referred to as *data sparsity,* and it significantly impacts the data mining process. The direct use of a quantitative data mining algorithm is often unlikely to work with sparse data without appropriate modifications. The sparsity also affects how the data are represented. For example, while it is possible to use the representation suggested in Definition 1.3.1, this is not a practical approach. Most values of x_i^j in Definition 1.3.1 are 0 for the case of text data. Therefore, it is inefficient to explicitly maintain a d-dimensional representation in which most values are 0. A bag-of-words representation is used containing only the words in the document. In addition, the frequencies of these words are explicitly maintained. This approach is typically more efficient. Because of data sparsity issues, text data are often processed with specialized methods. Therefore, text mining is often studied as a separate subtopic within data mining. Text mining methods are discussed in Chap. 13.

1.3.2 Dependency-Oriented Data

Most of the aforementioned discussion in this chapter is about the multidimensional scenario, where it is assumed that the data records can be treated independently of one another. In practice, the different data values may be (implicitly) related to each other temporally, spatially, or through explicit network relationship links between the data items. The knowledge about *preexisting* dependencies greatly changes the data mining process because data mining is all about finding relationships between data items. The presence of preexisting dependencies therefore changes the *expected* relationships in the data, and what may be considered *interesting* from the perspective of these expected relationships. Several types of dependencies may exist that may be either *implicit* or *explicit*:

1. *Implicit dependencies:* In this case, the dependencies between data items are not explicitly specified but are known to "typically" exist in that domain. For example, consecutive temperature values collected by a sensor are likely to be extremely similar to one another. Therefore, if the temperature value recorded by a sensor at a particular time is significantly different from that recorded at the next time instant then this is extremely unusual and may be interesting for the data mining process. This is different from multidimensional data sets where each data record is treated as an independent entity.

2. *Explicit dependencies:* This typically refers to graph or network data in which edges are used to specify explicit relationships. Graphs are a very powerful abstraction that are often used as an intermediate representation to solve data mining problems in the context of other data types.

In this section, the different dependency-oriented data types will be discussed in detail.

1.3.2.1 Time-Series Data

Time-series data contain values that are typically generated by continuous measurement over time. For example, an environmental sensor will measure the temperature continuously, whereas an electrocardiogram (ECG) will measure the parameters of a subject's heart rhythm. Such data typically have *implicit* dependencies built into the values received over time. For example, the adjacent values recorded by a temperature sensor will usually vary smoothly over time, and this factor needs to be explicitly used in the data mining process.

The nature of the temporal dependency may vary significantly with the application. For example, some forms of sensor readings may show periodic patterns of the measured attribute over time. An important aspect of time-series mining is the extraction of such dependencies in the data. To formalize the issue of dependencies caused by temporal correlation, the attributes are classified into two types:

1. *Contextual attributes:* These are the attributes that define the *context* on the basis of which the implicit dependencies occur in the data. For example, in the case of sensor data, the time stamp at which the reading is measured may be considered the contextual attribute. Sometimes, the time stamp is not explicitly used, but a position index is used. While the time-series data type contains only one contextual attribute, other data types may have more than one contextual attribute. A specific example is *spatial data*, which will be discussed later in this chapter.

2. *Behavioral attributes:* These represent the values that are measured in a particular context. In the sensor example, the temperature is the behavioral attribute value. It is possible to have more than one behavioral attribute. For example, if multiple sensors record readings at synchronized time stamps, then it results in a multidimensional time-series data set.

The contextual attributes typically have a strong impact on the dependencies between the behavioral attribute values in the data. Formally, time-series data are defined as follows:

Definition 1.3.2 (Multivariate Time-Series Data) *A time series of length n and dimensionality d contains d numeric features at each of n time stamps $t_1 \dots t_n$. Each time-stamp contains a component for each of the d series. Therefore, the set of values received at time stamp t_i is $\overline{Y_i} = (y_i^1 \dots y_i^d)$. The value of the jth series at time stamp t_i is y_i^j.*

For example, consider the case where two sensors at a particular location monitor the temperature and pressure every second for a minute. This corresponds to a multidimensional series with $d = 2$ and $n = 60$. In some cases, the time stamps $t_1 \dots t_n$ may be replaced by index values from 1 through n, especially when the time-stamp values are equally spaced apart.

Time-series data are relatively common in many sensor applications, forecasting, and financial market analysis. Methods for analyzing time series are discussed in Chap. 14.

1.3.2.2 Discrete Sequences and Strings

Discrete sequences can be considered the categorical analog of time-series data. As in the case of time-series data, the contextual attribute is a time stamp or a position index in the ordering. The behavioral attribute is a categorical value. Therefore, discrete sequence data are defined in a similar way to time-series data.

Definition 1.3.3 (Multivariate Discrete Sequence Data) *A discrete sequence of length n and dimensionality d contains d discrete feature values at each of n different time stamps $t_1 \dots t_n$. Each of the n components $\overline{Y_i}$ contains d discrete behavioral attributes $(y_i^1 \dots y_i^d)$, collected at the ith time-stamp.*

For example, consider a sequence of Web accesses, in which the Web page address and the originating IP address of the request are collected for 100 different accesses. This represents a discrete sequence of length $n = 100$ and dimensionality $d = 2$. A particularly common case in sequence data is the *univariate* scenario, in which the value of d is 1. Such sequence data are also referred to as *strings*.

It should be noted that the aforementioned definition is almost identical to the time-series case, with the main difference being that discrete sequences contain categorical attributes. In theory, it is possible to have series that are mixed between categorical and numerical data. Another important variation is the case where a sequence does not contain categorical attributes, but a *set* of any number of unordered categorical values. For example, supermarket transactions may contain a sequence of sets of items. Each set may contain any number of items. Such setwise sequences are not really multivariate sequences, but are univariate sequences, in which each element of the sequence is a *set* as opposed to a unit element. Thus, discrete sequences can be defined in a wider variety of ways, as compared to time-series data because of the ability to define sets on discrete elements.

In some cases, the contextual attribute may not refer to time explicitly, but it might be a position based on physical placement. This is the case for biological sequence data. In such cases, the time stamp may be replaced by an index representing the position of the value in the string, counting the leftmost position as 1. Some examples of common scenarios in which sequence data may arise are as follows:

- *Event logs:* A wide variety of computer systems, Web servers, and Web applications create event logs on the basis of user activity. An example of an event log is a sequence of user actions at a financial Web site:

  ```
  Login Password Login Password Login Password ....
  ```

  This particular sequence may represent a scenario where a user is attempting to break into a password-protected system, and it may be interesting from the perspective of anomaly detection.

- *Biological data:* In this case, the sequences may correspond to strings of nucleotides or amino acids. The ordering of such units provides information about the characteristics of protein function. Therefore, the data mining process can be used to determine interesting patterns that are reflective of different biological properties.

Discrete sequences are often more challenging for mining algorithms because they do not have the smooth value continuity of time-series data. Methods for sequence mining are discussed in Chap. 15.

1.3.2.3 Spatial Data

In spatial data, many nonspatial attributes (e.g., temperature, pressure, image pixel color intensity) are measured at spatial locations. For example, sea-surface temperatures are often collected by meteorologists to forecast the occurrence of hurricanes. In such cases, the spatial coordinates correspond to contextual attributes, whereas attributes such as the temperature correspond to the behavioral attributes. Typically, there are two spatial attributes. As in the case of time-series data, it is also possible to have multiple behavioral attributes. For example, in the sea-surface temperature application, one might also measure other behavioral attributes such as the pressure.

Definition 1.3.4 (Spatial Data) *A d-dimensional spatial data record contains d behavioral attributes and one or more contextual attributes containing the spatial location. Therefore, a d-dimensional spatial data set is a set of d dimensional records* $\overline{X_1} \dots \overline{X_n}$, *together with a set of n locations* $L_1 \dots L_n$, *such that the record* $\overline{X_i}$ *is associated with the location* L_i.

The aforementioned definition provides broad flexibility in terms of how record $\overline{X_i}$ and location L_i may be defined. For example, the behavioral attributes in record $\overline{X_i}$ may be numeric or categorical, or a mixture of the two. In the meteorological application, $\overline{X_i}$ may contain the temperature and pressure attributes at location L_i. Furthermore, L_i may be specified in terms of precise spatial coordinates, such as latitude and longitude, or in terms of a logical location, such as the city or state.

Spatial data mining is closely related to time-series data mining, in that the behavioral attributes in most commonly studied spatial applications are continuous, although some applications may use categorical attributes as well. Therefore, value continuity is observed across contiguous spatial locations, just as value continuity is observed across contiguous time stamps in time-series data.

Spatiotemporal Data

A particular form of spatial data is spatiotemporal data, which contains both spatial and temporal attributes. The precise nature of the data also depends on which of the attributes are contextual and which are behavioral. Two kinds of spatiotemporal data are most common:

1. *Both spatial and temporal attributes are contextual:* This kind of data can be viewed as a direct generalization of both spatial data and temporal data. This kind of data is particularly useful when the spatial and temporal dynamics of particular behavioral attributes are measured simultaneously. For example, consider the case where the variations in the sea-surface temperature need to be measured over time. In such cases, the temperature is the behavioral attribute, whereas the spatial and temporal attributes are contextual.

2. *The temporal attribute is contextual, whereas the spatial attributes are behavioral:* Strictly speaking, this kind of data can also be considered time-series data. However, the spatial nature of the behavioral attributes also provides better interpretability and more focused analysis in many scenarios. The most common form of this data arises in the context of *trajectory analysis.*

It should be pointed out that any 2- or 3-dimensional time-series data can be mapped onto trajectories. This is a useful transformation because it implies that trajectory mining algorithms can also be used for 2- or 3-dimensional time-series data. For example, the *Intel Research Berkeley data set* [556] contains readings from a variety of sensors. An example of a pair of readings from a temperature and voltage sensor are illustrated in Figs. 1.2a and b, respectively. The corresponding temperature–voltage trajectory is illustrated in Fig. 1.2c. Methods for spatial and spatiotemporal data mining are discussed in Chap. 16.

1.3.2.4 Network and Graph Data

In network and graph data, the data values may correspond to nodes in the network, whereas the relationships among the data values may correspond to the edges in the network. In some cases, attributes may be associated with nodes in the network. Although it is also possible to associate attributes with edges in the network, it is much less common to do so.

Definition 1.3.5 (Network Data) *A network $G = (N, A)$ contains a set of nodes N and a set of edges A, where the edges in A represent the relationships between the nodes. In*

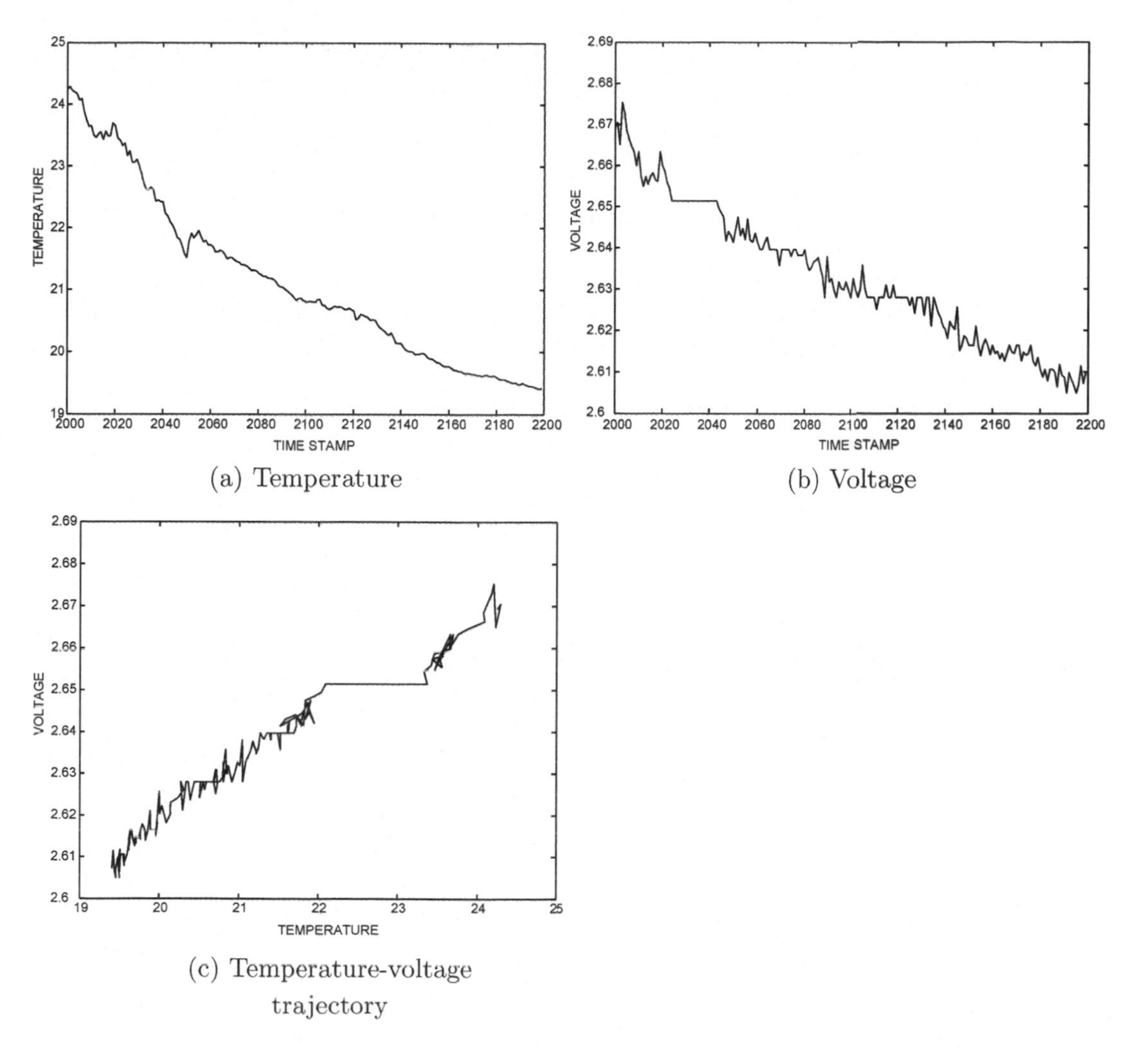

(a) Temperature

(b) Voltage

(c) Temperature-voltage trajectory

Figure 1.2: Mapping of multivariate time series to trajectory data

some cases, an attribute set $\overline{X_i}$ *may be associated with node* i*, or an attribute set* $\overline{Y_{ij}}$ *may be associated with edge* (i, j).

The edge (i, j) may be directed or undirected, depending on the application at hand. For example, the Web graph may contain directed edges corresponding to directions of hyperlinks between pages, whereas friendships in the Facebook social network are undirected.

A second class of graph mining problems is that of a database containing many small graphs such as chemical compounds. The challenges in these two classes of problems are very different. Some examples of data that are represented as graphs are as follows:

- *Web graph:* The nodes correspond to the Web pages, and the edges correspond to hyperlinks. The nodes have text attributes corresponding to the content in the page.
- *Social networks:* In this case, the nodes correspond to social network actors, whereas the edges correspond to friendship links. The nodes may have attributes corresponding to social page content. In some specialized forms of social networks, such as email or

chat-messenger networks, the edges may have content associated with them. This content corresponds to the communication between the different nodes.

- *Chemical compound databases:* In this case, the nodes correspond to the elements and the edges correspond to the chemical bonds between the elements. The structures in these chemical compounds are very useful for identifying important reactive and pharmacological properties of these compounds.

Network data are a very general representation and can be used for solving many similarity-based applications on other data types. For example, multidimensional data may be converted to network data by creating a node for each record in the database, and representing similarities between nodes by edges. Such a representation is used quite often for many similarity-based data mining applications, such as clustering. It is possible to use community detection algorithms to determine clusters in the network data and then map them back to multidimensional data. Some spectral clustering methods, discussed in Chap. 19, are based on this principle. This generality of network data comes at a price. The development of mining algorithms for network data is generally more difficult. Methods for mining network data are discussed in Chaps. 17, 18, and 19.

1.4 The Major Building Blocks: A Bird's Eye View

As discussed in the introduction Sect. 1.1, four problems in data mining are considered fundamental to the mining process. These problems correspond to clustering, classification, association pattern mining, and outlier detection, and they are encountered repeatedly in the context of many data mining applications. What makes these problems so special? Why are they encountered repeatedly? To answer these questions, one must understand the nature of the typical relationships that data scientists often try to extract from the data.

Consider a multidimensional database $\mathcal{D}$ with n records, and d attributes. Such a database $\mathcal{D}$ may be represented as an $n \times d$ matrix D, in which each row corresponds to one record and each column corresponds to a dimension. We generally refer to this matrix as the *data matrix*. This book will use the notation of a data matrix D, and a database $\mathcal{D}$ interchangeably. Broadly speaking, data mining is all about finding summary relationships between the entries in the data matrix that are either unusually frequent or unusually infrequent. Relationships between data items are one of two kinds:

- *Relationships between columns:* In this case, the frequent or infrequent relationships between the values in a particular row are determined. This maps into either the positive or negative association pattern mining problem, though the former is more commonly studied. In some cases, one particular column of the matrix is considered more important than other columns because it represents a target attribute of the data mining analyst. In such cases, one tries to determine how the relationships in the other columns relate to this special column. Such relationships can be used to predict the value of this special column, when the value of that special column is unknown. This problem is referred to as *data classification.* A mining process is referred to as *supervised* when it is based on treating a particular attribute as special and predicting it.

- *Relationships between rows:* In these cases, the goal is to determine subsets of rows, in which the values in the corresponding columns are related. In cases where these subsets are similar, the corresponding problem is referred to as *clustering.* On the other hand, when the entries in a row are very different from the corresponding entries in other rows, then the corresponding row becomes interesting as an unusual data point, or as an *anomaly*. This problem is referred to as *outlier analysis*. Interestingly, the clustering problem is closely related to that of classification, in that the latter can be considered a supervised version of the former. The discrete values of a special column in the data correspond to the group identifiers of different *desired* or *supervised* groups of application-specific similar records in the data. For example, when the special column corresponds to whether or not a customer is interested in a particular product, this represents the two groups in the data that one is interested in *learning*, with the use of *supervision*. The term "supervision" refers to the fact that the special column is used to direct the data mining process in an application-specific way, just as a teacher may supervise his or her student toward a specific goal.

Thus, these four problems are important because they seem to cover an exhaustive range of scenarios representing different kinds of positive, negative, supervised, or unsupervised relationships between the entries of the data matrix. These problems are also related to one another in a variety of ways. For example, association patterns may be considered indirect representations of (overlapping) clusters, where each pattern corresponds to a cluster of data points of which it is a subset.

It should be pointed out that the aforementioned discussion assumes the (most commonly encountered) multidimensional data type, although these problems continue to retain their relative importance for more complex data types. However, the more complex data types have a wider variety of problem formulations associated with them because of their greater complexity. This issue will be discussed in detail later in this section.

It has consistently been observed that many application scenarios determine such relationships between rows and columns of the data matrix as an intermediate step. This is the reason that a good understanding of these building-block problems is so important for the data mining process. Therefore, the first part of this book will focus on these problems in detail before generalizing to complex scenarios.

1.4.1 Association Pattern Mining

In its most primitive form, the association pattern mining problem is defined in the context of *sparse binary databases*, where the data matrix contains only 0/1 entries, and most entries take on the value of 0. Most customer transaction databases are of this type. For example, if each column in the data matrix corresponds to an item, and a customer transaction represents a row, the (i, j)th entry is 1, if customer transaction i contains item j as one of the items that was bought. A particularly commonly studied version of this problem is the frequent pattern mining problem or, more generally, the association pattern mining problem. In terms of the binary data matrix, the frequent pattern mining problem may be formally defined as follows:

Definition 1.4.1 (Frequent Pattern Mining) *Given a binary $n \times d$ data matrix D, determine all subsets of columns such that all the values in these columns take on the value of 1 for at least a fraction s of the rows in the matrix. The relative frequency of a pattern is referred to as its support. The fraction s is referred to as the minimum support.*

Patterns that satisfy the minimum support requirement are often referred to as *frequent patterns*, or *frequent itemsets*. Frequent patterns represent an important class of association patterns. Many other definitions of relevant association patterns are possible that do not use absolute frequencies but use other statistical quantifications such as the χ^2 measure. These measures often lead to generation of more *interesting* rules from a statistical perspective. Nevertheless, this particular definition of association pattern mining has become the most popular one in the literature because of the ease in developing algorithms for it. This book therefore refers to this problem as *association pattern mining* as opposed to *frequent pattern mining*.

For example, if the columns of the data matrix D corresponding to *Bread*, *Butter*, and *Milk* take on the value of 1 together frequently in a customer transaction database, then it implies that these items are often bought together. This is very useful information for the merchant from the perspective of physical placement of the items in the store, or from the perspective of product promotions. Association pattern mining is not restricted to the case of binary data and can be easily generalized to quantitative and numeric attributes by using appropriate data transformations, which will be discussed in Chap. 4.

Association pattern mining was originally proposed in the context of *association rule mining*, where an additional step was included based on a measure known as the *confidence* of the rule. For example, consider two sets of items A and B. The confidence of the rule $A \Rightarrow B$ is defined as the fraction of transactions containing A, which also contain B. In other words, the confidence is obtained by dividing the support of the pattern $A \cup B$ with the support of pattern A. A combination of support and confidence is used to define association rules.

Definition 1.4.2 (Association Rules) *Let A and B be two sets of items. The rule $A \Rightarrow B$ is said to be valid at support level s and confidence level c, if the following two conditions are satisfied:*

1. *The support of the item set A is at least s.*
2. *The confidence of $A \Rightarrow B$ is at least c.*

By incorporating supervision in association rule mining algorithms, it is possible to provide solutions for the classification problem. Many variations of association pattern mining are also related to clustering and outlier analysis. This is a natural consequence of the fact that horizontal and vertical analysis of the data matrix are often related to one another. In fact, many variations of the association pattern mining problem are used as a subroutine to solve the clustering, outlier analysis, and classification problems. These issues will be discussed in Chaps. 4 and 5.

1.4.2 Data Clustering

A rather broad and informal definition of the clustering problem is as follows:

Definition 1.4.3 (Data Clustering) *Given a data matrix D (database $\mathcal{D}$), partition its rows (records) into sets $\mathcal{C}_1 \dots \mathcal{C}_k$, such that the rows (records) in each cluster are "similar" to one another.*

We have intentionally provided an informal definition here because clustering allows a wide variety of definitions of similarity, some of which are not cleanly defined in closed form by a similarity function. A clustering problem can often be defined as an optimization problem, in which the variables of the optimization problem represent cluster memberships of data points, and the objective function maximizes a concrete mathematical quantification of intragroup similarity in terms of these variables.

An important part of the clustering process is the design of an appropriate similarity function for the computation process. Clearly, the computation of similarity depends heavily on the underlying data type. The issue of similarity computation will be discussed in detail in Chap. 3. Some examples of relevant applications are as follows:

- *Customer segmentation:* In many applications, it is desirable to determine customers that are similar to one another in the context of a variety of product promotion tasks. The segmentation phase plays an important role in this process.

- *Data summarization:* Because clusters can be considered similar groups of records, these similar groups can be used to create a summary of the data.

- *Application to other data mining problems:* Because clustering is considered an unsupervised version of classification, it is often used as a building block to solve the latter. Furthermore, this problem is also used in the context of the outlier analysis problem, as discussed below.

The data clustering problem is discussed in detail in Chaps. 6 and 7.

1.4.3 Outlier Detection

An outlier is a data point that is significantly different from the remaining data. Hawkins formally defined [259] the concept of an outlier as follows:
"*An outlier is an observation that deviates so much from the other observations as to arouse suspicions that it was generated by a different mechanism.*"

Outliers are also referred to as *abnormalities*, *discordants*, *deviants*, or *anomalies* in the data mining and statistics literature. In most applications, the data are created by one or more generating processes that can either reflect activity in the system or observations collected about entities. When the generating process behaves in an unusual way, it results in the creation of outliers. Therefore, an outlier often contains useful information about abnormal characteristics of the systems and entities that impact the data-generation process. The recognition of such unusual characteristics provides useful application-specific insights. The outlier detection problem is informally defined in terms of the data matrix as follows:

Definition 1.4.4 (Outlier Detection) *Given a data matrix D, determine the rows of the data matrix that are very different from the remaining rows in the matrix.*

The outlier detection problem is related to the clustering problem by complementarity. This is because outliers correspond to dissimilar data points from the main groups in the data. On the other hand, the main groups in the data are clusters. In fact, a simple methodology to determine outliers uses clustering as an intermediate step. Some examples of relevant applications are as follows:

- *Intrusion-detection systems:* In many networked computer systems, different kinds of data are collected about the operating system calls, network traffic, or other activity in the system. These data may show unusual behavior because of malicious activity. The detection of such activity is referred to as intrusion detection.

- *Credit card fraud:* Unauthorized use of credit cards may show different patterns, such as a buying spree from geographically obscure locations. Such patterns may show up as outliers in credit card transaction data.

- *Interesting sensor events:* Sensors are often used to track various environmental and location parameters in many real applications. The sudden changes in the underlying patterns may represent events of interest. Event detection is one of the primary motivating applications in the field of sensor networks.

- *Medical diagnosis:* In many medical applications, the data are collected from a variety of devices such as magnetic resonance imaging (MRI), positron emission tomography (PET) scans, or electrocardiogram (ECG) time series. Unusual patterns in such data typically reflect disease conditions.

- *Law enforcement:* Outlier detection finds numerous applications in law enforcement, especially in cases where unusual patterns can only be discovered over time through multiple actions of an entity. The identification of fraud in financial transactions, trading activity, or insurance claims typically requires the determination of unusual patterns in the data generated by the actions of the criminal entity.

- *Earth science:* A significant amount of spatiotemporal data about weather patterns, climate changes, or land-cover patterns is collected through a variety of mechanisms such as satellites or remote sensing. Anomalies in such data provide significant insights about hidden human or environmental trends that may have caused such anomalies.

The outlier detection problem is studied in detail in Chaps. 8 and 9.

1.4.4 Data Classification

Many data mining problems are directed toward a specialized goal that is sometimes represented by the value of a particular feature in the data. This particular feature is referred to as the *class label*. Therefore, such problems are *supervised*, wherein the relationships of the remaining features in the data with respect to this special feature are *learned*. The data used to learn these relationships is referred to as the *training data*. The *learned model* may then be used to determine the estimated class labels for records, where the label is missing.

For example, in a target marketing application, each record may be tagged by a particular *label* that represents the interest (or lack of it) of the customer toward a particular product. The labels associated with customers may have been derived from the previous buying behavior of the customer. In addition, a set of features corresponding the customer demographics may also be available. The goal is to predict whether or not a customer, whose buying behavior is unknown, will be interested in a particular product by relating the demographic features to the class label. Therefore, a *training model* is constructed, which is then used to predict class labels. The classification problem is informally defined as follows:

Definition 1.4.5 (Data Classification) *Given an $n \times d$ training data matrix D (database $\mathcal{D}$), and a class label value in $\{1 \dots k\}$ associated with each of the n rows in D (records in $\mathcal{D}$), create a training model $\mathcal{M}$, which can be used to predict the class label of a d-dimensional record $\overline{Y} \notin \mathcal{D}$.*

The record whose class label is unknown is referred to as the *test record*. It is interesting to examine the relationship between the clustering and the classification problems. In the case of the clustering problem, the data are partitioned into k groups on the basis of similarity. In the case of the classification problem, a (test) record is also categorized into one of k groups, except that this is achieved by learning a model from a training database $\mathcal{D}$, rather than on the basis of similarity. In other words, the supervision from the training data redefines the notion of a group of "similar" records. Therefore, from a learning perspective, clustering is often referred to as *unsupervised learning* (because of the lack of a special training database to "teach" the model about the notion of an appropriate grouping), whereas the classification problem is referred to as *supervised learning.*

The classification problem is related to association pattern mining, in the sense that the latter problem is often used to solve the former. This is because if the entire training database (including the class label) is treated as an $n \times (d+1)$ matrix, then frequent patterns containing the class label in this matrix provide useful hints about the correlations of other features to the class label. In fact, many forms of classifiers, known as *rule-based classifiers*, are based on this broader principle.

The classification problem can be mapped to a specific version of the outlier detection problem, by incorporating supervision in the latter. While the outlier detection problem is assumed to be unsupervised by default, many variations of the problem are either partially or fully supervised. In supervised outlier detection, some examples of outliers are available. Thus, such data records are tagged to belong to a *rare class*, whereas the remaining data records belong to the *normal class*. Thus, the supervised outlier detection problem maps to a binary classification problem, with the caveat that the class labels are highly *imbalanced.*

The incorporation of supervision makes the classification problem unique in terms of its *direct* application specificity due to its use of application-specific class labels. Compared to the other major data mining problems, the classification problem is relatively self-contained. For example, the clustering and frequent pattern mining problem are more often used as intermediate steps in larger application frameworks. Even the outlier analysis problem is sometimes used in an exploratory way. On the other hand, the classification problem is often used directly as a stand-alone tool in many applications. Some examples of applications where the classification problem is used are as follows:

- *Target marketing:* Features about customers are related to their buying behavior with the use of a training model.

- *Intrusion detection:* The sequences of customer activity in a computer system may be used to predict the possibility of intrusions.

- *Supervised anomaly detection:* The rare class may be differentiated from the normal class when previous examples of outliers are available.

The data classification problem is discussed in detail in Chaps. 10 and 11.

1.4.5 Impact of Complex Data Types on Problem Definitions

The specific data type has a profound impact on the kinds of problems that may be defined. In particular, in dependency-oriented data types, the dependencies often play a critical role in the problem definition, the solution, or both. This is because the contextual attributes and dependencies are often fundamental to how the data may be evaluated. Furthermore, because complex data types are much richer, they allow the formulation of novel problem definitions that may not even exist in the context of multidimensional data. A tabular summary of the different variations of data mining problems for dependency-oriented data types is provided in Table 1.2. In the following, a brief review will be provided as to how the different problem definitions are affected by data type.

Table 1.2: Some examples of variation in problem definition with data type

Problem	Time series	Spatial	Sequence	Networks
Patterns	Motif-mining Periodic pattern	Colocation patterns	Sequential patterns Periodic Sequence	Structural patterns
	Trajectory patterns			
Clustering	Shape clusters	Spatial clusters	Sequence clusters	Community detection
	Trajectory clusters			
Outliers	Position outlier Shape outlier	Position outlier Shape outlier	Position outlier Combination outlier	Node outlier Linkage outlier Community outliers
	Trajectory outliers			
Classification	Position classification Shape classification	Position classification Shape classification	Position classification Sequence classification	Collective classification Graph classification
	Trajectory classification			

1.4.5.1 Pattern Mining with Complex Data Types

The association pattern mining problem generally determines the patterns from the underlying data in the form of sets; however, this is not the case when dependencies are present in the data. This is because the dependencies and relationships often impose ordering among data items, and the direct use of frequent pattern mining methods fails to recognize the relationships among the different data values. For example, when a larger number of time series are made available, they can be used to determine different kinds of *temporally* frequent patterns, in which a temporal ordering is imposed on the items in the pattern. Furthermore, because of the presence of the additional contextual attribute representing time, temporal patterns may be defined in a much richer way than a set-based pattern as in association pattern mining. The patterns may be temporally contiguous, as in *time-series motifs*, or they may be periodic, as in *periodic patterns*. Some of these methods for temporal pattern mining will be discussed in Chap. 14. A similar analogy exists for the case of discrete sequence mining, except that the individual pattern constituents are categorical, as opposed to continuous. It is also possible to define 2-dimensional motifs for the spatial scenario, and such a formulation is useful for image processing. Finally, structural patterns are commonly defined in networks that correspond to frequent subgraphs in the data. Thus, the dependencies between the nodes are included within the definition of the patterns.

1.4.5.2 Clustering with Complex Data Types

The techniques used for clustering are also affected significantly by the underlying data type. Most importantly, the similarity function is significantly affected by the data type. For example, in the case of time series, sequential, or graph data, the similarity between a pair of time series cannot be easily defined by using straightforward metrics such as the Euclidean metric. Rather, it is necessary to use other kinds of metrics, such as the edit distance or structural similarity. In the context of spatial data, trajectory clustering is particularly useful in finding the relevant patterns for mobile data, or for multivariate time series. For network data, the clustering problem discovers densely connected groups of nodes, and is also referred to as *community detection.*

1.4.5.3 Outlier Detection with Complex Data Types

Dependencies can be used to define expected values of data items. Deviations from these expected values are outliers. For example, a sudden jump in the value of a time series will result in a position outlier at the specific spot at which the jump occurs. The idea in these methods is to use *prediction-based techniques* to forecast the value at that position. Significant deviation from the prediction is reported as a *position outlier.* Such outliers can be defined in the context of time-series, spatial, and sequential data, where significant deviations from the corresponding neighborhoods can be detected using autoregressive, Markovian, or other models. In the context of graph data, outliers may correspond to unusual properties of nodes, edges, or entire subgraphs. Thus, the complex data types show significant richness in terms of how outliers may be defined.

1.4.5.4 Classification with Complex Data Types

The classification problem also shows a significant amount of variation in the different complex data types. For example, class labels can be attached to specific positions in a series, or they can be attached to the entire series. When the class labels are attached to a specific position in the series, this can be used to perform supervised event detection, where the first occurrence of an event-specific label (e.g., the breakdown of a machine as suggested by the underlying temperature and pressure sensor) of a particular series represents the occurrence of the event. For the case of network data, the labels may be attached to individual nodes in a very large network, or to entire graphs in a collection of multiple graphs. The former case corresponds to the classification of nodes in a social network, and is also referred to as *collective classification.* The latter case corresponds to the chemical compound classification problem, in which labels are attached to compounds on the basis of their chemical properties.

1.5 Scalability Issues and the Streaming Scenario

Scalability is an important concern in many data mining applications due to the increasing sizes of the data in modern-day applications. Broadly speaking, there are two important scenarios for scalability:

1. The data are stored on one or more machines, but it is too large to process efficiently. For example, it is easy to design efficient algorithms in cases where the entire data can be maintained in main memory. When the data are stored on disk, it is important to be design the algorithms in such a way that random access to the disk is minimized. For very large data sets, big data frameworks, such as MapReduce, may need to be used. This book will touch upon this kind of scalability at the level of disk-resident processing, where needed.

2. The data are generated continuously over time in high volume, and it is not practical to store it entirely. This scenario is that of *data streams,* in which the data need to be processed with the use of an online approach.

The latter scenario requires some further exposition. The streaming scenario has become increasingly popular because of advances in data collection technology that can collect large amounts of data over time. For example, simple transactions of everyday life such as using a credit card or the phone may lead to automated data collection. In such cases, the volume of the data is so large that it may be impractical to store directly. Rather, all algorithms must be executed in a single pass over the data. The major challenges that arise in the context of data stream processing are as follows:

1. *One-pass constraint:* The algorithm needs to process the entire data set in one pass. In other words, after a data item has been processed and the relevant summary insights have been gleaned, the raw item is discarded and is no longer available for processing. The amount of data that may be processed at a given time depends on the storage available for retaining segments of the data.

2. *Concept drift:* In most applications, the data distribution changes over time. For example, the pattern of sales in a given hour of a day may not be similar to that at another hour of the day. This leads to changes in the output of the mining algorithms as well.

It is often challenging to design algorithms for such scenarios because of the varying rates at which the patterns in the data may change over time and the continuously evolving patterns in the underlying data. Methods for stream mining are addressed in Chap. 12.

1.6 A Stroll Through Some Application Scenarios

In this section, some common application scenarios will be discussed. The goal is to illustrate the wide diversity of problems and applications, and how they might map onto some of the building blocks discussed in this chapter.

1.6.1 Store Product Placement

The application scenario may be stated as follows:

Application 1.6.1 (Store Product Placement) *A merchant has a set of d products together with previous transactions from the customers containing baskets of items bought together. The merchant would like to know how to place the product on the shelves to increase the likelihood that items that are frequently bought together are placed on adjacent shelves.*

This problem is closely related to frequent pattern mining because the analyst can use the frequent pattern mining problem to determine groups of items that are frequently bought together at a particular support level. An important point to note here is that the determination of the frequent patterns, while providing useful insights, does not provide the merchant with precise guidance in terms of how the products may be placed on the different shelves. This situation is quite common in data mining. The building block problems often do not directly solve the problem at hand. In this particular case, the merchant may choose from a variety of heuristic ideas in terms of how the products may be stocked on the different shelves. For example, the merchant may already have an existing placement, and may use the frequent patterns to create a numerical score for the quality of the placement. This placement can be successively optimized by making incremental changes to the current placement. With an appropriate initialization methodology, the frequent pattern mining approach can be leveraged as a very useful subroutine for the problem. These parts of data mining are often application-specific and show such wide variations across different domains that they can only be learned through practical experience.

1.6.2 Customer Recommendations

This is a very commonly encountered problem in the data mining literature. Many variations of this problem exist, depending on the kind of input data available to that application. In the following, we will examine a particular instantiation of the recommendation problem and a straw-man solution.

Application 1.6.2 (Product Recommendations) *A merchant has an $n \times d$ binary matrix D representing the buying behavior of n customers across d items. It is assumed that the matrix is sparse, and therefore each customer may have bought only a few items. It is desirable to use the product associations to make recommendations to customers.*

This problem is a simple version of the collaborative filtering problem that is widely studied in the data mining and recommendation literature. There are literally hundreds of solutions to the vanilla version of this problem, and we provide three sample examples of varying complexity below:

1. A simple solution is to use association rule mining at particular levels of support and confidence. For a particular customer, the relevant rules are those in which all items in the left-hand side were previously bought by this customer. Items that appear frequently on the right-hand side of the relevant rules are reported.

2. The previous solution does not use the similarity across different customers to make recommendations. A second solution is to determine the most similar rows to a target customer, and then recommend the most common item occurring in these similar rows.

3. A final solution is to use clustering to create segments of similar customers. Within each similar segment, association pattern mining may be used to make recommendations.

Thus, there can be multiple ways of solving a particular problem corresponding to different analytical paths. These different paths may use different kinds of building blocks, which are all useful in different parts of the data mining process.

1.6.3 Medical Diagnosis

Medical diagnosis has become a common application in the context of data mining. The data types in medical diagnosis tend to be complex, and may correspond to image, time-series, or discrete sequence data. Thus, dependency-oriented data types tend to be rather common in medical diagnosis applications. A particular case is that of ECG readings from heart patients.

Application 1.6.3 (Medical ECG Diagnosis) *Consider a set of ECG time series that are collected from different patients. It is desirable to determine the anomalous series from this set.*

This application can be mapped to different problems, depending upon the nature of the input data available. For example, consider the case where no previous examples of anomalous ECG series are available. In such cases, the problem can be mapped to the outlier detection problem. A time series that differs significantly from the remaining series in the data may be considered an outlier. However, the solution methodology changes significantly if previous examples of normal and anomalous series are available. In such cases, the problem maps to a classification problem on time-series data. Furthermore, the class labels are likely to be imbalanced because the number of abnormal series are usually far fewer than the number of normal series.

1.6.4 Web Log Anomalies

Web logs are commonly collected at the hosts of different Web sites. Such logs can be used to detect unusual, suspicious, or malicious activity at the site. Financial institutions regularly analyze the logs at their site to detect intrusion attempts.

Application 1.6.4 (Web Log Anomalies) *A set of Web logs is available. It is desired to determine the anomalous sequences from the Web logs.*

Because the data are typically available in the form of raw logs, a significant amount of data cleaning is required. First, the raw logs need to be transformed into sequences of symbols. These sequences may then need to be decomposed into smaller windows to analyze the sequences at a particular level of granularity. Anomalous sequences may be determined by using a sequence clustering algorithm, and then determining the sequences that do not lie in these clusters [5]. If it is desired to find specific positions that correspond to anomalies, then more sophisticated methods such as Markovian models may be used to determine the anomalies [5].

As in the previous case, the analytical phase of this problem can be modeled differently, depending on whether or not examples of Web log anomalies are available. If no previous examples of Web log anomalies are available, then this problem maps to the unsupervised temporal outlier detection problem. Numerous methods for solving the unsupervised case for the temporal outlier detection problem are introduced in [5]. The topic is also briefly discussed in Chaps. 14 and 15 of this book. On the other hand, when examples of previous anomalies are available, then the problem maps to the rare class-detection problem. This problem is discussed in [5] as well, and in Chap. 11 of this book.

1.7 Summary

Data mining is a complex and multistage process. These different stages are data collection, preprocessing, and analysis. The data preprocessing phase is highly application-specific because the different formats of the data require different algorithms to be applied to them. The processing phase may include data integration, cleaning, and feature extraction. In some cases, feature selection may also be used to sharpen the data representation. After the data have been converted to a convenient format, a variety of analytical algorithms can be used.

A number of data mining building blocks are often used repeatedly in a wide variety of application scenarios. These correspond to the frequent pattern mining, clustering, outlier analysis, and classification problems, respectively. The final design of a solution for a particular data mining problem is dependent on the skill of the analyst in mapping the application to the different building blocks, or in using novel algorithms for a specific application. This book will introduce the fundamentals required for gaining such analytical skills.

1.8 Bibliographic Notes

The problem of data mining is generally studied by multiple research communities corresponding to statistics, data mining, and machine learning. These communities are highly overlapping and often share many researchers in common. The machine learning and statistics communities generally approach data mining from a theoretical and statistical perspective. Some good books written in this context may be found in [95, 256, 389]. However, because the machine learning community is generally focused on supervised learning methods, these books are mostly focused on the classification scenario. More general data mining books, which are written from a broader perspective, may be found in [250, 485, 536]. Because the data mining process often has to interact with databases, a number of relevant database textbooks [434, 194] provide knowledge about data representation and integration issues.

A number of books have also been written on each of the major areas of data mining. The frequent pattern mining problem and its variations have been covered in detail in [34]. Numerous books have been written on the topic of data clustering. A well-known data clustering book [284] discusses the classical techniques from the literature. Another book [219] discusses the more recent methods for data clustering, although the material is somewhat basic. The most recent book [32] in the literature provides a very comprehensive overview of the different data clustering algorithms. The problem of data classification has been addressed in the standard machine learning books [95, 256, 389]. The classification problem has also been studied extensively by the pattern recognition community [189]. More recent surveys on the topic may be found in [33]. The problem of outlier detection has been studied in detail in [89, 259]. These books are, however, written from a statistical perspective and do not address the problem from the perspective of the computer science community. The problem has been addressed from the perspective of the computer science community in [5].

1.9 Exercises

1. An analyst collects surveys from different participants about their likes and dislikes. Subsequently, the analyst uploads the data to a database, corrects erroneous or missing entries, and designs a recommendation algorithm on this basis. Which of the following actions represent data collection, data preprocessing, and data analysis? (a) Conducting surveys and uploading to database, (b) correcting missing entries, (c) designing a recommendation algorithm.

2. What is the data type of each of the following kinds of attributes (a) *Age*, (b) *Salary*, (c) *ZIP code*, (d) *State of residence*, (e) *Height*, (f) *Weight*?

3. An analyst obtains medical notes from a physician for data mining purposes, and then transforms them into a table containing the medicines prescribed for each patient. What is the data type of (a) the original data, and (b) the transformed data? (c) What is the process of transforming the data to the new format called?

4. An analyst sets up a sensor network in order to measure the temperature of different locations over a period. What is the data type of the data collected?

5. The same analyst as discussed in Exercise 4 above finds another database from a different source containing pressure readings. She decides to create a single database

containing her own readings and the pressure readings. What is the process of creating such a single database called?

6. An analyst processes Web logs in order to create records with the ordering information for Web page accesses from different users. What is the type of this data?

7. Consider a data object corresponding to a set of nucleotides arranged in a certain order. What is this type of data?

8. It is desired to partition customers into similar groups on the basis of their demographic profile. Which data mining problem is best suited to this task?

9. Suppose in Exercise 8, the merchant already knows for *some* of the customers whether or not they have bought widgets. Which data mining problem would be suited to the task of identifying groups among the remaining customers, who *might* buy widgets in the future?

10. Suppose in Exercise 9, the merchant also has information for other items bought by the customers (beyond widgets). Which data mining problem would be best suited to finding sets of items that are often bought together with widgets?

11. Suppose that a small number of customers lie about their demographic profile, and this results in a mismatch between the buying behavior and the demographic profile, as suggested by comparison with the remaining data. Which data mining problem would be best suited to finding such customers?

Chapter 2

Data Preparation

"Success depends upon previous preparation, and without such preparation there is sure to be failure."—Confucius

2.1 Introduction

The raw format of real data is usually widely variable. Many values may be missing, inconsistent across different data sources, and erroneous. For the analyst, this leads to numerous challenges in using the data effectively. For example, consider the case of evaluating the interests of consumers from their activity on a social media site. The analyst may first need to determine the types of activity that are valuable to the mining process. The activity might correspond to the interests entered by the user, the comments entered by the user, and the set of friendships of the user along with their interests. All these pieces of information are diverse and need to be collected from different databases within the social media site. Furthermore, some forms of data, such as raw logs, are often not directly usable because of their unstructured nature. In other words, useful features need to be extracted from these data sources. Therefore, a *data preparation phase* is needed.

The data preparation phase is a multistage process that comprises several individual steps, some or all of which may be used in a given application. These steps are as follows:

1. *Feature extraction and portability:* The raw data is often in a form that is not suitable for processing. Examples include raw logs, documents, semistructured data, and possibly other forms of heterogeneous data. In such cases, it may be desirable to derive meaningful features from the data. Generally, features with good semantic interpretability are more desirable because they simplify the ability of the analyst to understand intermediate results. Furthermore, they are usually better tied to the goals of the data mining application at hand. In some cases where the data is obtained from multiple sources, it needs to be integrated into a single database for processing. In addition, some algorithms may work only with a specific data type, whereas the data may contain heterogeneous types. In such cases, *data type portability* becomes

C. C. Aggarwal, *Data Mining: The Textbook*, DOI 10.1007/978-3-319-14142-8_2

important where attributes of one type are transformed to another. This results in a more homogeneous data set that can be processed by existing algorithms.

2. *Data cleaning:* In the data cleaning phase, missing, erroneous, and inconsistent entries are removed from the data. In addition, some missing entries may also be estimated by a process known as *imputation.*

3. *Data reduction, selection, and transformation:* In this phase, the size of the data is reduced through data subset selection, feature subset selection, or data transformation. The gains obtained in this phase are twofold. First, when the size of the data is reduced, the algorithms are generally more efficient. Second, if irrelevant features or irrelevant records are removed, the quality of the data mining process is improved. The first goal is achieved by generic sampling and dimensionality reduction techniques. To achieve the second goal, a highly problem-specific approach must be used for feature selection. For example, a feature selection approach that works well for clustering may not work well for classification.

Some forms of feature selection are tightly integrated with the problem at hand. Later chapters on specific problems such as clustering and classification will contain detailed discussions on feature selection.

This chapter is organized as follows. The feature extraction phase is discussed in Sect. 2.2. The data cleaning phase is covered in Sect. 2.3. The data reduction phase is explained in Sect. 2.4. A summary is given in Sect. 2.5.

2.2 Feature Extraction and Portability

The first phase of the data mining process is creating a set of features that the analyst can work with. In cases where the data is in raw and unstructured form (e.g., raw text, sensor signals), the relevant features need to be extracted for processing. In other cases where a heterogeneous mixture of features is available in different forms, an "off-the-shelf" analytical approach is often not available to process such data. In such cases, it may be desirable to transform the data into a uniform representation for processing. This is referred to as *data type porting.*

2.2.1 Feature Extraction

The first phase of feature extraction is a crucial one, though it is very application specific. In some cases, feature extraction is closely related to the concept of data type portability, where low-level features of one type may be transformed to higher-level features of another type. The nature of feature extraction depends on the domain from which the data is drawn:

1. *Sensor data:* Sensor data is often collected as large volumes of low-level signals, which are massive. The low-level signals are sometimes converted to higher-level features using wavelet or Fourier transforms. In other cases, the time series is used directly after some cleaning. The field of signal processing has an extensive literature devoted to such methods. These technologies are also useful for porting time-series data to multidimensional data.

2. *Image data:* In its most primitive form, image data are represented as pixels. At a slightly higher level, color histograms can be used to represent the features in different segments of an image. More recently, the use of *visual words* has become more

popular. This is a semantically rich representation that is similar to document data. One challenge in image processing is that the data are generally very high dimensional. Thus, feature extraction can be performed at different levels, depending on the application at hand.

3. *Web logs:* Web logs are typically represented as text strings in a prespecified format. Because the fields in these logs are clearly specified and separated, it is relatively easy to convert Web access logs into a multidimensional representation of (the relevant) categorical and numeric attributes.

4. *Network traffic:* In many intrusion-detection applications, the characteristics of the network packets are used to analyze intrusions or other interesting activity. Depending on the underlying application, a variety of features may be extracted from these packets, such as the number of bytes transferred, the network protocol used, and so on.

5. *Document data:* Document data is often available in raw and unstructured form, and the data may contain rich linguistic relations between different entities. One approach is to remove stop words, stem the data, and use a bag-of-words representation. Other methods use *entity extraction* to determine linguistic relationships.

   Named-entity recognition is an important subtask of information extraction. This approach locates and classifies atomic elements in text into predefined expressions of names of persons, organizations, locations, actions, numeric quantities, and so on. Clearly, the ability to identify such atomic elements is very useful because they can be used to understand the structure of sentences and complex events. Such an approach can also be used to populate a more conventional database of relational elements or as a sequence of atomic entities, which is more easily analyzed. For example, consider the following sentence.

   ```
   Bill Clinton lives in Chappaqua.
   ```

   Here, "Bill Clinton" is the name of a person, and "Chappaqua" is the name of a place. The word "lives" denotes an action. Each type of entity may have a different significance to the data mining process depending on the application at hand. For example, if a data mining application is mainly concerned with mentions of specific locations, then the word "Chappaqua" needs to be extracted.

   Popular techniques for named entity recognition include linguistic grammar-based techniques and statistical models. The use of grammar rules is typically very effective, but it requires work by experienced computational linguists. On the other hand, statistical models require a significant amount of training data. The techniques designed are very often domain-specific. The area of named entity recognition is vast in its own right, which is outside the scope of this book. The reader is referred to [400] for a detailed discussion of different methods for entity recognition.

Feature extraction is an art form that is highly dependent on the skill of the analyst to choose the features and their representation that are best suited to the task at hand. While this particular aspect of data analysis typically belongs to the domain expert, it is perhaps the most important one. If the correct features are not extracted, the analysis can only be as good as the available data.

2.2.2 Data Type Portability

Data type portability is a crucial element of the data mining process because the data is often heterogeneous, and may contain multiple types. For example, a demographic data set may contain both numeric and mixed attributes. A time-series data set collected from an electrocardiogram (ECG) sensor may have numerous other meta-information and text attributes associated with it. This creates a bewildering situation for an analyst who is now faced with the difficult challenge of designing an algorithm with an arbitrary combination of data types. The mixing of data types also restricts the ability of the analyst to use off-the-shelf tools for processing. Note that porting data types does lose representational accuracy and expressiveness in some cases. Ideally, it is best to customize the algorithm to the particular combination of data types to optimize results. This is, however, time-consuming and sometimes impractical.

This section will describe methods for converting between various data types. Because the numeric data type is the simplest and most widely studied one for data mining algorithms, it is particularly useful to focus on how different data types may be converted to it. However, other forms of conversion are also useful in many scenarios. For example, for similarity-based algorithms, it is possible to convert virtually any data type to a graph and apply graph-based algorithms to this representation. The following discussion, summarized in Table 2.1, will discuss various ways of transforming data across different types.

2.2.2.1 Numeric to Categorical Data: Discretization

The most commonly used conversion is from the numeric to the categorical data type. This process is known as *discretization*. The process of discretization divides the ranges of the numeric attribute into ϕ ranges. Then, the attribute is assumed to contain ϕ different categorical labeled values from 1 to ϕ, depending on the range in which the original attribute lies. For example, consider the age attribute. One could create ranges $[0, 10]$, $[11, 20]$, $[21, 30]$, and so on. The *symbolic* value for any record in the range $[11, 20]$ is "2" and the symbolic value for a record in the range $[21, 30]$ is "3". Because these are symbolic values, no ordering is assumed between the values "2" and "3". Furthermore, variations within a range are not distinguishable after discretization. Thus, the discretization process does lose some information for the mining process. However, for some applications, this loss of information is not too debilitating. One challenge with discretization is that the data may be nonuniformly distributed across the different intervals. For example, for the case of the salary attribute, a large subset of the population may be grouped in the $[40,000,\ 80,000]$ range, but very few will be grouped in the $[1,040,000,\ 1,080,000]$ range. Note that both ranges have the same size. Thus, the use of ranges of equal size may not be very helpful in discriminating between different data segments. On the other hand, many attributes, such as age, are not as nonuniformly distributed, and therefore ranges of equal size may work reasonably well. The discretization process can be performed in a variety of ways depending on application-specific goals:

1. *Equi-width ranges:* In this case, each range $[a, b]$ is chosen in such a way that $b - a$ is the same for each range. This approach has the drawback that it will not work for data sets that are distributed nonuniformly across the different ranges. To determine the actual values of the ranges, the minimum and maximum values of each attribute are determined. This range $[min, max]$ is then divided into ϕ ranges of equal length.

2. *Equi-log ranges:* Each range $[a, b]$ is chosen in such a way that $\log(b) - \log(a)$ has the same value. This kinds of range selection has the effect of geometrically increasing

Table 2.1: Portability of different data types

Source data type	Destination data type	Methods
Numeric	Categorical	Discretization
Categorical	Numeric	Binarization
Text	Numeric	Latent semantic analysis (*LSA*)
Time series	Discrete sequence	*SAX*
Time series	Numeric multidimensional	*DWT*, *DFT*
Discrete sequence	Numeric multidimensional	*DWT*, *DFT*
Spatial	Numeric multidimensional	2-d *DWT*
Graphs	Numeric multidimensional	*MDS*, spectral
Any type	Graphs	Similarity graph (Restricted applicability)

ranges $[a, a \cdot \alpha]$, $[a \cdot \alpha, a \cdot \alpha^2]$, and so on, for some $\alpha > 1$. This kind of range may be useful when the attribute shows an exponential distribution across a range. In fact, if the attribute frequency distribution for an attribute can be modeled in functional form, then a natural approach would be to select ranges $[a, b]$ such that $f(b) - f(a)$ is the same for some function $f(\cdot)$. The idea is to select this function $f(\cdot)$ in such a way that each range contains an approximately similar number of records. However, in most cases, it is hard to find such a function $f(\cdot)$ in closed form.

3. *Equi-depth ranges:* In this case, the ranges are selected so that each range has an equal number of records. The idea is to provide the same level of granularity to each range. An attribute can be divided into equi-depth ranges by first sorting it, and then selecting the division points on the sorted attribute value, such that each range contains an equal number of records.

The process of discretization can also be used to convert time-series data to discrete sequence data.

2.2.2.2 Categorical to Numeric Data: Binarization

In some cases, it is desirable to use numeric data mining algorithms on categorical data. Because binary data is a special form of both numeric and categorical data, it is possible to convert the categorical attributes to binary form and then use numeric algorithms on the binarized data. If a categorical attribute has ϕ different values, then ϕ different binary attributes are created. Each binary attribute corresponds to one possible value of the categorical attribute. Therefore, exactly one of the ϕ attributes takes on the value of 1, and the remaining take on the value of 0.

2.2.2.3 Text to Numeric Data

Although the vector-space representation of text can be considered a sparse numeric data set with very high dimensionality, this special numeric representation is not very amenable to conventional data mining algorithms. For example, one typically uses specialized similarity functions, such as the cosine, rather than the Euclidean distance for text data. This is the reason that text mining is a distinct area in its own right with its own family of specialized algorithms. Nevertheless, it is possible to convert a text collection into a form that is more amenable to the use of mining algorithms for numeric data. The first step is to use *latent semantic analysis (LSA)* to transform the text collection to a nonsparse representation with lower dimensionality. Furthermore, after transformation, each document $\overline{X} = (x_1 \ldots x_d)$ needs to be scaled to $\frac{1}{\sqrt{\sum_{i=1}^{d} x_i^2}}(x_1 \ldots x_d)$. This scaling is necessary to ensure that documents of varying length are treated in a uniform way. After this scaling, traditional numeric measures, such as the Euclidean distance, work more effectively. *LSA* is discussed in Sect. 2.4.3.3 of this chapter. Note that *LSA* is rarely used in conjunction with this kind of scaling. Rather, traditional text mining algorithms are directly applied to the reduced representation obtained from *LSA*.

2.2.2.4 Time Series to Discrete Sequence Data

Time-series data can be converted to discrete sequence data using an approach known as *symbolic aggregate approximation (SAX)*. This method comprises two steps:

1. *Window-based averaging:* The series is divided into windows of length w, and the average time-series value over each window is computed.

2. *Value-based discretization:* The (already averaged) time-series values are discretized into a smaller number of approximately *equi-depth* intervals. This is identical to the equi-depth discretization of numeric attributes that was discussed earlier. The idea is to ensure that each symbol has an approximately equal frequency in the time series. The interval boundaries are constructed by assuming that the time-series values are distributed with a Gaussian assumption. The mean and standard deviation of the (windowed) time-series values are estimated in the data-driven manner to instantiate the parameters of the Gaussian distribution. The quantiles of the Gaussian distribution are used to determine the boundaries of the intervals. This is more efficient than sorting all the data values to determine quantiles, and it may be a more practical approach for a long (or streaming) time series. The values are discretized into a small number (typically 3 to 10) of intervals for the best results. Each such equi-depth interval is mapped to a symbolic value. This creates a symbolic representation of the time series, which is essentially a discrete sequence.

Thus, *SAX* might be viewed as an equi-depth discretization approach after window-based averaging.

2.2.2.5 Time Series to Numeric Data

This particular transformation is very useful because it enables the use of multidimensional algorithms for time-series data. A common method used for this conversion is the discrete wavelet transform (*DWT*). The wavelet transform converts the time series data to multidimensional data, as a set of coefficients that represent averaged differences between different portions of the series. If desired, a subset of the largest coefficients may be used to reduce the data size. This approach will be discussed in Sect. 2.4.4.1 on data reduction. An alternative method, known as the discrete Fourier transform (*DFT*), is discussed in Sect. 14.2.4.2 of Chap. 14. The common property of these transforms is that the various coefficients are no longer as dependency oriented as the original time-series values.

2.2.2.6 Discrete Sequence to Numeric Data

This transformation can be performed in two steps. The first step is to convert the discrete sequence to a *set* of (binary) time series, where the number of time series in this set is equal to the number of distinct symbols. The second step is to map each of these time series into a multidimensional vector using the wavelet transform. Finally, the features from the different series are combined to create a single multidimensional record.

To convert a sequence to a binary time series, one can create a binary string in which the value denotes whether or not a particular symbol is present at a position. For example, consider the following nucleotide sequence, which is drawn on four symbols:

```
ACACACTGTGACTG
```

This series can be converted into the following set of four binary time series corresponding to the symbols A, C, T, and G, respectively:

```
10101000001000
01010100000100
00000010100010
00000001010001
```

A wavelet transformation can be applied to each of these series to create a multidimensional set of features. The features from the four different series can be appended to create a single numeric multidimensional record.

2.2.2.7 Spatial to Numeric Data

Spatial data can be converted to numeric data by using the same approach that was used for time-series data. The main difference is that there are now two contextual attributes (instead of one). This requires modification of the wavelet transformation method. Section 2.4.4.1 will briefly discuss how the one-dimensional wavelet approach can be generalized when there are two contextual attributes. The approach is fairly general and can be used for any number of contextual attributes.

2.2.2.8 Graphs to Numeric Data

Graphs can be converted to numeric data with the use of methods such as multidimensional scaling (*MDS*) and spectral transformations. This approach works for those applications where the edges are weighted, and represent similarity or distance relationships between nodes. The general approach of *MDS* can achieve this goal, and it is discussed in Sect. 2.4.4.2. A spectral approach can also be used to convert a graph into a multidimensional representation. This is also a dimensionality reduction scheme that converts the structural information into a multidimensional representation. This approach will be discussed in Sect. 2.4.4.3.

2.2.2.9 Any Type to Graphs for Similarity-Based Applications

Many applications are based on the notion of similarity. For example, the clustering problem is defined as the creation of groups of similar objects, whereas the outlier detection problem is defined as one in which a subset of objects differing significantly from the remaining objects are identified. Many forms of classification models, such as nearest neighbor classifiers, are also dependent on the notion of similarity. The notion of pairwise similarity can be best captured with the use of a *neighborhood graph.* For a given set of data objects $\mathcal{O} = \{O_1 \dots O_n\}$, a neighborhood graph is defined as follows:

1. A single node is defined for each object in $\mathcal{O}$. This is defined by the node set N, containing n nodes where the node i corresponds to the object O_i.

2. An edge exists between O_i and O_j, if the distance $d(O_i, O_j)$ is less than a particular threshold ϵ. Alternatively, the k-nearest neighbors of each node may be used. Because the k-nearest neighbor relationship is not symmetric, this results in a directed graph. The directions on the edges are ignored, and the parallel edges are removed. The weight w_{ij} of the edge (i, j) is equal to a kernelized function of the distance between the objects O_i and O_j, so that larger weights indicate greater similarity. An example is the *heat kernel*:
$$w_{ij} = e^{-d(O_i,O_j)^2/t^2} \tag{2.1}$$
Here, t is a user-defined parameter.

A wide variety of data mining algorithms are available for network data. All these methods can also be used on the similarity graph. Note that the similarity graph can be crisply defined for data objects of any type, as long as an appropriate distance function can be defined. This is the reason that distance function design is so important for virtually any data type. The issue of distance function design will be addressed in Chap. 3. Note that this approach is useful only for applications that are based on the notion of similarity or distances. Nevertheless, many data mining problems are directed or indirectly related to notions of similarity and distances.

2.3 Data Cleaning

The data cleaning process is important because of the errors associated with the data collection process. Several sources of missing entries and errors may arise during the data collection process. Some examples are as follows:

1. Some data collection technologies, such as sensors, are inherently inaccurate because of the hardware limitations associated with collection and transmission. Sometimes sensors may drop readings because of hardware failure or battery exhaustion.

2. Data collected using scanning technologies may have errors associated with it because optical character recognition techniques are far from perfect. Furthermore, speech-to-text data is also prone to errors.

3. Users may not want to specify their information for privacy reasons, or they may specify incorrect values intentionally. For example, it has often been observed that users sometimes specify their birthday incorrectly on automated registration sites such as those of social networks. In some cases, users may choose to leave several fields empty.

4. A significant amount of data is created manually. Manual errors are common during data entry.

5. The entity in charge of data collection may not collect certain fields for some records, if it is too costly. Therefore, records may be incompletely specified.

The aforementioned issues may be a significant source of inaccuracy for data mining applications. Methods are needed to remove or correct missing and erroneous entries from the data. There are several important aspects of data cleaning:

1. *Handling missing entries:* Many entries in the data may remain unspecified because of weaknesses in data collection or the inherent nature of the data. Such missing entries may need to be estimated. The process of estimating missing entries is also referred to as *imputation.*

2. *Handling incorrect entries:* In cases where the same information is available from multiple sources, *inconsistencies* may be detected. Such inconsistencies can be removed as a part of the analytical process. Another method for detecting the incorrect entries is to use domain-specific knowledge about what is already known about the data. For example, if a person's height is listed as 6 m, it is most likely incorrect. More generally, data points that are inconsistent with the remaining data distribution are often noisy. Such data points are referred to as *outliers.* It is, however, dangerous to assume that such data points are always caused by errors. For example, a record representing credit card fraud is likely to be inconsistent with respect to the patterns in most of the (normal) data but should not be removed as "incorrect" data.

3. *Scaling and normalization:* The data may often be expressed in very different scales (e.g., age and salary). This may result in some features being inadvertently weighted too much so that the other features are implicitly ignored. Therefore, it is important to normalize the different features.

The following sections will discuss each of these aspects of data cleaning.

2.3.1 Handling Missing Entries

Missing entries are common in databases where the data collection methods are imperfect. For example, user surveys are often unable to collect responses to all questions. In cases where data contribution is voluntary, the data is almost always incompletely specified. Three classes of techniques are used to handle missing entries:

1. Any data record containing a missing entry may be eliminated entirely. However, this approach may not be practical when most of the records contain missing entries.

2. The missing values may be estimated or imputed. However, errors created by the imputation process may affect the results of the data mining algorithm.

3. The analytical phase is designed in such a way that it can work with missing values. Many data mining methods are inherently designed to work robustly with missing values. This approach is usually the most desirable because it avoids the additional biases inherent in the imputation process.

The problem of estimating missing entries is directly related to the classification problem. In the classification problem, a single attribute is treated specially, and the other features are used to estimate its value. In this case, the missing value can occur on any feature, and therefore the problem is more challenging, although it is fundamentally not different. Many of the methods discussed in Chaps. 10 and 11 for classification can also be used for missing value estimation. In addition, the matrix completion methods discussed in Sect. 18.5 of Chap. 18 may also be used.

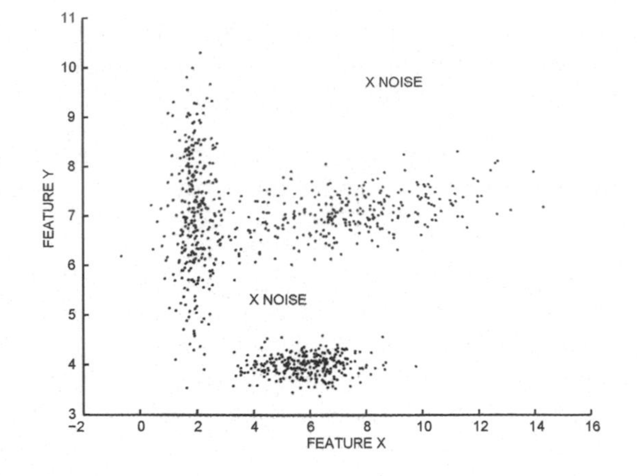

Figure 2.1: Finding noise by data-centric methods

In the case of dependency-oriented data, such as time series or spatial data, missing value estimation is much simpler. In this case, the behavioral attribute values of contextually nearby records are used for the imputation process. For example, in a time-series data set, the average of the values at the time stamp just before or after the missing attribute may be used for estimation. Alternatively, the behavioral values at the last n time-series data stamps can be linearly interpolated to determine the missing value. For the case of spatial data, the estimation process is quite similar, where the average of values at neighboring spatial locations may be used.

2.3.2 Handling Incorrect and Inconsistent Entries

The key methods that are used for removing or correcting the incorrect and inconsistent entries are as follows:

1. *Inconsistency detection:* This is typically done when the data is available from different sources in different formats. For example, a person's name may be spelled out in full in one source, whereas the other source may only contain the initials and a last name. In such cases, the key issues are duplicate detection and inconsistency detection. These topics are studied under the general umbrella of data integration within the database field.

2. *Domain knowledge:* A significant amount of domain knowledge is often available in terms of the ranges of the attributes or rules that specify the relationships across different attributes. For example, if the country field is "United States," then the city field cannot be "Shanghai." Many *data scrubbing* and *data auditing* tools have been developed that use such domain knowledge and constraints to detect incorrect entries.

3. *Data-centric methods:* In these cases, the *statistical behavior* of the data is used to detect outliers. For example, the two isolated data points in Fig. 2.1 marked as "noise" are outliers. These isolated points might have arisen because of errors in the data collection process. However, this may not always be the case because the anomalies may be the result of interesting behavior of the underlying system. Therefore, any detected outlier may need to be manually examined before it is discarded. The use of

data-centric methods for cleaning can sometimes be dangerous because they can result in the removal of useful knowledge from the underlying system. The outlier detection problem is an important analytical technique in its own right, and is discussed in detail in Chaps. 8 and 9.

The methods for addressing erroneous and inconsistent entries are generally highly domain specific.

2.3.3 Scaling and Normalization

In many scenarios, the different features represent different scales of reference and may therefore not be comparable to one another. For example, an attribute such as age is drawn on a very different scale than an attribute such as salary. The latter attribute is typically orders of magnitude larger than the former. As a result, any aggregate function computed on the different features (e.g., Euclidean distances) will be dominated by the attribute of larger magnitude.

To address this problem, it is common to use *standardization*. Consider the case where the jth attribute has mean μ_j and standard deviation σ_j. Then, the jth attribute value x_i^j of the ith record $\overline{X_i}$ may be normalized as follows:

$$z_i^j = \frac{x_i^j - \mu_j}{\sigma_j} \tag{2.2}$$

The vast majority of the normalized values will typically lie in the range $[-3, 3]$ under the normal distribution assumption.

A second approach uses *min-max scaling* to map all attributes to the range $[0, 1]$. Let min_j and max_j represent the minimum and maximum values of attribute j. Then, the jth attribute value x_i^j of the ith record $\overline{X_i}$ may be scaled as follows:

$$y_i^j = \frac{x_i^j - min_j}{max_j - min_j} \tag{2.3}$$

This approach is not effective when the maximum and minimum values are extreme value outliers because of some mistake in data collection. For example, consider the age attribute where a mistake in data collection caused an additional zero to be appended to an age, resulting in an age value of 800 years instead of 80. In this case, most of the scaled data along the age attribute will be in the range $[0, 0.1]$, as a result of which this attribute may be de-emphasized. Standardization is more robust to such scenarios.

2.4 Data Reduction and Transformation

The goal of data reduction is to represent it more compactly. When the data size is smaller, it is much easier to apply sophisticated and computationally expensive algorithms. The reduction of the data may be in terms of the number of rows (records) or in terms of the number of columns (dimensions). Data reduction does result in some loss of information. The use of a more sophisticated algorithm may sometimes compensate for the loss in information resulting from data reduction. Different types of data reduction are used in various applications:

1. *Data sampling:* The records from the underlying data are sampled to create a much smaller database. Sampling is generally much harder in the streaming scenario where the sample needs to be dynamically maintained.

2. *Feature selection:* Only a subset of features from the underlying data is used in the analytical process. Typically, these subsets are chosen in an application-specific way. For example, a feature selection method that works well for clustering may not work well for classification and vice versa. Therefore, this section will discuss the issue of feature subsetting only in a limited way and defer a more detailed discussion to later chapters.

3. *Data reduction with axis rotation:* The correlations in the data are leveraged to represent it in a smaller number of dimensions. Examples of such data reduction methods include principal component analysis (PCA), singular value decomposition (SVD), or latent semantic analysis (LSA) for the text domain.

4. *Data reduction with type transformation:* This form of data reduction is closely related to data type portability. For example, time series are converted to multidimensional data of a smaller size and lower complexity by discrete wavelet transformations. Similarly, graphs can be converted to multidimensional representations by using embedding techniques.

Each of the aforementioned aspects will be discussed in different segments of this section.

2.4.1 Sampling

The main advantage of sampling is that it is simple, intuitive, and relatively easy to implement. The type of sampling used may vary with the application at hand.

2.4.1.1 Sampling for Static Data

It is much simpler to sample data when the entire data is already available, and therefore the number of base data points is known in advance. In the unbiased sampling approach, a predefined fraction f of the data points is selected and retained for analysis. This is extremely simple to implement, and can be achieved in two different ways, depending upon whether or not replacement is used.

In sampling without replacement from a data set $\mathcal{D}$ with n records, a total of $\lceil n \cdot f \rceil$ records are randomly picked from the data. Thus, no duplicates are included in the sample, unless the original data set $\mathcal{D}$ also contains duplicates. In sampling with replacement from a data set $\mathcal{D}$ with n records, the records are sampled *sequentially and independently* from the *entire* data set $\mathcal{D}$ for a total of $\lceil n \cdot f \rceil$ times. Thus, duplicates are possible because the same record may be included in the sample over sequential selections. Generally, most applications do not use replacement because unnecessary duplicates can be a nuisance for some data mining applications, such as outlier detection. Some other specialized forms of sampling are as follows:

1. *Biased sampling:* In biased sampling, some parts of the data are intentionally emphasized because of their greater importance to the analysis. A classical example is that of temporal-decay bias where more recent records have a larger chance of being included in the sample, and stale records have a lower chance of being included. In exponential-decay bias, the probability $p(\overline{X})$ of sampling a data record $\overline{X}$, which was generated

δt time units ago, is proportional to an exponential decay function value regulated by the decay parameter λ:

$$p(\overline{X}) \propto e^{-\lambda \cdot \delta t} \tag{2.4}$$

Here e is the base of the natural logarithm. By using different values of λ, the impact of temporal decay can be regulated appropriately.

2. *Stratified sampling:* In some data sets, important parts of the data may not be sufficiently represented by sampling because of their rarity. A stratified sample, therefore, first partitions the data into a set of desired strata, and then independently samples from each of these strata based on predefined proportions in an application-specific way.

   For example, consider a survey that measures the economic diversity of the lifestyles of different individuals in the population. Even a sample of 1 million participants may not capture a billionaire because of their relative rarity. However, a stratified sample (by income) will independently sample a predefined fraction of participants from each income group to ensure greater robustness in analysis.

Numerous other forms of biased sampling are possible. For example, in density-biased sampling, points in higher-density regions are weighted less to ensure greater representativeness of the rare regions in the sample.

2.4.1.2 Reservoir Sampling for Data Streams

A particularly interesting form of sampling is that of *reservoir sampling* for data streams. In reservoir sampling, a sample of k points is *dynamically* maintained from a data stream. Recall that a stream is of an extremely large volume, and therefore one cannot store it on a disk to sample it. Therefore, for each incoming data point in the stream, one must use a set of efficiently implementable operations to *maintain* the sample.

In the static case, the probability of including a data point in the sample is k/n where k is the sample size, and n is the number of points in the "data set." In this case, the "data set" is not static and cannot be stored on disk. Furthermore, the value of n is constantly increasing as more points arrive and previous data points (outside the sample) have already been discarded. Thus, the sampling approach works with *incomplete knowledge* about the previous history of the stream at any given moment in time. In other words, for each incoming data point in the stream, we need to *dynamically* make two simple *admission control* decisions:

1. What sampling rule should be used to decide whether to include the newly incoming data point in the sample?

2. What rule should be used to decide how to eject a data point from the sample to "make room" for the newly inserted data point?

Fortunately, it is relatively simple to design an algorithm for reservoir sampling in data streams [498]. For a reservoir of size k, the first k data points in the stream are used to initialize the reservoir. Subsequently, for the nth incoming stream data point, the following two admission control decisions are applied:

1. Insert the nth incoming stream data point into the reservoir with probability k/n.

2. If the newly incoming data point was inserted, then eject one of the old k data points at random to make room for the newly arriving point.

It can be shown that the aforementioned rule maintains an unbiased reservoir sample from the data stream.

Lemma 2.4.1 *After n stream points have arrived, the probability of any stream point being included in the reservoir is the same, and is equal to k/n.*

Proof: This result is easy to show by induction. At initialization of the first k data points, the theorem is trivially true. Let us (inductively) assume that it is also true after $(n-1)$ data points have been received, and therefore the probability of each point being included in the reservoir is $k/(n-1)$. The probability of the arriving point being included in the stream is k/n, and therefore the lemma holds true for the arriving data point. It remains to prove the result for the remaining points in the data stream. There are two *disjoint* case events that can arise for an incoming data point, and the final probability of a point being included in the reservoir is the sum of these two cases:

I: The incoming data point is not inserted into the reservoir. The probability of this is $(n-k)/n$. Because the original probability of any point being included in the reservoir by the inductive assumption, is $k/(n-1)$, the overall probability of a point being included in the reservoir *and* Case I event, is the multiplicative value of $p_1 = \frac{k(n-k)}{n(n-1)}$.

II: The incoming data point is inserted into the reservoir. The probability of Case II is equal to insertion probability k/n of incoming data points. Subsequently, existing reservoir points are retained with probability $(k-1)/k$ because exactly one of them is ejected. Because the inductive assumption implies that any of the earlier points in the data stream was originally present in the reservoir with probability $k/(n-1)$, it implies that the probability of a point being included in the reservoir *and* Case II event is given by the product p_2 of the three aforementioned probabilities:

$$p_2 = \left(\frac{k}{n}\right)\left(\frac{k-1}{k}\right)\left(\frac{k}{n-1}\right) = \frac{k(k-1)}{n(n-1)} \tag{2.5}$$

Therefore, the total probability of a stream point being retained in the reservoir after the nth data point arrival is given by the sum of p_1 and p_2. It can be shown that this is equal to k/n. ■

It is possible to extend reservoir sampling to cases where temporal bias is present in the data stream. In particular, the case of exponential bias has been addressed in [35].

2.4.2 Feature Subset Selection

A second method for data preprocessing is feature subset selection. Some features can be discarded when they are known to be irrelevant. Which features are relevant? Clearly, this decision depends on the application at hand. There are two primary types of feature selection:

1. *Unsupervised feature selection:* This corresponds to the removal of noisy and redundant attributes from the data. Unsupervised feature selection is best defined in terms of its impact on clustering applications, though the applicability is much broader. It is difficult to comprehensively describe such feature selection methods without using the clustering problem as a proper context. Therefore, a discussion of methods for unsupervised feature selection is deferred to Chap. 6 on data clustering.

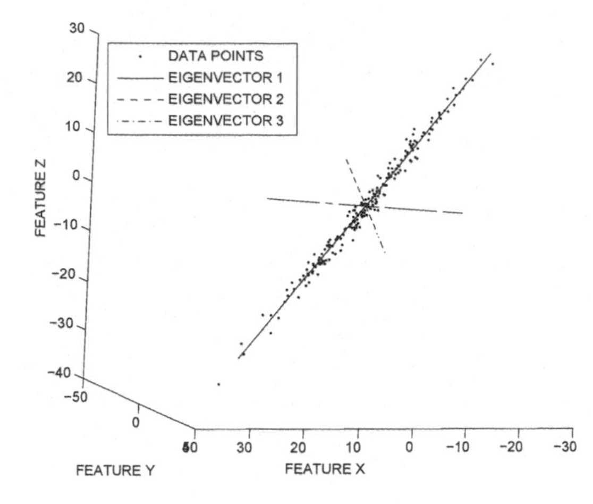

Figure 2.2: Highly correlated data represented in a small number of dimensions in an axis system that is rotated appropriately

2. *Supervised feature selection:* This type of feature selection is relevant to the problem of data classification. In this case, only the features that can predict the class attribute effectively are the most relevant. Such feature selection methods are often closely integrated with analytical methods for classification. A detailed discussion is deferred to Chap. 10 on data classification.

Feature selection is an important part of the data mining process because it defines the quality of the input data.

2.4.3 Dimensionality Reduction with Axis Rotation

In real data sets, a significant number of correlations exist among different attributes. In some cases, hard constraints or rules between attributes may uniquely define some attributes in terms of others. For example, the date of birth of an individual (represented quantitatively) is perfectly correlated with his or her age. In most cases, the correlations may not be quite as perfect, but significant dependencies may still exist among the different features. Unfortunately, real data sets contain many such redundancies that escape the attention of the analyst during the initial phase of data creation. These correlations and constraints correspond to implicit redundancies because they imply that knowledge of some subsets of the dimensions can be used to predict the values of the other dimensions. For example, consider the 3-dimensional data set illustrated in Fig. 2.2. In this case, if the axis is rotated to the orientation illustrated in the figure, the correlations and redundancies in the newly transformed feature values are removed. As a result of this redundancy removal, the entire data can be (approximately) represented along a 1-dimensional line. Thus, the *intrinsic dimensionality* of this 3-dimensional data set is 1. The other two axes correspond to the low-variance dimensions. If the data is represented as coordinates in the new axis system illustrated in Fig. 2.2, then the coordinate values along these low-variance dimensions will not vary much. Therefore, after the axis system has been rotated, these dimensions can be removed without much information loss.

A natural question arises as to how the correlation-removing axis system such as that in Fig. 2.2 may be determined in an automated way. Two natural methods to achieve this goal are those of *principal component analysis (PCA)* and *singular value decomposition (SVD)*. These two methods, while not exactly identical at the definition level, are closely related. Although the notion of principal component analysis is intuitively easier to understand, *SVD* is a more general framework and can be used to perform *PCA* as a special case.

2.4.3.1 Principal Component Analysis

PCA is generally applied after subtracting the mean of the data set from each data point. However, it is also possible to use it without mean centering, as long as the mean of the data is separately stored. This operation is referred to as *mean centering*, and it results in a data set centered at the origin. The goal of *PCA* is to rotate the data into an axis-system where the greatest amount of variance is captured in a small number of dimensions. It is intuitively evident from the example of Fig. 2.2 that such an axis system is affected by the correlations between attributes. An important observation, which we will show below, is that the variance of a data set along a particular direction can be expressed directly in terms of its covariance matrix.

Let C be the $d \times d$ symmetric covariance matrix of the $n \times d$ data matrix D. Thus, the (i, j)th entry c_{ij} of C denotes the covariance between the ith and jth columns (dimensions) of the data matrix D. Let μ_i represent the mean along the ith dimension. Specifically, if x_k^m be the mth dimension of the kth record, then the value of the covariance entry c_{ij} is as follows:

$$c_{ij} = \frac{\sum_{k=1}^{n} x_k^i x_k^j}{n} - \mu_i \mu_j \quad \forall i, j \in \{1 \dots d\} \tag{2.6}$$

Let $\overline{\mu} = (\mu_1 \dots \mu_d)$ is the d-dimensional row vector representing the means along the different dimensions. Then, the aforementioned $d \times d$ computations of Eq. 2.6 for different values of i and j can be expressed compactly in $d \times d$ matrix form as follows:

$$C = \frac{D^T D}{n} - \overline{\mu}^T \overline{\mu} \tag{2.7}$$

Note that the d diagonal entries of the matrix C correspond to the d variances. The covariance matrix C is positive semi-definite, because it can be shown that for any d-dimensional column vector $\overline{v}$, the value of $\overline{v}^T C \overline{v}$ is equal to the variance of the 1-dimensional projection $D\overline{v}$ of the data set D on $\overline{v}$.

$$\overline{v}^T C \overline{v} = \frac{(D\overline{v})^T D\overline{v}}{n} - (\overline{\mu}\,\overline{v})^2 = \text{Variance of 1-dimensional points in } D\overline{v} \geq 0 \tag{2.8}$$

In fact, the goal of *PCA* is to successively determine orthonormal vectors $\overline{v}$ maximizing $\overline{v}^T C \overline{v}$. How can one determine such directions? Because the covariance matrix is symmetric and positive semidefinite, it can be diagonalized as follows:

$$C = P \Lambda P^T \tag{2.9}$$

The columns of the matrix P contain the orthonormal eigenvectors of C, and Λ is a diagonal matrix containing the nonnegative eigenvalues. The entry Λ_{ii} is the eigenvalue corresponding to the ith eigenvector (or column) of the matrix P. These eigenvectors represent successive orthogonal solutions[1] to the aforementioned optimization model maximizing the variance $\overline{v}^T C \overline{v}$ along the unit direction $\overline{v}$.

[1] Setting the gradient of the Lagrangian relaxation $\overline{v}^T C \overline{v} - \lambda(||\overline{v}||^2 - 1)$ to 0 is equivalent to the eigenvector condition $C\overline{v} - \lambda\overline{v} = 0$. The variance along an eigenvector is $\overline{v}^T C \overline{v} = \overline{v}^T \lambda \overline{v} = \lambda$. Therefore, one should include the orthonormal eigenvectors in decreasing order of eigenvalue λ to maximize preserved variance in reduced subspace.

An interesting property of this diagonalization is that both the eigenvectors and eigenvalues have a geometric interpretation in terms of the underlying data distribution. Specifically, if the axis system of data representation is rotated to the orthonormal set of eigenvectors in the columns of P, then it can be shown that all $\binom{d}{2}$ covariances of the newly transformed feature values are zero. In other words, *the greatest variance-preserving directions are also the correlation-removing directions.* Furthermore, the eigenvalues represent the variances of the data along the corresponding eigenvectors. In fact, the diagonal matrix Λ is the new covariance matrix after axis rotation. Therefore, eigenvectors with large eigenvalues preserve greater variance, and are also referred to as *principal components.* Because of the nature of the optimization formulation used to derive this transformation, a new axis system containing only the eigenvectors with the largest eigenvalues is optimized to *retaining the maximum variance in a fixed number of dimensions.* For example, the scatter plot of Fig. 2.2 illustrates the various eigenvectors, and it is evident that the eigenvector with the largest variance is all that is needed to create a variance-preserving representation. It generally suffices to retain only a small number of eigenvectors with large eigenvalues.

Without loss of generality, it can be assumed that the columns of P (and corresponding diagonal matrix Λ) are arranged from left to right in such a way that they correspond to decreasing eigenvalues. Then, the transformed data matrix D' in the new coordinate system after axis rotation to the orthonormal columns of P can be algebraically computed as the following linear transformation:

$$D' = DP \tag{2.10}$$

While the transformed data matrix D' is also of size $n \times d$, only its first (leftmost) $k \ll d$ columns will show significant variation in values. Each of the remaining $(d-k)$ columns of D' will be approximately equal to the mean of the data in the rotated axis system. For mean-centered data, the values of these $(d-k)$ columns will be almost 0. Therefore, the dimensionality of the data can be reduced, and only the first k columns of the transformed data matrix D' may need to be retained[2] for representation purposes. Furthermore, it can be confirmed that the covariance matrix of the transformed data $D' = DP$ is the diagonal matrix Λ by applying the covariance definition of Eq. 2.7 to DP (transformed data) and $\overline{\mu}P$ (transformed mean) instead of D and $\overline{\mu}$, respectively. The resulting covariance matrix can be expressed in terms of the original covariance matrix C as P^TCP. Substituting $C = P\Lambda P^T$ from Eq. 2.9 shows equivalence because $P^TP = PP^T = I$. In other words, correlations have been removed from the transformed data because Λ is diagonal.

The variance of the data set defined by projections along top-k eigenvectors is equal to the sum of the k corresponding eigenvalues. In many applications, the eigenvalues show a precipitous drop-off after the first few values. For example, the behavior of the eigenvalues for the 279-dimensional *Arrythmia* data set from the *UCI Machine Learning Repository* [213] is illustrated in Fig. 2.3. Figure 2.3a shows the absolute magnitude of the eigenvalues in increasing order, whereas Fig. 2.3b shows the total amount of variance retained in the top-k eigenvalues. Figure 2.3b can be derived by using the cumulative sum of the smallest eigenvalues in Fig. 2.3a. It is interesting to note that the 215 smallest eigenvalues contain less than 1 % of the total variance in the data and can therefore be removed with little change to the results of similarity-based applications. Note that the *Arrythmia* data set is not a very strongly correlated data set along many pairs of dimensions. Yet, the dimensionality reduction is drastic because of the *cumulative* effect of the correlations across many dimensions.

---

[2]The means of the remaining columns also need be stored if the data set is not mean centered.

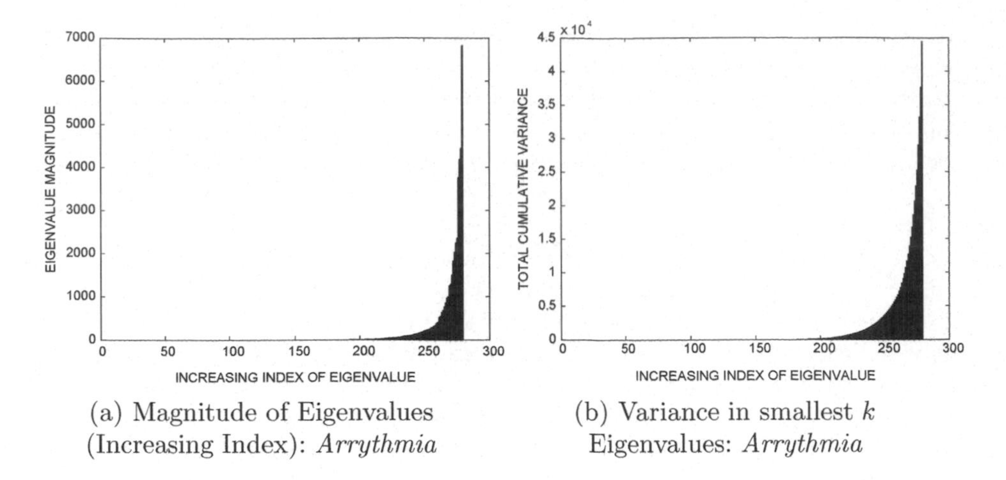

(a) Magnitude of Eigenvalues (Increasing Index): *Arrythmia*

(b) Variance in smallest k Eigenvalues: *Arrythmia*

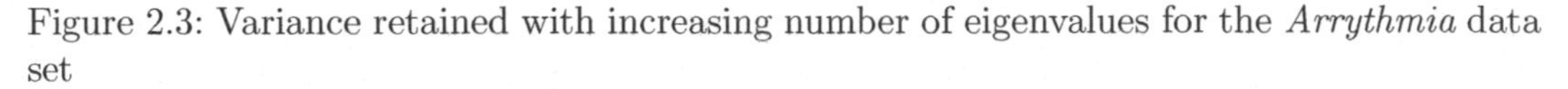

Figure 2.3: Variance retained with increasing number of eigenvalues for the *Arrythmia* data set

The eigenvectors of the matrix C may be determined by using any numerical method discussed in [295] or by an off-the-shelf eigenvector solver. *PCA* can be extended to discovering nonlinear embeddings with the use of a method known as the *kernel trick*. Refer to Sect. 10.6.4.1 of Chap. 10 for a brief description of kernel *PCA*.

2.4.3.2 Singular Value Decomposition

Singular value decomposition (*SVD*) is closely related to principal component analysis (*PCA*). However, these distinct methods are sometimes confused with one another because of the close relationship. Before beginning the discussion of *SVD*, we state how it is related to *PCA*. *SVD* is more general than *PCA* because it provides *two* sets of basis vectors instead of one. *SVD* provides basis vectors of both the rows and columns of the data matrix, whereas *PCA* only provides basis vectors of the rows of the data matrix. Furthermore, *SVD* provides the same basis as *PCA* for the rows of the data matrix in certain special cases:

SVD provides the same basis vectors and data transformation as PCA for data sets in which the mean of each attribute is 0.

The basis vectors of *PCA* are invariant to mean-translation, whereas those of *SVD* are not. When the data are not mean centered, the basis vectors of *SVD* and *PCA* will not be the same, and qualitatively different results may be obtained. *SVD* is often applied without mean centering to sparse nonnegative data such as document-term matrices. A formal way of defining *SVD* is as a decomposable product of (or *factorization* into) three matrices:

$$D = Q\Sigma P^T \tag{2.11}$$

Here, Q is an $n \times n$ matrix with orthonormal columns, which are the *left singular vectors*. Σ is an $n \times d$ diagonal matrix containing the *singular values*, which are always nonnegative and, by convention, arranged in nonincreasing order. Furthermore, P is a $d \times d$ matrix with orthonormal columns, which are the *right singular vectors*. Note that the diagonal matrix Σ is rectangular rather than square, but it is referred to as diagonal because only entries of the form Σ_{ii} are nonzero. It is a fundamental fact of linear algebra that such a decomposition always exists, and a proof may be found in [480]. The number of nonzero diagonal entries of Σ is equal to the rank of the matrix D, which is at most $\min\{n, d\}$. Furthermore, because of the orthonormality of the singular vectors, both $P^T P$ and $Q^T Q$ are identity matrices. We make the following observations:

1. The columns of matrix Q, which are also the left singular vectors, are the orthonormal eigenvectors of DD^T. This is because $DD^T = Q\Sigma(P^T P)\Sigma^T Q^T = Q\Sigma\Sigma^T Q^T$. Therefore, the square of the nonzero singular values, which are diagonal entries of the $n \times n$ diagonal matrix $\Sigma\Sigma^T$, represent the nonzero eigenvalues of DD^T.

2. The columns of matrix P, which are also the right singular vectors, are the orthonormal eigenvectors of $D^T D$. The square of the nonzero singular values, which are represented in diagonal entries of the $d \times d$ diagonal matrix $\Sigma^T\Sigma$, are the nonzero eigenvalues of $D^T D$. Note that the nonzero eigenvalues of DD^T and $D^T D$ are the same. The matrix P is particularly important because it provides the basis vectors, which are analogous to the eigenvectors of the covariance matrix in PCA.

3. Because the covariance matrix of mean-centered data is $\frac{D^T D}{n}$ (cf. Eq. 2.7) and the right singular vectors of SVD are eigenvectors of $D^T D$, it follows that the eigenvectors of PCA are the same as the right-singular vectors of SVD for mean-centered data. Furthermore, the squared singular values in SVD are n times the eigenvalues of PCA. This equivalence shows why SVD and PCA can provide the same transformation for mean-centered data.

4. Without loss of generality, it can be assumed that the diagonal entries of Σ are arranged in decreasing order, and the columns of matrix P and Q are also ordered accordingly. Let P_k and Q_k be the truncated $d \times k$ and $n \times k$ matrices obtained by selecting the first k columns of P and Q, respectively. Let Σ_k be the $k \times k$ square matrix containing the top k singular values. Then, the SVD factorization yields an *approximate* d-dimensional data representation of the original data set D:

$$D \approx Q_k \Sigma_k P_k^T \tag{2.12}$$

The columns of P_k represent a k-dimensional basis system for a reduced representation of the data set. The dimensionality reduced data set in this k-dimensional basis system is given by the $n \times k$ data set $D'_k = DP_k = Q_k\Sigma_k$, as in Eq. 2.10 of PCA. Each of the n rows of D'_k contain the k coordinates of each transformed data point in this new axis system. Typically, the value of k is much smaller than both n and d. Furthermore, unlike PCA, the rightmost $(d-k)$ columns of the full d-dimensional transformed data matrix $D' = DP$ will be approximately 0 (rather than the data mean), whether the data are mean centered or not. In general, PCA projects the data on a low-dimensional hyperplane passing through the data mean, whereas SVD projects the data on a low-dimensional hyperplane passing through the origin. PCA captures as much of the variance (or, squared Euclidean distance about the *mean*) of the data as possible, whereas SVD captures as much of the aggregate squared Euclidean distance about the *origin* as possible. This method of approximating a data matrix is referred to as *truncated* SVD.

In the following, we will show that truncated SVD maximizes the aggregate squared Euclidean distances (or *energy*) of the transformed data points about the origin. Let $\overline{v}$ be a

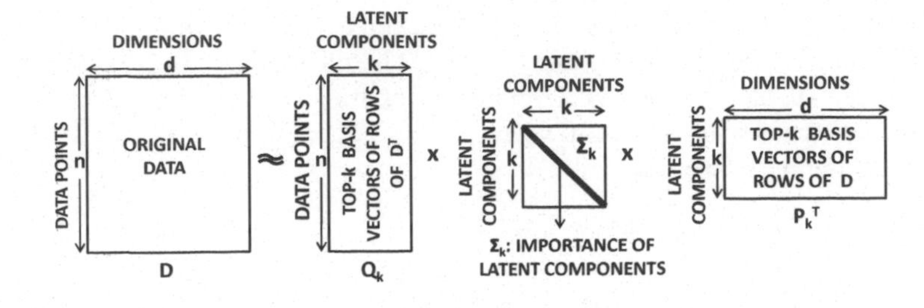

Figure 2.4: Complementary basis properties of matrix factorization in *SVD*

d-dimensional column vector and $D\overline{v}$ be the projection of the data set D on $\overline{v}$. Consider the problem of determining the *unit* vector $\overline{v}$ such that the sum of squared Euclidean distances $(D\overline{v})^T(D\overline{v})$ of the projected data points from the origin is maximized. Setting the gradient of the Lagrangian relaxation $\overline{v}^T D^T D\overline{v} - \lambda(||\overline{v}||^2 - 1)$ to 0 is equivalent to the eigenvector condition $D^T D\overline{v} - \lambda\overline{v} = 0$. Because the right singular vectors are eigenvectors of $D^T D$, it follows that the eigenvectors (right singular vectors) with the k largest eigenvalues (squared singular values) provide a basis that maximizes the preserved energy in the transformed and reduced data matrix $D'_k = DP_k = Q_k\Sigma_k$. Because the *energy*, which is the sum of squared Euclidean distances from the origin, is invariant to axis rotation, the energy in D'_k is the same as that in $D'_k P_k^T = Q_k\Sigma_k P_k^T$. Therefore, *$k$-rank SVD is a maximum energy-preserving factorization.* This result is known as the *Eckart–Young theorem.*

The total preserved energy of the projection $D\overline{v}$ of the data set D along unit right-singular vector $\overline{v}$ with singular value σ is given by $(D\overline{v})^T(D\overline{v})$, which can be simplified as follows:

$$(D\overline{v})^T(D\overline{v}) = \overline{v}^T(D^T D\overline{v}) = \overline{v}^T(\sigma^2\overline{v}) = \sigma^2$$

Because the energy is defined as a linearly separable sum along orthonormal directions, the preserved energy in the data projection along the top-k singular vectors is equal to the sum of the squares of the top-k singular values. Note that the total energy in the data set D is always equal to the sum of the squares of all the nonzero singular values. It can be shown that maximizing the preserved energy is the same as minimizing the squared error[3] (or *lost* energy) of the k-rank approximation. This is because the sum of the energy in the preserved subspace and the lost energy in the complementary (discarded) subspace is always a constant, which is equal to the energy in the original data set D.

When viewed purely in terms of eigenvector analysis, *SVD* provides two different perspectives for understanding the transformed and reduced data. The transformed data matrix can either be viewed as the *projection* DP_k of the data matrix D on the top k basis eigenvectors P_k of the $d \times d$ *scatter matrix* $D^T D$, or it can *directly* be viewed as the scaled eigenvectors $Q_k\Sigma_k = DP_k$ of the $n \times n$ *dot-product similarity matrix* DD^T. While it is generally computationally expensive to extract the eigenvectors of an $n \times n$ similarity matrix, such an approach also generalizes to nonlinear dimensionality reduction methods where notions of linear basis vectors do not exist in the original space. In such cases, the dot-product similarity matrix is replaced with a more complex similarity matrix in order to extract a nonlinear embedding (cf. Table 2.3).

SVD is more general than *PCA* and can be used to simultaneously determine a subset of k basis vectors for the data matrix and its transpose with the maximum energy. The latter can be useful in understanding complementary transformation properties of D^T.

[3]The squared error is the sum of squares of the entries in the error matrix $D - Q_k\Sigma_k P_k^T$.

The orthonormal columns of Q_k provide a k-dimensional basis system for (approximately) transforming "data points" corresponding to the rows of D^T, and the matrix $D^T Q_k = P_k \Sigma_k$ contains the corresponding coordinates. For example, in a user-item ratings matrix, one may wish to determine either a reduced representation of the users, or a reduced representation of the items. SVD provides the basis vectors for both reductions. Truncated SVD expresses the data in terms of k dominant *latent components.* The ith latent component is expressed in the ith basis vectors of both D and D^T, and its relative importance in the data is defined by the ith singular value. By decomposing the matrix product $Q_k \Sigma_k P_k^T$ into column vectors of Q_k and P_k (i.e., dominant basis vectors of D^T and D), the following additive sum of the k latent components can be obtained:

$$Q_k \Sigma_k P_k^T = \sum_{i=1}^{k} \overline{q_i} \sigma_i \overline{p_i}^T = \sum_{i=1}^{k} \sigma_i (\overline{q_i}\, \overline{p_i}^T) \tag{2.13}$$

Here $\overline{q_i}$ is the ith column of Q, $\overline{p_i}$ is the ith column of P, and σ_i is the ith diagonal entry of Σ. Each latent component $\sigma_i(\overline{q_i}\, \overline{p_i}^T)$ is an $n \times d$ matrix with rank 1 and energy σ_i^2. This decomposition is referred to as *spectral decomposition.* The relationships of the reduced basis vectors to SVD matrix factorization are illustrated in Fig. 2.4.

An example of a rank-2 truncated SVD of a toy 6×6 matrix is illustrated below:

$$D = \begin{pmatrix} 2 & 2 & 1 & 2 & 0 & 0 \\ 2 & 3 & 3 & 3 & 0 & 0 \\ 1 & 1 & 1 & 1 & 0 & 0 \\ 2 & 2 & 2 & 3 & 1 & 1 \\ 0 & 0 & 0 & 1 & 1 & 1 \\ 0 & 0 & 0 & 2 & 1 & 2 \end{pmatrix} \approx Q_2 \Sigma_2 P_2^T$$

$$\approx \begin{pmatrix} -0.41 & 0.17 \\ -0.65 & 0.31 \\ -0.23 & 0.13 \\ -0.56 & -0.20 \\ -0.10 & -0.46 \\ -0.19 & -0.78 \end{pmatrix} \begin{pmatrix} 8.4 & 0 \\ 0 & 3.3 \end{pmatrix} \begin{pmatrix} -0.41 & -0.49 & -0.44 & -0.61 & -0.10 & -0.12 \\ 0.21 & 0.31 & 0.26 & -0.37 & -0.44 & -0.68 \end{pmatrix}$$

$$= \begin{pmatrix} 1.55 & 1.87 & \underline{1.67} & 1.91 & 0.10 & 0.04 \\ 2.46 & 2.98 & 2.66 & 2.95 & 0.10 & -0.03 \\ 0.89 & 1.08 & 0.96 & 1.04 & 0.01 & -0.04 \\ 1.81 & 2.11 & 1.91 & 3.14 & 0.77 & 1.03 \\ 0.02 & -0.05 & -0.02 & 1.06 & 0.74 & 1.11 \\ 0.10 & -0.02 & 0.04 & 1.89 & 1.28 & 1.92 \end{pmatrix}$$

Note that the rank-2 matrix is a good approximation of the original matrix. The entry with the largest error is underlined in the final approximated matrix. Interestingly, this entry is also *inconsistent* with the structure of the remaining matrix in the *original* data (why?). Truncated SVD often tries to correct inconsistent entries, and this property is sometimes leveraged for noise reduction in error-prone data sets.

2.4.3.3 Latent Semantic Analysis

Latent semantic analysis (LSA) is an application of the SVD method to the text domain. In this case, the data matrix D is an $n \times d$ document-term matrix containing normalized word frequencies in the n documents, where d is the size of the lexicon. No mean centering is used, but the results are approximately the same as PCA because of the sparsity of D. The sparsity of D implies that most of the entries in D are 0, and the mean values of each column are much smaller than the nonzero values. In such scenarios, it can be shown that the covariance matrix is approximately proportional to $D^T D$. The sparsity of the data set also results in a low intrinsic dimensionality. Therefore, in the text domain, the reduction in dimensionality from LSA is rather drastic. For example, it is not uncommon to be able to represent a corpus drawn on a lexicon of 100,000 dimensions in fewer than 300 dimensions.

LSA is a classical example of how the "loss" of information from discarding some dimensions can actually result in an *improvement* in the quality of the data representation. The text domain suffers from two main problems corresponding to *synonymy* and *polysemy*. Synonymy refers to the fact that two words may have the same meaning. For example, the words "*comical*" and "*hilarious*" mean approximately the same thing. Polysemy refers to the fact that the same word may mean two different things. For example, the word "*jaguar*" could refer to a car or a cat. Typically, the significance of a word can only be understood in the context of other words in the document. This is a problem for similarity-based applications because the computation of similarity with the use of word frequencies may not be completely accurate. For example, two documents containing the words "*comical*" and "*hilarious,*" respectively, may not be deemed sufficiently similar in the original representation space. The two aforementioned issues are a direct result of synonymy and polysemy effects. The truncated representation after LSA typically removes the noise effects of synonymy and polysemy because the (high-energy) singular vectors represent the directions of correlation in the data, and the appropriate context of the word is implicitly represented along these directions. The variations because of individual differences in usage are implicitly encoded in the low-energy directions, which are truncated anyway. It has been observed that significant *qualitative* improvements [184, 416] for text applications may be achieved with the use of LSA. The improvement[4] is generally greater in terms of synonymy effects than polysemy. This noise-removing behavior of SVD has also been demonstrated in general multidimensional data sets [25].

2.4.3.4 Applications of PCA and SVD

Although PCA and SVD are primarily used for data reduction and compression, they have many other applications in data mining. Some examples are as follows:

1. *Noise reduction:* While removal of the smaller eigenvectors/singular vectors in PCA and SVD can lead to information loss, it can also lead to *improvement* in the quality of data representation in surprisingly many cases. The main reason is that the variations along the small eigenvectors are often the result of noise, and their removal is generally beneficial. An example is the application of LSA in the text domain where the removal of the smaller components leads to the enhancement of the semantic characteristics of text. SVD is also used for deblurring noisy images. These text- and image-specific results have also been shown to be true in arbitrary data domains [25]. Therefore, the data reduction is not just space efficient but actually provides *qualitative* benefits in many cases.

[4]Concepts that are not present predominantly in the collection will be ignored by truncation. Therefore, alternative meanings reflecting infrequent concepts in the collection will be ignored. While this has a robust effect on the *average*, it may not always be the correct or complete disambiguation of polysemous words.

2. *Data imputation: SVD* and *PCA* can be used for data imputation applications [23], such as collaborative filtering, because the *reduced* matrices Q_k, Σ_k, and P_k can be estimated for small values of k even from incomplete data matrices. Therefore, the entire matrix can be approximately reconstructed as $Q_k \Sigma_k P_k^T$. This application is discussed in Sect. 18.5 of Chap. 18.

3. *Linear equations:* Many data mining applications are optimization problems in which the solution is recast into a system of linear equations. For any linear system $A\overline{y} = 0$, any right singular vector of A with 0 singular value will satisfy the system of equations (see Exercise 14). Therefore, any linear combination of the 0 singular vectors will provide a solution.

4. *Matrix inversion: SVD* can be used for the inversion of a square $d \times d$ matrix D. Let the decomposition of D be given by $Q\Sigma P^T$. Then, the inverse of D is $D^{-1} = P\Sigma^{-1}Q^T$. Note that Σ^{-1} can be trivially computed from Σ by inverting its diagonal entries. The approach can also be generalized to the determination of the *Moore–Penrose pseudoinverse* D^+ of a rank-k matrix D by inverting only the nonzero diagonal entries of Σ. The approach can even be generalized to non-square matrices by performing the additional operation of transposing Σ. Such matrix inversion operations are required in many data mining applications such as least-squares regression (cf. Sect. 11.5 of Chap. 11) and social network analysis (cf. Chap. 19).

5. *Matrix algebra:* Many network mining applications require the application of algebraic operations such as the computation of the powers of a matrix. This is common in random-walk methods (cf. Chap. 19), where the kth powers of the symmetric adjacency matrix of an undirected network may need to be computed. Such symmetric adjacency matrices can be decomposed into the form $Q\Delta Q^T$. The kth power of this decomposition can be efficiently computed as $D^k = Q\Delta^k Q^T$. In fact, any polynomial function of the matrix can be computed efficiently.

SVD and *PCA* are extraordinarily useful because matrix and linear algebra operations are ubiquitous in data mining. *SVD* and *PCA* facilitate such matrix operations by providing convenient decompositions and basis representations. *SVD* has rightly been referred to [481] as "absolutely a high point of linear algebra."

2.4.4 Dimensionality Reduction with Type Transformation

In these methods, dimensionality reduction is coupled with type transformation. In most cases, the data is *transformed* from a more complex type to a less complex type, such as multidimensional data. Thus, these methods serve the dual purpose of data reduction and type portability. This section will study two such transformation methods:

1. *Time series to multidimensional:* A number of methods, such as the discrete Fourier transform and discrete wavelet transform are used. While these methods can also be viewed as a rotation of an axis system defined by the various time stamps of the contextual attribute, the data are no longer dependency oriented after the rotation. Therefore, the resulting data set can be processed in a similar way to multidimensional data. We will study the Haar wavelet transform because of its intuitive simplicity.

2. *Weighted graphs to multidimensional:* Multidimensional scaling and spectral methods are used to embed weighted graphs in multidimensional spaces, so that the similarity or distance values on the edges are captured by a multidimensional embedding.

Table 2.2: An example of wavelet coefficient computation

Granularity (order k)	Averages (Φ values)	DWT coefficients (ψ values)
$k = 4$	(8, 6, 2, 3, 4, 6, 6, 5)	–
$k = 3$	(7, 2.5, 5, 5.5)	(1, −0.5, −1, 0.5)
$k = 2$	(4.75, 5.25)	(2.25, −0.25)
$k = 1$	(5)	(−0.25)

This section will discuss each of these techniques.

2.4.4.1 Haar Wavelet Transform

Wavelets are a well-known technique that can be used for multigranularity decomposition and summarization of time-series data into the multidimensional representation. The *Haar* wavelet is a particularly popular form of wavelet decomposition because of its intuitive nature and ease of implementation. To understand the intuition behind wavelet decomposition, an example of sensor temperatures will be used.

Suppose that a sensor measured the temperatures over the course of 12 h from the morning until the evening. Assume that the sensor samples temperatures at the rate of 1 sample/s. Thus, over the course of a single day, a sensor will collect $12 \times 60 \times 60 = 43,200$ readings. Clearly, this will not scale well over many days and many sensors. An important observation is that many adjacent sensor readings will be very similar, causing this representation to be very wasteful. So, how can we represent this data approximately in a small amount of space? How can we determine the key regions where "variations" in readings occur, and store these variations instead of repeating values?

Suppose we only stored the average over the entire day. This provides some idea of the temperature but not much else about the variation over the day. Now, if the difference in average temperature between the first half and second half of the day is also stored, we can derive the averages for both the first and second half of the day from these two values. This principle can be applied recursively because the first half of the day can be divided into the first quarter of the day and the second quarter of the day. Thus, with four stored values, we can perfectly reconstruct the averages in four quarters of the day. This process can be applied recursively right down to the level of granularity of the sensor readings. These "difference values" are used to derive wavelet coefficients. Of course, we did not yet achieve any data *reduction* because the number of such coefficients can be shown to be exactly equal to the length of the original time series.

It is important to understand that large difference values tell us more about the *variations* in the temperature values than the small ones, and they are therefore more important to store. Therefore, larger coefficient values are stored after a normalization for the level of granularity. This normalization, which is discussed later, has a bias towards storing coefficients representing longer time scales because trends over longer periods of time are more informative for (global) series reconstruction.

More formally, the wavelet technique creates a decomposition of the time series into a set of coefficient-weighted wavelet basis vectors. Each of the coefficients represents the rough variation of the time series between the two halves of a particular time range. The

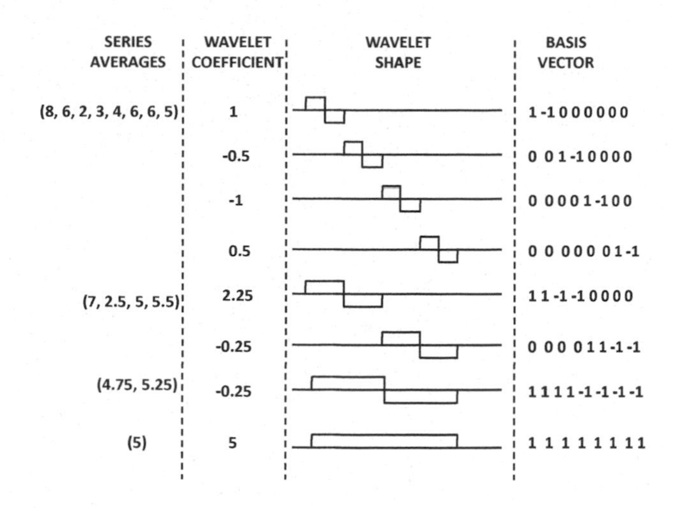

Figure 2.5: Illustration of the wavelet decomposition

wavelet basis vector is a time series that represents the temporal range of this variation in the form of a simple step function. The wavelet coefficients are of different *orders*, depending on the length of the time-series segment analyzed, which also represents the granularity of analysis. The higher-order coefficients represent the broad trends in the series because they correspond to larger ranges. The more localized trends are captured by the lower-order coefficients. Before providing a more notational description, a simple recursive description of wavelet decomposition of a time series segment S is provided below in two steps:

1. Report half the average difference of the behavioral attribute values between the first and second temporal halves of S as a wavelet coefficient.

2. Recursively apply this approach to first and second temporal halves of S.

At the end of the process, a reduction process is performed, where larger (normalized) coefficients are retained. This normalization step will be described in detail later.

A more formal and notation-intensive description will be provided at this point. For ease in discussion, assume that the length q of the series is a power of 2. For each value of $k \geq 1$, the Haar wavelet decomposition defines 2^{k-1} coefficients of order k. Each of these 2^{k-1} coefficients corresponds to a contiguous portion of the time series of length $q/2^{k-1}$. The ith of these 2^{k-1} coefficients corresponds to the segment in the series starting from position $(i-1) \cdot q/2^{k-1} + 1$ to the position $i \cdot q/2^{k-1}$. Let us denote this coefficient by ψ_k^i and the corresponding time-series segment by S_k^i. At the same time, let us define the average value of the first half of the S_k^i by a_k^i and that of the second half by b_k^i. Then, the value of ψ_k^i is given by $(a_k^i - b_k^i)/2$. More formally, if Φ_k^i denote the average value of the S_k^i, then the value of ψ_k^i can be defined recursively as follows:

$$\psi_k^i = (\Phi_{k+1}^{2 \cdot i-1} - \Phi_{k+1}^{2 \cdot i})/2 \tag{2.14}$$

The set of Haar coefficients is defined by all the coefficients of order 1 to $\log_2(q)$. In addition, the global average Φ_1^1 is required for the purpose of perfect reconstruction. The total number of coefficients is exactly equal to the length of the original series, and the dimensionality reduction is obtained by discarding the smaller (normalized) coefficients. This will be discussed later.

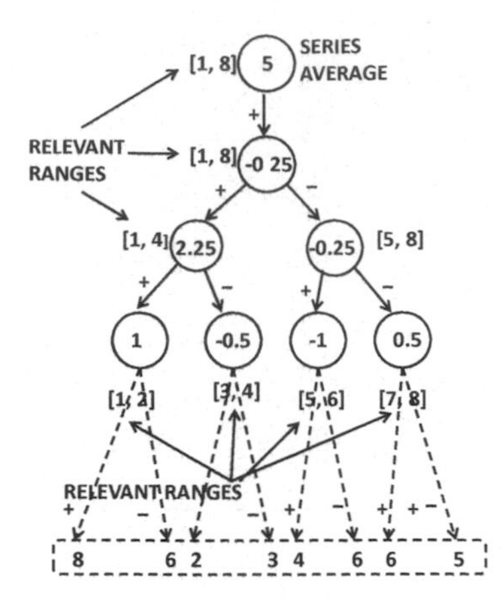

Figure 2.6: The error tree from the wavelet decomposition

The coefficients of different orders provide an understanding of the major trends in the data at a particular level of granularity. For example, the coefficient ψ_k^i is half the quantity by which the first half of the segment S_k^i is larger than the second half of the same segment. Because larger values of k correspond to geometrically reducing segment sizes, one can obtain an understanding of the basic trends at different levels of granularity. This definition of the Haar wavelet makes it very easy to compute by a sequence of averaging and differencing operations. Table 2.2 shows the computation of the wavelet coefficients for the sequence $(8, 6, 2, 3, 4, 6, 6, 5)$. This decomposition is illustrated in graphical form in Fig. 2.5. Note that each value in the original series can be represented as a sum of $\log_2(8) = 3$ wavelet coefficients with a positive or negative sign attached in front. In general, the entire decomposition may be represented as a tree of depth 3, which represents the hierarchical decomposition of the entire series. This is also referred to as the *error tree*. In Fig. 2.6, the error tree for the wavelet decomposition in Table 2.2 is illustrated. The nodes in the tree contain the values of the wavelet coefficients, except for a special *super-root* node that contains the series average.

The number of wavelet coefficients in this series is 8, which is also the length of the original series. The original series has been replicated just below the error tree in Fig. 2.6, and can be reconstructed by adding or subtracting the values in the nodes along the path leading to that value. Each coefficient in a node should be added, if we use the left branch below it to reach to the series values. Otherwise, it should be subtracted. This natural decomposition means that an entire contiguous range along the series can be reconstructed by using only the portion of the error tree which is relevant to it.

As in all dimensionality reduction methods, smaller coefficients are ignored. We will explain the process of discarding coefficients with the help of the notion of the *basis vectors* associated with each coefficient:

The wavelet representation is a decomposition of the original time series of length q into the weighted sum of a set of q "simpler" time series (or wavelets) that are orthogonal to one another. These "simpler" time series are the basis vectors, and the wavelet coefficients represent the weights of the different basis vectors in the decomposition.

Figure 2.5 shows these "simpler" time series along with their corresponding coefficients. The number of wavelet coefficients (and basis vectors) is equal to the length of the series q.

The length of the time series representing each basis vector is also q. Each basis vector has a $+1$ or -1 value in the contiguous time-series segment from which a particular coefficient was derived by a differencing operation. Otherwise, the value is 0 because the wavelet is not related to variations in that region of the time series. The first half of the nonzero segment of the basis vector is $+1$, and the second half is -1. This gives it the shape of a *wavelet* when it is plotted as a time series, and also reflects the differencing operation in the relevant time-series segment. Multiplying a basis vector with the coefficient has the effect of creating a weighted time series in which the difference between the first half and second half reflects the average difference between the corresponding segments in the original time series. Therefore, by adding up all these weighted wavelets over different levels of granularity in the error tree, it is possible to reconstruct the original series. The list of basis vectors in Fig. 2.5 are the rows of the following matrix:

$$\begin{pmatrix} 1 & -1 & 0 & 0 & 0 & 0 & 0 & 0 \\ 0 & 0 & 1 & -1 & 0 & 0 & 0 & 0 \\ 0 & 0 & 0 & 0 & 1 & -1 & 0 & 0 \\ 0 & 0 & 0 & 0 & 0 & 0 & 1 & -1 \\ 1 & 1 & -1 & -1 & 0 & 0 & 0 & 0 \\ 0 & 0 & 0 & 0 & 1 & 1 & -1 & -1 \\ 1 & 1 & 1 & 1 & -1 & -1 & -1 & -1 \\ 1 & 1 & 1 & 1 & 1 & 1 & 1 & 1 \end{pmatrix}$$

Note that the dot product of any pair of basis vectors is 0, and therefore these series are orthogonal to one another. The most detailed coefficients have only one $+1$ and one -1, whereas the most coarse coefficient has four $+1$ and -1 entries. In addition, the vector (11111111) is needed to represent the series average.

For a time series T, let $\overline{W_1} \dots \overline{W_q}$ be the corresponding basis vectors. Then, if $a_1 \dots a_q$ are the wavelet coefficients for the basis vectors $\overline{W_1} \dots \overline{W_q}$, the time series T can be represented as follows:

$$T = \sum_{i=1}^{q} a_i \overline{W_i} = \sum_{i=1}^{q} (a_i ||\overline{W_i}||) \frac{\overline{W_i}}{||\overline{W_i}||} \tag{2.15}$$

The coefficients represented in Fig. 2.5 are unnormalized because the underlying basis vectors do not have unit norm. While a_i is the unnormalized value from Fig. 2.5, the values $a_i||\overline{W_i}||$ represent normalized coefficients. The values of $||\overline{W_i}||$ are different for coefficients of different orders, and are equal to $\sqrt{2}$, $\sqrt{4}$, or $\sqrt{8}$ in this particular example. For example, in Fig. 2.5, the broadest level unnormalized coefficient is -0.25, whereas the corresponding normalized value is $-0.25\sqrt{8}$. After normalization, the basis vectors $\overline{W_1} \dots \overline{W_q}$ are orthonormal, and, therefore, the sum of the squares of the corresponding (normalized) coefficients is equal to the retained energy in the approximated time series. Because the normalized coefficients provide a new coordinate representation after axis rotation, Euclidean distances between time series are preserved in this new representation if coefficients are not dropped. It can be shown that by retaining the coefficients with the largest normalized values, the error loss from the wavelet representation is minimized.

The previous discussion focused on the approximation of a single time series. In practice, one might want to convert a database of N time series into N multidimensional vectors. When a database of multiple time series is available, then two strategies can be used:

1. The coefficient for the same basis vector is selected for each series to create a meaningful multidimensional database of low dimensionality. Therefore, the basis vectors

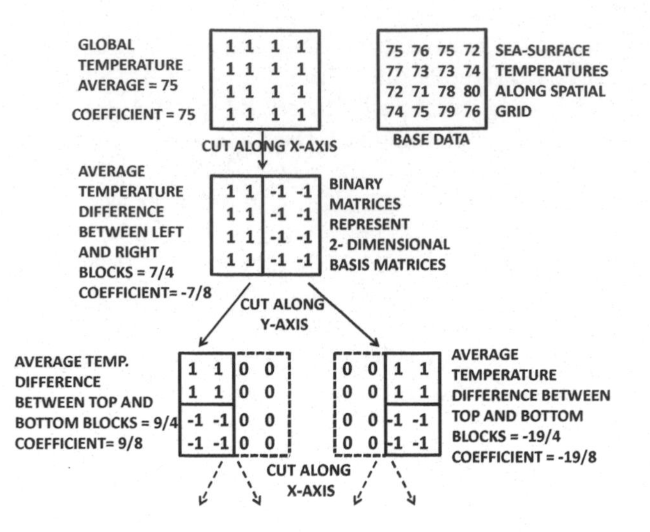

Figure 2.7: Illustration of the top levels of the wavelet decomposition for spatial data in a grid containing sea-surface temperatures

that have the largest *average* normalized coefficient across the N different series are selected.

2. The full dimensionality of the wavelet coefficient representation is retained. However, for each time series, the largest normalized coefficients (in magnitude) are selected individually. The remaining values are set to 0. This results in a sparse database of high dimensionality, in which many values are 0. A method such as *SVD* can be applied as a second step to further reduce the dimensionality. The second step of this approach has the disadvantage of losing interpretability of the features of the wavelet transform. Recall that the Haar wavelet is one of the few dimensionality reduction transforms where the coefficients do have some interpretability in terms of specific trends across particular time-series segments.

The wavelet decomposition method provides a natural method for dimensionality reduction (and data-type transformation) by retaining only a small number of coefficients.

Wavelet Decomposition with Multiple Contextual Attributes

Time-series data contain a single contextual attribute, corresponding to the time value. This helps in simplification of the wavelet decomposition. However, in some cases such as spatial data, there may be two contextual attributes corresponding to the X-coordinate and the Y-coordinate. For example, sea-surface temperatures are measured at spatial locations that are described with the use of two coordinates. How can wavelet decomposition be performed in such cases? In this section, a brief overview of the extension of wavelets to multiple contextual attributes is provided.

Assume that the spatial data is represented in the form of a 2-dimensional grid of size $q \times q$. Recall that in the 1-dimensional case, differencing operations were applied over contiguous segments of the time series by successive division of the time series in hierarchical fashion. The corresponding basis vectors have $+1$ and -1 at the relevant positions. The 2-dimensional case is completely analogous where contiguous *areas* of the spatial grid are used by successive divisions. These divisions are alternately performed along the different axes. The corresponding basis vectors are 2-dimensional *matrices* of size $q \times q$ that regulate how the differencing operations are performed.

An example of the strategy for 2-dimensional decomposition is illustrated in Fig. 2.7. Only the top two levels of the decomposition are illustrated in the figure. Here, a 4×4 grid of spatial temperatures is used as an example. The first division along the X-axis divides the spatial area into two blocks of size 4×2 each. The corresponding two-dimensional binary basis matrix is illustrated into the same figure. The next phase divides each of these 4×2 blocks into blocks of size 2×2 during the hierarchical decomposition process. As in the case of 1-dimensional time series, the wavelet coefficient is half the difference in the average temperatures between the two halves of the relevant block being decomposed. The alternating process of division along the X-axis and the Y-axis can be carried on to the individual data entries. This creates a hierarchical wavelet error tree, which has many similar properties to that created in the 1-dimensional case. The overall principles of this decomposition are almost identical to the 1-dimensional case, with the major difference in terms of how the cuts along different dimensions are performed by alternating at different levels. The approach can be extended to the case of $k > 2$ contextual attributes with the use of a round-robin rotation in the axes that are selected at different levels of the tree for the differencing operation.

2.4.4.2 Multidimensional Scaling

Graphs are a powerful mechanism for representing relationships between objects. In some data mining scenarios, the data type of an object may be very complex and heterogeneous such as a time series annotated with text and other numeric attributes. However, a crisp notion of distance between several pairs of data objects may be available based on application-specific goals. How can one visualize the inherent similarity between these objects? How can one visualize the "nearness" of two individuals connected in a social network? A natural way of doing so is the concept of multidimensional scaling (*MDS*). Although *MDS* was originally proposed in the context of spatial visualization of graph-structured distances, it has much broader applicability for embedding data objects of arbitrary types in multidimensional space. Such an approach also enables the use of multidimensional data mining algorithms on the embedded data.

For a graph with n nodes, let $\delta_{ij} = \delta_{ji}$ denote the specified distance between nodes i and j. It is assumed that all $\binom{n}{2}$ pairwise distances between nodes are specified. It is desired to map the n nodes to n different k-dimensional vectors denoted by $\overline{X_1} \ldots \overline{X_n}$, so that the distances in multidimensional space closely correspond to the $\binom{n}{2}$ distance values in the distance graph. In *MDS*, the k coordinates of each of these n points are treated as variables that need to be optimized, so that they can *fit* the current set of pairwise distances. Metric *MDS*, also referred to as classical *MDS*, attempts to solve the following optimization (minimization) problem:

$$O = \sum_{i,j:i<j} (||\overline{X_i} - \overline{X_j}|| - \delta_{ij})^2 \tag{2.16}$$

Here $|| \cdot ||$ represents Euclidean norm. In other words, each node is represented by a multidimensional data point, such that the Euclidean distances between these points reflect the graph distances as closely as possible. In other forms of *nonmetric MDS*, this objective function might be different. This optimization problem therefore has $n \cdot k$ variables, and it scales with the size of the data n and the desired dimensionality k of the embedding. The

Table 2.3: Scaled eigenvectors of various similarity matrices yield embeddings with different properties

Method	Relevant similarity matrix
PCA	Dot product matrix DD^T after mean centering D
SVD	Dot product matrix DD^T
Spectral embedding (Symmetric Version)	Sparsified/normalized similarity matrix $\Lambda^{-1/2}W\Lambda^{-1/2}$ (cf. Sect. 19.3.4 of Chap. 19)
MDS/ISOMAP	Similarity matrix derived from distance matrix Δ with cosine law $S = -\frac{1}{2}(I - \frac{U}{n})\Delta(I - \frac{U}{n})$
Kernel PCA	Centered kernel matrix $S = (I - \frac{U}{n})K(I - \frac{U}{n})$ (cf. Sect. 10.6.4.1 of Chap. 10)

objective function O of Eq. 2.16 is usually divided by $\sum_{i,j:i<j} \delta_{ij}^2$ to yield a value in $(0, 1)$. The square root of this value is referred to as *Kruskal stress*.

The basic assumption in classical *MDS* is that the distance matrix $\Delta = [\delta_{ij}^2]_{n\times n}$ is generated by computing pairwise Euclidean distances in some hypothetical data matrix D for which the entries and dimensionality are unknown. The matrix D can never be recovered completely in classical *MDS* because Euclidean distances are invariant to mean translation and axis rotation. The appropriate conventions for the data mean and axis orientation will be discussed later. While the optimization of Eq. 2.16 requires numerical techniques, a direct solution to classical *MDS* can be obtained by eigen decomposition *under the assumption that the specified distance matrix is Euclidean*:

1. Any pairwise (squared) distance matrix $\Delta = [\delta_{ij}^2]_{n\times n}$ can be converted into a symmetric dot-product matrix $S_{n\times n}$ with the help of the *cosine law* in Euclidean space. In particular, if $\overline{X_i}$ and $\overline{X_j}$ are the embedded representations of the ith and jth nodes, the dot product between $\overline{X_i}$ and $\overline{X_j}$ can be related to the distances as follows:

$$\overline{X_i} \cdot \overline{X_j} = -\frac{1}{2}\left[||\overline{X_i} - \overline{X_j}||^2 - (||\overline{X_i}||^2 + ||\overline{X_j}||^2)\right] \quad \forall i, j \in \{1 \dots n\} \tag{2.17}$$

For a *mean-centered* embedding, the value of $||\overline{X_i}||^2 + ||\overline{X_j}||^2$ can be expressed (see Exercise 9) in terms of the entries of the distance matrix Δ as follows:

$$||\overline{X_i}||^2 + ||\overline{X_j}||^2 = \frac{\sum_{p=1}^n ||\overline{X_i} - \overline{X_p}||^2}{n} + \frac{\sum_{q=1}^n ||\overline{X_j} - \overline{X_q}||^2}{n} - \frac{\sum_{p=1}^n \sum_{q=1}^n ||\overline{X_p} - \overline{X_q}||^2}{n^2} \tag{2.18}$$

A mean-centering assumption is necessary because the Euclidean distance is mean invariant, whereas the dot product is not. By substituting Eq. 2.18 in Eq. 2.17, it is possible to express the dot product $\overline{X_i} \cdot \overline{X_j}$ fully in terms of the entries of the distance matrix Δ. Because this condition is true for all possible values of i and j, we can conveniently express it in $n \times n$ matrix form. Let U be the $n \times n$ matrix of all 1s, and let I be the identity matrix. Then, our argument above shows that the dot-product matrix S is equal to $-\frac{1}{2}(I - \frac{U}{n})\Delta(I - \frac{U}{n})$. Under the Euclidean assumption, the matrix S is always positive semidefinite because it is equal to the $n \times n$ dot-product matrix DD^T of the *unobserved* data matrix D, which has unknown dimensionality. Therefore, it is desired to determine a high-quality factorization of S into the form $D_k D_k^T$, where D_k is an $n \times k$ matrix of dimensionality k.

2. Such a factorization can be obtained with eigen decomposition. Let $S \approx Q_k \Sigma_k^2 Q_k^T = (Q_k \Sigma_k)(Q_k \Sigma_k)^T$ represent the approximate diagonalization of S, where Q_k is an $n \times k$ matrix containing the largest k eigenvectors of S, and Σ_k^2 is a $k \times k$ diagonal matrix containing the eigenvalues. The embedded representation is given by $D_k = Q_k \Sigma_k$. Note that *SVD* also derives the optimal embedding as the scaled eigenvectors of the dot-product matrix of the original data. Therefore, the squared error of representation is minimized by this approach. This can also be shown to be equivalent to minimizing the Kruskal stress.

The optimal solution is not unique, because we can multiply $Q_k \Sigma_k$ with any $k \times k$ matrix with orthonormal columns, and the pairwise Euclidean distances will not be affected. In other words, any representation of $Q_k \Sigma_k$ in a rotated axis system is optimal as well. *MDS* finds an axis system like *PCA* in which the individual attributes are uncorrelated. In fact, if classical *MDS* is applied to a distance matrix Δ, which is constructed by computing the pairwise Euclidean distances in an actual data set, then it will yield the same embedding as the application of *PCA* on that data set. *MDS* is useful when such a data set is not available to begin with, and only the distance matrix Δ is available.

As in all dimensionality reduction methods, the value of the dimensionality k provides the trade-off between representation size and accuracy. Larger values of the dimensionality k will lead to lower stress. A larger number of data points typically requires a larger dimensionality of representation to achieve the same stress. The most crucial element is, however, the inherent structure of the distance matrix. For example, if a $10,000 \times 10,000$ distance matrix contains the pairwise driving distance between 10,000 cities, it can usually be approximated quite well with just a 2-dimensional representation. This is because driving distances are an approximation of Euclidean distances in 2-dimensional space. On the other hand, an arbitrary distance matrix may not be Euclidean and the distances may not even satisfy the triangle inequality. As a result, the matrix S might not be positive semidefinite. In such cases, it is sometimes still possible to use the metric assumption to obtain a high-quality embedding. Specifically, only those positive eigenvalues may be used, whose magnitude exceeds that of the most negative eigenvalue. This approach will work reasonably well if the negative eigenvalues have small magnitude.

MDS is commonly used in nonlinear dimensionality reduction methods such as *ISOMAP* (cf. Sect. 3.2.1.7 of Chap. 3). It is noteworthy that, in conventional *SVD*, the scaled eigenvectors of the $n \times n$ dot-product similarity matrix DD^T yield a low-dimensional embedded representation of D just as the eigenvectors of S yield the embedding in *MDS*. The eigen decomposition of similarity matrices is fundamental to many linear and nonlinear dimensionality reduction methods such as *PCA*, *SVD*, *ISOMAP*, *kernel PCA*, and spectral embedding. The specific properties of each embedding are a result of the choice of the similarity matrix and the scaling used on the resulting eigenvectors. Table 2.3 provides a preliminary comparison of these methods, although some of them are discussed in detail only in later chapters.

2.4.4.3 Spectral Transformation and Embedding of Graphs

Whereas *MDS* methods are designed for preserving global distances, spectral methods are designed for preserving local distances for applications such as clustering. Spectral methods work with *similarity graphs* in which the weights on the edges represent similarity rather than distances. When distance values are available they are converted to similarity values with kernel functions such as the heat kernel discussed earlier in this chapter. The notion of similarity is natural to many real Web, social, and information networks because of the notion of *homophily*. For example, consider a bibliographic network in which nodes correspond to authors, and the edges correspond to co-authorship relations. The weight of an edge represents the number of publications between authors and therefore represents one possible notion of similarity in author publications. Similarity graphs can also be constructed between arbitrary data types. For example, a set of n time series can be converted into a graph with n nodes, where a node represents each time series. The weight of an edge is equal to the similarity between the two nodes, and only edges with a "sufficient" level of similarity are retained. A discussion of the construction of the similarity graph is provided in Sect. 2.2.2.9. Therefore, if a similarity graph can be transformed to a multidimensional representation that preserves the similarity structure between nodes, it provides a transformation that can port virtually any data type to the easily usable multidimensional representation. The caveat here is that such a transformation can only be used for similarity-based applications such as clustering or nearest neighbor classification because the transformation is designed to preserve the *local* similarity structure. The local similarity structure of a data set is nevertheless fundamental to many data mining applications.

Let $G = (N, A)$ be an undirected graph with node set N and edge set A. It is assumed that the node set contains n nodes. A symmetric $n \times n$ weight matrix $W = [w_{ij}]$ represents the similarities between the different nodes. Unlike *MDS*, which works with a complete graph of *global* distances, this graph is generally a *sparsified* representation of the similarity of each object to its k *nearest objects* (cf. Sect. 2.2.2.9). The similarities to the remaining objects are not distinguished from one another and set to 0. This is because spectral methods preserve only the local similarity structure for applications such as clustering. All entries in this matrix are assumed to be nonnegative, and higher values indicate greater similarity. If an edge does not exist between a pair of nodes, then the corresponding entry is assumed to be 0. It is desired to embed the nodes of this graph into a k-dimensional space so that the similarity structure of the data is preserved.

First, let us discuss the much simpler problem of mapping the nodes onto a 1-dimensional space. The generalization to the k-dimensional case is relatively straightforward. We would like to map the nodes in N into a set of 1-dimensional real values $y_1 \ldots y_n$ on a line, so that the distances between these points reflect the edge connectivity among the nodes. Therefore, it is undesirable for nodes that are connected with high-weight edges, to be mapped onto distant points on this line. Therefore, we would like to determine values of y_i that minimize the following objective function O:

$$O = \sum_{i=1}^{n} \sum_{j=1}^{n} w_{ij} (y_i - y_j)^2 \tag{2.19}$$

This objective function penalizes the distances between y_i and y_j with weight proportional to w_{ij}. Therefore, when w_{ij} is very large (more similar nodes), the data points y_i and y_j will be more likely to be closer to one another in the embedded space. The objective function O can be rewritten in terms of the *Laplacian matrix* L of weight matrix W. The Laplacian matrix L is defined as $\Lambda - W$, where Λ is a diagonal matrix satisfying $\Lambda_{ii} = \sum_{j=1}^{n} w_{ij}$. Let the n-dimensional column vector of embedded values be denoted by $\overline{y} = (y_1 \ldots y_n)^T$. It can be shown after some algebraic simplification that the minimization objective function O can be rewritten in terms of the Laplacian matrix:

$$O = 2\overline{y}^T L \overline{y} \tag{2.20}$$

The matrix L is positive semidefinite with nonnegative eigenvalues because the sum-of-squares objective function O is always nonnegative. We need to incorporate a scaling constraint to ensure that the trivial value of $y_i = 0$ for all i, is not selected by the optimization solution. A possible scaling constraint is as follows:

$$\overline{y}^T \Lambda \overline{y} = 1 \tag{2.21}$$

The use of the matrix Λ in the constraint of Eq. 2.21 is essentially a normalization constraint, which is discussed in detail in Sect. 19.3.4 of Chap. 19.

It can be shown that the value of O is optimized by selecting $\overline{y}$ as the smallest eigenvector of the relationship $\Lambda^{-1} L \overline{y} = \lambda \overline{y}$. However, the smallest eigenvalue is always 0, and it corresponds to the trivial solution where the node embedding $\overline{y}$ is proportional to the vector containing only 1s. This trivial eigenvector is non-informative because it corresponds to an embedding in which every node is mapped to the same point. Therefore, it can be discarded, and it is not used in the analysis. The second-smallest eigenvector then provides an optimal solution that is more informative.

This solution can be generalized to finding an optimal k-dimensional embedding by determining successive directions corresponding to eigenvectors with increasing eigenvalues. After discarding the first trivial eigenvector $\overline{e_1}$ with eigenvalue $\lambda_1 = 0$, this results in a set of k eigenvectors $\overline{e_2}, \overline{e_3} \ldots \overline{e_{k+1}}$, with corresponding eigenvalues $\lambda_2 \leq \lambda_3 \leq \ldots \leq \lambda_{k+1}$. Each eigenvector is of length n and contains one coordinate value for each node. The ith value along the jth eigenvector represents the jth coordinate of the ith node. This creates an $n \times k$ matrix, corresponding to the k-dimensional embedding of the n nodes.

What do the small magnitude eigenvectors intuitively represent in the new transformed space? By using the ordering of the nodes along a small magnitude eigenvector to create a cut, the weight of the edges across the cut is likely to be small. Thus, this represents a cluster in the space of nodes. In practice, the k smallest eigenvectors (after ignoring the first) are selected to perform the reduction and create a k-dimensional embedding. This embedding typically contains an excellent representation of the underlying similarity structure of the nodes. The embedding can be used for virtually any similarity-based application, although the most common application of this approach is spectral clustering. Many variations of this approach exist in terms of how the Laplacian L is normalized, and in terms of how the final clusters are generated. The spectral clustering method will be discussed in detail in Sect. 19.3.4 of Chap. 19.

2.5 Summary

Data preparation is an important part of the data mining process because of the sensitivity of the analytical algorithms to the quality of the input data. The data mining process requires the collection of raw data from a variety of sources that may be in a form which is unsuitable for direct application of analytical algorithms. Therefore, numerous methods may need to be applied to extract features from the underlying data. The resulting data may have significant missing values, errors, inconsistencies, and redundancies. A variety of analytical methods and data scrubbing tools exist for imputing the missing entries or correcting inconsistencies in the data.

Another important issue is that of data heterogeneity. The analyst may be faced with a multitude of attributes that are distinct, and therefore the direct application of data mining algorithms may not be easy. Therefore, data type portability is important, wherein some subsets of attributes are converted to a predefined format. The multidimensional format is often preferred because of its simplicity. Virtually, any data type can be converted to multidimensional representation with the two-step process of constructing a similarity graph, followed by multidimensional embedding.

The data set may be very large, and it may be desirable to reduce its size both in terms of the number of rows and the number of dimensions. The reduction in terms of the number of rows is straightforward with the use of sampling. To reduce the number of columns in the data, either feature subset selection or data transformation may be used. In feature subset selection, only a smaller set of features is retained that is most suitable for analysis. These methods are closely related to analytical methods because the relevance of a feature may be application dependent. Therefore, the feature selection phase need to be tailored to the specific analytical method.

There are two types of feature transformation. In the first type, the axis system may be rotated to align with the correlations of the data and retain the directions with the greatest variance. The second type is applied to complex data types such as graphs and time series. In these methods, the size of the representation is reduced, and the data is also transformed to a multidimensional representation.

2.6 Bibliographic Notes

The problem of feature extraction is an important one for the data mining process but it is highly application specific. For example, the methods for extracting named entities from a document data set [400] are very different from those that extract features from an image data set [424]. An overview of some of the promising technologies for feature extraction in various domains may be found in [245].

After the features have been extracted from different sources, they need to be integrated into a single database. Numerous methods have been described in the conventional database literature for data integration [194, 434]. Subsequently, the data needs to be cleaned and missing entries need to be removed. A new field of probabilistic or uncertain data has emerged [18] that models uncertain and erroneous records in the form of probabilistic databases. This field is, however, still in the research stage and has not entered the mainstream of database applications. Most of the current methods either use tools for missing data analysis [71, 364] or more conventional data cleaning and data scrubbing tools [222, 433, 435].

After the data has been cleaned, its size needs to be reduced either in terms of numerosity or in terms of dimensionality. The most common and simple numerosity reduction method is sampling. Sampling methods can be used for either static data sets or dynamic data sets. Traditional methods for data sampling are discussed in [156]. The method of sampling has also been extended to data streams in the form of reservoir sampling [35, 498]. The work in [35] discusses the extension of reservoir sampling methods to the case where a biased sample needs to be created from the data stream.

Feature selection is an important aspect of the data mining process. The approach is often highly dependent on the particular data mining algorithm being used. For example, a feature selection method that works well for clustering may not work well for classification. Therefore, we have deferred the discussion of feature selection to the relevant chapters on the topic on clustering and classification in this book. Numerous books are available on the topic of feature selection [246, 366].

The two most common dimensionality reduction methods used for multidimensional data are *SVD* [480, 481] and *PCA* [295]. These methods have also been extended to text in the form of *LSA* [184, 416]. It has been shown in many domains [25, 184, 416] that the use of methods such as *SVD*, *LSA*, and *PCA* unexpectedly improves the quality of the underlying representation after performing the reduction. This improvement is because of reduction in noise effects by discarding the low-variance dimensions. Applications of *SVD* to data imputation are found in [23] and Chap. 18 of this book. Other methods for dimensionality reduction and transformation include Kalman filtering [260], Fastmap [202], and nonlinear methods such as *Laplacian eigenmaps* [90], *MDS* [328], and *ISOMAP* [490].

Many dimensionality reduction methods have also been proposed in recent years that simultaneously perform type transformation together with the reduction process. These include wavelet transformation [475] and graph embedding methods such as *ISOMAP* and *Laplacian eigenmaps* [90, 490]. A tutorial on spectral methods for graph embedding may be found in [371].

2.7 Exercises

1. Consider the time-series $(-3, -1, 1, 3, 5, 7, *)$. Here, a missing entry is denoted by $*$. What is the estimated value of the missing entry using linear interpolation on a window of size 3?

2. Suppose you had a bunch of text documents, and you wanted to determine all the personalities mentioned in these documents. What class of technologies would you use to achieve this goal?

3. Download the *Arrythmia* data set from the *UCI Machine Learning Repository* [213]. Normalize all records to a mean of 0 and a standard deviation of 1. Discretize each numerical attribute into (a) 10 equi-width ranges and (b) 10 equi-depth ranges.

4. Suppose that you had a set of arbitrary objects of different types representing different characteristics of widgets. A domain expert gave you the similarity value between every pair of objects. How would you convert these objects into a multidimensional data set for clustering?

5. Suppose that you had a data set, such that each data point corresponds to sea-surface temperatures over a square mile of resolution 10×10. In other words, each data record contains a 10×10 grid of temperature values with spatial locations. You also have some text associated with each 10×10 grid. How would you convert this data into a multidimensional data set?

6. Suppose that you had a set of discrete biological protein sequences that are annotated with text describing the properties of the protein. How would you create a multidimensional representation from this heterogeneous data set?

7. Download the *Musk* data set from the *UCI Machine Learning Repository* [213]. Apply *PCA* to the data set, and report the eigenvectors and eigenvalues.

8. Repeat the previous exercise using *SVD*.

9. For a mean-centered data set with points $\overline{X_1} \dots \overline{X_n}$, show that the following is true:

$$||\overline{X_i}||^2 + ||\overline{X_j}||^2 = \frac{\sum_{p=1}^n ||\overline{X_i} - \overline{X_p}||^2}{n} + \frac{\sum_{q=1}^n ||\overline{X_j} - \overline{X_q}||^2}{n} - \frac{\sum_{p=1}^n \sum_{q=1}^n ||\overline{X_p} - \overline{X_q}||^2}{n^2} \qquad (2.22)$$

10. Consider the time series 1, 1, 3, 3, 3, 3, 1, 1. Perform wavelet decomposition on the time series. How many coefficients of the series are nonzero?

11. Download the *Intel Research Berkeley data set.* Apply a wavelet transformation to the temperature values in the first sensor.

12. Treat each quantitative variable in the *KDD CUP 1999 Network Intrusion Data Set* from the *UCI Machine Learning Repository* [213] as a time series. Perform the wavelet decomposition of this time series.

13. Create samples of size $n = 1, 10, 100, 1000, 10000$ records from the data set of the previous exercise, and determine the average value e_i of each quantitative column i using the sample. Let μ_i and σ_i be the global mean and standard deviation over the entire data set. Compute the number of standard deviations z_i by which e_i varies from μ_i.

$$z_i = \frac{|e_i - \mu_i|}{\sigma_i}$$

How does z_i vary with n?

14. Show that any right singular vector $\overline{y}$ of A with 0 singular value satisfies $A\overline{y} = 0$.

15. Show that the diagonalization of a square matrix is a specialized variation of *SVD*.

Chapter 3

Similarity and Distances

"*Love is the power to see similarity in the dissimilar.*"—Theodor Adorno

3.1 Introduction

Many data mining applications require the determination of similar or dissimilar objects, patterns, attributes, and events in the data. In other words, a methodical way of quantifying similarity between data objects is required. Virtually all data mining problems, such as clustering, outlier detection, and classification, require the computation of similarity. A formal statement of the problem of similarity or distance quantification is as follows:

Given two objects O_1 and O_2, determine a value of the similarity $Sim(O_1, O_2)$ (or distance $Dist(O_1, O_2)$) between the two objects.

In similarity functions, larger values imply greater similarity, whereas in distance functions, smaller values imply greater similarity. In some domains, such as spatial data, it is more natural to talk about distance functions, whereas in other domains, such as text, it is more natural to talk about similarity functions. Nevertheless, the principles involved in the design of such functions are generally invariant across different data domains. This chapter will, therefore, use either of the terms "distance function" and "similarity function," depending on the domain at hand. Similarity and distance functions are often expressed in closed form (e.g., Euclidean distance), but in some domains, such as time-series data, they are defined algorithmically and cannot be expressed in closed form.

Distance functions are fundamental to the effective design of data mining algorithms, because a poor choice in this respect may be very detrimental to the quality of the results. Sometimes, data analysts use the Euclidean function as a "black box" without much thought about the overall impact of such a choice. It is not uncommon for an inexperienced analyst to invest significant effort in the algorithmic design of a data mining problem, while treating the distance function subroutine as an afterthought. This is a mistake. As this chapter will elucidate, poor choices of the distance function can sometimes be disastrously misleading

C. C. Aggarwal, *Data Mining: The Textbook*, DOI 10.1007/978-3-319-14142-8_3

depending on the application domain. Good distance function design is also crucial for type portability. As discussed in Sect. 2.4.4.3 of Chap. 2, spectral embedding can be used to convert a similarity graph constructed on any data type into multidimensional data.

Distance functions are highly sensitive to the data distribution, dimensionality, and data type. In some data types, such as multidimensional data, it is much simpler to define and compute distance functions than in other types such as time-series data. In some cases, user intentions (or *training feedback* on object pairs) are available to *supervise* the distance function design. Although this chapter will primarily focus on unsupervised methods, we will also briefly touch on the broader principles of using supervised methods.

This chapter is organized as follows. Section 3.2 studies distance functions for multidimensional data. This includes quantitative, categorical, and mixed attribute data. Similarity measures for text, binary, and set data are discussed in Sect. 3.3. Temporal data is discussed in Sect. 3.4. Distance functions for graph data are addressed in Sect. 3.5. A discussion of supervised similarity will be provided in Sect. 3.6. Section 3.7 gives a summary.

3.2 Multidimensional Data

Although multidimensional data are the simplest form of data, there is significant diversity in distance function design across different attribute types such as categorical or quantitative data. This section will therefore study each of these types separately.

3.2.1 Quantitative Data

The most common distance function for quantitative data is the L_p-norm. The L_p-norm between two data points $\overline{X} = (x_1 \dots x_d)$ and $\overline{Y} = (y_1 \dots y_d)$ is defined as follows:

$$Dist(\overline{X}, \overline{Y}) = \left(\sum_{i=1}^{d} |x_i - y_i|^p \right)^{1/p}. \tag{3.1}$$

Two special cases of the L_p-norm are the *Euclidean* ($p = 2$) and the *Manhattan* ($p = 1$) metrics. These special cases derive their intuition from spatial applications where they have clear physical interpretability. The Euclidean distance is the straight-line distance between two data points. The Manhattan distance is the "city block" driving distance in a region in which the streets are arranged as a rectangular grid, such as the Manhattan Island of New York City.

A nice property of the Euclidean distance is that it is rotation-invariant because the straight-line distance between two data points does not change with the orientation of the axis system. This property also means that transformations, such as *PCA*, *SVD*, or the wavelet transformation for time series (discussed in Chap. 2), can be used on the data without affecting[1] the distance. Another interesting special case is that obtained by setting $p = \infty$. The result of this computation is to select the dimension for which the two objects are the most distant from one another and report the absolute value of this distance. All other features are ignored.a

The L_p-norm is one of the most popular distance functions used by data mining analysts. One of the reasons for its popularity is the natural intuitive appeal and interpretability of L_1- and L_2-norms in spatial applications. The intuitive interpretability of these distances does not, however, mean that they are the most relevant ones, especially for the high-dimensional case. In fact, these distance functions may not work very well when the data

[1]The distances are affected after dimensions are dropped. However, the transformation itself does not impact distances.

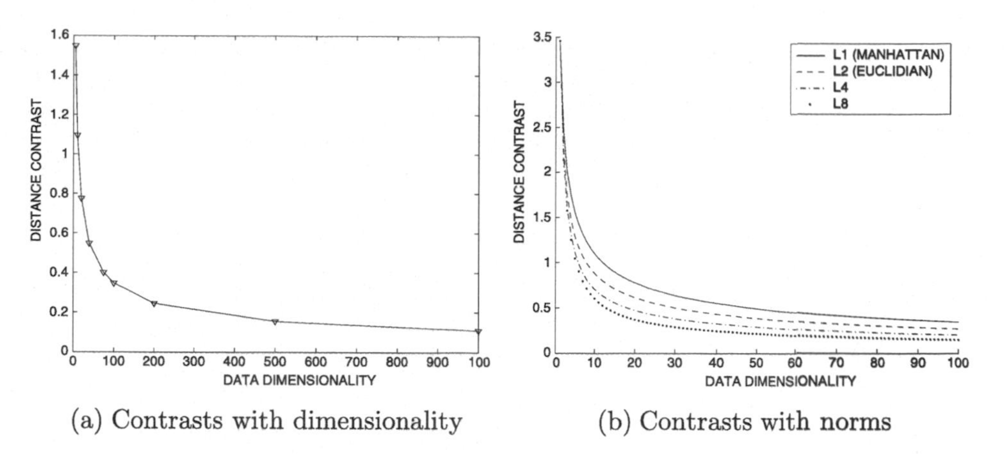

(a) Contrasts with dimensionality (b) Contrasts with norms

Figure 3.1: Reduction in distance contrasts with increasing dimensionality and norms

are high dimensional because of the varying impact of data sparsity, distribution, noise, and feature relevance. This chapter will discuss these broader principles in the context of distance function design.

3.2.1.1 Impact of Domain-Specific Relevance

In some cases, an analyst may know which features are more important than others for a particular application. For example, for a credit-scoring application, an attribute such as salary is much more relevant to the design of the distance function than an attribute such as gender, though both may have some impact. In such cases, the analyst may choose to weight the features differently if domain specific knowledge about the relative importance of different features is available. This is often a heuristic process based on experience and skill. The generalized L_p-distance is most suitable for this case and is defined in a similar way to the L_p-norm, except that a coefficient a_i is associated with the ith feature. This coefficient is used to weight the corresponding feature component in the L_p-norm:

$$Dist(\overline{X}, \overline{Y}) = \left(\sum_{i=1}^{d} a_i \cdot |x_i - y_i|^p \right)^{1/p}. \quad (3.2)$$

This distance is also referred to as the generalized *Minkowski distance*. In many cases, such domain knowledge is not available. Therefore, the L_p-norm may be used as a default option. Unfortunately, without knowledge about the most relevant features, the L_p-norm is susceptible to some undesirable effects of increasing dimensionality, as discussed subsequently.

3.2.1.2 Impact of High Dimensionality

Many distance-based data mining applications lose their effectiveness as the dimensionality of the data increases. For example, a distance-based clustering algorithm may group unrelated data points because the distance function may poorly reflect the intrinsic semantic distances between data points with increasing dimensionality. As a result, distance-based models of clustering, classification, and outlier detection are often *qualitatively* ineffective. This phenomenon is referred to as the "curse of dimensionality," a term first coined by Richard Bellman.

To better understand the impact of the dimensionality curse on distances, let us examine a unit cube of dimensionality d that is fully located in the nonnegative quadrant, with one corner at the origin $\overline{O}$. What is the Manhattan distance of the corner of this cube (say, at the origin) to a randomly chosen point $\overline{X}$ inside the cube? In this case, because one end point is the origin, and all coordinates are nonnegative, the Manhattan distance will sum up the coordinates of $\overline{X}$ over the different dimensions. Each of these coordinates is uniformly distributed in $[0, 1]$. Therefore, if Y_i represents the uniformly distributed random variable in $[0, 1]$, it follows that the Manhattan distance is as follows:

$$Dist(\overline{O}, \overline{X}) = \sum_{i=1}^{d} (Y_i - 0). \tag{3.3}$$

The result is a random variable with a mean of $\mu = d/2$ and a standard deviation of $\sigma = \sqrt{d/12}$. For large values of d, it can be shown by the law of large numbers that the vast majority of randomly chosen points inside the cube will lie in the range $[D_{min}, D_{max}] = [\mu - 3\sigma, \mu + 3\sigma]$. Therefore, most of the points in the cube lie within a distance range of $D_{max} - D_{min} = 6\sigma = \sqrt{3d}$ from the origin. Note that the expected Manhattan distance grows with dimensionality at a rate that is linearly proportional to d. Therefore, the *ratio* of the variation in the distances to the absolute values that is referred to as $Contrast(d)$, is given by:

$$\text{Contrast}(d) = \frac{D_{max} - D_{min}}{\mu} = \sqrt{12/d}. \tag{3.4}$$

This ratio can be interpreted as the distance *contrast* between the different data points, in terms of how different the minimum and maximum distances from the origin might be considered. Because the contrast reduces with $\sqrt{d}$, it means that there is virtually no contrast with increasing dimensionality. Lower contrasts are obviously not desirable because it means that the data mining algorithm will score the distances between all pairs of data points in approximately the same way and will not discriminate well between different pairs of objects with varying levels of semantic relationships. The variation in contrast with increasing dimensionality is shown in Fig. 3.1a. This behavior is, in fact, observed for all L_p-norms at different values of p, though with varying severity. These differences in severity will be explored in a later section. Clearly, with increasing dimensionality, a direct use of the L_p-norm may not be effective.

3.2.1.3 Impact of Locally Irrelevant Features

A more fundamental way of exploring the effects of high dimensionality is by examining the impact of irrelevant features. This is because many features are likely to be irrelevant in a typical high-dimensional data set. Consider, for example, a set of medical records, containing patients with diverse medical conditions and very extensive quantitative measurements about various aspects of an individual's medical history. For a cluster containing diabetic patients, certain attributes such as the blood glucose level are more important for the distance computation. On the other hand, for a cluster containing epileptic patients, a different set of features will be more important. The additive effects of the natural variations in the many attribute values may be quite significant. A distance metric such as the Euclidean metric may unnecessarily contribute a high value from the more noisy components because of its square-sum approach. The key point to understand here is that the precise features that are relevant to the distance computation may sometimes be sensitive to the particular pair of objects that are being compared. This problem cannot be solved by global feature

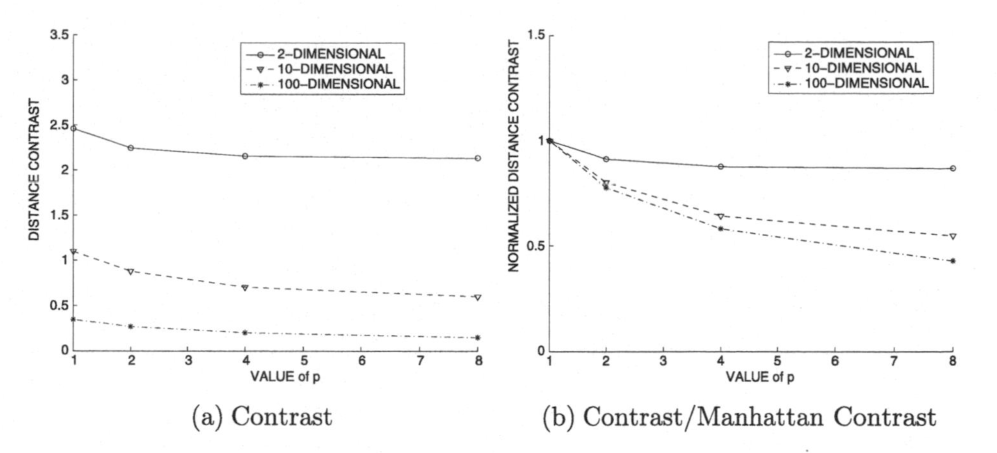

(a) Contrast

(b) Contrast/Manhattan Contrast

Figure 3.2: Impact of p on contrast

subset selection during preprocessing, because the relevance of features is *locally* determined by the pair of objects that are being considered. Globally, all features may be relevant.

When many features are irrelevant, the additive noise effects of the irrelevant features can sometimes be reflected in the concentration of the distances. In any case, such irrelevant features will almost always result in errors in distance computation. Because high-dimensional data sets are often likely to contain diverse features, many of which are irrelevant, the additive effect with the use of a sum-of-squares approach, such as the L_2-norm, can be very detrimental.

3.2.1.4 Impact of Different L_p-Norms

Different L_p-norms do not behave in a similar way either in terms of the impact of irrelevant features or the distance contrast. Consider the extreme case when $p = \infty$. This translates to using only the dimension where the two objects are the most *dissimilar*. Very often, this may be the impact of the natural variations in an irrelevant attribute that is not too useful for a similarity-based application. In fact, for a 1000-dimensional application, if two objects have similar values on 999 attributes, such objects should be considered *very* similar. However, a single irrelevant attribute on which the two objects are very different will throw off the distance value in the case of the L_∞ metric. In other words, local similarity properties of the data are de-emphasized by L_∞. Clearly, this is not desirable.

This behavior is generally true for larger values of p, where the irrelevant attributes are emphasized. In fact, it can also be shown that distance contrasts are also poorer for larger values of p for certain data distributions. In Fig. 3.1b, the distance contrasts have been illustrated for different values of p for the L_p-norm over different dimensionalities. The figure is constructed using the same approach as Fig. 3.1a. While all L_p-norms degrade with increasing dimensionality, the degradation is much faster for the plots representing larger values of p. This trend can be understood better from Fig. 3.2 where the value of p is used on the X-axis. In Fig. 3.2a, the contrast is illustrated with different values of p for data of different dimensionalities. Figure 3.2b is derived from Fig. 3.2a, except that the results show the fraction of the Manhattan performance achieved by higher order norms. It is evident that the *rate* of degradation with increasing p is higher when the dimensionality of the data is large. For 2-dimensional data, there is very little degradation. This is the reason that the value of p matters less in lower dimensional applications.

This argument has been used to propose the concept of fractional metrics, for which $p \in (0, 1)$. Such fractional metrics can provide more effective results for the high-dimensional case. As a rule of thumb, the larger the dimensionality, the lower the value of p. However, no exact rule exists on the precise choice of p because dimensionality is not the only factor in determining the proper value of p. The precise choice of p should be selected in an application-specific way, with the use of benchmarking. The bibliographic notes contain discussions on the use of fractional metrics.

3.2.1.5 Match-Based Similarity Computation

Because it is desirable to select locally relevant features for distance computation, a question arises as to how this can be achieved in a meaningful and practical way for data mining applications. A simple approach that is based on the cumulative evidence of matching many attribute values has been shown to be effective in many scenarios. This approach is also relatively easy to implement efficiently.

A broader principle that seems to work well for high-dimensional data is that the impact of the noisy variation along individual attributes needs to be de-emphasized while counting the cumulative match across many dimensions. Of course, such an approach poses challenges for low-dimensional data, because the cumulative impact of matching cannot be counted in a statistically robust way with a small number of dimensions. Therefore, an approach is needed that can automatically adjust to the dimensionality of the data.

With increasing dimensionality, a record is likely to contain both relevant and irrelevant features. A pair of semantically similar objects may contain feature values that are dissimilar (at the level of one standard deviation along that dimension) because of the noisy variations in irrelevant features. Conversely, a pair of objects are unlikely to have similar values across many attributes, just by chance, unless these attributes were relevant. Interestingly, the Euclidean metric (and L_p-norm in general) achieves exactly the opposite effect by using the squared sum of the difference in attribute values. As a result, the "noise" components from the irrelevant attributes dominate the computation and mask the similarity effects of a large number of relevant attributes. The L_∞-norm provides an extreme example of this effect where the dimension with the largest distance value is used. In high-dimensional domains such as text, similarity functions such as the cosine measure (discussed in Sect. 3.3), tend to emphasize the cumulative effect of matches on many attribute values rather than large distances along individual attributes. This general principle can also be used for quantitative data.

One way of de-emphasizing precise levels of dissimilarity is to use *proximity thresholding* in a dimensionality-sensitive way. To perform proximity thresholding, the data are discretized into equidepth buckets. Each dimension is divided into k_d equidepth buckets, containing a fraction $1/k_d$ of the records. The number of buckets, k_d, is dependent on the data dimensionality d.

Let $\overline{X} = (x_1 \ldots x_d)$ and $\overline{Y} = (y_1 \ldots y_d)$ be two d-dimensional records. Then, for dimension i, if both x_i and y_i belong to the same bucket, the two records are said to be in proximity on dimension i. The subset of dimensions on which $\overline{X}$ and $\overline{Y}$ map to the same bucket is referred to as the proximity set, and it is denoted by $\mathcal{S}(\overline{X}, \overline{Y}, k_d)$. Furthermore, for each dimension $i \in \mathcal{S}(\overline{X}, \overline{Y}, k_d)$, let m_i and n_i be the upper and lower bounds of the bucket in dimension i, in which the two records are proximate to one another. Then, the

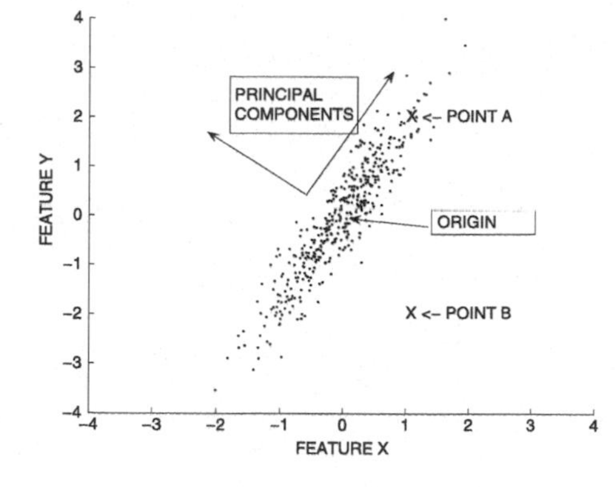

Figure 3.3: Global data distributions impact distance computations

similarity $PSelect(\overline{X}, \overline{Y}, k_d)$ is defined as follows:

$$PSelect(\overline{X}, \overline{Y}, k_d) = \left[\sum_{i \in \mathcal{S}(\overline{X}, \overline{Y}, k_d)} \left(1 - \frac{|x_i - y_i|}{m_i - n_i} \right)^p \right]^{1/p}. \tag{3.5}$$

The value of the aforementioned expression will vary between 0 and $|\mathcal{S}(\overline{X}, \overline{Y}, k_d)|$ because each individual expression in the summation lies between 0 and 1. This is a *similarity* function because larger values imply greater similarity.

The aforementioned similarity function guarantees a nonzero similarity component only for dimensions mapping to the same bucket. The use of equidepth partitions ensures that the probability of two records sharing a bucket for a particular dimension is given by $1/k_d$. Thus, on average, the aforementioned summation is likely to have d/k_d nonzero components. For more similar records, the number of such components will be greater, and each individual component is also likely to contribute more to the similarity value. The degree of dissimilarity on the distant dimensions is ignored by this approach because it is often dominated by noise. It has been shown theoretically [7] that picking $k_d \propto d$ achieves a constant level of contrast in high-dimensional space for certain data distributions. High values of k_d result in more stringent quality bounds for each dimension. These results suggest that in high-dimensional space, it is better to aim for higher quality bounds for each dimension, so that a smaller percentage (not number) of retained dimensions are used in similarity computation. An interesting aspect of this distance function is the nature of its sensitivity to data dimensionality. The choice of k_d with respect to d ensures that for low-dimensional applications, it bears some resemblance to the L_p-norm by using most of the dimensions; whereas for high-dimensional applications, it behaves similar to text domain-like similarity functions by using similarity on matching attributes. The distance function has also been shown to be more effective for a prototypical nearest-neighbor classification application.

3.2.1.6 Impact of Data Distribution

The L_p-norm depends only on the two data points in its argument and is invariant to the global statistics of the remaining data points. Should distances depend on the underlying data distribution of the remaining points in the data set? The answer is yes. To illustrate this point, consider the distribution illustrated in Fig. 3.3 that is centered at the origin. In

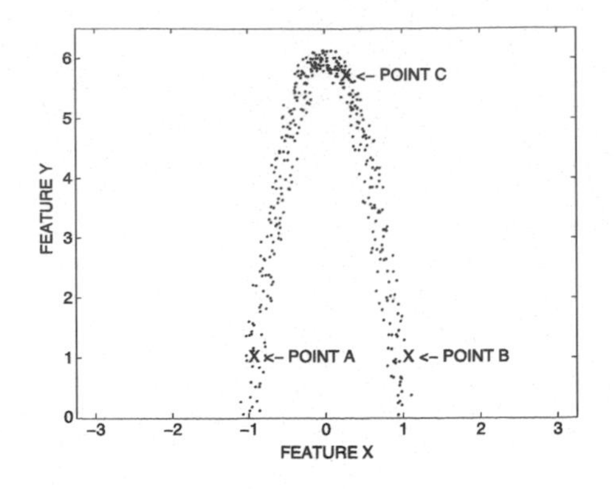

Figure 3.4: Impact of nonlinear distributions on distance computations

addition, two data points A= $(1, 2)$ and B= $(1, -2)$ are marked in the figure. Clearly, A and B are equidistant from the origin according to any L_p-norm. However, a question arises, as to whether A and B should truly be considered equidistant from the origin O. This is because the straight line from O to A is aligned with a *high-variance* direction in the data, and statistically, it is more *likely* for data points to be further away in this direction. On the other hand, many segments of the path from O to B are sparsely populated, and the corresponding direction is a *low-variance* direction. Statistically, it is much less likely for B to be so far away from O along this direction. Therefore, the distance from O to A *ought* to be less than that of O to B.

The *Mahalanobis distance* is based on this general principle. Let Σ be its $d \times d$ covariance matrix of the data set. In this case, the (i, j)th entry of the covariance matrix is equal to the covariance between the dimensions i and j. Then, the Mahalanobis distance $Maha(\overline{X}, \overline{Y})$ between two d-dimensional data points $\overline{X}$ and $\overline{Y}$ is as follows:

$$Maha(\overline{X}, \overline{Y}) = \sqrt{(\overline{X} - \overline{Y})\Sigma^{-1}(\overline{X} - \overline{Y})^T}.$$

A different way of understanding the Mahalanobis distance is in terms of principal component analysis (*PCA*). The Mahalanobis distance is similar to the Euclidean distance, except that it normalizes the data on the basis of the interattribute correlations. For example, if the axis system were to be rotated to the principal directions of the data (shown in Fig. 3.3), then the data would have no (second order) interattribute correlations. The Mahalanobis distance is equivalent to the Euclidean distance in such a transformed (axes-rotated) data set *after* dividing each of the transformed coordinate values by the standard deviation of the data along that direction. As a result, the data point B will have a larger distance from the origin than data point A in Fig. 3.3.

3.2.1.7 Nonlinear Distributions: ISOMAP

We now examine the case in which the data contain nonlinear distributions of arbitrary shape. For example, consider the global distribution illustrated in Fig. 3.4. Among the three data points A, B, and C, which pair are the closest to one another? At first sight, it would seem that data points A and B are the closest on the basis of Euclidean distance. However, the global data distribution tells us otherwise. One way of understanding distances is as the shortest length of the path from one data point to another, when using only point-to-point jumps from data points to one of their k-nearest neighbors based on a standard metric

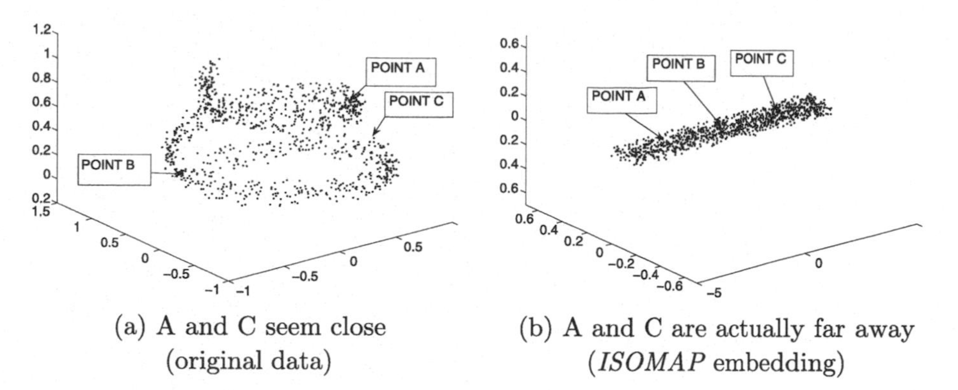

(a) A and C seem close (original data)

(b) A and C are actually far away (*ISOMAP* embedding)

Figure 3.5: Impact of *ISOMAP* embedding on distances

such as the Euclidean measure. The intuitive rationale for this is that only *short* point-to-point jumps can accurately measure minor changes in the generative process for that point. Therefore, the overall sum of the point-to-point jumps reflects the aggregate change (distance) from one point to another (distant) point more accurately than a straight-line distance between the points. Such distances are referred to as *geodesic distances.* In the case of Fig. 3.4, the only way to walk from A to B with short point-to-point jumps is to walk along the entire elliptical shape of the data distribution while passing C along the way. Therefore, A and B are actually the *farthest* pair of data points (from A, B, and C) on this basis! The implicit assumption is that nonlinear distributions are *locally* Euclidean but are *globally* far from Euclidean.

Such distances can be computed by using an approach that is derived from a nonlinear dimensionality reduction and embedding method, known as *ISOMAP*. The approach consists of two steps:

1. Compute the k-nearest neighbors of each point. Construct a weighted graph G with nodes representing data points, and edge weights (costs) representing distances of these k-nearest neighbors.

2. For any pair of points $\overline{X}$ and $\overline{Y}$, report $Dist(\overline{X}, \overline{Y})$ as the shortest path between the corresponding nodes in G.

These two steps are already able to compute the distances without explicitly performing dimensionality reduction. However, an additional step of embedding the data into a multidimensional space makes *repeated* distance computations between many pairs of points much faster, while losing some accuracy. Such an embedding also allows the use of algorithms that work naturally on numeric multidimensional data with predefined distance metrics.

This is achieved by using the all-pairs shortest-path problem to construct the full set of distances between any pair of nodes in G. Subsequently, *multidimensional scaling (MDS)* (cf. Sect. 2.4.4.2 of Chap. 2) is applied to embed the data into a lower dimensional space. The overall effect of the approach is to "straighten out" the nonlinear shape of Fig. 3.4 and embed it into a space where the data are aligned along a flat strip. In fact, a 1-dimensional representation can approximate the data after this transformation. Furthermore, in this new space, a distance function such as the Euclidean metric will work very well as long as metric *MDS* was used in the final phase. A 3-dimensional example is illustrated in Fig. 3.5a, in which the data are arranged along a spiral. In this figure, data points A and C seem much

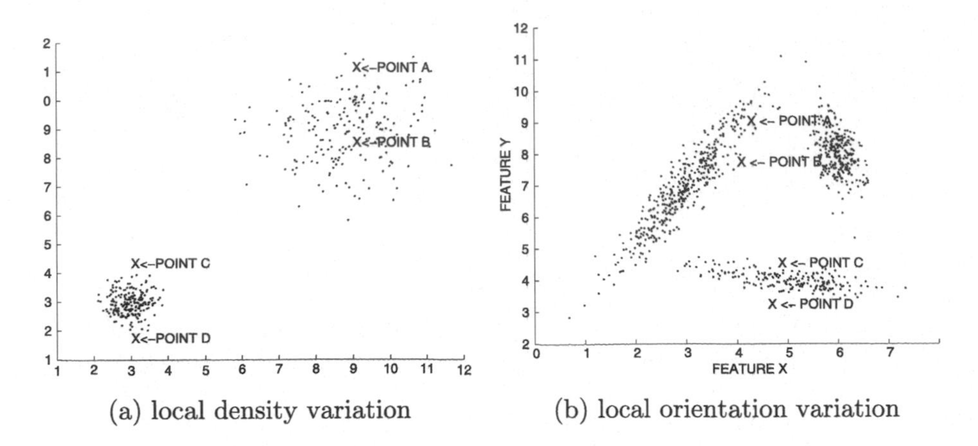

(a) local density variation (b) local orientation variation

Figure 3.6: Impact of local distributions on distance computations

closer to each other than data point B. However, in the *ISOMAP* embedding of Fig. 3.5b, the data point B is much closer to each of A and C. This example shows the drastic effect of data distributions on distance computation.

In general, high-dimensional data are aligned along nonlinear low-dimensional shapes, which are also referred to as *manifolds*. These manifolds can be "flattened out" to a new representation where metric distances can be used effectively. Thus, this is a data transformation method that facilitates the use of standard metrics. The major computational challenge is in performing the dimensionality reduction. However, after the one-time preprocessing cost has been paid for, repeated distance computations can be implemented efficiently.

Nonlinear embeddings can also be achieved with extensions of *PCA*. *PCA* can be extended to discovering nonlinear embeddings with the use of a method known as the *kernel trick*. Refer to Sect. 10.6.4.1 of Chap. 10 for a brief description of kernel *PCA*.

3.2.1.8 Impact of Local Data Distribution

The discussion so far addresses the impact of global distributions on the distance computations. However, the distribution of the data varies significantly with locality. This variation may be of two types. For example, the absolute density of the data may vary significantly with data locality, or the shape of clusters may vary with locality. The first type of variation is illustrated in Fig. 3.6a, which has two clusters containing the same number of points, but one of them is denser than the other. Even though the absolute distance between (A, B) is identical to that between (C, D), the distance between C and D should be considered greater on the basis of the *local* data distribution. In other words, C and D are much farther away in the *context* of what their local distributions look like. This problem is often encountered in many distance-based methods such as outlier detection. It has been shown that methods that adjust for the local variations in the distances typically perform much better than those that do not adjust for local variations. One of the most well-known methods for outlier detection, known as Local Outlier Factor (*LOF*), is based on this principle.

A second example is illustrated in Fig. 3.6b, which illustrates the impact of varying local orientation of the clusters. Here, the distance between (A, B) is identical to that between (C, D) using the Euclidean metric. However, the local clusters in each region show very different orientation. The high-variance axis of the cluster of data points relevant to (A, B) is aligned along the path from A to B. This is not true for (C, D). As a result, the intrinsic distance between C and D is much greater than that between A and B. For example, if the *local* Mahalanobis distance is computed using the relevant cluster covariance statistics, then the distance between C and D will evaluate to a larger value than that between A and B.

Shared Nearest-Neighbor Similarity: The first problem can be at least partially alleviated with the use of a shared nearest-neighbor similarity. In this approach, the k-nearest neighbors of each data point are computed in a preprocessing phase. The shared nearest-neighbor similarity is equal to the number of common neighbors between the two data points. This metric is locally sensitive because it depends on the number of common *neighbors*, and not on the absolute values of the distances. In dense regions, the k-nearest neighbor distances will be small, and therefore data points need to be closer together to have a larger number of shared nearest neighbors. Shared nearest-neighbor methods can be used to define a *similarity* graph on the underlying data points in which pairs of data points with at least one shared neighbor have an edge between them. Similarity graph-based methods are almost always locality sensitive because of their local focus on the k-nearest neighbor distribution.

Generic Methods: In generic local distance computation methods, the idea is to divide the space into a set of local regions. The distances are then adjusted in each region using the local statistics of this region. Therefore, the broad approach is as follows:

1. Partition the data into a set of local regions.

2. For any pair of objects, determine the most relevant region for the pair, and compute the pairwise distances using the local statistics of that region. For example, the local Mahalanobis distance may be used in each local region.

A variety of clustering methods are used for partitioning the data into local regions. In cases where each of the objects in the pair belongs to a different region, either the global distribution may be used, or the average may be computed using both local regions. Another problem is that the first step of the algorithm (partitioning process) itself requires a notion of distances for clustering. This makes the solution circular, and calls for an iterative solution. Although a detailed discussion of these methods is beyond the scope of this book, the bibliographic notes at the end of this chapter provide a number of pointers.

3.2.1.9 Computational Considerations

A major consideration in the design of distance functions is the computational complexity. This is because distance function computation is often embedded as a subroutine that is used repeatedly in the application at hand. If the subroutine is not efficiently implementable, the applicability becomes more restricted. For example, methods such as *ISOMAP* are computationally expensive and hard to implement for very large data sets because these methods scale with at least the square of the data size. However, they do have the merit that a one-time transformation can create a representation that can be used efficiently by data mining algorithms. Distance functions are executed repeatedly, whereas the preprocessing is performed only once. Therefore, it is definitely advantageous to use a preprocessing-intensive approach as long as it speeds up later computations. For many applications, sophisticated methods such as *ISOMAP* may be too expensive even for one-time analysis. For such cases, one of the earlier methods discussed in this chapter may need to be used. Among the methods discussed in this section, carefully chosen L_p-norms and match-based techniques are the fastest methods for large-scale applications.

3.2.2 Categorical Data

Distance functions are naturally computed as functions of value differences along dimensions in numeric data, which is *ordered*. However, no ordering exists among the discrete values of categorical data. How can distances be computed? One possibility is to transform the categorical data to numeric data with the use of the binarization approach discussed in Sect. 2.2.2.2 of Chap. 2. Because the binary vector is likely to be sparse (many zero values), similarity functions can be adapted from other sparse domains such as text. For the case of categorical data, it is more common to work with similarity functions rather than distance functions because discrete values can be matched more naturally.

Consider two records $\overline{X} = (x_1 \dots x_d)$ and $\overline{Y} = (y_1 \dots y_d)$. The simplest possible similarity between the records $\overline{X}$ and $\overline{Y}$ is the sum of the similarities on the individual attribute values. In other words, if $S(x_i, y_i)$ is the similarity between the attributes values x_i and y_i, then the overall similarity is defined as follows:

$$Sim(\overline{X}, \overline{Y}) = \sum_{i=1}^{d} S(x_i, y_i).$$

Therefore, the choice of $S(x_i, y_i)$ defines the overall similarity function.

The simplest possible choice is to set $S(x_i, y_i)$ to 1 when $x_i = y_i$ and 0 otherwise. This is also referred to as the *overlap* measure. The major drawback of this measure is that it does not account for the relative frequencies among the different attributes. For example, consider a categorical attribute in which the attribute value is "Normal" for 99 % of the records, and either "Cancer" or "Diabetes" for the remaining records. Clearly, if two records have a "Normal" value for this variable, this does not provide statistically significant information about the similarity, because the majority of pairs are likely to show that pattern just by chance. However, if the two records have a matching "Cancer" or "Diabetes" value for this variable, it provides significant statistical evidence of similarity. This argument is similar to that made earlier about the importance of the global data distribution. Similarities or differences that are unusual are statistically more significant than those that are common.

In the context of categorical data, the *aggregate statistical properties* of the data set should be used in computing similarity. This is similar to how the Mahalanobis distance was used to compute similarity more accurately with the use of global statistics. The idea is that matches on unusual values of a categorical attribute should be weighted more heavily than values that appear frequently. This also forms the underlying principle of many common normalization techniques that are used in domains such as text. An example, which is discussed in the next section, is the use of *inverse document frequency (IDF)* in the information retrieval domain. An analogous measure for categorical data will be introduced here.

The *inverse occurrence frequency* is a generalization of the simple matching measure. This measure weights the similarity between the matching attributes of two records by an inverse function of the frequency of the matched value. Thus, when $x_i = y_i$, the similarity $S(x_i, y_i)$ is equal to the inverse weighted frequency, and 0 otherwise. Let $p_k(x)$ be the fraction of records in which the kth attribute takes on the value of x in the data set. In other words, when $x_i = y_i$, the value of $S(x_i, y_i)$ is $1/p_k(x_i)^2$ and 0 otherwise.

$$S(x_i, y_i) = \begin{cases} 1/p_k(x_i)^2 & \text{if } x_i = y_i \\ 0 & \text{otherwise} \end{cases} \tag{3.6}$$

A related measure is the *Goodall* measure. As in the case of the inverse occurrence frequency, a higher similarity value is assigned to a match when the value is infrequent. In a simple variant of this measure [104], the similarity on the kth attribute is defined as $1 - p_k(x_i)^2$, when $x_i = y_i$, and 0 otherwise.

$$S(x_i, y_i) = \begin{cases} 1 - p_k(x_i)^2 & \text{if } x_i = y_i \\ 0 & \text{otherwise} \end{cases} \tag{3.7}$$

The bibliographic notes contain pointers to various similarity measures for categorical data.

3.2.3 Mixed Quantitative and Categorical Data

It is fairly straightforward to generalize the approach to mixed data by adding the weights of the numeric and quantitative components. The main challenge is in deciding how to assign the weights of the quantitative and categorical components. For example, consider two records $\overline{X} = (\overline{X_n}, \overline{X_c})$ and $\overline{Y} = (\overline{Y_n}, \overline{Y_c})$ where $\overline{X_n}$, $\overline{Y_n}$ are the subsets of numerical attributes and $\overline{X_c}$, $\overline{Y_c}$ are the subsets of categorical attributes. Then, the overall similarity between $\overline{X}$ and $\overline{Y}$ is defined as follows:

$$Sim(\overline{X}, \overline{Y}) = \lambda \cdot NumSim(\overline{X_n}, \overline{Y_n}) + (1 - \lambda) \cdot CatSim(\overline{X_c}, \overline{Y_c}). \tag{3.8}$$

The parameter λ regulates the relative importance of the categorical and numerical attributes. The choice of λ is a difficult one. In the absence of domain knowledge about the relative importance of attributes, a natural choice is to use a value of λ that is equal to the fraction of numerical attributes in the data. Furthermore, the proximity in numerical data is often computed with the use of distance functions rather than similarity functions. However, distance values can be converted to similarity values as well. For a distance value of *dist*, a common approach is to use a kernel mapping that yields [104] the similarity value of $1/(1 + dist)$.

Further normalization is required to meaningfully compare the similarity value components on the numerical and categorical attributes that may be on completely different scales. One way of achieving this goal is to determine the standard deviations in the similarity values over the two domains with the use of sample pairs of records. Each component of the similarity value (numerical or categorical) is divided by its standard deviation. Therefore, if σ_c and σ_n are the standard deviations of the similarity values in the categorical and numerical components, then Eq. 3.8 needs to be modified as follows:

$$Sim(\overline{X}, \overline{Y}) = \lambda \cdot NumSim(\overline{X_n}, \overline{Y_n})/\sigma_n + (1 - \lambda) \cdot CatSim(\overline{X_c}, \overline{Y_c})/\sigma_c. \tag{3.9}$$

By performing this normalization, the value of λ becomes more meaningful, as a true *relative weight* between the two components. By default, this weight can be set to be proportional to the number of attributes in each component unless specific domain knowledge is available about the relative importance of attributes.

3.3 Text Similarity Measures

Strictly speaking, text can be considered quantitative multidimensional data when it is treated as a bag of words. The frequency of each word can be treated as a quantitative attribute, and the base lexicon can be treated as the full set of attributes. However, the structure of text is *sparse* in which most attributes take on 0 values. Furthermore, all word frequencies are nonnegative. This special structure of text has important implications for similarity computation and other mining algorithms. Measures such as the L_p-norm do not adjust well to the varying length of the different documents in the collection. For example, the L_2-distance between two long documents will almost always be larger than that between two short documents even if the two long documents have many words in common, and the short documents are completely disjoint. How can one normalize for such irregularities? One way of doing so is by using the cosine measure. The cosine measure computes the *angle* between the two documents, which is insensitive to the absolute length of the document. Let $\overline{X} = (x_1 \dots x_d)$ and $\overline{Y} = (y_1 \dots y_d)$ be two documents on a lexicon of size d. Then, the cosine measure $\cos(\overline{X}, \overline{Y})$ between $\overline{X}$ and $\overline{Y}$ can be defined as follows:

$$\cos(\overline{X}, \overline{Y}) = \frac{\sum_{i=1}^{d} x_i \cdot y_i}{\sqrt{\sum_{i=1}^{d} x_i^2} \cdot \sqrt{\sum_{i=1}^{d} y_i^2}}. \tag{3.10}$$

The aforementioned measure simply uses the raw frequencies between attributes. However, as in other data types, it is possible to use global statistical measures to improve the similarity computation. For example, if two documents match on an uncommon word, it is more indicative of similarity than the case where two documents match on a word that occurs very commonly. The *inverse document frequency* id_i, which is a decreasing function of the number of documents n_i in which the ith word occurs, is commonly used for normalization:

$$id_i = \log(n/n_i). \tag{3.11}$$

Here, the number of documents in the collection is denoted by n. Another common adjustment is to ensure that the excessive presence of single word does not throw off the similarity measure. A damping function $f(\cdot)$, such as the square root or the logarithm, is optionally applied to the frequencies before similarity computation.

$$f(x_i) = \sqrt{x_i}$$
$$f(x_i) = \log(x_i)$$

In many cases, the damping function is not used, which is equivalent to setting $f(x_i)$ to x_i. Therefore, the *normalized* frequency $h(x_i)$ for the ith word may be defined as follows:

$$h(x_i) = f(x_i) \cdot id_i. \tag{3.12}$$

Then, the cosine measure is defined as in Eq. 3.10, except that the normalized frequencies of the words are used:

$$\cos(\overline{X}, \overline{Y}) = \frac{\sum_{i=1}^{d} h(x_i) \cdot h(y_i)}{\sqrt{\sum_{i=1}^{d} h(x_i)^2} \cdot \sqrt{\sum_{i=1}^{d} h(y_i)^2}}. \tag{3.13}$$

Another measure that is less commonly used for text is the *Jaccard coefficient* $J(\overline{X}, \overline{Y})$:

$$J(\overline{X}, \overline{Y}) = \frac{\sum_{i=1}^{d} h(x_i) \cdot h(y_i)}{\sum_{i=1}^{d} h(x_i)^2 + \sum_{i=1}^{d} h(y_i)^2 - \sum_{i=1}^{d} h(x_i) \cdot h(y_i)}. \tag{3.14}$$

The Jaccard coefficient is rarely used for the text domain, but it is used commonly for sparse binary data sets.

3.3.1 Binary and Set Data

Binary multidimensional data are a representation of set-based data, where a value of 1 indicates the presence of an element in a set. Binary data occur commonly in market-basket domains in which transactions contain information corresponding to whether or not an item is present in a transaction. It can be considered a special case of text data in which word frequencies are either 0 or 1. If S_X and S_Y are two sets with binary representations $\overline{X}$ and $\overline{Y}$, then it can be shown that applying Eq. 3.14 to the raw binary representation of the two sets is equivalent to:

$$J(\overline{X}, \overline{Y}) = \frac{\sum_{i=1}^{d} x_i \cdot y_i}{\sum_{i=1}^{d} x_i^2 + \sum_{i=1}^{d} y_i^2 - \sum_{i=1}^{d} x_i \cdot y_i} = \frac{|S_X \cap S_Y|}{|S_X \cup S_Y|}. \tag{3.15}$$

This is a particularly intuitive measure because it carefully accounts for the number of common and disjoint elements in the two sets.

3.4 Temporal Similarity Measures

Temporal data contain a single contextual attribute representing time and one or more behavioral attributes that measure the properties varying along a particular time period. Temporal data may be represented as continuous time series, or as discrete sequences, depending on the application domain. The latter representation may be viewed as the discrete version of the former. It should be pointed out that discrete sequence data are not always temporal because the contextual attribute may represent placement. This is typically the case in biological sequence data. Discrete sequences are also sometimes referred to as *strings*. Many of the similarity measures used for time series and discrete sequences can be reused across either domain, though some of the measures are more suited to one of the domains. Therefore, this section will address both data types, and each similarity measure will be discussed in a subsection on either continuous series or discrete series, based on its most common use. For some measures, the usage is common across both data types.

3.4.1 Time-Series Similarity Measures

The design of time-series similarity measures is highly application specific. For example, the simplest possible similarity measure between two time series of equal length is the Euclidean metric. Although such a metric may work well in many scenarios, it does not account for several distortion factors that are common in many applications. Some of these factors are as follows:

1. *Behavioral attribute scaling and translation:* In many applications, the different time series may not be drawn on the same scales. For example, the time series representing various stocks prices may show similar patterns of movements, but the absolute values may be very different both in terms of the mean and the standard deviation. For example, the share prices of several different hypothetical stock tickers are illustrated in Fig. 3.7. All three series show similar patterns but with different scaling and some random variations. Clearly, they show similar patterns but cannot be meaningfully compared if the absolute values of the series are used.

2. *Temporal (contextual) attribute translation:* In some applications, such as real-time analysis of financial markets, the different time series may represent the same periods

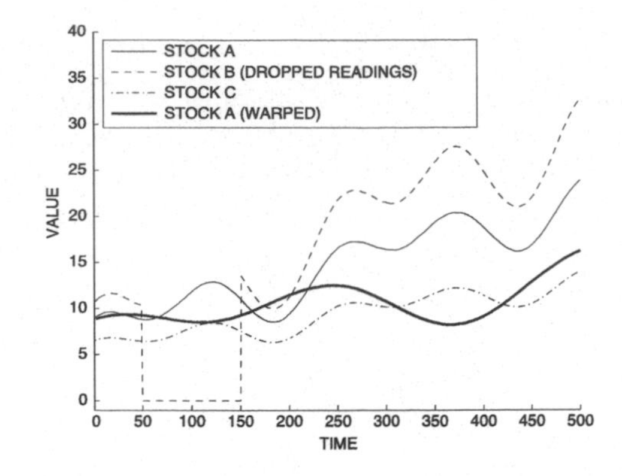

Figure 3.7: Impact of scaling, translation, and noise

in time. In other applications, such as the analysis of the time series obtained from medical measurements, the absolute time stamp of when the reading was taken is not important. In such cases, the temporal attribute value needs to be shifted in at least one of the time series to allow more effective matching.

3. *Temporal (contextual) attribute scaling:* In this case, the series may need to be stretched or compressed along the temporal axis to allow more effective matching. This is referred to as *time warping.* An additional complication is that different temporal segments of the series may need to be warped differently to allow for better matching. In Fig. 3.7, the simplest case of warping is shown where the entire set of values for stock A has been stretched. In general, the time warping can be more complex where different windows in the same series may be stretched or compressed differently. This is referred to as *dynamic* time warping (DTW).

4. *Noncontiguity in matching:* Long time series may have noisy segments that do not match very well with one another. For example, one of the series in Fig. 3.7 has a window of dropped readings because of data collection limitations. This is common in sensor data. The distance function may need to be robust to such noise.

Some of these issues can be addressed by attribute normalization during preprocessing.

3.4.1.1 Impact of Behavioral Attribute Normalization

The translation and scaling issues are often easier to address for the behavioral attributes as compared to contextual attributes, because they can be addressed by normalization during preprocessing:

1. *Behavioral attribute translation:* The behavioral attribute is mean centered during preprocessing.

2. *Behavioral attribute scaling:* The standard deviation of the behavioral attribute is scaled to 1 unit.

It is important to remember that these normalization issues may not be relevant to every application. Some applications may require only translation, only scaling, or neither of the two. Other applications may require both. In fact, in some cases, the wrong choice of

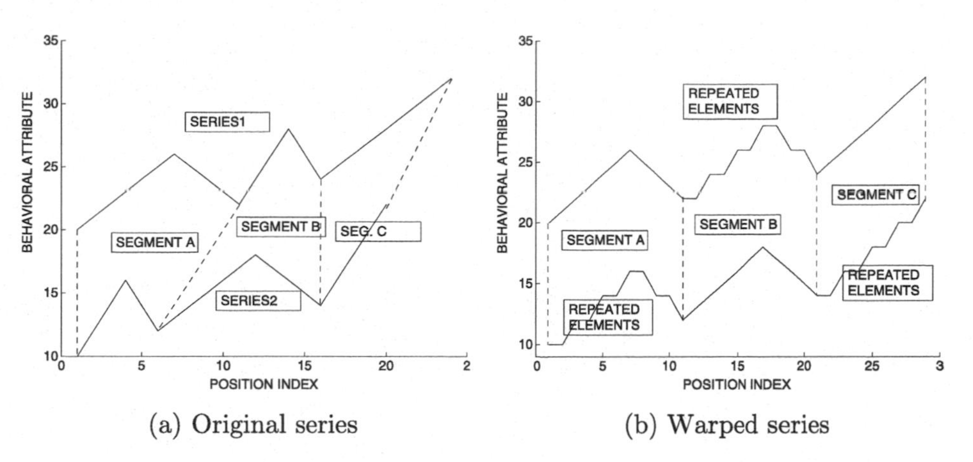

(a) Original series

(b) Warped series

Figure 3.8: Illustration of dynamic time warping by repeating elements

normalization may have detrimental effects on the interpretability of the results. Therefore, an analyst needs to judiciously select a normalization approach depending on application-specific needs.

3.4.1.2 L_p-Norm

The L_p-norm may be defined for two series $\overline{X} = (x_1 \ldots x_n)$ and $\overline{Y} = (y_1 \ldots y_n)$. This measure treats a time series as a multidimensional data point in which each time stamp is a dimension.

$$Dist(\overline{X}, \overline{Y}) = \left(\sum_{i=1}^{n} |x_i - y_i|^p \right)^{1/p} \tag{3.16}$$

The L_p-norm can also be applied to wavelet transformations of the time series. In the special case where $p = 2$, accurate distance computations are obtained with the wavelet representation, if most of the larger wavelet coefficients are retained in the representation. In fact, it can be shown that if no wavelet coefficients are removed, then the distances are identical between the two representations. This is because wavelet transformations can be viewed as a rotation of an axis system in which each dimension represents a time stamp. Euclidean metrics are invariant to axis rotation. The major problem with L_p-norms is that they are designed for time series of equal length and cannot address distortions on the temporal (contextual) attributes.

3.4.1.3 Dynamic Time Warping Distance

DTW stretches the series along the time axis in a varying (or *dynamic*) way over different portions to enable more effective matching. An example of warping is illustrated in Fig. 3.8a, where the two series have very similar shape in segments A, B, and C, but specific segments in each series need to be stretched appropriately to enable better matching. The DTW measure has been adapted from the field of speech recognition, where time warping was deemed necessary to match different speaking speeds. DTW can be used either for time-series or sequence data, because it addresses only the issue of contextual attribute scaling, and it is unrelated to the nature of the behavioral attribute. The following description is a generic one, which can be used either for time-series or sequence data.

The L_p-metric can only be defined between two time series of equal length. However, DTW, by its very nature, allows the measurement of distances between two series of *different* lengths. In the L_p distance, a one-to-one mapping exists between the time stamps of the two time series. However, in DTW, a many-to-one mapping is allowed to account for the time warping. This many-to-one mapping can be thought of in terms of repeating some of the elements in carefully chosen segments of either of the two time series. This can be used to artificially create two series of the same length that have a one-to-one mapping between them. The distances can be measured on the resulting warped series using any distance measure such as the L_p-norm. For example, in Fig. 3.8b, some elements in a few segments of either series are repeated to create a one-to-one mapping between the two series. Note that the two series now look much more similar than the two series in Fig. 3.8a. Of course, this repeating can be done in many different ways, and the goal is to perform it in an *optimal* way to minimize the DTW distance. The optimal choice of warping is determined using dynamic programming.

To understand how DTW generalizes a one-to-one distance metric such as the L_p-norm, consider the L_1 (Manhattan) metric $M(\overline{X_i}, \overline{Y_i})$, computed on the first i elements of two time series $\overline{X} = (x_1 \dots x_n)$ and $\overline{Y} = (y_1 \dots y_n)$ of equal length. The value of $M(\overline{X_i}, \overline{Y_i})$ can be written *recursively* as follows:

$$M(\overline{X_i}, \overline{Y_i}) = |x_i - y_i| + M(\overline{X_{i-1}}, \overline{Y_{i-1}}). \tag{3.17}$$

Note that the indices of *both* series are reduced by 1 in the right-hand side because of the one-to-one matching. In DTW, both indices need not reduce by 1 unit because a many-to-one mapping is allowed. Rather, any one or both indices may reduce by 1, depending on the *best match* between the two time series (or sequences). The index that did *not* reduce by 1 corresponds to the repeated element. The choice of index reduction is naturally defined, recursively, as an optimization over the various options.

Let $DTW(i, j)$ be the optimal distance between the first i and first j elements of two time series $\overline{X} = (x_1 \dots x_m)$ and $\overline{Y} = (y_1 \dots y_n)$, respectively. Note that the two time series are of lengths m and n, which may not be the same. Then, the value of $DTW(i, j)$ is defined recursively as follows:

$$DTW(i,j) = distance(x_i, y_j) + \min \begin{cases} DTW(i, j-1) & \text{repeat } x_i \\ DTW(i-1, j) & \text{repeat } y_j \\ DTW(i-1, j-1) & \text{repeat neither} \end{cases}. \tag{3.18}$$

The value of $distance(x_i, y_j)$ may be defined in a variety of ways, depending on the application domain. For example, for continuous time series, it may be defined as $|x_i - y_i|^p$, or by a distance that accounts for (behavioral attribute) scaling and translation. For discrete sequences, it may be defined using a categorical measure. The DTW approach is primarily focused on warping the *contextual* attribute, and has little to do with the nature of the behavioral attribute or distance function. Because of this fact, time warping can easily be extended to multiple behavioral attributes by simply using the distances along multiple attributes in the recursion.

Equation 3.18 yields a natural iterative approach. The approach starts by initializing $DTW(0, 0)$ to 0, $DTW(0, j)$ to ∞ for $j \in \{1 \dots n\}$, and $DTW(i, 0)$ to ∞ for $i \in \{1 \dots m\}$. The algorithm computes $DTW(i, j)$ by repeatedly executing Eq. 3.18 with increasing index values of i and j. This can be achieved by a simple nested loop in which the indices i and j increase from 1 to m and 1 to n, respectively:

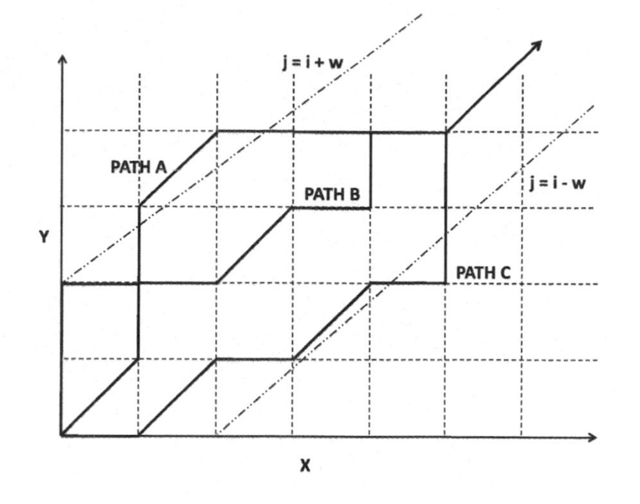

Figure 3.9: Illustration of warping paths

```
for i = 1 to m
   for j = 1 to n
      compute DTW(i, j) using Eq. 3.18
```

The aforementioned code snippet is a nonrecursive and iterative approach. It is also possible to implement a recursive computer program by directly using Eq. 3.18. Therefore, the approach requires the computation of all values of $DTW(i, j)$ for every $i \in [1, m]$ and every $j \in [1, n]$. This is a $m \times n$ grid of values, and therefore the approach may require $O(m \cdot n)$ iterations, where m and n are lengths of the series.

The optimal warping can be understood as an *optimal path* through different values of i and j in the $m \times n$ grid of values, as illustrated in Fig. 3.9. Three possible paths, denoted by A, B, and C, are shown in the figure. These paths only move to the right (increasing i and repeating y_j), upward (increasing j and repeating x_i), or both (repeating neither).

A number of practical constraints are often added to the DTW computation. One commonly used constraint is the *window constraint* that imposes a minimum level w of positional alignment between matched elements. The window constraint requires that $DTW(i, j)$ be computed only when $|i - j| \leq w$. Otherwise, the value may be set to ∞ by default. For example, the paths B and C in Fig. 3.9 no longer need to be computed. This saves the computation of many values in the dynamic programming recursion. Correspondingly, the computations in the inner variable j of the nested loop above can be saved by constraining the index j, so that it is never more than w units apart from the outer loop variable i. Therefore, the inner loop index j is varied from $\max\{0, i - w\}$ to $\min\{n, i + w\}$.

The *DTW* distance can be extended to multiple behavioral attributes easily, if it is assumed that the different behavioral attributes have the same time warping. In this case, the recursion is unchanged, and the only difference is that $distance(\overline{x_i}, \overline{y_j})$ is computed using a vector-based distance measure. We have used a bar on $\overline{x_i}$ and $\overline{y_j}$ to denote that these are vectors of multiple behavioral attributes. This multivariate extension is discussed in Sect. 16.3.4.1 of Chap. 16 for measuring distances between 2-dimensional trajectories.

3.4.1.4 Window-Based Methods

The example in Fig. 3.7 illustrates a case where dropped readings may cause a gap in the matching. Window-based schemes attempt to decompose the two series into windows and then "stitch" together the similarity measure. The intuition here is that if two series have many contiguous matching segments, they should be considered similar. For long time series, a global match becomes increasingly unlikely. The only reasonable choice is the use of windows for measurement of segment-wise similarity.

Consider two time series $\overline{X}$ and $\overline{Y}$, and let $\overline{X_1} \ldots \overline{X_r}$ and $\overline{Y_1} \ldots \overline{Y_r}$ be temporally ordered and nonoverlapping windows extracted from the respective series. Note that some windows from the base series may not be included in these segments at all. These correspond to the noise segments that are dropped. Then, the overall similarity between $\overline{X}$ and $\overline{Y}$ can be computed as follows:

$$Sim(\overline{X}, \overline{Y}) = \sum_{i=1}^{r} Match(\overline{X_i}, \overline{Y_i}). \tag{3.19}$$

A variety of measures discussed in this section may be used to instantiate the value of $Match(\overline{X_i}, \overline{Y_i})$. It is tricky to determine the proper value of $Match(\overline{X_i}, \overline{Y_i})$ because a contiguous match along a long window is more unusual than many short segments of the same length. The proper choice of $Match(\overline{X_i}, \overline{Y_i})$ may depend on the application at hand. Another problem is that the optimal decomposition of the series into windows may be a difficult task. These methods are not discussed in detail here, but the interested reader is referred to the bibliographic notes for pointers to relevant methods.

3.4.2 Discrete Sequence Similarity Measures

Discrete sequence similarity measures are based on the same general principles as time-series similarity measures. As in the case of time-series data, discrete sequence data may or may not have a one-to-one mapping between the positions. When a one-to-one mapping does exist, many of the multidimensional categorical distance measures can be adapted to this domain, just as the L_p-norm can be adapted to continuous time series. However, the application domains of discrete sequence data are most often such that a one-to-one mapping does not exist. Aside from the DTW approach, a number of other dynamic programming methods are commonly used.

3.4.2.1 Edit Distance

The edit distance defines the distance between two strings as the *least* amount of "effort" (or *cost*) required to transform one sequence into another by using a series of transformation operations, referred to as "edits." The edit distance is also referred to as the *Levenshtein* distance. The edit operations include the use of symbol insertions, deletions, and replacements with specific costs. In many models, replacements are assumed to have higher cost than insertions or deletions, though insertions and deletions are usually assumed to have the same cost. Consider the sequences *ababababab* and *bababababa*, which are drawn on the alphabet $\{a, b\}$. The first string can be transformed to the second in several ways. For example, if every alphabet in the first string was replaced by the other alphabet, it would result in the second string. The cost of doing so is that of ten replacements. However, a more cost-efficient way of achieving the same goal is to delete the leftmost element of the string, and insert the symbol "a" as the rightmost element. The cost of this sequence of operations is only one insertion and one deletion. The edit distance is defined as the optimal cost to transform one string to another with a sequence of insertions, deletions, and replacements. The computation of the optimal cost requires a dynamic programming recursion.

For two sequences $\overline{X} = (x_1 \dots x_m)$ and $\overline{Y} = (y_1 \dots y_n)$, let the edits be performed on sequence $\overline{X}$ to transform to $\overline{Y}$. Note that this distance function is asymmetric because of the directionality to the edit. For example, $Edit(\overline{X}, \overline{Y})$ may not be the same as $Edit(\overline{Y}, \overline{X})$ if the insertion and deletion costs are not identical. In practice, however, the insertion and deletion costs are assumed to be the same.

Let I_{ij} be a binary indicator that is 0 when the ith symbol of $\overline{X}$ and jth symbols of $\overline{Y}$ are the same. Otherwise, the value of this indicator is 1. Then, consider the first i symbols of $\overline{X}$ and the first j symbols of $\overline{Y}$. Assume that these segments are represented by $\overline{X_i}$ and $\overline{Y_j}$, respectively. Let $Edit(i, j)$ represent the optimal matching cost between these segments. The goal is to determine what operation to perform on the last element of $\overline{X_i}$ so that it either matches an element in $\overline{Y_j}$, or it is deleted. Three possibilities arise:

1. The last element of $\overline{X_i}$ is deleted, and the cost of this is $[Edit(i-1, j) + \text{Deletion Cost}]$. The last element of the truncated segment $\overline{X_{i-1}}$ may or may not match the last element of $\overline{Y_j}$ at this point.

2. An element is inserted at the end of $\overline{X_i}$ to match the last element of Y_j, and the cost of this is $[Edit(i, j-1) + \text{Insertion Cost}]$. The indices of the edit term $Edit(i, j-1)$ reflect the fact that the matched elements of both series can now be removed.

3. The last element of $\overline{X_i}$ is flipped to that of $\overline{Y_j}$ *if it is different*, and the cost of this is $[Edit(i-1, j-1) + I_{ij} \cdot (\text{Replacement Cost})]$. In cases where the last elements are the same, the additional replacement cost is not incurred, but progress is nevertheless made in matching. This is because the matched elements (x_i, y_j) of both series need not be considered further, and residual matching cost is $Edit(i-1, j-1)$.

Clearly, it is desirable to pick the minimum of these costs for the optimal matching. Therefore, the optimal matching is defined by the following recursion:

$$Edit(i, j) = \min \begin{cases} Edit(i-1, j) + \text{Deletion Cost} \\ Edit(i, j-1) + \text{Insertion Cost} \\ Edit(i-1, j-1) + I_{ij} \cdot (\text{Replacement Cost}) \end{cases} . \tag{3.20}$$

Furthermore, the bottom of the recursion also needs to be set up. The value of $Edit(i, 0)$ is equal to the cost of i deletions for any value of i, and that of $Edit(0, j)$ is equal to the cost of j insertions for any value of j. This nicely sets up the dynamic programming approach. It is possible to write the corresponding computer program either as a nonrecursive nested loop (as in *DTW*) or as a recursive computer program that directly uses the aforementioned cases.

The aforementioned discussion assumes general insertion, deletion, and replacement costs. In practice, however, the insertion and deletion costs are usually assumed to be the same. In such a case, the edit function is symmetric because it does not matter which of the two strings is edited to the other. For any sequence of edits from one string to the other, a reverse sequence of edits, with the same cost, will exist from the other string to the first.

The edit distance can be extended to numeric data by changing the primitive operations of *insert*, *delete*, and *replace* to transformation rules that are designed for time series. Such transformation rules can include making basic changes to the shape of the time series in window segments. This is more complex because it requires one to design the base set of allowed time-series shape transformations and their costs. Such an approach has not found much popularity for time-series distance computation.

3.4.2.2 Longest Common Subsequence

A *subsequence* of a sequence is a set of symbols drawn from the sequence in the same order as the original sequence. A subsequence is different from a *substring* in that the values of the subsequence need not be contiguous, whereas the values in the substring need to be contiguous. Consider the sequences $agbfcgdhei$ and $afbgchdiei$. In this case, ei is a substring of both sequences and also a subsequence. However, $abcde$ and fgi are subsequences of both strings but not substrings. Clearly, subsequences of longer length are indicative of a greater level of matching between the strings. Unlike the edit distance, the longest common subsequence (*LCSS*) is a similarity function because higher values indicate greater similarity. The number of possible subsequences is exponentially related to the length of a string. However, the LCSS can be computed in polynomial time with a dynamic programming approach.

For two sequences $\overline{X} = (x_1 \ldots x_m)$ and $\overline{Y} = (y_1 \ldots y_n)$, consider the first i symbols of $\overline{X}$ and the first j symbols of $\overline{Y}$. Assume that these segments are represented by $\overline{X_i}$ and $\overline{Y_j}$, respectively. Let $LCSS(i, j)$ represent the optimal LCSS values between these segments. The goal here is to either match the last element of $\overline{X_i}$ and $\overline{Y_j}$, or delete the last element in one of the two sequences. Two possibilities arise:

1. The last element of $\overline{X_i}$ matches $\overline{Y_j}$, in which case, it cannot hurt to instantiate the matching on the last element and then delete the last element of both sequences. The similarity value $LCSS(i, j)$ can be expressed recursively as this is $LCSS(i-1, j-1)+1$.

2. The last element does not match. In such a case, the last element of at least one of the two strings needs to be deleted under the assumption that it cannot occur in the matching. In this case, the value of $LCSS(i, j)$ is either $LCSS(i, j-1)$ or $LCSS(i-1, j)$, depending on which string is selected for deletion.

Therefore, the optimal matching can be expressed by enumerating these cases:

$$LCSS(i, j) = \max \begin{cases} LCSS(i-1, j-1) + 1 & \text{only if } x_i = y_j \\ LCSS(i-1, j) & \text{otherwise (no match on } x_i) \\ LCSS(i, j-1) & \text{otherwise (no match on } y_j) \end{cases} . \tag{3.21}$$

Furthermore, the boundary conditions need to be set up. The values of $LCSS(i, 0)$ and $LCSS(0, j)$ are always equal to 0 for any value of i and j. As in the case of the DTW and edit-distance computations, a nested loop can be set up to compute the final value. A recursive computer program can also be implemented that uses the aforementioned recursive relationship. Although the *LCSS* approach is defined for a discrete sequence, it can also be applied to a continuous time series after discretizing the time-series values into a sequence of categorical values. Alternatively, one can discretize the time-series *movement* between two contiguous time stamps. The particular choice of discretization depends on the goals of the application at hand.

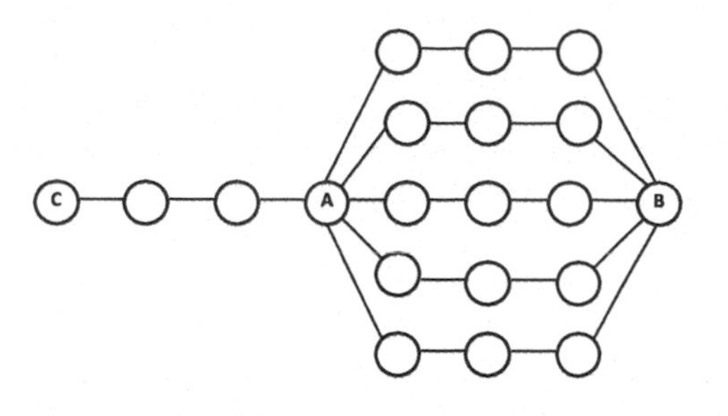

Figure 3.10: Shortest path versus homophily

3.5 Graph Similarity Measures

The similarity in graphs can be measured in different ways, depending on whether the similarity is being measured between two graphs, or between two nodes in a single graph. For simplicity, undirected networks are assumed, though the measures can be easily generalized to directed networks.

3.5.1 Similarity between Two Nodes in a Single Graph

Let $G = (N, A)$ be an undirected network with node set N and edge set A. In some domains, costs are associated with nodes, whereas in others, weights are associated with nodes. For example, in domains such as bibliographic networks, the edges are naturally weighted, and in road networks, the edges naturally have costs. Typically, *distance* functions work with costs, whereas *similarity* functions work with weights. Therefore, it may be assumed that either the cost c_{ij}, or the weight w_{ij} of the edge (i, j) is specified. It is often possible to convert costs into weights (and vice versa) using simple heuristic kernel functions that are chosen in an application-specific way. An example is the heat kernel $K(x) = e^{-x^2/t^2}$.

It is desired to measure the similarity between any pair of nodes i and j. The principle of similarity between two nodes in a single graph is based on the concept of *homophily* in real networks. The principle of homophily is that nodes are typically more similar in a network when they are connected to one another with edges. This is common in many domains such as the Web and social networks. Therefore, nodes that are connected via *short paths* and *many paths* should be considered more similar. The latter criterion is closely related to the concept of connectivity between nodes. The first criterion is relatively easy to implement with the use of the shortest-path algorithm in networks.

3.5.1.1 Structural Distance-Based Measure

The goal here is to measure the distances from any source node s to any other node in the network. Let $SP(s, j)$ be the shortest-path distance from source node s to any node j. The value of $SP(s, j)$ is initialized to 0 for $j = s$ and ∞ otherwise. Then, the distance computation of s to all other nodes in the network may be summarized in a single step that is performed exactly once for each node in the network in a certain order:

- Among all nodes not examined so far, select the node i with the smallest value of $SP(s, i)$ and update the distance labels of each of its neighbors j as follows:

$$SP(s, j) = \min\{SP(s, j), SP(s, i) + c_{ij}\}. \tag{3.22}$$

This is the essence of the well-known Dijkstra algorithm. This approach is linear in the number of edges in the network, because it examines each node and its incident edges exactly once. The approach provides the distances from a single node to all other nodes in a single pass. The final value of $SP(s, j)$ provides a quantification of the structural distance between node s and node j. Structural distance-based measures do not leverage the multiplicity in paths between a pair of nodes because they focus only on the raw structural distances.

3.5.1.2 Random Walk-Based Similarity

The structural measure of the previous section does not work well when pairs of nodes have varying numbers of paths between them. For example, in Fig. 3.10, the shortest-path length between nodes A and B is 4, whereas that between A and C is 3. Yet, node B should be considered more similar to A because the two nodes are more tightly connected with a *multiplicity* of paths. The idea of random walk-based similarity is based on this principle.

In random walk-based similarity, the approach is as follows: Imagine a random walk that starts at source node s, and proceeds to an adjacent node with weighted probability proportional to w_{ij}. Furthermore, at any given node, it is allowed to "jump back" to the source node s with a probability referred to as the *restart probability.* This will result in a probability distribution that is heavily biased toward the source node s. Nodes that are more similar to s will have higher probability of visits. Such an approach will adjust very well to the scenario illustrated in Fig. 3.10 because the walk will visit B more frequently.

The intuition here is the following: If you were lost in a road network and drove randomly, while taking turns randomly, which location are you more *likely* to reach? You are more likely to reach a location that is close by *and* can be reached in multiple ways. The random-walk measure therefore provides a result that is different from that of the shortest-path measure because it also accounts for multiplicity in paths during similarity computation.

This similarity computation is closely related to concept of *PageRank*, which is used to rank pages on the Web by search engines. The corresponding modification for measuring similarity between nodes is also referred to as *personalized PageRank*, and a symmetric variant is referred to as *SimRank*. This chapter will not discuss the details of *PageRank* and *SimRank* computation, because it requires more background on the notion of ranking. Refer to Sect. 18.4 of Chap. 18, which provides a more complete discussion.

3.5.2 Similarity Between Two Graphs

In many applications, multiple graphs are available, and it is sometimes necessary to determine the distances between multiple graphs. A complicating factor in similarity computation is that many nodes may have the same label, which makes them indistinguishable. Such cases arise often in domains such as chemical compound analysis. Chemical compounds can be represented as graphs where nodes are elements, and bonds are edges. Because an element may be repeated in a molecule, the labels on the nodes are not distinct. Determining a similarity measure on graphs is extremely challenging in this scenario, because even the very special case of determining whether the two graphs are identical is hard. The latter problem is referred to as the *graph isomorphism* problem, and is known to the NP-hard [221]. Numerous measures, such as the graph-edit distance and substructure-based similarity, have been proposed to address this very difficult case. The core idea in each of these methods is as follows:

1. *Maximum common subgraph distance:* When two graphs contain a large subgraph in common, they are generally considered more similar. The maximum common subgraph problem and the related distance functions are addressed in Sect. 17.2 of Chap. 17.

2. *Substructure-based similarity:* Although it is difficult to match two large graphs, it is much easier to match smaller substructures. The core idea is to count the frequently occurring substructures between the two graphs and report it as a similarity measure. This can be considered the graph analog of subsequence-based similarity in strings. Substructure-based similarity measures are discussed in detail in Sect. 17.3 of Chap. 17.

3. *Graph-edit distance:* This distance measure is analogous to the string-edit distance and is defined as the number of edits required to transform one graph to the other. Because graph matching is a hard problem, this measure is difficult to implement for large graphs. The graph-edit distance is discussed in detail in Sect. 17.2.3.2 of Chap. 17.

4. *Graph kernels:* Numerous kernel functions have been defined to measure similarity between graphs, such as the shortest-path kernel and the random-walk kernel. This topic is discussed in detail in Sect. 17.3.3 of Chap. 17.

These methods are quite complex and require a greater background in the area of graphs. Therefore, the discussion of these measures is deferred to Chap. 17 of this book.

3.6 Supervised Similarity Functions

The previous sections discussed similarity measures that do not require any understanding of user intentions. In practice, the relevance of a feature or the choice of distance function heavily depends on the domain at hand. For example, for an image data set, should the color feature or the texture feature be weighted more heavily? These aspects cannot be modeled by a distance function without taking the user intentions into account. Unsupervised measures, such as the L_p-norm, treat all features equally, and have little intrinsic understanding of the end user's *semantic* notion of similarity. The only way to incorporate this information into the similarity function is to use explicit feedback about the similarity and dissimilarity of objects. For example, the feedback can be expressed as the following sets of object pairs:

$$\mathcal{S} = \{(O_i, O_j) : O_i \text{ is similar to } O_j\}$$
$$\mathcal{D} = \{(O_i, O_j) : O_i \text{ is dissimilar to } O_j\}.$$

How can this information be leveraged to improve the computation of similarity? Many specialized methods have been designed for supervised similarity computation. A common approach is to assume a specific closed form of the similarity function for which the parameters need to be learned. An example is the weighted L_p-norm in Sect. 3.2.1.1, where the parameters represented by Θ correspond to the feature weights $(a_1 \dots a_d)$. Therefore, the first step is to create a distance function $f(O_i, O_j, \Theta)$, where Θ is a set of unknown weights. Assume that higher values of the function indicate greater dissimilarity. Therefore, this is a distance function, rather than a similarity function. Then, it is desirable to determine the parameters Θ, so that the following conditions are satisfied as closely as possible:

$$f(O_i, O_j, \Theta) = \begin{cases} 0 & \text{if } (O_i, O_j) \in \mathcal{S} \\ 1 & \text{if } (O_i, O_j) \in \mathcal{D} \end{cases}. \tag{3.23}$$

This can be expressed as a least squares optimization problem over Θ, with the following error E:

$$E = \sum_{(O_i,O_j)\in \mathcal{S}} (f(O_i, O_j, \Theta) - 0)^2 + \sum_{(O_i,O_j)\in \mathcal{D}} (f(O_i, O_j, \Theta) - 1)^2. \tag{3.24}$$

This objective function can be optimized with respect to Θ with the use of any off-the-shelf optimization solver. If desired, the additional constraint $\Theta \geq 0$ can be added where appropriate. For example, when Θ represents the feature weights $(a_1 \ldots a_d)$ in the Minkowski metric, it is natural to make the assumption of nonnegativity of the coefficients. Such a constrained optimization problem can be easily solved using many nonlinear optimization methods. The use of a closed form such as $f(O_i, O_j, \Theta)$ ensures that the function $f(O_i, O_j, \Theta)$ can be computed efficiently after the one-time cost of computing the parameters Θ.

Where possible, user feedback should be used to improve the quality of the distance function. The problem of learning distance functions can be modeled more generally as that of classification. The classification problem will be studied in detail in Chaps. 10 and 11. Supervised distance function design with the use of Fisher's method is also discussed in detail in the section on instance-based learning in Chap. 10.

3.7 Summary

The problem of distance function design is a crucial one in the context of data mining applications. This is because many data mining algorithms use the distance function as a key subroutine, and the design of the function directly impacts the quality of the results. Distance functions are highly sensitive to the type of the data, the dimensionality of the data, and the global and local nature of the data distribution.

The L_p-norm is the most common distance function used for multidimensional data. This distance function does not seem to work well with increasing dimensionality. Higher values of p work particularly poorly with increasing dimensionality. In some cases, it has been shown that fractional metrics are particularly effective when p is chosen in the range $(0, 1)$. Numerous proximity-based measures have also been shown to work effectively with increasing dimensionality.

The data distribution also has an impact on the distance function design. The simplest possible distance function that uses global distributions is the Mahalanobis metric. This metric is a generalization of the Euclidean measure, and stretches the distance values along the principal components according to their variance. A more sophisticated approach, referred to as *ISOMAP*, uses nonlinear embeddings to account for the impact of nonlinear data distributions. Local normalization can often provide more effective measures when the distribution of the data is heterogeneous.

Other data types such as categorical data, text, temporal, and graph data present further challenges. The determination of time-series and discrete-sequence similarity measures is closely related because the latter can be considered the categorical version of the former. The main problem is that two similar time series may exhibit different scaling of their behavioral and contextual attributes. This needs to be accounted for with the use of different normalization functions for the behavioral attribute, and the use of warping functions for the contextual attribute. For the case of discrete sequence data, many distance and similarity functions, such as the edit distance and the LCSS, are commonly used.

Because distance functions are often intended to model user notions of similarity, feedback should be used, where possible, to improve the distance function design. This feedback can be used within the context of a parameterized model to learn the optimal parameters that are consistent with the user-provided feedback.

3.8 Bibliographic Notes

The problem of similarity computation has been studied extensively by data mining researchers and practitioners in recent years. The issues with high-dimensional data were explored in [17, 88, 266]. In the work of [88], the impact of the distance concentration effects on high-dimensional computation was analyzed. The work in [266] showed the relative advantages of picking distance functions that are locality sensitive. The work also showed the advantages of the Manhattan metric over the Euclidean metric. Fractional metrics were proposed in [17] and generally provide more accurate results than the Manhattan and Euclidean metric. The *ISOMAP* method discussed in this chapter was proposed in [490]. Numerous local methods are also possible for distance function computation. An example of an effective local method is the instance-based method proposed in [543].

Similarity in categorical data was explored extensively in [104]. In this work, a number of similarity measures were analyzed, and how they apply to the outlier detection problem was tested. The Goodall measure is introduced in [232]. The work in [122] uses information theoretic measures for computation of similarity. Most of the measures discussed in this chapter do not distinguish between mismatches on an attribute. However, a number of methods proposed in [74, 363, 473] distinguish between mismatches on an attribute value. The premise is that infrequent attribute values are statistically expected to be more different than frequent attribute values. Thus, in these methods, $S(x_i, y_i)$ is not always set to 0 (or the same value) when x_i and y_i are different. A local similarity measure is presented in [182]. Text similarity measures have been studied extensively in the information retrieval literature [441].

The area of time-series similarity measures is a rich one, and a significant number of algorithms have been designed in this context. An excellent tutorial on the topic may be found in [241]. The use of wavelets for similarity computation in time series is discussed in [130]. While DTW has been used extensively in the context of speech recognition, its use in data mining applications was first proposed by [87]. Subsequently, it has been used extensively [526] for similarity-based applications in data mining. The major challenge in data mining applications is its computationally intensive nature. Numerous methods [307] have been proposed in the time series data mining literature to speed up DTW. A fast method for computing a lower bound on DTW was proposed in [308], and how this can be used for exact indexing was shown. A window-based approach for computing similarity in sequences with noise, scaling, and translation was proposed in [53]. Methods for similarity search in multivariate time series and sequences were proposed in [499, 500]. The edit distance has been used extensively in biological data for computing similarity between sequences [244]. The use of transformation rules for time-series similarity has been studied in [283, 432]. Such rules can be used to create edit distance-like measures for continuous time series. Methods for the string-edit distance are proposed in [438]. It has been shown in [141], how the L_p-norm may be combined with the edit distance. Algorithms for the LCSS problem may be found in [77, 92, 270, 280]. A survey of these algorithms is available in [92]. A variety of other measures for time series and sequence similarity are discussed in [32].

Numerous methods are available for similarity search in graphs. A variety of efficient shortest-path algorithms for finding distances between nodes may be found in [62]. The page rank algorithm is discussed in the Web mining book [357]. The NP-hardness of the graph isomorphism problem, and other closely related problems to the edit distance are discussed in [221]. The relationship between the maximum common subgraph problem and the graph-edit distance problem has been studied in [119, 120]. The problem of substructure similarity search, and the use of substructures for similarity search have been addressed in [520, 521]. A notion of mutation distance has been proposed in [522] to measure the distances between graphs. A method that uses the frequent substructures of a graph for similarity computation in clustering is proposed in [42]. A survey on graph-matching techniques may be found in [26].

User supervision has been studied extensively in the context of distance function learning. One of the earliest methods that parameterizes the weights of the L_p-norm was proposed in [15]. The problem of distance function learning has been formally related to that of classification and has been studied recently in great detail. A survey that covers the important topics in distance function learning is provided in [33].

3.9 Exercises

1. Compute the L_p-norm between $(1, 2)$ and $(3, 4)$ for $p = 1, 2, \infty$.

2. Show that the Mahalanobis distance between two data points is equivalent to the Euclidean distance on a transformed data set, where the transformation is performed by representing the data along the principal components, and dividing by the standard deviation of each component.

3. Download the *Ionosphere data set* from the *UCI Machine Learning Repository* [213], and compute the L_p distance between all pairs of data points, for $p = 1, 2$, and ∞. Compute the contrast measure on the data set for the different norms. Repeat the exercise after sampling the first r dimensions, where r varies from 1 to the full dimensionality of the data.

4. Compute the match-based similarity, cosine similarity, and the Jaccard coefficient, between the two sets $\{A, B, C\}$ and $\{A, C, D, E\}$.

5. Let $\overline{X}$ and $\overline{Y}$ be two data points. Show that the cosine angle between the vectors $\overline{X}$ and $\overline{Y}$ is given by:

$$cosine(\overline{X}, \overline{Y}) = \frac{||\overline{X}||^2 + ||\overline{Y}||^2 - ||\overline{X} - \overline{Y}||^2}{2||\overline{X}||||\overline{Y}||}. \tag{3.25}$$

6. Download the *KDD Cup Network Intrusion Data Set* for the *UCI Machine Learning Repository* [213]. Create a data set containing only the categorical attributes. Compute the nearest neighbor for each data point using the (a) match measure, and (b) inverse occurrence frequency measure. Compute the number of cases where there is a match on the class label.

7. Repeat Exercise 6 using only the quantitative attributes of the data set, and using the L_p-norm for values of $p = 1, 2, \infty$.

8. Repeat Exercise 6 using all attributes in the data set. Use the mixed-attribute function, and different combinations of the categorical and quantitative distance functions of Exercises 6 and 7.

9. Write a computer program to compute the edit distance.

10. Write a computer program to compute the *LCSS* distance.

11. Write a computer program to compute the *DTW* distance.

12. Assume that $Edit(\overline{X}, \overline{Y})$ represents the cost of transforming the string $\overline{X}$ to $\overline{Y}$. Show that $Edit(\overline{X}, \overline{Y})$ and $Edit(\overline{Y}, \overline{X})$ are the same, as long as the insertion and deletion costs are the same.

13. Compute the edit distance, and *LCSS* similarity between: (a) *ababcabc* and *babcbc* and (b) *cbacbacba* and *acbacbacb*. For the edit distance, assume equal cost of insertion, deletion, or replacement.

14. Show that $Edit(i, j)$, $LCSS(i, j)$, and $DTW(i, j)$ are all monotonic functions in i and j.

15. Compute the cosine measure using the raw frequencies between the following two sentences:

    (a) "The sly fox jumped over the lazy dog."

    (b) "The dog jumped at the intruder."

16. Suppose that insertion and deletion costs are 1, and replacement costs are 2 units for the edit distance. Show that the optimal edit distance between two strings can be computed only with insertion and deletion operations. Under the aforementioned cost assumptions, show that the optimal edit distance can be expressed as a function of the optimal *LCSS* distance and the lengths of the two strings.

Chapter 4

Association Pattern Mining

"The pattern of the prodigal is: rebellion, ruin, repentance, reconciliation, restoration."—Edwin Louis Cole

4.1 Introduction

The classical problem of association pattern mining is defined in the context of supermarket data containing sets of items bought by customers, which are referred to as *transactions*. The goal is to determine *associations* between groups of items bought by customers, which can intuitively be viewed as k-way correlations between items. The most popular model for association pattern mining uses the frequencies of sets of items as the quantification of the level of association. The discovered sets of items are referred to as *large itemsets*, *frequent itemsets*, or *frequent patterns*. The association pattern mining problem has a wide variety of applications:

1. *Supermarket data:* The supermarket application was the original motivating scenario in which the association pattern mining problem was proposed. This is also the reason that the term *itemset* is used to refer to a frequent pattern in the context of supermarket *items* bought by a customer. The determination of frequent itemsets provides useful insights about target marketing and shelf placement of the items.

2. *Text mining:* Because text data is often represented in the bag-of-words model, frequent pattern mining can help in identifying co-occurring terms and keywords. Such co-occurring terms have numerous text-mining applications.

3. *Generalization to dependency-oriented data types:* The original frequent pattern mining model has been generalized to many dependency-oriented data types, such as time-series data, sequential data, spatial data, and graph data, with a few modifications. Such models are useful in applications such as Web log analysis, software bug detection, and spatiotemporal event detection.

C. C. Aggarwal, *Data Mining: The Textbook*, DOI 10.1007/978-3-319-14142-8_4

4. *Other major data mining problems:* Frequent pattern mining can be used as a subroutine to provide effective solutions to many data mining problems such as clustering, classification, and outlier analysis.

Because the frequent pattern mining problem was originally proposed in the context of market basket data, a significant amount of terminology used to describe both the data (e.g., *transactions*) and the output (e.g., *itemsets*) is borrowed from the supermarket analogy. From an application-neutral perspective, a frequent pattern may be defined as a frequent subset, defined on the universe of all possible sets. Nevertheless, because the market basket terminology has been used popularly, this chapter will be consistent with it.

Frequent itemsets can be used to generate *association rules* of the form $X \Rightarrow Y$, where X and Y are sets of items. A famous example of an association rule, which has now become part[1] of the data mining folklore, is $\{Beer\} \Rightarrow \{Diapers\}$. This rule suggests that buying beer makes it more likely that diapers will also be bought. Thus, there is a certain directionality to the implication that is quantified as a conditional probability. Association rules are particularly useful for a variety of target market applications. For example, if a supermarket owner discovers that $\{Eggs, Milk\} \Rightarrow \{Yogurt\}$ is an association rule, he or she can promote yogurt to customers who often buy eggs and milk. Alternatively, the supermarket owner may place yogurt on shelves that are located in proximity to eggs and milk.

The frequency-based model for association pattern mining is very popular because of its simplicity. However, the raw frequency of a pattern is not quite the same as the statistical significance of the underlying correlations. Therefore, numerous models for frequent pattern mining have been proposed that are based on statistical significance. This chapter will also explore some of these alternative models, which are also referred to as *interesting patterns*.

This chapter is organized as follows. Section 4.2 introduces the basic model for association pattern mining. The generation of association rules from frequent itemsets is discussed in Sect. 4.3. A variety of algorithms for frequent pattern mining are discussed in Sect. 4.4. This includes the *Apriori* algorithm, a number of enumeration tree algorithms, and a suffix-based recursive approach. Methods for finding interesting frequent patterns are discussed in Sect. 4.5. Meta-algorithms for frequent pattern mining are discussed in Sect. 4.6. Section 4.7 discusses the conclusions and summary.

4.2 The Frequent Pattern Mining Model

The problem of association pattern mining is naturally defined on unordered set-wise data. It is assumed that the database $\mathcal{T}$ contains a set of n transactions, denoted by $T_1 \dots T_n$. Each transaction T_i is drawn on the universe of items U and can also be represented as a multidimensional record of dimensionality, $d = |U|$, containing only binary attributes. Each binary attribute in this record represents a particular item. The value of an attribute in this record is 1 if that item is present in the transaction, and 0 otherwise. In practical settings, the universe of items U is very large compared to the typical number of items in each transaction T_i. For example, a supermarket database may have tens of thousands of items, and a single transaction will typically contain less than 50 items. This property is often leveraged in the design of frequent pattern mining algorithms.

An *itemset* is a set of items. A k-itemset is an itemset that contains exactly k items. In other words, a k-itemset is a set of items of cardinality k. The fraction of transactions

[1]This rule was derived in some early publications on supermarket data. No assertion is made here about the likelihood of such a rule appearing in an arbitrary supermarket data set.

Table 4.1: Example of a snapshot of a market basket data set

tid	Set of items	Binary representation
1	$\{Bread, Butter, Milk\}$	110010
2	$\{Eggs, Milk, Yogurt\}$	000111
3	$\{Bread, Cheese, Eggs, Milk\}$	101110
4	$\{Eggs, Milk, Yogurt\}$	000111
5	$\{Cheese, Milk, Yogurt\}$	001011

in $T_1 \ldots T_n$ in which an itemset occurs as a subset provides a crisp quantification of its frequency. This frequency is also known as the *support*.

Definition 4.2.1 (Support) *The support of an itemset I is defined as the fraction of the transactions in the database $\mathcal{T} = \{T_1 \ldots T_n\}$ that contain I as a subset.*

The support of an itemset I is denoted by $sup(I)$. Clearly, items that are correlated will frequently occur together in transactions. Such itemsets will have high support. Therefore, the frequent pattern mining problem is that of determining itemsets that have the requisite level of *minimum support*.

Definition 4.2.2 (Frequent Itemset Mining) *Given a set of transactions $\mathcal{T} = \{T_1 \ldots T_n\}$, where each transaction T_i is a subset of items from U, determine all itemsets I that occur as a subset of at least a predefined fraction minsup of the transactions in $\mathcal{T}$.*

The predefined fraction *minsup* is referred to as the minimum support. While the default convention in this book is to assume that *minsup* refers to a fractional relative value, it is also sometimes specified as an absolute integer value in terms of the raw number of transactions. This chapter will always assume the convention of a relative value, unless specified otherwise. Frequent patterns are also referred to as frequent itemsets, or large itemsets. This book will use these terms interchangeably.

The unique identifier of a transaction is referred to as a *transaction identifier*, or *tid* for short. The frequent itemset mining problem may also be stated more generally in set-wise form.

Definition 4.2.3 (Frequent Itemset Mining: Set-wise Definition) *Given a set of sets $\mathcal{T} = \{T_1 \ldots T_n\}$, where each element of the set T_i is drawn on the universe of elements U, determine all sets I that occur as a subset of at least a predefined fraction minsup of the sets in $\mathcal{T}$.*

As discussed in Chap. 1, binary multidimensional data and set data are equivalent. This equivalence is because each multidimensional attribute can represent a set element (or item). A value of 1 for a multidimensional attribute corresponds to inclusion in the set (or transaction). Therefore, a transaction data set (or set of sets) can also be represented as a multidimensional binary database whose dimensionality is equal to the number of items.

Consider the transactions illustrated in Table 4.1. Each transaction is associated with a unique transaction identifier in the leftmost column, and contains a baskets of items that were bought together at the same time. The right column in Table 4.1 contains the binary multidimensional representation of the corresponding basket. The attributes of this binary representation are arranged in the order $\{Bread, Butter, Cheese, Eggs, Milk, Yogurt\}$. In this database of 5 transactions, the *support* of $\{Bread, Milk\}$ is $2/5 = 0.4$ because both items in this basket occur in 2 out of a total of 5 transactions. Similarly, the support of $\{Cheese, Yogurt\}$ is 0.2 because it appears in only the last transaction. Therefore, if the minimum support is set to 0.3, then the itemset $\{Bread, Milk\}$ will be reported but not the itemset $\{Cheese, Yogurt\}$.

The number of frequent itemsets is generally very sensitive to the minimum support level. Consider the case where a minimum support level of 0.3 is used. Each of the items $Bread$, $Milk$, $Eggs$, $Cheese$, and $Yogurt$ occur in more than 2 transactions, and can therefore be considered frequent items at a minimum support level of 0.3. These items are frequent 1-itemsets. In fact, the only item that is not frequent at a support level of 0.3 is $Butter$. Furthermore, the frequent 2-itemsets at a minimum support level of 0.3 are $\{Bread, Milk\}$, $\{Eggs, Milk\}$, $\{Cheese, Milk\}$, $\{Eggs, Yogurt\}$, and $\{Milk, Yogurt\}$. The only 3-itemset reported at a support level of 0.3 is $\{Eggs, Milk, Yogurt\}$. On the other hand, if the minimum support level is set to 0.2, it corresponds to an absolute support value of only 1. In such a case, every subset of every transaction will be reported. Therefore, the use of lower minimum support levels yields a larger number of frequent patterns. On the other hand, if the support level is too high, then no frequent patterns will be found. Therefore, an appropriate choice of the support level is crucial for discovering a set of frequent patterns with meaningful size.

When an itemset I is contained in a transaction, all its subsets will also be contained in the transaction. Therefore, the support of any subset J of I will always be at least equal to that of I. This property is referred to as the *support monotonicity property*.

Property 4.2.1 (Support Monotonicity Property) *The support of every subset J of I is at least equal to that of the support of itemset I.*

$$sup(J) \geq sup(I) \quad \forall J \subseteq I \tag{4.1}$$

The monotonicity property of support implies that every subset of a frequent itemset will also be frequent. This is referred to as the *downward closure property*.

Property 4.2.2 (Downward Closure Property) *Every subset of a frequent itemset is also frequent.*

The downward closure property of frequent patterns is algorithmically very convenient because it provides an important constraint on the inherent structure of frequent patterns. This constraint is often leveraged by frequent pattern mining algorithms to prune the search process and achieve greater efficiency. Furthermore, the downward closure property can be used to create concise representations of frequent patterns, wherein only the *maximal* frequent subsets are retained.

Definition 4.2.4 (Maximal Frequent Itemsets) *A frequent itemset is maximal at a given minimum support level minsup, if it is frequent, and no superset of it is frequent.*

In the example of Table 4.1, the itemset $\{Eggs, Milk, Yogurt\}$ is a maximal frequent itemset at a minimum support level of 0.3. However, the itemset $\{Eggs, Milk\}$ is not maximal because it has a superset that is also frequent. Furthermore, the set of *maximal* frequent patterns at a minimum support level of 0.3 is $\{Bread, Milk\}$, $\{Cheese, Milk\}$, and $\{Eggs, Milk, Yogurt\}$. Thus, there are only 3 maximal frequent itemsets, whereas the number of frequent itemsets in the entire transaction database is 11. All frequent itemsets can be derived from the maximal patterns by enumerating the subsets of the maximal frequent

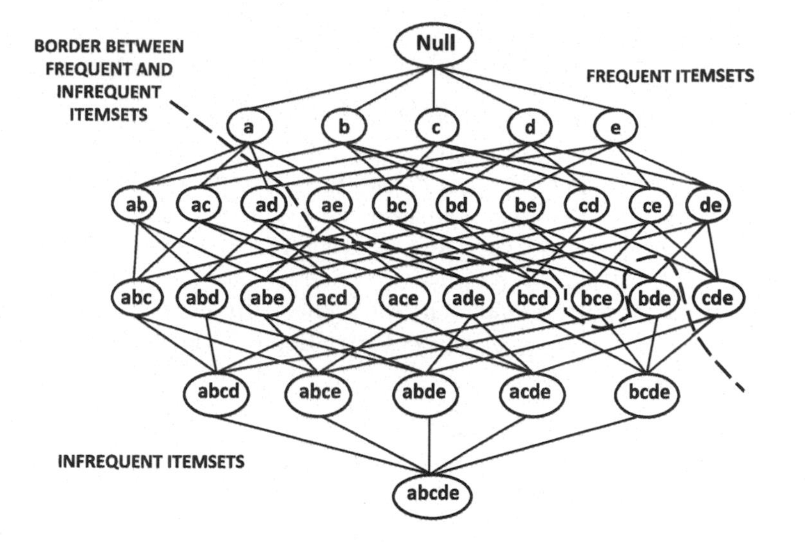

Figure 4.1: The itemset lattice

patterns. Therefore, the maximal patterns can be considered condensed representations of the frequent patterns. However, this condensed representation does not retain information about the support values of the subsets. For example, the support of $\{Eggs, Milk, Yogurt\}$ is 0.4, but it does not provide any information about the support of $\{Eggs, Milk\}$, which is 0.6. A different condensed representation, referred to as *closed frequent itemsets*, is able to retain support information as well. The notion of closed frequent itemsets will be studied in detail in Chap. 5.

An interesting property of itemsets is that they can be conceptually arranged in the form of a *lattice of itemsets*. This lattice contains one node for each of the $2^{|U|}$ sets drawn from the universe of items U. An edge exists between a pair of nodes, if the corresponding sets differ by exactly one item. An example of an itemset lattice of size $2^5 = 32$ on a universe of 5 items is illustrated in Fig. 4.1. The lattice represents the search space of frequent patterns. All frequent pattern mining algorithms, implicitly or explicitly, traverse this search space to determine the frequent patterns.

The lattice is separated into frequent and infrequent itemsets by a *border*, which is illustrated by a dashed line in Fig. 4.1. All itemsets above this border are frequent, whereas those below the border are infrequent. Note that all maximal frequent itemsets are adjacent to this border of itemsets. Furthermore, any valid border representing a true division between frequent and infrequent itemsets will always respect the downward closure property.

4.3 Association Rule Generation Framework

Frequent itemsets can be used to generate *association rules*, with the use of a measure known as the *confidence*. The confidence of a rule $X \Rightarrow Y$ is the conditional probability that a transaction contains the set of items Y, given that it contains the set X. This probability is estimated by dividing the support of itemset $X \cup Y$ with that of itemset X.

Definition 4.3.1 (Confidence) *Let X and Y be two sets of items. The confidence $conf(X \cup Y)$ of the rule $X \cup Y$ is the conditional probability of $X \cup Y$ occurring in a* *transaction, given that the transaction contains X. Therefore, the confidence $conf(X \Rightarrow Y)$ is defined as follows:*

$$conf(X \Rightarrow Y) = \frac{sup(X \cup Y)}{sup(X)}. \tag{4.2}$$

The itemsets X and Y are said to be the *antecedent* and the *consequent* of the rule, respectively. In the case of Table 4.1, the support of $\{Eggs, Milk\}$ is 0.6, whereas the support of $\{Eggs, Milk, Yogurt\}$ is 0.4. Therefore, the confidence of the rule $\{Eggs, Milk\} \Rightarrow \{Yogurt\}$ is $(0.4/0.6) = 2/3$.

As in the case of support, a minimum confidence threshold *minconf* can be used to generate the most relevant association rules. Association rules are defined using both *support* and *confidence* criteria.

Definition 4.3.2 (Association Rules) *Let X and Y be two sets of items. Then, the rule $X \Rightarrow Y$ is said to be an association rule at a minimum support of minsup and minimum confidence of minconf, if it satisfies both the following criteria:*

1. *The support of the itemset $X \cup Y$ is at least minsup.*

2. *The confidence of the rule $X \Rightarrow Y$ is at least minconf.*

The first criterion ensures that a sufficient number of transactions are relevant to the rule; therefore, it has the required critical mass for it to be considered relevant to the application at hand. The second criterion ensures that the rule has sufficient strength in terms of conditional probabilities. Thus, the two measures quantify different aspects of the association rule.

The overall framework for association rule generation uses two phases. These phases correspond to the two criteria in Definition 4.3.2, representing the support and confidence constraints.

1. In the first phase, all the frequent itemsets are generated at the minimum support of *minsup*.

2. In the second phase, the association rules are generated from the frequent itemsets at the minimum confidence level of *minconf*.

The first phase is more computationally intensive and is, therefore, the more interesting part of the process. The second phase is relatively straightforward. Therefore, the discussion of the first phase will be deferred to the remaining portion of this chapter, and a quick discussion of the (more straightforward) second phase is provided here.

Assume that a set of frequent itemsets $\mathcal{F}$ is provided. For each itemset $I \in \mathcal{F}$, a simple way of generating the rules would be to partition the set I into all possible combinations of sets X and $Y = I - X$, such that $I = X \cup Y$. The confidence of each rule $X \Rightarrow Y$ can then be determined, and it can be retained if it satisfies the minimum confidence requirement. Association rules also satisfy a confidence monotonicity property.

Property 4.3.1 (Confidence Monotonicity) *Let X_1, X_2, and I be itemsets such that $X_1 \subset X_2 \subset I$. Then the confidence of $X_2 \Rightarrow I - X_2$ is at least that of $X_1 \Rightarrow I - X_1$.*

$$conf(X_2 \Rightarrow I - X_2) \geq conf(X_1 \Rightarrow I - X_1) \tag{4.3}$$

This property follows directly from definition of confidence and the property of support monotonicity. Consider the rules $\{Bread\} \Rightarrow \{Butter, Milk\}$ and $\{Bread, Butter\} \Rightarrow \{Milk\}$. The second rule is redundant with respect to the first because it will have the same support, but a confidence that is no less than the first. Because of confidence monotonicity, it is possible to report only the non-redundant rules. This issue is discussed in detail in the next chapter.

4.4 Frequent Itemset Mining Algorithms

In this section, a number of popular algorithms for frequent itemset generation will be discussed. Because there are a large number of frequent itemset mining algorithms, the focus of the chapter will be to discuss specific algorithms in detail to introduce the reader to the key tricks in algorithmic design. These tricks are often reusable across different algorithms because the same *enumeration tree framework* is used by virtually all frequent pattern mining algorithms.

4.4.1 Brute Force Algorithms

For a universe of items U, there are a total of $2^{|U|} - 1$ distinct subsets, excluding the empty set. All 2^5 subsets for a universe of 5 items are illustrated in Fig. 4.1. Therefore, one possibility would be to generate all these *candidate* itemsets, and count their support against the transaction database $\mathcal{T}$. In the frequent itemset mining literature, the term *candidate itemsets* is commonly used to refer to itemsets that might *possibly* be frequent (or *candidates* for being frequent). These candidates need to be verified against the transaction database by *support counting*. To count the support of an itemset, we would need to check whether a given itemset I is a subset of each transaction $T_i \in \mathcal{T}$. Such an exhaustive approach is likely to be impractical, when the universe of items U is large. Consider the case where $d = |U| = 1000$. In that case, there are a total of $2^{1000} > 10^{300}$ candidates. To put this number in perspective, if the fastest computer available today were somehow able to process one candidate in one elementary machine cycle, then the time required to process all candidates would be hundreds of orders of magnitude greater than the age of the universe. Therefore, this is not a practical solution.

Of course, one can make the brute-force approach faster by observing that no $(k+1)$-patterns are frequent if no k-patterns are frequent. This observation follows directly from the downward closure property. Therefore, one can enumerate and count the support of all the patterns with increasing length. In other words, one can enumerate and count the support of all patterns containing one item, two items, and so on, until for a certain length l, none of the candidates of length l turn out to be frequent. For sparse transaction databases, the value of l is typically very small compared to $|U|$. At this point, one can terminate. This is a significant improvement over the previous approach because it requires the enumeration of $\sum_{i=1}^{l} \binom{|U|}{i} \ll 2^{|U|}$ candidates. Because the longest frequent itemset is of *much* smaller length than $|U|$ in sparse transaction databases, this approach is orders of magnitude faster. However, the resulting computational complexity is still not satisfactory for large values of U. For example, when $|U| = 1000$ and $l = 10$, the value of $\sum_{i=1}^{10} \binom{|U|}{i}$ is of the order of 10^{23}. This value is still quite large and outside reasonable computational capabilities available today.

One observation is that even a very minor and rather blunt application of the downward closure property made the algorithm hundreds of orders of magnitude faster. Many of the fast algorithms for itemset generation use the downward closure property in a more refined way, both to generate the candidates and to prune them before counting. Algorithms for frequent pattern mining search the lattice of possibilities (or candidates) for frequent patterns (see Fig. 4.1) and use the transaction database to count the support of candidates in this lattice. Better efficiencies can be achieved in a frequent pattern mining algorithm by using one or more of the following approaches:

1. Reducing the size of the explored search space (lattice of Fig. 4.1) by pruning candidate *itemsets* (lattice nodes) using tricks, such as the *downward closure* property.

2. Counting the support of each candidate more efficiently by pruning *transactions* that are known to be irrelevant for counting a candidate itemset.

3. Using compact data structures to represent either candidates or transaction databases that support efficient counting.

The first algorithm that used an effective pruning of the search space with the use of the downward closure property was the *Apriori* algorithm.

4.4.2 The Apriori Algorithm

The *Apriori* algorithm uses the downward closure property in order to prune the candidate search space. The downward closure property imposes a clear structure on the set of frequent patterns. In particular, information about the *infrequency* of itemsets can be leveraged to generate the superset candidates more carefully. Thus, if an itemset is infrequent, there is little point in counting the support of its superset candidates. This is useful for avoiding wasteful counting of support levels of itemsets that are known not to be frequent. The *Apriori* algorithm generates candidates with smaller length k first and counts their supports before generating candidates of length $(k+1)$. The resulting frequent k-itemsets are used to restrict the number of $(k+1)$-candidates with the downward closure property. Candidate generation and support counting of patterns with increasing length is interleaved in *Apriori*. Because the counting of candidate supports is the most expensive part of the frequent pattern generation process, it is extremely important to keep the number of candidates low.

For ease in description of the algorithm, it will be assumed that the items in U have a lexicographic ordering, and therefore an itemset $\{a, b, c, d\}$ can be treated as a (lexicographically ordered) string *abcd* of items. This can be used to impose an ordering among itemsets (patterns), which is the same as the order in which the corresponding strings would appear in a dictionary.

The *Apriori* algorithm starts by counting the supports of the individual items to generate the frequent 1-itemsets. The 1-itemsets are combined to create candidate 2-itemsets, whose support is counted. The frequent 2-itemsets are retained. In general, the frequent itemsets of length k are used to generate the candidates of length $(k+1)$ for increasing values of k. Algorithms that count the support of candidates with increasing length are referred to as *level-wise* algorithms. Let $\mathcal{F}_k$ denote the set of frequent k-itemsets, and $\mathcal{C}_k$ denote the set of candidate k-itemsets. The core of the approach is to iteratively generate the $(k+1)$-candidates $\mathcal{C}_{k+1}$ from frequent k-itemsets in $\mathcal{F}_k$ already found by the algorithm. The frequencies of these $(k+1)$-candidates are counted with respect to the transaction database. While generating the $(k+1)$-candidates, the search space may be pruned by checking whether all k-subsets of $\mathcal{C}_{k+1}$ are included in $\mathcal{F}_k$. So, how does one generate the relevant $(k+1)$-candidates in $\mathcal{C}_{k+1}$ from frequent k-patterns in $\mathcal{F}_k$?

If a pair of itemsets X and Y in $\mathcal{F}_k$ have $(k-1)$ items in common, then a join between them using the $(k-1)$ common items will create a candidate itemset of size $(k+1)$. For example, the two 3-itemsets $\{a, b, c\}$ (or *abc* for short) and $\{a, b, d\}$ (or *abd* for short), when

```
Algorithm Apriori(Transactions: T, Minimum Support: minsup)
begin
  k = 1;
  F_1 = { All Frequent 1-itemsets };
  while F_k is not empty do begin
    Generate C_{k+1} by joining itemset-pairs in F_k;
    Prune itemsets from C_{k+1} that violate downward closure;
    Determine F_{k+1} by support counting on (C_{k+1}, T) and retaining
          itemsets from C_{k+1} with support at least minsup;
    k = k + 1;
  end;
  return(∪_{i=1}^{k} F_i);
end
```

Figure 4.2: The *Apriori* algorithm

joined together on the two common items a and b, will yield the candidate 4-itemset $abcd$. Of course, it is possible to join other frequent patterns to create the same candidate. One might also join abc and bcd to achieve the same result. Suppose that all four of the 3-subsets of $abcd$ are present in the set of frequent 3-itemsets. One can create the candidate 4-itemset in $\binom{4}{2} = 6$ different ways. To avoid redundancy in candidate generation, the convention is to impose a lexicographic ordering on the items and use the first $(k-1)$ items of the itemset for the join. Thus, in this case, the only way to generate $abcd$ would be to join using the first two items a and b. Therefore, the itemsets abc and abd would need to be joined to create $abcd$. Note that, if either of abc and abd are *not* frequent, then $abcd$ will *not* be generated as a candidate using this join approach. Furthermore, in such a case, it is assured that $abcd$ will not be frequent because of the downward closure property of frequent itemsets. Thus, the downward closure property ensures that the candidate set generated using this approach does not miss any itemset that is truly frequent. As we will see later, this *non-repetitive* and *exhaustive* way of generating candidates can be interpreted in the context of a conceptual hierarchy of the patterns known as the *enumeration tree*. Another point to note is that the joins can usually be performed very efficiently. This efficiency is because, if the set $\mathcal{F}_k$ is sorted in lexicographic (dictionary) order, all itemsets with a common set of items in the first $k-1$ positions will appear contiguously, allowing them to be located easily.

A *level-wise pruning trick* can be used to further reduce the size of the $(k+1)$-candidate set. All the k-subsets (i.e., subsets of cardinality k) of an itemset $I \in \mathcal{C}_{k+1}$ need to be present in $\mathcal{F}_k$ because of the downward closure property. Otherwise, it is guaranteed that the itemset I is not frequent. Therefore, it is checked whether all k-subsets of each itemset $I \in \mathcal{C}_{k+1}$ are present in $\mathcal{F}_k$. If this is not the case, then such itemsets I are removed from $\mathcal{C}_{k+1}$.

After the candidate itemsets $\mathcal{C}_{k+1}$ of size $(k+1)$ have been generated, their support can be determined by counting the number of occurrences of each candidate in the transaction database $\mathcal{T}$. Only the candidate itemsets that have the required minimum support are retained to create the set of $(k+1)$-frequent itemsets $\mathcal{F}_{k+1} \subseteq \mathcal{C}_{k+1}$. In the event that the set $\mathcal{F}_{k+1}$ is empty, the algorithm terminates. At termination, the union $\cup_{i=1}^{k} \mathcal{F}_i$ of the frequent patterns of different sizes is reported as the final output of the algorithm.

The overall algorithm is illustrated in Fig. 4.2. The heart of the algorithm is an iterative loop that generates $(k+1)$-candidates from frequent k-patterns for successively higher values of k and counts them. The three main operations of the algorithm are candidate generation, pruning, and support counting. Of these, the support counting process is the most expensive one because it depends on the size of the transaction database $\mathcal{T}$. The level-wise approach ensures that the algorithm is relatively efficient at least from a disk-access cost perspective. This is because each set of candidates in $\mathcal{C}_{k+1}$ can be counted in a single pass over the data without the need for random disk accesses. The number of passes over the data is, therefore, equal to the cardinality of the longest frequent itemset in the data. Nevertheless, the counting procedure is still quite expensive especially if one were to use the naive approach of checking whether each itemset is a subset of a transaction. Therefore, efficient support counting procedures are necessary.

4.4.2.1 Efficient Support Counting

To perform support counting, *Apriori* needs to *efficiently* examined whether each candidate itemset is present in a transaction. This is achieved with the use of a data structure known as the *hash tree*. The hash tree is used to carefully organize the candidate patterns in $\mathcal{C}_{k+1}$ for more efficient counting. Assume that the items in the transactions and the candidate itemsets are sorted lexicographically. A hash tree is a tree with a fixed degree of the internal nodes. Each internal node is associated with a random hash function that maps to the index of the different children of that node in the tree. A leaf node of the hash tree contains a list of lexicographically sorted itemsets, whereas an interior node contains a hash table. Every itemset in $\mathcal{C}_{k+1}$ is contained in exactly one leaf node of the hash tree. The hash functions in the interior nodes are used to decide which candidate itemset belongs to which leaf node with the use of a methodology described below.

It may be assumed that all interior nodes use the same hash function $f(\cdot)$ that maps to $[0 \ldots h-1]$. The value of h is also the branching degree of the hash tree. A candidate itemset in $\mathcal{C}_{k+1}$ is mapped to a leaf node of the tree by defining a path from the root to the leaf node with the use of these hash functions at the internal nodes. Assume that the root of the hash tree is level 1, and all successive levels below it increase by 1. As before, assume that the items in the candidates and transactions are arranged in lexicographically sorted order. At an interior node in level i, a hash function is applied to the ith item of a candidate itemset $I \in \mathcal{C}_{k+1}$ to decide which branch of the hash tree to follow for the candidate itemset. The tree is constructed recursively in top-down fashion, and a minimum threshold is imposed on the number of candidates in the leaf node to decide where to terminate the hash tree extension. The candidate itemsets in the leaf node are stored in sorted order.

To perform the counting, all possible candidate k-itemsets in $\mathcal{C}_{k+1}$ that are subsets of a transaction $T_j \in \mathcal{T}$ are discovered in a single exploration of the hash tree. To achieve this goal, all possible paths in the hash tree, whose leaves *might* contain subset itemsets of the transaction T_j, are discovered using a recursive traversal. The selection of the relevant leaf nodes is performed by recursive traversal as follows. At the root node, all branches are followed such that any of the items in the transaction T_j hash to one of the branches. At a given interior node, if the ith item of the transaction T_j was last hashed (at the parent node), then all items following it in the transaction are hashed to determine the possible children to follow. Thus, by following all these paths, the relevant leaf nodes in the tree are determined. The candidates in the leaf node are stored in sorted order and can be compared efficiently to the transaction T_j to determine whether they are relevant. This process is repeated for each transaction to determine the final support count of each itemset in $\mathcal{C}_{k+1}$.

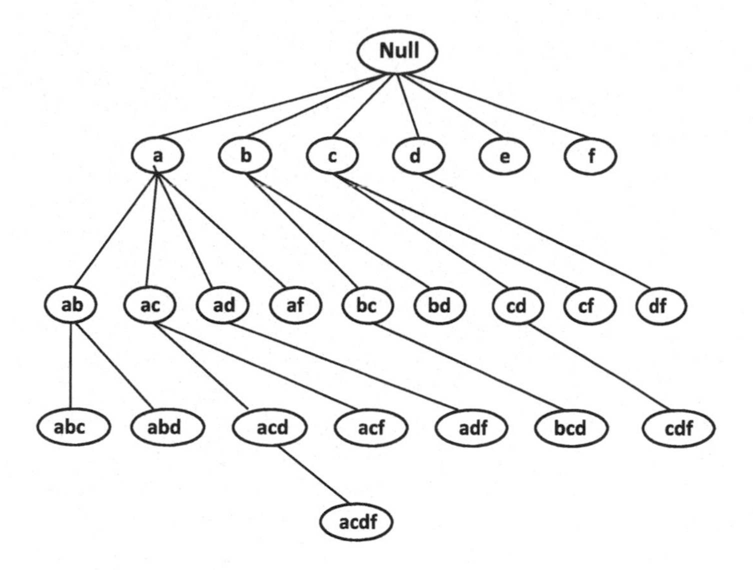

Figure 4.3: The lexicographic or enumeration tree of frequent itemsets

4.4.3 Enumeration-Tree Algorithms

These algorithms are based on set enumeration concepts, in which the different candidate itemsets are generated in a tree-like structure known as the *enumeration tree*, which is a subgraph of the lattice of itemsets introduced in Fig. 4.1. This tree-like structure is also referred to as a lexicographic tree because it is dependent on an upfront lexicographic ordering among the items. The candidate patterns are generated by growing this lexicographic tree. This tree can be grown in a wide variety of different strategies to achieve different trade-offs between storage, disk access costs, and computational efficiency. Because most of the discussion in this section will use this structure as a base for algorithmic development, this concept will be discussed in detail here. The main characteristic of the enumeration tree (or lexicographic tree) is that it provides an abstract hierarchical representation of the itemsets. This representation is leveraged by frequent pattern mining algorithms for systematic exploration of the candidate patterns in a non-repetitive way. The final output of these algorithms can also be viewed as an enumeration tree structure that is defined only on the frequent itemsets. The enumeration tree is defined on the frequent itemsets in the following way:

1. A node exists in the tree corresponding to each frequent itemset. The root of the tree corresponds to the *null* itemset.

2. Let $I = \{i_1, \ldots i_k\}$ be a frequent itemset, where $i_1, i_2 \ldots i_k$ are listed in lexicographic order. The parent of the node I is the itemset $\{i_1, \ldots i_{k-1}\}$. Thus, the child of a node can only be extended with items occurring lexicographically *after* all items occurring in that node. The enumeration tree can also be viewed as a *prefix* tree on the lexicographically ordered string representation of the itemsets.

This definition of an ancestral relationship naturally creates a tree structure on the nodes, which is rooted at the *null* node. An example of the frequent portion of the enumeration tree is illustrated in Fig. 4.3. An item that is used to extend a node to its (frequent) child in the enumeration tree is referred to as a *frequent tree extension*, or simply a tree extension. In the example of Fig. 4.3, the frequent tree extensions of node a are b, c, d, and f, because these items extend node a to the frequent itemsets ab, ac, ad, and af, respectively. The lattice provides many paths to extend the *null* itemset to a node, whereas an enumeration tree provides only one path. For example, itemset ab can be extended either in the order $a \rightarrow ab$, or in the order $b \rightarrow ab$ in the lattice. However, only the former is possible in the enumeration tree after the lexicographic ordering has been fixed. Thus, the lexicographic ordering imposes a strictly hierarchical structure on the itemsets. This hierarchical structure enables *systematic* and *non-redundant* exploration of the itemset search space by algorithms that generate candidates by extending frequent itemsets with one item at a time. The enumeration tree can be constructed in many ways with different lexicographic orderings of items. The impact of this ordering will be discussed later.

Most of the enumeration tree algorithms work by growing this enumeration tree of frequent itemsets with a predefined strategy. First, the root node of the tree is extended by finding the frequent 1-items. Then, these nodes may be extended to create *candidates*. These are checked against the transaction database to determine the ones that are frequent. The enumeration tree framework provides an order and structure to the frequent itemset discovery, which can be leveraged to improve the counting and pruning process of candidates. In the following discussion, the terms "node" and "itemset" will be used interchangeably. Therefore, the notation P will be used to denote both an itemset, and its corresponding node in the enumeration tree.

So, how can candidates nodes be generated in a systematic way from the frequent nodes in the enumeration tree that have already been discovered? For an item i to be considered a candidate for extending a frequent node P to $P \cup \{i\}$, it must also be a frequent extension of the parent Q of P. This is because of the downward closure property, and it can be used to systematically define the candidate extensions of a node P after the frequent extensions of its parent Q have been determined. Let $F(Q)$ represent the frequent lexicographic tree extensions of node Q. Let $i \in F(Q)$ be the frequent extension item that extends frequent node Q to frequent node $P = Q \cup \{i\}$. Let $C(P)$ denote the subset of items from $F(Q)$ occurring lexicographically *after* the item i used to extend node Q to node P. The set $C(P)$ defines the *candidate extension items* of node P, which are defined as items that can be appended at the end of P to create candidate itemsets. This provides a systematic methodology to generate candidate children of node P. As we will see in Sect. 4.4.3.1, the resulting candidates are identical to those generated by *Apriori* joins. Note that the relationship $F(P) \subseteq C(P) \subset F(Q)$ is always true. The value of $F(P)$ in Fig. 4.3, when $P = ab$, is $\{c, d\}$. The value of $C(P)$ for $P = ab$ is $\{c, d, f\}$ because these are frequent extensions of parent itemset $Q = \{a\}$ of P occurring lexicographically after the item b. Note that the set of *candidate* extensions $C(ab)$ also contains the (infrequent) item f that the set of *frequent* extensions $F(ab)$ does not. Such infrequent item extensions correspond to *failed* candidate tests in all enumeration tree algorithms. Note that the infrequent itemset abf is not included in the *frequent* itemset tree of Fig. 4.3. It is also possible to create an enumeration tree structure on the *candidate* itemsets, which contains an additional layer of infrequent candidate extensions of the nodes in Fig. 4.3. Such a tree would contain abf.

Enumeration tree algorithms iteratively grow the enumeration tree $\mathcal{ET}$ of frequent patterns. A very generic description of this iterative step, which is executed repeatedly to extend the enumeration tree $\mathcal{ET}$, is as follows:

Select one or more nodes $\mathcal{P}$ in $\mathcal{ET}$;
Determine candidate extensions $C(P)$ for each such node $P \in \mathcal{P}$;
Count support of generated candidates;
Add frequent candidates to $\mathcal{ET}$ (tree growth);

Algorithm *GenericEnumerationTree*(Transactions: $\mathcal{T}$,
Minimum Support: *minsup*)
begin
Initialize enumeration tree $\mathcal{ET}$ to single *Null* node;
while any node in $\mathcal{ET}$ has not been examined **do begin**
Select one of more unexamined nodes $\mathcal{P}$ from $\mathcal{ET}$ for examination;
Generate candidates extensions $C(P)$ of each node $P \in \mathcal{P}$;
Determine frequent extensions $F(P) \subseteq C(P)$ for each $P \in \mathcal{P}$ with support counting;
Extend each node $P \in \mathcal{P}$ in $\mathcal{ET}$ with its frequent extensions in $F(P)$;
end
return enumeration tree $\mathcal{ET}$;
end

Figure 4.4: Generic enumeration-tree growth with unspecified growth strategy and counting method

This approach is continued until none of the nodes can be extended any further. At this point, the algorithm terminates. A more detailed description is provided in Fig. 4.4. Interestingly, almost all frequent pattern mining algorithms can be viewed as variations and extensions of this simple enumeration-tree framework. Within this broader framework, a wide variability exists both in terms of the growth strategy of the tree and the specific data structures used for support counting. Therefore, the description of Fig. 4.4 is very generic because none of these aspects are specified. The different choices of growth strategy and counting methodology provide different trade-offs between efficiency, space-requirements, and disk access costs. For example, in breadth-first strategies, the node set $\mathcal{P}$ selected in an iteration of Fig. 4.4 corresponds to all nodes at one level of the tree. This approach may be more relevant for disk-resident databases because all nodes at a single level of the tree can be extended during one counting pass on the transaction database. Depth-first strategies select a single node at the deepest level to create $\mathcal{P}$. These strategies may have better ability to explore the tree deeply and discover long frequent patterns early. The early discovery of longer patterns is especially useful for computational efficiency in maximal pattern mining and for better memory management in certain classes of *projection-based* algorithms.

Because the counting approach is the most expensive part, the different techniques attempt to use growth strategies that optimize the work done during counting. Furthermore, it is crucial for the counting data structures to be efficient. This section will explore some of the common algorithms, data structures, and pruning strategies that leverage the enumeration-tree structure in the counting process. Interestingly, the enumeration-tree framework is so general that even the *Apriori* algorithm can be interpreted within this framework, although the concept of an enumeration tree was not used when *Apriori* was proposed.

4.4.3.1 Enumeration-Tree-Based Interpretation of Apriori

The *Apriori* algorithm can be viewed as the level-wise construction of the enumeration tree in breadth-first manner. The *Apriori* join for generating candidate $(k+1)$-itemsets is performed in a non-redundant way by using only the *first* $(k-1)$ items from two frequent k-itemsets. This is equivalent to joining all pairs of *immediate siblings* at the kth level of the enumeration tree. For example, the children of ab in Fig. 4.3 may be obtained by joining

ab with all its frequent siblings (other children of node a) that occur lexicographically later than it. In other words, the join operation of node P with its lexicographically later frequent siblings produces the candidates corresponding to the extension of P with each of its candidate tree-extensions $C(P)$. In fact, the candidate extensions $C(P)$ for all nodes P at a given level of the tree can be *exhaustively* and *non-repetitively* generated by using joins between all pairs of frequent *siblings* at that level. The *Apriori* pruning trick then discards some of the enumeration tree nodes because they are guaranteed not to be frequent. A single pass over the transaction database is used to count the support of these candidate extensions, and generate the *frequent* extensions $F(P) \subseteq C(P)$ for each node P in the level being extended. The approach terminates when the tree cannot be grown further in a particular pass over the database. Thus, the join operation of *Apriori* has a direct interpretation in terms of the enumeration tree, and the *Apriori* algorithm implicitly extends the enumeration tree in a level-wise fashion with the use of joins.

4.4.3.2 TreeProjection and DepthProject

TreeProjection is a family of methods that uses recursive *projections* of the transactions down the enumeration tree structure. The goal of these recursive projections is to reuse the counting work that has already been done at a given node of the enumeration tree at its descendent nodes. This reduces the overall counting effort by orders of magnitude. *TreeProjection* is a general framework that shows how to use database projection in the context of a variety of different strategies for construction of the enumeration tree, such as breadth-first, depth-first, or a combination of the two. The *DepthProject* approach is a specific instantiation of this framework with the depth-first strategy. Different strategies have different trade-offs between the memory requirements and disk-access costs.

The main observation in projection-based methods is that if a transaction does not contain the itemset corresponding to an enumeration-tree node, then this transaction will not be relevant for counting at any descendent (superset itemset) of that node. Therefore, when counting is done at an enumeration-tree node, the information about irrelevant transactions should somehow be preserved for counting at its descendent nodes. This is achieved with the notion of *projected databases*. Each projected transaction database is specific to an enumeration-tree node. Transactions that do not contain the itemset P are not included in the projected databases at node P and its descendants. This results in a significant reduction in the number of projected transactions. Furthermore, only the candidate extension items of P, denoted by $C(P)$, are relevant for counting at any of the subtrees rooted at node P. Therefore, the projected database at node P can be expressed only in terms of the items in $C(P)$. The size of $C(P)$ is much smaller than the universe of items, and therefore the projected database contains a smaller number of items per transaction with increasing size of P. We denote the projected database at node P by $\mathcal{T}(P)$. For example, consider the node $P = ab$ in Fig. 4.3, in which the candidate items for extending ab are $C(P) = \{c, d, f\}$. Then, the transaction $abcfg$ maps to the projected transaction cf in $\mathcal{T}(P)$. On the other hand, the transaction $acfg$ is not even present in $\mathcal{T}(P)$ because $P = ab$ is not a subset of $acfg$. The special case $\mathcal{T}(Null) = \mathcal{T}$ corresponds to the top level of the enumeration tree and is equal to the full transaction database. In fact, the subproblem at node P with transaction database $\mathcal{T}(P)$ is structurally identical to the top-level problem, except that it is a much smaller problem focused on determining frequent patterns with a prefix of P. Therefore, the frequent node P in the enumeration tree can be extended further by counting the support of individual items in $C(P)$ using the relatively small database $\mathcal{T}(P)$. This

```
Algorithm ProjectedEnumerationTree(Transactions: T,
            Minimum Support: minsup)
begin
  Initialize enumeration tree ET to a single (Null, T) root node;
  while any node in ET has not been examined do begin
    Select an unexamined node (P, T(P)) from ET for examination;
    Generate candidates item extensions C(P) of node (P, T(P));
    Determine frequent item extensions F(P) ⊆ C(P) by support counting
        of individual items in smaller projected database T(P);
    Remove infrequent items in T(P);
    for each frequent item extension i ∈ F(P) do begin
      Generate T(P ∪ {i}) from T(P);
      Add (P ∪ {i}, T(P ∪ {i})) as child of P in ET;
    end
  end
  return enumeration tree ET;
end
```

Figure 4.5: Generic enumeration-tree growth with unspecified growth strategy and database projections

results in a simplified and efficient counting process of candidate 1-item *extensions* rather than *itemsets*.

The enumeration tree can be grown with a variety of strategies such as the breadth-first or depth-first strategies. At each node, the counting is performed with the use of the projected database rather than the entire transaction database, and a further reduced and projected transaction database is propagated to the children of P. At each level of the hierarchical projection down the enumeration tree, the number of items and the number of transactions in the projected database are reduced. The basic idea is that $\mathcal{T}(P)$ contains the minimal portion of the transaction database that is *relevant* for counting the subtree rooted at P, based on the removal of irrelevant transactions and items by the counting process that has already been performed at higher levels of the tree. By *recursively* projecting the transaction database down the enumeration tree, this counting work is reused. We refer to this approach as *projection-based reuse* of counting effort.

The generic enumeration-tree algorithm with hierarchical projections is illustrated in Fig. 4.5. This generic algorithm does not assume any specific exploration strategy, and is quite similar to the generic enumeration-tree pseudocode shown in Fig. 4.4. There are two differences between the pseudocodes.

1. For simplicity of notation, we have shown the exploration of a single node P at one time in Fig. 4.5, rather than a group of nodes $\mathcal{P}$ (as in Fig. 4.4). However, the pseudocode shown in Fig. 4.5 can easily be rewritten for a group of nodes $\mathcal{P}$. Therefore, this is not a significant difference.

2. The key difference is that the projected database $\mathcal{T}(P)$ is used to count support at node P. Each node in the enumeration tree is now represented by the itemset and projected database pair $(P, \mathcal{T}(P))$. This is a very important difference because $\mathcal{T}(P)$ is much smaller than the original database. Therefore, a significant amount of information gained by counting the supports of ancestors of node P, is preserved in $\mathcal{T}(P)$. Furthermore, one only needs to count the support of single item *extensions* of node P in $\mathcal{T}(P)$ (rather than entire itemsets) in order to grow the subtree at P further.

The enumeration tree can be constructed in many different ways depending on the lexicographic ordering of items. How should the items be ordered? The structure of the enumeration tree has a built-in bias towards creating unbalanced trees in which the lexicographically smaller items have more descendants. For example, in Fig. 4.3, node a has many more descendants than node f. Therefore, ordering the items from least support to greatest support ensures that the computationally heavier branches of the enumeration tree have fewer relevant transactions. This is helpful in maximizing the selectivity of projections and ensuring better efficiency.

The strategy used for selection of the node P defines the order in which the nodes of the enumeration tree are materialized. This strategy has a direct impact on memory management because projected databases, which are no longer required for future computation, can be deleted. In depth-first strategies, the lexicographically smallest unexamined node P is selected for extension. In this case, one only needs to maintain projected databases along the current path of the enumeration tree being explored. In breadth-first strategies, an entire group of nodes $\mathcal{P}$ corresponding to all patterns of a particular size are grown first. In such cases, the projected databases need to be simultaneously maintained along the full breadth of the enumeration tree $\mathcal{ET}$ at the two current levels involved in the growth process. Although it may be possible to perform the projection on such a large number of nodes for smaller transaction databases, some modifications to the basic framework of Fig. 4.5 are needed for the general case of larger databases.

In particular, breadth-first variations of the *TreeProjection* framework perform hierarchical projections on the fly during counting from their ancestor nodes. The depth-first variations of *TreeProjection*, such as *DepthProject*, achieve full projection-based reuse because the projected transactions can be consistently maintained at each materialized node along the relatively small path of the enumeration tree from the root to the current node. The breadth-first variations do have the merit that they can optimize disk-access costs for arbitrarily large databases at the expense of losing some of the power of projection-based reuse. As will be discussed later, all (full) projection-based reuse methods face memory-management challenges with increasing database size. These additional memory requirements can be viewed as the price for persistently storing the relevant work done in earlier iterations in the indirect form of projected databases. There is usually a different trade-off between disk-access costs and memory/computational requirements in various strategies, which is exploited by the *TreeProjection* framework. The bibliographic notes contain pointers to specific details of these optimized variations of *TreeProjection*.

Optimized counting at deeper level nodes: The projection-based approach enables specialized counting techniques at deeper level nodes near the leaves of the enumeration tree. These specialized counting methods can provide the counts of *all* the itemsets in a lower-level subtree in the time required to *scan* the projected database. Because such nodes are more numerous, this can lead to large computational improvements.

What is the point at which such counting methods can be used? When the number of frequent extensions $F(P)$ of a node P falls below a threshold t such that 2^t fits in memory, an approach known as *bucketing* can be used. To obtain the best computational results, the value of t used should be such that 2^t is much smaller than the number of transactions in the projected database. This can occur only when there are many repeated transactions in the projected database.

A two-phase approach is used. In the first phase, the count of each distinct transaction in the projected database is determined. This can be accomplished easily by maintaining $2^{|F(P)|}$ buckets or counters, scanning the transactions one by one, and adding counts to the buckets. This phase can be completed in a simple scan of the small (projected) database of transactions. Of course, this process only provides transaction counts and not itemset counts.

In the second phase, the transaction frequency counts can be further aggregated in a systematic way to create itemset frequency counts. Conceptually, the process of aggregating projected transaction counts is similar to arranging all the $2^{|F(P)|}$ possibilities in the form of a lattice, as illustrated in Fig. 4.1. The counts of the lattice nodes, which are computed in the first phase, are aggregated up the lattice structure by adding the count of immediate supersets to their subsets. For small values of $|F(P)|$, such as 10, this phase is not the limiting computational factor, and the overall time is dominated by that required to scan the projected database in the first phase. An efficient implementation of the second phase is discussed in detail below.

Consider a string composed of 0, 1, and $*$ that refers to an itemset in which the positions with 0 and 1 are fixed to those values (corresponding to presence or absence of items), whereas a position with a $*$ is a "don't care." Thus, all transactions can be expressed in terms of 0 and 1 in their binary representation. On the other hand, all itemsets can be expressed in terms of 1 and $*$ because itemsets are traditionally defined with respect to presence of items and ambiguity with respect to absence. Consider, for example, the case when $|F(P)| = 4$, and there are four items, numbered $\{1, 2, 3, 4\}$. An itemset containing items 2 and 4 is denoted by $*1*1$. We start with the information on $2^4 = 16$ bitstrings that are composed 0 and 1. These represent all possible distinct transactions. The algorithm aggregates the counts in $|F(P)|$ iterations. The count for a string with a "*" in a particular position may be obtained by adding the counts for the strings with a 0 and 1 in those positions. For example, the count for the string *1*1 may be expressed as the sum of the counts of the strings 01*1 and 11*1. The positions may be processed in any order, although the simplest approach is to aggregate them from the least significant to the most significant.

A simple pseudocode to perform the aggregation is described below. In this pseudocode, the initial value of $bucket[i]$ is equal to the count of the transaction corresponding to the bitstring representation of integer i. The final value of $bucket[i]$ is one in which the transaction count has been converted to an itemset count by successive aggregation. In other words, the 0s in the bitstring are replaced by "don't cares."

for $i := 1$ **to** k **do begin**
  for $j := 1$ **to** 2^k **do begin**
    if the ith bit of bitstring representation
      of j is 0 **then** $bucket[j] = bucket[j] + bucket[j + 2^{i-1}]$;
  endfor
endfor

An example of bucketing for $|F(P)| = 4$ is illustrated in Fig. 4.6. The bucketing trick is performed commonly at lower nodes of the tree because the value of $|F(P)|$ falls drastically at the lower levels. Because the nodes at the lower levels dominate the total number of nodes in the enumeration-tree structure, the impact of bucketing can be very significant.

Optimizations for maximal pattern mining: The *DepthProject* method, which is a depth-first variant of the approach, is particularly adaptable for maximal pattern discovery. In this case, the enumeration tree is explored in depth-first order to maximize the advantages of pruning the search space of regions containing only non-maximal patterns. The order of construction of the enumeration tree is important in the particular case of maximal frequent

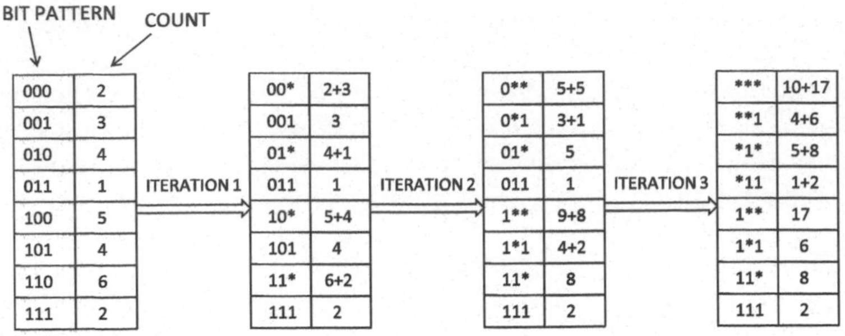

BIT PATTERN COUNT

BIT PATTERN	COUNT		ITERATION 1			ITERATION 2			ITERATION 3	
000	2		00*	2+3		0**	5+5		***	10+17
001	3		001	3		0*1	3+1		**1	4+6
010	4		01*	4+1		01*	5		*1*	5+8
011	1		011	1		011	1		*11	1+2
100	5		10*	5+4		1**	9+8		1**	17
101	4		101	4		1*1	4+2		1*1	6
110	6		11*	6+2		11*	8		11*	8
111	2		111	2		111	2		111	2

Figure 4.6: Performing the second phase of bucketing

pattern mining because certain kinds of non-maximal search-space pruning are optimized with the depth-first order. The notion of *lookaheads* is one such optimization.

Let $C(P)$ be the set of candidate item extensions of node P. Before support counting, it is tested whether $P \cup C(P)$ is a subset of a frequent pattern that has already been found. If such is indeed the case, then the pattern $P \cup C(P)$ is a non-maximal frequent pattern, and the entire subtree (of the enumeration tree) rooted at P can be pruned. This kind of pruning is referred to as *superset-based pruning.* When P cannot be pruned, the supports of its candidate extensions need to be determined. During this support counting, the support of $P \cup C(P)$ is counted along with the individual item extensions of P. If $P \cup C(P)$ is found to be frequent, then it eliminates any further work of counting the support of (non-maximal) nodes in the subtree rooted at node P.

While lookaheads can also be used with breadth-first algorithms, they are more effective with a depth-first strategy. In depth-first methods, longer patterns tend to be found first, and are, therefore, already available in the frequent set for superset-based pruning. For example, consider a frequent pattern of length 20 with 2^{20} subsets. In a depth-first strategy, it can be shown that the pattern of length 20 will be discovered after exploring only 19 of its immediate prefixes. On the other hand, a breadth-first method may remain trapped by discovery of shorter patterns. Therefore, the longer patterns become available very early in depth-first methods such as *DepthProject* to prune large portions of the enumeration tree with superset-based pruning.

4.4.3.3 Vertical Counting Methods

The *Partition* [446] and *Monet* [273] methods pioneered the concept of *vertical database representations* of the transaction database $\mathcal{T}$. In the *vertical representation*, each item is associated with a list of its transaction identifiers (*tids*). It can also be thought of as using the transpose of the binary transaction data matrix representing the transactions so that columns are transformed to rows. These rows are used as the new "records." Each item, thus, has a *tid* list of identifiers of transactions containing it. For example, the vertical representation of the database of Table 4.1 is illustrated in Table 4.2. Note that the binary matrix in Table 4.2 is the transpose of that in Table 4.1.

The intersection of two item *tid* lists yields a new *tid* list whose length is equal to the support of that 2-itemset. Further intersection of the resulting *tid* list with that of another item yields the support of 3-itemsets. For example, the intersection of the *tid* lists of *Milk* and *Yogurt* yields $\{2, 4, 5\}$ with length 3. Further intersection of the *tid* list of $\{Milk, Yogurt\}$ with that of *Eggs* yields the *tid* list $\{2, 4\}$ of length 2. This means that the support of

Table 4.2: Vertical representation of market basket data set

Item	Set of tids	Binary representation
Bread	$\{1,3\}$	10100
Butter	$\{1\}$	10000
Cheese	$\{3,5\}$	00101
Eggs	$\{2,3,4\}$	01110
Milk	$\{1,2,3,4,5\}$	11111
Yogurt	$\{2,4,5\}$	01011

$\{Milk, Yogurt\}$ is $3/5 = 0.6$ and that of $\{Milk, Eggs, Yogurt\}$ is $2/5 = 0.4$. Note that one can also intersect the smaller *tid* lists of $\{Milk, Yogurt\}$ and $\{Milk, Eggs\}$ to achieve the same result. For a pair of k-itemsets that join to create a $(k+1)$-itemset, it is possible to intersect the *tid* lists of the k-itemset pair to obtain the *tid*-list of the resulting $(k+1)$-itemset. Intersecting *tid* lists of k-itemsets is preferable to intersecting *tid* lists of 1-itemsets because the *tid* lists of k-itemsets are typically smaller than those of 1-itemsets, which makes intersection faster. Such an approach is referred to as *recursive tid* list intersection. This insightful notion of recursive *tid* list intersection was introduced[2] by the *Monet* [273] and *Partition* [446] algorithms. The *Partition* framework [446] proposed a vertical version of the *Apriori* algorithm with *tid* list intersection. The pseudocode of this vertical version of the *Apriori* algorithm is illustrated in Fig. 4.7. The only difference from the horizontal *Apriori* algorithm is the use of recursive *tid* list intersections for counting. While the vertical *Apriori* algorithm is computationally more efficient than horizontal *Apriori*, it is memory-intensive because of the need to store *tid* lists with each itemset. Memory requirements can be reduced with the use of a *partitioned ensemble* in which the database is divided into smaller chunks which are independently processed. This approach reduces the memory requirements at the expense of running-time overheads in terms of postprocessing, and it is discussed in Sect. 4.6.2. For smaller databases, no partitioning needs to be applied. In such cases, the vertical *Apriori* algorithm of Fig. 4.7 is also referred to as *Partition-1*, and it is the progenitor of all modern vertical pattern mining algorithms.

The vertical database representation can, in fact, be used in almost any enumeration-tree algorithm with a growth strategy that is different from the breadth-first method. As in the case of the vertical *Apriori* algorithm, the *tid* lists can be stored with the itemsets (nodes) during the growth of the tree. If the *tid* list of any node P is known, it can be intersected with the *tid* list of a sibling node to determine the support count (and *tid* list) of the corresponding extension of P. This provides an efficient way of performing the counting. By varying the strategy of growing the tree, the memory overhead of storing the *tid* lists can be reduced but not the number of operations. For example, while both breadth-first and depth-first strategies will require exactly the same *tid* list intersections for a particular pair of nodes, the depth-first strategy will have a smaller memory footprint because the *tid* lists need to be stored only at the nodes on the tree-path being explored and their immediate siblings. Reducing the memory footprint is, nevertheless, important because it increases the size of the database that can be processed entirely in core.

Subsequently, many algorithms, such as *Eclat* and *VIPER*, adopted *Partition*'s recursive *tid* list intersection approach. *Eclat* is a lattice-partitioned memory-optimization of the algo-

[2]Strictly speaking, *Monet* is the name of the vertical database, on top of which this (unnamed) algorithm was built.

```
Algorithm VerticalApriori(Transactions: T, Minimum Support: minsup)
begin
  k = 1;
  F_1 = { All Frequent 1-itemsets };
  Construct vertical tid lists of each frequent item;
  while F_k is not empty do begin
    Generate C_{k+1} by joining itemset-pairs in F_k;
    Prune itemsets from C_{k+1} that violate downward closure;
    Generate tid list of each candidate itemset in C_{k+1} by intersecting
       tid lists of the itemset-pair in F_k that was used to create it;
    Determine supports of itemsets in C_{k+1} using lengths of their tid lists;
    F_{k+1}= Frequent itemsets of C_{k+1} together with their tid lists;
    k = k + 1;
  end;
  return(∪_{i=1}^{k} F_i);
end
```

Figure 4.7: The vertical *Apriori* algorithm of Savasere et al. [446]

rithm in Fig. 4.7. In *Eclat* [537], an independent *Apriori*-like breadth-first strategy is used on each of the sublattices of itemsets with a common prefix. These groups of itemsets are referred to as equivalence classes. Such an approach can reduce the memory requirements by partitioning the candidate space into groups that are processed independently in conjunction with the relevant vertical lists of their prefixes. This kind of candidate partitioning is similar to parallel versions of *Apriori*, such as the *Candidate Distribution* algorithm [54]. Instead of using the candidate partitioning to distribute various sublattices to different processors, the *Eclat* approach sequentially processes the sublattices one after another to reduce peak memory requirements. Therefore, *Eclat* can avoid the postprocessing overheads associated with Savasere et al.'s *data* partitioning approach, if the database is too large to be processed in core by *Partition-1*, but small enough to be processed in core by *Eclat*. In such cases, *Eclat* is faster than *Partition*. Note that the number of computational operations for support counting in *Partition-1* is fundamentally no different from that of *Eclat* because the *tid* list intersections between any pair of itemsets remain the same. Furthermore, *Eclat* implicitly assumes an upper bound on the database size. This is because it assumes that multiple *tid* lists, each of size at least a fraction *minsup* of the number of database records, fit in main memory. The cumulative memory overhead of the multiple *tid* lists always scales proportionally with database size, whereas the memory overhead of the ensemble-based *Partition* algorithm is independent of database size.

4.4.4 Recursive Suffix-Based Pattern Growth Methods

Enumeration trees are constructed by extending *prefixes* of itemsets that are expressed in a lexicographic order. It is also possible to express some classes of itemset exploration methods recursively with *suffix*-based exploration. Although recursive pattern-growth is often understood as a completely different class of methods, it can be viewed as a special case of the generic enumeration-tree algorithm presented in the previous section. This relationship between recursive pattern-growth methods and enumeration-tree methods will be explored in greater detail in Sect. 4.4.4.5.

Recursive suffix-based pattern growth methods are generally understood in the context of the well-known FP-Tree data structure. While the FP-Tree provides a space- and time-efficient way to implement the recursive pattern exploration, these methods can also be implemented with the use of arrays and pointers. This section will present the recursive pattern growth approach in a simple way without introducing any specific data structure. We also present a number of simplified implementations[3] with various data structures to facilitate better understanding. The idea is to move from the simple to the complex by providing a top-down data structure-agnostic presentation, rather than a tightly integrated presentation with the commonly used FP-Tree data structure. This approach provides a clear understanding of how the search space of patterns is explored and the relational with conventional enumeration tree algorithms.

Consider the transaction database $\mathcal{T}$ which is expressed in terms of only frequent 1-items. It is assumed that a counting pass has already been performed on $\mathcal{T}$ to remove the infrequent items and count the supports of the items. Therefore, the input to the recursive procedure described here is slightly different from the other algorithms discussed in this chapter in which this database pass has not been performed. The items in the database are ordered with decreasing support. This lexicographic ordering is used to define the ordering of items within itemsets and transactions. This ordering is also used to define the notion of prefixes and suffixes of itemsets and transactions. The input to the algorithm is the transaction database $\mathcal{T}$ (expressed in terms of frequent 1-items), a current frequent itemset suffix P, and the minimum support *minsup*. The goal of a recursive call to the algorithm is to determine all the frequent patterns that have the suffix P. Therefore, at the top-level recursive call of the algorithm, the suffix P is empty. At deeper-level recursive calls, the suffix P is not empty. The assumption for deeper-level calls is that $\mathcal{T}$ contains only those transactions from the original database that include the itemset P. Furthermore, each transaction in $\mathcal{T}$ is represented using only those frequent extension items of P that are lexicographically smaller than all items of P. Therefore $\mathcal{T}$ is a *conditional transaction set*, or *projected database* with respect to suffix P. This suffix-based projection is similar to the prefix-based projection in *TreeProjection* and *DepthProject*.

In any given recursive call, the first step is to construct the itemset $P_i = \{i\} \cup P$ by concatenating each item i in the transaction database $\mathcal{T}$ to the beginning of suffix P, and reporting it as frequent. The itemset P_i is frequent because $\mathcal{T}$ is defined in terms of frequent items of the projected database of suffix P. For each item i, it is desired to further extend P_i by using a recursive call with the projected database of the (newly extended) frequent suffix P_i. The projected database for extended suffix P_i is denoted by $\mathcal{T}_i$, and it is created as follows. The first step is to extract all transactions from $\mathcal{T}$ that contain the item i. Because it is desired to extend the suffix P_i backwards, all items that are lexicographically greater than or equal to i are removed from the extracted transactions in $\mathcal{T}_i$. In other words, the part of the transaction occurring lexicographically after (and including) i is not relevant for counting frequent patterns ending in P_i. The frequency of each item in $\mathcal{T}_i$ is counted, and the infrequent items are removed.

It is easy to see that the transaction set $\mathcal{T}_i$ is sufficient to generate all the frequent patterns with P_i as a suffix. The problem of finding all frequent patterns ending in P_i using the transaction set $\mathcal{T}_i$ is an identical but smaller problem than the original one on $\mathcal{T}$. Therefore, the original procedure is called recursively with the smaller projected database $\mathcal{T}_i$ and extended suffix P_i. This procedure is repeated for each item i in $\mathcal{T}$.

[3]Variations of these strategies are actually used in some implementations of these methods. We stress that the simplified versions are not optimized for efficiency but are provided for clarity.

Algorithm *RecursiveSuffixGrowth*(Transactions in terms of frequent 1-items: $\mathcal{T}$,
Minimum Support: *minsup*, Current Suffix: P)
begin
  for each item i in $\mathcal{T}$ **do begin**
    report itemset $P_i = \{i\} \cup P$ as frequent;
    Extract all transactions $\mathcal{T}_i$ from $\mathcal{T}$ containing item i;
    Remove all items from $\mathcal{T}_i$ that are lexicographically $\geq i$;
    Remove all infrequent items from $\mathcal{T}_i$;
    if $(\mathcal{T}_i \neq \phi)$ **then** *RecursiveSuffixGrowth*$(\mathcal{T}_i, minsup, P_i)$;
  end
end

Figure 4.8: Generic recursive suffix growth on transaction database expressed in terms of frequent 1-items

The projected transaction set $\mathcal{T}_i$ will become successively smaller at deeper levels of the recursion in terms of the number of items and the number of transactions. As the number of transactions reduces, all items in it will eventually fall below the minimum support, and the resulting projected database (constructed on only the frequent items) will be empty. In such cases, a recursive call with $\mathcal{T}_i$ is not initiated; therefore, this branch of the recursion is not explored. For some data structures, such as the FP-Tree, it is possible to impose stronger boundary conditions to terminate the recursion even earlier. This boundary condition will be discussed in a later section.

The overall recursive approach is presented in Fig. 4.8. While the parameter *minsup* has always been assumed to be a (relative) fractional value in this chapter, it is assumed to be an absolute integer support value in this section and in Fig. 4.8. This deviation from the usual convention ensures consistency of the minimum support value across different recursive calls in which the size of the conditional transaction database reduces.

4.4.4.1 Implementation with Arrays but No Pointers

So, how can the projected database $\mathcal{T}$ be decomposed into the *conditional transaction sets* $\mathcal{T}_1 \dots \mathcal{T}_d$, corresponding to d different 1-item suffixes? The simplest solution is to use arrays. In this solution, the original transaction database $\mathcal{T}$ and the conditional transaction sets $\mathcal{T}_1 \dots \mathcal{T}_d$ can be represented in arrays. The transaction database $\mathcal{T}$ may be scanned within the "**for**" loop of Fig. 4.8, and the set $\mathcal{T}_i$ is created from $\mathcal{T}$. The infrequent items from $\mathcal{T}_i$ are removed within the loop. However, it is expensive and wasteful to repeatedly scan the database $\mathcal{T}$ inside a "**for**" loop. One alternative is to extract all projections $\mathcal{T}_i$ of $\mathcal{T}$ corresponding to the different suffix items simultaneously in a single scan of the database just before the "**for**" loop is initiated. On the other hand, the simultaneous creation of many such item-specific projected data sets can be memory-intensive. One way of obtaining an excellent trade-off between computational and storage requirements is by using pointers. This approach is discussed in the next section.

4.4.4.2 Implementation with Pointers but No FP-Tree

The array-based solution either needs to repeatedly scan the database $\mathcal{T}$ or simultaneously create many smaller item-specific databases in a single pass. Typically, the latter achieves

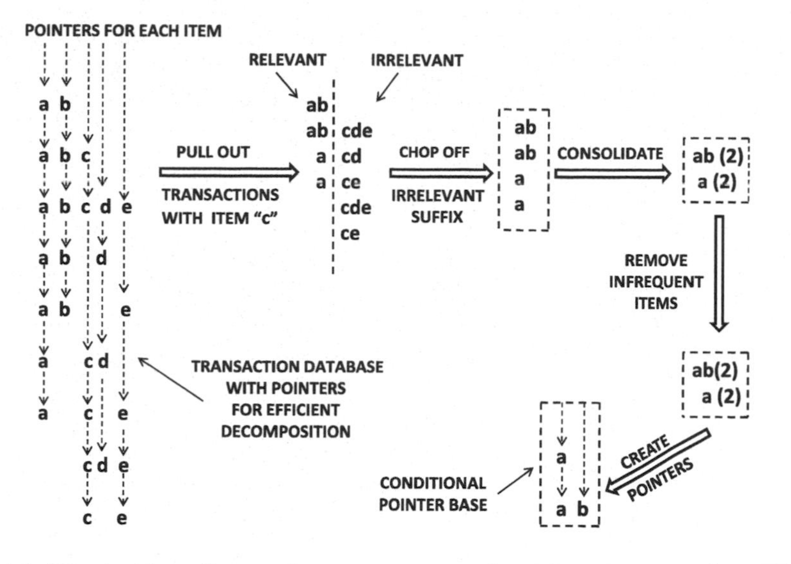

Figure 4.9: Illustration of recursive pattern growth with pointers and no FP-Tree

better efficiency but is more memory-intensive. One simple solution to this dilemma is to set up a data structure in the form of pointers in the first pass, which *implicitly* stores the decomposition of $\mathcal{T}$ into different item-specific data sets at a lower memory cost. This data structure is set up at the time that infrequent items are removed from the transaction database $\mathcal{T}$, and then utilized for extracting different conditional transaction sets $\mathcal{T}_i$ from $\mathcal{T}$. For each item i in $\mathcal{T}$, a pointer threads through the transactions containing that item in lexicographically sorted (dictionary) order. In other words, after arranging the database $\mathcal{T}$ in lexicographically sorted order, each item i in each transaction has a pointer to the same item i in the next transaction that contains it. Because a pointer is required at each item in each transaction, the storage overhead in this case is proportional to that of the original transaction database $\mathcal{T}$. An additional optimization is to consolidate repeated transactions and store them with their counts. An example of a sample database with nine transactions on the five items $\{a, b, c, d, e\}$ is illustrated in Fig. 4.9. It is clear from the figure that there are five sets of pointers, one for each item in the database.

After the pointers have been set up, $\mathcal{T}_i$ is extracted by just "chasing" the pointer thread for item i. The time for doing this is proportional to the number of transactions in $\mathcal{T}_i$. The infrequent items in $\mathcal{T}_i$ are removed, and the pointers for the conditional transaction data need to be reconstructed to create a *conditional pointer base* which is basically the conditional transaction set augmented with pointers. The modified pseudocode with the use of pointers is illustrated in Fig. 4.10. Note that the only difference between the pseudocode of Figs. 4.8 and 4.10 is the setting up of pointers after extraction of conditional transaction sets and the use of these pointers to efficiently extract the conditional transaction data sets $\mathcal{T}_i$. A recursive call is initiated at the next level with the extended suffix $P_i = \{i\} \cup P$, and conditional database $\mathcal{T}_i$.

To illustrate how $\mathcal{T}_i$ can be extracted, an example of a transaction database with 5 items and 9 transactions is illustrated in Fig. 4.9. For simplicity, we use a (raw) minimum support value of 1. The transactions corresponding to the item c are extracted, and the irrelevant suffix including and after item c are removed for further recursive calls. Note that this leads to shorter transactions, some of which are repeated. As a result, the conditional database

```
Algorithm RecursiveGrowthPointers(Transactions in terms of frequent 1-items: T,
        Minimum Support: minsup, Current Suffix: P)
begin
  for each item i in T do begin
    report itemset P_i = {i} ∪ P as frequent;
    Use pointers to extract all transactions T_i
        from T containing item i;
    Remove all items from T_i that are lexicographically ≥ i;
    Remove all infrequent items from T_i;
    Set up pointers for T_i;
    if (T_i ≠ φ) then RecursiveGrowthPointers(T_i, minsup, P_i);
  end
end
```

Figure 4.10: Generic recursive suffix growth with pointers

for $\mathcal{T}_i$ contains only two distinct transactions after consolidation. The infrequent items from this conditional database need to be removed. No items are removed at a minimum support of 1. Note that if the minimum support had been 3, then the item b would have been removed. The pointers for the new conditional transaction set do need to be set up again because they will be different for the conditional transaction database than in the original transactions. Unlike the pseudocode of Fig. 4.8, an additional step of setting up pointers is included in the pseudocode of Fig. 4.10.

The pointers provide an efficient way to extract the conditional transaction database. Of course, the price for this is that the pointers are a space overhead, with size exactly proportional to the original transaction database $\mathcal{T}$. Consolidating repeated transactions does save some space. The FP-Tree, which will be discussed in the next section, takes this approach one step further by consolidating not only repeated transactions, but also repeated *prefixes* of transactions with the use of a trie data structure. This representation reduces the space-overhead by consolidating prefixes of the transaction database.

4.4.4.3 Implementation with Pointers and FP-Tree

The FP-Tree is designed with the primary goal of space efficiency of the projected database. The FP-Tree is a trie data structure representation of the conditional transaction database by consolidating the prefixes. This trie replaces the array-based implementation of the previous sections, but it retains the pointers. The path from the root to the leaf in the trie represents a (possibly repeated) transaction in the database. The path from the root to an internal node may represent either a transaction or the prefix of a transaction in the database. Each internal node is associated with a count representing the number of transactions in the original database that contain the prefix corresponding to the path from the root to that node. The count on a leaf represents the number of repeated instances of the transaction defined by the path from the root to that leaf. Thus, the FP-Tree maintains all counts of all the repeated transactions as well as their prefixes in the database. As in a standard trie data-structure, the prefixes are sorted in dictionary order. The lexicographic ordering of items is from the most frequent to the least frequent to maximize the advantages of prefix-based compression. This ordering also provides excellent selectivity in reducing the size of various conditional transaction sets in a balanced way. An example of the FP-Tree

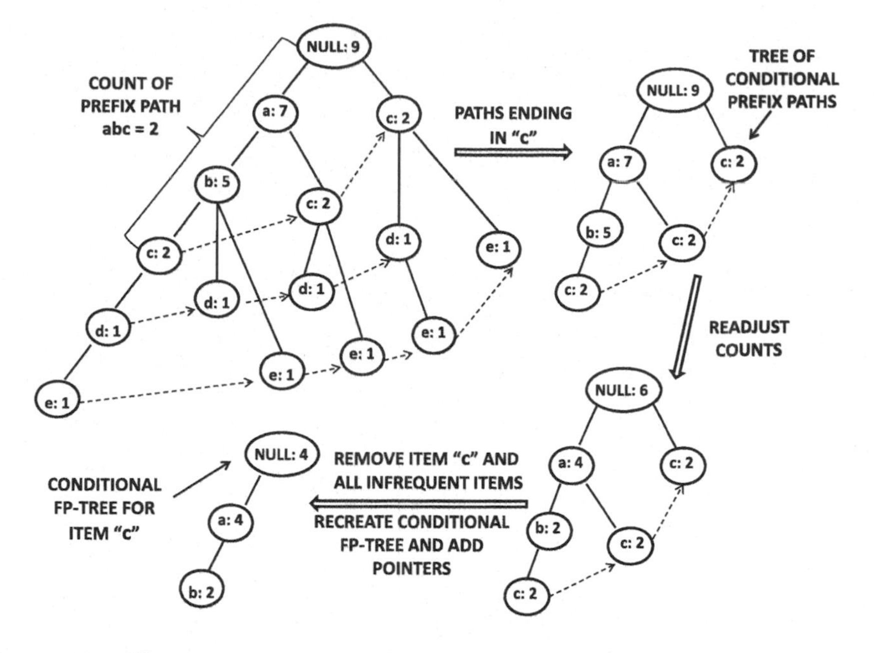

Figure 4.11: Illustration of recursive pattern growth with pointers and FP-Tree

data structure for the same database (as the previous example of Fig. 4.9) is shown in Fig. 4.11. In the example, the number "2" associated with the leftmost item c in the FP-Tree, represents the count of prefix path abc, as illustrated in Fig. 4.11.

The initial FP-Tree $\mathcal{FPT}$ can be constructed as follows. First the infrequent items in the database are removed. The resulting transactions are then successively inserted into the trie. The counts on the overlapping nodes are incremented by 1 when the prefix of the inserted transaction overlaps with an existing path in the trie. For the non-overlapping portion of the transaction, a new path needs to be created containing this portion. The newly created nodes are assigned a count of 1. This process of insertion is identical to that of trie creation, except that counts are also associated with nodes. The resulting tree is a compressed representation because common items in the prefixes of multiple transactions are represented by a single node.

The pointers can be constructed in an analogous way to the simpler array data structure of the previous section. The pointer for each item points to the next occurrence of the same item in the trie. Because a trie stores the transactions in dictionary order, it is easy to create pointers threading each of the items. However, the number of pointers is smaller, because many nodes have been consolidated. As an illustrative example, one can examine the relationship between the array-based data structure of Fig. 4.9, and the FP-Tree in Fig. 4.11. The difference is that the prefixes of the arrays in Fig. 4.9 are consolidated and compressed into a trie in Fig. 4.11.

The conditional FP-Tree $\mathcal{FPT}_i$ (representing the conditional database $\mathcal{T}_i$) needs to be extracted and reorganized for each item $i \in \mathcal{FPT}$. This extraction is required to initiate recursive calls with conditional FP-Trees. As in the case of the simple pointer-based structure of the previous section, it is possible to use the pointers of an item to extract the subset of the projected database containing that item. The following steps need to be performed for extraction of the conditional FP-Tree of item i:

1. The pointers for item i are chased to extract the *tree of conditional prefix paths* for the item. These are the paths from the item to the root. The remaining branches are pruned.

2. The counts of the nodes in the tree of prefix-paths are adjusted to account for the pruned branches. The counts can be adjusted by aggregating the counts on the leaves upwards.

3. The frequency of each item is counted by aggregating the counts over all occurrences of that item in the tree of prefix paths. The items that do not meet the minimum support requirement are removed from the prefix paths. Furthermore, the last item i is also removed from each prefix path. The resulting conditional FP-Tree might have a completely different organization than the extracted tree of prefix-paths because of the removal of infrequent items. Therefore, the conditional FP-Tree may need to be recreated by reinserting the conditional prefix paths obtained after removing infrequent items. The pointers for the conditional FP-Tree need to be reconstructed as well.

Consider the example in Fig. 4.11 which is the same data set as in Fig. 4.9. As in Fig. 4.9, it is possible to follow the pointers for item c in Fig. 4.11 to extract a tree of conditional prefix paths (shown in Fig. 4.11). The counts on many nodes in the tree of conditional prefix paths need to be reduced because many branches from the original FP-Tree (that do not contain the item c) are not included. These reduced counts can be determined by aggregating the counts on the leaves upwards. After removing the item c and infrequent items, two frequency-annotated conditional prefix paths $ab(2)$ and $a(2)$ are obtained, which are identical to the two projected and consolidated transactions of Fig. 4.9. The conditional FP-tree is then constructed for item c by reinserting these two conditional prefix paths into a new conditional FP-Tree. Again, this conditional FP-Tree is a trie representation of the conditional pointer base of Fig. 4.9. In this case, there are no infrequent items because a minimum support of 1 is used. If a minimum support of 3 had been used, then the item b would have to be removed. The resulting conditional FP-Tree is used in the next level recursive call. After extracting the conditional FP-Tree $\mathcal{FPT}_i$, it is checked whether it is empty. An empty conditional FP-Tree could occur when there are no frequent items in the extracted tree of conditional prefix paths. If the tree is not empty, then the next level recursive call is initiated with suffix $P_i = \{i\} \cup P$, and the conditional FP-Tree $\mathcal{FPT}_i$.

The use of the FP-Tree allows an additional optimization in the form of a boundary condition for quickly extracting frequent patterns at deeper levels of the recursion. In particular, it is checked whether all the nodes of the FP-Tree lie on a single path. In such a case, the frequent patterns can be directly extracted from this path by extracting all combinations of nodes on this path together with the aggregated support counts. For example, in the case of Fig. 4.11, all nodes on the conditional FP-Tree lie on a single path. Therefore, in the next recursive call, the bottom of the recursion will be reached. The pseudocode for *FP-growth* is illustrated in Fig. 4.12. This pseudocode is similar to the pointer-based pseudocode of Fig. 4.10, except that a compressed FP-Tree is used.

4.4.4.4 Trade-offs with Different Data Structures

The main advantage of an FP-Tree over pointer-based implementation is one of space compression. The FP-Tree requires less space than pointer-based implementation because of trie-based compression, although it might require more space than an array-based implementation because of the pointer overhead. The precise space requirements depend on the

Algorithm *FP-growth*(FP-Tree of frequent items: $\mathcal{FPT}$, Minimum Support: *minsup*,
Current Suffix: P)
begin
  if $\mathcal{FPT}$ is a single path
    then determine all combinations C of nodes on the
      path, and report $C \cup P$ as frequent;
  else (Case when $\mathcal{FPT}$ is not a single path)
  for each item i in $\mathcal{FPT}$ **do begin**
    report itemset $P_i = \{i\} \cup P$ as frequent;
    Use pointers to extract conditional prefix paths
      from $\mathcal{FPT}$ containing item i;
    Readjust counts of prefix paths and remove i;
    Remove infrequent items from prefix paths and reconstruct
      conditional FP-Tree $\mathcal{FPT}_i$;
    if $(\mathcal{FPT}_i \neq \phi)$ **then** *FP-growth*$(\mathcal{FPT}_i, minsup, P_i)$;
  end
end

Figure 4.12: The *FP-growth* algorithm with an FP-Tree representation of the transaction database expressed in terms of frequent 1-items

level of consolidation at higher level nodes in the trie-like FP-Tree structure for a particular data set. Different data structures may be more suitable for different data sets.

Because projected databases are repeatedly constructed and scanned during recursive calls, it is crucial to maintain them in main memory. Otherwise, drastic disk-access costs will be incurred by the potentially exponential number of recursive calls. The sizes of the projected databases increase with the original database size. For certain kinds of databases with limited consolidation of repeated transactions, the number of *distinct* transactions in the projected database will always be approximately proportional to the number of transactions in the original database, where the proportionality factor f is equal to the (fractional) minimum support. For databases that are larger than a factor $1/f$ of the main memory availability, projected databases may not fit in main memory either. Therefore, the limiting factor on the use of the approach is the size of the original transaction database. This issue is specific to almost all projection-based methods and vertical counting methods. Memory is always at a premium in such methods and therefore it is crucial for projected transaction data structures to be designed as compactly as possible. As we will discuss later, the *Partition* framework of Savasere et al. [446] provides a partial solution to this issue at the expense of running time.

4.4.4.5 Relationship Between FP-Growth and Enumeration-Tree Methods

FP-growth is popularly believed to be radically different from enumeration-tree methods. This is, in part, because *FP-growth* was originally presented as a method that extracts frequent patterns without candidate generation. However, such an exposition provides an incomplete understanding of how the search space of patterns is explored. *FP-growth* is an instantiation of enumeration-tree methods. All enumeration-tree methods generate candidate extensions to grow the tree. In the following, we will show the equivalence between enumeration-tree methods and *FP-growth*.

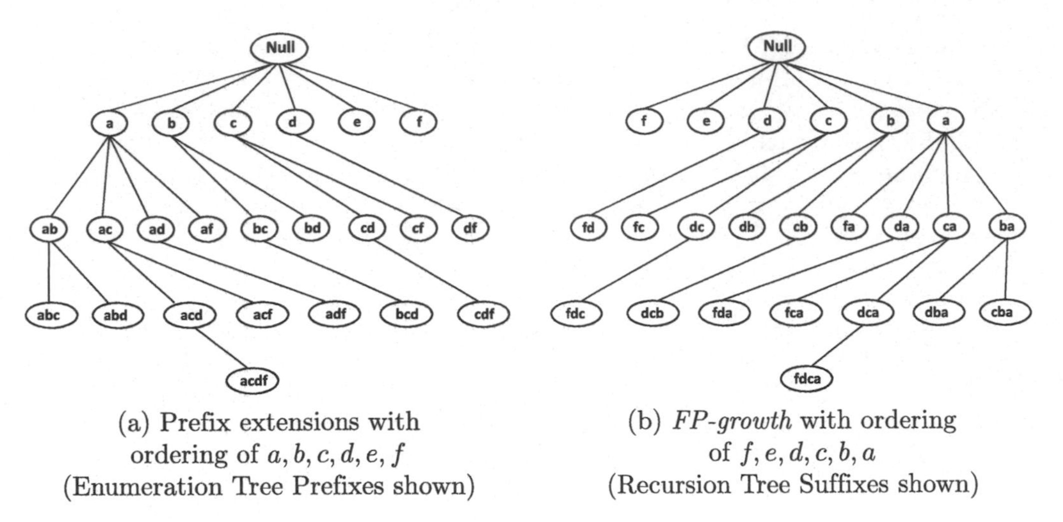

(a) Prefix extensions with ordering of a, b, c, d, e, f (Enumeration Tree Prefixes shown)

(b) *FP-growth* with ordering of f, e, d, c, b, a (Recursion Tree Suffixes shown)

Figure 4.13: Enumeration trees are identical to *FP-growth* recursion trees with reverse lexicographic ordering

FP-growth is a recursive algorithm that extends suffixes of frequent patterns. Any recursive approach has a tree-structure associated with it that is referred to as its *recursion tree*, and a dynamic *recursion stack* that stores the recursion variables on the current path of the recursion tree during execution. Therefore, it is instructive to examine the suffix-based recursion tree created by the *FP-growth* algorithm, and compare it with the classical prefix-based enumeration tree used by enumeration-tree algorithms.

In Fig. 4.13a, the enumeration tree from the earlier example of Fig. 4.3 has been replicated. This tree of frequent patterns is counted by all enumeration-tree algorithms along with a single layer of infrequent candidate extensions of this tree corresponding to failed candidate tests. Each call of *FP-growth* discovers the set of frequent patterns extending a particular suffix of items, just as each branch of an enumeration tree explores the itemsets for a particular prefix. So, what is the hierarchical recursive relationship among the suffixes whose conditional pattern bases are explored? First, we need to decide on an ordering of items. Because the recursion is performed on suffixes and enumeration trees are constructed on prefixes, the *opposite* ordering $\{f, e, d, c, b, a\}$ is assumed to adjust for the different convention in the two methods. Indeed, most enumeration-tree methods order items from the least frequent to the most frequent, whereas *FP-growth* does the reverse. The corresponding recursion tree of *FP-growth*, when the 1-itemsets are ordered from left to right in dictionary order, is illustrated in Fig. 4.13b. The trees in Figs 4.13a and 4.13b are identical, with the only difference being that they are drawn differently, to account for the opposite lexicographic ordering. The *FP-growth* recursion tree on the reverse lexicographic ordering has an identical structure to the traditional enumeration tree on the prefixes. During any given recursive call of *FP-growth*, the current (recursion) stack of suffix items is the path in the enumeration tree that is currently being explored. This enumeration tree is explored in depth-first order by *FP-growth* because of its recursive nature.

Traditional enumeration-tree methods typically count the support of a single layer of infrequent extensions of the frequent patterns in the enumeration-tree, as (failed) candidates, to rule them out. Therefore, it is instructive to explore whether *FP-growth* avoids counting these infrequent candidates. Note that when conditional transaction databases $\mathcal{FPT}_i$ are created (see Fig. 4.12), infrequent items must be removed from them. This requires the counting of the support of these (implicitly failed) *candidate extensions*. In a traditional candidate generate-and-test algorithm, the frequent candidate extensions would be reported immediately after the counting step as a successful candidate test. However, in *FP-growth*, these frequent extensions are encoded back into the conditional transaction database $\mathcal{FPT}_i$, and the reporting is delayed to the next level recursive call. In the next level recursive call, these frequent extensions are then extracted from $\mathcal{FPT}_i$ and reported. The counting and removal of infrequent items from conditional transaction sets is an implicit candidate evaluation and testing step. The number of such failed candidate tests[4] in *FP-growth* is exactly equal to that of enumeration-tree algorithms, such as *Apriori* (without the level-wise pruning step). This equality follows directly from the relationship of all these algorithms to how they explore the enumeration tree and rule out infrequent portions. All pattern-growth methods, including *FP-growth*, should be considered enumeration-tree methods, as should *Apriori*. Whereas traditional enumeration trees are constructed on prefixes, the (implicit) *FP-growth* enumeration trees are constructed using suffixes. This is a difference only in the item-ordering convention.

The depth-first strategy is the approach of choice in database projection methods because it is more memory-efficient to maintain the conditional transaction sets along a (relatively small) depth of the enumeration (recursion) tree rather than along the (much larger) breadth of the enumeration tree. As discussed in the previous section, memory management becomes a problem even with the depth-first strategy beyond a certain database size. However, the specific strategy used for tree exploration does not have any impact on the size of the enumeration tree (or candidates) explored over the course of the entire algorithm execution. The only difference is that breadth-first methods process candidates in large batches based on pattern size, whereas depth-first methods process candidates in smaller batches of immediate siblings in the enumeration tree. From this perspective, *FP-growth* cannot avoid the exponential candidate search space exploration required by enumeration-tree methods, such as *Apriori*.

Whereas methods such as *Apriori* can also be interpreted as counting methods on an enumeration-tree of exactly the same size as the recursion tree of *FP-growth*, the counting work done at the higher levels of the enumeration tree is lost. This loss is because the counting is done from scratch at each level in *Apriori* with the entire transaction database rather than a projected database that remembers and *reuses* the work done at the higher levels of the tree. Projection-based reuse is also utilized by Savasere et al.'s vertical counting methods [446] and *DepthProject*. The use of a pointer-trie combination data structure for projected transaction representation is the primary difference of *FP-growth* from other projection-based methods. In the context of depth-first exploration, these methods can be understood either as divide-and-conquer strategies or as projection-based reuse strategies. The notion of projection-based reuse is more general because it applies to both the breadth-first and depth-first versions of the algorithm, and it provides a clearer picture of how computational savings are achieved by avoiding wasteful and repetitive counting. Projection-based reuse enables the efficient testing of candidate *item extensions* in a restricted portion of the database rather than the testing of candidate *itemsets* in the full database. Therefore, the efficiencies in *FP-growth* are a result of more efficient counting *per candidate* and not because of fewer candidates. The only differences in search space size between various methods are

[4] An *ad hoc* pruning optimization in *FP-growth* terminates the recursion when all nodes in the FP-Tree lie on a single path. This pruning optimization reduces the number of successful candidate tests but not the number of failed candidate tests. Failed candidate tests often dominate successful candidate tests in real data sets.

the result of *ad hoc* pruning optimizations, such as level-wise pruning in *Apriori*, bucketing in the *DepthProject* algorithm, and the single-path boundary condition of *FP-growth*.

The bookkeeping of the projected transaction sets can be done differently with the use of different data structures, such as arrays, pointers, or a pointer-trie combination. Many different data structure variations are explored in different projection algorithms, such as *TreeProjection*, *DepthProject*, *FP-growth*, and *H-Mine* [419]. Each data structure is associated with a different set of efficiencies and overheads.

In conclusion, the enumeration tree[5] is the most general framework to describe all previous frequent pattern mining algorithms. This is because the enumeration tree is a subgraph of the lattice (candidate space) and it provides a way to explore the candidate patterns in a systematic and non-redundant way. The support testing of the frequent portion of the enumeration tree along with a single layer of infrequent candidate extensions of these nodes is fundamental to all frequent itemset mining algorithms for ruling in and ruling out *possible* (or *candidate*) frequent patterns. Any algorithm, such as *FP-growth*, which uses the enumeration tree to rule in and rule out possible extensions of frequent patterns with support counting, is a candidate generate-and-test algorithm.

4.5 Alternative Models: Interesting Patterns

The traditional model for frequent itemset generation has found widespread popularity and acceptance because of its simplicity. The simplicity of using raw frequency counts for the support, and that of using the conditional probabilities for the confidence is very appealing. Furthermore, the downward closure property of frequent itemsets enables the design of efficient algorithms for frequent itemset mining. This algorithmic convenience does not, however, mean that the patterns found are always significant from an *application-specific* perspective. Raw frequencies of itemsets do not always correspond to the most *interesting* patterns.

For example, consider the transaction database illustrated in Fig. 4.1. In this database, *all* the transactions contain the item $Milk$. Therefore, the item $Milk$ can be appended to *any* set of items, without changing its frequency. However, this does not mean that $Milk$ is truly associated with any set of items. Furthermore, for any set of items X, the association rule $X \Rightarrow \{Milk\}$ has 100% confidence. However, it would not make sense for the supermarket merchant to assume that the basket of items X is *discriminatively* indicative of $Milk$. Herein lies the limitation of the traditional support-confidence model.

Sometimes, it is also desirable to design measures that can adjust to the skew in the individual item support values. This adjustment is especially important for negative pattern mining. For example, the support of the pair of items $\{Milk, Butter\}$ is very different from that of $\{\neg Milk, \neg Butter\}$. Here, $\neg$ indicates negation. On the other hand, it can be argued that the statistical coefficient of correlation is exactly the same in both cases. Therefore, the measure should quantify the association between both pairs in exactly the same way. Clearly, such measures are important for negative pattern mining. Measures that satisfy this property are said to satisfy the *bit symmetric* property because values of 0 in the binary matrix are treated in a similar way to values of 1.

[5] *FP-growth* has been presented in a separate section from enumeration tree methods only because it uses a different convention of constructing *suffix*-based enumeration trees. It is not necessary to distinguish "pattern growth" methods from "candidate-based" methods to meaningfully categorize various frequent pattern mining methods. Enumeration tree methods are best categorized on the basis of their (i) tree exploration strategy, (ii) projection-based reuse properties, and (iii) relevant data structures.

Although it is possible to quantify the affinity of sets of items in ways that are statistically more robust than the support-confidence framework, the major computational problem faced by most such *interestingness-based models* is that the downward closure property is generally not satisfied. This makes algorithmic development rather difficult on the exponentially large search space of patterns. In some cases, the measure is defined only for the special case of 2-itemsets. In other cases, it is possible to design more efficient algorithms. The following contains a discussion of some of these models.

4.5.1 Statistical Coefficient of Correlation

A natural statistical measure is the Pearson coefficient of correlation between a pair of items. The Pearson coefficient of correlation between a pair of random variables X and Y is defined as follows:

$$\rho = \frac{E[X \cdot Y] - E[X] \cdot E[Y]}{\sigma(X) \cdot \sigma(Y)}. \tag{4.4}$$

In the case of market basket data, X and Y are binary variables whose values reflect presence or absence of items. The notation $E[X]$ denotes the expectation of X, and $\sigma(X)$ denotes the standard deviation of X. Then, if $sup(i)$ and $sup(j)$ are the relative supports of individual items, and $sup(\{i,j\}$ is the relative support of itemset $\{i,j\}$, then the overall correlation can be estimated from the data as follows:

$$\rho_{ij} = \frac{sup(\{i,j\}) - sup(i) \cdot sup(j)}{\sqrt{sup(i) \cdot sup(j) \cdot (1 - sup(i)) \cdot (1 - sup(j))}}. \tag{4.5}$$

The coefficient of correlation always lies in the range $[-1,1]$, where the value of $+1$ indicates perfect positive correlation, and the value of -1 indicates perfect negative correlation. A value near 0 indicates weakly correlated data. This measure satisfies the bit symmetric property. While the coefficient of correlation is statistically considered the most robust way of measuring correlations, it is often intuitively hard to interpret when dealing with items of varying but low support values.

4.5.2 χ^2 Measure

The χ^2 measure is another bit-symmetric measure that treats the presence and absence of items in a similar way. Note that for a set of k binary random variables (items), denoted by X, there are 2^k-possible states representing presence or absence of different items of X in the transaction. For example, for $k = 2$ items $\{Bread, Butter\}$, the 2^2 states are $\{Bread, Butter\}$, $\{Bread, \neg Butter\}$, $\{\neg Bread, Butter\}$, and $\{\neg Bread, \neg Butter\}$. The expected fractional presence of each of these combinations can be quantified as the product of the supports of the states (presence or absence) of the individual items. For a given data set, the *observed* value of the support of a state may vary significantly from the expected value of the support. Let O_i and E_i be the observed and expected values of the absolute support of state i. For example, the expected support E_i of $\{Bread, \neg Butter\}$ is given by the total number of transactions multiplied by each of the fractional supports of *Bread* and $\neg$*Butter*, respectively. Then, the χ^2-measure for set of items X is defined as follows:

$$\chi^2(X) = \sum_{i=1}^{2^{|X|}} \frac{(O_i - E_i)^2}{E_i}. \tag{4.6}$$

For example, when $X = \{Bread, Butter\}$, one would need to perform the summation in Eq. 4.6 over the $2^2 = 4$ states corresponding to $\{Bread, Butter\}$, $\{Bread, \neg Butter\}$, $\{\neg Bread, Butter\}$, and $\{\neg Bread, \neg Butter\}$. A value that is close to 0 indicates statistical independence among the items. Larger values of this quantity indicate greater dependence between the variables. However, large χ^2 values do not reveal whether the dependence between items is positive or negative. This is because the χ^2 test measures dependence between variables, rather than the nature of the correlation between the specific states of these variables.

The χ^2 measure is bit-symmetric because it treats the presence and absence of items in a similar way. The χ^2-test satisfies the *upward closure property* because of which an efficient algorithm can be devised for discovering interesting k-patterns. On the other hand, the computational complexity of the measure in Eq. 4.6 increases exponentially with $|X|$.

4.5.3 Interest Ratio

The interest ratio is a simple and intuitively interpretable measure. The interest ratio of a set of items $\{i_1 \ldots i_k\}$ is denoted as $I(\{i_1, \ldots i_k\})$, and is defined as follows:

$$I(\{i_1 \ldots i_k\}) = \frac{sup(\{i_1 \ldots i_k\})}{\prod_{j=1}^{k} sup(i_j)}. \tag{4.7}$$

When the items are statistically independent, the joint support in the numerator will be equal to the product of the supports in the denominator. Therefore, an interest ratio of 1 is the break-even point. A value greater than 1 indicates that the variables are positively correlated, whereas a ratio of less than 1 is indicative of negative correlation.

When some items are extremely rare, the interest ratio can be misleading. For example, if an item occurs in only a single transaction in a large transaction database, each item that co-occurs with it in that transaction can be paired with it to create a 2-itemset with a very high interest ratio. This is statistically misleading. Furthermore, because the interest ratio does not satisfy the downward closure property, it is difficult to design efficient algorithms for computing it.

4.5.4 Symmetric Confidence Measures

The traditional confidence measure is *asymmetric* between the antecedent and consequent. However, the support measure is symmetric. Symmetric confidence measures can be used to replace the support-confidence framework with a single measure. Let X and Y be two 1-itemsets. Symmetric confidence measures can be derived as a function of the confidence of $X \Rightarrow Y$ and the confidence of $Y \Rightarrow X$. The various symmetric confidence measures can be any one of the minimum, average, or maximum of these two confidence values. The minimum is not desirable when either X or Y is very infrequent, causing the combined measure to be too low. The maximum is not desirable when either X or Y is very frequent, causing the combined measure to be too high. The average provides the most robust trade-off in many scenarios. The measures can be generalized to k-itemsets by using all k possible individual items in the consequent for computation. Interestingly, the *geometric* mean of the two confidences evaluates to the cosine measure, which is discussed below. The computational problem with symmetric confidence measures is that the relevant itemsets satisfying a specific threshold on the measure do not satisfy the downward closure property.

4.5.5 Cosine Coefficient on Columns

The cosine coefficient is usually applied to the rows to determine the similarity among transactions. However, it can also be applied to the columns, to determine the similarity between items. The cosine coefficient is best computed using the vertical *tid* list representation on the corresponding binary vectors. The cosine value on the binary vectors computes to the following:

$$cosine(i,j) = \frac{sup(\{i,j\})}{\sqrt{sup(i)} \cdot \sqrt{sup(j)}}. \quad (4.8)$$

The numerator can be evaluated as the length of the intersection of the *tid* lists of items i and j. The cosine measure can be viewed as the geometric mean of the confidences of the rules $\{i\} \Rightarrow \{j\}$ and $\{j\} \Rightarrow \{i\}$. Therefore, the cosine is a kind of symmetric confidence measure.

4.5.6 Jaccard Coefficient and the Min-hash Trick

The Jaccard coefficient was introduced in Chap. 3 to measure similarity between sets. The *tid* lists on a column can be viewed as a set, and the Jaccard coefficient between two *tid* lists can be used to compute the similarity. Let S_1 and S_2 be two sets. As discussed in Chap. 3, the Jaccard coefficient $J(S_1, S_2)$ between the two sets can be computed as follows:

$$J(S_1, S_2) = \frac{|S_1 \cap S_2|}{|S_1 \cup S_2|}. \quad (4.9)$$

The Jaccard coefficient can easily be generalized to multiway sets, as follows:

$$J(S_1 \dots S_k) = \frac{|\cap S_i|}{|\cup S_i|}. \quad (4.10)$$

When the sets $S_1 \dots S_k$ correspond to the *tid* lists of k items, the intersection and union of the *tid* lists can be used to determine the numerator and denominator of the aforementioned expression. This provides the Jaccard-based significance for that k-itemset. It is possible to use a minimum threshold on the Jaccard coefficient to determine all the relevant itemsets.

A nice property of Jaccard-based significance is that it satisfies the set-wise monotonicity property. The k-way Jaccard coefficient $J(S_1 \dots S_k)$ is always no smaller than the $(k+1)$-way Jaccard coefficient $J(S_1 \dots S_{k+1})$. This is because the numerator of the Jaccard coefficient is monotonically non-increasing with increasing values of k (similar to support), whereas the denominator is monotonically non-decreasing. Therefore, the Jaccard coefficient cannot increase with increasing values of k. Therefore, when a minimum threshold is used on the Jaccard-based significance of an itemset, the resulting itemsets satisfy the downward closure property, as well. This means that most of the traditional algorithms, such as *Apriori* and enumeration tree methods, can be generalized to the Jaccard coefficient quite easily.

It is possible to use sampling to speed up the computation of the Jaccard coefficient further, and transform it to a standard frequent pattern mining problem. This kind of sampling uses hash functions to simulate sorted samples of the data. So, how can the Jaccard coefficient be computed using sorted sampling? Let D be the $n \times d$ binary data matrix representing the n rows and d columns. Without loss of generality, consider the case when the Jaccard coefficient needs to be computed on the first k columns. Suppose one were to sort the rows in D, and pick the first row in which *at least* one of the first k columns in this row has a value of 1 in this column. Then, it is easy to see that the probability of the event that *all* the k columns have a value of 1 is equal to the k-way Jaccard coefficient. If one were to sort the rows multiple times, it is possible to estimate this probability as the fraction of sorts over which the event of all k columns taking on unit values occurs. Of course, it is rather inefficient to do it in this way because every sort requires a pass over the database. Furthermore, this approach can only estimate the Jaccard coefficient for a *particular* set of k columns, and it does not *discover* all the k-itemsets that satisfy the minimum criterion on the Jaccard coefficient.

The *min-hash* trick can be used to efficiently perform the sorts in an implicit way and transform to a concise sampled representation on which traditional frequent pattern mining algorithms can be applied to discover combinations satisfying the Jaccard threshold. The basic idea is as follows. A random hash function $h(\cdot)$ is applied to each *tid*. For each column of binary values, the *tid*, with the smallest hash function value, is selected among all entries *that have a unit value in that column.* This results in a vector of d different *tids*. What is the probability that the *tids* in the first k columns are the same? It is easy to see that this is equal to the Jaccard coefficient because the hashing process simulates the sort, and reports the *index* of the first non-zero element in the binary matrix. Therefore, by using independent hash functions to create multiple *samples*, it is possible to *estimate* the Jaccard coefficient. It is possible to repeat this process with r different hash functions, to create r different samples. Note that the r hash-functions can be applied simultaneously in a single pass over the transaction database. This creates a $r \times d$ categorical data matrix of *tids*. By determining the subsets of columns where the *tid* value is the same with support equal to a minimum support value, it is possible to estimate all sets of k-items whose Jaccard coefficient is at least equal to the minimum support value. This is a standard frequent pattern mining problem, except that it is defined on categorical values instead of a binary data matrix.

One way of transforming this $r \times d$ categorical data matrix to a binary matrix is to pull out the column identifiers where the *tids* are the same from each row and create a new transaction of column-identifier "items." Thus, a single row from the $r \times d$ matrix will map to multiple transactions. The resulting transaction data set can be represented by a new binary matrix D'. Any off-the-shelf frequent pattern mining algorithm can be applied to this binary matrix to discover relevant column-identifier combinations. The advantage of an off-the-shelf approach is that many efficient algorithms for the conventional frequent pattern mining model are available. It can be shown that the accuracy of the approach increases exponentially fast with the number of data samples.

4.5.7 Collective Strength

The collective strength of an itemset is defined in terms of its *violation rate*. An itemset I is said to be in *violation* of a transaction, if some of the items are present in the transaction, and others are not. The *violation rate* $v(I)$ of an itemset I is the fraction of violations of the itemset I over all transactions. The *collective strength* $C(I)$ of an itemset I is defined in terms of the violation rate as follows:

$$C(I) = \frac{1 - v(I)}{1 - E[v(I)]} \cdot \frac{E[v(I)]}{v(I)}. \tag{4.11}$$

The collective strength is a number between 0 to ∞. A value of 0 indicates a perfect negative correlation, whereas a value of ∞ indicates a perfectly positive correlation. The value of 1 is the break-even point. The expected value of $v(I)$ is calculated assuming statistical independence of the individual items. No violation occurs when all items in I are included in transaction, *or* when no items in I are included in a transaction. Therefore, if p_i is the fraction of transactions in which the item i occurs, we have:

$$E[v(I)] = 1 - \prod_{i \in I} p_i - \prod_{i \in I} (1 - p_i). \tag{4.12}$$

Intuitively, if the violation of an itemset in a transaction is a "bad event" from the perspective of trying to establish a high correlation among items, then $v(I)$ is the fraction of bad events, and $(1 - v(I))$ is the fraction of "good events." Therefore, collective strength may be understood as follows:

$$C(I) = \frac{\text{Good Events}}{\text{E[Good Events]}} \cdot \frac{\text{E[Bad Events]}}{\text{Bad Events}}. \tag{4.13}$$

The concept of collective-strength may be strengthened to *strongly collective* itemsets.

Definition 4.5.1 *An itemset I is denoted to be strongly collective at level s, if it satisfies the following properties:*

1. *The collective strength $C(I)$ of the itemset I is at least s.*
2. **Closure property:** *The collective strength $C(J)$ of every subset J of I is at least s.*

It is necessary to force the closure property to ensure that unrelated items may not be present in an itemset. Consider, for example, the case when itemset I_1 is $\{Milk, Bread\}$ and itemset I_2 is $\{Diaper, Beer\}$. If I_1 and I_2 each have a high collective strength, then it may often be the case that the itemset $I_1 \cup I_2$ may also have a high collective strength, even though items such as milk and beer may be independent. Because of the closure property of this definition, it is possible to design an *Apriori*-like algorithm for the problem.

4.5.8 Relationship to Negative Pattern Mining

In many applications, it is desirable to determine patterns between items *or their absence*. Negative pattern mining requires the use of *bit-symmetric* measures that treat the presence or absence of an item evenly. The traditional support-confidence measure is not designed for finding such patterns. Measures such as the statistical coefficient of correlation, χ^2 measure, and collective strength are better suited for finding such positive or negative correlations between items. However, many of these measures are hard to use in practice because they do not satisfy the downward closure property. The multiway Jaccard coefficient and collective strength are among the few measures that do satisfy the downward closure property.

4.6 Useful Meta-algorithms

A number of meta-algorithms can be used to obtain different insights from pattern mining. A *meta-algorithm* is defined as an algorithm that uses a particular algorithm as a subroutine, either to make the original algorithm more efficient (e.g., by sampling), or to gain new insights. Two types of meta-algorithms are most common in pattern mining. The first type uses sampling to improve the efficiency of association pattern mining algorithms. The second uses preprocessing and postprocessing subroutines to apply the algorithm to other scenarios. For example, after using these wrappers, standard frequent pattern mining algorithms can be applied to quantitative or categorical data.

4.6.1 Sampling Methods

When the transaction database is very large, it cannot be stored in main memory. This makes the application of frequent pattern mining algorithms more challenging. This is because such databases are typically stored on disk, and only level-wise algorithms may be used. Many depth-first algorithms on the enumeration tree may be challenged by these scenarios because they require random access to the transactions. This is inefficient for disk-resident data. As discussed earlier, such depth-first algorithms are usually the most efficient for *memory-resident* data. By sampling, it is possible to apply many of these algorithms in an efficient way, with only limited loss in accuracy. When a standard itemset mining algorithm is applied to sampled data, it will encounter two main challenges:

1. *False positives:* These are patterns that meet the support threshold on the sample but not on the base data.

2. *False negatives:* These are patterns that do not meet the support threshold on the sample, but meet the threshold on the data.

False positives are easier to address than false negatives because the former can be removed by scanning the disk-resident database only once. However, to address false negatives, one needs to reduce the support thresholds. By reducing support thresholds, it is possible to probabilistically guarantee the level of loss for specific thresholds. Pointers to these probabilistic guarantees may be found in the bibliographic notes. Reducing the support thresholds too much will lead to many spurious itemsets and increase the work in the postprocessing phase. Typically, the number of false positives increases rapidly with small changes in support levels.

4.6.2 Data Partitioned Ensembles

One approach that can guarantee no false positives and no false negatives, is the use of partitioned ensembles by the *Partition* algorithm [446]. This approach may be used either for reduction of disk-access costs or for reduction of memory requirements of projection-based algorithms. In partitioned ensembles, the transaction database is partitioned into k disjoint segments, each of which is main-memory resident. The frequent itemset mining algorithm is independently applied to each of these k different segments with the required minimum support level. An important property is that every frequent pattern *must* appear in at least one of the segments. Otherwise, its cumulative support across different segments will not meet the minimum support requirement. Therefore, the union of the frequent itemset generated from different segments provides a superset of the frequent patterns. In other words, the union contains false positives but no false negatives. A postprocessing phase of support counting can be applied to this superset to remove the false positives. This approach is particularly useful for memory-intensive projection-based algorithms when the projected databases do not fit in main memory. In the original *Partition* algorithm, the data structure used to perform projection-based reuse was the vertical *tid* list. While partitioning is almost always necessary for memory-based implementations of projection-based algorithms in databases of arbitrarily large size, the cost of postprocessing overhead can sometimes be significant. Therefore, one should use the minimum number of partitions based on the available memory. Although *Partition* is well known mostly for its ensemble approach, an even more significant but unrecognized contribution of the method was to propose the notion of vertical lists. The approach is credited with recognizing the projection-based reuse properties of recursive *tid* list intersections.

4.6.3 Generalization to Other Data Types

The generalization to other data types is quite straightforward with the use of type-transformation methods discussed in Chap. 2.

4.6.3.1 Quantitative Data

In many applications, it is desirable to discover quantitative association rules when some of the attributes take on quantitative values. Many online merchants collect profile information, such as age, which have numeric values. For example, in supermarket applications, it may be desirable to relate demographic information to item attributes in the data. An example of such a rule is as follows:

$$(Age = 90) \Rightarrow Checkers.$$

This rule may not have sufficient support if the transactions do not contain enough individuals of that age. However, the rule may be relevant to the broader age group. Therefore, one possibility is to create a rule that groups the different ages into one range:

$$Age[85, 95] \Rightarrow Checkers.$$

This rule will have the required level of minimum support. In general, for quantitative association rule mining, the quantitative attributes are discretized and converted to binary form. Thus, the entire data set (including the item attributes) can be represented as a binary matrix. A challenge with the use of such an approach is that the appropriate level of discretization is often hard to know *a priori*. A standard association rule mining algorithm may be applied to this representation. Furthermore, rules on adjacent ranges can be merged to create summarized rules on larger ranges.

4.6.3.2 Categorical Data

Categorical data is common in many application domains. For example, attributes such as the gender and ZIP code are typical categorical. In other cases, the quantitative and categorical data may be mixed. An example of a rule with mixed attributes is as follows:

$$(Gender = Male), \quad Age[20, 30] \Rightarrow Basketball.$$

Categorical data can be transformed to binary values with the use of the binarization approach discussed in Chap. 2. For each categorical attribute value, a single binary value is used to indicate the presence or absence of the item. This can be used to determine the association rules. In some cases, when domain knowledge is available, clusters on categorical values on may used as binary attributes. For example, the ZIP codes may be clustered by geography into k clusters, and then these k clusters may be treated as binary attributes.

4.7 Summary

The problem of association rule mining is used to identify relationships between different attributes. Association rules are typically generated using a two-phase framework. In the first phase, all the patterns that satisfy the minimum support requirement are determined. In the second phase, rules that satisfy the minimum confidence requirement are generated from the patterns.

The *Apriori* algorithm is one of the earliest and most well known methods for frequent pattern mining. In this algorithm, candidate patterns are generated with the use of joins between frequent patterns. Subsequently, a number of enumeration-tree algorithms were proposed for frequent pattern mining techniques. Many of these methods use projections to count the support of transactions in the database more efficiently. The traditional support-confidence framework has the shortcoming that it is not based on robust statistical measures. Many of the patterns generated are not interesting. Therefore, a number of interest measures have been proposed for determining more relevant patterns.

A number of sampling methods have been designed for improving the efficiency of frequent pattern mining. Sampling methods result in both false positives and false negatives, though the former can be addressed by postprocessing. A partitioned sample ensemble is also able to avoid false negatives. Association rules can be determined in quantitative and categorical data with the use of type transformations.

4.8 Bibliographic Notes

The problem of frequent pattern mining was first proposed in [55]. The *Apriori* algorithm discussed in this chapter was first proposed in [56], and an enhanced variant of the approach was proposed in [57]. Maximal and non-maximal frequent pattern mining algorithms are usually different from one another primarily in terms of additional pruning steps in the former. The *MaxMiner* algorithm used superset-based non-maximality pruning [82] for more efficient counting. However, the exploration is in breadth-first order, to reduce the number of passes over the data. The *DepthProject* algorithm recognized that superset-based non-maximality pruning is more effective with a depth-first approach.

The *FP-growth* [252] and *DepthProject* [3, 4] methods independently proposed the notion of projection-based reuse in the horizontal database layout. A variety of different data structures are used by different projection-based reuse algorithms such as *TreeProjection* [3], *DepthProject* [4], *FP-growth* [252], and *H-Mine* [419]. A method, known as *Opportune-Project* [361], chooses opportunistically between array-based and tree-based structures to represent the projected transactions. The *TreeProjection* framework also recognized that breadth-first and depth-first strategies have different trade-offs. Breadth-first variations of *TreeProjection* sacrifice some of the power of projection-based reuse to enable fewer disk-based passes on arbitrarily large data sets. Depth-first variations of *TreeProjection*, such as *DepthProject*, achieve full projection-based reuse but the projected databases need to be consistently maintained in main memory. A book and a survey on frequent pattern mining methods may be found in [34] and [253], respectively.

The use of the vertical representation for frequent pattern mining was independently pioneered by Holsheimer et al. [273] and Savasere et al. [446]. These works introduced the clever insight that *recursive tid* list intersections provide significant computational savings in support counting because k-itemsets have shorter *tid* lists than those of $(k-1)$-itemsets or individual items. The vertical *Apriori* algorithm is based on an ensemble component of the *Partition* framework [446]. Although the use of vertical lists by this algorithm was mentioned [537, 534, 465] in the earliest vertical pattern mining papers, some of the contributions of the *Partition* algorithm and their relationship to the subsequent work seem to have remained unrecognized by the research community over the years. Savasere et al.'s *Apriori*-like algorithm, in fact, formed the basis for all vertical algorithms such as *Eclat* [534] and *VIPER* [465]. *Eclat* is described as a breadth-first algorithm in the book by Han et al. [250], and as a depth-first algorithm in the book by Zaki et al. [536]. A careful examination of the

Eclat paper [537] reveals that it is a memory optimization of the breadth-first approach by Savasere et al. [446]. The main contribution of *Eclat* is a memory optimization of the individual ensemble component of Savasere et al.'s algorithm with lattice partitioning (instead of data partitioning), thereby increasing the maximum size of the databases that can be processed in memory without the computational overhead of data-partitioned postprocessing. The number of computational operations for support counting in a *single* component version of *Partition* is fundamentally no different from that of *Eclat*. The *Eclat* algorithm partitions the lattice based on common prefixes, calling them equivalence classes, and then uses a breadth-first approach [537] over each of these smaller sublattices in main memory. This type of lattice partitioning was adopted from parallel versions of *Apriori*, such as the *Candidate Distribution* algorithm [54], where a similar choice exists between lattice partitioning and data partitioning. Because a breadth-first approach is used for search on each sublattice, such an approach has significantly higher memory requirements than a pure depth-first approach. As stated in [534], *Eclat* explicitly *decouples* the lattice decomposition phase from the pattern search phase. This is different from a pure depth-first strategy in which both are tightly integrated. Depth-first algorithms do not require an explicitly decoupled approach for reduction of memory requirements. Therefore, the lattice-partitioning in *Eclat*, which was motivated by the *Candidate Distribution* algorithm [54], seems to have been specifically designed with a breadth-first approach in mind for the second (pattern search) phase. Both the conference [537] and journal versions [534] of the *Eclat* algorithm state that a breadth-first (bottom-up) procedure is used in the second phase for all experiments. *FP-growth* [252] and *DepthProject* [4] were independently proposed as the first depth-first algorithms for frequent pattern mining. *MAFIA* was the first vertical method to use a pure depth-first approach [123]. Other later variations of vertical algorithms, such as *GenMax* and *dEclat* [233, 538], also incorporated the depth-first approach. The notion of *diffsets* [538, 233], which uses incremental vertical lists along the enumeration tree hierarchy, was also proposed in these algorithms. The approach provides memory and efficiency advantages for certain types of data sets.

Numerous measures for finding interesting frequent patterns have been proposed. The χ^2 measure was one of the first such tests, and was discussed in [113]. This measure satisfies the *upward closure property*. Therefore, efficient pattern mining algorithms can be devised. The use of the min-hashing technique for determining interesting patterns without support counting was discussed in [180]. The impact of skews in the support of individual items has been addressed in [517]. An affinity-based algorithm for mining interesting patterns in data with skews has been proposed in the same work. A common scenario in which there is significant skew in support distributions is that of mining negative association rules [447]. The collective strength model was proposed in [16], and a level-wise algorithm for finding all strongly collective itemsets was discussed in the same work. The collective strength model can also discover negative associations from the data. The work in [486] addresses the problem of selecting the right measure for finding interesting association rules.

Sampling is a popular approach for finding frequent patterns in an efficient way with memory-resident algorithms. The first sampling approach was discussed in [493], and theoretical bounds were presented. The work in [446] enables the application of memory-based frequent pattern mining algorithms on large data sets by using ensembles on data partitions. The problem of finding quantitative association rules, and different kinds of patterns from quantitative data is discussed in [476]. The *CLIQUE* algorithm can also be considered an association pattern mining algorithm on quantitative data [58].

4.9 Exercises

1. Consider the transaction database in the table below:

tid	Items
1	a, b, c, d
2	b, c, e, f
3	a, d, e, f
4	a, e, f
5	b, d, f

   Determine the absolute support of itemsets $\{a, e, f\}$, and $\{d, f\}$. Convert the absolute support to the relative support.

2. For the database in Exercise 1, compute all frequent patterns at absolute minimum support values of 2, 3, and 4.

3. For the database in Exercise 1, determine all the maximal frequent patterns at absolute minimum support values of 2, 3, and 4.

4. Represent the database of Exercise 1 in vertical format.

5. Consider the transaction database in the table below:

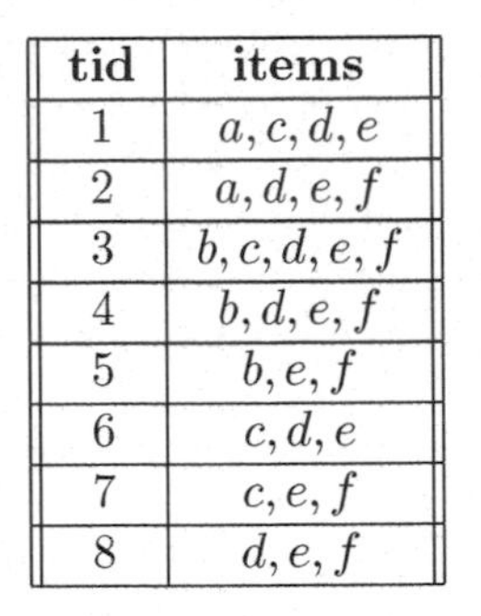

tid	items
1	a, c, d, e
2	a, d, e, f
3	b, c, d, e, f
4	b, d, e, f
5	b, e, f
6	c, d, e
7	c, e, f
8	d, e, f

   Determine all frequent patterns and maximal patterns at support levels of 3, 4, and 5.

6. Represent the transaction database of Exercise 5 in vertical format.

7. Determine the confidence of the rules $\{a\} \Rightarrow \{f\}$, and $\{a, e\} \Rightarrow \{f\}$ for the transaction database in Exercise 1.

8. Determine the confidence of the rules $\{a\} \Rightarrow \{f\}$, and $\{a, e\} \Rightarrow \{f\}$ for the transaction database in Exercise 5.

9. Show the candidate itemsets and the frequent itemsets in each level-wise pass of the *Apriori* algorithm in Exercise 1. Assume an absolute minimum support level of 2.

10. Show the candidate itemsets and the frequent itemsets in each level-wise pass of the *Apriori* algorithm in Exercise 5. Assume an absolute minimum support level of 3.

11. Show the prefix-based enumeration tree of frequent itemsets, for the data set of Exercise 1 at an absolute minimum support level of 2. Assume a lexicographic ordering of a, b, c, d, e, f. Construct the tree for the reverse lexicographic ordering.

12. Show the prefix-based enumeration tree of frequent itemsets, for the data set in Exercise (5), at an absolute minimum support of 3. Assume a lexicographic ordering of a, b, c, d, e, f. Construct the tree for the reverse lexicographic ordering.

13. Show the frequent suffixes generated in the recursion tree of the generic pattern growth method for the data set and support level in Exercise 9. Assume the lexicographic ordering of a, b, c, d, e, f, and f, e, d, c, b, a. How do these trees compare with those generated in Exercise 11?

14. Show the frequent suffixes generated in the recursion tree of the generic pattern growth method for the data set and support level in Exercise 10. Assume the lexicographic ordering of a, b, c, d, e, f, and f, e, d, c, b, a. How do these trees compare with those generated in Exercise 12?

15. Construct a prefix-based FP-Tree for the lexicographic ordering a, b, c, d, e, f for the data set in Exercise 1. Create the same tree for the reverse lexicographic ordering.

16. Construct a prefix-based FP-Tree for the lexicographic ordering a, b, c, d, e, f for the data set in Exercise 5. Create the same tree for the reverse lexicographic ordering.

17. The pruning approach in *Apriori* was inherently designed for a breadth-first strategy because all frequent k-itemsets are generated before $(k+1)$-itemsets. Discuss how one might implement such a pruning strategy with a depth-first algorithm.

18. Implement the pattern growth algorithm with the use of (a) an array-based data structure, (b) a pointer-based data structure with no FP-Tree, and (c) a pointer-based data structure with FP-Tree.

19. Implement Exercise 18(c) by growing patterns from prefixes and the FP-Tree on suffixes.

20. For the itemset $\{d, f\}$ and the data set of Exercise 1, compute the (a) statistical correlation coefficient, (b) interest ratio, (c) cosine coefficient, and (d) Jaccard coefficient.

21. For the itemset $\{d, f\}$ and the data set of Exercise 1, compute the (a) statistical correlation coefficient, (b) interest ratio, (c) cosine coefficient, and (d) Jaccard coefficient.

22. Discuss the similarities and differences between *TreeProjection*, *DepthProject*, *VerticalApriori*, and *FP-growth*.

Chapter 5

Association Pattern Mining: Advanced Concepts

"Each child is an adventure into a better life—an opportunity to change the old pattern and make it new."—Hubert H. Humphrey

5.1 Introduction

Association pattern mining algorithms often discover a large number of patterns, and it is difficult to use this large output for application-specific tasks. One reason for this is that a vast majority of the discovered associations may be uninteresting or redundant for a specific application. This chapter discusses a number of advanced methods that are designed to make association pattern mining more application-sensitive:

1. *Summarization:* The output of association pattern mining is typically very large. For an end-user, a smaller set of discovered itemsets is much easier to understand and assimilate. This chapter will introduce a number of summarization methods such as finding maximal itemsets, closed itemsets, or nonredundant rules.

2. *Querying:* When a large number of itemsets are available, the users may wish to query them for smaller summaries. This chapter will discuss a number of specialized summarization methods that are query friendly. The idea is to use a two-phase approach in which the data is preprocessed to create a summary. This summary is then queried.

3. *Constraint incorporation:* In many real scenarios, one may wish to incorporate application-specific constraints into the itemset generation process. Although a constraint-based algorithm may not always provide online responses, it does allow for the use of much lower support-levels for mining, than a two-phase "preprocess-once query-many" approach.

These topics are all related to the extraction of interesting summary information from itemsets in different ways. For example, compressed representations of itemsets are very useful

C. C. Aggarwal, *Data Mining: The Textbook*, DOI 10.1007/978-3-319-14142-8_5

Table 5.1: Example of a snapshot of a market basket data set (Replicated from Table 4.1 of Chap. 4)

tid	Set of items
1	$\{Bread, Butter, Milk\}$
2	$\{Eggs, Milk, Yogurt\}$
3	$\{Bread, Cheese, Eggs, Milk\}$
4	$\{Eggs, Milk, Yogurt\}$
5	$\{Cheese, Milk, Yogurt\}$

for querying. A query-friendly compression scheme is very different from a summarization scheme that is designed to assure nonredundancy. Similarly, there are fewer constrained itemsets than unconstrained itemsets. However, the shrinkage of the discovered itemsets is because of the constraints rather than a compression or summarization scheme. This chapter will also discuss a number of useful applications of association pattern mining.

This chapter is organized as follows. The problem of pattern summarization is addressed in Sect. 5.2. A discussion of querying methods for pattern mining is provided in Sect. 5.3. Section 5.4 discusses numerous applications of frequent pattern mining. The conclusions are discussed in Sect. 5.5.

5.2 Pattern Summarization

Frequent itemset mining algorithms often discover a large number of patterns. The size of the output creates challenges for users to assimilate the results and make meaningful inferences. An important observation is that the vast majority of the generated patterns are often redundant. This is because of the downward closure property, which ensures that all subsets of a frequent itemset are also frequent. There are different kinds of compact representations in frequent pattern mining that retain different levels of knowledge about the true set of frequent patterns and their support values. The most well-known representations are those of *maximal* frequent itemsets, *closed* frequent itemsets, and other approximate representations. These representations vary in the degree of information loss in the summarized representation. Closed representations are fully lossless with respect to the support and membership of itemsets. Maximal representations are lossy with respect to the support but lossless with respect to membership of itemsets. Approximate condensed representations are lossy with respect to both but often provide the best practical alternative in application-driven scenarios.

5.2.1 Maximal Patterns

The concept of maximal itemsets was discussed briefly in the previous chapter. For convenience, the definition of maximal itemsets is restated here:

Definition 5.2.1 (Maximal Frequent Itemset) *A frequent itemset is maximal at a given minimum support level minsup if it is frequent and no superset of it is frequent.*

For example, consider the example of Table 5.1, which is replicated from the example of Table 4.1 in the previous chapter. It is evident that the itemset $\{Eggs, Milk, Yogurt\}$ is frequent at a minimum support level of 2 and is also maximal. The support of proper subsets of a maximal itemset is always equal to, or strictly larger than the latter because of the support-monotonicity property. For example, the support of $\{Eggs, Milk\}$, which is a proper subset of the itemset $\{Eggs, Milk, Yogurt\}$, is 3. Therefore, one strategy for summarization is to mine only the maximal itemsets. The remaining itemsets are derived as subsets of the maximal itemsets.

Although all the itemsets can be derived from the maximal itemsets with the subsetting approach, their support values cannot be derived. Therefore, maximal itemsets are lossy because they do not retain information about the support values. To provide a lossless representation in terms of the support values, the notion of *closed itemset mining* is used. This concept will be discussed in the next section.

A trivial way to find all the maximal itemsets would be to use any frequent itemset mining algorithm to find *all* itemsets. Then, only the maximal ones can be retained in a postprocessing phase by examining itemsets in decreasing order of length, and removing proper subsets. This process is continued until all itemsets have either been examined or removed. The itemsets that have not been removed at termination are the maximal ones. However, this approach is an inefficient solution. When the itemsets are very long, the number of maximal frequent itemsets may be orders of magnitude smaller than the number of frequent itemsets. In such cases, it may make sense to design algorithms that can directly prune parts of the search space of patterns during frequent itemset discovery. Most of the tree-enumeration methods can be modified with the concept of *lookaheads* to prune the search space of patterns. This notion is discussed in the previous chapter in the context of the *DepthProject* algorithm.

Although the notion of lookaheads is described in the Chap. 4, it is repeated here for completeness. Let P be a frequent pattern in the enumeration tree of itemsets, and $F(P)$ represent the set of candidate extensions of P in the enumeration tree. Then, if $P \cup F(P)$ is a subset of a frequent pattern that has already been found, then it implies that the entire enumeration tree rooted at P is frequent and can, therefore, be removed from further consideration. In the event that the subtree is not pruned, the candidate extensions of P need to be counted. During counting, the support of $P \cup F(P)$ is counted at the same time that the supports of single-item candidate extensions of P are counted. If $P \cup F(P)$ is frequent then the subtree rooted at P can be pruned as well. The former kind of *subset-based* pruning approach is particularly effective with depth-first methods. This is because maximal patterns are found much earlier with a depth-first strategy than with a breadth-first strategy. For a maximal pattern of length k, the depth-first approach discovers it after exploring only $(k-1)$ of its prefixes, rather than the 2^k possibilities. This maximal pattern then becomes available for subset-based pruning. The remaining subtrees containing subsets of $P \cup F(P)$ are then pruned. The superior lookahead-pruning of depth-first methods was first noted in the context of the *DepthProject* algorithm.

The pruning approach provides a smaller set of patterns that includes *all* maximal patterns but may also include some nonmaximal patterns despite the pruning. Therefore, the approach discussed above may be applied to remove these nonmaximal patterns. Refer to the bibliographic notes for pointers to various maximal frequent pattern mining algorithms.

5.2.2 Closed Patterns

A simple definition of a closed pattern, or closed itemset, is as follows:

Definition 5.2.2 (Closed Itemsets) *An itemset X is closed, if none of its supersets have exactly the same support count as X.*

Closed frequent pattern mining algorithms require itemsets to be closed in addition to being frequent. So why are closed itemsets important? Consider a closed itemset X, and the set $\mathcal{S}(X)$ of itemsets which are subsets of X, and which have the same support as X. The only itemset from $\mathcal{S}(X)$ that will be returned by a closed frequent itemset mining algorithm, will be X. The itemsets contained in $\mathcal{S}(X)$ may be referred to as the *equi-support* subsets of X. An important observation is as follows:

Observation 5.2.1 *Let X be a closed itemset, and $\mathcal{S}(X)$ be its equi-support subsets. For any itemset $Y \in \mathcal{S}(X)$, the set of transactions $\mathcal{T}(Y)$ containing Y is exactly the same. Furthermore, there is no itemset Z outside $\mathcal{S}(X)$ such that the set of transactions in $\mathcal{T}(Z)$ is the same as $\mathcal{T}(X)$.*

This observation follows from the downward closed property of frequent itemsets. For any proper subset Y of X, the set of transactions $\mathcal{T}(Y)$ is always a superset of $\mathcal{T}(X)$. However, if the support values of X and Y are the same, then $\mathcal{T}(X)$ and $\mathcal{T}(Y)$ are the same, as well. Furthermore, if any itemset $Z \notin \mathcal{S}(X)$ yields $\mathcal{T}(Z) = \mathcal{T}(X)$, then the support of $Z \cup X$ must be the same as that of X. Because Z is not a subset of X, $Z \cup X$ must be a proper superset of X. This would lead to a contradiction with the assumption that X is closed.

It is important to understand that the itemset X encodes information about *all* the nonredundant counting information needed with respect to any itemset in $\mathcal{S}(X)$. *Every itemset in $\mathcal{S}(X)$ describes the same set of transactions, and therefore, it suffices to keep the single representative itemset.* The maximal itemset X from $\mathcal{S}(X)$ is retained. It should be pointed out that Definition 5.2.2 is a simplification of a more formal definition that is based on the use of a set-closure operator. The formal definition with the use of a set-closure operator is directly based on Observation 5.2.1 (which was derived here from the simplified definition). The informal approach used by this chapter is designed for better understanding. The frequent closed itemset mining problem is defined below.

Definition 5.2.3 (Closed Frequent Itemsets) *An itemset X is a closed frequent itemset at minimum support minsup, if it is both closed and frequent.*

The set of closed itemsets can be discovered in two ways:

1. The set of frequent itemsets at any given minimum support level may be determined, and the closed frequent itemsets can be derived from this set.

2. Algorithms can be designed to directly find the closed frequent patterns during the process of frequent pattern discovery.

While the second class of algorithms is beyond the scope of this book, a brief description of the first approach for finding all the closed itemsets will be provided here. The reader is referred to the bibliographic notes for algorithms of the second type.

A simple approach for finding frequent closed itemsets is to first partition all the frequent itemsets into equi-support groups. The maximal itemsets from each equi-support group may be reported. Consider a set of frequent patterns $\mathcal{F}$, from which the closed frequent patterns need to be determined. The frequent patterns in $\mathcal{F}$ are processed in increasing order of support and either ruled in or ruled out, depending on whether or not they are closed. Note that an increasing support ordering also ensures that closed patterns are encountered earlier than their redundant subsets. Initially, all patterns are unmarked. When an unmarked pattern $X \in \mathcal{F}$ is processed (based on the increasing support order selection), it is added to the frequent closed set $\mathcal{CF}$. The proper subsets of X with the same support cannot be closed. Therefore, all the proper subsets of X with the same support are marked. To achieve this goal, the subset of the itemset lattice representing $\mathcal{F}$ can be traversed in depth-first or breadth-first order starting at X, and exploring subsets of X. Itemsets that are subsets of X are marked when they have the same support as X. The traversal process backtracks when an itemset is reached with strictly larger support, or the itemset has already been marked by the current or a previous traversal. After the traversal is complete, the next unmarked node is selected for further exploration and added to $\mathcal{CF}$. The entire process of marking nodes is repeated, starting from the pattern newly added to $\mathcal{CF}$. At the end of the process, the itemsets in $\mathcal{CF}$ represent the frequent closed patterns.

5.2.3 Approximate Frequent Patterns

Approximate frequent pattern mining schemes are almost always lossy schemes because they do not retain all the information about the itemsets. The approximation of the patterns may be performed in one of the following two ways:

1. *Description in terms of transactions:* The closure property provides a lossless description of the itemsets in terms of their membership in transactions. A generalization of this idea is to allow "almost" closures, where the closure property is not exactly satisfied but is approximately specified. Thus, a "play" is allowed in the support values of the closure definition.

2. *Description in terms of itemsets themselves:* In this case, the frequent itemsets are clustered, and representatives can be drawn from each cluster to provide a concise summary. In this case, the "play" is allowed in terms of the distances between the representatives and remaining itemsets.

These two types of descriptions yield different insights. One is defined in terms of *transaction membership*, whereas the other is defined in terms of the *structure of the itemset.* Note that among the subsets of a 10-itemset X, a 9-itemset may have a much higher support, but a 1-itemset may have exactly the same support as X. In the first definition, the 10-itemset and 1-itemset are "almost" redundant with respect to each other in terms of transaction membership. In the second definition, the 10-itemset and 9-itemset are almost redundant with respect to each other in terms of itemset structure. The following sections will introduce methods for discovering each of these kinds of itemsets.

5.2.3.1 Approximation in Terms of Transactions

The closure property describes itemsets in terms of transactions, and the equivalence of different itemsets with this criterion. The notion of "approximate closures" is a generalization of this criterion. There are multiple ways to define "approximate closure," and a simpler definition is introduced here for ease in understanding.

In the earlier case of exact closures, one chooses the maximal supersets at a particular support value. In approximate closures, one does not necessarily choose the maximal supersets at a particular support value but allows a "play" δ, within a range of supports. Therefore, all frequent itemsets $\mathcal{F}$ can be segmented into a disjoint set of k "almost equi-support" groups $\mathcal{F}_1 \ldots \mathcal{F}_k$, such that for any pair of itemsets X, Y within any group $\mathcal{F}_i$, the value of $|sup(X) - sup(Y)|$ is at most δ. From each group, $\mathcal{F}_i$, only the maximal frequent representatives are reported. Clearly, when δ is chosen to be 0, this is exactly the set of closed itemsets. If desired, the exact error value obtained by removing individual items from approximately closed itemsets is also stored. There is, of course, still some uncertainty in support values because the support values of itemsets obtained by removing *two* items cannot be *exactly* inferred from this additional data.

Note that the "almost equi-support" groups may be constructed in many different ways when $\delta > 0$. This is because the ranges of the "almost equi-support" groups need not exactly be δ but can be less than δ. Of course, a greedy way of choosing the ranges is to always pick the itemset with the lowest support and add δ to it to pick the upper end of the range. This process is repeated to construct all the ranges. Then, the frequent closed itemsets can be extracted on the basis of these ranges.

The algorithm for finding frequent "almost closed" itemsets is very similar to that of finding frequent closed itemsets. As in the previous case, one can partition the frequent itemsets into almost equi-support groups, and determine the maximal ones among them. A traversal algorithm in terms of the graph lattice is as follows.

The first step is to decide the different ranges of support for the "almost equi-support" groups. The itemsets in $\mathcal{F}$ are processed groupwise in increasing order of support ranges for the "almost equi-support" groups. Within a group, unmarked itemsets are processed in increasing order of support. When these nodes are examined they are added to the almost closed set $\mathcal{AC}$. When a pattern $X \in \mathcal{F}$ is examined, all its proper subsets within the same group are marked, unless they have already been marked. To achieve this goal, the subset of the itemset lattice representing $\mathcal{F}$ can be traversed in the same way as discussed in the previous case of (exactly) closed sets. This process is repeated with the next unmarked node. At the end of the process, the set $\mathcal{AC}$ contains the frequent "almost closed" patterns. A variety of other ways of defining "almost closed" itemsets are available in the literature. The bibliographic notes contain pointers to these methods.

5.2.3.2 Approximation in Terms of Itemsets

The approximation in terms of itemsets can also be defined in many different ways and is closely related to clustering. Conceptually, the goal is to create clusters from the set of frequent itemsets $calF$, and pick representatives $\mathcal{J} = J_1 \dots J_k$ from the clusters. Because clusters are always defined with respect to a distance function $Dist(X, Y)$ between itemsets X and Y, the notion of δ-approximate sets is also based on a distance function.

Definition 5.2.4 (δ-Approximate Sets) *The set of representatives $\mathcal{J} = \{J_1 \dots J_k\}$ is δ-approximate, if for each frequent pattern $X \in \mathcal{F}$, and each $J_i \in \mathcal{J}$, the following is true:*

$$Dist(X, J_i) \leq \delta \tag{5.1}$$

Any distance function for set-valued data, such as the Jaccard coefficient, may be used. Note that the cardinality of the set k defines the level of compression. Therefore, the goal is to determine the smallest value of k for a particular level of compression δ. This objective is closely related to the partition-based formulation of clustering, in which the value of k is fixed, and the average distance of the individual objects to their representatives are optimized. Conceptually, this process also creates a clustering on the frequent itemsets. The frequent itemsets can be either strictly partitioned to their closest representative, or they can be allowed to belong to multiple sets for which their distance to the closest representative is at most δ.

So, how can the optimal size of the representative set be determined? It turns out that a simple greedy solution is very effective in most scenarios. Let $\mathcal{C}(\mathcal{J}) \subseteq \mathcal{F}$ denote the set of frequent itemsets *covered* by the representatives in $\mathcal{J}$. An itemset in $\mathcal{F}$ is said to be covered by a representative in $\mathcal{J}$, if it lies within a distance of at most δ from at least one representative of $\mathcal{J}$. Clearly, it is desirable to determine $\mathcal{J}$ so that $\mathcal{C}(\mathcal{J}) = \mathcal{F}$ and the size of the set $\mathcal{J}$ is as small as possible.

The idea of the greedy algorithm is to start with $\mathcal{J} = \{\}$ and add the first element from $\mathcal{F}$ to $\mathcal{J}$ that covers the maximum number of itemsets in $\mathcal{F}$. The covered itemsets are then removed from $\mathcal{F}$. This process is repeated iteratively by greedily adding more elements to $\mathcal{J}$ to maximize coverage in the residual set $\mathcal{F}$. The process terminates when the set $\mathcal{F}$ is empty. It can be shown that the function $f(\mathcal{J}) = |\mathcal{C}(\mathcal{J})|$ satisfies the *submodularity* property with respect to the argument $\mathcal{J}$. In such cases, greedy algorithms are generally effective in practice. In fact, in a minor variation of this problem in which $|\mathcal{C}(J)|$ is directly optimized for fixed size of J, a theoretical bound can also be established on the quality achieved by the greedy algorithm. The reader is referred to the bibliographic notes for pointers on submodularity.

5.3 Pattern Querying

Although the compression approach provides a concise summary of the frequent itemsets, there may be scenarios in which users may wish to query the patterns with specific properties. The query responses provide the *relevant* sets of patterns in an application. This relevant set is usually much smaller than the full set of patterns. Some examples are as follows:

1. Report all the frequent patterns containing X that have a minimum support of *minsup*.
2. Report all the association rules containing X that have a minimum support of *minsup* and a minimum confidence of *minconf*.

One possibility is to exhaustively scan all the frequent itemsets and report the ones satisfying the user-specified constraints. This is, however, quite inefficient when the number of frequent patterns is large. There are two classes of methods that are frequently used for querying interesting subsets of patterns:

1. *Preprocess-once query-many paradigm:* The first approach is to mine all the itemsets at a low level of support and arrange them in the form of a hierarchical or lattice data structure. Because the first phase needs to be performed only once in offline fashion, sufficient computational resources may be available. Therefore, a low level of support is used to maximize the number of patterns preserved in the first phase. Many queries can be addressed in the second phase with the summary created in the first phase.
2. *Constraint-based pattern mining:* In this case, the user-specified constraints are pushed directly into the mining process. Although such an approach can be slower for each query, it allows the mining of patterns at much lower values of the support than is possible with the first approach. This is because the constraints can reduce the pattern sizes in the intermediate steps of the itemset discovery algorithm and can, therefore, enable the discovery of patterns at much lower values of the support than an (unconstrained) preprocessing phase.

In this section, both types of methods will be discussed.

5.3.1 Preprocess-once Query-many Paradigm

This particular paradigm is very effective for the case of simpler queries. In such cases, the key is to first determine all the frequent patterns at a very low value of the support. The resulting itemsets can then be arranged in the form of a data structure for querying. The simplest data structure is the itemset lattice, which can be considered a graph data structure for querying. However, itemsets can also be queried with the use of data structures

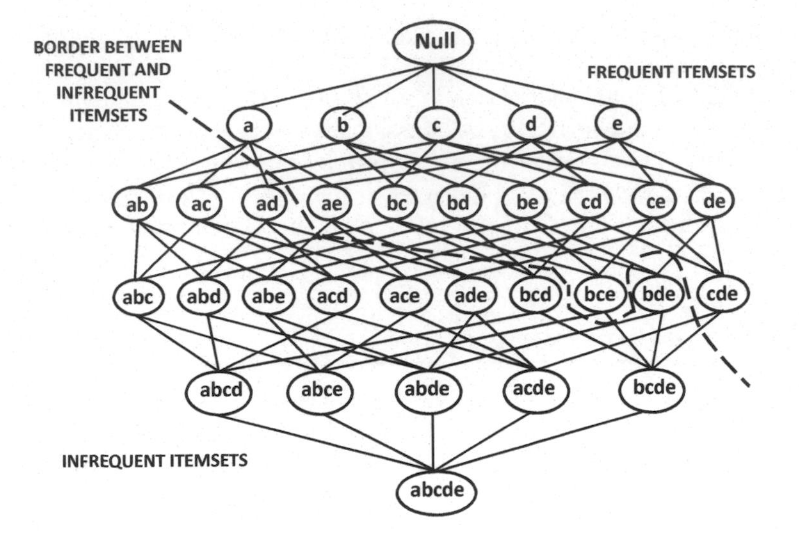

Figure 5.1: The itemset lattice (replicated from Fig. 4.1 of Chap. 4)

adapted from the information retrieval literature that use the bag-of-words representation. Both options will be explored in this chapter.

5.3.1.1 Leveraging the Itemset Lattice

As discussed in the previous chapter, the space of itemsets can be expressed as a lattice. For convenience, Fig. 4.1 of the previous chapter is replicated in Fig. 5.1. Itemsets above the dashed border are frequent, whereas itemsets below the border are infrequent.

In the preprocess-once query-many paradigm, the itemsets are mined at the lowest possible level of support s, so that a large frequent portion of the lattice (graph) of itemsets can be stored in main memory. This stage is a preprocessing phase; therefore, running time is not a primary consideration. The edges on the lattice are implemented as pointers for efficient traversal. In addition, a hash table maps the itemsets to the nodes in the graph. The lattice has a number of important properties, such as downward closure, which enable the discovery of nonredundant association rules and patterns.

This structure can effectively provide responses to many queries that are posed with support $minsup \geq s$. Some examples are as follows:

1. To determine all itemsets containing a set X at a particular level of *minsup*, one uses the hash table to map to the itemset X. Then, the lattice is traversed to determine the relevant supersets of X and report them. A similar approach can be used to determine all the frequent itemsets contained in X by using a traversal in the opposite direction.

2. It is possible to determine maximal itemsets directly during the traversal by identifying nodes that do not have edges to their immediate supersets at the user-specified minimum support level *minsup*.

3. It is possible to identify nodes within a specific hamming distance of X and a specified minimum support, by traversing the lattice structure both upward and downward from X for a prespecified number of steps.

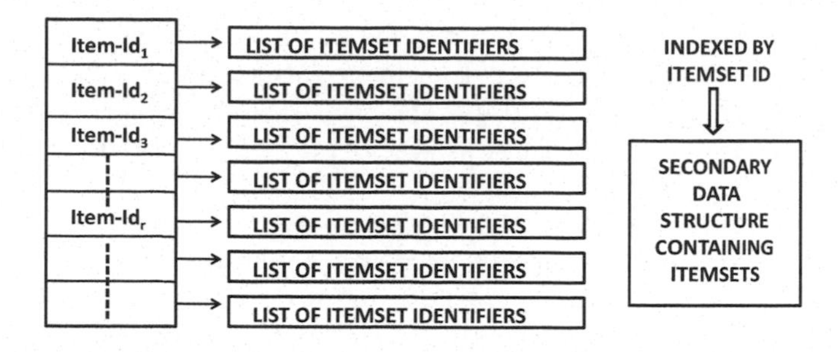

Figure 5.2: Illustration of the inverted lists

It is also possible to determine nonredundant rules with the use of this approach. For example, for any itemset $Y' \subseteq Y$, the rule $X \Rightarrow Y$ has a confidence and support that is no greater than that of the rule $X \Rightarrow Y'$. Therefore, the rule $X \Rightarrow Y'$ is redundant with respect to the rule $X \Rightarrow Y$. This is referred to as *strict redundancy.* Furthermore, for any itemset I, the rule $I - Y' \Rightarrow Y'$ is redundant with respect to the rule $I - Y \Rightarrow Y$ only in terms of the confidence. This is referred to as *simple redundancy.* The lattice structure provides an efficient way to identify such nonredundant rules in terms of both simple redundancy and strict redundancy. The reader is referred to the bibliographic notes for specific search strategies on finding such rules.

5.3.1.2 Leveraging Data Structures for Querying

In some cases, it is desirable to use disk-resident representations for querying. In such cases, the memory-based lattice traversal process is likely to be inefficient. The two most commonly used data structures are the *inverted index* and the *signature table.* The major drawback in using these data structures is that they do not allow an ordered exploration of the set of frequent patterns, as in the case of the lattice structure.

The data structures discussed in this section can be used for *either* transactions or itemsets. However, some of these data structures, such as signature tables, work particularly well for itemsets because they explicitly leverage correlations between itemsets for efficient indexing. Note that correlations are more significant in the case of itemsets than raw transactions. Both these data structures are described in some detail below.

Inverted Index: The inverted index is a data structure that is used for retrieving sparse set-valued data, such as the bag-of-words representation of text. Because frequent patterns are also sparse sets drawn over a much larger universe of items, they can be retrieved efficiently with an inverted index.

Each itemset is assigned a unique *itemset-id.* This can easily be generated with a hash function. This itemset-id is similar to the *tid* that is used to represent transactions. The itemsets themselves may be stored in a secondary data structure that is indexed by the itemset-id. This secondary data structure can be a hash table that is based on the same hash function used to create the itemset-id.

The inverted list contains a list for each item. Each item points to a list of itemset-ids. This list may be stored on disk. An example of an inverted list is illustrated in Fig. 5.2. The inverted representation is particularly useful for *inclusion* queries over small sets of items. Consider a query for all itemsets containing X, where X is a small set of items. The inverted lists for each item in X is stored on the disk. The intersection of these lists is determined.

This provides the relevant itemset-ids but not the itemsets. If desired, the relevant itemsets can be accessed from disk and reported. To achieve this goal, the secondary data structure on disk needs to be accessed with the use of the recovered itemset-ids. This is an additional overhead of the inverted data structure because it may require random access to disk. For large query responses, such an approach may not be practical.

While inverted lists are effective for inclusion queries over small sets of items, they are not quite as effective for similarity queries over longer itemsets. One issue with the inverted index is that it treats each item independently, and it does not leverage the significant correlations between the items in the itemset. Furthermore, the retrieval of the full itemsets is more challenging than that of only itemset-ids. For such cases, the *signature table* is the data structure of choice.

Signature Tables: Signature tables were originally designed for indexing market basket transactions. Because itemsets have the same set-wise data structure as transactions, they can be used in the context of signature tables. Signature tables are particularly useful for sparse binary data in which there are significant correlations among the different items. Because itemsets are inherently defined on the basis of correlations, and different itemsets have large overlaps among them, signature tables are particularly useful in such scenarios.

A *signature* is a set of items. The set of items U in the original data is partitioned into sets of K signatures $S_1 \ldots S_K$, such that $U = \cup_{i=1}^{K} S_i$. The value of K is referred to as the *signature cardinality*. An itemset X is said to *activate* a signature S_i at level r if and only if $|S_i \cap X| \geq r$. This level r is referred to as the *activation threshold*. In other words, the itemset needs to have a user-specified minimum number r of items in common with the signature to activate it.

The super-coordinate of an itemset exists in K-dimensional space, where K is the signature cardinality. Each dimension of the super-coordinate has a unique correspondence with a particular signature and vice versa. The value of this dimension is 0–1, which indicates whether or not the corresponding signature is activated by that itemset. Thus, if the items are partitioned into K signatures $\{S_1, \ldots S_K\}$, then there are 2^K possible super-coordinates. Each itemset maps on to a unique super-coordinate, as defined by the set of signatures activated by that itemset. If $S_{i_1}, S_{i_2}, \ldots S_{i_l}$ be the set of signatures which an itemset activates, then the super-coordinates of that itemset are defined by setting the $l \leq K$ dimensions $\{i_1, i_2, \ldots i_l\}$ in this super-coordinate to 1 and the remaining dimensions to 0. Thus, this approach creates a many-to-one mapping, in which multiple itemsets may map into the same super-coordinate. For highly correlated itemsets, only a small number of signatures will be activated by an itemset, provided that the partitioning of U into signatures is designed to ensure that each signature contains correlated items.

The signature table contains a set of 2^K entries. One entry in the signature table corresponds to each possible super-coordinate. This creates a strict partitioning of the itemsets on the basis of the mapping of itemsets to super-coordinates. This partitioning can be used for similarity search. The signature table can be stored in main memory because the number of distinct super-coordinates can be mapped to main memory when K is small. For example, when K is chosen to be 20, the number of super-coordinates is about a million. The actual itemsets that are indexed by each entry of the signature table are stored on disk. Each entry in the signature table points to a list of pages that contain the itemsets indexed by that super-coordinate. The signature table is illustrated in Fig. 5.3.

A signature can be understood as a small category of items from the universal set of items U. Thus, if the items in each signature are closely correlated, then an itemset is likely to activate a small number of signatures. These signatures provide an idea of the

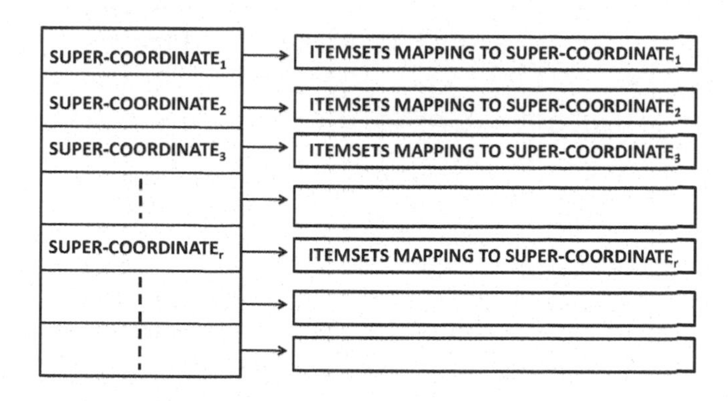

Figure 5.3: Illustration of the signature table

approximate pattern of buying behavior for that itemset. Thus, it is crucial to perform the clustering of the items into signatures so that two criteria are satisfied:

1. The items within a cluster S_i are correlated. This ensures a more *discriminative* mapping, which provides better indexing performance.

2. The aggregate support of the items within each cluster is similar. This is necessary to ensure that the signature table is balanced.

To construct the signature table, a graph is constructed so that each node of the graph corresponds to an item. For every pair of items that is frequent, an edge is added to the graph, and the weight of the edge is a function of the support of that pair of items. In addition, the weight of a node is the support of a particular item. It is desired to determine a clustering of this graph into K partitions so that the cumulative weights of edges across the partitions is as small as possible and the partitions are well balanced. Reducing the weights of edges across partitions ensures that correlated sets of items are grouped together. The partitions should be as well balanced as possible so that the itemsets mapping to each super-coordinate are as well balanced as possible. Thus, this approach transforms the items into a similarity graph, which can be clustered into partitions. A variety of clustering algorithms can be used to partition the graph into groups of items. Any of the graph clustering algorithms discussed in Chap. 19, such as *METIS*, may be used for this purpose. The bibliographic notes also contain pointers to some methods for signature table construction.

After the signatures have been determined, the partitions of the data may be defined by using the super-coordinates of the itemsets. Each itemset belongs to the partition that is defined by its super-coordinate. Unlike the case of inverted lists, the itemsets are *explicitly* stored within this list, rather than just their identifiers. This ensures that the secondary data structure does not need to be accessed to explicitly recover the itemsets. This is the reason that the signature table can be used to recover the itemsets themselves, rather than only the identifiers of the itemsets.

The signature table is capable of handling general similarity queries that cannot be efficiently addressed with inverted lists. Let x be the number of items in which an itemset matches with a target Q, and y be the number of items in which it differs with the target Q. The signature table is capable of handling similarity functions of the form $f(x, y)$ that satisfy the following two properties, for a fixed target record Q:

$$\frac{\Delta f(x,y)}{\Delta x} \geq 0 \quad (5.2)$$

$$\frac{\Delta f(x,y)}{\Delta y} \leq 0 \quad (5.3)$$

This is referred to as the *monotonicity* property. These intuitive conditions on the function ensure that it is an increasing function in the number of matches and decreasing in the hamming distance. While the match function and the hamming distance obviously satisfy these conditions, it can be shown that other functions for set-wise similarity, such as the cosine and the Jaccard coefficient, also satisfy them. For example, let P and Q be the sets of items in two itemsets, where Q is the target itemset. Then, the cosine function can be expressed as follows, in terms of x and y:

$$\begin{aligned} Cosine(P,Q) &= \frac{x}{\sqrt{|P|}\cdot\sqrt{|Q|}} \\ &= \frac{x}{\sqrt{(2\cdot x + y - |Q|)}\cdot\sqrt{|Q|}} \\ Jaccard(P,Q) &= \frac{x}{x+y} \end{aligned}$$

These functions are increasing in x and decreasing in y. These properties are important because they allow bounds to be computed on the similarity function in terms of bounds on the arguments. In other words, if γ is an upper bound on the value of x and θ is a lower bound on the value of y, then it can be shown that $f(\gamma, \theta)$ is an upper (optimistic) bound on the value of the function $f(x, y)$. This is useful for implementing a branch-and-bound method for similarity computation.

Let Q be the target itemset. Optimistic bounds on the match and hamming distance from Q to the itemsets within each super-coordinate are computed. These bounds can be shown to be a function of the target Q, the activation threshold, and the choice of signatures. The precise method for computing these bounds is described in the pointers found in the bibliographic notes. Let the optimistic bound on the match be x_i and that on distance be y_i for the ith super-coordinate. These are used to determine an optimistic bound on the similarity $f(x, y)$ between the target and any itemset indexed by the ith super-coordinate. Because of the monotonicity property, this optimistic bound for the ith super-coordinate is $B_i = f(x_i, y_i)$. The super-coordinates are sorted in decreasing (worsening) order of the optimistic bounds B_i. The similarity of Q to the itemsets that are pointed to by these super-coordinates is computed in this sorted order. The closest itemset found so far is dynamically maintained. The process terminates when the optimistic bound B_i to a super-coordinate is lower (worse) than the similarity value of the closest itemset found so far to the target. At this point, the closest itemset found so far is reported.

5.3.2 Pushing Constraints into Pattern Mining

The methods discussed so far in this chapter are designed for retrieval queries with item-specific constraints. In practice, however, the constraints may be much more general and cannot be easily addressed with any particular data structure. In such cases, the constraints may be need to be directly pushed into the mining process.

In all the previous methods, a preprocess-once query-many paradigm is used; therefore, the querying process is limited by the initial minimum support chosen during the preprocessing phase. Although such an approach has the advantage of providing online capabilities for query responses, it is sometimes not effective when the constraints result in removal of most of the itemsets. In such cases, a much lower level of minimum support may be required than could be reasonably selected during the initial preprocessing phase. The advantage of pushing the constraints into the mining process is that the constraints can be used to prune out many of the intermediate itemsets *during the execution* of the frequent pattern mining algorithms. This allows the use of much lower minimum support levels. The price for this flexibility is that the resulting algorithms can no longer be considered truly online algorithms when the data sets are very large.

Consider, for example, a scenario where the different items are tagged into different categories, such as snacks, dairy, baking products, and so on. It is desired to determine patterns, such that all items belong to the same category. Clearly, this is a constraint on the discovery of the underlying patterns. Although it is possible to first mine all the patterns, and then filter down to the relevant patterns, this is not an efficient solution. If the number of patterns mined during the preprocessing phase is no more than 10^6 and the level of selectivity of the constraint is more than 10^{-6}, then the final set returned may be empty, or too small.

Numerous methods have been developed in the literature to address such constraints directly in the mining process. These constraints are classified into different types, depending upon their impact on the mining algorithm. Some examples of well-known types of constraints, include *succinct*, *monotonic*, *antimonotonic*, and *convertible*. A detailed description of these methods is beyond the scope of this book. The bibliographic section contains pointers to many of these algorithms.

5.4 Putting Associations to Work: Applications

Association pattern mining has numerous applications in a wide variety of real scenarios. This section will discuss some of these applications briefly.

5.4.1 Relationship to Other Data Mining Problems

The association model is intimately related to other data mining problems such as classification, clustering, and outlier detection. Association patterns can be used to provide effective solutions to these data mining problems. This section will explore these relationships briefly. Many of the relevant algorithms are also discussed in the chapters on these different data mining problems.

5.4.1.1 Application to Classification

The association pattern mining problem is closely related to that of classification. *Rule-based classifiers* are closely related to association-rule mining. These types of classifiers are discussed in detail in Sect. 10.4 of Chap. 10, and a brief overview is provided here.

Consider the rule $X \Rightarrow Y$, where X is the antecedent and Y is the consequent. In associative classification, the consequent Y is a single item corresponding to the class variable, and the antecedent contains the feature variables. These rules are mined from the training data. Typically, the rules are not determined with the traditional support and confidence measures. Rather, the most *discriminative* rules with respect to the different classes need to be determined. For example, consider an itemset X and two classes c_1 and c_2. Intuitively, the itemset X is discriminative between the two classes, if the absolute difference in the confidence of the rules $X \Rightarrow c_1$ and $X \Rightarrow c_2$ is as large as possible. Therefore, the mining process should determine such discriminative rules.

Interestingly, it has been discovered, that even a relatively straightforward modification of the association framework to the classification problem is quite effective. An example of such a classifier is the CBA framework for Classification Based on Associations. More details on rule-based classifiers are discussed in Sect. 10.4 of Chap. 10.

5.4.1.2 Application to Clustering

Because association patterns determine highly correlated subsets of attributes, they can be applied to quantitative data after discretization to determine dense regions in the data. The *CLIQUE* algorithm, discussed in Sect. 7.4.1 of Chap. 7, uses discretization to transform quantitative data into binary attributes. Association patterns are discovered on the transformed data. The data points that overlap with these regions are reported as subspace clusters. This approach, of course, reports clusters that are highly overlapping with one another. Nevertheless, the resulting groups correspond to the dense regions in the data, which provide significant insights about the underlying clusters.

5.4.1.3 Applications to Outlier Detection

Association pattern mining has also been used to determine outliers in market basket data. The key idea here is that the outliers are defined as transactions that are not "covered" by most of the association patterns in the data. A transaction is said to be covered by an association pattern when the corresponding association pattern is contained in the transaction. This approach is particularly useful in scenarios where the data is high dimensional and traditional distance-based algorithms cannot be easily used. Because transaction data is inherently high dimensional, such an approach is particularly effective. This approach is discussed in detail in Sect. 9.2.3 of Chap. 9.

5.4.2 Market Basket Analysis

The prototypical problem for which the association rule mining problem was first proposed is that of market basket analysis. In this problem, it is desired to determine rules relating buying behavior of customers. The knowledge of such rules can be very useful for a retailer. For example, if an association rule reveals that the sale of beer implies a sale of diapers, then a merchant may use this information to optimize his or her shelf placement and promotion decisions. In particular, rules that are interesting or unexpected are the most informative for market basket analysis. Many of the traditional and alternative models for market basket analysis are focused on such decisions.

5.4.3 Demographic and Profile Analysis

A closely related problem is that of using demographic profiles to make recommendations. An example is the rule discussed in Sect. 4.6.3 of Chap. 4.

$$Age[85, 95] \Rightarrow Checkers$$

Other demographic attributes, such as the gender or the ZIP code, can be used to determine more refined rules. Such rules are referred to as *profile association rules*. Profile association rules are very useful for target marketing decisions because they can be used to identify relevant population segments for specific products. Profile association rules can be viewed in a similar way to classification rules, except that the antecedent of the rule typically identifies a profile segment, and the consequent identifies a population segment for target marketing.

5.4.4 Recommendations and Collaborative Filtering

Both the aforementioned applications are closely related to the generic problem of recommendation analysis and collaborative filtering. In collaborative filtering, the idea is to make recommendations to users on the basis of the buying behavior of other similar users. In this context, *localized pattern mining* is particularly useful. In localized pattern mining, the idea is to cluster the data into segments, and then determine the patterns in these segments. The patterns from each segment are typically more resistant to noise from the global data distribution and provide a clearer idea of the patterns within like-minded customers. For example, in a movie recommendation system, a particular pattern for movie titles, such as $\{\text{Gladiator}, \text{Nero}, \text{Julius Caesar}\}$, may not have sufficient support on a global basis. However, within like-minded customers, who are interested in historical movies, such a pattern may have sufficient support. This approach is used in applications such as collaborative filtering. The problem of localized pattern mining is much more challenging because of the need to simultaneously determine the clustered segments and the association rules. The bibliographic section contains pointers to such localized pattern mining methods. Collaborative filtering is discussed in detail in Sect. 18.5 of Chap. 18.

5.4.5 Web Log Analysis

Web log analysis is a common scenario for pattern mining methods. For example, the set of pages accessed during a session is no different than a market-basket data set containing transactions. When a set of Web pages is accessed together frequently, this provides useful insights about correlations in user behavior with respect to Web pages. Such insights can be leveraged by site-administrators to improve the structure of the Web site. For example, if a pair of Web pages are frequently accessed together in a session but are not currently linked together, it may be helpful to add a link between them. The most sophisticated forms of Web log analysis typically work with the temporal aspects of logs, beyond the set-wise framework of frequent itemset mining. These methods will be discussed in detail in Chaps. 15 and 18.

5.4.6 Bioinformatics

Many new technologies in bioinformatics, such as microarray and mass spectrometry technologies, allow the collection of different kinds of very high-dimensional data sets. A classical example of this kind of data is gene-expression data, which can be expressed as an $n \times d$ matrix, where the number of columns d is very large compared with typical market basket applications. It is not uncommon for a microarray application to contain a hundred thousand columns. The discovery of frequent patterns in such data has numerous applications in the discovery of key biological properties that are encoded by these data sets. For such cases, long pattern mining methods, such as maximal and closed pattern mining are very useful. In fact, a number of methods, discussed in the bibliographic notes, have specifically been designed for such data sets.

5.4.7 Other Applications for Complex Data Types

Frequent pattern mining algorithms have been generalized to more complex data types such as temporal data, spatial data, and graph data. This book contains different chapters for these complex data types. A brief discussion of these more complex applications is provided here:

1. *Temporal Web log analytics:* The use of temporal information from Web logs greatly enriches the analysis process. For example, certain patterns of accesses may occur frequently in the logs and these can be used to build event prediction models in cases where future events may be predicted from the current pattern of events.

2. *Spatial co-location patterns:* Spatial co-location patterns provide useful insights about the spatial correlations among different individuals. Frequent pattern mining algorithms have been generalized to such scenarios. Refer to Chap. 16.

3. *Chemical and biological graph applications:* In many real scenarios, such as chemical and biological compounds, the determination of structural patterns provides insights about the properties of these molecules. Such patterns are also used to create classification models. These methods are discussed in Chap. 17.

4. *Software bug analysis:* The structure of computer programs can often be represented as call graphs. The analysis of the frequent patterns in the call graphs and key deviations from these patterns provides insights about the bugs in the underlying software.

Many of the aforementioned applications will be discussed in later chapters of this book.

5.5 Summary

In order to use frequent patterns effectively in data-driven applications, it is crucial to create concise summaries of the underlying patterns. This is because the number of returned patterns may be very large and difficult to interpret. Numerous methods have been designed to create a compressed summary of the frequent patterns. Maximal patterns provide a concise summary but are lossy in terms of the support of the underlying patterns. They can often be determined effectively by incorporating different kinds of pruning strategies in frequent pattern mining algorithms.

Closed patterns provide a lossless description of the underlying frequent itemsets. On the other hand, the compression obtained from closed patterns is not quite as significant as that obtained from the use of maximal patterns. The concept of "almost closed" itemsets allows good compression, but there is some degree of information loss in the process. A different way of compressing itemsets is to cluster itemsets so that all itemsets can be expressed within a prespecified distance of particular representatives.

Query processing of itemsets is important in the context of many applications. For example, the itemset lattice can be used to resolve simple queries on itemsets. In some cases, the lattice may not fit in main memory. For these cases, it may be desirable to use disk resident data structures such as the inverted index or the signature table. In cases where the constraints are arbitrary or have a high level of selectivity, it may be desirable to push the constraints directly into the mining process.

Frequent pattern mining has many applications, including its use as a subroutine for other data mining problems. Other applications include market basket analysis, profile analysis, recommendations, Web log analysis, spatial data, and chemical data. Many of these applications are discussed in later chapters of this book.

5.6 Bibliographic Notes

The first algorithm for maximal pattern mining was proposed in [82]. Subsequently, the *DepthProject* [4] and *GenMax* [233] algorithms were also designed for maximal pattern mining. *DepthProject* showed that the depth-first method has several advantages for determining maximal patterns. Vertical bitmaps were used in *MAFIA* [123] to compress the sizes of the underlying *tid* lists. The problem of closed pattern mining was first proposed in [417] in which an *Apriori*-based algorithm, known as *A-Close*, was presented. Subsequently, numerous algorithms such as *CLOSET* [420], *CLOSET+* [504], and *CHARM* [539] were proposed for closed frequent pattern mining. The last of these algorithms uses the vertical data format to mine long patterns in a more efficient way. For the case of very high-dimensional data sets, closed pattern mining algorithms were proposed in the form of *CARPENTER* and *COBBLER*, respectively [413, 415]. Another method, known as pattern-fusion [553], fuses the different pattern segments together to create a long pattern.

The work in [125] shows how to use deduction rules to construct a minimal representation for all frequent itemsets. An excellent survey on condensed representations of frequent itemsets may be found in [126]. Numerous methods have subsequently been proposed to approximate closures in the form of δ-freesets [107]. Information-theoretic methods for itemset compression have been discussed in [470].

The use of clustering-based methods for compression focuses on the itemsets rather than the transactions. The work in [515] clusters the patterns on the basis of their similarity and frequency to create a condensed representation of the patterns. The submodularity property used in the greedy algorithm for finding the best set of covering itemsets is discussed in [403].

The algorithm for using the itemset lattice for interactive rule exploration is discussed in [37]. The concepts of simple redundancy and strict redundancy are also discussed in this work. This method was also generalized to the case of profile association rules [38]. The inverted index, presented in this chapter, may be found in [441]. A discussion of a market basket-specific implementation, together with the signature table, may be found in [41]. A compact disk structure for storing and querying frequent itemsets has been studied in [359].

A variety of constraint-based methods have been developed for pattern mining. Succinct constraints are the easiest to address because they can be pushed directly into data selection. *Monotonic constraints* need to be checked only once to restrict pattern growth [406, 332], whereas *antimonotonic constraints* need to be pushed deep into the pattern mining process. Another form of pattern mining, known as *convertible* constraints [422], can be addressed by sorting items in ascending or descending order for restraining pattern growth.

The *CLIQUE* algorithm [58] shows how association pattern mining methods may be used for clustering algorithms. The *CBA* algorithm for rule-based classification is discussed in [358]. A survey on rule-based classification methods may be found in [115]. The frequent pattern mining problem has also been used for outlier detection in very long transactions [263]. Frequent pattern mining has also been used in the field of bioinformatics [413, 415]. The determination of localized associations [27] is very useful for the problem of recommendations and collaborative filtering. Methods for mining long frequent patterns in the context of bioinformatics applications may be found in [413, 415, 553]. Association rules can also be used to discover spatial co-location patterns [388]. A detailed discussion of frequent pattern mining methods for graph applications, such as software bug analysis, and chemical and biological data, is provided in Aggarwal and Wang [26].

5.7 Exercises

1. Consider the transaction database in the table below:

tid	items
1	a, b, c, d
2	b, c, e, f
3	a, d, e, f
4	a, e, f
5	b, d, f

   Determine all maximal patterns in this transaction database at support levels of 2, 3, and 4.

2. Write a program to determine the set of maximal patterns, from a set of frequent patterns.

3. For the transaction database of Exercise 1, determine all the closed patterns at support levels of 2, 3, and 4.

4. Write a computer program to determine the set of closed frequent patterns, from a set of frequent patterns.

5. Consider the transaction database in the table below:

tid	items
1	a, c, d, e
2	a, d, e, f
3	b, c, d, e, f
4	b, d, e, f
5	b, e, f
6	c, d, e
7	c, e, f
8	d, e, f

   Determine all frequent maximal and closed patterns at support levels of 3, 4, and 5.

6. Write a computer program to implement the greedy algorithm for finding a representative itemset from a group of itemsets.

7. Write a computer program to implement an inverted index on a set of market baskets. Implement a query to retrieve all itemsets containing a particular set of items.

8. Write a computer program to implement a signature table on a set of market baskets. Implement a query to retrieve the closest market basket to a target basket on the basis of the cosine similarity.

Chapter 6

Cluster Analysis

"In order to be an immaculate member of a flock of sheep, one must, above all, be a sheep oneself." —Albert Einstein

6.1 Introduction

Many applications require the partitioning of data points into intuitively similar groups. The partitioning of a large number of data points into a smaller number of groups helps greatly in summarizing the data and understanding it for a variety of data mining applications. An informal and intuitive definition of clustering is as follows:

Given a set of data points, partition them into groups containing very similar data points.

This is a very rough and intuitive definition because it does not state much about the different ways in which the problem can be formulated, such as the number of groups, or the objective criteria for similarity. Nevertheless, this simple description serves as the basis for a number of models that are specifically tailored for different applications. Some examples of such applications are as follows:

- *Data summarization:* At the broadest level, the clustering problem may be considered as a form of data summarization. As data mining is all about extracting summary information (or concise *insights*) from data, the clustering process is often the first step in many data mining algorithms. In fact, many applications use the summarization property of cluster analysis in one form or the other.

- *Customer segmentation:* It is often desired to analyze the common behaviors of groups of similar customers. This is achieved by *customer segmentation*. An example of an application of customer segmentation is *collaborative filtering*, in which the stated or derived preferences of a similar group of customers are used to make product recommendations within the group.

C. C. Aggarwal, *Data Mining: The Textbook*, DOI 10.1007/978-3-319-14142-8_6

- *Social network analysis:* In the case of network data, nodes that are tightly clustered together by linkage relationships are often similar groups of friends, or *communities.* The problem of community detection is one of the most widely studied in social network analysis, because a broader understanding of human behaviors is obtained from an analysis of community group dynamics.

- *Relationship to other data mining problems:* Due to the summarized representation it provides, the clustering problem is useful for enabling other data mining problems. For example, clustering is often used as a preprocessing step in many classification and outlier detection models.

A wide variety of models have been developed for cluster analysis. These different models may work better in different scenarios and data types. A problem, which is encountered by many clustering algorithms, is that many features may be noisy or uninformative for cluster analysis. Such features need to be removed from the analysis early in the clustering process. This problem is referred to as *feature selection.* This chapter will also study feature-selection algorithms for clustering.

In this chapter and the next, the study of clustering will be restricted to simpler multidimensional data types, such as numeric or discrete data. More complex data types, such as temporal or network data, will be studied in later chapters. The key models differ primarily in terms of how similarity is defined within the groups of data. In some cases, similarity is defined explicitly with an appropriate distance measure, whereas in other cases, it is defined implicitly with a probabilistic mixture model or a density-based model. In addition, certain scenarios for cluster analysis, such as high-dimensional or very large-scale data sets, pose special challenges. These issues will be discussed in the next chapter.

This chapter is organized as follows. The problem of feature selection is studied in Sect. 6.2. Representative-based algorithms are addressed in Sect. 6.3. Hierarchical clustering algorithms are discussed in Sect. 6.4. Probabilistic and model-based methods for data clustering are addressed in Sect. 6.5. Density-based methods are presented in Sect. 6.6. Graph-based clustering techniques are presented in Sect. 6.7. Section 6.8 presents the nonnegative matrix factorization method for data clustering. The problem of cluster validity is discussed in Sect. 6.9. Finally, the chapter is summarized in Sect. 6.10.

6.2 Feature Selection for Clustering

The key goal of feature selection is to remove the noisy attributes that do not cluster well. Feature selection is generally more difficult for *unsupervised* problems, such as clustering, where external validation criteria, such as labels, are not available for feature selection. Intuitively, the problem of feature selection is intimately related to that of determining the inherent *clustering tendency* of a set of features. Feature selection methods determine subsets of features that maximize the underlying clustering tendency. There are two primary classes of models for performing feature selection:

1. *Filter models:* In this case, a score is associated with each feature with the use of a similarity-based criterion. This criterion is essentially a *filter* that provides a crisp condition for feature removal. Data points that do not meet the required score are removed from consideration. In some cases, these models may quantify the quality of a subset of features as a *combination*, rather than a single feature. Such models are more powerful because they implicitly take into account the *incremental* impact of adding a feature to others.

2. *Wrapper models:* In this case, a clustering algorithm is used to evaluate the quality of a subset of features. This is then used to refine the subset of features on which the clustering is performed. This is a naturally iterative approach in which a good choice of features depends on the clusters and vice versa. The features selected will typically be at least somewhat dependent on the particular methodology used for clustering. Although this may seem like a disadvantage, the fact is that different clustering methods may work better with different sets of features. Therefore, this methodology can also optimize the feature selection to the specific clustering technique. On the other hand, the inherent informativeness of the specific features may sometimes not be reflected by this approach due to the impact of the specific clustering methodology.

A major distinction between filter and wrapper models is that the former can be performed purely as a preprocessing phase, whereas the latter is integrated directly into the clustering process. In the following sections, a number of filter and wrapper models will be discussed.

6.2.1 Filter Models

In filter models, a specific criterion is used to evaluate the impact of specific features, or subsets of features, on the clustering tendency of the data set. The following will introduce many of the commonly used criteria.

6.2.1.1 Term Strength

Term strength is suitable for sparse domains such as text data. In such domains, it is more meaningful to talk about presence or absence of nonzero values on the attributes (words), rather than distances. Furthermore, it is more meaningful to use similarity functions rather than distance functions. In this approach, pairs of documents are sampled, but a random ordering is imposed between the pair. The term strength is defined as the fraction of similar document pairs (with similarity *greater* than β), in which the term occurs in both the documents, conditional on the fact that it appears in the first. In other words, for any term t, and document pair $(\overline{X}, \overline{Y})$ that have been deemed to be sufficiently similar, the term strength is defined as follows:

$$\text{Term Strength} = P(t \in \overline{Y} | t \in \overline{X}). \tag{6.1}$$

If desired, term strength can also be generalized to multidimensional data by discretizing the quantitative attributes into binary values. Other analogous measures use the correlations between the overall distances and attribute-wise distances to model relevance.

6.2.1.2 Predictive Attribute Dependence

The intuitive motivation of this measure is that correlated features will always result in better clusters than uncorrelated features. When an attribute is relevant, other attributes can be used to predict the value of this attribute. A classification (or regression modeling) algorithm can be used to evaluate this predictiveness. If the attribute is numeric, then a regression modeling algorithm is used. Otherwise, a classification algorithm is used. The overall approach for quantifying the relevance of an attribute i is as follows:

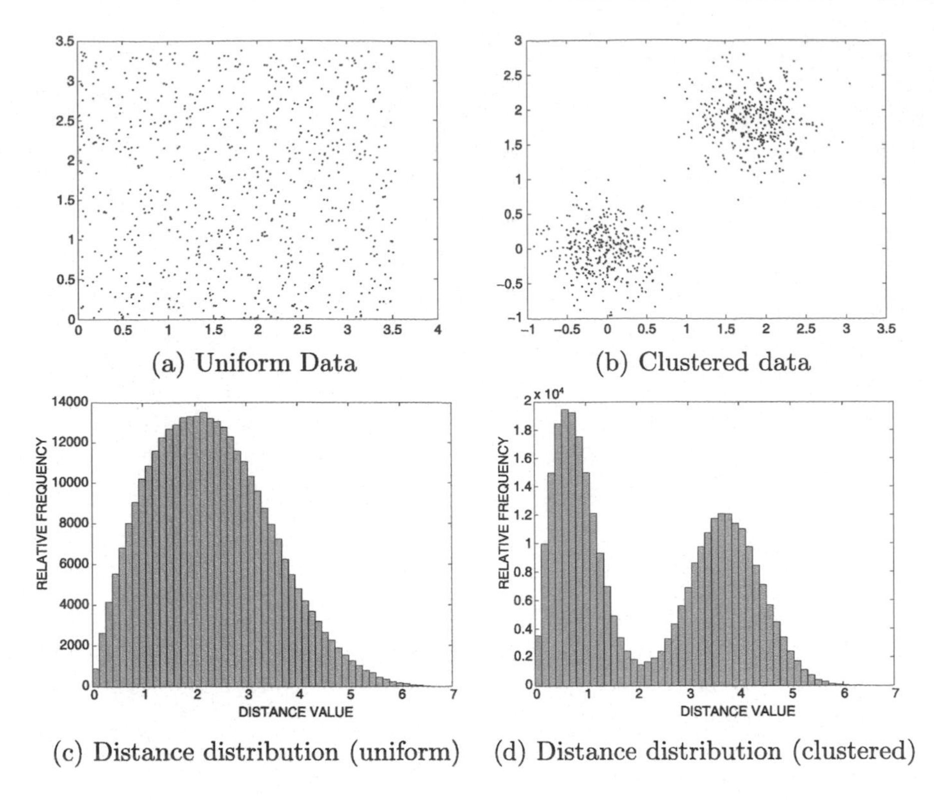

(a) Uniform Data (b) Clustered data

(c) Distance distribution (uniform) (d) Distance distribution (clustered)

Figure 6.1: Impact of clustered data on distance distribution entropy

1. Use a classification algorithm on all attributes, except attribute i, to predict the value of attribute i, while treating it as an artificial class variable.

2. Report the classification accuracy as the relevance of attribute i.

Any reasonable classification algorithm can be used, although a nearest neighbor classifier is desirable because of its natural connections with similarity computation and clustering. Classification algorithms are discussed in Chap. 10.

6.2.1.3 Entropy

The basic idea behind these methods is that highly clustered data reflects some of its clustering characteristics on the underlying distance distributions. To illustrate this point, two different data distributions are illustrated in Figures 6.1a and b, respectively. The first plot depicts uniformly distributed data, whereas the second one depicts data with two clusters. In Figures 6.1c and d, the distribution of the pairwise point-to-point distances is illustrated for the two cases. It is evident that the distance distribution for uniform data is arranged in the form of a bell curve, whereas that for clustered data has two different peaks corresponding to the intercluster distributions and intracluster distributions, respectively. The number of such peaks will typically increase with the number of clusters. The goal of entropy-based measures is to quantify the "shape" of this distance distribution *on a given subset of features*, and then pick the subset where the distribution shows behavior that is more similar to the case of Fig. 6.1b. Therefore, such algorithms typically require a systematic way to search for the appropriate combination of features, in addition to quantifying the distance-based entropy. So how can the distance-based entropy be quantified on a particular subset of attributes?

A natural way of quantifying the entropy is to directly use the probability distribution on the data points and quantify the entropy using these values. Consider a k-dimensional subset of features. The first step is to discretize the data into a set of multidimensional grid regions using ϕ grid regions for each dimension. This results in $m = \phi^k$ grid ranges that are indexed from 1 through m. The value of m is approximately the same across all the evaluated feature subsets by selecting $\phi = \lceil m^{1/k} \rceil$. If p_i is the fraction of data points in grid region i, then the probability-based entropy E is defined as follows:

$$E = -\sum_{i=1}^{m} [p_i \log(p_i) + (1 - p_i)\log(1 - p_i)]. \tag{6.2}$$

A uniform distribution with poor clustering behavior has high entropy, whereas clustered data has lower entropy. Therefore, the entropy measure provides feedback about the clustering quality of a subset of features.

Although the aforementioned quantification can be used directly, the probability density p_i of grid region i is sometimes hard to accurately estimate from high-dimensional data. This is because the grid regions are multidimensional, and they become increasingly sparse in high dimensionality. It is also hard to fix the number of grid regions m over feature subsets of varying dimensionality k because the value of $\phi = \lceil m^{1/k} \rceil$ is rounded up to an integer value. Therefore, an alternative is to compute the entropy on the 1-dimensional point-to-point distance distribution on a sample of the data. This is the same as the distributions shown in Fig. 6.1. The value of p_i then represents the fraction of *distances* in the ith 1-dimensional discretized range. Although this approach does not fully address the challenges of high dimensionality, it is usually a better option for data of modest dimensionality. For example, if the entropy is computed on the histograms in Figs. 6.1c and d, then this will distinguish between the two distributions well. A heuristic approximation on the basis of the raw distances is also often used. Refer to the bibliographic notes.

To determine the subset of features, for which the entropy E is minimized, a variety of search strategies are used. For example, starting from the full set of features, a simple greedy approach may be used to drop the feature that leads to the greatest reduction in the entropy. Features are repeatedly dropped greedily until the incremental reduction is not significant, or the entropy increases. Some enhancements of this basic approach, both in terms of the quantification measure and the search strategy, are discussed in the bibliographic section.

6.2.1.4 Hopkins Statistic

The Hopkins statistic is often used to measure the clustering tendency of a data set, although it can also be applied to a particular subset of attributes. The resulting measure can then be used in conjunction with a feature search algorithm, such as the greedy method discussed in the previous subsection.

Let $\mathcal{D}$ be the data set whose clustering tendency needs to be evaluated. A sample S of r synthetic data points is randomly generated in the domain of the data space. At the same time, a sample R of r data points is selected from $\mathcal{D}$. Let $\alpha_1 \ldots \alpha_r$ be the distances of the data points in the sample $R \subseteq D$ to their nearest neighbors within the original database $\mathcal{D}$. Similarly, let $\beta_1 \ldots \beta_r$ be the distances of the data points in the synthetic sample S to their nearest neighbors within $\mathcal{D}$. Then, the Hopkins statistic H is defined as follows:

$$H = \frac{\sum_{i=1}^{r} \beta_i}{\sum_{i=1}^{r} (\alpha_i + \beta_i)}. \tag{6.3}$$

The Hopkins statistic will be in the range $(0, 1)$. Uniformly distributed data will have a Hopkins statistic of 0.5 because the values of α_i and β_i will be similar. On the other hand, the values of α_i will typically be much lower than β_i for the clustered data. This will result in a value of the Hopkins statistic that is closer to 1. Therefore, a high value of the Hopkins statistic H is indicative of highly clustered data points.

One observation is that the approach uses random sampling, and therefore the measure will vary across different random samples. If desired, the random sampling can be repeated over multiple trials. A statistical tail confidence test can be employed to determine the level of confidence at which the Hopkins statistic is greater than 0.5. For feature selection, the average value of the statistic over multiple trials can be used. This statistic can be used to evaluate the quality of any particular subset of attributes to evaluate the clustering tendency of that subset. This criterion can be used in conjunction with a greedy approach to discover the relevant subset of features. The greedy approach is similar to that discussed in the case of the distance-based entropy method.

6.2.2 Wrapper Models

Wrapper models use an *internal cluster validity criterion* in conjunction with a clustering algorithm that is applied to an appropriate subset of features. Cluster validity criteria are used to evaluate the quality of clustering and are discussed in detail in Sect. 6.9. The idea is to use a clustering algorithm with a subset of features, and then evaluate the quality of this clustering with a cluster validity criterion. Therefore, the search space of different subsets of features need to be explored to determine the optimum combination of features. As the search space of subsets of features is exponentially related to the dimensionality, a greedy algorithm may be used to successively drop features that result in the greatest improvement of the cluster validity criterion. The major drawback of this approach is that it is sensitive to the choice of the validity criterion. As you will learn in this chapter, cluster validity criteria are far from perfect. Furthermore, the approach can be computationally expensive.

Another simpler methodology is to select individual features with a feature selection criterion that is borrowed from that used in classification algorithms. In this case, the features are evaluated individually, rather than collectively, as a subset. The clustering approach artificially creates a set of labels L, corresponding to the cluster identifiers of the individual data points. A feature selection criterion may be borrowed from the classification literature with the use of the labels in L. This criterion is used to identify the most discriminative features:

1. Use a clustering algorithm on the current subset of selected features F, in order to fix cluster labels L for the data points.

2. Use any *supervised* criterion to quantify the quality of the individual features with respect to labels L. Select the top-k features on the basis of this quantification.

There is considerable flexibility in the aforementioned framework, where different kinds of clustering algorithms and feature selection criteria are used in each of the aforementioned steps. A variety of supervised criteria can be used, such as the *class-based entropy* or the

Algorithm *GenericRepresentative*(Database: $\mathcal{D}$, Number of Representatives: k)
begin
Initialize representative set S;
repeat
Create clusters $(\mathcal{C}_1 \ldots \mathcal{C}_k)$ by assigning each
point in $\mathcal{D}$ to closest representative in S
using the distance function $Dist(\cdot,\cdot)$;
Recreate set S by determining one representative $\overline{Y_j}$ for
each $\mathcal{C}_j$ that minimizes $\sum_{\overline{X_i} \in \mathcal{C}_j} Dist(\overline{X_i}, \overline{Y_j})$;
until convergence;
return $(\mathcal{C}_1 \ldots \mathcal{C}_k)$;
end

Figure 6.2: Generic representative algorithm with unspecified distance function

Fisher score (cf. Sect. 10.2 of Chap. 10). The Fisher score, discussed in Sect. 10.2.1.3 of Chap. 10, measures the ratio of the intercluster variance to the intracluster variance on any particular attribute. Furthermore, it is possible to apply this two-step procedure iteratively. However, some modifications to the first step are required. Instead of selecting the top-k features, the weights of the top-k features are set to 1, and the remainder are set to $\alpha < 1$. Here, α is a user-specified parameter. In the final step, the top-k features are selected.

Wrapper models are often combined with filter models to create *hybrid models* for better efficiency. In this case, candidate feature subsets are constructed with the use of filter models. Then, the quality of each candidate feature subset is evaluated with a clustering algorithm. The evaluation can be performed either with a cluster validity criterion or with the use of a classification algorithm on the resulting cluster labels. The best candidate feature subset is selected. Hybrid models provide better accuracy than filter models and are more efficient than wrapper models.

6.3 Representative-Based Algorithms

Representative-based algorithms are the simplest of all clustering algorithms because they rely directly on intuitive notions of distance (or similarity) to cluster data points. In representative-based algorithms, the clusters are created in one shot, and hierarchical relationships do not exist among different clusters. This is typically done with the use of a set of partitioning *representatives.* The partitioning representatives may either be created as a function of the data points in the clusters (e.g., the mean) or may be selected from the existing data points in the cluster. The main insight of these methods is that the discovery of high-quality clusters in the data is equivalent to discovering a high-quality set of representatives. Once the representatives have been determined, a distance function can be used to assign the data points to their closest representatives.

Typically, it is assumed that the number of clusters, denoted by k, is specified by the user. Consider a data set $\mathcal{D}$ containing n data points denoted by $\overline{X_1} \ldots \overline{X_n}$ in d-dimensional space. The goal is to determine k representatives $\overline{Y_1} \ldots \overline{Y_k}$ that minimize the following objective function O:

$$O = \sum_{i=1}^{n} \left[\min_j Dist(\overline{X_i}, \overline{Y_j}) \right]. \tag{6.4}$$

In other words, the sum of the distances of the different data points to their closest representatives needs to be minimized. Note that the assignment of data points to representatives depends on the choice of the representatives $\overline{Y_1} \dots \overline{Y_k}$. In some variations of representative algorithms, such as k-medoid algorithms, it is assumed that the representatives $\overline{Y_1} \dots \overline{Y_k}$ are drawn from the original database $\mathcal{D}$, although this will obviously not provide an optimal solution. In general, the discussion in this section will not automatically assume that the representatives are drawn from the original database $\mathcal{D}$, unless specified otherwise.

One observation about the formulation of Eq. 6.4 is that the representatives $\overline{Y_1} \dots \overline{Y_k}$ and the optimal assignment of data points to representatives are unknown a priori, but they depend on each other in a circular way. For example, if the optimal representatives are known, then the optimal assignment is easy to determine, and vice versa. Such optimization problems are solved with the use of an iterative approach where candidate representatives and candidate assignments are used to improve each other. Therefore, the generic k-representatives approach starts by initializing the k representatives S with the use of a straightforward heuristic (such as random sampling from the original data), and then refines the representatives and the clustering assignment, iteratively, as follows:

- (Assign step) Assign each data point to its closest representative in S using distance function $Dist(\cdot, \cdot)$, and denote the corresponding clusters by $\mathcal{C}_1 \dots \mathcal{C}_k$.
- (Optimize step) Determine the optimal representative $\overline{Y_j}$ for each cluster $\mathcal{C}_j$ that minimizes its *local* objective function $\sum_{\overline{X_i} \in \mathcal{C}_j} \left[Dist(\overline{X_i}, \overline{Y_j}) \right]$.

It will be evident later in this chapter that this two-step procedure is very closely related to generative models of cluster analysis in the form of *expectation-maximization* algorithms. The second step of *local* optimization is simplified by this two-step iterative approach, because it no longer depends on an unknown assignment of data points to clusters as in the global optimization problem of Eq. 6.4. Typically, the optimized representative can be shown to be some central measure of the data points in the jth cluster $\mathcal{C}_j$, and the precise measure depends on the choice of the distance function $Dist(\overline{X_i}, \overline{Y_j})$. In particular, for the case of the Euclidean distance and cosine similarity functions, it can be shown that the optimal centralized representative of each cluster is its mean. However, different distance functions may lead to a slightly different type of centralized representative, and these lead to different variations of this broader approach, such as the k-means and k-medians algorithms. Thus, the k-representative approach defines a *family* of algorithms, in which minor changes to the basic framework allow the use of different distance criteria. These different criteria will be discussed below. The generic framework for representative-based algorithms with an unspecified distance function is illustrated in the pseudocode of Fig. 6.2. The idea is to improve the objective function over multiple iterations. Typically, the increase is significant in early iterations, but it slows down in later iterations. When the improvement in the objective function in an iteration is less than a user-defined threshold, the algorithm may be allowed to terminate. The primary computational bottleneck of the approach is the assignment step where the distances need to be computed between all point-representative pairs. The time complexity of each iteration is $O(k \cdot n \cdot d)$ for a data set of size n and dimensionality d. The algorithm typically terminates in a small constant number of iterations.

The inner workings of the k-representatives algorithm are illustrated with an example in Fig. 6.3, where the data contains three natural clusters, denoted by A, B, and C. For illustration, it is assumed that the input k to the algorithm is the same as the number of natural clusters in the data, which, in this case, is 3. The Euclidean distance function

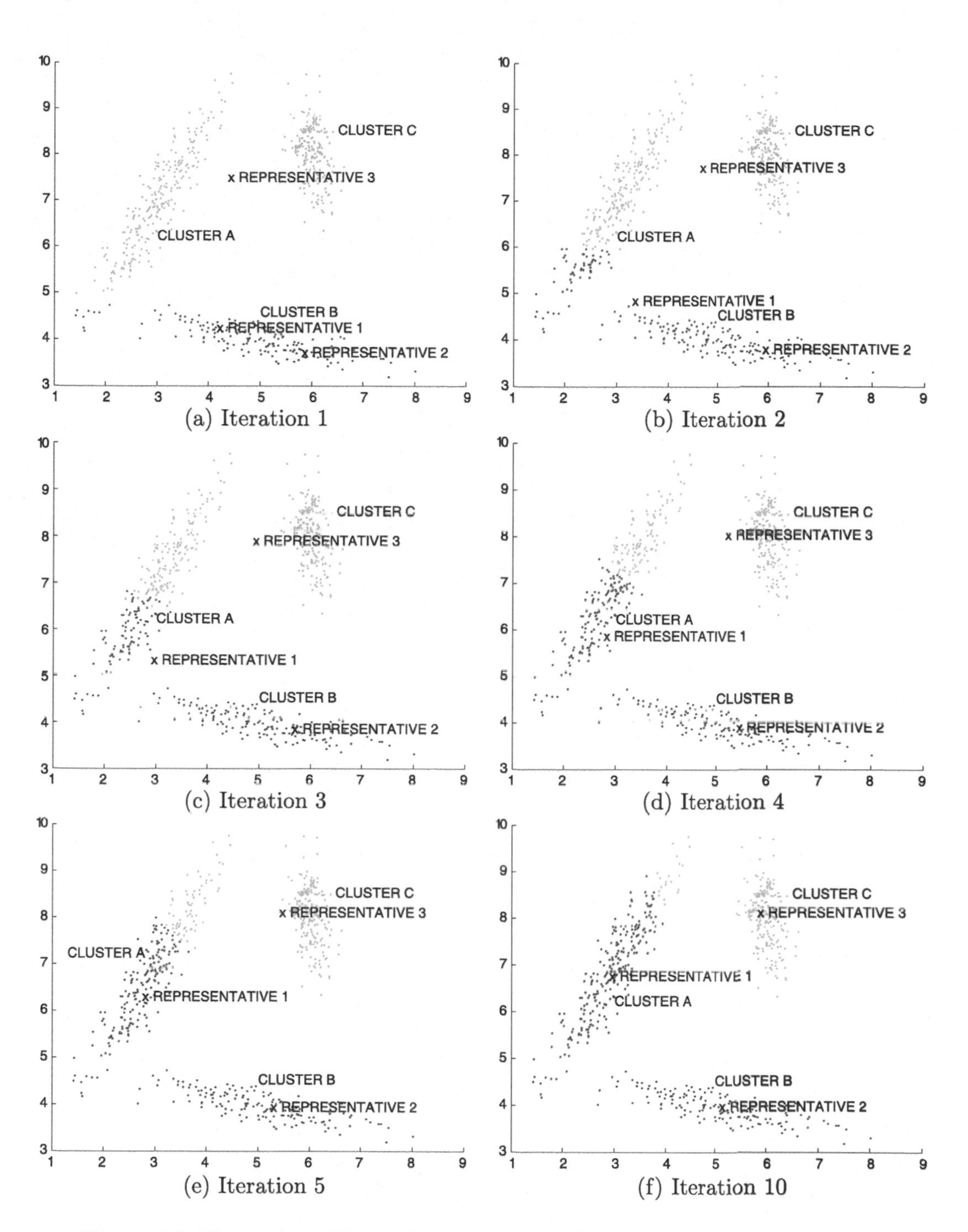

Figure 6.3: Illustration of k-representative algorithm with random initialization

is used, and therefore the "re-centering" step uses the mean of the cluster. The initial set of representatives (or seeds) is chosen randomly from the data space. This leads to a particularly bad initialization, where two of the representatives are close to cluster B, and one of them lies somewhere midway between clusters A and C. As a result, the cluster B is initially split up by the "sphere of influence" of two representatives, whereas most of the points in clusters A and C are assigned to a single representative in the first assignment step. This situation is illustrated in Fig. 6.3a. However, because each representative is assigned a different number of data points from the different clusters, the representatives drift in subsequent iterations to one of the unique clusters. For example, representative 1 steadily drifts toward cluster A, and representative 3 steadily drifts toward cluster C. At the same time, representative 2 becomes a better centralized representative of cluster B. As a result, cluster B is no longer split up among different representatives by the end of iteration 10 (Fig. 6.3f). An interesting observation is that even though the initialization was so poor, it required only 10 iterations for the k-representatives approach to create a reasonable clustering of the data. In practice, this is generally true of k-representative methods, which converge relatively fast toward a good clustering of the data points. However, it is possible for k-means to converge to suboptimal solutions, especially when an outlier data point is selected as an initial representative for the algorithm. In such a case, one of the clusters may contain a singleton point that is not representative of the data set, or it may contain two merged clusters. The handling of such cases is discussed in the section on implementation issues. In the following section, some special cases and variations of this framework will be discussed. Most of the variations of the k-representative framework are defined by the choice of the distance function $Dist(\overline{X_i}, \overline{Y_j})$ between the data points $\overline{X_i}$ and the representatives $\overline{Y_j}$. Each of these choices results in a different type of centralized representative of a cluster.

6.3.1 The k-Means Algorithm

In the k-means algorithm, the sum of the squares of the Euclidean distances of data points to their closest representatives is used to quantify the objective function of the clustering. Therefore, we have:

$$Dist(\overline{X_i}, \overline{Y_j}) = ||\overline{X_i} - \overline{Y_j}||_2^2. \tag{6.5}$$

Here, $|| \cdot ||_p$ represents the L_p-norm. The expression $Dist(\overline{X_i}, \overline{Y_j})$ can be viewed as the squared error of approximating a data point with its closest representative. Thus, the overall objective minimizes the sum of square errors over different data points. This is also sometimes referred to as SSE. In such a case, it can be shown[1] that the *optimal representative $\overline{Y_j}$ for each of the "optimize" iterative steps is the mean of the data points in cluster $\mathcal{C}_j$*. Thus, the only difference between the generic pseudocode of Fig. 6.2 and a k-means pseudocode is the specific instantiation of the distance function $Dist(\cdot, \cdot)$, and the choice of the representative as the local mean of its cluster.

An interesting variation of the k-means algorithm is to use the *local* Mahalanobis distance for assignment of data points to clusters. This distance function is discussed in Sect. 3.2.1.6 of Chap. 3. Each cluster $\mathcal{C}_j$ has its $d \times d$ own covariance matrix Σ_j, which can be computed using the data points assigned to that cluster in the previous iteration. The squared Mahalanobis distance between data point $\overline{X_i}$ and representative $\overline{Y_j}$ with a covariance matrix Σ_j is defined

[1] For a *fixed* cluster assignment $\mathcal{C}_1 \ldots \mathcal{C}_k$, the gradient of the clustering objective function $\sum_{j=1}^{k} \sum_{\overline{X_i} \in \mathcal{C}_j} ||\overline{X_i} - \overline{Y_j}||^2$ with respect to $\overline{Y_j}$ is $2 \sum_{\overline{X_i} \in \mathcal{C}_j} (\overline{X_i} - \overline{Y_j})$. Setting the gradient to 0 yields the mean of cluster $\mathcal{C}_j$ as the optimum value of $\overline{Y_j}$. Note that the other clusters do not contribute to the gradient, and, therefore, the approach effectively optimizes the local clustering objective function for $\mathcal{C}_j$.

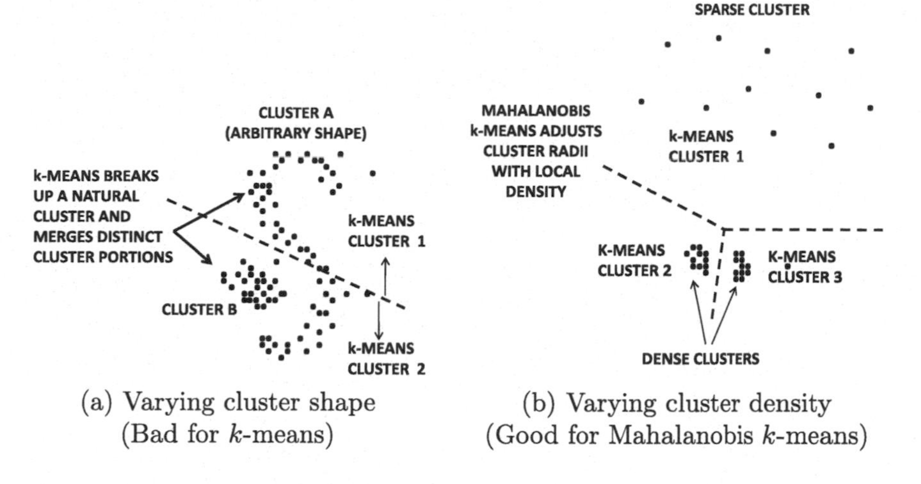

(a) Varying cluster shape (Bad for k-means)

(b) Varying cluster density (Good for Mahalanobis k-means)

Figure 6.4: Strengths and weaknesses of k-means

as follows:

$$Dist(\overline{X_i}, \overline{Y_j}) = (\overline{X_i} - \overline{Y_j})\Sigma_j^{-1}(\overline{X_i} - \overline{Y_j})^T. \tag{6.6}$$

The use of the Mahalanobis distance is generally helpful when the clusters are elliptically elongated along certain directions, as in the case of Fig. 6.3. The factor Σ_j^{-1} also provides local density normalization, which is helpful in data sets with varying local density. The resulting algorithm is referred to as the *Mahalanobis k-means* algorithm.

The k-means algorithm does not work well when the clusters are of arbitrary shape. An example is illustrated in Fig. 6.4a, in which cluster A has a nonconvex shape. The k-means algorithm breaks it up into two parts, and also merges one of these parts with cluster B. Such situations are common in k-means, because it is biased toward finding spherical clusters. Even the Mahalanobis k-means algorithm does not work well in this scenario in spite of its ability to adjust for the elongation of clusters. On the other hand, the Mahalanobis k-means algorithm can adjust well to varying cluster density, as illustrated in Fig. 6.4b. This is because the Mahalanobis method normalizes local distances with the use of a cluster-specific covariance matrix. The data set of Fig. 6.4b cannot be effectively clustered by many density-based algorithms, which are designed to discover arbitrarily shaped clusters (cf. Sect. 6.6). Therefore, different algorithms are suitable in different application settings.

6.3.2 The Kernel k-Means Algorithm

The k-means algorithm can be extended to discovering clusters of arbitrary shape with the use of a method known as the *kernel trick*. The basic idea is to implicitly transform the data so that arbitrarily shaped clusters map to Euclidean clusters in the new space. Refer to Sect. 10.6.4.1 of Chap. 10 for a brief description of the kernel k-means algorithm. The main problem with the kernel k-means algorithm is that the complexity of computing the kernel matrix alone is quadratically related to the number of data points. Such an approach can effectively discover the arbitrarily shaped clusters of Fig. 6.4a.

```
Algorithm GenericMedoids(Database: D, Number of Representatives: k)
begin
 Initialize representative set S by selecting from D;
 repeat
   Create clusters (C_1 ... C_k) by assigning
      each point in D to closest representative in S
      using the distance function Dist(·,·);
   Determine a pair X_i ∈ D and Y_j ∈ S such that
     replacing Y_j ∈ S with X_i leads to the
     greatest possible improvement in objective function;
   Perform the exchange between X_i and Y_j only
     if improvement is positive;
 until no improvement in current iteration;
 return (C_1 ... C_k);
end
```

Figure 6.5: Generic k-medoids algorithm with unspecified hill-climbing strategy

6.3.3 The k-Medians Algorithm

In the k-medians algorithm, the Manhattan distance is used as the objective function of choice. Therefore, the distance function $Dist(\overline{X_i}, \overline{Y_j})$ is defined as follows:

$$Dist(\overline{X_i}, \overline{Y_j}) = ||X_i - Y_j||_1. \tag{6.7}$$

In such a case, it can be shown that the optimal representative $\overline{Y_j}$ is the median of the data points along each dimension in cluster $\mathcal{C}_j$. This is because the point that has the minimum sum of L_1-distances to a set of points distributed on a line is the median of that set. The proof of this result is simple. The definition of a median can be used to show that a perturbation of ϵ in either direction from the median cannot strictly reduce the sum of L_1-distances. This implies that the median optimizes the sum of the L_1-distances to the data points in the set.

As the median is chosen independently along each dimension, the resulting d-dimensional representative will (typically) not belong to the original data set $\mathcal{D}$. The k-medians approach is sometimes confused with the k-medoids approach, which chooses these representatives from the original database $\mathcal{D}$. In this case, the only difference between the generic pseudocode of Fig. 6.2, and a k-medians variation would be to instantiate the distance function to the Manhattan distance and use the representative as the local median of the cluster (independently along each dimension). The k-medians approach generally selects cluster representatives in a more robust way than k-means, because the median is not as sensitive to the presence of outliers in the cluster as the mean.

6.3.4 The k-Medoids Algorithm

Although the k-medoids algorithm also uses the notion of representatives, its algorithmic structure is different from the generic k-representatives algorithm of Fig. 6.2. The clustering objective function is, however, of the same form as the k-representatives algorithm. The main distinguishing feature of the k-medoids algorithm is that the representatives are always selected from the database $\mathcal{D}$, and this difference necessitates changes to the basic structure of the k-representatives algorithm.

A question arises as to why it is sometimes desirable to select the representatives from $\mathcal{D}$. There are two reasons for this. One reason is that the representative of a k-means cluster may be distorted by outliers in that cluster. In such cases, it is possible for the representative to be located in an empty region which is unrepresentative of most of the data points in that cluster. Such representatives may result in partial merging of different clusters, which is clearly undesirable. This problem can, however, be partially resolved with careful outlier handling and the use of outlier-robust variations such as the k-medians algorithm. The second reason is that it is sometimes difficult to compute the optimal central representative of a set of data points of a complex data type. For example, if the k-representatives clustering algorithm were to be applied on a set of time series of *varying lengths*, then how should the central representatives be defined as a function of these heterogeneous time-series? In such cases, selecting representatives from the original data set may be very helpful. As long as a representative object is selected from each cluster, the approach will provide reasonably high quality results. Therefore, a key property of the k-medoids algorithm is that it can be defined virtually on any data type, as long as an appropriate similarity or distance function can be defined on the data type. Therefore, k-medoids methods directly relate the problem of distance function design to clustering.

The k-medoids approach uses a generic hill-climbing strategy, in which the representative set S is initialized to a set of points from the original database $\mathcal{D}$. Subsequently, this set S is iteratively improved by exchanging a single point from set S with a data point selected from the database $\mathcal{D}$. This iterative exchange can be viewed as a hill-climbing strategy, because the set S implicitly defines a solution to the clustering problem, and each exchange can be viewed as a hill-climbing step. So what should be the criteria for the exchange, and when should one terminate?

Clearly, in order for the clustering algorithm to be successful, the hill-climbing approach should at least improve the objective function of the problem to some extent. Several choices arise in terms of how the exchange can be performed:

1. One can try *all* $|S| \cdot |\mathcal{D}|$ possibilities for replacing a representative in S with a data point in $\mathcal{D}$ and then select the best one. However, this is extremely expensive because the computation of the incremental objective function change for *each* of the $|S| \cdot |\mathcal{D}|$ alternatives will require time proportional to the original database size.

2. A simpler solution is to use a randomly selected set of r pairs $(\overline{X_i}, \overline{Y_j})$ for possible exchange, where $\overline{X_i}$ is selected from the database $\mathcal{D}$, and $\overline{Y_j}$ is selected from the representative set S. The best of these r pairs is used for the exchange.

The second solution requires time proportional to r times the database size but is usually practically implementable for databases of modest size. The solution is said to have converged when the objective function does not improve, or if the average objective function improvement is below a user-specified threshold in the previous iteration. The k-medoids approach is generally much slower than the k-means method but has greater applicability to different data types. The next chapter will introduce the *CLARANS* algorithm, which is a scalable version of the k-medoids framework.

Practical and Implementation Issues

A number of practical issues arise in the proper implementation of all representative-based algorithms, such as the k-means, k-medians, and k-medoids algorithms. These issues relate to the initialization criteria, the choice of the number of clusters k, and the presence of outliers.

The simplest initialization criteria is to either select points randomly from the *domain of the data space*, or to sample the original database $\mathcal{D}$. Sampling the original database $\mathcal{D}$ is generally superior to sampling the data space, because it leads to better statistical representatives of the underlying data. The k-representatives algorithm seems to be surprisingly robust to the choice of initialization, though it is possible for the algorithm to create suboptimal clusters. One possible solution is to sample more data points from $\mathcal{D}$ than the required number k, and use a more expensive hierarchical agglomerative clustering approach to create k robust centroids. Because these centroids are more representative of the database $\mathcal{D}$, this provides a better starting point for the algorithm.

A very simple approach, which seems to work surprisingly well, is to select the initial representatives as centroids of m randomly chosen samples of points for some user-selected parameter m. This will ensure that the initial centroids are not too biased by any particular outlier. Furthermore, while all these centroid representatives will be approximately equal to the mean of the data, they will typically be slightly biased toward one cluster or another because of random variations across different samples. Subsequent iterations of k-means will eventually associate each of these representatives with a cluster.

The presence of outliers will typically have a detrimental impact on such algorithms. This can happen in cases where the initialization procedure selects an outlier as one of the initial centers. Although a k-medoids algorithm will typically discard an outlier representative during an iterative exchange, a k-center approach can become stuck with a singleton cluster or an empty cluster in subsequent iterations. In such cases, one solution is to add an additional step in the iterative portion of the algorithm that discards centers with very small clusters and replaces them with randomly chosen points from the data.

The number of clusters k is a parameter used by this approach. Section 6.9.1.1 on cluster validity provides an approximate method for selecting the number of clusters k. As discussed in Sect. 6.9.1.1, this approach is far from perfect. The number of natural clusters is often difficult to determine using automated methods. Because the number of natural clusters is not known a priori, it may sometimes be desirable to use a larger value of k than the analyst's "guess" about the true natural number of clusters in the data. This will result in the splitting of some of the data clusters into multiple representatives, but it is less likely for clusters to be incorrectly merged. As a postprocessing step, it may be possible to merge some of the clusters based on the intercluster distances. Some hybrid agglomerative and partitioning algorithms include a merging step within the k-representative procedure. Refer to the bibliographic notes for references to these algorithms.

6.4 Hierarchical Clustering Algorithms

Hierarchical algorithms typically cluster the data with distances. However, the use of distance functions is not compulsory. Many hierarchical algorithms use other clustering methods, such as density- or graph-based methods, as a subroutine for constructing the hierarchy.

So why are hierarchical clustering methods useful from an application-centric point of view? One major reason is that different levels of clustering granularity provide different application-specific insights. This provides a *taxonomy* of clusters, which may be browsed for semantic insights. As a specific example, consider the taxonomy[2] of Web pages created by the well-known *Open Directory Project (ODP)*. In this case, the clustering has been

[2] `http://www.dmoz.org`

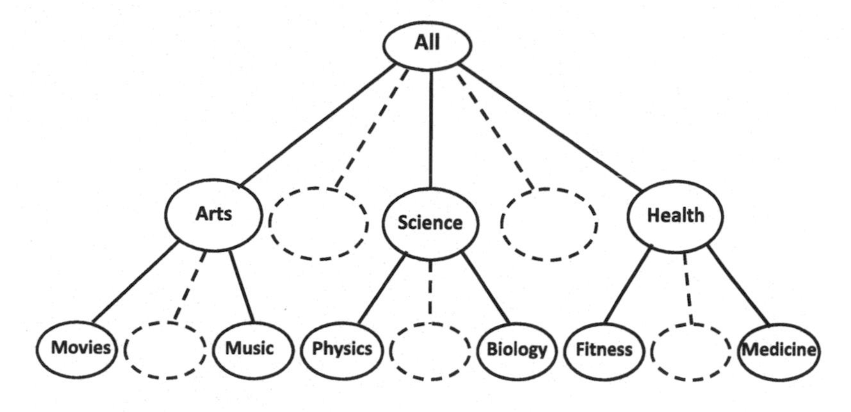

Figure 6.6: Multigranularity insights from hierarchical clustering

created by a manual volunteer effort, but it nevertheless provides a good understanding of the multigranularity insights that may be obtained with such an approach. A small portion of the hierarchical organization is illustrated in Fig. 6.6. At the highest level, the Web pages are organized into topics such as arts, science, health, and so on. At the next level, the topic of science is organized into subtopics, such as biology and physics, whereas the topic of health is divided into topics such as fitness and medicine. This organization makes manual browsing very convenient for a user, especially when the content of the clusters can be described in a semantically comprehensible way. In other cases, such hierarchical organizations can be used by indexing algorithms. Furthermore, such methods can sometimes also be used for creating better "flat" clusters. Some agglomerative hierarchical methods and divisive methods, such as bisecting k-means, can provide better quality clusters than partitioning methods such as k-means, albeit at a higher computational cost.

There are two types of hierarchical algorithms, depending on how the hierarchical tree of clusters is constructed:

1. *Bottom-up (agglomerative) methods:* The individual data points are successively agglomerated into higher-level clusters. The main variation among the different methods is in the choice of objective function used to decide the merging of the clusters.

2. *Top-down (divisive) methods:* A top-down approach is used to successively partition the data points into a tree-like structure. A flat clustering algorithm may be used for the partitioning in a given step. Such an approach provides tremendous flexibility in terms of choosing the trade-off between the balance in the tree structure and the balance in the number of data points in each node. For example, a tree-growth strategy that splits the heaviest node will result in leaf nodes with a similar number of data points in them. On the other hand, a tree-growth strategy that constructs a balanced tree structure with the same number of children at each node will lead to leaf nodes with varying numbers of data points.

In the following sections, both types of hierarchical methods will be discussed.

6.4.1 Bottom-Up Agglomerative Methods

In bottom-up methods, the data points are successively agglomerated into higher level clusters. The algorithm starts with individual data points in their own clusters and successively

Algorithm *AgglomerativeMerge*(Data: $\mathcal{D}$)
begin
  Initialize $n \times n$ distance matrix M using $\mathcal{D}$;
  repeat
    Pick closest pair of clusters i and j using M;
    Merge clusters i and j;
    Delete rows/columns i and j from M and create
      a new row and column for newly merged cluster;
    Update the entries of new row and column of M;
  until termination criterion;
  return current merged cluster set;
end

Figure 6.7: Generic agglomerative merging algorithm with unspecified merging criterion

agglomerates them into higher level clusters. In each iteration, two clusters are selected that are deemed to be as close as possible. These clusters are merged and replaced with a newly created merged cluster. Thus, each merging step reduces the number of clusters by 1. Therefore, a method needs to be designed for measuring proximity between clusters containing multiple data points, so that they may be merged. It is in this choice of computing the distances between clusters, that most of the variations among different methods arise.

Let n be the number of data points in the d-dimensional database $\mathcal{D}$, and $n_t = n - t$ be the number of clusters after t agglomerations. At any given point, the method maintains an $n_t \times n_t$ distance matrix M between the current *clusters* in the data. The precise methodology for computing and maintaining this distance matrix will be described later. In any given iteration of the algorithm, the (nondiagonal) entry in the distance matrix with the least distance is selected, and the corresponding clusters are merged. This merging will require the distance matrix to be updated to a smaller $(n_t - 1) \times (n_t - 1)$ matrix. The dimensionality reduces by 1 because the rows and columns for the two merged clusters need to be deleted, and a new row and column of distances, corresponding to the newly created cluster, needs to be added to the matrix. This corresponds to the newly created cluster in the data. The algorithm for determining the values of this newly created row and column depends on the cluster-to-cluster distance computation in the merging procedure and will be described later. The incremental update process of the distance matrix is a more efficient option than that of computing all distances from scratch. It is, of course, assumed that sufficient memory is available to maintain the distance matrix. If this is not the case, then the distance matrix will need to be fully recomputed in each iteration, and such agglomerative methods become less attractive. For termination, either a maximum threshold can be used on the distances between two merged clusters or a minimum threshold can be used on the number of clusters at termination. The former criterion is designed to automatically determine the natural number of clusters in the data but has the disadvantage of requiring the specification of a quality threshold that is hard to guess intuitively. The latter criterion has the advantage of being intuitively interpretable in terms of the number of clusters in the data. The order of merging naturally creates a hierarchical tree-like structure illustrating the relationship between different clusters, which is referred to as a *dendrogram*. An example of a dendrogram on successive merges on six data points, denoted by A, B, C, D, E, and F, is illustrated in Fig. 6.8a.

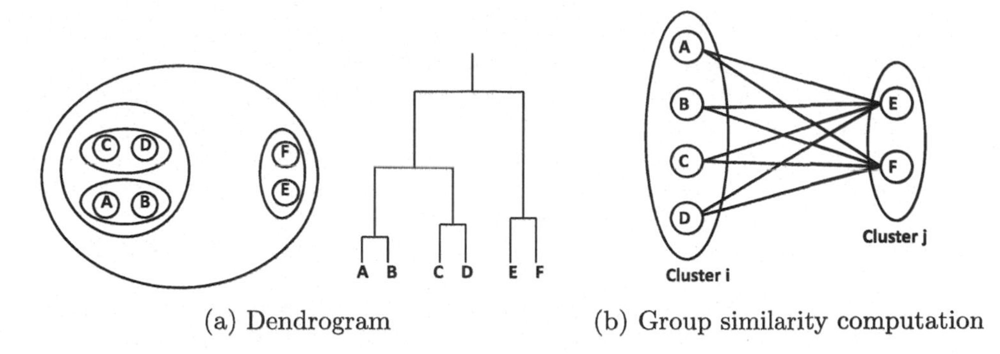

(a) Dendrogram (b) Group similarity computation

Figure 6.8: Illustration of hierarchical clustering steps

The generic agglomerative procedure with an unspecified merging criterion is illustrated in Fig. 6.7. The distances are encoded in the $n_t \times n_t$ distance matrix M. This matrix provides the pairwise cluster distances computed with the use of the merging criterion. The different choices for the merging criteria will be described later. The merging of two clusters corresponding to rows (columns) i and j in the matrix M requires the computation of some measure of distances between their constituent objects. For two clusters containing m_i and m_j objects, respectively, there are $m_i \cdot m_j$ pairs of distances between constituent objects. For example, in Fig. 6.8b, there are $2 \times 4 = 8$ pairs of distances between the constituent objects, which are illustrated by the corresponding edges. The overall distance between the two clusters needs to be computed as a function of these $m_i \cdot m_j$ pairs. In the following, different ways of computing the distances will be discussed.

6.4.1.1 Group-Based Statistics

The following discussion assumes that the indices of the two clusters to be merged are denoted by i and j, respectively. In group-based criteria, the distance between two groups of objects is computed as a function of the $m_i \cdot m_j$ pairs of distances among the constituent objects. The different ways of computing distances between two groups of objects are as follows:

1. *Best (single) linkage:* In this case, the distance is equal to the minimum distance between all $m_i \cdot m_j$ pairs of objects. This corresponds to the closest pair of objects between the two groups. After performing the merge, the matrix M of pairwise distances needs to be updated. The ith and jth rows and columns are deleted and replaced with a single row and column representing the merged cluster. The new row (column) can be computed using the minimum of the values in the previously deleted pair of rows (columns) in M. This is because the distance of the other clusters to the merged cluster is the minimum of their distances to the individual clusters in the best-linkage scenario. For any other cluster $k \neq i, j$, this is equal to $\min\{M_{ik}, M_{jk}\}$ (for rows) and $\min\{M_{ki}, M_{kj}\}$ (for columns). The indices of the rows and columns are then updated to account for the deletion of the two clusters and their replacement with a new one. The best linkage approach is one of the instantiations of agglomerative methods that is very good at discovering clusters of arbitrary shape. This is because the data points in clusters of arbitrary shape can be successively merged with chains of data point pairs at small pairwise distances to each other. On the other hand, such chaining may also inappropriately merge distinct clusters when it results from noisy points.

2. *Worst (complete) linkage:* In this case, the distance between two groups of objects is equal to the maximum distance between all $m_i \cdot m_j$ pairs of objects in the two groups. This corresponds to the farthest pair in the two groups. Correspondingly, the matrix M is updated using the maximum values of the rows (columns) in this case. For any value of $k \neq i, j$, this is equal to $\max\{M_{ik}, M_{jk}\}$ (for rows), and $\max\{M_{ki}, M_{kj}\}$ (for columns). The worst-linkage criterion implicitly attempts to minimize the maximum diameter of a cluster, as defined by the largest distance between any pair of points in the cluster. This method is also referred to as the *complete linkage* method.

3. *Group-average linkage:* In this case, the distance between two groups of objects is equal to the average distance between all $m_i \cdot m_j$ pairs of objects in the groups. To compute the row (column) for the merged cluster in M, a weighted average of the ith and jth rows (columns) in the matrix M is used. For any value of $k \neq i, j$, this is equal to $\frac{m_i \cdot M_{ik} + m_j \cdot M_{jk}}{m_i + m_j}$ (for rows), and $\frac{m_i \cdot M_{ki} + m_j \cdot M_{kj}}{m_i + m_j}$ (for columns).

4. *Closest centroid:* In this case, the closest centroids are merged in each iteration. This approach is not desirable, however, because the centroids lose information about the relative spreads of the different clusters. For example, such a method will not discriminate between merging pairs of clusters of varying sizes, as long as their centroid pairs are at the same distance. Typically, there is a bias toward merging pairs of larger clusters because centroids of larger clusters are statistically more likely to be closer to each other.

5. *Variance-based criterion:* This criterion minimizes the *change* in the objective function (such as cluster variance) as a result of the merging. Merging always results in a worsening of the clustering objective function value because of the loss of granularity. It is desired to merge clusters where the change (degradation) in the objective function as a result of merging is as little as possible. To achieve this goal, the zeroth, first, and second order *moment statistics* are maintained with each cluster. The average squared error SE_i of the ith cluster can be computed as a function of the number m_i of points in the cluster (zeroth-order moment), the sum F_{ir} of the data points in the cluster i along each dimension r (first-order moment), and the squared sum S_{ir} of the data points in the cluster i across each dimension r (second-order moment) according to the following relationship;

$$SE_i = \sum_{r=1}^{d} (S_{ir}/m_i - F_{ir}^2/m_i^2). \tag{6.8}$$

This relationship can be shown using the basic definition of variance and is used by many clustering algorithms such as *BIRCH* (cf. Chap. 7). Therefore, for each cluster, one only needs to maintain these cluster-specific statistics. Such statistics are easy to maintain across merges because the moment statistics of a merge of the two clusters i and j can be computed easily as the sum of their moment statistics. Let $SE_{i \cup j}$ denote the variance of a potential merge between the two clusters i and j. Therefore, the change in variance on executing a merge of clusters i and j is as follows:

$$\Delta SE_{i \cup j} = SE_{i \cup j} - SE_i - SE_j. \tag{6.9}$$

This change can be shown to always be a positive quantity. The cluster pair with the smallest increase in variance because of the merge is selected as the relevant pair to

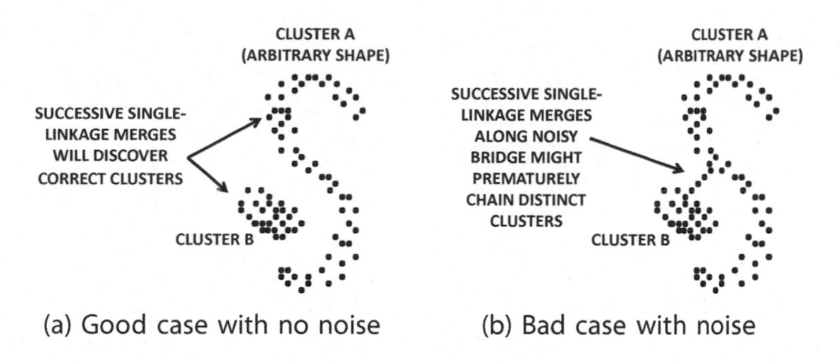

Figure 6.9: Good and bad cases for single-linkage clustering

be merged. As before, a matrix M of pairwise values of $\Delta SE_{i\cup j}$ is maintained along with moment statistics. After each merge of the ith and jth clusters, the ith and jth rows and columns of M are deleted and a new column for the merged cluster is added. The kth row (column) entry ($k \neq i, j$) in M of this new column is equal to $SE_{i\cup j\cup k} - SE_{i\cup j} - SE_k$. These values are computed using the cluster moment statistics. After computing the new row and column, the indices of the matrix M are updated to account for its reduction in size.

6. *Ward's method:* Instead of using the change in variance, one might also use the (unscaled) sum of squared error as the merging criterion. This is equivalent to setting the RHS of Eq. 6.8 to $\sum_{r=1}^{d}(m_i S_{ir} - F_{ir}^2)$. Surprisingly, this approach is a variant of the centroid method. The objective function for merging is obtained by multiplying the (squared) Euclidean distance between centroids with the harmonic mean of the number of points in each of the pair. Because larger clusters are penalized by this additional factor, the approach performs more effectively than the centroid method.

The various criteria have different advantages and disadvantages. For example, the single linkage method is able to successively merge chains of closely related points to discover clusters of arbitrary shape. However, this property can also (inappropriately) merge two unrelated clusters, when the chaining is caused by noisy points between two clusters. Examples of good and bad cases for single-linkage clustering are illustrated in Figs. 6.9a and b, respectively. Therefore, the behavior of single-linkage methods depends on the impact and relative presence of noisy data points. Interestingly, the well-known *DBSCAN* algorithm (cf. Sect. 6.6.2) can be viewed as a robust variant of single-linkage methods, and it can therefore find arbitrarily shaped clusters. The *DBSCAN* algorithm excludes the noisy points between clusters from the merging process to avoid undesirable chaining effects.

The complete (worst-case) linkage method attempts to minimize the maximum distance between any pair of points in a cluster. This quantification can implicitly be viewed as an approximation of the diameter of a cluster. Because of its focus on minimizing the diameter, it will try to create clusters so that all of them have a similar diameter. However, if some of the natural clusters in the data are larger than others, then the approach will break up the larger clusters. It will also be biased toward creating clusters of spherical shape irrespective of the underlying data distribution. Another problem with the complete linkage method is that it gives too much importance to data points at the noisy fringes of a cluster because of its focus on the maximum distance between any pair of points in the cluster. The group-average, variance, and Ward's methods are more robust to noise due to the use of multiple linkages in the distance computation.

The agglomerative method requires the maintenance of a heap of sorted distances to efficiently determine the minimum distance value in the matrix. The initial distance matrix computation requires $O(n^2 \cdot d)$ time, and the maintenance of a sorted heap data structure requires $O(n^2 \cdot \log(n))$ time over the course of the algorithm because there will be a total of $O(n^2)$ additions and deletions into the heap. Therefore, the overall running time is $O(n^2 \cdot d + n^2 \cdot \log(n))$. The required space for the distance matrix is $O(n^2)$. The space-requirement is particularly problematic for large data sets. In such cases, a similarity matrix M cannot be incrementally maintained, and the time complexity of many hierarchical methods will increase dramatically to $O(n^3 \cdot d)$. This increase occurs because the similarity computations between clusters need to be performed explicitly at the time of the merging. Nevertheless, it is possible to speed up the algorithm in such cases by approximating the merging criterion. The *CURE* method, discussed in Sect. 7.3.3 of Chap. 7, provides a scalable single-linkage implementation of hierarchical methods and can discover clusters of arbitrary shape. This improvement is achieved by using carefully chosen representative points from clusters to approximately compute the single-linkage criterion.

Practical Considerations

Agglomerative hierarchical methods naturally lead to a binary tree of clusters. It is generally difficult to control the structure of the hierarchical tree with bottom-up methods as compared to the top-down methods. Therefore, in cases where a taxonomy of a specific structure is desired, bottom-up methods are less desirable.

A problem with hierarchical methods is that they are sensitive to a small number of mistakes made during the merging process. For example, if an incorrect merging decision is made at some stage because of the presence of noise in the data set, then there is no way to undo it, and the mistake may further propagate in successive merges. In fact, some variants of hierarchical clustering, such as single-linkage methods, are notorious for successively merging neighboring clusters because of the presence of a small number of noisy points. Nevertheless, there are numerous ways to reduce these effects by treating noisy data points specially.

Agglomerative methods can become impractical from a *space- and time-efficiency* perspective for larger data sets. Therefore, these methods are often combined with sampling and other partitioning methods to efficiently provide solutions of high quality.

6.4.2 Top-Down Divisive Methods

Although bottom-up agglomerative methods are typically distance-based methods, top-down hierarchical methods can be viewed as general-purpose meta-algorithms that can use almost any clustering algorithm as a subroutine. Because of the top-down approach, greater control is achieved on the global structure of the tree in terms of its degree and balance between different branches.

The overall approach for top-down clustering uses a general-purpose flat-clustering algorithm $\mathcal{A}$ as a subroutine. The algorithm initializes the tree at the root node containing all the data points. In each iteration, the data set at a particular node of the current tree is split into multiple nodes (clusters). By changing the criterion for node selection, one can create trees balanced by height or trees balanced by the number of clusters. If the algorithm $\mathcal{A}$ is randomized, such as the k-means algorithm (with random seeds), it is possible to use multiple trials of the same algorithm at a particular node and select the best one. The generic pseudocode for a top-down divisive strategy is illustrated in Fig. 6.10. The algo-

Algorithm *GenericTopDownClustering*(Data: $\mathcal{D}$, Flat Algorithm: $\mathcal{A}$)
begin
  Initialize tree $\mathcal{T}$ to root containing $\mathcal{D}$;
  repeat
    Select a leaf node L in $\mathcal{T}$ based on pre-defined criterion;
    Use algorithm $\mathcal{A}$ to split L into $L_1 \dots L_k$;
    Add $L_1 \dots L_k$ as children of L in $\mathcal{T}$;
  until termination criterion;
end

Figure 6.10: Generic top-down meta-algorithm for clustering

rithm recursively splits nodes with a top-down approach until either a certain height of the tree is achieved or each node contains fewer than a predefined number of data objects. A wide variety of algorithms can be designed with different instantiations of the algorithm $\mathcal{A}$ and growth strategy. Note that the algorithm $\mathcal{A}$ can be any arbitrary clustering algorithm, and not just a distance-based algorithm.

6.4.2.1 Bisecting k-Means

The bisecting k-means algorithm is a top-down hierarchical clustering algorithm in which each node is split into exactly two children with a 2-means algorithm. To split a node into two children, several randomized trial runs of the split are used, and the split that has the best impact on the overall clustering objective is used. Several variants of this approach use different growth strategies for selecting the node to be split. For example, the heaviest node may be split first, or the node with the smallest distance from the root may be split first. These different choices lead to balancing either the cluster weights or the tree height.

6.5 Probabilistic Model-Based Algorithms

Most clustering algorithms discussed in this book are *hard* clustering algorithms in which each data point is deterministically assigned to a particular cluster. Probabilistic model-based algorithms are *soft* algorithms in which each data point may have a nonzero assignment probability to many (typically all) clusters. A soft solution to a clustering problem may be converted to a hard solution by assigning a data point to a cluster with respect to which it has the largest assignment probability.

The broad principle of a mixture-based *generative* model is to assume that the data was generated from a mixture of k distributions with probability distributions $\mathcal{G}_1 \dots \mathcal{G}_k$. Each distribution $\mathcal{G}_i$ represents a cluster and is also referred to as a *mixture component*. Each data point $\overline{X_i}$, where $i \in \{1 \dots n\}$, is generated by this mixture model as follows:

1. Select a mixture component with prior probability $\alpha_i = P(\mathcal{G}_i)$, where $i \in \{1 \dots k\}$. Assume that the rth one is selected.

2. Generate a data point from $\mathcal{G}_r$.

This generative model will be denoted by $\mathcal{M}$. The different prior probabilities α_i and the parameters of the different distributions $\mathcal{G}_r$ are not known in advance. Each distribution $\mathcal{G}_i$ is often assumed to be the Gaussian, although any arbitrary (and different) family of distributions may be assumed for each $\mathcal{G}_i$. The choice of distribution $\mathcal{G}_i$ is important because it reflects the user's a priori understanding about the distribution and shape of the individual clusters (mixture components). The parameters of the distribution of each mixture component, such as its mean and variance, need to be estimated from the data, so that the overall data has the maximum likelihood of being *generated* by the model. This is achieved with the *expectation-maximization (EM)* algorithm. The parameters of the different mixture components can be used to describe the clusters. For example, the estimation of the mean of each Gaussian component is analogous to determine the mean of each cluster center in a k-representative algorithm. After the parameters of the mixture components have been estimated, the *posterior* generative (or assignment) probabilities of data points with respect to each mixture component (cluster) can be determined.

Assume that the probability density function of mixture component $\mathcal{G}_i$ is denoted by $f^i(\cdot)$. The probability (density function) of the data point $\overline{X_j}$ being generated by the model is given by the weighted sum of the probability densities over different mixture components, where the weight is the prior probability $\alpha_i = P(\mathcal{G}_i)$ of the mixture components:

$$f^{point}(\overline{X_j}|\mathcal{M}) = \sum_{i=1}^{k} \alpha_i \cdot f^i(\overline{X_j}). \tag{6.10}$$

Then, for a data set $\mathcal{D}$ containing n data points, denoted by $\overline{X_1} \ldots \overline{X_n}$, the probability density of the data set being generated by the model $\mathcal{M}$ is the product of all the point-specific probability densities:

$$f^{data}(\mathcal{D}|\mathcal{M}) = \prod_{j=1}^{n} f^{point}(\overline{X_j}|\mathcal{M}). \tag{6.11}$$

The log-likelihood fit $\mathcal{L}(\mathcal{D}|\mathcal{M})$ of the data set $\mathcal{D}$ with respect to model $\mathcal{M}$ is the logarithm of the aforementioned expression and can be (more conveniently) represented as a sum of values over different data points. The log-likelihood fit is preferred for computational reasons.

$$\mathcal{L}(\mathcal{D}|\mathcal{M}) = \log(\prod_{j=1}^{n} f^{point}(\overline{X_j}|\mathcal{M})) = \sum_{j=1}^{n} \log(\sum_{i=1}^{k} \alpha_i f^i(\overline{X_j})). \tag{6.12}$$

This log-likelihood fit needs to maximized to determine the model parameters. A salient observation is that if the probabilities of data points being generated from different clusters were known, then it becomes relatively easy to determine the optimal model parameters separately for each component of the mixture. At the same time, the probabilities of data points being generated from different components are dependent on these optimal model parameters. This circularity is reminiscent of a similar circularity in optimizing the objective function of partitioning algorithms in Sect. 6.3. In that case, the knowledge of a *hard* assignment of data points to clusters provides the ability to determine optimal cluster representatives locally for each cluster. In this case, the knowledge of a *soft* assignment provides the ability to estimate the optimal (maximum likelihood) model parameters *locally* for each cluster. This naturally suggests an iterative EM algorithm, in which the model parameters and probabilistic assignments are iteratively estimated from one another.

Let Θ be a vector, representing the *entire set* of parameters describing all components of the mixture model. For example, in the case of the Gaussian mixture model, Θ contains all the component mixture means, variances, covariances, and the *prior* generative probabilities $\alpha_1 \ldots \alpha_k$. Then, the EM algorithm starts with an initial set of values of Θ (possibly corresponding to random assignments of data points to mixture components), and proceeds as follows:

1. (E-step) Given the current value of the parameters in Θ, estimate the *posterior* probability $P(\mathcal{G}_i|\overline{X_j}, \Theta)$ of the component $\mathcal{G}_i$ having been selected in the generative process, given that we have observed data point $\overline{X_j}$. The quantity $P(\mathcal{G}_i|\overline{X_j}, \Theta)$ is also the soft cluster assignment probability that we are trying to estimate. This step is executed for each data point $\overline{X_j}$ and mixture component $\mathcal{G}_i$.

2. (M-step) Given the current probabilities of assignments of data points to clusters, use the maximum likelihood approach to determine the values of all the parameters in Θ that maximize the log-likelihood fit on the basis of current assignments.

The two steps are executed repeatedly in order to improve the maximum likelihood criterion. The algorithm is said to converge when the objective function does not improve significantly in a certain number of iterations. The details of the E-step and the M-step will now be explained.

The E-step uses the currently available model parameters to compute the probability density of the data point $\overline{X_j}$ being generated by each component of the mixture. This probability density is used to compute the Bayes probability that the data point $\overline{X_j}$ was generated by component $\mathcal{G}_i$ (with model parameters fixed to the current set of the parameters Θ):

$$P(\mathcal{G}_i|\overline{X_j}, \Theta) = \frac{P(\mathcal{G}_i) \cdot P(\overline{X_j}|\mathcal{G}_i, \Theta)}{\sum_{r=1}^{k} P(\mathcal{G}_r) \cdot P(\overline{X_j}|\mathcal{G}_r, \Theta)} = \frac{\alpha_i \cdot f^{i,\Theta}(\overline{X_j})}{\sum_{r=1}^{k} \alpha_r \cdot f^{r,\Theta}(\overline{X_j})}. \tag{6.13}$$

As you will learn in Chap. 10 on classification, Eq. 6.13 is exactly the mechanism with which a Bayes classifier assigns previously unseen data points to categories (classes). A superscript Θ has been added to the probability density functions to denote the fact that they are evaluated for current model parameters Θ.

The M-step requires the optimization of the parameters for each probability distribution under the assumption that the E-step has provided the "correct" soft assignment. To optimize the fit, the partial derivative of the log-likelihood fit with respect to corresponding model parameters needs to be computed and set to zero. Without specifically describing the details of these algebraic steps, the values of the model parameters that are computed as a result of the optimization are described here.

The value of each α_i is estimated as the current weighted fraction of points assigned to cluster i, where a weight of $P(\mathcal{G}_i|\overline{X_j}, \Theta)$ is associated with data point $\overline{X_j}$. Therefore, we have:

$$\alpha_i = P(\mathcal{G}_i) = \frac{\sum_{j=1}^{n} P(\mathcal{G}_i|\overline{X_j}, \Theta)}{n}. \tag{6.14}$$

In practice, in order to obtain more robust results for smaller data sets, the expected number of data points belonging to each cluster in the numerator is augmented by 1, and the total number of points in the denominator is $n + k$. Therefore, the estimated value is as follows:

$$\alpha_i = \frac{1 + \sum_{j=1}^{n} P(\mathcal{G}_i|\overline{X_j}, \Theta)}{k + n}. \tag{6.15}$$

This approach is also referred to as *Laplacian smoothing.*

To determine the other parameters for component i, the value of $P(\mathcal{G}_i|\overline{X_j}, \Theta)$ is treated as a weight of that data point. Consider a Gaussian mixture model in d dimensions, in which the distribution of the ith component is defined as follows:

$$f^{i,\Theta}(\overline{X_j}) = \frac{1}{\sqrt{|\Sigma_i|}(2 \cdot \pi)^{(d/2)}} e^{-\frac{1}{2}(\overline{X_j} - \overline{\mu_i})\Sigma_i^{-1}(\overline{X_j} - \overline{\mu_i})}. \tag{6.16}$$

Here, $\overline{\mu_i}$ is the d-dimensional mean vector of the ith Gaussian component, and Σ_i is the $d \times d$ covariance matrix of the generalized Gaussian distribution of the ith component. The notation $|\Sigma_i|$ denotes the determinant of the covariance matrix. It can be shown[3] that the maximum-likelihood estimation of $\overline{\mu_i}$ and Σ_i yields the (probabilistically weighted) means and covariance matrix of the data points in that component. These probabilistic weights were derived from the assignment probabilities in the E-step. Interestingly, this is exactly how the representatives and covariance matrices of the Mahalanobis k-means approach are derived in Sect. 6.3. The only difference was that the data points were not weighted because hard assignments were used by the deterministic k-means algorithm. Note that the term in the exponent of the Gaussian distribution is the square of the Mahalanobis distance.

The E-step and the M-step can be iteratively executed to convergence to determine the optimal parameter set Θ. At the end of the process, a probabilistic model is obtained that describes the entire data set in terms of a generative model. The model also provides soft assignment probabilities $P(\mathcal{G}_i|\overline{X_j}, \Theta)$ of the data points, on the basis of the final execution of the E-step.

In practice, to minimize the number of estimated parameters, the non-diagonal entries of Σ_i are often set to 0. In such cases, the determinant of Σ_i simplifies to the product of the variances along the individual dimensions. This is equivalent to using the square of the *Minkowski* distance in the exponent. If all diagonal entries are further constrained to have the same value, then it is equivalent to using the Euclidean distance, and all components of the mixture will have spherical clusters. Thus, different choices and complexities of mixture model distributions provide different levels of flexibility in representing the probability distribution of each component.

This two-phase iterative approach is similar to representative-based algorithms. The E-step can be viewed as a soft version of the *assign* step in distance-based partitioning algorithms. The M-step is reminiscent of the *optimize* step, in which optimal component-specific parameters are learned on the basis of the fixed assignment. The distance term in the exponent of the probability distribution provides the natural connection between probabilistic and distance-based algorithms. This connection is discussed in the next section.

6.5.1 Relationship of EM to k-means and Other Representative Methods

The EM algorithm provides an extremely flexible framework for probabilistic clustering, and certain special cases can be viewed as soft versions of distance-based clustering methods. As a specific example, consider the case where all a priori generative probabilities α_i are fixed to $1/k$ as a part of the model setting. Furthermore, *all* components of the mixture have the same radius σ along all directions, and the mean of the jth cluster is assumed to be $\overline{Y_j}$. Thus, the only parameters to be learned are σ, and $\overline{Y_1} \dots \overline{Y_k}$. In that case, the jth component of the mixture has the following distribution:

$$f^{j,\Theta}(\overline{X_i}) = \frac{1}{(\sigma\sqrt{2 \cdot \pi})^d} e^{-\left(\frac{||\overline{X_i}-\overline{Y_j}||^2}{2\sigma^2}\right)}. \tag{6.17}$$

This model assumes that all mixture components have the same radius σ, and the cluster in each component is spherical. Note that the exponent in the distribution is the scaled square

[3]This is achieved by setting the partial derivative of $\mathcal{L}(\mathcal{D}|\mathcal{M})$ (see Eq. 6.12) with respect to each parameter in $\overline{\mu_i}$ and Σ to 0.

of the Euclidean distance. How do the E-step and M-step compare to the assignment and re-centering steps of the k-means algorithm?

1. (E-step) Each data point i has a probability belonging to cluster j, which is proportional to the scaled and exponentiated Euclidean distance to each representative $\overline{Y_j}$. In the k-means algorithm, this is done in a hard way, by picking the *best* Euclidean distance to any representative $\overline{Y_j}$.
2. (M-step) The center $\overline{Y_j}$ is the weighted mean over all the data points where the weight is defined by the probability of assignment to cluster j. The hard version of this is used in k-means, where each data point is either assigned to a cluster or not assigned to a cluster (i.e., analogous to 0-1 probabilities).

When the mixture distribution is defined with more general forms of the Gaussian distribution, the corresponding k-representative algorithm is the Mahalanobis k-means algorithm. It is noteworthy that the exponent of the general Gaussian distribution is the Mahalanobis distance. This implies that *special cases of the EM algorithm are equivalent to a soft version of the k-means algorithm*, where the exponentiated k-representative distances are used to define soft EM assignment probabilities.

The E-step is structurally similar to the *Assign* step, and the M-step is similar to the *Optimize* step in k-representative algorithms. Many mixture component distributions can be expressed in the form $K_1 \cdot e^{-K_2 \cdot Dist(\overline{X_i}, \overline{Y_j})}$, where K_1 and K_2 are regulated by distribution parameters. The log-likelihood of such an exponentiated distribution directly maps to an additive distance term $Dist(\overline{X_i}, \overline{Y_j})$ in the M-step objective function, which is structurally identical to the corresponding additive optimization term in k-representative methods. For many EM models with mixture probability distributions of the form $K_1 \cdot e^{-K_2 \cdot Dist(\overline{X_i}, \overline{Y_j})}$, a corresponding k-representative algorithm can be defined with a distance function $Dist(\overline{X_i}, \overline{Y_j})$.

Practical Considerations

The major practical consideration in mixture modeling is the level of the desired flexibility in defining the mixture components. For example, when each mixture component is defined as a generalized Gaussian, it is more effective at finding clusters of arbitrary shape and orientation. On the other hand, this requires the learning of a larger number of parameters, such as a $d \times d$ covariance matrix Σ_j. When the amount of data available is small, such an approach will not work very well because of *overfitting*. Overfitting refers to the situation where the parameters learned on a small sample of the true generative model are not reflective of this model because of the noisy variations within the data. Furthermore, as in k-means algorithms, the EM-algorithm can converge to a local optimum.

At the other extreme end, one can pick a spherical Gaussian where each component of the mixture has an identical radius, and also fix the a priori generative probability α_i to $1/k$. In this case, the EM model will work quite effectively even on very small data sets, because only a single parameter needs to be learned by the algorithm. However, if the different clusters have different shapes, sizes, and orientations, the clustering will be poor even on a large data set. The general rule of thumb is to tailor the model complexity to the available data size. Larger data sets allow more complex models. In some cases, an analyst may have domain knowledge about the distribution of data points in clusters. In these scenarios, the best option is to select the mixture components on the basis of this domain knowledge.

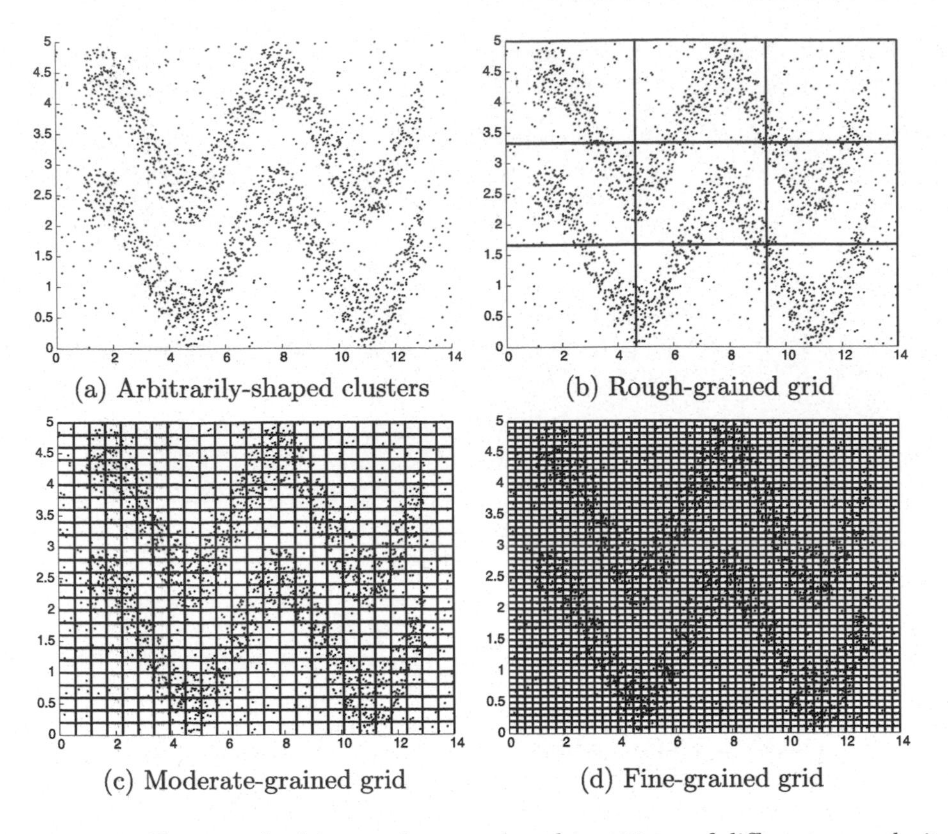

(a) Arbitrarily-shaped clusters

(b) Rough-grained grid

(c) Moderate-grained grid

(d) Fine-grained grid

Figure 6.11: Clusters of arbitrary shape and grid partitions of different granularity

6.6 Grid-Based and Density-Based Algorithms

One of the major problems with distance-based and probabilistic methods is that the shape of the underlying clusters is already defined implicitly by the underlying distance function or probability distribution. For example, a k-means algorithm implicitly assumes a spherical shape for the cluster. Similarly, an EM algorithm with the generalized Gaussian assumes elliptical clusters. In practice, the clusters may be hard to model with a prototypical shape implied by a distance function or probability distribution. To understand this point, consider the clusters illustrated in Fig. 6.11a. It is evident that there are two clusters of sinusoidal shape in the data. However, virtually any choice of representatives in a k-means algorithm will result in the representatives of one of the clusters pulling away data points from the other.

Density-based algorithms are very helpful in such scenarios. The core idea in such algorithms is to first identify fine-grained dense regions in the data. These form the "building blocks" for constructing the arbitrarily-shaped clusters. These can also be considered pseudo-data points that need to be re-clustered together carefully into groups of arbitrary shape. Thus, most density-based methods can be considered two-level hierarchical algorithms. Because there are a fewer building blocks in the second phase, as compared to the number of data points in the first phase, it is possible to organize them together into complex shapes using more detailed analysis. This detailed analysis (or postprocessing) phase is conceptually similar to a single-linkage agglomerative algorithm, which is usually better tailored to determining arbitrarily-shaped clusters from a small number of (pseudo)-data points. Many variations of this broader principle exist, depending on the particular type of building blocks that are chosen. For example, in grid-based methods, the fine-grained clusters are grid-like *regions* in the data space. When pre-selected data points in dense regions are clustered with a single-linkage method, the approach is referred to as *DBSCAN*. Other more sophisticated density-based methods, such as *DENCLUE*, use gradient ascent on the kernel-density estimates to create the building blocks.

6.6.1 Grid-Based Methods

In this technique, the data is discretized into p intervals that are typically equi-width intervals. Other variations such as equi-depth intervals are possible, though they are often not used in order to retain the intuitive notion of density. For a d-dimensional data set, this leads to a total of p^d hyper-cubes in the underlying data. Examples of grids of different granularity with $p = 3, 25$, and 80 are illustrated in Figures 6.11b, c, and d, respectively. The resulting hyper-cubes (rectangles in Fig. 6.11) are the building blocks in terms of which the clustering is defined. A density threshold τ is used to determine the subset of the p^d hyper-cubes that are dense. In most real data sets, an arbitrarily shaped cluster will result in multiple dense regions that are connected together by a side or at least a corner. Therefore, two grid regions are said to be *adjacently connected*, if they share a side in common. A weaker version of this definition considers two regions to be adjacently connected if they share a *corner* in common. Many grid-clustering algorithms use the strong definition of adjacent connectivity, where a side is used instead of a corner. In general, for data points in k-dimensional space, two k-dimensional cubes may be defined as adjacent, if they have share a surface of dimensionality at least r, for some user-defined parameter $r < k$.

This directly adjacent connectivity can be generalized to indirect density connectivity between grid regions that are not immediately adjacent to one another. Two grid regions are density connected, if a path can be found from one grid to the other containing only a sequence of adjacently connected grid regions. The goal of grid-based clustering is to determine the connected regions created by such grid cells. It is easy to determine such connected grid regions by using a graph-based model on the grids. Each *dense* grid node is associated with a node in the graph, and each edge represents adjacent connectivity. The connected components in the graph may be determined by using breadth-first or depth-first traversal on the graph, starting from nodes in different components. The data points in these connected components are reported as the final clusters. An example of the construction of the clusters of arbitrary shape from the building blocks is illustrated in Fig. 6.13. Note that the corners of the clusters found are artificially rectangular, which is one of the limitations of grid-based methods. The generic pseudocode for the grid-based approach is discussed in Fig. 6.12.

One desirable property of grid-based (and most other density-based) algorithms is that the number of data clusters is not pre-defined in advance, as in k-means algorithms. Rather, the goal is to return the natural clusters in the data together with their corresponding shapes. On the other hand, two different parameters need to be defined corresponding to the number of grid ranges p and the density threshold τ. The correct choice of these parameters is often difficult and semantically un-intuitive to guess. An inaccurate choice can lead to unintended consequences:

1. When the number of grid ranges selected is too small, as in Fig. 6.11b, the data points from multiple clusters will be present in the same grid region. This will result in the

Algorithm *GenericGrid*(Data: $\mathcal{D}$, Ranges: p, Density: τ)
begin
  Discretize each dimension of data $\mathcal{D}$ into p ranges;
  Determine dense grid cells at density level τ;
  Create graph in which dense grids are connected if they are adjacent;
  Determine connected components of graph;
  return points in each connected component as a cluster;
end

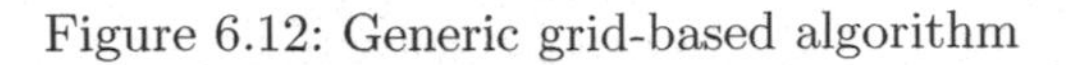
Figure 6.12: Generic grid-based algorithm

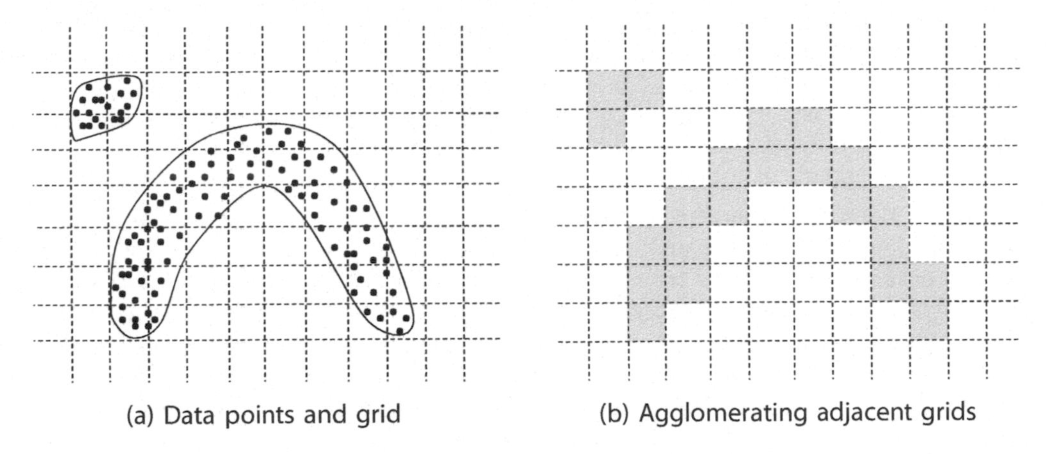
(a) Data points and grid (b) Agglomerating adjacent grids

Figure 6.13: Agglomerating adjacent grids

undesirable merging of clusters. When the number of grid ranges selected is too large, as in Fig. 6.11d, this will result in many empty grid cells even within the clusters. As a result, natural clusters in the data may be disconnected by the algorithm. A larger number of grid ranges also leads to computational challenges because of the increasing number of grid cells.

2. The choice of the density threshold has a similar effect on the clustering. For example, when the density threshold τ is too low, all clusters, including the ambient noise, will be merged into a single large cluster. On the other hand, an unnecessarily high density can partially or entirely miss a cluster.

The two drawbacks discussed above are serious ones, especially when there are significant variations in the cluster size and density over different local regions.

Practical Issues

Grid-based methods do not require the specification of the number of clusters, and also do not assume any specific shape for the clusters. However, this comes at the expense of having to specify a density parameter τ, which is not always intuitive from an analytical perspective. The choice of grid resolution is also challenging because it is not clear how it can be related to the density τ. As will be evident later, this is much easier with *DBSCAN*,

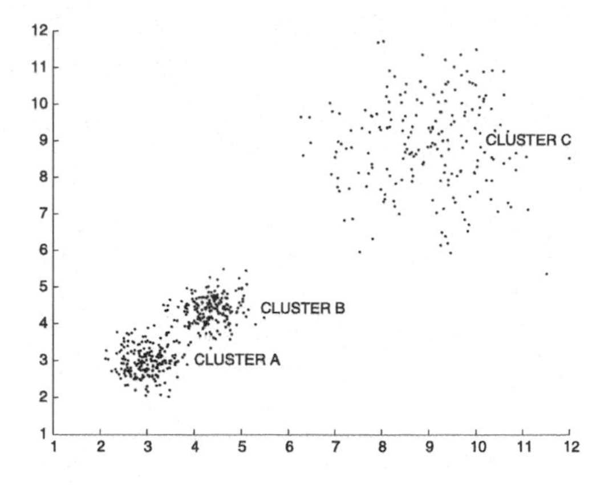

Figure 6.14: Impact of local distributions on density-based methods

where the resolution of the density-based approach is more easily related to the specified density threshold.

A major challenge with many density-based methods, including grid-based methods, is that they use a single density parameter τ globally. However, the clusters in the underlying data may have varying density, as illustrated in Fig. 6.14. In this particular case, if the density threshold is selected to be too high, then cluster C may be missed. On the other hand, if the density threshold is selected to be too low, then clusters A and B may be merged artificially. In such cases, distance-based algorithms, such as k-means, may be more effective than a density-based approach. This problem is not specific to the grid-based method but is generally encountered by all density-based methods.

The use of rectangular grid regions is an approximation of this class of methods. This approximation degrades with increasing dimensionality because high-dimensional rectangular regions are poor approximations of the underlying clusters. Furthermore, grid-based methods become computationally infeasible in high dimensions because the number of grid cells increase exponentially with the underlying data dimensionality.

6.6.2 DBSCAN

The *DBSCAN* approach works on a very similar principle as grid-based methods. However, unlike grid-based methods, the density characteristics of data points are used to merge them into clusters. Therefore, the individual data points *in dense regions* are used as building blocks after classifying them on the basis of their density.

The density of a data point is defined by the number of points that lie within a radius *Eps* of that point (including the point itself). The densities of these spherical regions are used to classify the data points into *core*, *border*, or *noise* points. These notions are defined as follows:

1. *Core point:* A data point is defined as a *core* point, if it contains[4] at least τ data points.

2. *Border point:* A data point is defined as a *border* point, if it contains less than τ points, but it also contains at least one core point within a radius *Eps*.

[4]The parameter *MinPts* is used in the original *DBSCAN* description. However, the notation τ is used here to retain consistency with the grid-clustering description.

Algorithm *DBSCAN*(Data: $\mathcal{D}$, Radius: *Eps*, Density: τ)
begin
  Determine core, border and noise points of $\mathcal{D}$ at level (Eps, τ);
  Create graph in which core points are connected
    if they are within *Eps* of one another;
  Determine connected components in graph;
  Assign each border point to connected component
    with which it is best connected;
  return points in each connected component as a cluster;
end

Figure 6.15: Basic *DBSCAN* algorithm

3. *Noise point:* A data point that is neither a core point nor a border point is defined as a *noise* point.

Examples of core points, border points, and noise points are illustrated in Fig. 6.16 for $\tau = 10$. The data point A is a core point because it contains 10 data points within the illustrated radius *Eps*. On the other hand, data point B contains only 6 points within a radius of *Eps*, but it contains the core point A. Therefore, it is a border point. The data point C is a noise point because it contains only 4 points within a radius of *Eps*, and it does not contain any core point.

After the core, border, and noise points have been determined, the *DBSCAN* clustering algorithm proceeds as follows. First, a connectivity graph is constructed with respect to the core points, in which each node corresponds to a core point, and an edge is added between a pair of core points, if and only if they are within a distance of *Eps* from one another. Note that the graph is constructed on the data *points* rather than on partitioned *regions*, as in grid-based algorithms. All connected components of this graph are identified. These correspond to the clusters constructed on the core points. The border points are then assigned to the cluster with which they have the highest level of connectivity. The resulting groups are reported as clusters and noise points are reported as outliers. The basic *DBSCAN* algorithm is illustrated in Fig. 6.15. It is noteworthy that the first step of graph-based clustering is identical to a single-linkage agglomerative clustering algorithm with termination-criterion of *Eps*-distance, *which is applied only to the core points.* Therefore, the *DBSCAN* algorithm may be viewed as an enhancement of single-linkage agglomerative clustering algorithms by treating marginal (border) and noisy points specially. This special treatment can reduce the outlier-sensitive chaining characteristics of single-linkage algorithms without losing the ability to create clusters of arbitrary shape. For example, in the pathological case of Fig. 6.9(b), the bridge of noisy data points will not be used in the agglomerative process if *Eps* and τ are selected appropriately. In such cases, *DBSCAN* will discover the correct clusters in spite of the noise in the data.

Practical Issues

The *DBSCAN* approach is very similar to grid-based methods, except that it uses circular regions as building blocks. The use of circular regions generally provides a smoother contour to the discovered clusters. Nevertheless, at more detailed levels of granularity, the two methods will tend to become similar. The strengths and weaknesses of *DBSCAN* are also

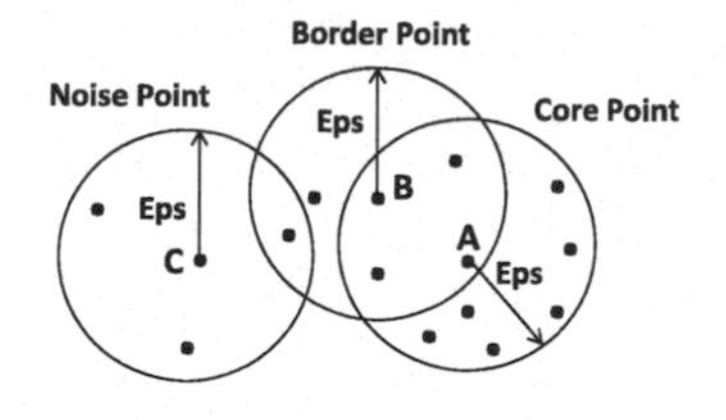

Figure 6.16: Examples of core, border, and noise points

similar to those of grid-based methods. The *DBSCAN* method can discover clusters of arbitrary shape, and it does not require the number of clusters as an input parameter. As in the case of grid-based methods, it is susceptible to variations in the local cluster density. For example, in Figs. 6.4b and 6.14, *DBSCAN* will either not discover the sparse cluster, or it might merge the two dense clusters. In such cases, algorithms such as Mahalanobis k-means are more effective because of their ability to normalize the distances with local density. On the other hand, *DBSCAN* will be able to effectively discover the clusters of Fig. 6.4a, which is not possible with the Mahalanobis k-means method.

The major time complexity of *DBSCAN* is in finding the neighbors of the different data points within a distance of Eps. For a database of size n, the time complexity can be $O(n^2)$ in the worst case. However, for some special cases, the use of a spatial index for finding the nearest neighbors can reduce this to approximately $O(n \cdot \log(n))$ distance computations. The $O(\log(n))$ query performance is realized only for low-dimensional data, in which nearest neighbor indexes work well. In general, grid-based methods are more efficient because they partition the *space*, rather than opting for the more computationally intensive approach of finding the nearest neighbors.

The parameters τ and Eps are related to one another in an intuitive way, which is useful for parameter setting. In particular, after the value of τ has been set by the user, the value of Eps can be determined in a data-driven way. The idea is to use a value of Eps that can capture most of the data points in clusters as core points. This can be achieved as follows. For each data point, its τ-nearest neighbor distance is determined. Typically, the vast majority of the data points inside clusters will have a small value of the τ-nearest neighbor distance. However, the value of the τ-nearest neighbor often increases suddenly for a small number of noisy points (or points at the fringes of clusters). Therefore, the key is to identify the tail of the distribution of τ-nearest neighbor distances. Statistical tests, such as the Z-value test, can be used in order to determine the value of Eps at which the τ-nearest neighbor distance starts increasing abruptly. This value of the τ-nearest neighbor distance at this cutoff point provides a suitable value of Eps.

Algorithm *DENCLUE*(Data: $\mathcal{D}$, Density: τ)
begin
  Determine density attractor of each data point in $\mathcal{D}$ with
    gradient-ascent of Equation 6.20;
  Create clusters of data points that converge to the same
    density attractor;
  Discard clusters whose density attractors have density less
    than τ and report as outliers;
  Merge clusters whose density attractors are connected with
    a path of density at least τ;
  return points in each cluster;
end

Figure 6.17: Basic *DENCLUE* algorithm

6.6.3 DENCLUE

The *DENCLUE* algorithm is based on firm statistical foundations that are rooted in kernel-density estimation. Kernel-density estimation can be used to create a smooth profile of the density distribution. In kernel-density estimation, the density $f(\overline{X})$ at coordinate $\overline{X}$ is defined as a sum of the influence (kernel) functions $K(\cdot)$ over the n different data points in the database $\mathcal{D}$:

$$f(\overline{X}) = \frac{1}{n}\sum_{i=1}^{n} K(\overline{X} - \overline{X_i}). \tag{6.18}$$

A wide variety of kernel functions may be used, and a common choice is the Gaussian kernel. For a d-dimensional data set, the Gaussian kernel is defined as follows:

$$K(\overline{X} - \overline{X_i}) = \left(\frac{1}{h\sqrt{2\pi}}\right)^d e^{-\frac{||\overline{X}-\overline{X_i}||^2}{2\cdot h^2}}. \tag{6.19}$$

The term $||\overline{X} - \overline{X_i}||$ represents the Euclidean distance between these d-dimensional data points. Intuitively, the effect of kernel-density estimation is to replace each discrete data point with a smooth "bump," and the density at a point is the sum of these "bumps." This results in a smooth profile of the data in which the random artifacts of the data are suppressed, and a smooth estimate of the density is obtained. Here, h represents the bandwidth of the estimation that regulates the smoothness of the estimation. Large values of the bandwidth h smooth out the noisy artifacts but may also lose some detail about the distribution. In practice, the value of h is chosen heuristically in a data-driven manner. An example of a kernel-density estimate in a data set with three natural clusters is illustrated in Fig. 6.18.

The goal is to determine clusters by using a density threshold τ that intersects this smooth density profile. Examples are illustrated in Figs. 6.18 and 6.19. The data points that lie in each (arbitrarily shaped) connected contour of this intersection will belong to the corresponding cluster. Some of the border data points of a cluster that lie just outside this contour may also be included because of the way in which data points are associated with clusters with the use of a hill-climbing approach. The choice of the density threshold will impact the number of clusters in the data. For example, in Fig. 6.18, a low-density threshold is used, and therefore two distinct clusters are merged. As a result, the approach will report

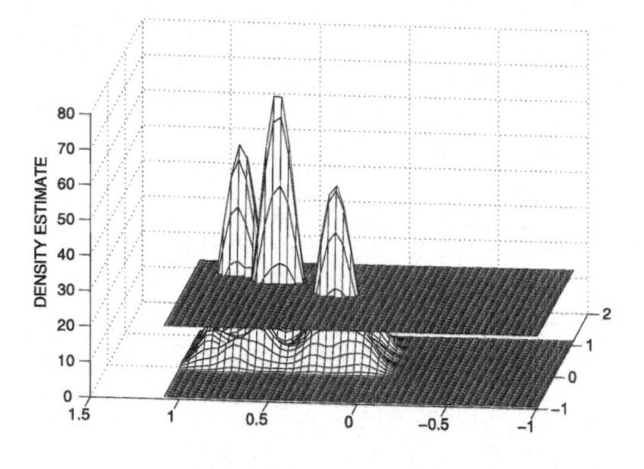

Figure 6.18: Density-based profile with lower density threshold

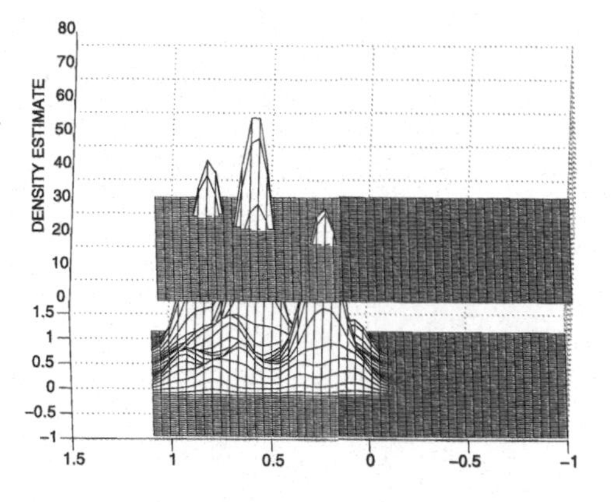

Figure 6.19: Density-based profile with higher density threshold

only two clusters. In Fig. 6.19, a higher density threshold is used, and therefore the approach will report three clusters. Note that, if the density threshold is increased further, one or more of the clusters will be completely missed. Such a cluster, whose peak density is lower than the user-specified threshold, is considered a noise cluster, and not reported by the *DENCLUE* algorithm.

The *DENCLUE* algorithm uses the notion of *density attractors* to partition data points into clusters. The idea is to treat each local peak of the density distribution as a density attractor, and associate each data point with its relevant peak by hill climbing toward its relevant peak. The different peaks that are connected by a path of density at least τ are then merged. For example, in each of Figs. 6.18 and 6.19, there are three density attractors. However, for the density threshold of Fig 6.18, only two clusters will be discovered because of the merging of a pair of peaks.

The *DENCLUE* algorithm uses an iterative gradient ascent approach in which each data point $\overline{X} \in \mathcal{D}$ is iteratively updated by using the gradient of the density profile with respect to $\overline{X}$. Let $\overline{X^{(t)}}$ be the updated value of $\overline{X}$ in the tth iteration. The value of $\overline{X^{(t)}}$ is updated as follows:

$$\overline{X^{(t+1)}} = \overline{X^{(t)}} + \alpha \nabla f(\overline{X^{(t)}}). \tag{6.20}$$

Here, $\nabla f(\overline{X^{(t)}})$ denotes the d-dimensional vector of partial derivatives of the kernel density with respect to each coordinate, and α is the step size. The data points are continually updated using the aforementioned rule, until they converge to a local optimum, which will always be one of the density attractors. Therefore, multiple data points may converge to the same density attractor. This creates an implicit clustering of the points, corresponding to the different density attractors (or local peaks). The density at each attractor is computed according to Eq. 6.18. Those attractors whose density does not meet the user-specified threshold τ are excluded because they are deemed to be small "noise" clusters. Furthermore, any pair of clusters whose density attractors are connected to each other by a path of density at least τ will be merged. This step addresses the merging of multiple density peaks, as illustrated in Fig. 6.18, and is analogous to the postprocessing step used in grid-based methods and *DBSCAN*. The overall *DENCLUE* algorithm is illustrated in Fig. 6.17.

One advantage of kernel-density estimation is that the gradient values $\nabla f(\overline{X})$ can be computed easily using the gradient of the constituent kernel-density values:

$$\nabla f(\overline{X}) = \frac{1}{n} \sum_{i=1}^{n} \nabla K(\overline{X} - \overline{X_i}). \tag{6.21}$$

The precise value of the gradient will depend on the choice of kernel function, though the differences across different choices are often not significant when the number of data points is large. In the particular case of the Gaussian kernel, the gradient can be shown to take on the following special form because of the presence of the negative squared distance in the exponent:

$$\nabla K(\overline{X} - \overline{X_i}) \propto (\overline{X_i} - \overline{X}) K(\overline{X} - \overline{X_i}). \tag{6.22}$$

This is because the derivative of an exponential function is itself, and the gradient of the negative squared distance is proportional to $(\overline{X_i} - \overline{X})$. The gradient of the kernel is the product of these two terms. Note that the constant of proportionality in Eq. 6.22 is irrelevant because it is indirectly included in the step size α of the gradient-ascent method.

A different way of determining the local optimum is by setting the gradient $\nabla f(\overline{X})$ to 0 as the optimization condition for $f(\overline{X})$, and solving the resulting system of equations using an iterative method, but using different starting points corresponding to the various data points. For example, by setting the gradient in Eq. 6.21 for the Gaussian kernel to 0, we obtain the following by substituting Eq. 6.22 in Eq. 6.21:

$$\sum_{i=1}^{n} \overline{X} K(\overline{X} - \overline{X_i}) = \sum_{i=1}^{n} \overline{X_i} K(\overline{X} - \overline{X_i}). \tag{6.23}$$

This is a nonlinear system of equations in terms of the d coordinates of $\overline{X}$ and it will have multiple solutions corresponding to different density peaks (or local optima). Such systems of equations can be solved numerically using iterative update methods and the choice of the starting point will yield different peaks. When a particular data point is used as the starting point in the iterations, it will always reach its density attractor. Therefore, one obtains the following modified update rule instead of the gradient ascent method:

$$\overline{X^{(t+1)}} = \frac{\sum_{i=1}^{n} \overline{X_i} K(\overline{X^{(t)}} - \overline{X_i})}{\sum_{i=1}^{n} K(\overline{X^{(t)}} - \overline{X_i})}. \tag{6.24}$$

This update rule replaces Eq. 6.20 and has a much faster rate of convergence. Interestingly, this update rule is widely known as the *mean-shift* method. Thus, there are interesting connections between *DENCLUE* and the mean-shift method. The bibliographic notes contain pointers to this optimized method and the mean-shift method.

The approach requires the computation of the density at each data point, which is $O(n)$. Therefore, the overall computational complexity is $O(n^2)$. This computational complexity can be reduced by observing that the density of a data point is mostly influenced only by its neighboring data points, and that the influence of distant data points is relatively small for exponential kernels such as the Gaussian kernel. In such cases, the data is discretized into grids, and the density of a point is computed only on the basis of the data points inside its grid and immediately neighboring grids. Because the grids can be efficiently accessed with the use of an index structure, this implementation is more efficient. Interestingly, the clustering of the *DBSCAN* method can be shown to be a special case of *DENCLUE* by using a binary kernel function that takes on the value of 1 within a radius of *Eps* of a point, and 0 otherwise.

Practical Issues

The *DENCLUE* method can be more effective than other density-based methods, when the number of data points is relatively small, and, therefore, a smooth estimate provides a more accurate understanding of the density distribution. *DENCLUE* is also able to handle data points at the borders of clusters in a more elegant way by using density attractors to attract relevant data points from the fringes of the cluster, even if they have density less than τ. Small groups of noisy data points will be appropriately discarded if their density attractor does not meet the user-specified density threshold τ. The approach also shares many advantages of other density-based algorithms. For example, the approach is able to discover arbitrarily shaped clusters, and it does not require the specification of the number of clusters. On the other hand, as in all density-based methods, it requires the specification of density threshold τ, which is difficult to determine in many real applications. As discussed earlier in the context of Fig. 6.14, local variations of density can be a significant challenge for any density-based algorithm. However, by varying the density threshold τ, it is possible to create a hierarchical dendrogram of clusters. For example, the two different values of τ in Figs. 6.18 and 6.19 will create a natural hierarchical arrangement of the clusters.

6.7 Graph-Based Algorithms

Graph-based methods provide a general *meta-framework*, in which data of virtually any type can be clustered. As discussed in Chap. 2, data of virtually any type can be converted to similarity graphs for analysis. This transformation is the key that allows the implicit clustering of any data type by performing the clustering on the corresponding transformed graph.

This transformation will be revisited in the following discussion. The notion of pairwise similarity is defined with the use of a *neighborhood graph*. Consider a set of data objects $\mathcal{O} = \{O_1 \dots O_n\}$, on which a neighborhood graph can be defined. Note that these objects can be of any type, such as time series or discrete sequences. The main constraint is that it should be possible to define a distance function on these objects. The neighborhood graph is constructed as follows:

1. A single node is defined for each object in $\mathcal{O}$. This is defined by the node set N, containing n nodes, where the node i corresponds to the object O_i.

2. An edge exists between O_i and O_j, if the distance $d(O_i, O_j)$ is less than a particular threshold ϵ. A better approach is to compute the k-nearest neighbors of both O_i and O_j, and add an edge when either one is a k-nearest neighbor of the other. The weight w_{ij} of the edge (i, j) is equal to a kernelized function of the distance between the objects O_i and O_j, so that larger weights indicate greater similarity. An example is the *heat kernel*, which is defined in terms of a parameter t:

$$w_{ij} = e^{-d(O_i, O_j)^2/t^2}. \tag{6.25}$$

For multidimensional data, the Euclidean distance is typically used to instantiate $d(O_i, O_j)$.

3. (Optional step) This step can be helpful for reducing the impact of local density variations such as those discussed in Fig. 6.14. Note that the quantity $deg(i) = \sum_{r=1}^{n} w_{ir}$ can be viewed as a proxy for the local kernel-density estimate near object O_i. Each

Algorithm *GraphMetaFramework*(Data: $\mathcal{D}$)
begin
  Construct the neighborhood graph G on $\mathcal{D}$;
  Determine clusters (communities) on the nodes in G;
  return clusters corresponding to the node partitions;
end

Figure 6.20: Generic graph-based meta-algorithm

edge weight w_{ij} is normalized by dividing it with $\sqrt{deg(i) \cdot deg(j)}$. Such an approach ensures that the clustering is performed after normalization of the similarity values with local densities. This step is not essential when algorithms such as normalized spectral clustering are used for finally clustering nodes in the neighborhood graph. This is because spectral clustering methods perform a similar normalization under the covers.

After the neighborhood graph has been constructed, *any* network clustering or community detection algorithm (cf. Sect. 19.3 of Chap. 19) can be used to cluster the nodes in the neighborhood graph. The clusters on the nodes can be used to map back to clusters on the original data objects. The spectral clustering method, which is a specific instantiation of the final node clustering step, is discussed in some detail below. However, the graph-based approach should be treated as a generic meta-algorithm that can use any community detection algorithm in the final node clustering step. The overall meta-algorithm for graph-based clustering is provided in Fig. 6.20.

Let $G = (N, A)$ be the undirected graph with node set N and edge set A, which is created by the aforementioned neighborhood-based transformation. A symmetric $n \times n$ weight matrix W defines the corresponding node similarities, based on the specific choice of neighborhood transformation, as in Eq. 6.25. All entries in this matrix are assumed to be non-negative, and higher values indicate greater similarity. If an edge does not exist between a pair of nodes, then the corresponding entry is assumed to be 0. It is desired to embed the nodes of this graph into a k-dimensional space, so that the similarity structure of the data is approximately preserved for the clustering process. This embedding is then used for a second phase of clustering.

First, let us discuss the much simpler problem of mapping the nodes into a 1-dimensional space. The generalization to the k-dimensional case is relatively straightforward. We would like to map the nodes in N into a set of 1-dimensional real values $y_1 \dots y_n$ on a line, so that the distances between these points reflect the connectivity among the nodes. It is undesirable for nodes that are connected with high-weight edges to be mapped onto distant points on this line. Therefore, we would like to determine values of y_i that minimize the following objective function O:

$$O = \sum_{i=1}^{n} \sum_{j=1}^{n} w_{ij} \cdot (y_i - y_j)^2. \tag{6.26}$$

This objective function penalizes the distances between y_i and y_j with weight proportional to w_{ij}. Therefore, when w_{ij} is very large (more similar nodes), the data points y_i and y_j will be more likely to be closer to one another in the embedded space. The objective function O can be rewritten in terms of the *Laplacian matrix* L of the weight matrix $W = [w_{ij}]$.

The Laplacian matrix L is defined as $\Lambda - W$, where Λ is a diagonal matrix satisfying $\Lambda_{ii} = \sum_{j=1}^{n} w_{ij}$. Let the n-dimensional column vector of embedded values be denoted by $\overline{y} = (y_1 \ldots y_n)^T$. It can be shown after some algebraic simplification that the objective function O can be rewritten in terms of the Laplacian matrix:

$$O = 2\overline{y}^T L \overline{y}. \quad (6.27)$$

The Laplacian matrix L is positive semi-definite with non-negative eigenvalues because the sum-of-squares objective function O is always non-negative. We need to incorporate a scaling constraint to ensure that the trivial value of $y_i = 0$ for all i is not selected by the optimization solution. A possible scaling constraint is as follows:

$$\overline{y}^T \Lambda \overline{y} = 1. \quad (6.28)$$

The presence of Λ in the constraint ensures better local normalization of the embedding. It can be shown using constrained optimization techniques, that the optimal solution for $\overline{y}$ that minimizes the objective function O is equal to the smallest eigenvector of $\Lambda^{-1}L$, satisfying the relationship $\Lambda^{-1}L\overline{y} = \lambda\overline{y}$. Here, λ is an eigenvalue. However, the smallest eigenvalue of $\Lambda^{-1}L$ is always 0, and it corresponds to the trivial solution where $\overline{y}$ is proportional to the vector containing only 1s. This trivial eigenvector is non-informative because it embeds every node to the same point on the line. Therefore, it can be discarded, and it is not used in the analysis. The second-smallest eigenvector then provides an optimal solution that is more informative.

This optimization formulation and the corresponding solution can be generalized to finding an optimal k-dimensional embedding. This is achieved by determining eigenvectors of $\Lambda^{-1}L$ with successively increasing eigenvalues. After discarding the first trivial eigenvector $\overline{e_1}$ with eigenvalue $\lambda_1 = 0$, this results in a set of k eigenvectors $\overline{e_2}, \overline{e_3} \ldots \overline{e_{k+1}}$, with corresponding eigenvalues $\lambda_2 \leq \lambda_3 \leq \ldots \leq \lambda_{k+1}$. Each eigenvector is an n-dimensional vector and is scaled to unit norm. The ith component of the jth eigenvector represents the jth coordinate of the ith data point. Because a total of k eigenvectors were selected, this approach creates an $n \times k$ matrix, corresponding to a new k-dimensional representation of each of the n data points. A k-means clustering algorithm can then be applied to the transformed representation.

Why is the transformed representation more suitable for an off-the-shelf k-means algorithm than the original data? It is important to note that the spherical clusters naturally found by the Euclidean-based k-means in the new embedded space may correspond to arbitrarily shaped clusters in the original space. As discussed in the next section, this behavior is a direct result of the way in which the similarity graph and objective function O are defined. This is also one of the main advantages of using a transformation to similarity graphs. For example, if the approach is applied to the arbitrarily shaped clusters in Fig. 6.11, the similarity graph will be such that a k-means algorithm on the transformed data (or a community detection algorithm on the similarity graph) will typically result in the correct arbitrarily-shaped clusters in the original space. Many variations of the spectral approach are discussed in detail in Sect. 19.3.4 of Chap. 19.

6.7.1 Properties of Graph-Based Algorithms

One interesting property of graph-based algorithms is that clusters of arbitrary shape can be discovered with the approach. This is because the neighborhood graph encodes the relevant *local* distances (or k-nearest neighbors), and therefore the communities in the induced

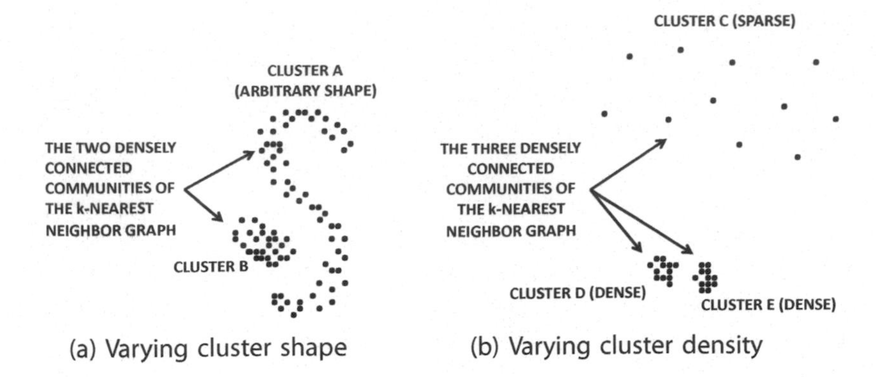

Figure 6.21: The merits of the k-nearest neighbor graph for handling clusters of varying shape and density

neighborhood graph are implicitly determined by agglomerating locally dense regions. As discussed in the previous section on density-based clustering, the agglomeration of locally dense regions corresponds to arbitrarily shaped clusters. For example, in Fig. 6.21a, the data points in the arbitrarily shaped cluster A will be densely connected to one another in the k-nearest neighbor graph, but they will not be significantly connected to data points in cluster B. As a result, any community detection algorithm will be able to discover the two clusters A and B on the graph representation.

Graph-based methods are also able to adjust much better to local variations in data density (see Fig. 6.14) when they use the k-nearest neighbors to construct the neighborhood graph rather than an absolute distance threshold. This is because the k-nearest neighbors of a node are chosen on the basis of *relative* comparison of distances within the locality of a data point whether they are large or small. For example, in Fig. 6.21b, even though clusters D and E are closer to each other than any pair of data points in sparse cluster C, all three clusters should be considered distinct clusters. Interestingly, a k-nearest neighbor graph will not create too many cross-connections between these clusters for small values of k. Therefore, all three clusters will be found by a community detection algorithm on the k-nearest neighbor graph in spite of their varying density. Therefore, graph-based methods can provide better results than algorithms such as *DBSCAN* because of their ability to adjust to varying local density *in addition to* their ability to discover arbitrarily shaped clusters. This desirable property of k-nearest neighbor graph algorithms is not restricted to the use of spectral clustering methods in the final phase. Many other graph-based algorithms have also been shown to discover arbitrarily shaped clusters in a locality-sensitive way. These desirable properties are therefore embedded within the k-nearest neighbor graph representation and are generalizable[5] to other data mining problems such as outlier analysis. Note that the locality-sensitivity of the shared nearest neighbor similarity function (cf. Sect. 3.2.1.8 of Chap. 3) is also due to the same reason. The locality-sensitivity of many classical clustering algorithms, such as k-medoids, bottom-up algorithms, and *DBSCAN*, can be improved by incorporating graph-based similarity functions such as the shared nearest neighbor method.

On the other hand, high computational costs are the major drawback of graph-based algorithms. It is often expensive to apply the approach to an $n \times n$ matrix of similarities. Nevertheless, because similarity graphs are sparse, many recent community detection methods can exploit this sparsity to provide more efficient solutions.

[5] See [257], which is a graph-based alternative to the *LOF* algorithm for locality-sensitive outlier analysis.

6.8 Non-negative Matrix Factorization

Nonnegative matrix factorization (*NMF*) is a dimensionality reduction method that is tailored to clustering. In other words, it embeds the data into a latent space that makes it more amenable to clustering. This approach is suitable for data matrices that are *non-negative* and *sparse*. For example, the $n \times d$ document-term matrix in text applications always contains non-negative entries. Furthermore, because most word frequencies are zero, this matrix is also sparse.

Nonnegative matrix factorization creates a new *basis system* for data representation, as in all dimensionality reduction methods. However, a distinguishing feature of *NMF* compared to many other dimensionality reduction methods is that the basis system does not necessarily contain orthonormal vectors. Furthermore, the basis system of vectors and the coordinates of the data records in this system are non-negative. The non-negativity of the representation is highly interpretable and well-suited for clustering. Therefore, non-negative matrix factorization is one of the dimensionality reduction methods that serves the dual purpose of enabling data clustering.

Consider the common use-case of *NMF* in the text domain, where the $n \times d$ data matrix D is a document-term matrix. In other words, there are n documents defined on a lexicon of size d. *NMF* transforms the data to a reduced k-dimensional basis system, in which each basis vector is a topic. Each such basis vector is a vector of nonnegatively weighted words that define that topic. Each document has a non-negative coordinate with respect to each basis vector. Therefore, the cluster membership of a document may be determined by examining the largest coordinate of the document along any of the k vectors. This provides the "topic" to which the document is most related and therefore defines its cluster. An alternative way of performing the clustering is to apply another clustering method such as k-means on the transformed representation. Because the transformed representation better discriminates between the clusters, the k-means approach will be more effective. The expression of each document as an additive and non-negative combination of the underlying topics also provides *semantic interpretability* to this representation. This is why the non-negativity of matrix factorization is so desirable.

So how are the basis system and the coordinate system determined? The non-negative matrix factorization method attempts to determine the matrices U and V that minimize the following objective function:

$$J = \frac{1}{2}||D - UV^T||^2. \tag{6.29}$$

Here, $|| \cdot ||^2$ represents the (squared) Frobenius norm, which is the sum of the squares of all the elements in the matrix, U is an $n \times k$ non-negative matrix, and V is a $d \times k$ non-negative matrix. The value of k is the dimensionality of the embedding. The matrix U provides the new k-dimensional coordinates of the rows of D in the transformed basis system, and the matrix V provides the basis vectors in terms of the original lexicon. Specifically, the rows of U provide the k-dimensional coordinates for each of the n documents, and the columns of V provide the k d-dimensional basis vectors.

What is the significance of the aforementioned optimization problem? Note that by minimizing J, the goal is to factorize the document-term matrix D as follows:

$$D \approx UV^T. \tag{6.30}$$

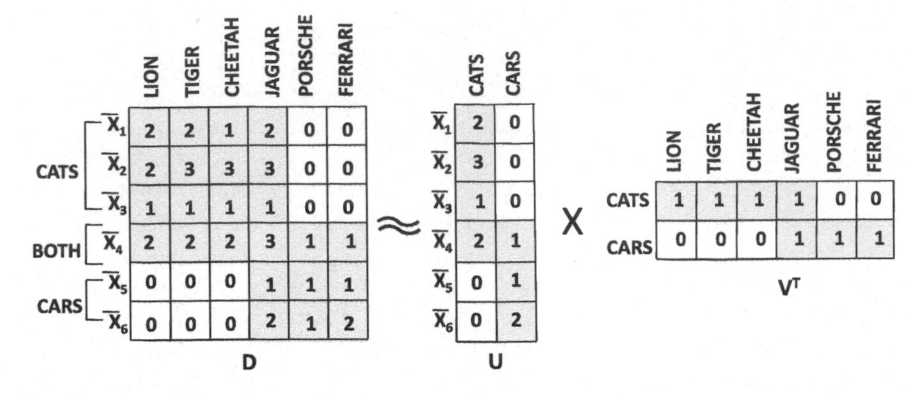

Figure 6.22: An example of non-negative matrix factorization

For each *row* $\overline{X_i}$ of D (document vector), and each k-dimensional row $\overline{Y_i}$ of U (transformed document vector), the aforementioned equation can be rewritten as follows:

$$\overline{X_i} \approx \overline{Y_i} V^T. \tag{6.31}$$

This is exactly in the same form as any standard dimensionality reduction method, where the columns of V provide the basis space and row-vector $\overline{Y_i}$ represents the reduced coordinates. In other words, the document vector $\overline{X_i}$ can be rewritten as an approximate (non-negative) linear combination of the k basis vectors. The value of k is typically small compared to the full dimensionality because the column vectors of V discover the latent structure in the data. Furthermore, the non-negativity of the matrices U and V ensures that the documents are expressed as a non-negative combination of the key concepts (or, clustered regions) in the term-based feature space.

An example of *NMF* for a toy 6×6 document-term matrix D is illustrated in Fig. 6.22. The rows correspond to 6 documents $\{\overline{X_1} \ldots \overline{X_6}\}$ and the 6 words correspond to columns. The matrix entries correspond to word frequencies in the documents. The documents $\{\overline{X_1}, \overline{X_2}, \overline{X_3}\}$ are related to cats, the documents $\{\overline{X_5}, \overline{X_6}\}$ are related to cars, and the document $\overline{X_4}$ is related to both. Thus, there are two natural clusters in the data, and the matrix is correspondingly factorized into two matrices U and V^T with rank $k = 2$. An approximately optimal factorization, with each entry rounded to the nearest integer, is illustrated in Fig. 6.22. Note that most of the entries in the factorized matrices will not be exactly 0 in a real-world example, but many of them might be close to 0, and almost all will be non-integer values. It is evident that the columns and rows, respectively, of U and V map to either the car or the cat cluster in the data. The 6×2 matrix U provides information about the relationships of 6 documents to 2 clusters, whereas the 6×2 matrix V provides information about the corresponding relationships of 6 words to 2 clusters. Each document can be assigned to the cluster for which it has the largest coordinate in U.

The rank-k matrix factorization UV^T can be decomposed into k components by expressing the matrix product in terms of the k columns $\overline{U_i}$ and $\overline{V_i}$, respectively, of U and V:

$$UV^T = \sum_{i=1}^{k} \overline{U_i}\, \overline{V_i}^T. \tag{6.32}$$

Each $n \times d$ matrix $\overline{U_i}\, \overline{V_i}^T$ is rank-1 matrix, which corresponds to a latent component in the data. Because of the interpretable nature of non-negative decomposition, it is easy

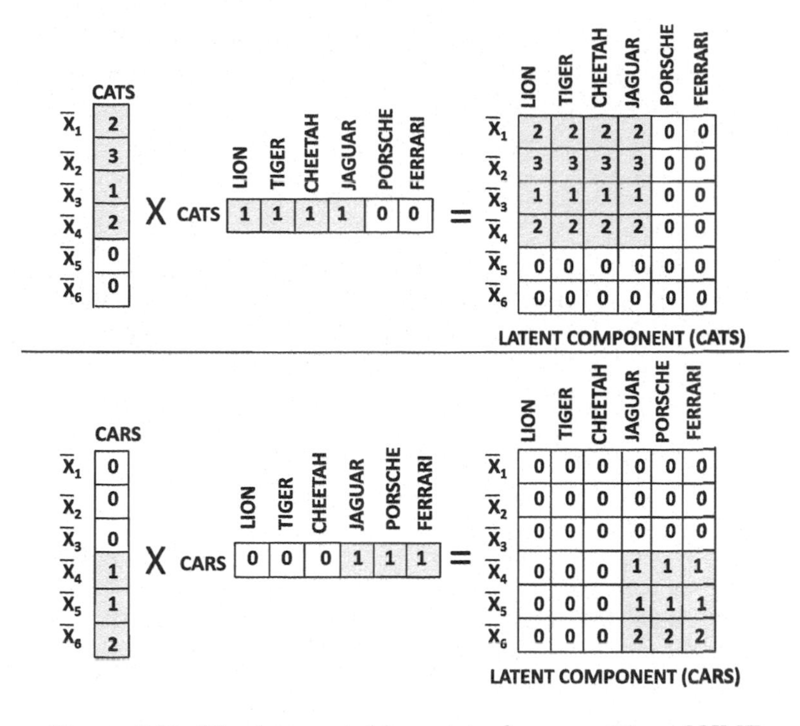

Figure 6.23: The interpretable matrix decomposition of NMF

to map these latent components to clusters. For example, the two latent components of the aforementioned example corresponding to cats and cars, respectively, are illustrated in Fig. 6.23.

It remains to be explained how the aforementioned optimization problem for J is solved. The squared norm of any matrix Q can be expressed as the trace of the matrix QQ^T. Therefore, the objective function J can be expressed as follows:

$$J = \frac{1}{2} tr\left[(D - UV^T)(D - UV^T)^T\right] \tag{6.33}$$

$$= \frac{1}{2}\left[tr(DD^T) - tr(DVU^T) - tr(UV^TD^T) + tr(UV^TVU^T)\right] \tag{6.34}$$

This is an optimization problem with respect to the matrices $U = [u_{ij}]$ and $V = [v_{ij}]$. Therefore, the matrix entries u_{ij} and v_{ij} are the optimization variables. In addition, the constraints $u_{ij} \geq 0$ and $v_{ij} \geq 0$ ensure non-negativity. This is a typical constrained non-linear optimization problem and can be solved using the *Lagrangian relaxation*, which relaxes these non-negativity constraints and replaces them in the objective function with constraint-violation penalties. The *Lagrange parameters* are the multipliers of these new penalty terms. Let $P_\alpha = [\alpha_{ij}]_{n\times k}$ and $P_\beta = [\beta_{ij}]_{d\times k}$ be matrices with the same dimensions as U and V, respectively. The elements of the matrices P_α and P_β are the corresponding Lagrange multipliers for the non-negativity conditions on the different elements of U and V, respectively. Furthermore, note that $tr(P_\alpha U^T)$ is equal to $\sum_{i,j} \alpha_{ij} u_{ij}$, and $tr(P_\beta V^T)$ is equal to $\sum_{i,j} \beta_{ij} v_{ij}$. These correspond to the Lagrangian penalties for the non-negativity constraints on U and V, respectively. Then, the augmented objective function with constraint penalties can be expressed as follows:

$$L = J + tr(P_\alpha U^T) + tr(P_\beta V^T). \tag{6.35}$$

To optimize this problem, the partial derivative of L with respect to U and V are computed and set to 0. Matrix calculus on the trace-based objective function yields the following:

$$\frac{\partial L}{\partial U} = -DV + UV^TV + P_\alpha = 0 \tag{6.36}$$

$$\frac{\partial L}{\partial V} = -D^TU + VU^TU + P_\beta = 0 \tag{6.37}$$

The aforementioned expressions provide two *matrices* of constraints. The (i, j)th entry of the above (two matrices of) conditions correspond to the partial derivatives of L with respect to u_{ij} and v_{ij}, respectively. These constraints are multiplied by u_{ij} and v_{ij}, respectively. By using the Kuhn-Tucker optimality conditions $\alpha_{ij}u_{ij} = 0$ and $\beta_{ij}v_{ij} = 0$, the (i, j)th pair of constraints can be written as follows:

$$(DV)_{ij}u_{ij} - (UV^TV)_{ij}u_{ij} = 0 \quad \forall i \in \{1 \dots n\}, \forall j \in \{1 \dots k\} \tag{6.38}$$

$$(D^TU)_{ij}v_{ij} - (VU^TU)_{ij}v_{ij} = 0 \quad \forall i \in \{1 \dots d\}, \forall j \in \{1 \dots k\} \tag{6.39}$$

These conditions are independent of P_α and P_β, and they provide a system of equations in terms of the entries of U and V. Such systems of equations are often solved using iterative methods. It can be shown that this particular system can be solved by using the following multiplicative update rules for u_{ij} and v_{ij}, respectively:

$$u_{ij} = \frac{(DV)_{ij}u_{ij}}{(UV^TV)_{ij}} \quad \forall i \in \{1 \dots n\}, \forall j \in \{1 \dots k\} \tag{6.40}$$

$$v_{ij} = \frac{(D^TU)_{ij}v_{ij}}{(VU^TU)_{ij}} \quad \forall i \in \{1 \dots d\}, \forall j \in \{1 \dots k\} \tag{6.41}$$

The entries of U and V are initialized to random values in $(0, 1)$, and the iterations are executed to convergence.

One interesting observation about the matrix factorization technique is that it can also be used to determine word-clusters instead of document clusters. Just as the columns of V provide a basis that can be used to discover document clusters, one can use the columns of U to discover a basis that corresponds to word clusters. Thus, this approach provides complementary insights into spaces where the dimensionality is very large.

6.8.1 Comparison with Singular Value Decomposition

Singular value decomposition (cf. Sect. 2.4.3.2 of Chap. 2) is a matrix factorization method. *SVD* factorizes the data matrix into three matrices instead of two. Equation 2.12 of Chap. 2 is replicated here:

$$D \approx Q_k \Sigma_k P_k^T. \tag{6.42}$$

It is instructive to compare this factorization to that of Eq. 6.30 for non-negative matrix factorization. The $n \times k$ matrix $Q_k\Sigma_k$ is analogous to the $n \times k$ matrix U in non-negative matrix factorization. The $d \times k$ matrix P_k is analogous to the $d \times k$ matrix V in matrix factorization. Both representations minimize the squared-error of data representation. The main differences between *SVD* and *NMF* arise from the different constraints in the corresponding optimization formulations. *SVD* can be viewed as a matrix-factorization in which the objective function is the same, but the optimization formulation imposes *orthogonality* constraints on the basis vectors rather than non-negativity constraints. Many other kinds of constraints can be used to design different forms of matrix factorization. Furthermore, one can change the objective function to be optimized. For example, *PLSA* (cf. Sect. 13.4 of Chap. 13) interprets the non-negative elements of the (scaled) matrix as probabilities and maximizes the likelihood estimate of a generative model with respect to the observed matrix elements. The different variations of matrix factorization provide different types of utility in various applications:

1. The latent factors in *NMF* are more easily interpretable for clustering applications, because of non-negativity. For example, in application domains such as text clustering, each of the k columns in U and V can be associated with document clusters and word clusters, respectively. The magnitudes of the non-negative (transformed) coordinates reflect which concepts are strongly expressed in a document. This "additive parts" representation of *NMF* is highly interpretable, especially in domains such as text, in which the features have semantic meaning. This is not possible with *SVD* in which transformed coordinate values and basis vector components may be negative. This is also the reason that *NMF* transformations are more useful than those of *SVD* for clustering. Similarly, the probabilistic forms of non-negative matrix factorization, such as *PLSA*, are also used commonly for clustering. It is instructive to compare the example of Fig. 6.22, with the *SVD* of the same matrix at the end of Sect. 2.4.3.2 in Chap. 2. Note that the *NMF* factorization is more easily interpretable.

2. Unlike *SVD*, the k latent factors of *NMF* are not orthogonal to one another. This is a disadvantage of *NMF* because orthogonality of the axis-system allows intuitive interpretations of the data transformation as an axis-rotation. It is easy to project *out-of-sample* data points (i.e., data points not included in D) on an orthonormal basis system. Furthermore, distance computations between transformed data points are more meaningful in *SVD*.

3. The addition of a constraint, such as non-negativity, to any optimization problem usually reduces the quality of the solution found. However, the addition of orthogonality constraints, as in *SVD*, do not affect the *theoretical* global optimum of the *unconstrained* matrix factorization formulation (see Exercise 13). Therefore, *SVD* provides better rank-k approximations than *NMF*. Furthermore, it is much easier *in practice* to determine the global optimum of *SVD*, as compared to unconstrained matrix factorization for matrices that are completely specified. Thus, *SVD* provides one of the alternate global optima of unconstrained matrix factorization, which is computationally easy to determine.

4. *SVD* is generally hard to implement for incomplete data matrices as compared to many other variations of matrix factorization. This is relevant in recommender systems where rating matrices are incomplete. The use of latent factor models for recommendations is discussed in Sect. 18.5.5 of Chap. 18.

Thus, *SVD* and *NMF* have different advantages and disadvantages and may be more suitable for different applications.

6.9 Cluster Validation

After a clustering of the data has been determined, it is important to evaluate its quality. This problem is referred to as *cluster validation*. Cluster validation is often difficult in real data sets because the problem is defined in an unsupervised way. Therefore, no external validation criteria may be available to evaluate a clustering. Thus, a number of *internal* criteria may be defined to validate the quality of a clustering. The major problem with internal criteria is that they may be biased toward one algorithm or the other, depending on how they are defined. In some cases, external validation criteria may be available when a test data set is synthetically generated, and therefore the true (ground-truth) clusters are known. Alternatively, for real data sets, the class labels, if available, may be used as proxies for the cluster identifiers. In such cases, the evaluation is more effective. Such criteria are referred to as *external validation criteria*.

6.9.1 Internal Validation Criteria

Internal validation criteria are used when no external criteria are available to evaluate the quality of a clustering. In most cases, the criteria used to validate the quality of the algorithm are borrowed directly from the objective function, which is optimized by a particular clustering model. For example, virtually any of the objective functions in the k-representatives, EM algorithms, and agglomerative methods could be used for validation purposes. The problem with the use of these criteria is obvious in comparing algorithms with disparate methodologies. A validation criterion will always favor a clustering algorithm that uses a similar kind of objective function for its optimization. Nevertheless, in the absence of external validation criteria, this is the best that one can hope to achieve. Such criteria can also be effective in comparing two algorithms using the same broad approach. The commonly used internal evaluation criteria are as follows:

1. *Sum of square distances to centroids:* In this case, the centroids of the different clusters are determined, and the sum of squared (SSQ) distances are reported as the corresponding objective function. Smaller values of this measure are indicative of better cluster quality. This measure is obviously more optimized to distance-based algorithms, such as k-means, as opposed to a density-based method, such as *DBSCAN*. Another problem with SSQ is that the absolute distances provide no meaningful information to the user about the quality of the underlying clusters.

2. *Intracluster to intercluster distance ratio:* This measure is more detailed than the SSQ measure. The idea is to sample r pairs of data points from the underlying data. Of these, let P be the set of pairs that belong to the same cluster found by the algorithm. The remaining pairs are denoted by set Q. The average intercluster distance and intracluster distance are defined as follows:

$$Intra = \sum_{(\overline{X_i}, \overline{X_j}) \in P} dist(\overline{X_i}, \overline{X_j})/|P| \tag{6.43}$$

$$Inter = \sum_{(\overline{X_i}, \overline{X_j}) \in Q} dist(\overline{X_i}, \overline{X_j})/|Q|. \tag{6.44}$$

   Then the ratio of the average intracluster distance to the intercluster distance is given by $Intra/Inter$. Small values of this measure indicate better clustering behavior.

3. *Silhouette coefficient:* Let $Davg_i^{in}$ be the average distance of $\overline{X_i}$ to data points *within* the cluster of $\overline{X_i}$. The average distance of data point $\overline{X_i}$ to the points in each cluster (other than its own) is also computed. Let $Dmin_i^{out}$ represent the minimum of these

(average) distances, over the other clusters. Then, the silhouette coefficient S_i *specific to the ith object*, is as follows:

$$S_i = \frac{Dmin_i^{out} - Davg_i^{in}}{\max\{Dmin_i^{out}, Davg_i^{in}\}}. \tag{6.45}$$

The overall silhouette coefficient is the average of the data point-specific coefficients. The silhouette coefficient will be drawn from the range $(-1, 1)$. Large positive values indicate highly separated clustering, and negative values are indicative of some level of "mixing" of data points from different clusters. This is because $Dmin_i^{out}$ will be less than $Davg_i^{in}$ only in cases where data point $\overline{X_i}$ is closer to at least one other cluster than its own cluster. One advantage of this coefficient is that the absolute values provide a good intuitive feel of the quality of the clustering.

4. *Probabilistic measure:* In this case, the goal is to use a mixture model to estimate the quality of a particular clustering. The centroid of each mixture component is assumed to be the centroid of each discovered cluster, and the other parameters of each component (such as the covariance matrix) are computed from the discovered clustering using a method similar to the M-step of EM algorithms. The overall log-likelihood of the measure is reported. Such a measure is useful when it is known from domain-specific knowledge that the clusters *ought* to have a specific shape, as is suggested by the distribution of each component in the mixture.

The major problem with internal measures is that they are heavily biased toward particular clustering algorithms. For example, a distance-based measure, such as the silhouette coefficient, will not work well for clusters of arbitrary shape. Consider the case of the clustering in Fig. 6.11. In this case, some of the *point-specific* coefficients might have a negative value for the correct clustering. Even the overall silhouette coefficient for the correct clustering might not be as high as an incorrect k-means clustering, which mixes points from different clusters. This is because the clusters in Fig. 6.11 are of arbitrary shape that do not conform to the quality metrics of distance-based measures. On the other hand, if a density-based criterion were designed, it would also be biased toward density-based algorithms. The major problem in relative comparison of different methodologies with internal criteria is that all criteria attempt to define a "prototype" model for goodness. The quality measure very often only tells us *how well the prototype validation model matches the model used for discovering clusters*, rather than anything intrinsic about the underlying clustering. This can be viewed as a form of *overfitting*, which significantly affects such evaluations. At the very least, this phenomenon creates uncertainty about the reliability of the evaluation, which defeats the purpose of evaluation in the first place. This problem is fundamental to the unsupervised nature of data clustering, and there are no completely satisfactory solutions to this issue.

Internal validation measures do have utility in some practical scenarios. For example, they can be used to compare clusterings by a similar class of algorithms, or different runs of the same algorithm. Finally, these measures are also sensitive to the number of clusters found by the algorithm. For example, two different clusterings cannot be compared on a particular criterion when the number of clusters determined by different algorithms is different. A fine-grained clustering will typically be associated with superior values of many internal qualitative measures. Therefore, these measures should be used with great caution, because of their tendency to favor specific algorithms, or different settings of the same algorithm. Keep in mind that clustering is an *unsupervised* problem, which, by definition, implies that there is no well-defined notion of a "correct" model of clustering in the absence of external criteria.

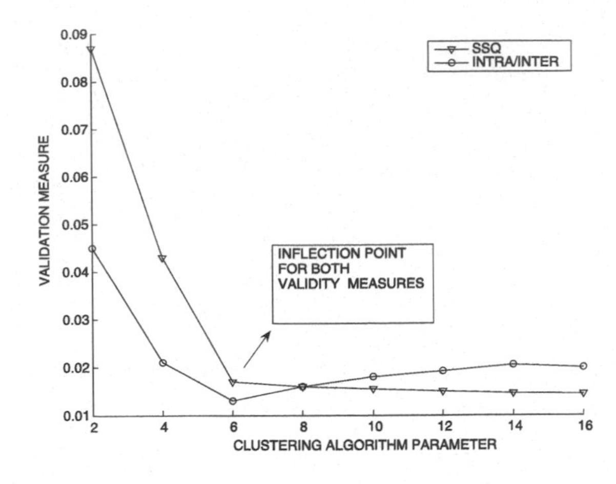

Figure 6.24: Inflection points in validity measures for parameter tuning

6.9.1.1 Parameter Tuning with Internal Measures

All clustering algorithms use a number of parameters as input, such as the number of clusters or the density. Although internal measures are inherently flawed, a limited amount of parameter tuning can be performed with these measures. The idea here is that the variation in the validity measure may show an inflection point (or "elbow") at the correct choice of parameter. Of course, because these measures are flawed to begin with, such techniques should be used with great caution. Furthermore, the shape of the inflection point may vary significantly with the nature of the parameter being tuned, and the validation measure being used. Consider the case of k-means clustering where the parameter being tuned is the number of clusters k. In such a case, the SSQ measure will always reduce with the number of clusters, though it will reduce at a sharply lower rate after the inflection point. On the other hand, for a measure such as the ratio of the intra-cluster to inter-cluster distance, the measure will reduce until the inflection point and then may increase slightly. An example of these two kinds of inflections are illustrated in Fig. 6.24. The X-axis indicates the parameter being tuned (number of clusters), and the Y-axis illustrates the (relative) values of the validation measures. In many cases, if the validation model does not reflect either the natural shape of the clusters in the data, or the algorithmic model used to create the clusters very well, such inflection points may either be misleading, or not even be observed. However, plots such as those illustrated in Fig. 6.24 can be used in conjunction with visual inspection of the scatter plot of the data and the algorithm partitioning to determine the correct number of clusters in many cases. Such tuning techniques with internal measures should be used as an informal rule of thumb, rather than as a strict criterion.

6.9.2 External Validation Criteria

Such criteria are used when ground truth is available about the true clusters in the underlying data. In general, this is not possible in most real data sets. However, when synthetic data is generated from known benchmarks, it is possible to associate cluster identifiers with the generated records. In the context of real data sets, these goals can be *approximately* achieved with the use of class labels when they are available. The major risk with the use of class labels is that these labels are based on application-specific properties of that data set and may not reflect the natural clusters in the underlying data. Nevertheless, such criteria

Cluster Indices	1	2	3	4
1	97	0	2	1
2	5	191	1	3
3	4	3	87	6
4	0	0	5	195

Figure 6.25: Confusion matrix for a clustering of good quality

Cluster Indices	1	2	3	4
1	33	30	17	20
2	51	101	24	24
3	24	23	31	22
4	46	40	44	70

Figure 6.26: Confusion matrix for a clustering of poor quality

are still preferable to internal methods because they can usually avoid *consistent* bias in evaluations, when used over multiple data sets. In the following discussion, the term "class labels" will be used interchangeably to refer to either cluster identifiers in a synthetic data set or class labels in a real data set.

One of the problems is that the number of natural clusters in the data may not reflect the number of class labels (or cluster identifiers). The number of class labels is denoted by k_t, which represents the true or ground-truth number of clusters. The number of clusters determined by the algorithm is denoted by k_d. In some settings, the number of true clusters k_t is equal to the number of algorithm-determined clusters k_d, though this is often not the case. In cases where $k_d = k_t$, it is particularly helpful to create a *confusion matrix*, which relates the mapping of the true clusters to those determined by the algorithm. Each row i corresponds to the class label (ground-truth cluster) i, and each column j corresponds to the points in *algorithm-determined* cluster j. Therefore, the (i, j)th entry of this matrix is equal to the number of data points in the true cluster i, which are mapped to the algorithm-determined cluster j. The sum of the values across a particular row i will always be the same across different clustering algorithms because it reflects the size of ground-truth cluster i in the data set.

When the clustering is of high quality, it is usually possible to permute the rows and columns of this confusion matrix, so that only the diagonal entries are large. On the other hand, when the clustering is of poor quality, the entries across the matrix will be more evenly distributed. Two examples of confusion matrices are illustrated in Figs. 6.25 and 6.26, respectively. The first clustering is obviously of much better quality than the second.

The confusion matrix provides an intuitive method to visually assess the clustering. However, for larger confusion matrices, this may not be a practical solution. Furthermore, while confusion matrices can also be created for cases where $k_d \neq k_t$, it is much harder to assess the quality of a particular clustering by visual inspection. Therefore, it is important to design hard measures to evaluate the overall quality of the confusion matrix. Two commonly used measures are the *cluster purity*, and *class-based Gini index*. Let m_{ij} represent the number of data points from class (ground-truth cluster) i that are mapped to (algorithm-determined) cluster j. Here, i is drawn from the range $[1, k_t]$, and j is drawn from the range $[1, k_d]$. Also assume that the number of data points in true cluster i are denoted by N_i, and the number of data points in algorithm-determined cluster j are denoted by M_j. Therefore, the number of data points in different clusters can be related as follows:

$$N_i = \sum_{j=1}^{k_d} m_{ij} \qquad \forall i = 1 \ldots k_t \tag{6.46}$$

$$M_j = \sum_{i=1}^{k_t} m_{ij} \qquad \forall j = 1 \ldots k_d \tag{6.47}$$

A high-quality algorithm-determined cluster j should contain data points that are largely dominated by a single class. Therefore, for a given algorithm-determined cluster j, the number of data points P_j in its *dominant class* is equal to the maximum of the values of m_{ij} over different values of ground truth cluster i:

$$P_j = \max_i m_{ij}. \tag{6.48}$$

A high-quality clustering will result in values of $P_j \leq M_j$, which are very close to M_j. Then, the overall purity is given by the following:

$$\text{Purity} = \frac{\sum_{j=1}^{k_d} P_j}{\sum_{j=1}^{k_d} M_j}. \tag{6.49}$$

High values of the purity are desirable. The cluster purity can be computed in two different ways. The method discussed above computes the purity of each algorithm-determined cluster (with respect to ground-truth clusters), and then computes the aggregate purity on this basis. The second way can compute the purity of each ground-truth cluster with respect to the algorithm-determined clusters. The two methods will not lead to the same results, especially when the values of k_d and k_t are significantly different. The mean of the two values may also be used as a single measure in such cases. The first of these measures, according to Eq. 6.49, is the easiest to intuitively interpret, and it is therefore the most popular.

One of the major problems with the purity-based measure is that it only accounts for the dominant label in the cluster and ignores the distribution of the remaining points. For example, a cluster that contains data points predominantly drawn from two classes, is better than one in which the data points belong to many different classes, even if the cluster purity is the same. To account for the variation across the different classes, the Gini index may be used. This measure is closely related to the notion of entropy, and it measures the level of *inequality* (or confusion) in the distribution of the entries in a row (or column) of the confusion matrix. As in the case of the purity measure, it can be computed with a row-wise method or a column-wise method, and it will evaluate to different values. Here the column-wise method is described. The Gini index G_j for column (algorithm-determined cluster) j is defined as follows:

$$G_j = 1 - \sum_{i=1}^{k_t} \left(\frac{m_{ij}}{M_j} \right)^2. \tag{6.50}$$

The value of G_j will be close to 0 when the entries in a column of a confusion matrix are skewed, as in the case of Fig. 6.25. When the entries are evenly distributed, the value will be close to $1 - 1/k_t$, which is also the upper bound on this value. The average Gini coefficient is the weighted average of these different column-wise values where the weight of G_j is M_j:

$$G_{average} = \frac{\sum_{j=1}^{k_d} G_j \cdot M_j}{\sum_{j=1}^{k_d} M_j}. \tag{6.51}$$

Low values of the Gini index are desirable. The notion of the Gini index is closely related to the notion of entropy E_j (of algorithm-determined cluster j), which measures the same intuitive characteristics of the data:

$$E_j = - \sum_{i=1}^{k_t} \left(\frac{m_{ij}}{M_j} \right) \cdot \log \left(\frac{m_{ij}}{M_j} \right). \tag{6.52}$$

Lower values of the entropy are indicative of a higher quality clustering. The overall entropy is computed in a similar way to the Gini index, with the use of cluster specific entropies.

$$E_{average} = \frac{\sum_{j=1}^{k_d} E_j \cdot M_j}{\sum_{j=1}^{k_d} M_j}. \quad (6.53)$$

Finally, a pairwise precision and pairwise recall measure can be used to evaluate the quality of a clustering. To compute this measure, all pairs of data points within the same algorithm-determined cluster are generated. The fraction of pairs which belong to the same ground-truth clusters is the precision. To determine the recall, pairs of points within the same ground-truth clusters are sampled, and the fraction that appear in the same algorithm-determined cluster are computed. A unified measure is the *Fowlkes-Mallows* measure, which reports the geometric mean of the precision and recall.

6.9.3 General Comments

Although cluster validation is a widely studied problem in the clustering literature, most methods for cluster validation are rather imperfect. Internal measures are imperfect because they are typically biased toward one algorithm or the other. External measures are imperfect because they work with class labels that may not reflect the true clusters in the data. Even when synthetic data is generated, the method of generation will implicitly favor one algorithm or the other. These challenges arise because clustering is an *unsupervised* problem, and it is notoriously difficult to validate the quality of such algorithms. Often, the only true measure of clustering quality is its ability to meet the goals of a specific application.

6.10 Summary

A wide variety of algorithms have been designed for the problem of data clustering, such as representative-based methods, hierarchical methods, probabilistic methods, density-based methods, graph-based methods, and matrix factorization-based methods. All methods typically require the algorithm to specify some parameters, such as the number of clusters, the density, or the rank of the matrix factorization. Representative-based methods, and probabilistic methods restrict the shape of the clusters but adjust better to varying cluster density. On the other hand, agglomerative and density-based methods adjust better to the shape of the clusters but do not adjust to varying density of the clusters. Graph-based methods provide the best adjustment to varying shape and density but are typically more expensive to implement. The problem of cluster validation is a notoriously difficult one for unsupervised problems, such as clustering. Although external and internal validation criteria are available for the clustering, they are often biased toward different algorithms, or may not accurately reflect the internal clusters in the underlying data. Such measures should be used with caution.

6.11 Bibliographic Notes

The problem of clustering has been widely studied in the data mining and machine learning literature. The classical books [74, 284, 303] discuss most of the traditional clustering methods. These books present many of the classical algorithms, such as the partitioning and hierarchical algorithms, in great detail. Another book [219] discusses more recent methods for data clustering. An excellent survey on data clustering may be found in [285]. The most recent book [32] in the literature provides a very comprehensive overview of the different data clustering algorithms. A detailed discussion on feature selection methods is provided in [366]. The distance-based entropy measure is discussed in [169]. Various validity measures derived from spectral clustering and the cluster scatter matrix can be used for feature selection [262, 350, 550]. The second chapter in the clustering book [32] provides a detailed review of feature selection methods.

A classical survey [285] provides an excellent review of k-means algorithms. The problem of refining the initial data points for k-means type algorithms is discussed in [108]. The problem of discovering the correct number of clusters in a k-means algorithm is addressed in [423]. Other notable criteria for representative algorithms include the use of Bregman divergences [79].

The three main density-based algorithms presented in this chapter are *STING* [506], *DBSCAN* [197], and *DENCLUE* [267]. The faster update rule for *DENCLUE* appears in [269]. The faster update rule was independently discovered earlier in [148, 159] as mean-shift clustering. Among the grid-based algorithms, the most common ones include *WaveCluster* [464] and *MAFIA* [231]. The incremental version of *DBSCAN* is addressed in [198]. The *OPTICS* algorithm [76] performs density-based clustering based on ordering of the data points. It is also useful for hierarchical clustering and visualization. Another variation of the *DBSCAN* algorithm is the *GDBSCAN* method [444] that can work with more general kinds of data.

One of the most well-known graph-based algorithms is the *Chameleon* algorithm [300]. Shared nearest neighbor algorithms [195], are inherently graph-based algorithms, and adjust well to the varying density in different data localities. A well-known top-down hierarchical multilevel clustering algorithm is the *METIS* algorithm [301]. An excellent survey on spectral clustering methods may be found in [371]. Matrix factorization and its variations [288, 440, 456] are closely related to spectral clustering [185]. Methods for community detection in graphs are discussed in [212]. Any of these methods can be used for the last phase of graph-based clustering algorithms. Cluster validity methods are discussed in [247, 248]. In addition, the problem of cluster validity is studied in detail in [32].

6.12 Exercises

1. Consider the 1-dimensional data set with 10 data points $\{1, 2, 3, \ldots 10\}$. Show three iterations of the k-means algorithms when $k = 2$, and the random seeds are initialized to $\{1, 2\}$.

2. Repeat Exercise 1 with an initial seed set of $\{2, 9\}$. How did the different choice of the seed set affect the quality of the results?

3. Write a computer program to implement the k-representative algorithm. Use a modular program structure, in which the distance function and centroid determination are separate subroutines. Instantiate these subroutines to the cases of (i) the k-means algorithm, and (ii) the k-medians algorithm.

4. Implement the Mahalanobis k-means algorithm.

5. Consider the 1-dimensional data set $\{1 \ldots 10\}$. Apply a hierarchical agglomerative approach, with the use of minimum, maximum, and group average criteria for merging. Show the first six merges.

6. Write a computer program to implement a hierarchical merging algorithm with the single-linkage merging criterion.

7. Write a computer program to implement the EM algorithm, in which there are two spherical Gaussian clusters with the same radius. Download the *Ionosphere* data set from the *UCI Machine Learning Repository* [213]. Apply the algorithm to the data set (with randomly chosen centers), and record the centroid of the Gaussian in each iteration. Now apply the k-means algorithm implemented in Exercise 3, with the same set of initial seeds as Gaussian centroids. How do the centroids in the two algorithms compare over the different iterations?

8. Implement the computer program of Exercise 7 with a general Gaussian distribution, rather than a spherical Gaussian.

9. Consider a 1-dimensional data set with three natural clusters. The first cluster contains the consecutive integers $\{1 \dots 5\}$. The second cluster contains the consecutive integers $\{8 \dots 12\}$. The third cluster contains the data points $\{24, 28, 32, 36, 40\}$. Apply a k-means algorithm with initial centers of 1, 11, and 28. Does the algorithm determine the correct clusters?

10. If the initial centers are changed to 1, 2, and 3, does the algorithm discover the correct clusters? What does this tell you?

11. Use the data set of Exercise 9 to show how hierarchical algorithms are sensitive to local density variations.

12. Use the data set of Exercise 9 to show how grid-based algorithms are sensitive to local density variations.

13. It is a fundamental fact of linear algebra that any rank-k matrix has a singular value decomposition in which exactly k singular values are non-zero. Use this result to show that the lowest error of rank-k approximation in *SVD* is the same as that of unconstrained matrix factorization in which basis vectors are not constrained to be orthogonal. Assume that the Frobenius norm of the error matrix is used in both cases to compute the approximation error.

14. Suppose that you constructed a k-nearest neighbor similarity graph from a data set with weights on edges. Describe the bottom-up single-linkage algorithm in terms of the similarity graph.

15. Suppose that a shared nearest neighbor similarity function (see Chap. 3) is used in conjunction with the k-medoids algorithm to discover k clusters from n data points. The number of nearest neighbors used to define shared nearest neighbor similarity is m. Describe how a reasonable value of m may be selected in terms of k and n, so as to not result in poor algorithm performance.

16. Suppose that matrix factorization is used to approximately represent a data matrix D as $D \approx D' = UV^T$. Show that one or more of the rows/columns of U and V can be multiplied with constant factors, so as represent $D' = UV^T$ in an infinite number of different ways. What would be a reasonable choice of U and V among these solutions?

17. Explain how each of the internal validity criteria is biased toward one of the algorithms.

18. Suppose that you generate a synthetic data set containing arbitrarily oriented Gaussian clusters. How well does the SSQ criterion reflect the quality of the clusters?

19. Which algorithms will perform best for the method of synthetic data generation in Exercise 18?

Chapter 7

Cluster Analysis: Advanced Concepts

"The crowd is just as important as the group. It takes everything to make it work."—Levon Helm

7.1 Introduction

In the previous chapter, the basic data clustering methods were introduced. In this chapter, several advanced clustering scenarios will be studied, such as the impact of the size, dimensionality, or type of the underlying data. In addition, it is possible to obtain significant insights with the use of advanced supervision methods, or with the use of ensemble-based algorithms. In particular, two important aspects of clustering algorithms will be addressed:

1. *Difficult clustering scenarios:* Many data clustering scenarios are more challenging. These include the clustering of categorical data, high-dimensional data, and massive data. Discrete data are difficult to cluster because of the challenges in distance computation, and in appropriately defining a "central" cluster representative from a set of categorical data points. In the high-dimensional case, many irrelevant dimensions may cause challenges for the clustering process. Finally, massive data sets are more difficult for clustering due to scalability issues.

2. *Advanced insights:* Because the clustering problem is an unsupervised one, it is often difficult to evaluate the quality of the underlying clusters in a meaningful way. This weakness of cluster validity methods was discussed in the previous chapter. Many alternative clusterings may exist, and it may be difficult to evaluate their relative quality. There are many ways of improving application-specific relevance and robustness by using external supervision, human supervision, or meta-algorithms such as ensemble clustering that combine multiple clusterings of the data.

The difficult clustering scenarios are typically caused by particular aspects of the data that make the analysis more challenging. These aspects are as follows:

C. C. Aggarwal, *Data Mining: The Textbook*, DOI 10.1007/978-3-319-14142-8_7

1. *Categorical data clustering:* Categorical data sets are more challenging for clustering because the notion of similarity is harder to define in such scenarios. Furthermore, many intermediate steps in clustering algorithms, such as the determination of the mean of a cluster, are not quite as naturally defined for categorical data as for numeric data.

2. *Scalable clustering:* Many clustering algorithms require multiple passes over the data. This can create a challenge when the data are very large and resides on disk.

3. *High-dimensional clustering:* As discussed in Sect. 3.2.1.2 of Chap. 3, the computation of similarity between high-dimensional data points often does not reflect the intrinsic distance because of many irrelevant attributes and concentration effects. Therefore, many methods have been designed that use projections to determine the clusters in relevant subsets of dimensions.

Because clustering is an unsupervised problem, the quality of the clusters may be difficult to evaluate in many real scenarios. Furthermore, when the data are noisy, the quality may also be poor. Therefore, a variety of methods are used to either supervise the clustering, or gain advanced insights from the clustering process. These methods are as follows:

1. *Semisupervised clustering:* In some cases, partial information may be available about the underlying clusters. This information may be available in the form of labels or other external feedback. Such information can be used to greatly improve the clustering quality.

2. *Interactive and visual clustering:* In these cases, feedback from the user may be utilized to improve the quality of the clustering. In the case of clustering, this feedback is typically achieved with the help of visual interaction. For example, an interactive approach may explore the data in different subspace projections and isolate the most relevant clusters.

3. *Ensemble clustering:* As discussed in the previous chapter, the different models for clustering may produce clusters that are very different from one another. Which of these clusterings is the best solution? Often, there is no single answer to this question. Rather the knowledge from multiple models may be combined to gain a more unified insight from the clustering process. Ensemble clustering can be viewed as a meta-algorithm, which is used to gain more significant insights from multiple models.

This chapter is organized as follows: Section 7.2 discusses algorithms for clustering categorical data. Scalable clustering algorithms are discussed in Sect. 7.3. High-dimensional algorithms are addressed in Sect. 7.4. Semisupervised clustering algorithms are discussed in Sect. 7.5. Interactive and visual clustering algorithms are discussed in Sect. 7.6. Ensemble clustering methods are presented in Sect. 7.7. Section 7.8 discusses the different applications of data clustering. Section 7.9 provides a summary.

7.2 Clustering Categorical Data

The problem of categorical (or discrete) data clustering is challenging because most of the primitive operations in data clustering, such as distance computation, representative determination, and density estimation, are naturally designed for numeric data. A salient observation is that categorical data can always be converted to binary data with the use of

Table 7.1: Example of a 2-dimensional categorical data cluster

Data	(Color, Shape)
1	(Blue, Square)
2	(Red, Circle)
3	(Green, Cube)
4	(Blue, Cube)
5	(Green, Square)
6	(Red, Circle)
7	(Blue, Square)
8	(Green, Cube)
9	(Blue, Circle)
10	(Green, Cube)

Table 7.2: Mean histogram and modes for categorical data cluster

Attribute	Histogram	Mode
Color	Blue= 0.4 Green = 0.4 Red = 0.2	Blue *or* Green
Shape	Cube = 0.4 Square = 0.3 Circle = 0.3	Cube

the binarization process discussed in Chap. 2. It is often easier to work with binary data because it is also a special case of numeric data. However, in such cases, the algorithms need to be tailored to binary data.

This chapter will discuss a wide variety of algorithms for clustering categorical data. The specific challenges associated with applying the various classical methods to categorical data will be addressed in detail along with the required modifications.

7.2.1 Representative-Based Algorithms

The centroid-based representative algorithms, such as k-means, require the repeated determination of centroids of clusters, and the determination of similarity between the centroids and the original data points. As discussed in Sect. 6.3 of the previous chapter, these algorithms iteratively determine the centroids of clusters, and then assign data points to their closest centroid. At a higher level, these steps remain the same for categorical data. However, the specifics of both steps are affected by the categorical data representation as follows:

1. *Centroid of a categorical data set:* All representative-based algorithms require the determination of a central representative of a set of objects. In the case of numerical data, this is achieved very naturally by averaging. However, for categorical data, the equivalent centroid is a probability histogram of values on *each attribute.* For each attribute i, and possible value v_j, the histogram value p_{ij} represents the fraction of the number of objects in the cluster for which attribute i takes on value v_j. Therefore, for a d-dimensional data set, the centroid of a cluster of points is a set of d different histograms, representing the probability distribution of categorical values of each attribute in the cluster. If n_i is the number of distinct values of attribute i, then such an approach will require $O(n_i)$ space to represent the centroid of the ith attribute. A cluster of 2-dimensional data points with attributes *Color* and *Shape* is illustrated in Table 7.1. The corresponding histograms for the *Color* and *Shape* attributes are illustrated in Table 7.2. Note that the probability values over a particular attribute always sum to one unit.

2. *Calculating similarity to centroids:* A variety of similarity functions between a pair of categorical records are introduced in Sect. 3.2.2 of Chap. 3. The simplest of these is match-based similarity. However, in this case, the goal is to determine the similarity

between a probability histogram (corresponding to a representative) and a categorical attribute value. If the attribute i takes on the value v_j for a particular data record, then the analogous match-based similarity is its histogram-based probability p_{ij}. These probabilities are summed up over the different attributes to determine the total similarity. Each data record is assigned to the centroid with the greatest similarity.

The other steps of the k-means algorithm remain the same as for the case of numeric data. The effectiveness of a k-means algorithm is highly dependent on the distribution of the attribute values in the underlying data. For example, if the attribute values are highly skewed, as in the case of market basket data, the histogram-based variation of the match-based measure may perform poorly. This is because this measure treats all attribute values evenly, however, rare attribute values should be treated with greater importance in such cases. This can be achieved by a prekshiprocessing phase that assigns a weight to each categorical attribute *value*, which is the inverse of its global frequency. Therefore, the categorical data records now have weights associated with each attribute. The presence of these weights will affect both probability histogram generation and match-based similarity computation.

7.2.1.1 k-Modes Clustering

In k-modes clustering, each attribute value for a representative is chosen as the mode of the categorical values for that attribute in the cluster. The *mode* of a set of categorical values is the value with the maximum frequency in the set. The modes of each attribute for the cluster of ten points in Table 7.1 are illustrated in Table 7.2. Intuitively, this corresponds to the categorical value v_j for each attribute i for which the frequency histogram has the largest value of p_{ij}. The mode of an attribute may not be unique if two categorical values have the same frequency. In the case of Table 7.2, two possible values of the mode are $(Blue, Cube)$, and $(Green, Cube)$. Any of these could be used as the representative, if a random tie-breaking criterion is used. The mode-based representative may not be drawn from the original data set because the mode of each attribute is determined independently. Therefore, the particular combination of d-dimensional modes obtained for the representative may not belong to the original data. One advantage of the mode-based approach is that the representative is also a categorical data record, rather than a histogram. Therefore, it is easier to use a richer set of similarity functions for computing the distances between data points and their modes. For example, the inverse occurrence frequency-based similarity function, described in Chap. 3, may be used to normalize for the skew in the attribute values. On the other hand, when the attribute values in a categorical data set are naturally skewed, as in market basket data, the use of modes may not be informative. For example, for a market basket data set, all item attributes for the representative point may be set to the value of 0 because of the natural sparsity of the data set. Nevertheless, for cases where the attribute values are more evenly distributed, the k-modes approach can be used effectively. One way of making the k-modes algorithm work well in cases where the attribute values are distributed unevenly, is by dividing the cluster-specific frequency of an attribute by its (global) occurrence frequency to determine a *normalized* frequency. This essentially corrects for the differential global distribution of different attribute values. The modes of this normalized frequency are used. The most commonly used similarity function is the match-based similarity metric, discussed in Sect. 3.2.2 of Chap. 3. However, for biased categorical data distributions, the inverse occurrence frequency should be used for normalizing the similarity function, as discussed in Chap. 3. This can be achieved indirectly by weighting each attribute of each data point with the inverse occurrence frequency of the corresponding attribute *value*. With normalized modes and weights associated with each attribute of each data point, the straightforward match-based similarity computation will provide effective results.

7.2.1.2 k-Medoids Clustering

The medoid-based clustering algorithms are easier to generalize to categorical data sets because the representative data point is chosen from the input database. The broad description of the medoids approach remains the same as that described in Sect. 6.3.4 of the previous chapter. The only difference is in terms of how the similarity is computed between a pair of categorical data points, as compared to numeric data. Any of the similarity functions discussed in Sect. 3.2.2 of Chap. 3 can be used for this purpose. As in the case of k-modes clustering, because the representative is also a categorical data point (as opposed to a histogram), it is easier to directly use the categorical similarity functions of Chap. 3. These include the use of inverse occurrence frequency-based similarity functions that normalize for the skew across different attribute values.

7.2.2 Hierarchical Algorithms

Hierarchical algorithms are discussed in Sect. 6.4 of Chap. 6. Agglomerative bottom-up algorithms have been used successfully for categorical data. The approach in Sect. 6.4 has been described in a general way with a distance matrix of values. As long as a distance (or similarity) matrix can be defined for the case of categorical attributes, most of the algorithms discussed in the previous chapter can be easily applied to this case. An interesting hierarchical algorithm that works well for categorical data is *ROCK*.

7.2.2.1 ROCK

The *ROCK* (ROBust Clustering using linKs) algorithm is based on an agglomerative bottom-up approach in which the clusters are merged on the basis of a similarity criterion. The *ROCK* algorithm uses a criterion that is based on the shared nearest-neighbor metric. Because agglomerative methods are somewhat expensive, the *ROCK* method applies the approach to only a sample of data points to discover prototype clusters. The remaining data points are assigned to one of these prototype clusters in a final pass.

The first step of the *ROCK* algorithm is to convert the categorical data to a binary representation using the binarization approach introduced in Chap. 2. For each value v_j of categorical attribute i, a new pseudo-item is created, which has a value of 1, only if attribute i takes on the value v_j. Therefore, if the ith attribute in a d-dimensional categorical data set has n_i different values, such an approach will create a binary data set with $\sum_{i=1}^{d} n_i$ binary attributes. When the value of each n_i is high, this binary data set will be sparse, and it will resemble a market basket data set. Thus, each data record can be treated as a binary transaction, or a set of items. The similarity between the two transactions is computed with the use of the Jaccard coefficient between the corresponding sets:

$$Sim(T_i, T_j) = \frac{|T_i \cap T_j|}{|T_i \cup T_j|}. \tag{7.1}$$

Subsequently, two data points T_i and T_j are defined to be *neighbors*, if the similarity $Sim(T_i, T_j)$ between them is greater than a threshold θ. Thus, the concept of neighbors implicitly defines a graph structure on the data items, where the nodes correspond to the data items, and the links correspond to the neighborhood relations. The notation $Link(T_i, T_j)$ denotes a shared nearest-neighbor similarity function, which is equal to the number of shared nearest neighbors between T_i and T_j.

The similarity function $Link(T_i, T_j)$ provides a merging criterion for agglomerative algorithms. The algorithm starts with each data point (from the initially chosen sample) in its own cluster and then hierarchically merges clusters based on a similarity criterion between clusters. Intuitively, two clusters $\mathcal{C}_1$ and $\mathcal{C}_2$ should be merged, if the cumulative number of shared nearest neighbors between objects in $\mathcal{C}_1$ and $\mathcal{C}_2$ is large. Therefore, it is possible to generalize the notion of link-based similarity using clusters as arguments, as opposed to individual data points:

$$GroupLink(\mathcal{C}_i, \mathcal{C}_j) = \sum_{T_u \in \mathcal{C}_i, T_v \in \mathcal{C}_j} Link(T_u, T_v). \tag{7.2}$$

Note that this criterion has a slight resemblance to the group-average linkage criterion discussed in the previous chapter. However, this measure is not yet normalized because the *expected* number of cross-links between larger clusters is greater. Therefore, one must normalize by the expected number of cross-links between a pair of clusters to ensure that the merging of larger clusters is not unreasonably favored. Therefore, the normalized linkage criterion $V(\mathcal{C}_i, \mathcal{C}_j)$ is as follows:

$$V(\mathcal{C}_i, \mathcal{C}_j) = \frac{GroupLink(\mathcal{C}_i, \mathcal{C}_j)}{E[CrossLinks(\mathcal{C}_i, \mathcal{C}_j)]}. \tag{7.3}$$

The expected number of cross-links between $\mathcal{C}_i$ and $\mathcal{C}_j$ can be computed as function of the expected number of intracluster links $Intra(\cdot)$ in individual clusters as follows:

$$E[CrossLinks(\mathcal{C}_i, \mathcal{C}_j)] = E[Intra(\mathcal{C}_i \cup \mathcal{C}_j)] - E[Intra(\mathcal{C}_i)] - E[Intra(\mathcal{C}_j)]. \tag{7.4}$$

The expected number of intracluster links is specific to a single cluster and is more easily estimated as a function of cluster size q_i and θ. The number of intracluster links in a cluster containing q_i data points is heuristically estimated by the *ROCK* algorithm as $q_i^{1+2 \cdot f(\theta)}$. Here, the function $f(\theta)$ is a property of both the data set, and the kind of clusters that one is interested in. The value of $f(\theta)$ is heuristically defined as follows:

$$f(\theta) = \frac{1 - \theta}{1 + \theta}. \tag{7.5}$$

Therefore, by substituting the expected number of cross-links in Eq. 7.3, one obtains the following merging criterion $V(\mathcal{C}_i, \mathcal{C}_j)$:

$$V(\mathcal{C}_i, \mathcal{C}_j) = \frac{GroupLink(\mathcal{C}_i, \mathcal{C}_j)}{(q_i + q_j)^{1+2 \cdot f(\theta)} - q_i^{1+2 \cdot f(\theta)} - q_j^{1+2 \cdot f(\theta)}}. \tag{7.6}$$

The denominator explicitly normalizes for the sizes of the clusters being merged by penalizing larger clusters. The goal of this kind of normalization is to prevent the imbalanced preference toward successively merging only large clusters.

The merges are successively performed until a total of k clusters remain in the data. Because the agglomerative approach is applied only to a sample of the data, it remains to assign the remaining data points to one of the clusters. This can be achieved by assigning each disk-resident data point to the cluster with which it has the greatest similarity. This similarity is computed using the same quality criterion in Eq. 7.6 as was used for cluster–cluster merges. In this case, similarity is computed between clusters and individual data points by treating each data point as a singleton cluster.

7.2.3 Probabilistic Algorithms

The probabilistic approach to data clustering is introduced in Sect. 6.5 of Chap. 6. Generative models can be generalized to virtually any data type as long as an appropriate generating probability distribution can be defined for each mixture component. This provides unprecedented flexibility in adapting probabilistic clustering algorithms to various data types. After the mixture distribution model has been defined, the E- and M-steps need to be defined for the corresponding expectation–maximization (EM) approach. The main difference from numeric clustering is that the soft assignment process in the E-step, and the parameter estimation process in the M-step will depend on the relevant probability distribution model for the corresponding data type.

Let the k components of the mixture be denoted by $\mathcal{G}_1 \dots \mathcal{G}_k$. Then, the generative process for each point in the data set $\mathcal{D}$ uses the following two steps:

1. Select a mixture component with prior probability α_i, where $i \in \{1 \dots k\}$.

2. If the mth component of the mixture was selected in the first step, then generate a data point from $\mathcal{G}_m$.

The values of α_i denote the prior probabilities $P(\mathcal{G}_i)$, which need to be estimated along with other model parameters in a data-driven manner. The main difference from the numerical case is in the mathematical form of the generative model for the mth cluster (or mixture component) $\mathcal{G}_m$, which is now a discrete probability distribution rather than the probability density function used in the numeric case. This difference reflects the corresponding difference in data type. One reasonable choice for the discrete probability distribution of $\mathcal{G}_m$ is to assume that the jth categorical value of ith attribute is independently generated by mixture component (cluster) m with probability p_{ijm}. Consider a data point $\overline{X}$ containing the attribute value indices $j_1 \dots j_d$ for its d dimensions. In other words, the rth attribute takes on the j_rth possible categorical value. For convenience, the entire set of model parameters is denoted by the generic notation Θ. Then, the discrete probability distribution $g^{m,\Theta}(\overline{X})$ from cluster m is given by the following expression:

$$g^{m,\Theta}(\overline{X}) = \prod_{r=1}^{d} p_{rj_r m}. \tag{7.7}$$

The discrete probability distribution is $g^{m,\Theta}(\cdot)$, which is analogous to the continuous density function $f^{m,\Theta}(\cdot)$ of the EM model in the previous chapter. Correspondingly, the *posterior* probability $P(\mathcal{G}_m|\overline{X}, \Theta)$ of the component $\mathcal{G}_m$ having generated *observed* data point $\overline{X}$ may be estimated as follows:

$$P(\mathcal{G}_m|\overline{X_j}, \Theta) = \frac{\alpha_m \cdot g^{m,\Theta}(\overline{X})}{\sum_{r=1}^{k} \alpha_r \cdot g^{r,\Theta}(\overline{X})}. \tag{7.8}$$

This defines the E-step for categorical data, and it provides a soft assignment probability of the data point to a cluster.

After the soft assignment probability has been determined, the M-step applies maximum likelihood estimation to the *individual components* of the mixture to estimate the probability p_{ijm}. While estimating the parameters for cluster m, the *weight* of a record is assumed to be equal to its assignment probability $P(\mathcal{G}_m|\overline{X}, \Theta)$ to cluster m. For each cluster m, the *weighted* number w_{ijm} of data points for which attribute i takes on its jth possible categorical value is estimated. This is equal to the sum of the assignment probabilities (to cluster m) of data points that *do take on the jth value.* By dividing this value with the aggregate assignment probability of *all* data points to cluster m, the probability p_{ijm} may be estimated as follows:

$$p_{ijm} = \frac{w_{ijm}}{\sum_{\overline{X} \in \mathcal{D}} P(\mathcal{G}_m | \overline{X}, \Theta)}. \tag{7.9}$$

The parameter α_m is estimated as the average assignment probabilities of data points to cluster m. The aforementioned formulas for estimation may be derived from maximum likelihood estimation methods. Refer to the bibliographic notes for detailed derivations.

Sometimes, the estimation of Eq. 7.9 can be inaccurate because the available data may be limited, or particular values of categorical attributes may be rare. In such cases, some of the attribute values may not appear in a cluster (or $w_{ijm} \approx 0$). This can lead to poor parameter estimation, or *overfitting.* The *Laplacian smoothing* method is commonly used to address such ill-conditioned probabilities. This is achieved by adding a small positive value β to the estimated values of w_{ijm}, where β is the smoothing parameter. This will generally lead to more robust estimation. This type of smoothing is also applied in the estimation of the prior probabilities α_m when the data sets are very small. This completes the description of the M-step. As in the case of numerical data, the E-step and M-step are iterated to convergence.

7.2.4 Graph-Based Algorithms

Because graph-based methods are *meta-algorithms*, the broad description of these algorithms remains virtually the same for categorical data as for numeric data. Therefore, the approach described in Sect. 6.7 of the previous chapter applies to this case as well. The only difference is in terms of how the edges and values on the similarity graph are constructed. The first step is the determination of the k-nearest neighbors of each data record, and subsequent assignment of similarity values to edges. Any of the similarity functions described in Sect. 3.2.2 of Chap. 3 can be used to compute similarity values along the edges of the graph. These similarity measures could include the inverse occurrence frequency measure discussed in Chap. 3, which corrects for the natural skew in the different attribute values. As discussed in the previous chapter, one of the major advantages of graph-based algorithms is that they can be leveraged for virtually any kind of data type as long as a similarity function can be defined on that data type.

7.3 Scalable Data Clustering

In many applications, the size of the data is very large. Typically, the data cannot be stored in main memory, but it need to reside on disk. This is a significant challenge, because it imposes a constraint on the algorithmic design of clustering algorithms. This section will discuss the *CLARANS*, *BIRCH*, and *CURE* algorithms. These algorithms are all scalable implementations of one of the basic types of clustering algorithms discussed in the previous chapter. For example, the *CLARANS* approach is a scalable implementation of the k-medoids algorithm for clustering. The *BIRCH* algorithm is a top-down hierarchical generalization of the k-means algorithm. The *CURE* algorithm is a bottom-up agglomerative approach to clustering. These different algorithms inherit the advantages and disadvantages of the base classes of algorithms that they are generalized from. For example, while the *CLARANS* algorithm has the advantage of being more easily generalizable to different data types (beyond numeric data), it inherits the relatively high computational complexity of k-medoids methods. The *BIRCH* algorithm is much faster because it is based on the k-means methodology, and its hierarchical clustering structure can be tightly controlled because of its top-down partitioning approach. This can be useful for indexing applications. On the other hand, *BIRCH* is not designed for arbitrary data types or clusters of arbitrary shape. The *CURE* algorithm can determine clusters of arbitrary shape because of its bottom-up hierarchical approach. The choice of the most suitable algorithm depends on the application at hand. This section will provide an overview of these different methods.

7.3.1 CLARANS

The *CLARA* and *CLARANS* methods are two generalizations of the k-medoids approach to clustering. Readers are referred to Sect. 6.3.4 of the previous chapter for a description of the generic k-medoids approach. Recall that the k-medoids approach works with a set of representatives, and iteratively exchanges one of the medoids with a non-medoid in each iteration to improve the clustering quality. The generic k-medoids algorithm allows considerable flexibility in deciding how this exchange might be executed.

The *Clustering LARge Applications (CLARA)* method is based on a particular instantiation of the k-medoids method known as *Partitioning Around Medoids (PAM)*. In this method, to exchange a medoid with another non-medoid representative, all possible $k \cdot (n-k)$ pairs are tried for a possible exchange to improve the clustering objective function. The best improvement of these pairs is selected for an exchange. This exchange is performed until the algorithm converges to a locally optimal value. The exchange process requires $O(k \cdot n^2)$ distance computations. Therefore, each iteration requires $O(k \cdot n^2 \cdot d)$ time for a d-dimensional data set, which can be rather expensive. Because the complexity is largely dependent on the number of data points, it can be reduced by applying the algorithm to a smaller sample. Therefore, the *CLARA* approach applies *PAM* to a smaller sampled data set of size $f \cdot n$ to discover the medoids. The value of f is a sampling fraction, which is much smaller than 1. The remaining nonsampled data points are assigned to the optimal medoids discovered by applying *PAM* to the smaller sample. This overall approach is applied repeatedly over independently chosen samples of data points of the same size $f \cdot n$. The best clustering over these independently chosen samples is selected as the optimal solution. Because the complexity of each iteration is $O(k \cdot f^2 \cdot n^2 \cdot d + k \cdot (n-k))$, the approach may be orders of magnitude faster for small values of the sampling fraction f. The main problem with *CLARA* occurs when each of the preselected samples does not include good choices of medoids.

The *Clustering Large Applications based on RANdomized Search (CLARANS)* approach works with the full data set for the clustering in order to avoid the problem with preselected samples. The approach iteratively attempts exchanges between random medoids with random non-medoids. After a randomly chosen non-medoid is tried for an exchange with a randomly chosen medoid, it is checked if the quality improves. If the quality does improve, then this exchange is made final. Otherwise, the number of unsuccessful exchange attempts is counted. A local optimal solution is said to have been found when a user-specified number of unsuccessful attempts *MaxAttempt* have been reached. This entire process of finding the local optimum is repeated for a user-specified number of iterations, denoted by *MaxLocal*. The clustering objective function of each of these *MaxLocal* locally optimal solutions is evaluated. The best among these local optima is selected as the optimal solution. The advantage of *CLARANS* over *CLARA* is that a greater diversity of the search space is explored.

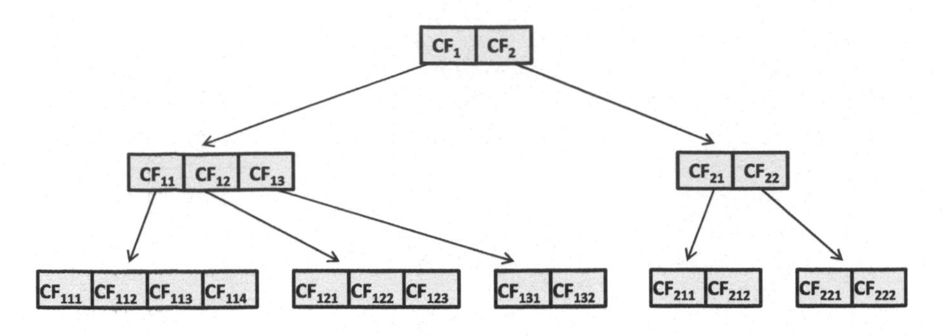

Figure 7.1: The CF-Tree

7.3.2 BIRCH

The *Balanced Iterative Reducing and Clustering using Hierarchies (BIRCH)* approach can be viewed as a combination of top-down hierarchical and k-means clustering. To achieve this goal, the approach introduces a hierarchical data structure, known as the CF-Tree. This is a height-balanced data structure organizing the clusters hierarchically. Each node has a branching factor of at most B, which corresponds to its (at most) B children subclusters. This structure shares a resemblance to the B-Tree data structure commonly used in database indexing. This is by design because the CF-Tree is inherently designed to support dynamic insertions into the hierarchical clustering structure. An example of the CF-Tree is illustrated in Fig. 7.1.

Each node contains a concise summary of each of the at most B subclusters that it points to. This concise summary of a cluster is referred to as its *cluster feature (CF)*, or *cluster feature vector*. The summary contains the triple $(\overline{SS}, \overline{LS}, m)$, where $\overline{SS}$ is a vector[1] containing the sum of the square of the points in the cluster (second-order moment), $\overline{LS}$ is a vector containing the linear sum of the points in the cluster (first-order moment), and m is the number of points in the cluster (zeroth-order moment). Thus, the size of the summary is $(2 \cdot d + 1)$ for a d-dimensional data set and is also referred to as a CF-vector. The cluster feature vector thus contains all moments of order at most 2. This summary has two very important properties:

1. Each cluster feature can be represented as a linear sum of the cluster features of the individual data points. Furthermore, the cluster feature of a parent node in the CF-Tree is the sum of the cluster features of its children. The cluster feature of a merged cluster can also be computed as the sum of the cluster features of the constituent clusters. Therefore, incremental updates of the cluster feature vector can be *efficiently* achieved by adding the cluster feature vector of a data point to that of the cluster.

2. The cluster features can be used to compute useful properties of a cluster, such as its radius and centroid. Note that these are the only two computations required by a centroid-based algorithm, such as k-means or *BIRCH*. These computations are discussed below.

To understand how the cluster feature can be used to measure the radius of a cluster, consider a set of data points denoted by $\overline{X_1} \ldots \overline{X_m}$, where $\overline{X_i} = (x_i^1 \ldots x_i^d)$. The mean and

[1] It is possible to store the sum of the values in $\overline{SS}$ across the d dimensions in lieu of $\overline{SS}$, without affecting the usability of the cluster feature. This would result in a cluster feature of size $(d+2)$ instead of $(2 \cdot d + 1)$.

variance of any set of points can be expressed in terms of the their first and second moments. It is easy to see that the centroid (vector) of the cluster is simply $\overline{LS}/m$. The variance of a random variable Z is defined to be $E[Z^2] - E[Z]^2$, where $E[\cdot]$ denotes expected values. Therefore, the variances along the ith dimension can be expressed as $SS_i/m - (LS_i/m)^2$. Here SS_i and LS_i represent the component of the corresponding moment vector along the ith dimension. The sum of these dimension-specific variances yields the variance of the entire cluster. Furthermore, the distance of any point to the centroid can be computed using the cluster feature by using the computed centroid $\overline{LS}/m$. Therefore, the cluster feature vector contains all the information needed to insert a data point into the CF-Tree.

Each leaf node in the CF-Tree has a diameter threshold T. The diameter[2] can be any spread measure of the cluster such as its radius or variance, as long as it can be computed directly from the cluster feature vector. The value of T regulates the granularity of the clustering, the height of the tree, and the aggregate number of clusters at the leaf nodes. Lower values of T will result in a larger number of fine-grained clusters. Because the CF-Tree is always assumed to be main-memory resident, the size of the data set will typically have a critical impact on the value of T. Smaller data sets will allow the use of a small threshold T, whereas a larger data set will require a larger value of the threshold T. Therefore, an incremental approach such as *BIRCH* gradually increases the value of T to balance the greater need for memory with increasing data size. In other words, the value of T may need to be increased whenever the tree can no longer be kept within main-memory availability.

The incremental insertion of a data point into the tree is performed with a top-down approach. Specifically, the closest centroid is selected at each level for insertion into the tree structure. This approach is similar to the insertion process in a traditional database index such as a B-Tree. The cluster feature vectors are updated along the corresponding path of the tree by simple addition. At the leaf node, the data point is inserted into its closest cluster only if the insertion does not increase the cluster diameter beyond the threshold T. Otherwise, a new cluster must be created containing only that data point. This new cluster is added to the leaf node, if it is not already full. If the leaf node is already full, then it needs to be split into two nodes. Therefore, the cluster feature entries in the old leaf node need to be assigned to one of the two new nodes. The two cluster features in the leaf node, whose centroids are the furthest apart, can serve as the seeds for the split. The remaining entries are assigned to the seed node to which they are closest. As a result, the branching factor of the parent node of the leaf increases by 1. Therefore, the split might result in the branching factor of the parent increasing beyond B. If this is the case, then the parent would need to be split as well in a similar way. Thus, the split may be propagated upward until the branching factors of all nodes are below B. If the split propagates all the way to the root node, then the height of the CF-Tree increases by 1.

These repeated splits may sometimes result in the tree running out of main memory. In such cases, the CF-Tree needs to be rebuilt by increasing the threshold T, and reinserting the old leaf nodes into a new tree with a higher threshold T. Typically, this reinsertion will result in the merging of some older clusters into larger clusters that meet the new modified threshold T. Therefore, the memory requirement of the new tree is lower. Because the old leaf nodes are reinserted with the use of cluster feature vectors, this step can be accomplished without reading the original database from disk. Note that the cluster feature

[2]The original *BIRCH* algorithm proposes to use the *pairwise* root mean square (RMS) distance between cluster data points as the diameter. This is one possible measure of the intracluster distance. This value can also be shown to be computable from the CF vector as $\sqrt{\frac{\sum_{i=1}^{d}(2 \cdot m \cdot SS_i - 2 \cdot LS_i^2)}{m \cdot (m-1)}}$.

vectors allow the computation of the diameters resulting from the merge of two clusters without using the original data points.

An optional cluster refinement phase can be used to group related clusters within the leaf nodes and remove small outlier clusters. This can be achieved with the use of an agglomerative hierarchical clustering algorithm. Many agglomerative merging criteria, such as the variance-based merging criterion (see Sect. 6.4.1 of Chap. 6), can be easily computed from the CF-vectors. Finally, an optional refinement step reassigns all data points to their closest center, as produced by the global clustering step. This requires an additional scan of the data. If desired, outliers can be removed during this phase.

The *BIRCH* algorithm is very fast because the basic approach (without refinement) requires only one scan over the data, and each insertion is an efficient operation, which resembles the insertion operation in a traditional index structure. It is also highly adaptive to the underlying main-memory requirements. However, it implicitly assumes a spherical shape of the underlying clusters.

7.3.3 CURE

The *Clustering Using REpresentatives (CURE)* algorithm is an agglomerative hierarchical algorithm. Recall from the discussion in Sect. 6.4.1 of Chap. 6 that single-linkage implementation of bottom-up hierarchical algorithms can discover clusters of arbitrary shape. As in all agglomerative methods, a current set of clusters is maintained, which are successively merged with one another, based on single-linkage distance between clusters. However, instead of directly computing distances between all pairs of points in the two clusters for agglomerative merging, the algorithm uses a set of representatives to achieve better efficiency. These representatives are carefully chosen to capture the shape of each of the current clusters, so that the ability of agglomerative methods to capture clusters of arbitrary shape is retained even with the use of a smaller number of representatives. The first representative is chosen to be a data point that is farthest from the center of the cluster, the second representative is farthest to the first, the third is chosen to be the one that has the largest distance to the closest of two representatives, and so on. In particular, the rth representative is a data point that has the largest distance to the closest of the current set of $(r - 1)$ representatives. As a result, the representatives tend to be arranged along the contours of the cluster. Typically, a small number of representatives (such as ten) are chosen from each cluster. This farthest distance approach does have the unfortunate effect of favoring selection of outliers. After the representatives have been selected, they are shrunk toward the cluster center to reduce the impact of outliers. This shrinking is performed by replacing a representative with a new synthetic data point on the line segment L joining the representative to the cluster center. The distance between the synthetic representative and the original representative is a fraction $\alpha \in (0, 1)$ of the length of line segment L. Shrinking is particularly useful in single-linkage implementations of agglomerative clustering because of the sensitivity of such methods to noisy representatives at the fringes of a cluster. Such noisy representatives may chain together unrelated clusters. Note that if the representatives are shrunk too far ($\alpha \approx 1$), the approach will reduce to centroid-based merging, which is also known to work poorly (see Sect. 6.4.1 of Chap. 6).

The clusters are merged using an agglomerative bottom-up approach. To perform the merging, the minimum distance between any pair of representative data points is used. This is the *single-linkage approach* of Sect. 6.4.1 in Chap. 6, which is most well suited to discovering clusters of arbitrary shape. By using a smaller number of representative data points, the *CURE* algorithm is able to significantly reduce the complexity of the merging criterion in agglomerative hierarchical algorithms. The merging can be performed until the number of remaining clusters is equal to k. The value of k is an input parameter specified by the user. *CURE* can handle outliers by periodically eliminating small clusters during the merging process. The idea here is that the clusters remain small because they contain mostly outliers.

To further improve the complexity, the *CURE* algorithm draws a random sample from the underlying data, and performs the clustering on this random sample. In a final phase of the algorithm, all the data points are assigned to one of the remaining clusters by choosing the cluster with the closest representative data point.

Larger sample sizes can be efficiently used with a partitioning trick. In this case, the sample is further divided into a set of p partitions. Each partition is hierarchically clustered until a desired number of clusters is reached, or some merging quality criterion is met. These intermediate clusters (across all partitions) are then reclustered together hierarchically to create the final set of k clusters from the sampled data. The final assignment phase is applied to the representatives of the resulting clusters. Therefore, the overall process may be described by the following steps:

1. Sample s points from the database $\mathcal{D}$ of size n.

2. Divide the sample s into p partitions of size s/p each.

3. Cluster each partition independently using the hierarchical merging to k' clusters in each partition. The overall number $k' \cdot p$ of clusters across all partitions is still larger than the user-desired target k.

4. Perform hierarchical clustering over the $k' \cdot p$ clusters derived across all partitions to the user-desired target k.

5. Assign each of the $(n - s)$ nonsample data points to the cluster containing the closest representative.

The *CURE* algorithm is able to discover clusters of arbitrary shape unlike other scalable methods such as *BIRCH* and *CLARANS*. Experimental results have shown that *CURE* is also faster than these methods.

7.4 High-Dimensional Clustering

High-dimensional data contain many irrelevant features that cause noise in the clustering process. The feature selection section of the previous chapter discussed how the irrelevant features may be removed to improve the quality of clustering. When a large number of features are irrelevant, the data cannot be separated into meaningful and cohesive clusters. This scenario is especially likely to occur when features are uncorrelated with one another. In such cases, the distances between all pairs of data points become very similar. The corresponding phenomenon is referred to as the *concentration* of distances.

The feature selection methods discussed in the previous chapter can reduce the detrimental impact of the irrelevant features. However, it may often not be possible to remove any particular set of features *a priori* when the optimum choice of features depends on the underlying data locality. Consider the case illustrated in Fig. 7.2a. In this case, cluster A exists in the XY-plane, whereas cluster B exists in the YZ-plane. Therefore, the feature relevance is *local*, and it is no longer possible to remove any feature *globally* without losing

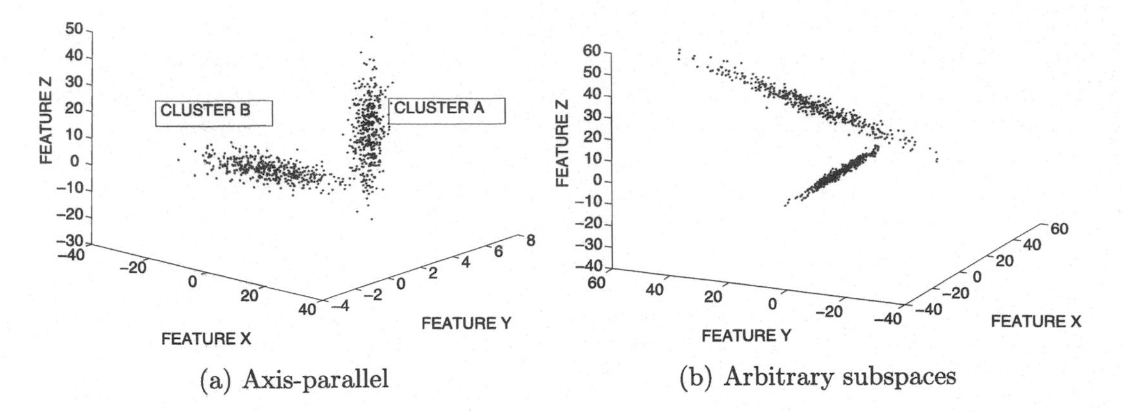

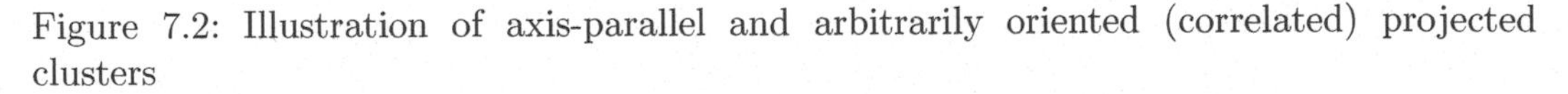
(a) Axis-parallel (b) Arbitrary subspaces

Figure 7.2: Illustration of axis-parallel and arbitrarily oriented (correlated) projected clusters

relevant features for some of the data localities. The concept of projected clustering was introduced to address this issue.

In conventional clustering methods, each cluster is a set of points. In *projected clustering*, each cluster is defined as a set of points *together with* a set of dimensions (or *subspace*). For example, the *projected* cluster A in Fig. 7.2a would be defined as its relevant set of points, *together* with the subspace corresponding to the X and Y dimensions. Similarly, the projected cluster B in Fig. 7.2a is defined as its relevant set of points, together with the subspace corresponding to the Y and Z dimensions. Therefore, a projected cluster is defined as the pair $(\mathcal{C}_i, \mathcal{E}_i)$, where $\mathcal{C}_i$ is a set of points, and the subspace $\mathcal{E}_i$ is the subspace defined by a set of dimensions.

An even more challenging situation is illustrated in Fig. 7.2b in which the clusters do not exist in axis-parallel subspaces, but they exist in arbitrarily oriented subspaces of the data. This problem is also a generalization of the principal component analysis (PCA) method discussed in Chap. 2, where a single global projection with the *largest* variance is found to retain the greatest information about the data. In this case, it is desired to retain the best local projections with the *least* variance to determine the subspaces in which each set of data points is tightly clustered. These types of clusters are referred to as *arbitrarily oriented projected clusters*, *generalized projected clusters*, or *correlation clusters*. Thus, the subspace $\mathcal{E}_i$ for each cluster $\mathcal{C}_i$ cannot be described in terms of the original set of dimensions. Furthermore, the orthogonal subspace to $\mathcal{E}_i$ provides the subspace for performing *local dimensionality reduction*. This is an interesting problem in its own right. Local dimensionality reduction provides enhanced reduction of data dimensionality because of the local selection of the subspaces for dimensionality reduction.

This problem has two different variations, which are referred to as *subspace clustering* and *projected clustering*, respectively.

1. *Subspace clustering:* In this case, overlaps are allowed among the points drawn from the different clusters. This problem is much closer to pattern mining, wherein the association patterns are mined from the numeric data after discretization. Each pattern therefore corresponds to a hypercube within a subspace of the numeric data, and the data points within this cube represent the subspace cluster. Typically, the number of subspace clusters mined can be very large, depending upon a user-defined parameter, known as the density threshold.

Algorithm *CLIQUE*(Data: $\mathcal{D}$, Ranges: p, Density: τ)
begin
Discretize each dimension of data set $\mathcal{D}$ into p ranges;
Determine dense combinations of grid cells at minimum support τ
using any frequent pattern mining algorithm;
Create graph in which dense grid combinations are
connected if they are adjacent;
Determine connected components of graph;
return (point set, subspace) pair for each connected component;
end

Figure 7.3: The *CLIQUE* algorithm

2. *Projected clustering:* In this case, no overlaps are allowed among the points drawn from the different clusters. This definition provides a concise summary of the data. Therefore, this model is much closer, in principle, to the original goals of the clustering framework of data summarization.

In this section, three different clustering algorithms will be described. The first of these is *CLIQUE*, which is a subspace clustering method. The other two are *PROCLUS* and *ORCLUS*, which are the first projected clustering methods proposed for the axis-parallel and the correlated versions of the problem, respectively.

7.4.1 CLIQUE

The *CLustering In QUEst (CLIQUE)* technique is a generalization of grid-based methods discussed in the previous chapter. The input to the method is the number of grid ranges p for each dimension, and the density τ. This density τ represents the minimum number of data points in a dense grid cell and can also be viewed as a minimum *support* requirement of the grid cell. As in all grid-based methods, the first phase of discretization is used to create a grid structure. In *full-dimensional* grid-based methods, the relevant dense regions are based on the intersection of the discretization ranges across *all* dimensions. The main difference of *CLIQUE* from these methods is that it is desired to determine the ranges only over a relevant *subset of* dimensions with density greater than τ. This is the same as the frequent pattern mining problem, where each discretized range is treated as an "item," and the support is set to τ. In the original *CLIQUE* algorithm, the *Apriori* method was used, though any other frequent pattern mining method could be used in principle. As in generic grid-based methods, the adjacent grid cells (defined on the same subspace) are put together. This process is also identical to the generic grid-based methods, except that two grids have to be defined on the same subspace for them to even be considered for adjacency. All the found patterns are returned together with the data points in them. The *CLIQUE* algorithm is illustrated in Fig. 7.3. An easily understandable description can also be generated for each set of k-dimensional connected grid regions by decomposing it into a *minimal* set of k-dimensional hypercubes. This problem is NP-hard. Refer to the bibliographic notes for efficient heuristics.

Strictly speaking, *CLIQUE* is a quantitative frequent pattern mining method rather than a clustering method. The output of *CLIQUE* can be very large and can sometimes be greater than the size of the data set, as is common in frequent pattern mining. Clustering and frequent pattern mining are related but different problems with different objectives. The primary goal of frequent pattern mining is that of finding dimension correlation, whereas the primary goal of clustering is summarization. From this semantic point of view, the approach does not seem to achieve the primary application-specific goal of data summarization. The worst-case complexity of the approach and the number of discovered patterns can be exponentially related to the number of dimensions. The approach may not terminate at low values of the density (support) threshold τ.

7.4.2 PROCLUS

The *PROjected CLUStering (PROCLUS)* algorithm uses a medoid-based approach to clustering. The algorithm proceeds in three phases: an initialization phase, an iterative phase, and a cluster refinement phase. The initialization phase selects a small candidate set M of medoids, which restricts the search space for hill climbing. In other words, the final medoid set will be a subset of the candidate set M. The iterative phase uses a medoid-based technique for hill climbing to better solutions until convergence. The final refinement phase assigns data points to the optimal medoids and removes outliers.

A small candidate set M of medoids is selected during initialization as follows:

1. A random sample M of data points of size proportional to the number of clusters k is picked. Let the size of this subset be denoted by $A \cdot k$, where A is a constant greater than 1.

2. A greedy method is used to further reduce the size of the set M to $B \cdot k$, where $A > B > 1$. Specifically, a farthest distance approach is applied, where points are iteratively selected by selecting the data point with the farthest distance to the closest of the previously selected points.

Although the selection of a small candidate medoid set M greatly reduces the complexity of the search space, it also tends to include many outliers because of its farthest distance approach. Nevertheless, the farthest distance approach ensures well-separated seeds, which also tend to separate out the clusters well.

The algorithm starts by choosing a random subset S of k medoids from M, and it progressively improves the quality of medoids by iteratively replacing the "bad" medoids in the current set with new points from M. The best set of medoids found so far is always stored in S_{best}. Each medoid in S is associated with a set of dimensions based on the statistical distribution of data points in its locality. This set of dimensions represents the subspace specific to the corresponding cluster. The algorithm determines a set of "bad" medoids in S_{best}, using an approach described later. These bad medoids are replaced with randomly selected replacement points from M and the impact on the objective function is measured. If the objective function improves, then the current best set of medoids S_{best} is updated to S. Otherwise, another randomly selected replacement set is tried for exchanging with the bad medoids in S_{best} in the next iteration. If the medoids in S_{best} do not improve for a predefined number of successive replacement attempts, then the algorithm terminates. All computations, such as the assignment and objective function computation, are executed in the subspace associated with each medoid. The overall algorithm is illustrated in Fig. 7.4. Next, we provide a detailed description of each of the aforementioned steps.

Determining projected dimensions for a medoid: The aforementioned approach requires the determination of the quality of a particular set of medoids. This requires the assignment of

Algorithm *PROCLUS*(Database: $\mathcal{D}$, Clusters: k, Dimensions: l)
begin
  Select candidate medoids $M \subseteq \mathcal{D}$ with a farthest distance approach;
  S = Random subset of M of size k;
  $BestObjective = \infty$;
  repeat
    Compute dimensions (subspace) associated with each medoid in S;
    Assign points in $\mathcal{D}$ to closest medoids in S using projected distance;
    $CurrentObjective$ = Mean projected distance of points to cluster centroids;
    if ($CurrentObjective < BestObjective$) **then begin**
      $S_{best} = S$;
      $BestObjective = CurrentObjective$;
    end;
    Recompute S by replacing bad medoids in S_{best} with random points from M;
  until termination criterion;
  Assign data points to medoids in S_{best} using refined subspace computations;
  return all cluster-subspace pairs;
end

Figure 7.4: The *PROCLUS* algorithm

data points to medoids by computing the distance of the data point to each medoid i in the subspace $\mathcal{E}_i$ relevant to the ith medoid. First, the *locality* of each medoid in S is defined. The locality of the medoid is defined as the set of data points that lies in a sphere of radius equal to the distance to the closest medoid. The (statistically normalized) average distance along each dimension from the medoid to the points in its locality is computed. Let r_{ij} be the average distance of the data points in the locality of medoid i to medoid i along dimension j. The mean $\mu_i = \sum_{j=1}^{d} r_{ij}/d$ and standard deviation $\sigma_i = \sqrt{\frac{\sum_{j=1}^{d}(r_{ij}-\mu_i)^2}{d-1}}$ of these distance values r_{ij} are computed, specific to each locality. This can then be converted into a statistically normalized value z_{ij}:

$$z_{ij} = \frac{r_{ij} - \mu_i}{\sigma_i}. \tag{7.10}$$

The reason for this locality-specific normalization is that different data localities have different natural sizes, and it is difficult to compare dimensions from different localities without normalization. Negative values of z_{ij} are particularly desirable because they suggest smaller average distances than expectation for a medoid-dimension pair. The basic idea is to select the smallest (most negative) $k \cdot l$ values of z_{ij} to determine the relevant cluster-specific dimensions. Note that this may result in the assignment of a different number of dimensions to the different clusters. The sum of the total number of dimensions associated with the different medoids must be equal to $k \cdot l$. An additional constraint is that the number of dimensions associated with a medoid must be at least 2. To achieve this, all the z_{ij} values are sorted in increasing order, and the two smallest ones are selected for each medoid i. Then, the remaining $k \cdot (l-2)$ medoid-dimension pairs are greedily selected as the smallest ones from the remaining values of z_{ij}.

Assignment of data points to clusters and cluster evaluation: Given the medoids and their associated sets of dimensions, the data points are assigned to the medoids using a single pass over the database. The distance of the data points to the medoids is computed using the Manhattan segmental distance. The *Manhattan segmental distance* is the same as the Manhattan distance, except that it is normalized for the varying number of dimensions associated with each medoid. To compute this distance, the Manhattan distance is computed using only the relevant set of dimensions, and then divided by the number of relevant dimensions. A data point is assigned to the medoid with which it has the least Manhattan segmental distance. After determining the clusters, the objective function of the clustering is evaluated as the average Manhattan segmental distance of data points to the *centroids* of their respective clusters. If the clustering objective improves, then S_{best} is updated.

Determination of bad medoids: The determination of "bad" medoids from S_{best} is performed as follows: The medoid of the cluster with the least number of points is bad. In addition, the medoid of any cluster with less than $(n/k) \cdot minDeviation$ points is bad, where $minDeviation$ is a constant smaller than 1. The typical value was set to 0.1. The assumption here is that bad medoids have small clusters either because they are outliers or because they share points with another cluster. The bad medoids are replaced with random points from the candidate medoid set M.

Refinement phase: After the best set of medoids is found, a final pass is performed over the data to improve the quality of the clustering. The dimensions associated with each medoid are computed differently than in the iterative phase. The main difference is that to analyze the dimensions associated with each medoid, the distribution of the points in the clusters at the end of the iterative phase is used, as opposed to the localities of the medoids. After the new dimensions have been computed, the points are reassigned to medoids based on the Manhattan segmental distance with respect to the new set of dimensions. Outliers are also handled during this final pass over the data. For each medoid i, its closest other medoid is computed using the Manhattan segmental distance in the relevant subspace of medoid i. The corresponding distance is referred to as its sphere of influence. If the Manhattan segmental distance of a data point to each medoid is greater than the latter's sphere of influence, then the data point is discarded as an outlier.

7.4.3 ORCLUS

The *arbitrarily ORiented projected CLUStering (ORCLUS)* algorithm finds clusters in arbitrarily oriented subspaces, as illustrated in Fig. 7.2b. Clearly, such clusters cannot be found by axis-parallel projected clustering. Such clusters are also referred to as *correlation clusters*. The algorithm uses the number of clusters k, and the dimensionality l of each subspace $\mathcal{E}_i$ as an input parameter. Therefore, the algorithm returns k different pairs $(\mathcal{C}_i, \mathcal{E}_i)$, where the cluster$\mathcal{C}_i$ is defined in the arbitrarily oriented subspace $\mathcal{E}_i$. In addition, the algorithm reports a set of outliers $\mathcal{O}$. This method is also referred to as *correlation clustering*. Another difference between the *PROCLUS* and *ORCLUS* models is the simplifying assumption in the latter that the dimensionality of each subspace is fixed to the same value l. In the former case, the value of l is simply the *average* dimensionality of the cluster-specific subspaces.

The *ORCLUS* algorithm uses a combination of hierarchical and k-means clustering in conjunction with subspace refinement. While hierarchical merging algorithms are generally more effective, they are expensive. Therefore, the algorithm uses hierarchical *representatives* that are successively merged. The algorithm starts with $k_c = k_0$ initial seeds, denoted by S.

Algorithm *ORCLUS*(Data: $\mathcal{D}$, Clusters: k, Dimensions: l)
begin
  Sample set S of $k_0 > k$ points from $\mathcal{D}$;
  $k_c = k_0$; $l_c = d$;
  Set each $\mathcal{E}_i$ to the full data dimensionality;
  $\alpha = 0.5$; $\beta = e^{-\log(d/l) \cdot \log(1/\alpha)/\log(k_0/k)}$;
  while $(k_c > k)$ **do**
  begin
    Assign each data point in $\mathcal{D}$ to closest seed in S using
      projected distance in $\mathcal{E}_i$ to create $\mathcal{C}_i$;
    Re-center each seed in S to centroid of cluster $\mathcal{C}_i$;
    Use PCA to determine subspace $\mathcal{E}_i$ associated with $\mathcal{C}_i$ by selecting
      smallest l_c eigenvectors of covariance matrix of $\mathcal{C}_i$;
    $k_c = \max\{k, k_c \cdot \alpha\}$; $l_c = \max\{l, l_c \cdot \beta\}$;
    Repeatedly merge clusters to reduce number of clusters to
      the new reduced value of k_c;
  end;
  Perform final assignment pass of points to clusters;
  return cluster-subspace pairs $(\mathcal{C}_i, \mathcal{E}_i)$ for each $i \in \{1 \dots k\}$;
end

Figure 7.5: The *ORCLUS* algorithm

The current number of seeds, k_c, are reduced over successive merging iterations. Methods from representative-based clustering are used to assign data points to these seeds, except that the distance of a data point to its seed is measured in its associated subspace $\mathcal{E}_i$. Initially, the current dimensionality, l_c, of each cluster is equal to the full data dimensionality. The value l_c is reduced gradually to the user-specified dimensionality l by successive reduction over different iterations. The idea behind this gradual reduction is that in the first few iterations, the clusters may not necessarily correspond very well to the natural lower dimensional subspace clusters in the data; so a larger subspace is retained to avoid loss of information. In later iterations, the clusters are more refined, and therefore subspaces of lower rank may be extracted.

The overall algorithm consists of a number of iterations, in each of which a sequence of merging operations is alternated with a k-means style assignment with projected distances. The number of current clusters is reduced by the factor $\alpha < 1$, and the dimensionality of current cluster $\mathcal{C}_i$ is reduced by $\beta < 1$ in a given iteration. The first few iterations correspond to a higher dimensionality, and each successive iteration continues to peel off more and more noisy subspaces for the different clusters. The values of α and β are related in such a way that the reduction from k_0 to k clusters occurs in the same number of iterations as the reduction from $l_0 = d$ to l dimensions. The value of α is 0.5, and the derived value of β is indicated in Fig. 7.5. The overall description of the algorithm is also illustrated in this figure.

The overall procedure uses the three alternating steps of assignment, subspace recomputation, and merging in each iteration. Therefore, the algorithm uses concepts from both hierarchical and k-means methods in conjunction with subspace refinement. The assignment step assigns each data point to its closest seed, by comparing the projected distance of a data point to the ith seed in S, using the subspace $\mathcal{E}_i$. After the assignment, all the seeds in S are re-centered to the centroids of the corresponding clusters. At this point, the subspace $\mathcal{E}_i$ of dimensionality l_c associated with each cluster $\mathcal{C}_i$ is computed. This is done by using PCA on cluster $\mathcal{C}_i$. The subspace $\mathcal{E}_i$ is defined by the l_c orthonormal eigenvectors of the covariance matrix of cluster $\mathcal{C}_i$ with the least eigenvalues. To perform the merging, the algorithm computes the projected energy (variance) of the union of the two clusters in the corresponding least spread subspace. The pair with the least energy is selected to perform the merging. Note that this is a subspace generalization of the variance criterion for hierarchical merging algorithms (see Sect. 6.4.1 of Chap. 6).

The algorithm terminates when the merging process over all the iterations has reduced the number of clusters to k. At this point, the dimensionality l_c of the subspace $\mathcal{E}_i$ associated with each cluster $\mathcal{C}_i$ is also equal to l. The algorithm performs one final pass over the database to assign data points to their closest seed based on the projected distance. Outliers are handled during the final phase. A data point is considered an outlier when its projected distance to the closest seed i is greater than the projected distance of other seeds to seed i in subspace $\mathcal{E}_i$.

A major computational challenge is that the merging technique requires the computation of the eigenvectors of the union of clusters, which can be expensive. To efficiently perform the merging, the *ORCLUS* approach extends the concept of cluster feature vectors from *BIRCH* to covariance matrices. The idea is to store not only the moments in the cluster feature vector but also the sum of the products of attribute values for each pair of dimensions. The covariance matrix can be computed from this extended cluster feature vector. This approach can be viewed as a covariance- and subspace-based generalization of the variance-based merging implementation of Sect. 6.4.1 in Chap. 6. For details on this optimization, the reader is referred to the bibliographic section.

Depending on the value of k_0 chosen, the time complexity is dominated by either the merges or the assignments. The merges require eigenvector computation, which can be expensive. With an efficient implementation based on cluster feature vectors, the merges can be implemented in $O(k_0^2 \cdot d \cdot (k_0 + d^2))$ time, whereas the assignment step always requires $O(k^0 \cdot n \cdot d)$ time. This can be made faster with the use of optimized eigenvector computations. For smaller values of k_0, the computational complexity of the method is closer to k-means, whereas for larger values of k_0, the complexity is closer to hierarchical methods. The *ORCLUS* algorithm does not assume the existence of an incrementally updatable similarity matrix, as is common with bottom-up hierarchical methods. At the expense of additional space, the maintenance of such a similarity matrix can reduce the $O(k_0^3 \cdot d)$ term to $O(k_0^2 \cdot \log(k_0) \cdot d)$.

7.5 Semisupervised Clustering

One of the challenges with clustering is that a wide variety of alternative solutions may be found by various algorithms. The quality of these alternative clusterings may be ranked differently by different internal validation criteria depending on the alignment between the clustering criterion and validation criterion. This is a major problem with any unsupervised algorithm. Semisupervision, therefore, relies on external *application-specific* criteria to guide the clustering process.

It is important to understand that different clusterings may not be equally useful from an application-specific perspective. The utility of a clustering result is, after all, based on the ability to use it effectively for a given application. One way of guiding the clustering results toward an application-specific goal is with the use of *supervision*. For example, consider the case where an analyst wishes to segment a set of documents approximately along the lines of the *Open Directory Project (ODP)*,[3] where users have already manually labeled documents into a set of predefined categories. One may want to use this directory only as soft guiding principle because the number of clusters and their topics in the analyst's collection may not always be exactly the same as in the *ODP* clusters. One way of incorporating supervision is to download example documents from each category of *ODP* and mix them with the documents that need to be clustered. This newly downloaded set of documents are labeled with their category and provide information about how the features are related to the different clusters (categories). The added set of labeled documents, therefore, provides *supervision* to the clustering process in the same way that a teacher guides his or her students toward a specific goal.

A different scenario is one in which it is known from background knowledge that certain documents should belong to the same class, and others should not. Correspondingly, two types of semisupervision are commonly used in clustering:

1. *Pointwise supervision:* Labels are associated with individual data points and provide information about the category (or cluster) of the object. This version of the problem is closely related to that of data classification.

2. *Pairwise supervision:* "Must-link" and "cannot-link" constraints are provided for the individual data points. This provides information about cases where pairs of objects are allowed to be in the same cluster or are forbidden to be in the same cluster, respectively. This form of supervision is also sometimes referred to as *constrained clustering*.

For each of these variations, a number of simple semisupervised clustering methods are described in the following sections.

7.5.1 Pointwise Supervision

Pointwise supervision is significantly easier to address than pairwise supervision because the labels associated with the data points can be used more naturally in conjunction with existing clustering algorithms. In *soft supervision*, the labels are used as guidance, but data points with different labels are allowed to mix. In *hard* supervision, data points with different labels are not allowed to mix. Some examples of different ways of modifying existing clustering algorithms are as follows:

1. *Semisupervised clustering by seeding:* In this case, the initial seeds for a k-means algorithm are chosen as data points of different labels. These are used to execute a standard k-means algorithm. The biased initialization has a significant impact on the final results, even when labeled data points are allowed to be assigned to a cluster whose initial seed had a different label (soft supervision). In hard supervision, clusters are explicitly associated with labels corresponding to their initial seeds. The assignment of *labeled* data points is constrained so that such points can be assigned to a cluster with the same label. In some cases, the weights of the unlabeled points are discounted while computing cluster centers to increase the impact of supervision. The second form of semisupervision is closely related to semisupervised classification, which is

[3] `http://www.dmoz.org/`.

discussed in Chap. 11. An EM algorithm, which performs semisupervised classification with labeled and unlabeled data, uses a similar approach. Refer to Sect. 11.6 of Chap. 11 for a discussion of this algorithm. For more robust initialization, an unsupervised clustering can be separately applied to each labeled data segment to create the seeds.

2. *EM algorithms:* Because the EM algorithm is a soft version of the k-means method, the changes required to EM methods are exactly identical to those in the case of k-means. The initialization of the EM algorithm is performed with mixtures centered at the labeled data points. In addition, for hard supervision, the posterior probabilities of labeled data points are always set to 0 for mixture components that do not belong to the same label. Furthermore, the unlabeled data points are discounted during computation of model parameters. This approach is discussed in detail in Sect. 11.6 of Chap. 11.

3. *Agglomerative algorithms:* Agglomerative algorithms can be generalized easily to the semisupervised case. In cases where the merging allows the mixing of different labels (soft supervision), the distance function between clusters during the clustering can incorporate the similarity in their class label distributions across the two components being merged by providing an extra credit to clusters with the same label. The amount of this credit regulates the level of supervision. Many different choices are also available to incorporate the supervision more strongly in the merging criterion. For example, the merging criterion may require that only clusters containing the same label are merged together (hard supervision).

4. *Graph-based algorithms:* Graph-based algorithms can be modified to work in the semisupervised scenario by changing the similarity graph to incorporate supervision. The edges joining data points with the same label have an extra weight of α. The value of α regulates the level of supervision. Increased values of α will be closer to hard supervision, whereas smaller values of α will be closer to soft supervision. All other steps in the clustering algorithm remain identical. A different form of graph-based supervision, known as collective classification, is used for the semisupervised classification problem (cf. Sect. 19.4 of Chap. 19).

Thus, pointwise supervision is easily incorporated in most clustering algorithms.

7.5.2 Pairwise Supervision

In pairwise supervision, "must-link" and "cannot-link" constraints are specified between pairs of objects. An immediate observation is that it is not necessary for a feasible and consistent solution to exist for an arbitrary set of constraints. Consider the case where three data points A, B, and C are such that (A, B), and (A, C) are both "must-link" pairs, whereas (B, C) is a "cannot-link" pair. It is evident that no feasible clustering can be found that satisfies all three constraints. The problem of finding clusters with pairwise constraints is generally more difficult than one in which pointwise constraints are specified. In cases where only "must-link" constraints are specified, the problem can be approximately reduced to the case of pointwise supervision.

The k-means algorithm can be modified to handle pairwise supervision quite easily. The basic idea is to start with an initial set of randomly chosen centroids. The data points are processed in a random order for assignment to the seeds. Each data point is assigned to its closest seed that does not violate any of the constraints implied by the assignments that have already been executed. In the event that the data point cannot be assigned to any cluster in a consistent way, the algorithm terminates. In this case, the clusters in the last iteration where a feasible solution was found are reported. In some cases, no feasible solution may be found even in the first iteration, depending on the choice of seeds. Therefore, the constrained k-means approach may be executed multiple times, and the best solution over these executions is reported. Numerous other methods are available in the literature, both in terms of the kinds of constraints that are specified, and in terms of the solution methodology. The bibliographic notes contain pointers to many of these methods.

7.6 Human and Visually Supervised Clustering

The previous section discussed ways of incorporating supervision in the *input data* in the form of constraints or labels. A different way of incorporating supervision is to use direct feedback from the user *during the clustering process*, based on an understandable summary of the clusters.

The core idea is that semantically meaningful clusters are often difficult to isolate using fully automated methods in which rigid mathematical formalizations are used as the only criteria. The utility of clusters is based on their application-specific usability, which is often semantically interpretable. In such cases, human involvement is necessary to incorporate the intuitive and semantically meaningful aspects during the cluster discovery process. Clustering is a problem that requires both the computational power of a computer and the intuitive understanding of a human. Therefore, a natural solution is to divide the clustering task in such a way that each entity performs the task that it is most well suited to. In the interactive approach, the computer performs the computationally intensive analysis, which is leveraged to provide the user with an intuitively understandable summary of the clustering structure. The user utilizes this summary to provide feedback about the key choices that should be made by a clustering algorithm. The result of this cooperative technique is a system that can perform the task of clustering better than either a human or a computer.

There are two natural ways of providing feedback during the clustering process:

1. *Semantic feedback as an intermediate process in standard clustering algorithms:* Such an approach is relevant in domains where the objects are semantically interpretable (e.g., documents or images), and the user provides feedback at specific stages in a clustering algorithm when critical choices are made. For example, in a k-means algorithm, a user may choose to drop some clusters during each iteration and manually specify new seeds reflecting uncovered segments of the data.

2. *Visual feedback in algorithms specifically designed for human–computer interaction:* In many high-dimensional data sets, the number of attributes is very large, and it is difficult to associate direct semantic interpretability with the objects. In such cases, the user must be provided visual representations of the clustering structure of the data in different subsets of attributes. The user may leverage these representatives to provide feedback to the clustering process. This approach can be viewed as an interactive version of projected clustering methods.

In the following section, each of these different types of algorithms will be addressed in detail.

7.6.1 Modifications of Existing Clustering Algorithms

Most clustering algorithms use a number of key decision steps in which choices need to be made, such as the choice of merges in a hierarchical clustering algorithm, or the resolution of close ties in assignment of data points to clusters. When these choices are made on the basis of stringent and predefined clustering criteria, the resulting clusters may not reflect the natural structure of clusters in the data. Therefore, the goal in this kind of approach is to present the user with a small number of *alternatives* corresponding to critical choices in the clustering process. Some examples of simple modifications of the existing clustering algorithms are as follows:

1. *Modifications to k-means and related methods:* A number of critical decision points in the k-means algorithm can be utilized to improve the clustering process. For example, after each iteration, representative data points from each cluster can be presented to the user. The user may choose to manually discard either clusters with very few data points or clusters that are closely related to others. The corresponding seeds may be dropped and replaced with randomly chosen seeds in each iteration. Such an approach works well when the representative data points presented to the user have clear semantic interpretability. This is true in many domains such as image data or document data.

2. *Modifications to hierarchical methods:* In the bottom-up hierarchical algorithms, the clusters are successively merged by selecting the closest pair for merging. The key here is that if a bottom-up algorithm makes an error in the merging process, the merging decision is final, resulting in a lower quality clustering. Therefore, one way of reducing such mistakes is to present the users with the top-ranking choices for the merge corresponding to a small number of different pairs of clusters. These choices can be made by the user on the basis of semantic interpretability.

It is important to point out that the key steps at which a user may provide the feedback depend on the level of semantic interpretability of the objects in the underlying clusters. In some cases, such semantic interpretability may not be available.

7.6.2 Visual Clustering

Visual clustering is particularly helpful in scenarios, such as high-dimensional data, where the semantic interpretability of the individual objects is low. In such cases, it is useful to visualize lower dimensional projections of the data to determine subspaces in which the data are clustered. The ability to discover such lower dimensional projections is based on a combination of the computational ability of a computer and the intuitive feedback of the user. *IPCLUS* is an approach that combines interactive projected clustering methods with visualization methods derived from density-based methods.

One challenge with high-dimensional clustering is that the density, distribution, and shapes of the clusters may be quite different in different data localities and subspaces. Furthermore, it may not be easy to decide the optimum density threshold at which to separate out the clusters in any particular subspace. This is a problem even for full-dimensional clustering algorithms where any particular density threshold[4] may either merge clusters or completely miss clusters. While subspace clustering methods such as *CLIQUE* address these issues by reporting a huge number of overlapping clusters, projected clustering methods such

[4]See discussion in Chap. 6 about Fig. 6.14.

```
Algorithm IPCLUS(Data Set: D, Polarization Points: k)
begin
  while not(termination_criterion) do
  begin
    Randomly sample k points Y_1 ... Y_k from D;
    Compute 2-dimensional subspace E polarized around Y_1 ... Y_k;
    Generate density profile in E and present to user;
    Record membership statistics of clusters based on
          user-specified density-based feedback;
  end;
  return consensus clusters from membership statistics;
end
```

Figure 7.6: The *IPCLUS* algorithm

as *PROCLUS* address these issues by making hard decisions about how the data should be most appropriately summarized. Clearly, such decisions can be made more effectively by interactive user exploration of these alternative *views* and creating a final *consensus* from these different views. The advantage of involving the user is the greater intuition available in terms of the quality of feedback provided to the clustering process. The result of this cooperative technique is a system that can perform the clustering task better than either a human or a computer.

The idea behind the *Interactive Projected CLUStering algorithm (IPCLUS)* is to provide the user with a set of meaningful visualizations in lower dimensional projections together with the ability to decide how to separate the clusters. The overall algorithm is illustrated in Fig. 7.6. The interactive projected clustering algorithm works in a series of iterations; in each, a projection is determined in which there are distinct sets of points that can be clearly distinguished from one another. Such projections are referred to as *well polarized*. In a well-polarized projection, it is easier for the user to clearly distinguish a set of clusters from the rest of the data. Examples of the data density distribution of a well-polarized projection and a poorly polarized projection are illustrated in Fig. 7.7a and b, respectively.

These polarized projections are determined by randomly selecting a set of k records from the database that are referred to as the *polarization anchors*. The number of polarization anchors k is one of the inputs to the algorithm. A 2-dimensional subspace of the data is determined such that the data are clustered around each of these polarization anchors. Specifically, a 2-dimensional subspace is selected so that the mean square radius of assignments of data points to the polarization points as anchors is minimized. Different projections are repeatedly determined with different sampled anchors in which the user can provide feedback. A consensus clustering is then determined from the different clusterings generated by the user over multiple subspace views of the data.

The polarization subspaces can be determined either in axis-parallel subspaces or arbitrary subspaces, although the former provides greater interpretability. The overall approach for determining polarization subspaces starts with the full dimensionality and iteratively reduces the dimensionality of the current subspace until a 2-dimensional subspace is obtained. This is achieved by iteratively assigning data points to their closest subspace-specific anchor points in each iteration, while discarding the most noisy (high variance) dimensions in each iteration about the polarization points. The dimensionality is reduced

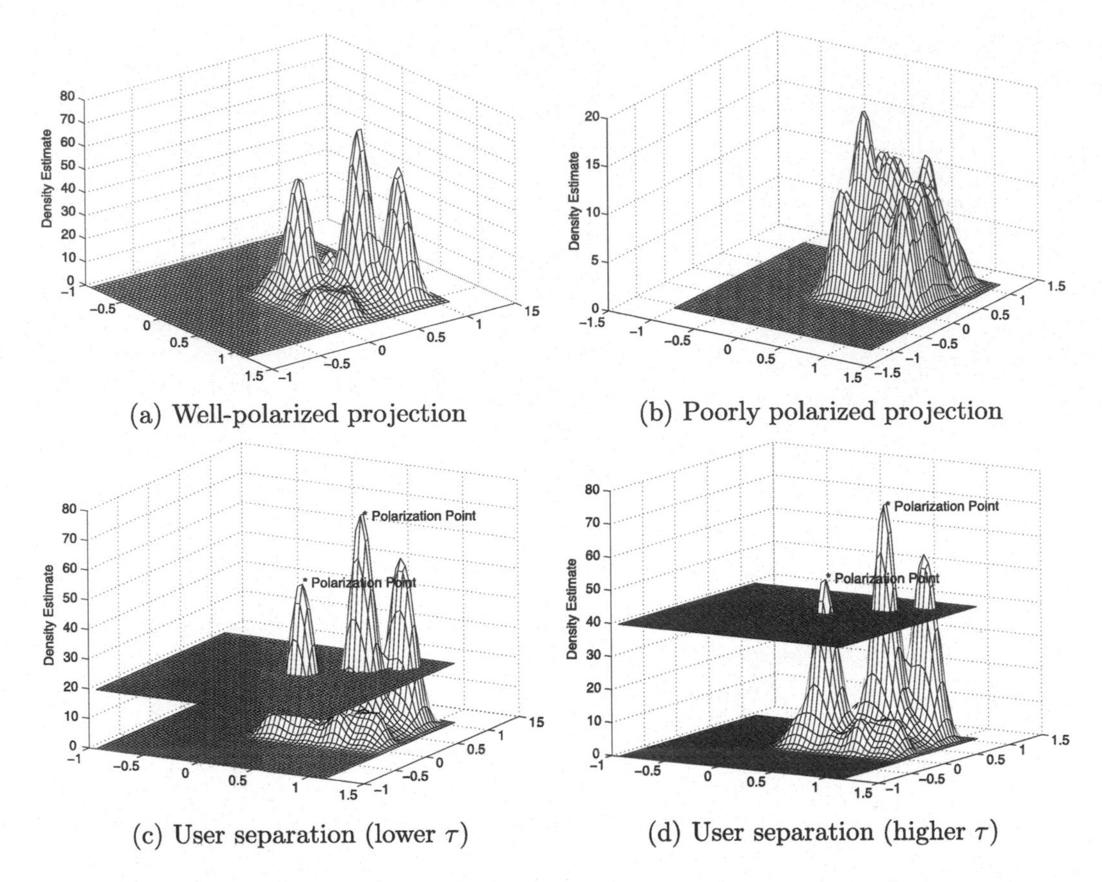

(a) Well-polarized projection

(b) Poorly polarized projection

(c) User separation (lower τ)

(d) User separation (higher τ)

Figure 7.7: Polarizations of different quality and flexibility of user interaction

by a constant factor of 2 in each iteration. Thus, this is a k-medoids type approach, which reduces the dimensionality of the subspaces for distance computation, but not the seeds in each iteration. This typically results in the discovery of a 2-dimensional subspace that is highly clustered around the polarization anchors. Of course, if the polarization anchors are poorly sampled, then this will result in poorly separated clusters. Nevertheless, repeated sampling of polarization points ensures that good subspaces will be selected in at least a few iterations.

After the projection subspace has been found, kernel density estimation techniques can be used to determine the data density at each point in a 2-dimensional grid of values in the relevant subspace. The density values at these grid points are used to create a surface plot. Examples of such plots are illustrated in Fig. 7.7. Because clusters correspond to dense regions in the data, they are represented by peaks in the density profile. To actually separate out the clusters, the user can visually specify density value thresholds that correspond to noise levels at which clusters can be separated from one another. Specifically, a cluster may be defined to be a connected region in the space with density above a certain noise threshold τ that is specified by the user. This cluster may be of arbitrary shape, and the points inside it can be determined. Note that when the density distribution varies significantly with locality, different numbers, shapes, and sizes of clusters will be discovered by different density thresholds. Examples of density thresholding are illustrated in Fig. 7.7c and 7.7d, where clusters of different numbers and shapes are discovered at different thresholds. It is in this step that the user intuition is very helpful, both in terms of deciding which polarized projections are most relevant, and in terms of deciding what density thresholds to specify. If desired, the user may discard a projection altogether or specify multiple thresholds in the same projection to discover clusters of different density in different localities. The specification of the density threshold τ need not be done directly by value. The density separator hyperplane can be visually superposed on the density profile with the help of a graphical interface.

Each feedback of the user results in the generation of connected sets of points within the density contours. These sets of points can be viewed as one or more binary "transactions" drawn on the "item" space of data points. The key is to determine the consensus clusters from these newly created transactions that encode user feedback. While the problem of finding consensus clusters from multiple clusterings will be discussed in detail in the next section, a very simple way of doing this is to use either frequent pattern mining (to find overlapping clusters) or a second level of clustering on the transactions to generate nonoverlapping clusters. Because this new set of transactions encodes the user preferences, the quality of the clusters found with such an approach will typically be quite high.

7.7 Cluster Ensembles

The previous section illustrated how different views of the data can lead to different solutions to the clustering problem. This notion is closely related to the concept of *multiview clustering* or *ensemble clustering*, which studies this issue from a broader perspective. It is evident from the discussion in this chapter and the previous one that clustering is an unsupervised problem with many alternative solutions. In spite of the availability of a large number of validation criteria, the ability to truly test the quality of a clustering algorithm remains elusive. The goal in ensemble clustering is to combine the results of many clustering models to create a more robust clustering. The idea is that no single model or criterion truly captures the optimal clustering, but an ensemble of models will provide a more robust solution.

Most ensemble models use the following two steps to generate the clustering solution:

1. Generate k different clusterings with the use of different models or data selection mechanisms. These represent the different ensemble *components*.

2. Combine the different results into a single and more robust clustering.

The following section provides a brief overview of the different ways in which the alternative clusterings can be constructed.

7.7.1 Selecting Different Ensemble Components

The different ensemble components can be selected in a wide variety of ways. They can be either *modelbased* or *data-selection* based. In model-based ensembles, the different components of the ensemble reflect different models, such as the use of different clustering models, different settings of the same model, or different clusterings provided by different runs of the same randomized algorithm. Some examples follow:

1. The different components can be a variety of models such as partitioning methods, hierarchical methods, and density-based methods. The qualitative differences between the models will be data set-specific.

2. The different components can correspond to different settings of the same algorithm. An example is the use of different initializations for algorithms such as k-means or EM, the use of different mixture models for EM, or the use of different parameter settings of the same algorithm, such as the choice of the density threshold in *DBSCAN*. An ensemble approach is useful because the optimum choice of parameter settings is also hard to determine in an unsupervised problem such as clustering.

3. The different components could be obtained from a single algorithm. For example, a 2-means clustering applied to the 1-dimensional embedding obtained from spectral clustering will yield a different clustering solution for *each* eigenvector. Therefore, the smallest k nontrivial eigenvectors will provide k different solutions that are often quite different as a result of the orthogonality of the eigenvectors.

A second way of selecting the different components of the ensemble is with the use of *data selection*. Data selection can be performed in two different ways:

1. *Point selection:* Different subsets of the data are selected, either via random sampling, or by systematic selection for the clustering process.

2. *Dimension selection:* Different subsets of dimensions are selected to perform the clustering. An example is the *IPCLUS* method discussed in the previous section.

After the individual ensemble components have been constructed, it is often a challenge to combine the results from these different components to create a *consensus* clustering.

7.7.2 Combining Different Ensemble Components

After the different clustering solutions have been obtained, it is desired to create a robust consensus from the different solutions. In the following section, some simple methods are described that use the base clusterings as input for the generation of the final set of clusters.

7.7.2.1 Hypergraph Partitioning Algorithm

Each object in the data is represented by a vertex. A cluster in any of the ensemble components is represented as a *hyperedge*. A hyperedge is a generalization of the notion of edge, because it connects more than two nodes in the form of a complete clique. Any off-the-shelf hypergraph clustering algorithm such as *HMETIS* [302] can be used to determine the optimal partitioning. Constraints are added to ensure a balanced partitioning. One major challenge with hypergraph partitioning is that a hyperedge can be "broken" by a partitioning in many different ways, not all of which are qualitatively equivalent. Most hypergraph partitioning algorithms use a constant penalty for breaking a hyperedge. This can sometimes be undesirable from a qualitative perspective.

7.7.2.2 Meta-clustering Algorithm

This is also a graph-based approach, except that vertices are associated with each cluster in the ensemble components. For example, if there are $k_1 \ldots k_r$ different clusters in each of the r ensemble components, then a total of $\sum_{i=1}^{r} k_i$ vertices will be created. Each vertex therefore represents a set of data objects. An edge is added between a pair of vertices if the Jaccard coefficient between the corresponding object sets is nonzero. The weight of the edge is equal to the Jaccard coefficient. This is therefore an r-partite graph because there are no edges between two vertices from the same ensemble component. A graph partitioning algorithm is applied to this graph to create the desired number of clusters. Each data point has r different instances corresponding to the different ensemble components. The distribution of the membership of different instances of the data point to the meta-partitions can be used to determine its meta-cluster membership, or soft assignment probability. Balancing constraints may be added to the meta-clustering phase to ensure that the resulting clusters are balanced.

7.8 Putting Clustering to Work: Applications

Clustering can be considered a specific type of data summarization where the summaries of the data points are constructed on the basis of similarity. Because summarization is a first step to many data mining applications, such summaries can be widely useful. This section will discuss the many applications of data clustering.

7.8.1 Applications to Other Data Mining Problems

Clustering is intimately related to other data mining problems and is used as a first summarization step in these cases. In particular, it is used quite often for the data mining problems of outlier analysis and classification. These specific applications are discussed below.

7.8.1.1 Data Summarization

Although many forms of data summarization, such as sampling, histograms, and wavelets are available for different kinds of data, clustering is the only natural form of summarization based on the notion of similarity. Because the notion of similarity is fundamental to many data mining applications, such summaries are very useful for similarity-based applications. Specific applications include recommendation analysis methods, such as collaborative filtering. This application is discussed later in this chapter, and in Chap. 18 on Web mining.

7.8.1.2 Outlier Analysis

Outliers are defined as data points that are generated by a different mechanism than the normal data points. This can be viewed as a complementary problem to clustering where the goal is to determine groups of closely related data points generated by the same mechanism. Therefore, outliers may be defined as data points that do not lie in any particular cluster. This is of course a simplistic abstraction but is nevertheless a powerful principle as a starting point. Sections 8.3 and 8.4 of Chap. 8 discuss how many algorithms for outlier analysis can be viewed as variations of clustering algorithms.

7.8.1.3 Classification

Many forms of clustering are used to improve the accuracy of classification methods. For example, nearest-neighbor classifiers report the class label of the closest set of training data points to a given test instance. Clustering can help speed up this process by replacing the data points with centroids of fine-grained clusters belonging to a particular class. In addition, semisupervised methods can also be used to perform categorization in many domains such as text. The bibliographic notes contain pointers to these methods.

7.8.1.4 Dimensionality Reduction

Clustering methods, such as nonnegative matrix factorization, are related to the problem of dimensionality reduction. In fact, the *dual* output of this algorithm is a set of concepts, together with a set of clusters. Another related approach is probabilistic latent semantic indexing, which is discussed in Chap. 13 on mining text data. These methods show the intimate relationship between clustering and dimensionality reduction and that common solutions can be exploited by both problems.

7.8.1.5 Similarity Search and Indexing

A hierarchical clustering such as CF-Tree can sometimes be used as an index, at least from a heuristic perspective. For any given target record, only the branches of the tree that are closest to the relevant clusters are searched, and the most relevant data points are returned. This can be useful in many scenarios where it is not practical to build exact indexes with guaranteed accuracy.

7.8.2 Customer Segmentation and Collaborative Filtering

In customer segmentation applications, similar customers are grouped together on the basis of the similarity of their profiles or other actions at a given site. Such segmentation methods are very useful in cases where the data analysis is naturally focused on similar segments of the data. A specific example is the case of *collaborative filtering* applications in which ratings are provided by different customers based on their items of interest. Similar customers are grouped together, and recommendations are made to the customers in a cluster on the basis of the distribution of ratings in a particular group.

7.8.3 Text Applications

Many Web portals need to organize the material at their Web sites on the basis of similarity in content. Text clustering methods can be useful for organization and browsing of text documents. Hierarchical clustering methods can be used to organize the documents in an exploration-friendly tree structure. Many hierarchical directories in Web sites are constructed with a combination of user labeling and semisupervised clustering methods. The semantic insights provided by hierarchical cluster organizations are very useful in many applications.

7.8.4 Multimedia Applications

With the increasing proliferation of electronic forms of multimedia data, such as images, photos, and music, numerous methods have been designed in the literature for finding clusters in such scenarios. Clusters of such multimedia data also provide the user the ability to search for relevant objects in social media Web sites containing this kind of data. This is because heuristic indexes can be constructed with the use of clustering methods. Such indexes are useful for effective retrieval.

7.8.5 Temporal and Sequence Applications

Many forms of temporal data, such as time-series data, and Web logs can be clustered for effective analysis. For example, clusters of sequences in a Web log provide insights about the normal patterns of users. This can be used to reorganize the site, or optimize its structure. In some cases, such information about normal patterns can be used to discover anomalies that do not conform to the normal patterns of interaction. A related domain is that of biological sequence data where clusters of sequences are related to their underlying biological properties.

7.8.6 Social Network Analysis

Clustering methods are useful for finding related communities of users in social-networking Web sites. This problem is known as *community detection*. Community detection has a wide variety of other applications in network science, such as anomaly detection, classification, influence analysis, and link prediction. These applications are discussed in detail in Chap. 19 on social network analysis.

7.9 Summary

This chapter discusses a number of advanced scenarios for cluster analysis. These scenarios include the clustering of advanced data types such as categorical data, large-scale data, and high-dimensional data. Many traditional clustering algorithms can be modified to work with categorical data by making changes to specific criteria, such as the similarity function or mixture model. Scalable algorithms require a change in algorithm design to reduce the number of passes over the data. High-dimensional data is the most difficult case because of the presence of many irrelevant features in the underlying data.

Because clustering algorithms yield many alternative solutions, supervision can help guide the cluster discovery process. This supervision can either be in the form of background knowledge or user interaction. In some cases, the alternative clusterings can be combined to create a *consensus* clustering that is more robust than the solution obtained from a single model.

7.10 Bibliographic Notes

The problem of clustering categorical data is closely related to that of finding suitable similarity measures [104, 182], because many clustering algorithms use similarity measures as a subroutine. The k-modes and a fuzzy version of the algorithm may be found in [135, 278]. Popular clustering algorithms include *ROCK* [238], *CACTUS* [220], *LIMBO* [75], and *STIRR* [229]. The three scalable clustering algorithms discussed in this book are *CLARANS* [407], *BIRCH* [549], and *CURE* [239]. The high-dimensional clustering algorithms discussed in this chapter include *CLIQUE* [58], *PROCLUS* [19], and *ORCLUS* [22]. Detailed surveys on many different types of categorical, scalable, and high-dimensional clustering algorithms may be found in [32].

Methods for semisupervised clustering with the use of seeding, constraints, metric learning, probabilistic learning, and graph-based learning are discussed in [80, 81, 94, 329]. The *IPCLUS* method presented in this chapter was first presented in [43]. Two other tools that are able to discover clusters by visualizing lower dimensional subspaces include *HD-Eye* [268] and *RNavGraph* [502]. The cluster ensemble framework was first proposed in [479]. The hypergraph partitioning algorithm *HMETIS*, which is used in ensemble clustering, was proposed in [302]. Subsequently, the utility of the method has also been demonstrated for high-dimensional data [205].

7.11 Exercises

1. Implement the k-modes algorithm. Download the *KDD CUP 1999 Network Intrusion Data Set* [213] from the *UCI Machine Learning Repository*, and apply the algorithm to the categorical attributes of the data set. Compute the cluster purity with respect to class labels.

2. Repeat the previous exercise with an implementation of the *ROCK* algorithm.

3. What changes would be required to the *BIRCH* algorithm to implement it with the use of the Mahalanobis distance, to compute distances between data points and centroids? The diameter of a cluster is computed as its RMS Mahalanobis radius.

4. Discuss the connection between high-dimensional clustering algorithms, such as *PROCLUS* and *ORCLUS*, and wrapper models for feature selection.

5. Show how to create an implementation of the cluster feature vector that allows the incremental computation of the covariance matrix of the cluster. Use this to create an incremental and scalable version of the Mahalanobis k-means algorithm.

6. Implement the k-means algorithm, with an option of selecting any of the points from the original data as seeds. Apply the approach to the quantitative attributes of the data set in Exercise 1, and select one data point from each class as a seed. Compute the cluster purity with respect to an implementation that uses random seeds.

7. Describe an automated way to determine whether a set of "must-link" and "cannot-link" constraints are consistent.

Chapter 8

Outlier Analysis

"*You are unique, and if that is not fulfilled, then something has been lost.*"—Martha Graham

8.1 Introduction

An outlier is a data point that is very different from most of the remaining data. Hawkins formally defined the notion of an outlier as follows:

"*An outlier is an observation which deviates so much from the other observations as to arouse suspicions that it was generated by a different mechanism.*"

Outliers can be viewed as a complementary concept to that of clusters. While clustering attempts to determine groups of data points that are similar, outliers are *individual* data points that are different from the remaining data. Outliers are also referred to as *abnormalities*, *discordants*, *deviants*, or *anomalies* in the data mining and statistics literature. Outliers have numerous applications in many data mining scenarios:

1. *Data cleaning:* Outliers often represent noise in the data. This noise may arise as a result of errors in the data collection process. Outlier detection methods are, therefore, useful for removing such noise.

2. *Credit card fraud:* Unusual patterns of credit card activity may often be a result of fraud. Because such patterns are much rarer than the normal patterns, they can be detected as outliers.

3. *Network intrusion detection:* The traffic on many networks can be considered as a stream of multidimensional records. Outliers are often defined as unusual records in this stream or unusual changes in the underlying trends.

C. C. Aggarwal, *Data Mining: The Textbook*, DOI 10.1007/978-3-319-14142-8_8

Most outlier detection methods create a model of normal patterns. Examples of such models include clustering, distance-based quantification, or dimensionality reduction. Outliers are defined as data points that do not naturally fit within this normal model. The "outlierness" of a data point is quantified by a numeric value, known as the outlier score. Consequently, most outlier detection algorithms produce an output that can be one of two types:

1. *Real-valued outlier score:* Such a score quantifies the tendency for a data point to be considered an outlier. Higher values of the score make it more (or, in some cases, less) likely that a given data point is an outlier. Some algorithms may even output a probability value quantifying the likelihood that a given data point is an outlier.

2. *Binary label:* A binary value is output, indicating whether or not a data point is an outlier. This type of output contains less information than the first one because a threshold can be imposed on the outlier scores to convert them into binary labels. However, the reverse is not possible. Therefore, outlier scores are more general than binary labels. Nevertheless, a binary score is required as the end result in most applications because it provides a crisp decision.

The generation of an outlier score requires the construction of a model of the normal patterns. In some cases, a model may be designed to produce *specialized* types of outliers based on a very restrictive model of normal patterns. Examples of such outliers are extreme values, and they are useful only for certain specific types of applications. In the following, some of the key models for outlier analysis are summarized. These will be discussed in more detail in later sections.

1. *Extreme values:* A data point is an extreme value, if it lies at one of the two ends of a probability distribution. Extreme values can also be defined equivalently for multidimensional data by using a multivariate probability distribution, instead of a univariate one. These are very *specialized* types of outliers but are useful in general outlier analysis because of their utility in converting scores to labels.

2. *Clustering models:* Clustering is considered a complementary problem to outlier analysis. The former problem looks for data points that occur together in a group, whereas the latter problem looks for data points that are isolated from groups. In fact, many clustering models determine outliers as a side-product of the algorithm. It is also possible to optimize clustering models to specifically detect outliers.

3. *Distance-based models:* In these cases, the k-nearest neighbor distribution of a data point is analyzed to determine whether it is an outlier. Intuitively, a data point is an outlier, if its k-nearest neighbor distance is much larger than that of other data points. Distance-based models can be considered a more fine-grained and instance-centered version of clustering models.

4. *Density-based models:* In these models, the local density of a data point is used to define its outlier score. Density-based models are intimately connected to distance-based models because the local density at a given data point is low only when its distance to its nearest neighbors is large.

5. *Probabilistic models:* Probabilistic algorithms for clustering are discussed in Chap. 6. Because outlier analysis can be considered a complementary problem to clustering, it is natural to use probabilistic models for outlier analysis as well. The steps are almost analogous to those of clustering algorithms, except that the EM algorithm is used for

clustering, and the probabilistic fit values are used to quantify the outlier scores of data points (instead of distance values).

6. *Information-theoretic models:* These models share an interesting relationship with other models. Most of the other methods fix the model of normal patterns and then quantify outliers in terms of deviations from the model. On the other hand, information-theoretic methods constrain the maximum deviation allowed from the normal model and then examine the difference in space requirements for constructing a model with or without a specific data point. If the difference is large, then this point is reported as an outlier.

In the following sections, these different types of models will be discussed in detail. Representative algorithms from each of these classes of algorithms will also be introduced.

It should be pointed out that this chapter defines outlier analysis as an *unsupervised problem* in which previous examples of anomalies and normal data points are not available. The supervised scenario, in which examples of previous anomalies are available, is a special case of the classification problem. That case will be discussed in detail in Chap. 11.

This chapter is organized as follows. Section 8.2 discusses methods for extreme value analysis. Probabilistic methods are introduced in Sect. 8.3. These can be viewed as modifications of EM-clustering methods that leverage the connections between the clustering and outlier analysis problem for detecting outliers. This issue is discussed more formally in Sect. 8.4. Distance-based models for outlier detection are discussed in Sect. 8.5. Density-based models are discussed in Sect. 8.6. Information-theoretic models are addressed in Sect. 8.7. The problem of cluster validity is discussed in Sect. 8.8. A summary is given in Sect. 8.9.

8.2 Extreme Value Analysis

Extreme value analysis is a very specific kind of outlier analysis where the data points at the outskirts of the data are reported as outliers. Such outliers correspond to the *statistical tails* of probability distributions. Statistical tails are more naturally defined for 1-dimensional distributions, although an analogous concept can be defined for the multidimensional case.

It is important to understand that extreme values are very specialized types of outliers; in other words, all extreme values are outliers, but the reverse may not be true. The traditional definition of outliers is based on Hawkins's definition of generative probabilities. For example, consider the 1-dimensional data set corresponding to $\{1, 3, 3, 3, 50, 97, 97, 97, 100\}$. Here, the values 1 and 100 may be considered extreme values. The value 50 is the mean of the data set and is therefore not an extreme value. However, it is the most *isolated* point in the data set and should, therefore, be considered an outlier from a generative perspective.

A similar argument applies to the case of multivariate data where the extreme values lie in the *multivariate tail area* of the distribution. It is more challenging to formally define the concept of multivariate tails, although the basic concept is analogous to that of univariate tails. Consider the example illustrated in Fig. 8.1. Here, data point A may be considered an extreme value, and also an outlier. However, data point B is also isolated, and should, therefore, be considered an outlier. However, it cannot be considered a multivariate extreme value.

Extreme value analysis has important applications in its own right, and, therefore, plays an integral role in outlier analysis. An example of an important application of extreme value analysis is that of converting outlier scores to binary labels by identifying those outlier scores that are extreme values. Multivariate extreme value analysis is often useful in multicriteria

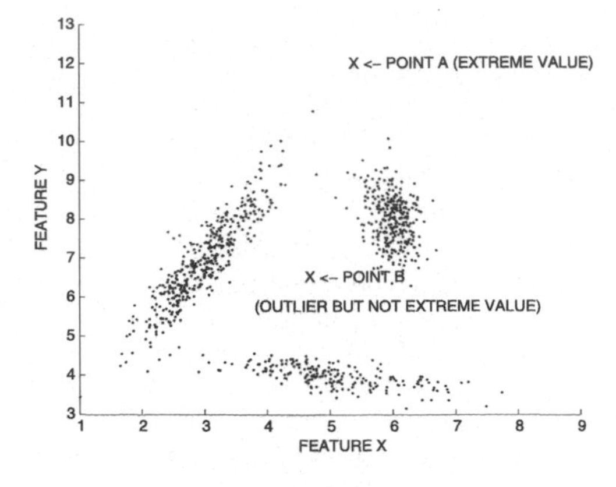

Figure 8.1: Multivariate extreme values

outlier-detection algorithms where it can be utilized to unify multiple outlier scores into a single value, and also generate a binary label as the output. For example, consider a meteorological application where outlier scores of spatial regions have been generated on the basis of analyzing their temperature and pressure variables independently. These evidences need to be unified into a single outlier score for the spatial region, or a binary label. Multivariate extreme value analysis is very useful in these scenarios. In the following discussion, methods for univariate and multivariate extreme value analysis will be discussed.

8.2.1 Univariate Extreme Value Analysis

Univariate extreme value analysis is intimately related to the notion of *statistical tail confidence tests.* Typically, statistical tail confidence tests assume that the 1-dimensional data are described by a specific distribution. These methods attempt to determine the fraction of the objects expected to be more extreme than the data point, based on these distribution assumptions. This quantification provides a level of confidence about whether or not a specific data point is an extreme value.

How is the "tail" of a distribution defined? For distributions that are not symmetric, it is often meaningful to talk about an upper tail and a lower tail, which may not have the same probability. The upper tail is defined as all extreme values larger than a particular threshold, and the lower tail is defined as all extreme values lower than a particular threshold. Consider the density distribution $f_X(x)$. In general, the tail may be defined as the two extreme regions of the distribution for which $f_X(x) \leq \theta$, for some user defined threshold θ. Examples of the lower tail and the upper tail for symmetric and asymmetric distributions are illustrated in Fig. 8.2a and b, respectively. As evident from Fig. 8.2b, the area in the upper tail and the lower tail of an asymmetric distribution may not be the same. Furthermore, some regions in the interior of the distribution of Fig. 8.2b have density below the density threshold θ, but are not extreme values because they do not lie in the tail of the distribution. The data points in this region may be considered outliers, but not extreme values. The areas inside the upper tail or lower tail in Fig. 8.2a and b represent the cumulative probability of these extreme regions. In symmetric probability distributions, the tail is defined in terms of this area, rather than a density threshold. However, the concept of density threshold is the defining characteristic of the tail, especially in the case of asymmetric univariate or multivariate

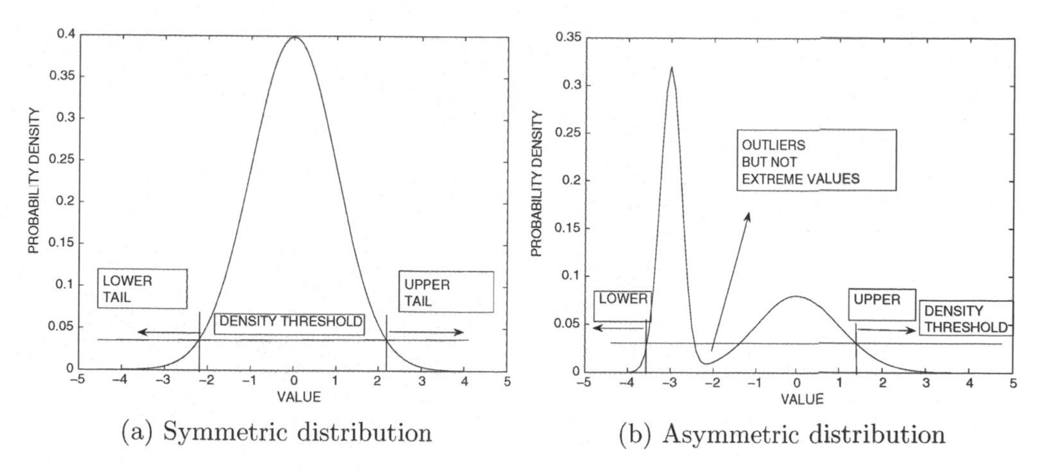

(a) Symmetric distribution (b) Asymmetric distribution

Figure 8.2: Tails of a symmetric and asymmetric distribution

distributions. Some asymmetric distributions, such as an exponential distribution, may not even have a tail at one end of the distribution.

A *model* distribution is selected for quantifying the tail probability. The most commonly used model is the normal distribution. The density function $f_X(x)$ of the normal distribution with mean μ and standard deviation σ is defined as follows:

$$f_X(x) = \frac{1}{\sigma \cdot \sqrt{2 \cdot \pi}} \cdot e^{\frac{-(x-\mu)^2}{2 \cdot \sigma^2}}. \tag{8.1}$$

A *standard* normal distribution is one in which the mean is 0, and the standard deviation σ is 1. In some application scenarios, the mean μ and standard deviation σ of the distribution may be known through prior domain knowledge. Alternatively, when a *large* number of data samples is available, the mean and standard deviation may be estimated very accurately. These can be used to compute the Z-value for a random variable. The Z-number z_i of an observed value x_i can be computed as follows:

$$z_i = (x_i - \mu)/\sigma. \tag{8.2}$$

Large positive values of z_i correspond to the upper tail, whereas large negative values correspond to the lower tail. The normal distribution can be expressed directly in terms of the Z-number because it corresponds to a scaled and translated random variable with a mean 0 and standard deviation of 1. The normal distribution of Eq. 8.3 can be written directly in terms of the Z-number, with the use of a *standard* normal distribution as follows:

$$f_X(z_i) = \frac{1}{\sigma \cdot \sqrt{2 \cdot \pi}} \cdot e^{\frac{-z_i^2}{2}}. \tag{8.3}$$

This implies that the cumulative normal distribution may be used to determine the area of the tail that is larger than z_i. As a rule of thumb, if the absolute values of the Z-number are greater than 3, the corresponding data points are considered extreme values. At this threshold, the cumulative area inside the tail can be shown to be less than 0.01 % for the normal distribution.

When a smaller number n of data samples is available for estimating the mean μ and standard deviations σ, the aforementioned methodology can be used with a minor modification. The value of z_i is computed as before, and the student t-distribution with n degrees

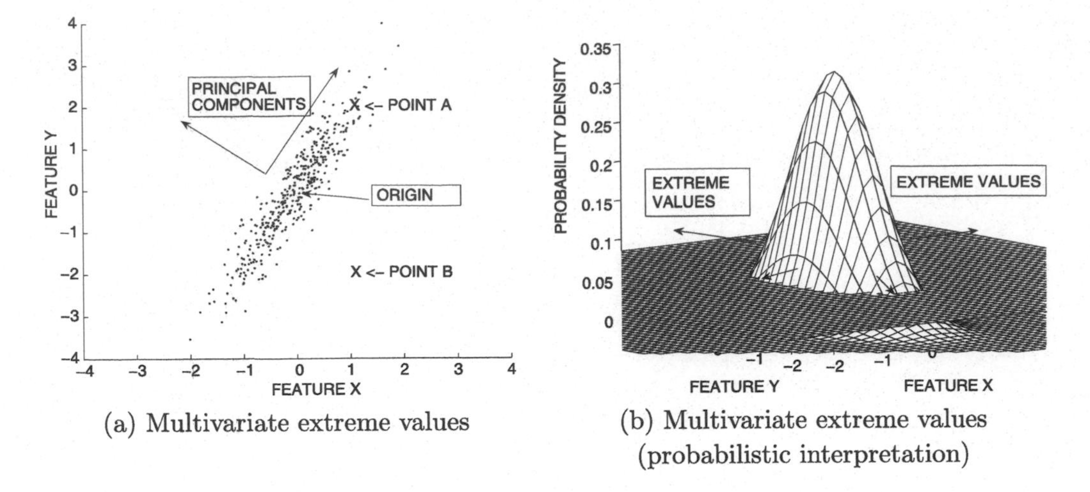

(a) Multivariate extreme values

(b) Multivariate extreme values (probabilistic interpretation)

Figure 8.3: Multivariate extreme values

of freedom is used to quantify the cumulative distribution in the tail instead of the normal distribution. Note that, when n is large, the t-distribution converges to the normal distribution.

8.2.2 Multivariate Extreme Values

Strictly speaking, tails are defined for univariate distributions. However, just as the univariate tails are defined as extreme regions with probability density less than a particular threshold, an analogous concept can also be defined for multivariate distributions. The concept is more complex than the univariate case and is defined for unimodal probability distributions with a single peak. As in the previous case, a multivariate Gaussian model is used, and the corresponding parameters are estimated in a data-driven manner. The implicit modeling assumption of multivariate extreme value analysis is that all data points are located in a probability distribution with a single peak (i.e., single Gaussian cluster), and data points in all directions that are as far away as possible from the center of the cluster should be considered extreme values.

Let $\overline{\mu}$ be the d-dimensional mean vector of a d-dimensional data set, and Σ be its $d \times d$ covariance matrix. Thus, the (i, j)th entry of the covariance matrix is equal to the covariance between the dimensions i and j. These represent the estimated parameters of the multivariate Gaussian distribution. Then, the probability distribution $f(\overline{X})$ for a d-dimensional data point $\overline{X}$ can be defined as follows:

$$f(\overline{X}) = \frac{1}{\sqrt{|\Sigma|} \cdot (2 \cdot \pi)^{(d/2)}} \cdot e^{-\frac{1}{2} \cdot (\overline{X} - \overline{\mu}) \Sigma^{-1} (\overline{X} - \overline{\mu})^T}. \tag{8.4}$$

The value of $|\Sigma|$ denotes the determinant of the covariance matrix. The term in the exponent is half the square of the *Mahalanobis distance* between data point $\overline{X}$ and the mean $\overline{\mu}$ of the data. In other words, if $Maha(\overline{X}, \overline{\mu}, \Sigma)$ represents the Mahalanobis distance between $\overline{X}$ and $\overline{\mu}$, with respect to the covariance matrix Σ, then the probability density function of the normal distribution is as follows:

$$f(\overline{X}) = \frac{1}{\sqrt{|\Sigma|} \cdot (2 \cdot \pi)^{(d/2)}} \cdot e^{-\frac{1}{2} \cdot Maha(\overline{X}, \overline{\mu}, \Sigma)^2}. \tag{8.5}$$

For the probability density to fall below a particular threshold, the Mahalanobis distance needs to be *larger* than a particular threshold. Thus, the Mahalanobis distance to the mean of the data can be used as an extreme-value score. The relevant extreme values are defined by the multidimensional region of the data for which the Mahalanobis distance to the mean is larger than a particular threshold. This region is illustrated in Fig. 8.3b. Therefore, the extreme value score for a data point can be reported as the Mahalanobis distance between that data point and the mean. Larger values imply more extreme behavior. In some cases, one might want a more intuitive probability measure. Correspondingly, the extreme value *probability* of a data point $\overline{X}$ is defined by the cumulative probability of the multidimensional region for which the Mahalanobis distance to the mean $\overline{\mu}$ of the data is greater than that between $\overline{X}$ and $\overline{\mu}$. How can one estimate this cumulative probability?

As discussed in Chap. 3, the Mahalanobis distance is similar to the Euclidean distance except that it standardizes the data along uncorrelated directions. For example, if the axis system of the data were to be rotated to the principal directions (shown in Fig. 8.3), then the transformed coordinates in this new axis system would have no interattribute correlations (i.e., a diagonal covariance matrix). The Mahalanobis distance is simply equal to the Euclidean distance in such a transformed (axes-rotated) data set *after* dividing each of the transformed coordinate values by the standard deviation along its direction. This approach provides a neat way to model the probability distribution of the Mahalanobis distance, and it also provides a concrete estimate of the cumulative probability in the multivariate tail.

Because of the scaling by the standard deviation, each of the independent components of the Mahalanobis distances along the principal correlation directions can be modeled as a 1-dimensional *standard* normal distribution with mean 0 and variance 1. The sum of the squares of d variables, drawn independently from standard normal distributions, will result in a variable drawn from an χ^2 distribution with d degrees of freedom. Therefore, the cumulative probability of the region of the χ^2 distribution with d degrees of freedom, for which the value is greater than $Maha(\overline{X}, \overline{\mu}, \Sigma)$, can be reported as the extreme value probability of $\overline{X}$. Smaller values of the probability imply greater likelihood of being an extreme value.

Intuitively, this approach models the data distribution along the various uncorrelated directions as statistically independent normal distributions and standardizes them so as to provide each such direction equal importance in the outlier score. In Fig. 8.3a, data point B can be more reasonably considered a multivariate extreme value than data point A, on the basis of the natural correlations in the data. On the other hand, the data point B is closer to the centroid of the data (than data point A) on the basis of Euclidean distance but not on the basis of the Mahalanobis distance. This shows the utility of the Mahalanobis distance in using the underlying statistical distribution of the data to infer the outlier behavior of the data points more effectively.

8.2.3 Depth-Based Methods

Depth-based methods are based on the general principle that the convex hull of a set of data points represents the pareto-optimal extremes of this set. A depth-based algorithm proceeds in an iterative fashion, where during the k-th iteration, all points at the corners of the convex hull of the data set are removed. The index of the iteration k also provides an outlier score where smaller values indicate a greater tendency for a data point to be an outlier. These steps are repeated until the data set is empty. The outlier score may be converted to a binary label by reporting all data points with depth at most r as outliers.

Algorithm *FindDepthOutliers*(Data Set: $\mathcal{D}$, Score Threshold: r)
begin
  $k = 1$;
  repeat
    Find set S of corners of convex hull of $\mathcal{D}$;
    Assign depth k to points in S;
    $\mathcal{D} = \mathcal{D} - S$;
    $k = k + 1$;
  until(D is empty);
  Report points with depth at most r as outliers;
end

Figure 8.4: Depth-based methods

The value of r may itself need to be determined by univariate extreme value analysis. The steps of the depth-based approach are illustrated in Fig. 8.4.

A pictorial illustration of the depth-based method is illustrated in Fig. 8.5. The process can be viewed as analogous to peeling the different layers of an onion (as shown in Fig. 8.5b) where the outermost layers define the outliers. Depth-based methods try to achieve the same goal as the multivariate method of the previous section, but it generally tends to be less effective both in terms of quality and computational efficiency. From a qualitative perspective, depth-based methods do not normalize for the characteristics of the statistical data distribution, as is the case for multivariate methods based on the Mahalanobis distance. All data points at the corners of a convex hull are treated equally. This is clearly not desirable, and the scores of many data points are indistinguishable because of ties. Furthermore, the fraction of data points at the corners of the convex hull generally increases with dimensionality. For very high dimensionality, it may not be uncommon for the majority of the data points to be located at the corners of the outermost convex hull. As a result, it is no longer possible to distinguish the outlier scores of different data points. The computational complexity of convex-hull methods increases significantly with dimensionality. The combination of qualitative and computational issues associated with this method make it a poor alternative to the multivariate methods based on the Mahalanobis distance.

8.3 Probabilistic Models

Probabilistic models are based on a generalization of the multivariate extreme values analysis methods discussed in Sect. 8.2.2. The Mahalanobis distance-based multivariate extreme value analysis method can be viewed as a Gaussian mixture model with a *single* component in the mixture. By generalizing this model to multiple mixture components, it is possible to determine general outliers, rather than multivariate extreme values. This idea is intimately related to the EM-clustering algorithm discussed in Sect. 6.5 of Chap. 6. At an intuitive level, data points that do not naturally fit any cluster in the probabilistic sense may be reported as outliers. The reader is referred to Sect. 6.5 of Chap. 6 for a more detailed discussion of the EM algorithm, though a brief outline is provided here for convenience.

The broad principle of a mixture-based generative model is to assume that the data were generated from a mixture of k distributions with the probability distributions $\mathcal{G}_1 \ldots \mathcal{G}_k$ based on the following process:

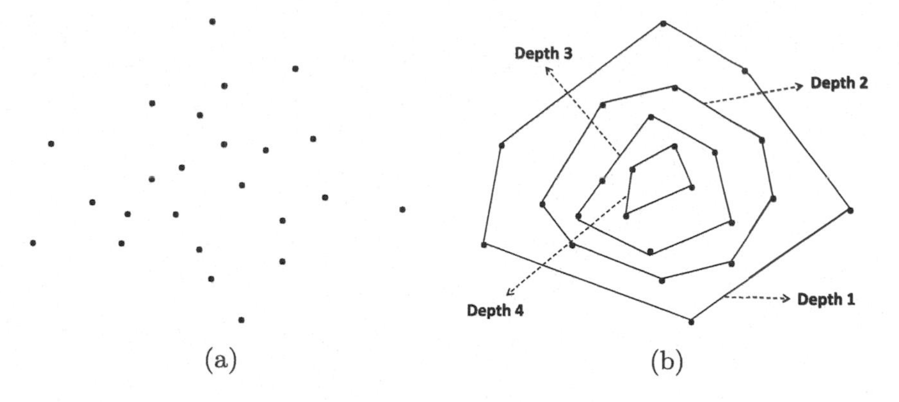

Figure 8.5: Depth-based outlier detection

1. Select a mixture component with prior probability α_i, where $i \in \{1 \ldots k\}$. Assume that the rth one is selected.

2. Generate a data point from $\mathcal{G}_r$.

This generative model will be denoted by $\mathcal{M}$, and it generates each point in the data set $\mathcal{D}$. The data set $\mathcal{D}$ is used to estimate the parameters of the model. Although it is natural to use Gaussians to represent each component of the mixture, other models may be used if desired. This flexibility is very useful to apply the approach to different data types. For example, in a categorical data set, a categorical probability distribution may be used for each mixture component instead of the Gaussian distribution. After the parameters of the model have been estimated, outliers are defined as those data points in $\mathcal{D}$ that are highly unlikely to be generated by this model. Note that such an assumption exactly reflects Hawkins's definition of outliers, as stated at the beginning of this chapter.

Next, we discuss the estimation of the various parameters of the model such as the estimation of different values of α_i and the parameters of the different distributions $\mathcal{G}_r$. The objective function of this estimation process is to ensure that the full data $\mathcal{D}$ has the maximum likelihood fit to the generative model. Assume that the density function of $\mathcal{G}_i$ is given by $f^i(\cdot)$. The probability (density function) of the data point $\overline{X_j}$ being generated by the model is given by the following:

$$f^{point}(\overline{X_j}|\mathcal{M}) = \sum_{i=1}^{k} \alpha_i \cdot f^i(\overline{X_j}). \tag{8.6}$$

Note that the density value $f^{point}(\overline{X_j}|\mathcal{M})$ provides an estimate of the outlier score of the data point. Data points that are outliers will naturally have low fit values. Examples of the relationship of the fit values to the outlier scores are illustrated in Fig. 8.6. Data points A and B will typically have very low fit to the mixture model and will be considered outliers because the data points A and B do not naturally belong to any of the mixture components. Data point C will have high fit to the mixture model and will, therefore, not be considered an outlier. The parameters of the model $\mathcal{M}$ are estimated using a maximum likelihood criterion, which is discussed below.

For data set $\mathcal{D}$ containing n data points, denoted by $\overline{X_1} \ldots \overline{X_n}$, the probability density of the data set being generated by model $\mathcal{M}$ is the product of the various point-specific

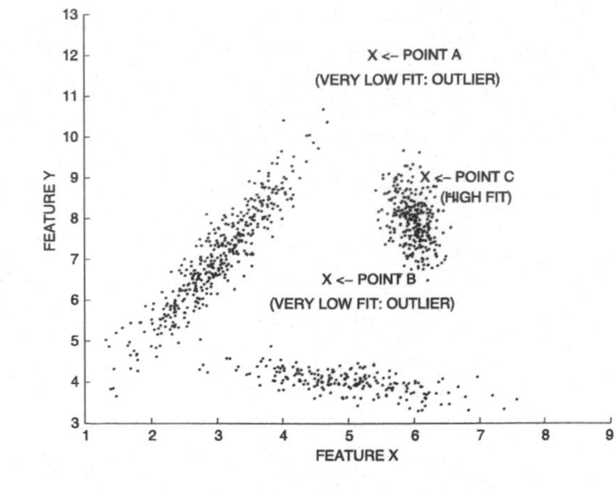

Figure 8.6: Likelihood fit values versus outlier scores

probability densities:

$$f^{data}(\mathcal{D}|\mathcal{M}) = \prod_{j=1}^{n} f^{point}(\overline{X_j}|\mathcal{M}). \tag{8.7}$$

The log-likelihood fit $\mathcal{L}(\mathcal{D}|\mathcal{M})$ of the data set $\mathcal{D}$ with respect to $\mathcal{M}$ is the logarithm of the aforementioned expression, and can be (more conveniently) represented as a sum of values over the different data points:

$$\mathcal{L}(\mathcal{D}|\mathcal{M}) = \log(\prod_{j=1}^{n} f^{point}(\overline{X_j}|\mathcal{M})) = \sum_{j=1}^{n} \log(\sum_{i=1}^{k} \alpha_i \cdot f^i(\overline{X_j})). \tag{8.8}$$

This log-likelihood fit needs to be optimized to determine the model parameters. This objective function maximizes the fit of the data points to the generative model. For this purpose, the EM algorithm discussed in Sect. 6.5 of Chap. 6 is used.

After the parameters of the model have been determined, the value of $f^{point}(\overline{X_j}|\mathcal{M})$ (or its logarithm) may be reported as the outlier score. The major advantage of such mixture models is that the mixture components can also incorporate domain knowledge about the shape of each individual mixture component. For example, if it is known that the data points in a particular cluster are correlated in a certain way, then this fact can be incorporated in the mixture model by fixing the appropriate parameters of the covariance matrix, and learning the remaining parameters. On the other hand, when the available data is limited, mixture models may overfit the data. This will cause data points that are truly outliers to be missed.

8.4 Clustering for Outlier Detection

The probabilistic algorithm of the previous section provides a preview of the relationship between clustering and outlier detection. Clustering is all about finding "crowds" of data points, whereas outlier analysis is all about finding data points that are far away from these crowds. Clustering and outlier detection, therefore, share a well-known complementary relationship. A simplistic view is that every data point is either a member of a cluster or an outlier. Clustering algorithms often have an "outlier handling" option that removes data

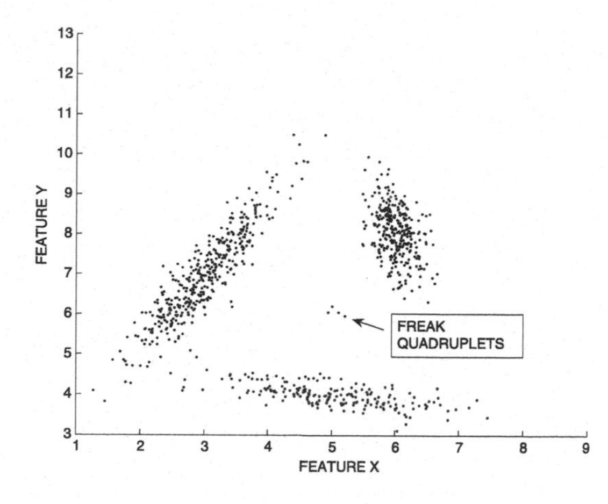

Figure 8.7: Small isolated groups of anomalies

points outside the clusters. The detection of outliers as a side-product of clustering methods is, however, not an appropriate approach because clustering algorithms are not optimized for outlier detection. Data points on the boundary regions of a cluster may also be considered weak outliers but are rarely useful in most application-specific scenarios.

Clustering models do have some advantages as well. Outliers often tend to occur in small clusters of their own. This is because the anomaly in the generating process may be repeated a few times. As a result, a small group of related outliers may be created. An example of a small set of isolated outliers is illustrated in Fig. 8.7. As will be discussed later, clustering methods are generally robust to such scenarios because such groups often do not have the critical mass required to form clusters of their own.

A simple way of defining the outlier score of a data point is to first cluster the data set and then use the raw distance of the data point to its closest cluster centroid. One can, however, do better when the clusters are elongated or have varying density over the data set. As discussed in Chap. 3, the local data distribution often distorts the distances, and, therefore, it is not optimal to use the raw distance. This broader principle is used in multivariate extreme value analysis where the *global* Mahalanobis distance defines outlier scores. In this case, the *local* Mahalanobis distance can be used with respect to the centroid of the closest cluster.

Consider a data set in which k clusters have been discovered with the use of a clustering algorithm. Assume that the rth cluster in d-dimensional space has a corresponding d-dimensional mean vector $\overline{\mu_r}$, and a $d \times d$ covariance matrix Σ_r. The (i, j)th entry of this matrix is the covariance between the dimensions i and j in that cluster. Then, the Mahalanobis distance $Maha(\overline{X}, \overline{\mu_r}, \Sigma_r)$ between a data point $\overline{X}$ and cluster centroid $\overline{\mu_r}$ is defined as follows:

$$Maha(\overline{X}, \overline{\mu_r}, \Sigma_r) = \sqrt{(\overline{X} - \overline{\mu_r})\Sigma_r^{-1}(\overline{X} - \overline{\mu_r})^T}. \quad (8.9)$$

This distance is reported as the outlier score. Larger values of the outlier score indicate a greater outlier tendency. After the outlier score has been determined, univariate extreme value analysis may be used to convert the scores to binary labels.

The justification for using the Mahalanobis distance is exactly analogous to the case of extreme value analysis of multivariate distances, as discussed in Sect. 8.2. The only difference is that the *local cluster-specific* Mahalanobis distances are more relevant to determination of general outliers, whereas global Mahalanobis distances are more relevant to determination of specific types of outliers, such as extreme values. The use of the local Mahalanobis distance also has an interesting connection to the likelihood fit criterion of EM algorithm where the (squared) Mahalanobis distance occurs in the exponent of each Gaussian mixture. Thus, the sum of the inverse exponentiated Mahalanobis distances of a data point to different mixture component means (cluster means) are used to determine the outlier score in the EM algorithm. Such a score can be viewed as a soft version of the score determined by hard clustering algorithms.

Clustering methods are based on global analysis. Therefore, small, closely related groups of data points will not form their own clusters in most cases. For example, the four isolated points in Fig. 8.7 will not typically be considered a cluster. Most clustering algorithms require a minimum critical mass for a set of data points to be considered a cluster. As a result, these points will have a high outlier score. This means that clustering methods are able to detect these small and closely related groups of data points meaningfully and report them as outliers. This is not the case for some of the density-based methods that are based purely on local analysis.

The major problem with clustering algorithms is that they are sometimes not able to properly distinguish between a data point that is *ambient noise* and a data point that is a *truly isolated anomaly.* Clearly, the latter is a much stronger anomaly than the former. Both these types of points will not reside in a cluster. Therefore, the distance to the closest cluster centroid will often not be very representative of their local distribution (or *instance-specific* distribution). In these cases, distance-based methods are more effective.

8.5 Distance-Based Outlier Detection

Because outliers are defined as data points that are far away from the "crowded regions" (or clusters) in the data, a natural and *instance-specific* way of defining an outlier is as follows:

The distance-based outlier score of an object O is its distance to its kth nearest neighbor.

The aforementioned definition, which uses the k-nearest neighbor distance, is the most common one. Other variations of this definition are sometimes used, such as the average distance to the k-nearest neighbors. The value of k is a user-defined parameter. Selecting a value of k larger than 1 helps identify isolated groups of outliers. For example, in Fig. 8.7, as long as k is fixed to any value larger than 3, all data points within the small groups of closely related points will have a high outlier score. Note that the target data point, for which the outlier score is computed, is itself not included among its k-nearest neighbors. This is done to avoid scenarios where a 1-nearest neighbor method will always yield an outlier score of 0.

Distance-based methods typically use a finer granularity of analysis than clustering methods and can therefore distinguish between ambient noise and truly isolated anomalies. This is because ambient noise will typically have a lower k-nearest neighbor distance than a truly isolated anomaly. This distinction is lost in clustering methods where the distance to the closest cluster centroid does not accurately reflect the *instance-specific* isolation of the underlying data point.

The price of this better granularity is higher computational complexity. Consider a data set $\mathcal{D}$ containing n data points. The determination of the k-nearest neighbor distance requires $O(n)$ time *for each data point*, when a sequential scan is used. Therefore, the determination of the outlier scores of all data points may require $O(n^2)$ time. This is clearly not a feasible option for very large data sets. Therefore, a variety of methods are used to speed up the computation:

1. *Index structures:* Index structures can be used to determine kth nearest neighbor distances efficiently. This is, however, not an option, if the data is high dimensional. In such cases, the effectiveness of index structures tends to degrade.

2. *Pruning tricks:* In most applications, the outlier scores of all the data points are not required. It may suffice to return binary labels for the top-r outliers, together with their scores. The outlier scores of the remaining data points are irrelevant. In such cases, it may be possible to terminate a k-nearest neighbor sequential scan for an outlier candidate when its current *upper bound estimate* on the k-nearest neighbor distance value falls below the rth best outlier score found so far. This is because such a candidate is guaranteed to be not among the top-r outliers. This methodology is referred to as the "early termination trick," and it is described in detail later in this section.

In some cases, it is also possible to combine the pruning approach with index structures.

8.5.1 Pruning Methods

Pruning methods are used only for the case where the top-r ranked outliers need to be returned, and the outlier scores of the remaining data points are irrelevant. Thus, pruning methods can be used only for the binary-decision version of the algorithm. The basic idea in pruning methods is to reduce the time required for the k-nearest neighbor distance computations by ruling out data points quickly that are obviously nonoutliers even with approximate computation.

8.5.1.1 Sampling Methods

The first step is to pick a sample $\mathcal{S}$ of size $s \ll n$ from the data $\mathcal{D}$, and compute all pairwise distances between the data points in sample $\mathcal{S}$ and those in database $\mathcal{D}$. There are a total of $n \cdot s$ such pairs. This process requires $O(n \cdot s) \ll O(n^2)$ distance computations. Thus, for each of the sampled points in $\mathcal{S}$, the k-nearest neighbor distance is already known exactly. The top rth ranked outlier in sample $\mathcal{S}$ is determined, where r is the number of outliers to be returned. The score of the rth rank outlier provides a *lower* bound[1] L on the rth ranked outlier score over the entire data set $\mathcal{D}$. For the data points in $\mathcal{D} - \mathcal{S}$, only an *upper bound* $V^k(\overline{X})$ on the k-nearest neighbor distance is known. This upper bound is equal to the k-nearest neighbor distance of each point in $\mathcal{D} - \mathcal{S}$ to the sample $\mathcal{S} \subset \mathcal{D}$. However, if this upper bound $V^k(\overline{X})$ is no larger than the lower bound L already determined, then such a data point $\overline{X} \in \mathcal{D} - \mathcal{S}$ can be excluded from further consideration as a top-r outlier. Typically, this will result in the removal of a large number of outlier candidates from $\mathcal{D} - \mathcal{S}$ immediately, as long as the underlying data set is clustered well. This is because most of the data points in clusters will be removed, as long as at least one point from each cluster is included in the sample $\mathcal{S}$, and at least r points in $\mathcal{S}$ are located in somewhat sparse regions. This can often be achieved with modest values of the sample size s in real-world data sets. After removing these data points from $\mathcal{D} - \mathcal{S}$, the remaining set of points is $\mathcal{R} \subseteq \mathcal{D} - \mathcal{S}$. The k-nearest neighbor approach can be applied to a much smaller set of candidates $\mathcal{R}$.

[1]Note that higher k-nearest neighbor distances indicate greater outlierness.

The top-r ranked outliers in $\mathcal{R} \cup \mathcal{S}$ are returned as the final output. Depending on the level of pruning already achieved, this can result in a very significant reduction in computational time, especially when $|\mathcal{R} \cup \mathcal{S}| \ll |\mathcal{D}|$.

8.5.1.2 Early Termination Trick with Nested Loops

The approach discussed in the previous section can be improved even further by speeding up the second phase of computing the k-nearest neighbor distances of each data point in $\mathcal{R}$. The idea is that the computation of the k-nearest neighbor distance of any data point $\overline{X} \in \mathcal{R}$ need not be followed through to termination once it has been determined that $\overline{X}$ cannot possibly be among the top-r outliers. In such cases, the scan of the database $\mathcal{D}$ for computation of the k-nearest neighbor of $\overline{X}$ can be terminated early.

Note that one already has an estimate (upper bound) $V^k(\overline{X})$ of the k-nearest neighbor distance of every $\overline{X} \in \mathcal{R}$, based on distances to sample $\mathcal{S}$. Furthermore, the k-nearest neighbor distance of the rth best outlier in $\mathcal{S}$ provides a lower bound on the "cut-off" required to make it to the top-r outliers. This lower-bound is denoted by L. This estimate $V^k(\overline{X})$ of the k-nearest neighbor distance of $\overline{X}$ is further tightened (reduced) as the database $\mathcal{D} - \mathcal{S}$ is scanned, and the distance of $\overline{X}$ is computed to each point in $\mathcal{D} - \mathcal{S}$. Because this running estimate $V^k(\overline{X})$ is always an upper bound on the true k-nearest neighbor distance of $\overline{X}$, the process of determining the k-nearest neighbor of $\overline{X}$ can be terminated as soon as $V^k(\overline{X})$ falls below the known lower bound L on the top-r outlier distance. This is referred to as *early termination* and provides significant computational savings. Then, the next data point in $\mathcal{R}$ can be processed. In cases where early termination is not achieved, the data point $\overline{X}$ will almost[2] always be among the top-r (current) outliers. Therefore, in this case, the lower bound L can be tightened (increased) as well, to the new rth best outlier score. This will result in even better pruning when the next data point from $\mathcal{R}$ is processed to determine its k-nearest neighbor distance value. To maximize the benefits of pruning, the data points in $\mathcal{R}$ should not be processed in arbitrary order. Rather, they should be processed in decreasing order of the initially sampled estimate $V^k(\cdot)$ of the k-nearest neighbor distances (based on $\mathcal{S}$). This ensures that the outliers in $\mathcal{R}$ are found early on, and the global bound L is tightened as fast as possible for even better pruning. Furthermore, in the inner loop, the data points $\overline{Y}$ in $\mathcal{D} - \mathcal{S}$ can be ordered in the opposite direction, based on *increasing* value of $V^k(\overline{Y})$. Doing so ensures that the k-nearest neighbor distances are updated as fast as possible, and the advantage of early termination is maximized. The nested loop approach can also be implemented without the first phase[3] of sampling, but such an approach will not have the advantage of proper ordering of the data points processed. Starting with an initial lower bound L on the rth best outlier score obtained from the sampling phase, the nested loop is executed as follows:

[2]We say "almost," because the very last distance computation for $\overline{X}$ may bring $V(\overline{X})$ below L. This scenario is unusual, but might occasionally occur.

[3]Most descriptions in the literature omit the first phase of sampling, which is very important for efficiency maximization. A number of implementations in time-series analysis [306] do order the data points more carefully but not with sampling.

for each $\overline{X} \in \mathcal{R}$ **do begin**
  for each $\overline{Y} \in \mathcal{D} - \mathcal{S}$ **do begin**
    Update current k-nearest neighbor distance estimate $V^k(\overline{X})$ by
      computing distance of $\overline{Y}$ to $\overline{X}$;
    if $V^k(\overline{X}) \leq L$ **then** terminate inner loop;
  endfor
  if $V^k(\overline{X}) > L$ **then**
    include $\overline{X}$ in current r best outliers and update L to
      the new rth best outlier score;
endfor

Note that the k-nearest neighbors of a data point $\overline{X}$ do not include the data point itself. Therefore, care must be taken in the nested loop structure to ignore the trivial cases where $\overline{X} = \overline{Y}$ while updating k-nearest neighbor distances.

8.5.2 Local Distance Correction Methods

Section 3.2.1.8 of Chap. 3 provides a detailed discussion of the impact of the local data distribution on distance computation. In particular, it is shown that straightforward measures, such as the Euclidean distance, do not reflect the *intrinsic* distances between data points when the density and shape of the clusters vary significantly with data locality. This principle was also used in Sect. 8.4 to justify the use of the local Mahalanobis distance for measuring the distances to cluster centroids, rather than the Euclidean distance. One of the earliest methods that recognized this principle in the context of varying data density was the *L*ocal *O*utlier *F*actor (LOF) method. A formal justification is based on the generative principles of data sets, but only an intuitive understanding will be provided here. It should be pointed out that the use of the Mahalanobis distance (instead of the Euclidean distance) for multivariate extreme value analysis (Sect. 8.2.2) is also based on generative principles of the *likelihood* of a data point conforming to the statistical properties of the underlying distribution. The main difference is that the analysis was *global* in that case, whereas the analysis is *local* in this case. The reader is also advised to revisit Sect. 3.2.1.8 of Chap. 3 for a discussion of the impact of data distributions on *intrinsic* distances between data points.

To motivate the principles of local distance correction in the context of outlier analysis, two examples will be used. One of these examples illustrates the impact of varying local distribution density, whereas another example illustrates the impact of varying local cluster shape. Both these aspects can be addressed with different kinds of local normalization of distance computations. In Fig. 8.8a, two different clusters have been shown, one of which is much sparser than the other. In this case, both data points A and B are clearly outliers. While the outlier B will be easily detected by most distance-based algorithms, a challenge arises in the detection of outlier A. This is because the nearest neighbor distance of many data points in the sparser cluster is at least as large as the nearest neighbor distance of outlier A. As a result, depending on the distance-threshold used, a k-nearest neighbor algorithm will either falsely report portions of the sparse cluster, or will completely miss outlier A. Simply speaking, the *ranking* of the outliers by distance-based algorithms is an incorrect one. This is because the true distance of points in cluster A should be computed in a *normalized* way, based on its *local* data distribution. This aspect is relevant to the discussion in Sect. 3.2.1.8 of Chap. 3 on the impact of local data distributions on distance function design, and it is important for many distance-based data mining problems. The key

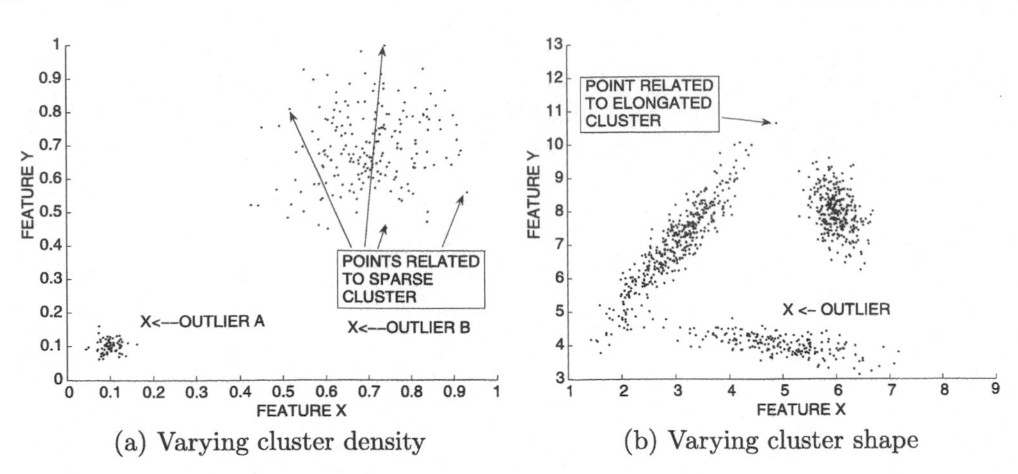

(a) Varying cluster density

(b) Varying cluster shape

Figure 8.8: Impact of local variations in data distribution on distance-based outlier detection

issue here is the generative principle, that data point A is much *less* likely to be generated by its closest (tightly knit) cluster than many slightly isolated data points belonging to the relatively diffuse cluster are likely to be generated by their cluster. Hawkins's definition of outliers, stated at the beginning of this chapter, was formulated on the basis of generative principles. It should be pointed out that the probabilistic EM algorithm of Sect. 8.3 does a much better job at recognizing these generative differences. However, the probabilistic EM method is often not used practically because of overfitting issues with smaller data sets. The *LOF* approach is the first method that recognized the importance of incorporating these generative principles in nonparametric distance-based algorithms.

This point can be emphasized further by examining clusters of different local shape and orientation in Fig. 8.8b. In this case, a distance-based algorithm will report one of the data points along the long axis of one of the elongated clusters, as the strongest outlier, if the 1-nearest neighbor distance is used. This data point is far *more* likely to be generated by its closest cluster, than the outlier marked by "X." However, the latter has a smaller 1-nearest neighbor distance. Therefore, the significant problem with distance-based algorithms is that they do not account for the local generative behavior of the underlying data. In this section, two methods will be discussed for addressing this issue. One of them is *LOF*, and the other is a direct generalization of the global Mahalanobis method for extreme value analysis. The first method can adjust for the generative variations illustrated in Fig. 8.8a, and the second method can adjust for the generative variations illustrated in Fig. 8.8b.

8.5.2.1 Local Outlier Factor (LOF)

The *Local Outlier Factor (LOF)* approach adjusts for local variations in cluster density by normalizing distances with the average point-specific distances in a data locality. It is often understood popularly as a density-based approach, although, in practice, it is a (normalized) distance-based approach where the normalization factor corresponds to the average local data density. This normalization is the key to addressing the challenges posed by the scenario of Fig. 8.8a.

For a given data point $\overline{X}$, let $V^k(\overline{X})$ be the distance to its k-nearest neighbor, and let $L_k(\overline{X})$ be the set of points within the k-nearest neighbor distance of $\overline{X}$. The set $L_k(\overline{X})$ will typically contain k points, but may sometimes contain more than k points because of ties in the k-nearest neighbor distance.

Then, the reachability distance $R_k(\overline{X}, \overline{Y})$ of object $\overline{X}$ with respect to $\overline{Y}$ is defined as the maximum of the distance $Dist(\overline{X}, \overline{Y})$, between the pair $(\overline{X}, \overline{Y})$ and the k-nearest neighbor distance of $\overline{Y}$.

$$R_k(\overline{X}, \overline{Y}) = \max\{Dist(\overline{X}, \overline{Y}), V^k(\overline{Y})\} \tag{8.10}$$

The reachability distance is not symmetric between $\overline{X}$ and $\overline{Y}$. Intuitively, when $\overline{Y}$ is in a dense region and the distance between $\overline{X}$ and $\overline{Y}$ is large, the reachability distance of $\overline{X}$ with respect to it is equal to the true distance $Dist(\overline{X}, \overline{Y})$. On the other hand, when the distances between $\overline{X}$ and $\overline{Y}$ are small, then the reachability distance is smoothed out by the k-nearest neighbor distance of $\overline{Y}$. The larger the value of k, the greater the smoothing. Correspondingly, the reachability distances with respect to different points will also become more similar. The reason for using this smoothing is that it makes the intermediate distance computations more stable. This is especially important when the distances between $\overline{X}$ and $\overline{Y}$ are small, and it will result in greater statistical fluctuations in the raw distances. At a conceptual level, it is possible to define a version of *LOF* directly in terms of raw distances, rather than reachability distances. However, such a version would be missing the stability provided by smoothing.

The *average reachability distance* $AR_k(\overline{X})$ of data point $\overline{X}$ with respect to its neighborhood $L_k(\overline{X})$ is defined as the average of its reachability distances to all objects in its neighborhood.

$$AR_k(\overline{X}) = \text{MEAN}_{\overline{Y} \in L_k(\overline{X})} R_k(\overline{X}, \overline{Y}) \tag{8.11}$$

Here, the MEAN function simply represents the mean value over the entire set $L_k(\overline{X})$. The *Local Outlier Factor* $LOF_k(\overline{X})$ is then equal to the mean ratio of $AR_k(\overline{X})$ to the corresponding values of all points in the k-neighborhood of $\overline{X}$.

$$LOF_k(\overline{X}) = \text{MEAN}_{\overline{Y} \in L_k(\overline{X})} \frac{AR_k(\overline{X})}{AR_k(\overline{Y})} \tag{8.12}$$

The use of distance ratios in the definition ensures that the local distance behavior is well accounted for in this definition. As a result, the *LOF* values for the objects in a cluster are often close to 1 when the data points in the cluster are homogeneously distributed. For example, in the case of Fig. 8.8a, the *LOF* values of data points in *both* clusters will be quite close to 1, even though the densities of the two clusters are different. On the other hand, the *LOF* values of *both* the outlying points will be much higher because they will be computed in terms of the ratios to the average neighbor reachability distances. In practice, the maximum value of $LOF_k(\overline{X})$ over a range of different values of k is used as the outlier score to determine the best size of the neighborhood.

One observation about the *LOF* method is that while it is popularly understood in the literature as a density-based approach, it can be more simply understood as a *relative* distance-based approach with smoothing. The smoothing is really a refinement to make distance computations more stable. The basic *LOF* method will work reasonably well on many data sets, even if the raw distances are used instead of reachability distances, for the aforementioned computations of Eq. 8.11.

The *LOF* method, therefore, has the ability to adjust well to regions of varying density because of this relative normalization in the denominator of each term of Eq. 8.12. In the original presentation of the *LOF* algorithm (see bibliographic notes), the *LOF* is defined in terms of a density variable. The density variable is loosely defined as the inverse of the average of the smoothed reachability distances. This is, of course, not a precise definition of density. Density is traditionally defined in terms of the number of data points within a specified area or volume. This book provides exactly the same definition of LOF but presents it slightly differently by omitting the intermediate density variable. This is done both for simplicity, and for a definition of LOF directly in terms of (normalized) distances. The real connection of LOF to data density lies in its insightful ability to *adjust* to varying data density with the use of *relative distances.* Therefore, this book has classified this approach as a (normalized) distance-based method, rather than as a density-based method.

8.5.2.2 Instance-Specific Mahalanobis Distance

The instance-specific Mahalanobis distance is designed for adjusting to varying *shapes* of the distributions in the locality of a particular data point, as illustrated in Fig. 8.8b. The Mahalanobis distance is directly related to shape of the data distribution, although it is traditionally used in the *global* sense. Of course, it is also possible to use the *local* Mahalanobis distance by using the covariance structure of the neighborhood of a data point.

The problem here is that the neighborhood of a data point is hard to define with the Euclidean distance when the shape of the neighborhood cluster is not spherical. For example, the use of the Euclidean distance to a data point is biased toward capturing the circular region around that point, rather than an elongated cluster. To address this issue, an *agglomerative* approach is used to determine the k-neighborhood $L_k(\overline{X})$ of a data point $\overline{X}$. First, data point $\overline{X}$ is added to $L_k(\overline{X})$. Then, data points are iteratively added to $L_k(\overline{X})$ that have the smallest distance to *their closest* point in $L_k(\overline{X})$. This approach can be viewed as a special case of single-linkage hierarchical clustering methods, where singleton points are merged with clusters. Single-linkage methods are well-known for creating clusters of arbitrary shape. Such an approach tends to "grow" the neighborhood with the same shape as the cluster. The mean $\overline{\mu_k(X)}$ and covariance matrix $\Sigma_k(\overline{X})$ of the neighborhood $L_k(\overline{X})$ are computed. Then, the *instance-specific Mahalanobis score* $LMaha_k(\overline{X})$ of a data point $\overline{X}$ provides its outlier score. This score is defined as the Mahalanobis distance of $\overline{X}$ to the mean $\overline{\mu_k(X)}$ of data points in $L_k(\overline{X})$.

$$LMaha_k(\overline{X}) = Maha(\overline{X}, \overline{\mu_k(X)}, \Sigma_k(\overline{X})) \tag{8.13}$$

The only difference between this computation and that of the global Mahalanobis distance for extreme value analysis is that the local neighborhood set $L_k(\overline{X})$ is used as the "relevant" data for comparison in the former. While the clustering approach of Sect. 8.4 does use a Mahalanobis metric on the local neighborhood, the computation is subtly different in this case. In the case of clustering-based outlier detection, a preprocessing approach *predefines* a limited number of clusters as the universe of possible neighborhoods. In this case, the neighborhood is constructed in an *instance-specific* way. Different points will have slightly different neighborhoods, and they may not neatly correspond to a predefined cluster. This additional granularity allows more refined analysis. At a conceptual level, this approach computes whether data point $\overline{X}$ can be regarded as an extreme value with respect to its local cluster. As in the case of the LOF method, the approach can be applied for different values of k, and the highest outlier score for each data point can be reported.

If this approach is applied to the example of Fig. 8.8b, the method will correctly determine the outlier because of the local Mahalanobis normalization with the appropriate (local) covariance matrix for each data point. No distance normalizations are necessary for varying data density (scenario of Fig. 8.8a) because the Mahalanobis distance already performs these local normalizations under the covers. Therefore, such a method can be used for the scenario of Fig. 8.8a as well. The reader is referred to the bibliographic notes for variations of *LOF* that use the concept of varying local cluster shapes with agglomerative neighborhood computation.

8.6 Density-Based Methods

Density-based methods are loosely based on similar principles as density-based clustering. The idea is to determine sparse regions in the underlying data in order to report outliers. Correspondingly, histogram-based, grid-based, or kernel density-based methods can be used. Histograms can be viewed as 1-dimensional special cases of grid-based methods. These methods have not, however, found significant popularity because of their difficulty in adjusting to variations of density in different data localities. The definition of density also becomes more challenging with increasing dimensionality. Nevertheless, these methods are used more frequently in the univariate case because of their natural probabilistic interpretation.

8.6.1 Histogram- and Grid-Based Techniques

Histograms are simple and easy to construct for univariate data, and are therefore used quite frequently in many application domains. In this case, the data is discretized into bins, and the frequency of each bin is estimated. Data points that lie in bins with very low frequency are reported as outliers. If a continuous outlier score is desired, then the number of *other* data points in the bin for data point $\overline{X}$ is reported as the outlier score for $\overline{X}$. Therefore, the count for a bin does not include the point itself in order to minimize overfitting for smaller bin widths or smaller number of data points. In other words, the outlier score for each data point is one less than its bin count.

In the context of multivariate data, a natural generalization is the use of a grid-structure. Each dimension is partitioned into p equi-width ranges. As in the previous case, the number of points in a particular grid region is reported as the outlier score. Data points that have density less than τ in any particular grid region are reported as outliers. The appropriate value of τ can be determined by using univariate extreme value analysis.

The major challenge with histogram-based techniques is that it is often hard to determine the optimal histogram width well. Histograms that are too wide, or too narrow, will not model the frequency distribution well. These are similar issues to those encountered with the use of grid-structures for clustering. When the bins are too narrow, the normal data points falling in these bins will be declared outliers. On the other hand, when the bins are too wide, anomalous data points and high-density regions may be merged into a single bin. Therefore, such anomalous data points may not be declared outliers.

A second issue with the use of histogram techniques is that they are too local in nature, and often do not take the global characteristics of the data into account. For example, for the case of Fig. 8.7, a multivariate grid-based approach may not be able to classify an isolated group of data points as outliers, unless the resolution of the grid structure is calibrated carefully. This is because the density of the grid only depends on the data points inside it, and an isolated group of points may create an artificially dense grid cell when the granularity of representation is high. Furthermore, when the density distribution varies significantly with data locality, grid-based methods may find it difficult to normalize for local variations in density.

Finally, histogram methods do not work very well in high dimensionality because of the sparsity of the grid structure with increasing dimensionality, unless the outlier score is computed with respect to carefully chosen lower dimensional projections. For example, a d-dimensional space will contain at least 2^d grid-cells, and, therefore, the number of data points expected to populate each cell reduces exponentially with increasing dimensionality. These problems with grid-based methods are well known, and are also frequently encountered in the context of other data mining applications such as clustering.

8.6.2 Kernel Density Estimation

Kernel density estimation methods are similar to histogram techniques in terms of building density profiles, though the major difference is that a smoother version of the density profile is constructed. In kernel density estimation, a continuous estimate of the density is generated at a given point. The value of the density at a given point is estimated as the sum of the smoothed values of kernel functions $K_h(\cdot)$ associated with each point in the data set. Each kernel function is associated with a kernel width h that determines the level of smoothing created by the function. The kernel estimation $f(\overline{X})$ based on n data points of dimensionality d, and kernel function $K_h(\cdot)$ is defined as follows:

$$f(\overline{X}) = \frac{1}{n} \cdot \sum_{i=1}^{n} K_h(\overline{X} - \overline{X_i}). \tag{8.14}$$

Thus, each discrete point $\overline{X_i}$ in the data set is replaced by a continuous function $K_h(\cdot)$ that peaks at $\overline{X_i}$ and has a variance determined by the smoothing parameter h. An example of such a distribution is the Gaussian kernel with width h.

$$K_h(\overline{X} - \overline{X_i}) = \left(\frac{1}{\sqrt{2\pi} \cdot h}\right)^d \cdot e^{-||\overline{X}-\overline{X_i}||^2/(2h^2)} \tag{8.15}$$

The estimation error is defined by the kernel width h, which is chosen in a data-driven manner. It has been shown that for most smooth functions $K_h(\cdot)$, when the number of data points goes to infinity, the estimate asymptotically converges to the true density value, provided that the width h is chosen appropriately. The density at each data point is computed without including the point itself in the density computation. The value of the density is reported as the outlier score. Low values of the density indicate greater tendency to be an outlier.

Density-based methods have similar challenges as histogram- and grid-based techniques. In particular, the use of a global kernel width h to estimate density may not work very well in cases where there are wide variations in local density, such as those in Figs. 8.7 and 8.8. This is because of the myopic nature of density-based methods, in which the variations in the density distribution are not well accounted for. Nevertheless, kernel-density-based methods can be better generalized to data with local variations, especially if the bandwidth is chosen locally. As in the case of grid-based methods, these techniques are not very effective for higher dimensionality. The reason is that the accuracy of the density estimation approach degrades with increasing dimensionality.

8.7 Information-Theoretic Models

Outliers are data points that do not naturally fit the remaining data distribution. Therefore, if a data set were to be somehow compressed with the use of the "normal" patterns in the data distribution, the outliers would increase the minimum code length required to describe it. For example, consider the following two strings:

```
ABABABABABABABABABABABABABABABABAB
ABABACABABABABABABABABABABABABABAB
```

The second string is of the same length as the first and is different at only a single position containing the unique symbol C. The first string can be described concisely as "AB 17 times." However, the second string has a single position corresponding to the symbol C. Therefore, the second string can no longer be described as concisely. In other words, the presence of the symbol C somewhere in the string increases its *minimum description length.* It is also easy to see that this symbol corresponds to an outlier. Information-theoretic models are based on this general principle because they measure the increase in model size required to describe the data as concisely as possible.

Information-theoretic models can be viewed as almost equivalent to conventional deviation-based models, except that the outlier score is defined by the model size for a fixed deviation, rather than the deviation for a fixed model. In conventional models, outliers are always defined on the basis of a "summary" model of normal patterns. When a data point deviates significantly from the estimations of the summary model, then this deviation value is reported as the outlier score. Clearly, a trade-off exists between the size of the summary model and the level of deviation. For example, if a clustering model is used, then a larger number of cluster centroids (model size) will result in lowering the maximum deviation of any data point (including the outlier) from its nearest centroid. Therefore, in conventional models, the same clustering is used to compute deviation values (scores) for the different data points. A slightly different way of computing the outlier score is to fix the maximum allowed deviation (instead of the number of cluster centroids) and compute the number of cluster centroids required to achieve the same level of deviation, with and without a particular data point. It is this *increase* that is reported as the outlier score in the information-theoretic version of the same model. The idea here is that each point can be estimated by its closest cluster centroid, and the cluster centroids serve as a "code-book" in terms of which the data is compressed in a lossy way.

Information-theoretic models can be viewed as a complementary version of conventional models where a different aspect of the space-deviation trade-off curve is examined. Virtually every conventional model can be converted into an information-theoretic version by examining the bi-criteria space-deviation trade-off in terms of space rather than deviation. The bibliographic notes will also provide specific examples of each of the cases below:

1. The probabilistic model of Sect. 8.3 models the normal patterns in terms of generative model parameters such as the mixture means and covariance matrices. The space required by the model is defined by its complexity (e.g., number of mixture components), and the deviation corresponds to the probabilistic fit. In an information-theoretic version of the model, the complementary approach is to examine the size of the model required to achieve a fixed level of fit.

2. A clustering or density-based summarization model describes a data set in terms of cluster descriptions, histograms or other summarized representations. The granularity of these representations (number of cluster centroids, or histogram bins) controls the space, whereas the error in approximating the data point with a central element of the cluster (bin) defines the deviation. In conventional models, the size of the model (number of bins or clusters) is fixed, whereas in the information-theoretic version, the

maximum allowed deviation is fixed, and the required model size is reported as the outlier score.

3. A frequent pattern mining model describes the data in terms of an underlying code-book of frequent patterns. The larger the size of the code-book (by using frequent patterns of lower support), the more accurately the data can be described. These models are particularly popular, and some pointers are provided in the bibliographic notes.

All these models represent the data approximately in terms of individual condensed components representing aggregate trends. In general, outliers increase the length of the description in terms of these condensed components to achieve the same level of approximation. For example, a data set with outliers will require a larger number of mixture parameters, clusters, or frequent patterns to achieve *the same level of approximation.* Therefore, in information-theoretic methods, the components of these summary models are loosely referred to as "code books." Outliers are defined as data points whose removal results in the *largest decrease* in description length for the same error. The actual construction of the coding is often heuristic, and is not very different from the summary models used in conventional outlier analysis. In some cases, the description length for a data set can be *estimated* without explicitly constructing a code book, or building a summary model. An example is that of the *entropy* of a data set, or the *Kolmogorov complexity* of a string. Readers are referred to the bibliographic notes for examples of such methods.

While information-theoretic models are approximately equivalent to conventional models in that they explore the same trade-off in a slightly different way, they do have an advantage in some cases. These are cases where an accurate summary model of the data is hard to *explicitly* construct, and measures such as the entropy or Kolmogorov complexity can be used to *estimate* the compressed space requirements of the data set *indirectly.* In such cases, information-theoretic methods can be useful. In cases where the summary models can be explicitly constructed, it is better to use conventional models because the outlier scores are directly optimized to point-specific deviations rather than the more blunt measure of differential space impact. The bibliographic notes provide specific examples of some of the aforementioned methods.

8.8 Outlier Validity

As in the case of clustering models, it is desirable to determine the validity of outliers determined by a particular algorithm. Although the relationship between clustering and outlier analysis is complementary, the measures for outlier validity cannot easily be designed in a similar complementary way. In fact, validity analysis is much harder in outlier detection than data clustering. The reasons for this are discussed in the next section.

8.8.1 Methodological Challenges

As in the case of data clustering, outlier analysis is an unsupervised problem. Unsupervised problems are hard to validate because of the lack of external criteria, unless such criteria are synthetically generated, or some rare aspects of real data sets are used as *proxies.* Therefore, a natural question arises, as to whether *internal criteria* can be defined for outlier validation, as is the case for data clustering.

However, internal criteria are rarely used in outlier analysis. While such criteria are well-known to be flawed even in the context of data clustering, these flaws become significant enough to make these criteria unusable for outlier analysis. The reader is advised to refer to Sect. 6.9.1 of Chap. 6 for a discussion of the challenges of internal cluster validity. Most of these challenges are related to the fact that cluster validity criteria are derived from the objective function criteria of clustering algorithms. Therefore, a particular validity measure will favor (or *overfit*) a clustering algorithm using a similar objective function criterion. These problems become magnified in outlier analysis because of the *small sample solution space*. A model only needs to be correct on a few outlier data points to be considered a good model. Therefore, the overfitting of internal validity criteria, which is significant even in clustering, becomes even more problematic in outlier analysis. As a specific example, if one used the k-nearest neighbor distance as an internal validity measure, then a pure distance-based outlier detector will always outperform a locally normalized detector such as *LOF*. This is, of course, not consistent with known experience in real settings, where *LOF* usually provides more meaningful results. One can try to reduce the overfitting effect by designing a validity measure which is different from the outlier detection models being compared. However, this is not a satisfactory solution because significant uncertainty always remains about the impact of hidden interrelationships between such measures and outlier detection models. The main problem with internal measures is that the relative bias in evaluation of various algorithms is *consistently* present, even when the data set is varied. A biased selection of internal measures can easily be abused in algorithm benchmarking.

Internal measures are almost never used in outlier analysis, although they are often used in clustering evaluation. Even in clustering, the use of internal validity measures is questionable in spite of its wider acceptance. Therefore, most of the validity measures used for outlier analysis are based on external measures such as the *Receiver Operating Characteristic* curve.

8.8.2 Receiver Operating Characteristic

Outlier detection algorithms are typically evaluated with the use of *external* measures where the *known* outlier labels from a synthetic data set or the *rare* class labels from a real data set are used as the ground-truth. This ground-truth is compared systematically with the outlier score to generate the final output. While such rare classes may not always reflect all the natural outliers in the data, the results are usually reasonably representative of algorithm quality, when evaluated over many data sets.

In outlier detection models, a threshold is typically used on the outlier score to generate a binary label. If the threshold is picked too restrictively to minimize the number of declared outliers then the algorithm will miss true outlier points (false-negatives). On the other hand, if the threshold is chosen in a more relaxed way, this will lead to too many false-positives. This leads to a trade-off between the false-positives and false-negatives. The problem is that the "correct" threshold to use is never known exactly in a real scenario. However, this entire trade-off curve can be generated, and various algorithms can be compared over the entire trade-off curve. One example of such a curve is the *Receiver Operating Characteristic (ROC)* curve.

For any given threshold t on the outlier score, the declared outlier set is denoted by $\mathcal{S}(t)$. As t changes, the size of $\mathcal{S}(t)$ changes as well. Let $\mathcal{G}$ represent the true set (ground-truth set) of outliers in the data set. The *true-positive rate*, which is also referred to as the *recall*, is defined as the percentage of *ground-truth* outliers that have been reported as outliers at threshold t.

$$TPR(t) = Recall(t) = 100 * \frac{|\mathcal{S}(t) \cap \mathcal{G}|}{|\mathcal{G}|}$$

Table 8.1: ROC construction with rank of ground-truth outliers

Algorithm	**Rank of ground-truth outliers**
Algorithm A	1, 5, 8, 15, 20
Algorithm B	3, 7, 11, 13, 15
Random Algorithm	17, 36, 45, 59, 66
Perfect Oracle	1, 2, 3, 4, 5

The false positive rate $FPR(t)$ is the percentage of the falsely reported positives out of the ground-truth negatives. Therefore, for a data set $\mathcal{D}$ with ground-truth positives $\mathcal{G}$, this measure is defined as follows:

$$FPR(t) = 100 * \frac{|\mathcal{S}(t) - \mathcal{G}|}{|\mathcal{D} - \mathcal{G}|}. \tag{8.16}$$

The ROC curve is defined by plotting the $FPR(t)$ on the X-axis, and $TPR(t)$ on the Y-axis for varying values of t. Note that the end points of the ROC curve are always at $(0, 0)$ and $(100, 100)$, and a random method is expected to exhibit performance along the diagonal line connecting these points. The *lift* obtained above this diagonal line provides an idea of the accuracy of the approach. The area under the ROC curve provides a concrete quantitative evaluation of the effectiveness of a particular method.

To illustrate the insights gained from these different graphical representations, consider an example of a data set with 100 points from which 5 points are outliers. Two algorithms, A and B, are applied to this data set that rank all data points from 1 to 100, with lower rank representing greater propensity to be an outlier. Thus, the true-positive rate and false-positive rate values can be generated by determining the ranks of the 5 ground-truth outlier points. In Table 8.1, some hypothetical ranks for the five ground-truth outliers have been illustrated for the different algorithms. In addition, the ranks of the ground-truth outliers for a random algorithm have been indicated. The random algorithm outputs a random outlier score for each data point. Similarly, the ranks for a "perfect oracle" algorithm, which ranks the correct top 5 points as outliers, have also been illustrated in the table. The corresponding ROC curves are illustrated in Fig. 8.9.

What do these curves really tell us? For cases in which one curve strictly dominates another, it is clear that the algorithm for the former curve is superior. For example, it is immediately evident that the oracle algorithm is superior to all algorithms, and the random algorithm is inferior to all the other algorithms. On the other hand, algorithms A and B show domination at different parts of the ROC curve. In such cases, it is hard to say that one algorithm is strictly superior. From Table 8.1, it is clear that Algorithm A, ranks three of the correct ground-truth outliers very highly, but the remaining two outliers are ranked poorly. In the case of Algorithm B, the highest ranked outliers are not as well ranked as the case of Algorithm A, though all five outliers are determined much earlier in terms of rank threshold. Correspondingly, Algorithm A dominates on the earlier part of the ROC curve whereas Algorithm B dominates on the later part. Some practitioners use the area under the ROC curve as a proxy for the overall effectiveness of the algorithm, though such a measure should be used very carefully because all parts of the ROC curve may not be equally important for different applications.

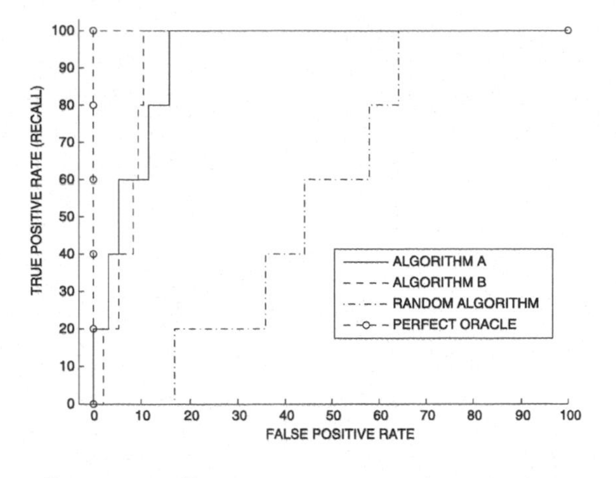

Figure 8.9: Receiver operating characteristic

8.8.3 Common Mistakes

A common mistake in benchmarking outlier analysis applications is that the area under the ROC curve is used repeatedly to tune the parameters of the outlier analysis algorithm. Note that such an approach implicitly uses the ground-truth labels for model construction, and it is, therefore, no longer an unsupervised algorithm. For problems such as clustering and outlier detection, it is not acceptable to use external labels in any way to tune the algorithm. In the particular case of outlier analysis, the accuracy can be *drastically* overestimated with such a tuning approach because the relative scores of a small number of outlier points have a very large influence on the ROC curve.

8.9 Summary

The problem of outlier analysis is an important one because of its applicability to a variety of problem domains. The common models in outlier detection include probabilistic models, clustering models, distance-based models, density-based models, and information-theoretic models. Among these, distance models are the most popular, but are computationally more expensive. A number of speed-up tricks have been proposed to make these models much faster. Local variations of distance-based models generally tend to be more effective because of their sensitivity to the generative aspects of the underlying data. Information-theoretic models are closely related to conventional models, and explore a different aspect of the space-deviation trade-off than conventional models.

Outlier validation is a difficult problem because of the challenges associated with the unsupervised nature of outlier detection, and the small sample-space problem. Typically, external validation criteria are used. The effectiveness of an outlier analysis algorithm is quantified with the use of the receiver operating characteristic curve that shows the trade-off between the false-positives and false-negatives for different thresholds on the outlier score. The area under this curve provides a quantitative evaluation of the outlier detection algorithm.

8.10 Bibliographic Notes

A number of books and surveys have been written on the problem of outlier analysis. The classical books [89, 259] in this area have mostly been written from the perspective of the statistics community. Most of these books were written before the wider adoption of database technology and are therefore not written from a computational perspective. More recently, this problem has been studied extensively by the computer science community. These works consider practical aspects of outlier detection corresponding to the cases where the data may be very large, or may have very high dimensionality. A recent book [5] also studies this area from the perspective of the computer science community. Numerous surveys have also been written that discuss the concept of outliers from different points of view, methodologies, or data types [61, 84, 131, 378, 380]. Among these, the survey by Chandola et al. [131] is the most recent and, arguably, the most comprehensive. It is an excellent review that covers the work on outlier detection quite broadly from the perspective of multiple communities.

The Z-value test is used commonly in the statistical literature, and numerous extensions, such as the t-value test are available [118]. While this test makes the normal distribution assumption for large data sets, it has been used fairly extensively as a good heuristic even for data distributions that do not satisfy the normal distribution assumption.

A variety of distance-based methods for outlier detection are proposed in [319, 436], and distance-correction methods for outlier detection are proposed in [109]. The determination of arbitrarily-shape clusters in the context of the *LOF* algorithm is explored in [487]. The agglomerative algorithm for discovering arbitrarily shaped neighborhoods, in the section on instance-specific Mahalanobis distance, is based on that approach. However, this method uses a connectivity-outlier factor, rather than the instance-specific Mahalanobis distance. The use of the Mahalanobis distance as a model for outlier detection was proposed in [468], though these methods are global, rather than local. A graph-based algorithm for local outlier detection is discussed in [257]. The *ORCLUS* algorithm also shows how to determine outliers in the presence of arbitrarily shaped clusters [22]. Methods for interpreting distance-based outliers were first proposed in [320].

A variety of information-theoretic methods for outlier detection are discussed in [68, 102, 160, 340, 472]. Many of these different models can be viewed in a complementary way to traditional models. For example, the work in [102] explores probabilistic methods in the context of information-theoretic models. The works in [68, 472] use code books of frequent-patterns for the modeling process. The connection between frequent patterns and compression has been explored in [470]. The use of measures such as entropy and Kolmogorov complexity for outlier analysis is explored in [340, 305]. The concept of coding complexity is explored in [129] in the context of set-based sequences.

Evaluation methods for outlier analysis are essentially identical to the techniques used in information retrieval for understanding precision-recall trade-offs, or in classification for ROC curve analysis. A detailed discussion may be found in [204].

8.11 Exercises

1. Suppose a particular random variable has mean 3 and standard deviation 2. Compute the Z-number for the values -1, 3. and 9. Which of these values can be considered the most extreme value?

2. Define the Mahalanobis-based extreme value measure when the d dimensions are statistically independent of one another, in terms of the dimension-specific standard deviations $\sigma_1 \dots \sigma_d$.

3. Consider the four 2-dimensional data points $(0, 0)$, $(0, 1)$, $(1, 0)$, and $(100, 100)$. Plot them using mathematical software such as MATLAB. Which data point visually seems like an extreme value? Which data point is reported by the Mahalanobis measure as the strongest extreme value? Which data points are reported by depth-based measure?

4. Implement the EM algorithm for clustering, and use it to implement a computation of the probabilistic outlier scores.

5. Implement the Mahalanobis k-means algorithm, and use it to implement a computation of the outlier score in terms of the local Mahalanobis distance to the closest cluster centroids.

6. Discuss the connection between the algorithms implemented in Exercises 4 and 5.

7. Discuss the advantages and disadvantages of clustering models over distance-based models.

8. Implement a naive distance-based outlier detection algorithm with no pruning.

9. What is the effect of the parameter k in k-nearest neighbor outlier detection? When do small values of k work well and when do larger values of k work well?

10. Design an outlier detection approach with the use of the *NMF* method of Chap. 6.

11. Discuss the relative effectiveness of pruning of distance-based algorithms in data sets which are (a) uniformly distributed, and (b) highly clustered with modest ambient noise and outliers.

12. Implement the *LOF*-algorithm for outlier detection.

13. Consider the set of 1-dimensional data points $\{1, 2, 2, 2, 2, 2, 6, 8, 10, 12, 14\}$. What are the data point(s) with the highest outlier score for a distance-based algorithm, using $k = 2$? What are the data points with highest outlier score using the *LOF* algorithm? Why the difference?

14. Implement the instance-specific Mahalanobis method for outlier detection.

15. Given a set of ground-truth labels, and outlier scores, implement a computer program to compute the ROC curve for a set of data points.

16. Use the objective function criteria of various outlier detection algorithms to design corresponding internal validity measures. Discuss the bias in these measures towards favoring specific algorithms.

17. Suppose that you construct a *directed* k-nearest neighbor graph from a data set. How can you use the degrees of the nodes to obtain an outlier score? What characteristics does this algorithm share with *LOF*?

Chapter 9

Outlier Analysis: Advanced Concepts

"If everyone is thinking alike, then somebody isn't thinking."—George S. Patton

9.1 Introduction

Many scenarios for outlier analysis cannot be addressed with the use of the techniques discussed in the previous chapter. For example, the data type has a critical impact on the outlier detection algorithm. In order to use an outlier detection algorithm on categorical data, it may be necessary to change the distance function or the family of distributions used in expectation–maximization (EM) algorithms. In many cases, these changes are exactly analogous to those required in the context of the clustering problem.

Other cases are more challenging. For example, when the data is very high dimensional, it is often difficult to apply outlier analysis because of the masking behavior of the noisy and irrelevant dimensions. In such cases, a new class of methods, referred to as ***subspace*** methods, needs to be used. In these methods, the outlier analysis is performed in lower dimensional projections of the data. In many cases, it is hard to discover these projections, and therefore results from multiple subspaces may need to be combined for better robustness.

The combination of results from multiple models is more generally referred to as *ensemble analysis*. Ensemble analysis is also used for other data mining problems such as clustering and classification. In principle, ensemble analysis in outlier detection is analogous to that in data clustering or classification. However, in the case of outlier detection, ensemble analysis is especially challenging. This chapter will study the following three classes of challenging problems in outlier analysis:

1. *Outlier detection in categorical data:* Because outlier models use notions such as nearest neighbor computation and clustering, these models need to be adjusted to the data type at hand. This chapter will address the changes required to handle categorical data types.

C. C. Aggarwal, *Data Mining: The Textbook*, DOI 10.1007/978-3-319-14142-8_9

2. *High-dimensional data:* This is a very challenging scenario for outlier detection because of the "curse-of-dimensionality." Many of the attributes are irrelevant and contribute to the errors in model construction. A common approach to address these issues is that of subspace outlier detection.

3. *Outlier ensembles:* In many cases, the robustness of an outlier detection algorithm can be improved with ensemble analysis. This chapter will study the fundamental principles of ensemble analysis for outlier detection.

Outlier analysis has numerous applications in a very wide variety of domains such as data cleaning, fraud detection, financial markets, intrusion detection, and law enforcement. This chapter will also study some of the more common applications of outlier analysis.

This chapter is organized as follows: Section 9.2 discusses outlier detection models for categorical data. The difficult case of high-dimensional data is discussed in Sect. 9.3. Outlier ensembles are studied in Sect. 9.4. A variety of applications of outlier detection are discussed in Sect. 9.5. Section 9.6 provides the summary.

9.2 Outlier Detection with Categorical Data

As in the case of other problems in data mining, the type of the underlying data has a significant impact on the specifics of the algorithm used for solving it. Outlier analysis is no exception. However, in the case of outlier detection, the changes required are relatively minor because, unlike clustering, many of the outlier detection algorithms (such as distance-based algorithms) use very simple definitions of outliers. These definitions can often be modified to work with categorical data with small modifications. In this section, some of the models discussed in the previous chapter will be revisited for categorical data.

9.2.1 Probabilistic Models

Probabilistic models can be modified easily to work with categorical data. A probabilistic model represents the data as a mixture of cluster components. Therefore, each component of the mixture needs to reflect a set of discrete attributes rather than numerical attributes. In other words, a generative mixture model of categorical data needs to be designed. Data points that do not fit this mixture model are reported as outliers.

The k components of the mixture model are denoted by $\mathcal{G}_1 \ldots \mathcal{G}_k$. The generative process uses the following two steps to generate each point in the d-dimensional data set $\mathcal{D}$:

1. Select a mixture component with prior probability α_i, where $i \in \{1 \ldots k\}$.

2. If the rth component of the mixture was selected in the first step, then generate a data point from $\mathcal{G}_r$.

The values of α_i denote the prior probabilities. An example of a model for the mixture component is one in which the jth value of the ith attribute is generated by cluster m with probability p_{ijm}. The set of all model parameters is collectively denoted by the notation Θ.

Consider a data point $\overline{X}$ containing the attribute value indices $j_1 \ldots j_d$ where the rth attribute takes on the value j_r. Then, the value of the generative probability $g^{m,\Theta}(\overline{X})$ of a data point from cluster m is given by the following expression:

$$g^{m,\Theta}(\overline{X}) = \prod_{r=1}^{d} p_{rj_rm}. \tag{9.1}$$

The fit probability of the data point to the rth component is given by $\alpha_r \cdot g^{r,\Theta}(\overline{X})$. Therefore, the sum of the fits over all components is given by $\sum_{r=1}^{k} \alpha_r \cdot g^{r,\Theta}(\overline{X})$. This fit represents the likelihood of the data point being generated by the model. Therefore, it is used as the outlier score. However, in order to compute this fit value, one needs to estimate the parameters Θ. This is achieved with the EM algorithm.

The assignment (or *posterior*) probability $P(\mathcal{G}_m|\overline{X}, \Theta)$ for the mth cluster may be estimated as follows:

$$P(\mathcal{G}_m|\overline{X}, \Theta) = \frac{\alpha_m \cdot g^{m,\Theta}(\overline{X})}{\sum_{r=1}^{k} \alpha_r \cdot g^{r,\Theta}(\overline{X})}. \quad (9.2)$$

This step provides a soft assignment probability of the data point to a cluster, and it corresponds to the E-step.

The soft-assignment probability is used to estimate the probability p_{ijm}. While estimating the parameters for cluster m, the *weight* of a data point is assumed to be equal to its assignment probability $P(\mathcal{G}_m|\overline{X}, \Theta)$ to cluster m. The value α_m is estimated to be the average assignment probability to cluster m over all data points. For each cluster m, the *weighted* number w_{ijm} of records for which the ith attribute takes on the jth possible discrete value in cluster m is estimated. The value of w_{ijm} is estimated as the sum of the posterior probabilities $P(\mathcal{G}_m|\overline{X}, \Theta)$ for all records $\overline{X}$ in which the ith attribute takes on the jth value. Then, the value of the probability p_{ijm} may be estimated as follows:

$$p_{ijm} = \frac{w_{ijm}}{\sum_{\overline{X} \in \mathcal{D}} P(\mathcal{G}_m|\overline{X}, \Theta)}. \quad (9.3)$$

When the number of data points is small, the estimation of Eq. 9.3 can be difficult to perform in practice. In such cases, some of the attribute values may not appear in a cluster (or $w_{ijm} \approx 0$). This situation can lead to poor parameter estimation, or *overfitting*. The *Laplacian smoothing* method is commonly used to address such ill-conditioned probabilities. Let m_i be the number of distinct attribute values of categorical attribute i. In Laplacian smoothing, a small value β is added to the numerator, and $m_i \cdot \beta$ is added to the denominator of Eq. 9.3. Here, β is a parameter that controls the level of smoothing. This form of smoothing is also sometimes applied in the estimation of the prior probabilities α_i when the data sets are very small. This completes the description of the M-step. As in the case of numerical data, the E- and M-steps are iterated to convergence. The maximum likelihood fit value is reported as the outlier score.

9.2.2 Clustering and Distance-Based Methods

Most of the clustering- and distance-based methods can be generalized easily from numerical to categorical data. Two main modifications required are as follows:

1. Categorical data requires specialized clustering methods that are typically different from those of numerical data. This is discussed in detail in Chap. 7. Any of these models can be used to create the initial set of clusters. If a distance- or similarity-based clustering algorithm is used, then the same distance or similarity function should be used to compute the distance (similarity) of the candidate point to the cluster centroids.

2. The choice of the similarity function is important in the context of categorical data, whether a centroid-based algorithm is used or a raw, distance-based algorithm is used. The distance functions in Sect. 3.2.2 of Chap. 3 can be very useful for this task. The pruning tricks for distance-based algorithms are agnostic to the choice of distance

function and can therefore be generalized to this case. Many of the local methods, such as *LOF*, can be generalized to this case as well with the use of this modified definition of distances.

Therefore, clustering and distance-based methods can be generalized to the scenario of categorical data with relatively modest modifications.

9.2.3 Binary and Set-Valued Data

Binary data are a special kind of categorical data, which occur quite frequently in many real scenarios. The chapters on frequent pattern mining were based on these kind of data. Furthermore, both categorical and numerical data can always be converted to binary data. One common characteristic of this domain is that, while the number of attributes is large, the number of *nonzero* values of the attribute is small in a typical transaction.

Frequent pattern mining is used as a subroutine for the problem of outlier detection in these cases. The basic idea is that frequent patterns are much less likely to occur in outlier transactions. Therefore, one possible measure is to use the sum of all the supports of frequent patterns occurring in a particular transaction. The total sum is normalized by dividing with the number of frequent patterns. This provides an outlier score for the pattern. Strictly speaking, the normalization can be omitted from the final score, because it is the same across all transactions.

Let $\mathcal{D}$ be a transaction database containing the transactions denoted by $T_1 \dots T_N$. Let $s(X, \mathcal{D})$ represent the support of itemset X in $\mathcal{D}$. Therefore, if $FPS(\mathcal{D}, s_m)$ represents the set of frequent patterns in the database $\mathcal{D}$ at minimum support level s_m, then, the frequent pattern outlier factor $FPOF(T_i)$ of a transaction $T_i \in \mathcal{D}$ at minimum support s_m is defined as follows:

$$FPOF(T_i) = \frac{\sum_{X \in FPS(\mathcal{D}, s_m), X \subseteq T_i} s(X, \mathcal{D})}{|FPS(\mathcal{D}, s_m)|}. \tag{9.4}$$

Intuitively, a transaction containing a large number of frequent patterns with high support will have a high value of $FPOF(T_i)$. Such a transaction is unlikely to be an outlier because it reflects the major patterns in the data. Therefore, lower scores indicate greater propensity to be an outlier.

Such an approach is analogous to nonmembership of data points in clusters to define outliers rather than determining the deviation or sparsity level of the transactions in a more direct way. The problem with this approach is that it may not be able to distinguish between truly isolated data points and ambient noise. This is because neither of these kinds of data points will be likely to contain many frequent patterns. As a result, such an approach may sometimes not be able to effectively determine the strongest anomalies in the data.

9.3 High-Dimensional Outlier Detection

High-dimensional outlier detection can be particularly challenging because of the varying importance of the different attributes with data locality. The idea is that the *causality* of an anomaly can be typically perceived in only a small subset of the dimensions. The remaining dimensions are irrelevant and only add noise to the anomaly-detection process. Furthermore, different subsets of dimensions may be relevant to different anomalies. As a result, full-dimensional analysis often does not properly expose the outliers in high-dimensional data.

This concept is best understood with a motivating example. In Fig. 9.1, four different 2-dimensional views of a hypothetical data set have been illustrated. Each of these views

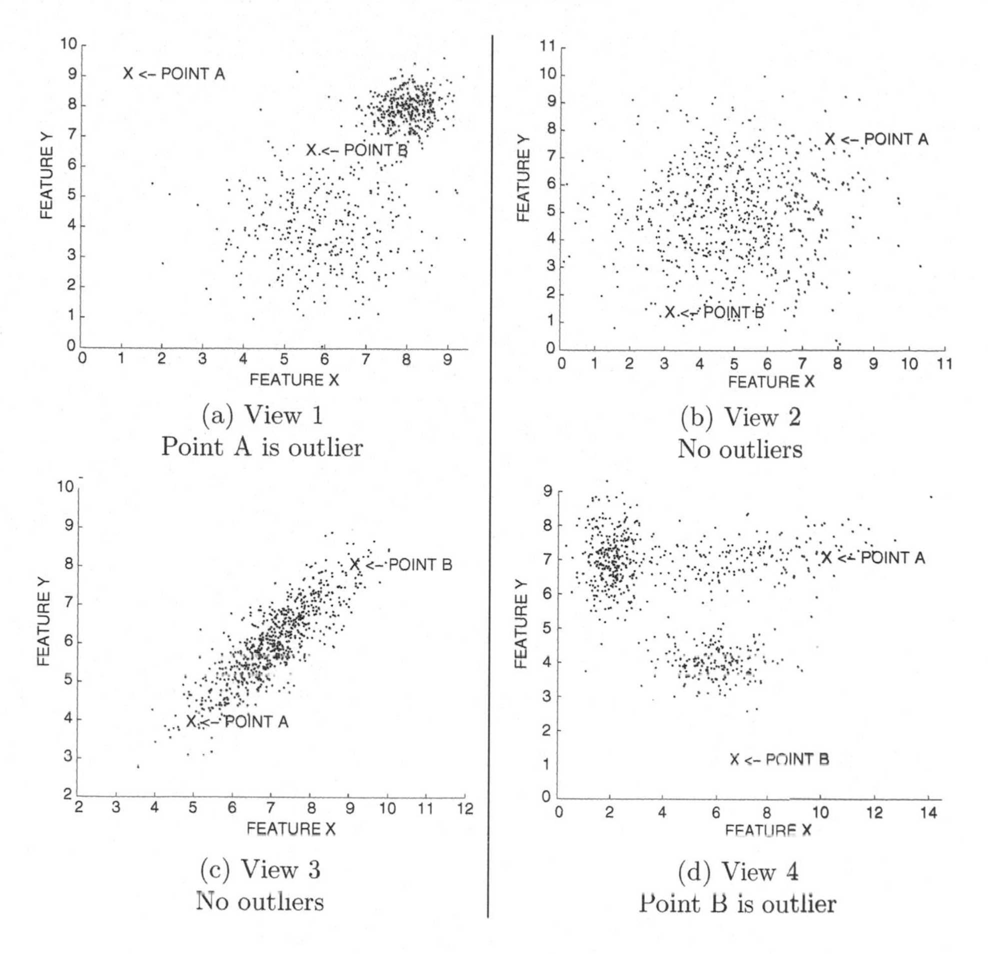

(a) View 1
Point A is outlier

(b) View 2
No outliers

(c) View 3
No outliers

(d) View 4
Point B is outlier

Figure 9.1: Impact of irrelevant attributes on outlier analysis

corresponds to a disjoint set of dimensions. As a result, these views look very different from one another. Data point A is exposed as an outlier in the first view of the data set, whereas point B is exposed as an outlier in the fourth view of the data set. However, data points A and B are not exposed as outliers in the second and third views of the data set. These views are therefore not very useful from the perspective of measuring the outlierness of *either* A or B. Furthermore, from the perspective of any *specific* data point (e.g., point A), three of the four views are irrelevant. Therefore, the outliers are lost in the random distributions within these views when the distance measurements are performed in *full* dimensionality. In many scenarios, the proportion of irrelevant views (features) may increase with dimensionality. In such cases, outliers are *lost* in low-dimensional subspaces of the data because of irrelevant attributes.

The physical interpretation of this situation is quite clear in many application-specific scenarios. For example, consider a credit card fraud application in which different features such as a customer's purchase location, frequency of purchase, and size of purchase are being tracked over time. In the case of one particular customer, the anomaly may be tracked by examining the location and purchase frequency attributes. In another anomalous customer, the size and timing of the purchase may be relevant. Therefore, all the features are useful from a *global* perspective but only a small subset of features is useful from a *local* perspective.

The major problem here is that the dilution effects of the vast number of "normally noisy" dimensions will make the detection of outliers difficult. In other words, outliers are lost in low-dimensional subspaces when full-dimensional analysis is used because of the *masking* and *dilution* effects of the noise in full-dimensional computations. A similar problem is also discussed in Chap. 7 in the context of data clustering.

In the case of data clustering, this problem is solved by defining subspace-specific clusters, or *projected* clusters. This approach also provides a natural path for outlier analysis in high dimensions. In other words, an outlier can now be defined by associating it with one or more subspaces that are *specific to that outlier.* While there is a clear analogy between the problems of subspace clustering and subspace outlier detection, the *difficulty levels* of the two problems are not even remotely similar.

Clustering is, after all, about the determination of *frequent groups* of data points, whereas outliers are about determination of *rare groups* of data points. As a rule, statistical learning methods find it much easier to determine frequent characteristics than rare characteristics of a data set. This problem is further magnified in high dimensionality. The number of possible subspaces of a d-dimensional data point is 2^d. Of these, only a small fraction will expose the outlier behavior of individual data points. In the case of clustering, dense subspaces can be easily determined by aggregate statistical analysis of the data points. This is not true of outlier detection, where the subspaces need to be explicitly explored in a way that is specific to the *individual* data points.

An effective outlier detection method would need to search the data points and dimensions in *an integrated way* to reveal the most relevant outliers. This is because different subsets of dimensions may be relevant to different outliers, as is evident from the example of Fig. 9.1. The integration of point and subspace exploration leads to a further expansion in the number of possibilities that need to be examined for outlier analysis. This chapter will explore two methods for subspace exploration, though many other methods are pointed out in the bibliographic notes. These methods are as follows:

1. *Grid-based rare subspace exploration:* In this case, rare subspaces of the data are explored after discretizing the data into a grid-like structure.

2. *Random subspace sampling:* In this case, subspaces of the data are sampled to discover the most relevant outliers.

The following subsections will discuss each of these methods in detail.

9.3.1 Grid-Based Rare Subspace Exploration

Projected outliers are determined by finding localized regions of the data in low-dimensional space that have abnormally low density. Because this is a density-based approach, a grid-based technique is the method of choice. To do so, nonempty grid-based subspaces need to be identified, whose density is very low. This is complementary to the definitions that were used for subspace clustering methods such as *CLIQUE.* In those cases, *frequent* subspaces were reported. It should be pointed out that the determination of frequent subspaces is a much simpler problem than the determination of rare subspaces, simply because there are many more rare subspaces than there are dense ones in a typical data set. This results in a combinatorial explosion of the number of possibilities, and level-wise algorithms, such as those used in *CLIQUE*, are no longer practical avenues for finding rare subspaces. The first step in all these models is to determine a proper statistical definition of rare lower dimensional projections.

9.3.1.1 Modeling Abnormal Lower Dimensional Projections

An abnormal lower dimensional projection is one in which the density of the data is exceptionally lower than the average. The concept of Z-number, introduced in the previous chapter, comes in very handy in this respect. The first step is to discretize the data. Each attribute of the data is divided into p ranges. These ranges are constructed on an *equidepth* basis. In other words, each range contains a fraction $f = 1/p$ of the records.

When k different intervals from different dimensions are selected, it creates a grid cell of dimensionality k. The expected fraction of the data records in this grid cell is equal to f^k, if the attributes are statistically independent. In practice, the data are not statistically independent, and, therefore, the distribution of the points in the cube differs significantly from the expected value. Deviations that correspond to rare regions in the data are of particular interest.

Let $\mathcal{D}$ be a database with n records, and dimensionality d. Under the independence assumption mentioned earlier, the presence or absence of any point in a k-dimensional cube is a Bernoulli random variable with probability f^k. The expected number and standard deviation of the points in a k-dimensional cube are given by $n \cdot f^k$ and $\sqrt{n \cdot f^k \cdot (1 - f^k)}$. When the value of n is large, the number of data points in a cube is a random variable that is approximated by a normal distribution, with the aforementioned mean and standard deviation.

Let $\mathcal{R}$ represent a cube in k-dimensional space, and $n_{\mathcal{R}}$ represent the number of data points inside this cube. Then, the sparsity coefficient $S(\mathcal{R})$ of the cube $\mathcal{R}$ can be computed as follows:

$$S(\mathcal{R}) = \frac{n_{\mathcal{R}} - n \cdot f^k}{\sqrt{n \cdot f^k \cdot (1 - f^k)}}. \tag{9.5}$$

A negative value of the sparsity coefficient indicates that the presence of data points in the cube is significantly lower than expected. Because $n_{\mathcal{R}}$ is assumed to fit a normal distribution, the normal distribution can be used to quantify the probabilistic level of significance of its deviation. This is only a heuristic approximation because the assumption of a normal distribution is generally not true in practice.

9.3.1.2 Grid Search for Subspace Outliers

As discussed earlier, level-wise algorithms are not practical for rare subspace discovery. Another challenge is that lower dimensional projections provide no information about the statistical behavior of combinations of dimensions, especially in the context of subspace analysis.

For example, consider a data set containing student scores in an examination. Each attribute represents a score in a particular subject. The scores in some of the subjects are likely to be highly correlated. For example, a student scoring well in a course on probability theory would likely also score well in a course on statistics. However, it would be extremely uncommon to find a student who scored well in one, but not the other. The problem here is that the *individual* dimensions provide no information about the combination of the dimensions. The rare nature of outliers makes such unpredictable scenarios common. This lack of predictability about the behavior of combinations of dimensions necessitates the use of evolutionary (genetic) algorithms to explore the search space.

Genetic algorithms mimic the process of biological evolution to solve optimization problems. Evolution is, after all, nature's great optimization experiment in which only the fittest individuals survive, thereby leading to more "optimal" organisms. Correspondingly, every solution to an optimization problem can be disguised as an individual in an evolutionary system. The measure of fitness of this "individual" is equal to the objective function value of the corresponding solution. The competing population with an individual is the group of other solutions to the optimization problem. In general, fitter organisms are more likely to survive and multiply in the population. This corresponds to the *selection* operator. The other two operators that are commonly used are *crossover* and *mutation*. Each feasible solution is encoded in the form of a string, and is considered the chromosome representation of the solution. This is also referred to as *encoding*.

Thus, each string is a solution that is associated with a particular objective function value. In genetic algorithms, this objective function value is also referred to as the *fitness function*. The idea here is that the selection operator should favor strings with better fitness (objective function value). This is similar to hill-climbing algorithms, except that genetic algorithms work with a population of solutions instead of a single one. Furthermore, instead of only checking the neighborhood (as in hill-climbing methods), genetic algorithms use crossover and mutation operators to explore a more complex concept of a neighborhood.

Therefore, evolutionary algorithms are set up to repeat the process of selection, crossover, and mutation to improve the fitness (objective) function value. As the process of evolution progresses, all the individuals in the population typically improve in fitness and also become more similar to each other. The convergence of a particular position in the string is defined as the stage at which a predefined fraction of the population has the same value for that gene. The population is said to have converged when all positions in the string representation have converged.

So, how does all this map to finding rare patterns? The relevant localized subspace patterns can be easily represented as strings of length d, where d denotes the dimensionality of the data. Each position in the string represents the index of an equi-depth range. Therefore, each position in the string can take on any value from 1 through p, where p is the granularity of the discretization. It can also take on the value $*$ ("don't care"), which indicates that the dimension is not included in the subspace corresponding to that string. The fitness of a string is based on the sparsity coefficient discussed earlier. Highly negative values of the objective function (sparsity coefficient) indicate greater fitness because one is searching for sparse subspaces. The only caveat is that the subspaces need to be nonempty to be considered fit. This is because empty subspaces are not useful for finding anomalous data points.

Consider a 4-dimensional problem in which the data have been discretized into ten ranges. Then, the string will be of length 4, and each position can take on one of 11 possible values (including the "*"). Therefore, there are a total of 11^4 strings, each of which corresponds to a subspace. For example, the string *2*6 is a 2-dimensional subspace on the second and fourth dimensions.

The evolutionary algorithm uses the dimensionality of the projection k as an input parameter. Therefore, for a d-dimensional data set, the string of length d will contain k specified position and $(d - k)$ "don't care" positions. The fitness for the corresponding solution may be computed using the sparsity coefficient discussed earlier. The evolutionary search technique starts with a population of Q random solutions and iteratively uses the processes of selection, crossover, and mutation to perform a combination of hill climbing, solution recombination, and random search over the space of possible projections. The process is continued until the population converges, based on the criterion discussed above.

At each stage of the algorithm, the m best projection solutions (most negative sparsity coefficients) are kept track of. At the end of the algorithm, these solutions are reported as the best projections in the data. The following operators are defined for selection, crossover, and mutation:

1. *Selection:* This step achieves the *hill climbing* required by the method, though quite differently from traditional hill-climbing methods. The copies of a solution are replicated by ordering them by rank and biasing them in the population in the favor of higher ranked solutions. This is referred to as *rank selection.* This results in a bias in favor of more optimal solutions.

2. *Crossover:* The crossover technique is key to the success of the algorithm because it implicitly defines the subspace exploration process. One solution is to use a uniform two-point crossover to create the recombinant children strings. It can be viewed as the process of combining the characteristics of two solutions to create two new recombinant solutions. Traditional hill-climbing methods only test an adjacent solution for a single string. The recombinant crossover approach examines a more complex neighborhood by combining the characteristics of *two* different strings, to yield two new neighborhood points.

   The two-point crossover mechanism works by determining a point in the string at random, called the crossover point, and exchanging the segments to the right of this point. This is equivalent to creating two new subspaces, by sampling subspaces from both solutions and combining them. However, such a blind recombination process may create poor solutions too often. Therefore, an optimized crossover mechanism is defined. In this mechanism, it is guaranteed that both children solutions correspond to a feasible k-dimensional projection, and the children typically have high fitness values. This is achieved by examining a subset of the different possibilities for recombination and picking the best among them.

3. *Mutation:* In this case, random positions in the string are flipped with a predefined mutation probability. Care must be taken to ensure that the dimensionality of the projection does not change after the flipping process. This is an exact analogue of traditional hill-climbing except that it is done to a population of solutions to ensure robustness. Thus, while genetic algorithms try to achieve the same goals as hill climbing, they do so in a different way to achieve better solutions.

At termination, the algorithm is followed by a postprocessing phase. In the postprocessing phase, all data points containing the abnormal projections are reported by the algorithm as the outliers. The approach also provides the relevant projections that provide the *causality* (or intensional knowledge) for the outlier behavior of a data point. Thus, this approach also has a high degree of interpretability in terms of providing the reasoning for *why* a data point should be considered an outlier.

9.3.2 Random Subspace Sampling

The determination of the rare subspaces is a very difficult task, as is evident from the unusual genetic algorithm designed discussed in the previous section. A different way of addressing the same problem is to explore many possible subspaces and examine if at least one of them contains outliers. One such well known approach is *feature bagging.* The broad approach is to repeatedly apply the following two steps:

1. Randomly, select between $(d/2)$ and d features from the underlying data set in iteration t to create a data set D_t in the tth iteration.

2. Apply the outlier detection algorithm O_t on the data set D_t to create score vectors S_t.

Virtually any algorithm O_t can be used to determine the outlier score, as long as the scores across different instantiations are comparable. From this perspective, the *LOF* algorithm is the ideal algorithm to use because of its normalized scores. At the end of the process, the outlier scores from the different algorithms need to be combined. Two distinct methods are used to combine the different subspaces:

1. *Breadth-first approach:* The ranking of the data points returned by the different algorithms is used for combination purposes. The top-ranked outliers over all the different algorithm executions are ranked first, followed by the second-ranked outliers (with repetitions removed), and so on. Minor variations could exist because of tie-breaking between the outliers within a particular rank. This is equivalent to using the best rank for each data point over all executions as the final outlier score.

2. *Cumulative sum approach:* The outlier scores over the different algorithm executions are summed up. The top-ranked outliers are reported on this basis.

At first sight, it seems that the random subspace sampling approach does not attempt to optimize the discovery of subspaces to finding rare instances at all. However, the idea here is that it is often hard to discover the rare subspaces anyway, even with the use of heuristic optimization methods. The robustness resulting from multiple subspace sampling is clearly a very desirable quality of the approach. Such methods are common in high-dimensional analysis where multiple subspaces are sampled for greater robustness. This is related to the field of *ensemble analysis*, which will be discussed in the next section.

9.4 Outlier Ensembles

Several algorithms in outlier analysis, such as the high-dimensional methods discussed in the previous section, combine the scores from different executions of outlier detection algorithms. These algorithms can be viewed as different forms of ensemble analysis. Some examples are enumerated below:

1. *Parameter tuning in LOF:* Parameter tuning in the *LOF* algorithm (cf. Sect. 8.5.2.1 of Chap. 8) can be viewed as a form of ensemble analysis. This is because the algorithm is executed over different values of the neighborhood size k, and the highest *LOF* score over each data point is selected. As a result, a more robust *ensemble* model is constructed. In fact, many parameter-tuning algorithms in outlier analysis can be viewed as ensemble methods.

2. *Random subspace sampling:* The random subspace sampling method applies the approach to multiple random subspaces of the data to determine the outlier scores as a combination function of the original scores. Even the evolutionary high-dimensional outlier detection algorithm can be shown to be an ensemble with a maximization combination function.

Ensemble analysis is a popular approach in many data mining problems such as clustering, classification, and outlier detection. On the other hand, ensemble analysis is not quite well studied in the context of outlier analysis. Furthermore, ensemble analysis is particularly important in the context of outlier analysis because of the rare nature of outliers, and a corresponding possibility of overfitting. A typical outlier ensemble contains a number of different components:

1. *Model components:* These are the individual methodologies or algorithms that are integrated to create an ensemble. For example, a random subspace sampling method combines many *LOF* algorithms that are each applied to different subspace projections.

2. *Normalization:* Different methods may create outlier scores on very different scales. In some cases, the scores may be in ascending order. In others, the scores may be in descending order. In such cases, normalization is important for meaningfully combining the scores, so that the scores from different components are roughly comparable.

3. *Model combination:* The combination process refers to the approach used to integrate the outlier score from different components. For example, in random subspace sampling, the cumulative outlier score over different ensemble components is reported. Other combination functions include the use of the maximum score, or the highest *rank* of the score among all ensembles. It is assumed that higher ranks imply greater propensity to be an outlier. Therefore, the highest rank is similar to the maximum outlier score, except that the rank is used instead of the raw score. The highest-rank combination function is also used by random subspace sampling.

Outlier ensemble methods can be categorized into different types, depending on the dependencies among different components and the process of selecting a specific model. The following section will study some of the common methods.

9.4.1 Categorization by Component Independence

This categorization examines whether or not the components are developed independently.

1. In *sequential ensembles*, a given algorithm or set of algorithms is applied sequentially, so that future applications of the algorithm are influenced by previous applications. This influence may be realized in terms of either modifications of the base data for analysis, or in terms of the specific choices of the algorithms. The final result is either a weighted combination of, or the final result of the last application of the outlier algorithm component. The typical scenario for sequential ensembles is that of *model refinement*, where successive iterations continue to improve a particular base model.

2. In *independent ensembles*, different algorithms, or different instantiations of the same algorithm, are *independently* applied to either the complete data or portions of the data. In other words, the various ensemble components are independent of the results of each other's executions.

In this section, both types of ensembles will be studied in detail.

9.4.1.1 Sequential Ensembles

In sequential ensembles, one or more outlier detection algorithms are applied sequentially to either all or portions of the data. The idea is that the result of a particular algorithmic execution may provide insights that may help refine future executions. Thus, depending upon the approach, either the data set or the algorithm may be changed in sequential executions. For example, consider the case where a clustering model is created for outlier detection. Because outliers interfere with the robust cluster generation, one possibility would be to apply the method to a successively refined data set after removing the obvious outliers through the insights gained in earlier iterations of the ensemble. Typically, the quality of the

```
Algorithm SequentialEnsemble(Data Set: D
    Base Algorithms: A_1 ... A_r)
begin
  j = 1;
  repeat
    Pick an algorithm Q_j ∈ {A_1 ... A_r} based
       on results from past executions;
    Create a new data set f_j(D) from D based
       on results from past executions;
    Apply Q_j to f_j(D);
    j = j + 1;
  until(termination);
  return outliers based on combinations of results
     from previous executions;
end
```

Figure 9.2: Sequential ensemble framework

outliers in later iterations will be better. This also facilitates a more robust outlier detection model. Thus, the sequential nature of the approach is used for successive refinement. If desired, this approach can either be applied for a fixed number of times or used to converge to a more robust solution. The broad framework of a sequential ensemble approach is provided in Fig. 9.2.

It is instructive to examine the execution of the algorithm in Fig. 9.2 in some detail. In each iteration, a successively refined algorithm may be used on a refined data set, based on the results from previous executions. The function $f_j(\cdot)$ is used to create a refinement of the data, which could correspond to data subset selection, attribute-subset selection, or a generic data transformation method. The generality of the aforementioned description ensures that many natural variations of the method can be explored with the use of this ensemble. For example, while the algorithm of Fig. 9.2 assumes that many different algorithms $\mathcal{A}_1 \dots \mathcal{A}_r$ are available, it is possible to select only one of them, and use it on successive modifications of the data. Sequential ensembles are often hard to use effectively in outlier analysis because of the lack of available groundtruth in interpreting the intermediate results. In many cases, the distribution of the outlier scores is used as a proxy for these insights.

9.4.1.2 Independent Ensembles

In independent ensembles, different instantiations of the algorithm or different portions of the data are executed *independently* for outlier analysis. Alternatively, the same algorithm may be applied, but with either a different initialization, parameter set, or even random seed in the case of a randomized algorithm. The *LOF* method, the high-dimensional evolutionary exploration method, and the random subspace sampling method discussed earlier are all examples of independent ensembles. Independent ensembles are more common in outlier analysis than sequential ensembles. In this case, the combination function requires careful normalization, especially if the different components of the ensemble are heterogeneous. A general-purpose description of independent ensemble algorithms is provided in the pseudocode description of Fig. 9.3.

```
Algorithm IndependentEnsemble(Data Set: D
    Base Algorithms: A_1 ... A_r)
begin
  j = 1;
  repeat
    Pick an algorithm Q_j ∈ {A_1 ... A_r};
    Create a new data set f_j(D) from D;
    Apply Q_j to f_j(D);
    j = j + 1;
  until(termination);
  return outliers based on combination of results
    from previous executions;
end
```

Figure 9.3: Independent ensemble framework

The broad principle of independent ensembles is that different ways of looking at the same problem provide more robust results that are not dependent on specific artifacts of a particular algorithm or data set. Independent ensembles are used commonly for parameter tuning of outlier detection algorithms. Another application is that of exploring outlier scores over multiple subspaces, and then providing the best result.

9.4.2 Categorization by Constituent Components

A second way of categorizing ensemble analysis algorithms is on the basis of their constituent components. In general, these two ways of categorization are orthogonal to one another, and an ensemble algorithm may be any of the four combinations created by these two forms of categorization.

Consider the case of parameter tuning in *LOF* and the case of subspace sampling in the feature bagging method. In the first case, each model is an application of the *LOF* model with a different parameter choice. Therefore, each component can itself be viewed as an outlier analysis model. On the other hand, in the random subspace method, the same algorithm is applied to a different selection (projection) of the data. In principle, it is possible to create an ensemble with *both* types of components, though this is rarely done in practice. Therefore, the categorization by component independence leads to either *model-centered* ensembles, or *data-centered* ensembles.

9.4.2.1 Model-Centered Ensembles

Model-centered ensembles combine the outlier scores from different models built on the same data set. The example of parameter tuning for *LOF* can be considered a model-centered ensemble. Thus, it is an *independent* ensemble based on one form or categorization, and a *model-centered* ensemble, based on another.

One advantage of using *LOF* in an ensemble algorithm is that the scores are roughly comparable to one another. This may not be true for an arbitrary algorithm. For example, if the raw k-nearest neighbor distance were used, the parameter tuning ensemble would always favor larger values of k when using a combination function that picks the maximum value of the outlier score. This is because the scores across different components would not be comparable to one another. Therefore, it is crucial to use normalization during the combination process. This is an issue that will be discussed in some detail in Sect. 9.4.3.

9.4.2.2 Data-Centered Ensembles

In data-centered ensembles, different parts, samples, or functions of the data are explored to perform the analysis. A function of the data could include either a sample of the data (horizontal sample) or a relevant subspace (vertical sample). The random subspace sampling approach of the previous section is an example of a data-centered ensemble. More general functions of the data are also possible, though are rarely used. Each function of the data may provide different insights about a specific part of the data. This is the key to the success of the approach. It should be pointed out that a data-centered ensemble may also be considered a model-centered ensemble by incorporating a preprocessing phase that generates a specific function of the data as a part of the model.

9.4.3 Normalization and Combination

The final stage of ensemble analysis is to put together the scores derived from the different models. The major challenge in model combination arises when the scores across different models are not comparable with one another. For example, in a model-centered ensemble, if the different components of the model are heterogeneous, the scores will not be comparable to one another. A k-nearest neighbor outlier score is not comparable to an LOF score. Therefore, *normalization* is important. In this context, univariate extreme value analysis is required to convert scores to normalized values. Two methods of varying levels of complexity are possible:

1. The univariate extreme value analysis methods in Sect. 8.2.1 of Chap. 8 may be used. In this case, a Z-number may be computed for each data point. While such a model makes the normal distribution approximation, it still provides better scores than using raw values.

2. If more refined scores are desired, and some insights are available about "typical" distributions of outlier scores, then the mixture model of Sect. 6.5 in Chap. 6 may be used to generate probabilistically interpretable fit values. The bibliographic notes provide a specific example of one such method.

Another problem is that the ordering of the outlier scores may vary with the outlier detection algorithm (ensemble component) at hand. In some algorithms, high scores indicate greater outlierness, whereas the reverse is true in other algorithms. In the former case, the Z-number of the outlier score is computed, whereas in the latter case, the negative of the Z-number is computed. These two values are on the same scale and more easily comparable.

The final step is to combine the scores from the different ensemble components. The method of combination may, in general, depend upon the composition of the ensemble. For example, in sequential ensembles, the final outlier score may be the score from the last execution of the ensemble. However, in general, the scores from the different components are combined together with the use of a combination function. In the following, the convention assumed is that higher (normalized) scores are indicative of greater abnormality. Two combination functions are particularly common.

1. *Maximum function:* The score is the *maximum* of the outlier scores from the different components.

2. *Average function:* The score is the *average* of the outlier scores from the different components.

Both the *LOF* method and the random subspace sampling method use the *maximum* function, either on the outlier scores or the ranks[1] of the outlier scores, to avoid dilution of the score from irrelevant models. The *LOF* research paper [109] provides a convincing argument as to why the maximum combination function has certain advantages. Although the average combination function will do better at discovering many "easy" outliers that are discoverable in many ensemble components, the maximum function will do better at finding well-hidden outliers. While there might be relatively fewer well-hidden outliers in a given data set, they are often the most interesting ones in outlier analysis. A common misconception[2] is that the maximum function might overestimate the absolute outlier scores, or that it might declare normal points as outliers because it computes the maximum score over many ensemble components. This is not an issue because outlier scores are *relative*, and the key is to make sure that the maximum is computed over an equal number of ensemble components for each data point. Absolute scores are irrelevant because outlier scores are comparable on a relative basis only over a fixed data set and not across multiple data sets. If desired, the combination scores can be standardized to zero mean and unit variance. The random subspace ensemble method has been implemented [334] with a rudimentary (rank-based) maximization and an average-based combination function as well. The experimental results show that the relative performance of the maximum and average combination functions is data specific. Therefore, either the maximum or average scores can achieve better performance, depending on the data set, but the maximum combination function will be consistently better at discovering well-hidden outliers. This is the reason that many methods such as *LOF* have advocated the use of the maximum combination function.

9.5 Putting Outliers to Work: Applications

The applications of outlier analysis are very diverse, and they extend to a variety of domains such as fault detection, intrusion detection, financial fraud, and Web log analytics. Many of these applications are defined for complex data types, and cannot be fully solved with the methodologies introduced in this chapter. Nevertheless, it will be evident from the discussion in later chapters that analogous methodologies can be defined for complex data types. In many cases, other data types can be converted to multidimensional data for analysis.

9.5.1 Quality Control and Fault Detection

Numerous applications arise in outlier analysis in the context of quality control and fault detection. Some of these applications typically require simple univariate extreme value analysis, whereas others require more complex methods. For example, anomalies in the manufacturing process may be detected by evaluating the number of defective units produced by each machine in a day. When the number of defective units is too large, it can be indicative of an anomaly. Univariate extreme value analysis is useful in such scenarios.

[1]In the case of ranks, if the maximum function is used, then outliers occurring early in the ranking are assigned larger rank values. Therefore, the most abnormal data point is assigned a score (rank) of n out of n data points.

[2]This is a common misunderstanding of the Bonferroni principle [343].

Other applications include the detection of faults in machine engines, where the engine measurements are tracked to determine faults. The system may be continuously monitored on a variety of parameters such as rotor speed, temperature, pressure, performance, and so on. It is desired to detect a fault in the engine system as soon as it occurs. Such applications are often temporal, and the outlier detection approach needs to be adapted to temporal data types. These methods will be discussed in detail in Chaps. 14 and 15.

9.5.2 Financial Fraud and Anomalous Events

Financial fraud is one of the more common applications of outlier analysis. Such outliers may arise in the context of credit card fraud, insurance transactions, and insider trading. A credit card company maintains the data corresponding to the card transactions by the different users. Each transaction contains a set of attributes corresponding to the user identifier, amount spent, geographical location, and so on. It is desirable to determine fraudulent transactions from the data. Typically, the fraudulent transactions often show up as unusual combinations of attributes. For example, high frequency transactions in a particular location may be more indicative of fraud. In such cases, subspace analysis can be very useful because the number of attributes tracked is very large, and only a particular subset of attributes may be relevant to a specific user. A similar argument applies to the case of related applications such as insurance fraud.

More complex temporal scenarios can be captured with the use of time-series data streams. An example is the case of financial markets, where the stock tickers correspond to the movements of different stocks. A sudden movement, or an anomalous crash, may be detected with the use of temporal outlier detection methods. Alternatively, time-series data may be transformed to multidimensional data with the use of the data portability methods discussed in Chap. 2. A particular example is wavelet transformation. The multidimensional outlier detection techniques discussed in this chapter can be applied to the transformed data.

9.5.3 Web Log Analytics

The user behavior at different Web sites is often tracked in an automated way. The anomalies in these behaviors may be determined with the use of Web log analytics. For example, consider a user trying to break into a password-protected Web site. The sequence of actions performed by the user is unusual, compared to the actions of the majority of users that are normal. The most effective methods for outlier detection work with optimized models for sequence data (see Chap. 15). Alternatively, sequence data can be transformed to multidimensional data, using a variation of the wavelet method, as discussed in Chap. 2. Anomalies can be detected on the transformed multidimensional data.

9.5.4 Intrusion Detection Applications

Intrusions correspond to different kinds of malicious security violations over a network or a computer system. Two common scenarios are *host-based intrusions*, and *network-based intrusions*. In host-based intrusions, the operating system call logs of a computer system are analyzed to determine anomalies. Such applications are typically discrete sequence mining applications that are not very different from Web log analytics. In *network-based intrusions*, the temporal relationships between the data values are much weaker, and the data can be treated as a stream of multidimensional data records. Such applications require streaming outlier detection methods, which are addressed in Chap. 12.

9.5.5 Biological and Medical Applications

Most of the data types produced in the biological data are complex data types. Such data types are studied in later chapters. Many diagnostic tools, such as sensor data and medical imaging, produce one or more complex data types. Some examples are as follows:

1. Many diagnostic tools used commonly in emergency rooms, such as electrocardiogram (ECG), are temporal sensor data. Unusual shapes in these readings may be used to make predictions.

2. Medical imaging applications are able to store 2-dimensional and 3-dimensional spatial representations of various tissues. Examples include magnetic resonance imaging (MRI) and computerized axial tomography (CAT) scans. These representations may be utilized to determine anomalous conditions.

3. Genetic data are represented in the form of discrete sequences. Unusual mutations are indicative of specific diseases, the determination of which are useful for diagnostic and research purposes.

Most of the aforementioned applications relate to the complex data types, and are discussed in detail later in this book.

9.5.6 Earth Science Applications

Anomaly detection is useful in detecting anomalies in earth science applications such as the unusual variations of temperature and pressure in the environment. These variations can be used to detect unusual changes in the climate, or important events, such as the detection of hurricanes. Another interesting application is that of determining land cover anomalies, where interesting changes in the forest cover patterns are determined with the use of outlier analysis methods. Such applications typically require the use of spatial outlier detection methods, which are discussed in Chap. 16.

9.6 Summary

Outlier detection methods can be generalized to categorical data with the use of similar methodologies that are used for cluster analysis. Typically, it requires a change in the mixture model for probabilistic models, and a change in the distance function for distance-based models. High-dimensional outlier detection is a particularly difficult case because of the large number of irrelevant attributes that interfere with the outlier detection process. Therefore, subspace methods need to be designed. Many of the subspace exploration methods use insights from multiple views of the data to determine outliers. Most high-dimensional methods are ensemble methods. Ensemble methods can be applied beyond high-dimensional scenarios to applications such as parameter tuning. Outlier analysis has numerous applications to diverse domains, such as fault detection, financial fraud, Web log analytics, medical applications, and earth science. Many of these applications are based on complex data types, which are discussed in later chapters.

9.7 Bibliographic Notes

A mixture model algorithm for outlier detection in categorical data is proposed in [518]. This algorithm is also able to address mixed data types with the use of a joint mixture model between quantitative and categorical attributes. Any of the categorical data clustering methods discussed in Chap. 7 can be applied to outlier analysis as well. Popular clustering algorithms include k-modes [135, 278], *ROCK* [238], *CACTUS* [220], *LIMBO* [75], and *STIRR* [229]. Distance-based outlier detection methods require the redesign of the distance function. Distance functions for categorical data are discussed in [104, 182]. In particular, the work in [104] explores categorical distance functions in the context of the outlier detection problem. A detailed description of outlier detection algorithms for categorical data may be found in [5].

Subspace outlier detection explores the effectiveness issue of outlier analysis, and was first proposed in [46]. In the context of high-dimensional data, there are two distinct lines of research, one of which investigates the *efficiency* of high-dimensional outlier detection [66, 501], and the other investigates the more fundamental issue of the *effectiveness* of high-dimensional outlier detection [46]. The masking behavior of the noisy and irrelevant dimensions was discussed by Aggarwal and Yu [46]. The efficiency-based methods often design more effective indexes, which are tuned toward determining nearest neighbors, and pruning more efficiently for distance-based algorithms. The random subspace sampling method discussed in this book was proposed in [334]. An isolation-forest approach was proposed in [365]. A number of ranking methods for subspace outlier exploration have been proposed in [396, 397]. In these methods, outliers are determined in multiple subspaces of the data. Different subspaces may provide information either about different outliers or about the same outliers. Therefore, the goal is to combine the information from these different subspaces in a robust way to report the final set of outliers. The *OUTRES* algorithm proposed in [396] uses recursive subspace exploration to determine all the subspaces relevant to a particular data point. The outlier scores from these different subspaces are combined to provide a final value. A more recent method for using multiple views of the data for subspace outlier detection is proposed in [397].

Recently, the problem of outlier detection has also been studied in the context of dynamic data and data streams. The *SPOT* approach was proposed in [546], which is able to determine projected outliers from high-dimensional data streams. This approach employs a window-based time model and decaying cell summaries to capture statistics from the data stream. A set of top sparse subspaces is obtained by a variety of supervised and unsupervised learning processes. These are used to detect the projected outliers. A multiobjective genetic algorithm is employed for finding outlying subspaces from training data. The problem of high-dimensional outlier detection has also been extended to other application-specific scenarios such as astronomical data [265] and transaction data [264]. A detailed description of the high-dimensional case for outlier detection may be found in [5].

The problem of outlier ensembles is generally less well developed in the context of outlier analysis, than in the context of problems such as clustering and classification. Many outlier ensemble methods, such the *LOF* method [109], do not explicitly state the ensemble component in their algorithms. The issue of score normalization has been studied in [223], and can be used for combining ensembles. A recent position paper has formalized the concept of outlier ensembles, and defined different categories of outlier ensembles [24]. Because outlier detection problems are evaluated in a similar way to classification problems, most classification ensemble algorithms, such as different variants of bagging/subsampling, will also improve outlier detection at least from a benchmarking perspective. While the results do reflect an improved quality of outliers in many cases, they should be interpreted with caution. Many recent subspace outlier detection methods [46, 396, 397] can also be considered ensemble methods. The first algorithm on high-dimensional outlier detection [46] may also be considered an ensemble method. A detailed description of different applications of outlier analysis may be found in the last chapter of [5].

9.8 Exercises

1. Suppose that algorithm A is designed for outlier detection in numeric data, whereas algorithm B is designed for outlier detection in categorical data. Show how you can use these algorithms to perform outlier detection in a mixed-attribute data set.

2. Design an algorithm for categorical outlier detection using the Mahalanobis distance. What are the advantages of such an approach?

3. Implement a distance-based outlier detection algorithm with the use of match-based similarity.

4. Design a feature bagging approach that uses arbitrary subspaces of the data rather than axis-parallel ones. Show how arbitrary subspaces may be efficiently sampled in a data distribution-sensitive way.

5. Compare and contrast multiview clustering with subspace ensembles in outlier detection.

6. Implement any two outlier detection algorithms of your choice. Convert the scores to Z-numbers. Combine the scores using the *max* function.

Chapter 10

Data Classification

"Science is the systematic classification of experience."—George Henry Lewes

10.1 Introduction

The classification problem is closely related to the clustering problem discussed in Chaps. 6 and 7. While the clustering problem is that of determining similar groups of data points, the classification problem is that of *learning* the structure of a data set of examples, *already partitioned into groups*, that are referred to as *categories* or *classes*. The learning of these categories is typically achieved with a *model*. This model is used to estimate the group identifiers (or *class labels*) of one or more previously unseen data examples with unknown labels. Therefore, one of the inputs to the classification problem is an example data set that has already been partitioned into different classes. This is referred to as the *training data*, and the group identifiers of these classes are referred to as class labels. In most cases, the class labels have a clear semantic interpretation in the context of a specific application, such as a group of customers interested in a specific product, or a group of data objects with a desired property of interest. The model learned is referred to as the *training model*. The previously unseen data points that need to be classified are collectively referred to as the *test data set*. The algorithm that creates the training model for prediction is also sometimes referred to as the *learner*.

Classification is, therefore, referred to as *supervised* learning because an example data set is used to learn the structure of the groups, just as a teacher supervises his or her students towards a specific goal. While the groups learned by a classification model may often be related to the similarity structure of the feature variables, as in clustering, this need not necessarily be the case. In classification, the example training data is paramount in providing the guidance of how groups are defined. Given a data set of test examples, the groups created by a classification model on the test examples will try to mirror the number and structure of the groups available in the example data set of training instances. Therefore, the classification problem may be intuitively stated as follows:

C. C. Aggarwal, *Data Mining: The Textbook*, DOI 10.1007/978-3-319-14142-8_10

Given a set of training data points, each of which is associated with a class label, determine the class label of one or more previously unseen test instances.

Most classification algorithms typically have two phases:

1. *Training phase:* In this phase, a training model is constructed from the training instances. Intuitively, this can be understood as a summary mathematical model of the labeled groups in the training data set.

2. *Testing phase:* In this phase, the training model is used to determine the class label (or group identifier) of one or more unseen test instances.

The classification problem is more powerful than clustering because, unlike clustering, it captures a *user-defined notion of grouping* from an example data set. Such an approach has almost direct applicability to a wide variety of problems, in which groups are defined naturally based on *external application-specific criteria*. Some examples are as follows:

1. *Customer target marketing:* In this case, the groups (or labels) correspond to the user interest in a particular product. For example, one group may correspond to customers interested in a product, and the other group may contain the remaining customers. In many cases, training examples of previous buying behavior are available. These can be used to provide examples of customers who may or may not be interested in a specific product. The feature variables may correspond to the demographic profiles of the customers. These training examples are used to learn whether or not a customer, with a known demographic profile, but unknown buying behavior, may be interested in a particular product.

2. *Medical disease management:* In recent years, the use of data mining methods in medical research has gained increasing traction. The features may be extracted from patient medical tests and treatments, and the class label may correspond to treatment outcomes. In these cases, it is desired to predict treatment outcomes with models constructed on the features.

3. *Document categorization and filtering:* Many applications, such as newswire services, require real-time classification of documents. These are used to organize the documents under specific topics in Web portals. Previous examples of documents from each topic may be available. The features correspond to the words in the document. The class labels correspond to the various topics, such as politics, sports, current events, and so on.

4. *Multimedia data analysis:* It is often desired to perform classification of large volumes of multimedia data such as photos, videos, audio, or other more complex multimedia data. Previous examples of particular activities of users associated with example videos may be available. These may be used to determine whether a particular video describes a specific activity. Therefore, this problem can be modeled as a binary classification problem containing two groups corresponding to the occurrence or nonoccurrence of a specific activity.

The applications of classification are diverse because of the ability to *learn by example.*

It is assumed that the training data set is denoted by $\mathcal{D}$ with n data points and d features, or dimensions. In addition, each of the data points in $\mathcal{D}$ is associated with a label drawn from $\{1 \dots k\}$. In some models, the label is assumed to be binary ($k = 2$) for simplicity. In the latter case, a commonly used convention is to assume that the labels are drawn from $\{-1, +1\}$. However, it is sometimes notationally convenient to assume that the labels are drawn from $\{0, 1\}$. This chapter will use either of these conventions depending on the classifier. A training model is constructed from $\mathcal{D}$, which is used to predict the label of unseen test instances. The output of a classification algorithm can be one of two types:

1. *Label prediction:* In this case, a label is predicted for each test instance.
2. *Numerical score:* In most cases, the learner assigns a score to each instance–label combination that measures the propensity of the instance to belong to a particular class. This score can be easily converted to a label prediction by using either the maximum value, or a cost-weighted maximum value of the numerical score across different classes. One advantage of using a score is that different test instances can be compared and ranked by their propensity to belong to a particular class. Such scores are particularly useful in situations where one of the classes is very rare, and a numerical score provides a way to determine the top *ranked* candidates belonging to that class.

A subtle but important distinction exists in the design process of these two types of models, especially when numerical scores are used for ranking different test instances. In the first model, the training model does not need to account for the relative classification propensity across different *test instances*. The model only needs to worry about the relative propensity towards different *labels* for a specific instance. The second model also needs to properly normalize the classification scores across different test instances so that they can be meaningfully compared for ranking. Minor variations of most classification models are able to handle either the labeling or the ranking scenario.

When the training data set is small, the performance of classification models is sometimes poor. In such cases, the model may describe the specific random characteristics of the training data set, and it may not *generalize* to the group structure of *previously unseen* test instances. In other words, such models might accurately predict the labels of instances used to construct them, but they perform poorly on unseen test instances. This phenomenon is referred to as *overfitting*. This issue will be revisited several times in this chapter and the next.

Various models have been designed for data classification. The most well-known ones include decision trees, rule-based classifiers, probabilistic models, instance-based classifiers, support vector machines, and neural networks. The modeling phase is often preceded by a feature selection phase to identify the most informative features for classification. Each of these methods will be addressed in this chapter.

This chapter is organized as follows. Section 10.2 introduces some of the common models used for feature selection. Decision trees are introduced in Sect. 10.3. Rule-based classifiers are introduced in Sect. 10.4. Section 10.5 discusses probabilistic models for data classification. Section 10.6 introduces support vector machines. Neural network classifiers are discussed in Sect. 10.7. Instance-based learning methods are explained in Sect. 10.8. Evaluation methods are discussed in Sect. 10.9. The summary is presented in Sect. 10.10.

10.2 Feature Selection for Classification

Feature selection is the first stage in the classification process. Real data may contain features of varying relevance for predicting class labels. For example, the gender of a person is less relevant for predicting a disease label such as "diabetes," as compared to his or her age. Irrelevant features will typically harm the accuracy of the classification model in addition to being a source of computational inefficiency. Therefore, the goal of feature selection algorithms is to select the most informative features with respect to the class label. Three primary types of methods are used for feature selection in classification.

1. *Filter models:* A crisp mathematical criterion is available to evaluate the quality of a feature or a subset of features. This criterion is then used to filter out irrelevant features.

2. *Wrapper models:* It is assumed that a classification algorithm is available to evaluate how well the algorithm performs with a particular subset of features. A feature search algorithm is then wrapped around this algorithm to determine the relevant set of features.

3. *Embedded models:* The solution to a classification model often contains useful hints about the most relevant features. Such features are isolated, and the classifier is retrained on the pruned features.

In the following discussion, each of these models will be explained in detail.

10.2.1 Filter Models

In filter models, a feature or a subset of features is evaluated with the use of a class-sensitive discriminative criterion. The advantage of evaluating a *group* of features at one time is that redundancies are well accounted for. Consider the case where two feature variables are perfectly correlated with one another, and therefore each can be predicted using the other. In such a case, it makes sense to use only one of these features because the other adds no *incremental* knowledge with respect to the first. However, such methods are often expensive because there are 2^d possible subsets of features on which a search may need to be performed. Therefore, in practice, most feature selection methods evaluate the features independently of one another and select the most discriminative ones.

Some feature selection methods, such as *linear discriminant analysis*, create a linear combination of the original features as a new set of features. Such analytical methods can be viewed either as stand-alone classifiers or as dimensionality reduction methods that are used *before* classification, depending on how they are used. These methods will also be discussed in this section.

10.2.1.1 Gini Index

The Gini index is commonly used to measure the discriminative power of a particular feature. Typically, it is used for categorical variables, but it can be generalized to numeric attributes by the process of discretization. Let $v_1 \dots v_r$ be the r possible values of a particular categorical attribute, and let p_j be the fraction of data points containing attribute value v_i that belong to the class $j \in \{1 \dots k\}$ for the attribute value v_i. Then, the Gini index $G(v_i)$ for the value v_i of a categorical attribute is defined as follows:

$$G(v_i) = 1 - \sum_{j=1}^{k} p_j^2. \tag{10.1}$$

When the different classes are distributed evenly for a particular attribute value, the value of the Gini index is $1 - 1/k$. On the other hand, if all data points for an attribute value

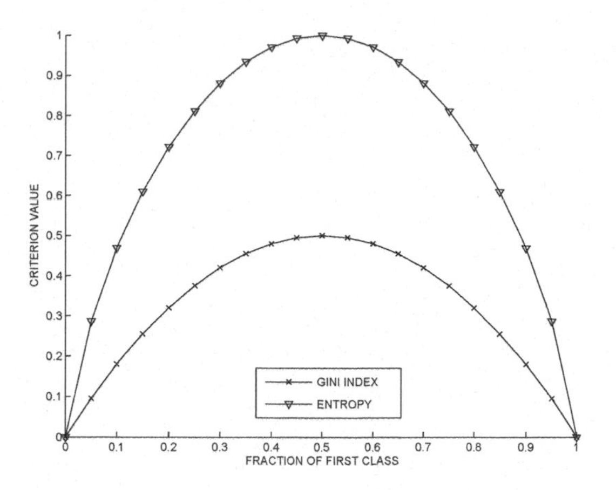

Figure 10.1: Variation of two feature selection criteria with class distribution skew

v_i belong to the same class, then the Gini index is 0. Therefore, lower values of the Gini index imply greater discrimination. An example of the Gini index for a two-class problem for varying values of p_1 is illustrated in Fig. 10.1. Note that the index takes on its maximum value at $p_1 = 0.5$.

The value-specific Gini index is converted into an attributewise Gini index. Let n_i be the number of data points that take on the value v_i for the attribute. Then, for a data set containing $\sum_{i=1}^{r} n_i = n$ data points, the overall Gini index G for the attribute is defined as the weighted average over the different attribute values as follows:

$$G = \sum_{i=1}^{r} n_i G(v_i)/n. \quad (10.2)$$

Lower values of the Gini index imply greater discriminative power. The Gini index is typically defined for a particular feature rather than a subset of features.

10.2.1.2 Entropy

The class-based entropy measure is related to notions of *information gain* resulting from fixing a specific attribute value. The entropy measure achieves a similar goal as the Gini index at an intuitive level, but it is based on sound information-theoretic principles. As before, let p_j be the fraction of data points belonging to the class j for attribute value v_i. Then, the class-based entropy $E(v_i)$ for the attribute value v_i is defined as follows:

$$E(v_i) = -\sum_{j=1}^{k} p_j \log_2(p_j). \quad (10.3)$$

The class-based entropy value lies in the interval $[0, \log_2(k)]$. Higher values of the entropy imply greater "mixing" of different classes. A value of 0 implies perfect separation, and, therefore, the largest possible discriminative power. An example of the entropy for a two-class problem with varying values of the probability p_1 is illustrated in Fig. 10.1. As in the case of the Gini index, the overall entropy E of an attribute is defined as the weighted average over the r different attribute values:

$$E = \sum_{i=1}^{r} n_i E(v_i)/n. \tag{10.4}$$

Here, n_i is the frequency of attribute value v_i.

10.2.1.3 Fisher Score

The Fisher score is naturally designed for numeric attributes to measure the ratio of the average interclass separation to the average intraclass separation. The larger the Fisher score, the greater the discriminatory power of the attribute. Let μ_j and σ_j, respectively, be the mean and standard deviation of data points belonging to class j for a particular feature, and let p_j be the fraction of data points belonging to class j. Let μ be the global mean of the data on the feature being evaluated. Then, the Fisher score F for that feature may be defined as the ratio of the interclass separation to intraclass separation:

$$F = \frac{\sum_{j=1}^{k} p_j (\mu_j - \mu)^2}{\sum_{j=1}^{k} p_j \sigma_j^2}. \tag{10.5}$$

The numerator quantifies the average interclass separation, whereas the denominator quantifies the average intraclass separation. The attributes with the largest value of the Fisher score may be selected for use with the classification algorithm.

10.2.1.4 Fisher's Linear Discriminant

Fisher's linear discriminant may be viewed as a generalization of the Fisher score in which newly created features correspond to linear combinations of the original features rather than a subset of the original features. This direction is designed to have a high level of discriminatory power with respect to the class labels. Fisher's discriminant can be viewed as a *supervised* dimensionality reduction method in contrast to *PCA*, which maximizes the preserved variance in the feature space but does not maximize the *class-specific* discrimination. For example, the most discriminating direction is aligned with the highest variance direction in Fig. 10.2a, but it is aligned with the lowest variance direction in Fig. 10.2b. In each case, if the data were to be projected along the most discriminating direction $\overline{W}$, then the *ratio* of interclass to intraclass separation is maximized. How can we determine such a d-dimensional vector $\overline{W}$?

The selection of a direction with high discriminative power is based on the same quantification as the Fisher score. Fisher's discriminant is naturally defined for the two-class scenario, although generalizations exist for the case of multiple classes. Let $\overline{\mu_0}$ and $\overline{\mu_1}$ be the d-dimensional row vectors representing the means of the data points in the two classes, and let Σ_0 and Σ_1 be the corresponding $d \times d$ covariance matrices in which the (i, j)th entry represents the covariance between dimensions i and j for that class. The fractional presence of the two classes are denoted by p_0 and p_1, respectively. Then, the equivalent Fisher score $FS(\overline{W})$ for a d-dimensional row vector $\overline{W}$ may be written in terms of *scatter* matrices, which are weighted versions of covariance matrices:

$$FS(\overline{W}) = \frac{\text{Between Class Scatter along } \overline{W}}{\text{Within Class Scatter along } \overline{W}} \propto \frac{(\overline{W} \cdot \overline{\mu_1} - \overline{W} \cdot \overline{\mu_0})^2}{p_0[\text{Variance(Class 0)}] + p_1[\text{Variance(Class 1)}]}$$

$$= \frac{\overline{W}[(\overline{\mu_1} - \overline{\mu_0})^T(\overline{\mu_1} - \overline{\mu_0})]\overline{W}^T}{p_0[\overline{W}\Sigma_0\overline{W}^T] + p_1[\overline{W}\Sigma_1\overline{W}^T]} = \frac{[\overline{W} \cdot (\overline{\mu_1} - \overline{\mu_0})]^2}{\overline{W}(p_0\Sigma_0 + p_1\Sigma_1)\overline{W}^T}.$$

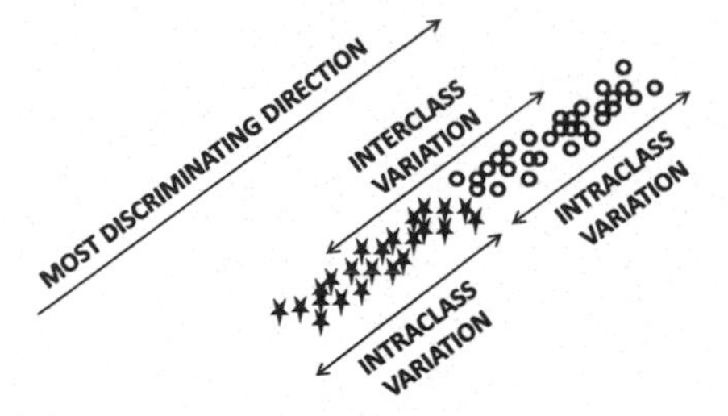

(a) Discriminating direction is aligned with high-variance direction

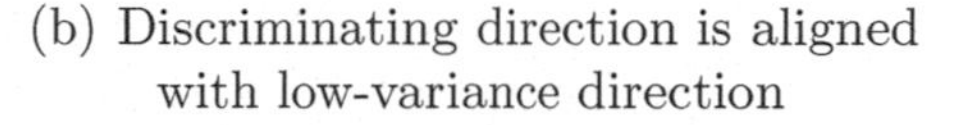

(b) Discriminating direction is aligned with low-variance direction

Figure 10.2: Impact of class distribution on Fisher's discriminating direction

Note that the quantity $\overline{W}\Sigma_i\overline{W}^T$ in one of the aforementioned expressions represents the variance of the projection of a data set along $\overline{W}$, whose covariance matrix is Σ_i. This result is derived in Sect. 2.4.3.1 of Chap. 2. The rank-1 matrix $S_b = [(\overline{\mu_1} - \overline{\mu_0})^T(\overline{\mu_1} - \overline{\mu_0})]$ is also referred[1] to as the (scaled) *between-class scatter-matrix* and the matrix $S_w = (p_0\Sigma_0 + p_1\Sigma_1)$ is the (scaled) *within-class scatter matrix*. The quantification $FS(\overline{W})$ is a direct generalization of the *axis-parallel* score in Eq. 10.5 to an arbitrary direction $\overline{W}$. The goal is to determine a direction $\overline{W}$ that maximizes the Fisher score. It can be shown[2] that the optimal direction $\overline{W^*}$, expressed as a row vector, is given by the following:

$$\overline{W^*} \propto (\overline{\mu_1} - \overline{\mu_0})(p_0\Sigma_0 + p_1\Sigma_1)^{-1}. \tag{10.6}$$

If desired, successive orthogonal directions may be determined by iteratively projecting the data into the orthogonal subspace to the optimal directions found so far, and determining the Fisher's discriminant in this reduced subspace. The final result is a new representation of lower dimensionality that is more discriminative than the original feature space. Interestingly, the matrix $S_w + p_0p_1S_b$ can be shown to be invariant to the values of the class labels of the data points (see Exercise 21), and it is equal to the covariance matrix of the data. Therefore, the top-k eigenvectors of $S_w + p_0p_1S_b$ yield the basis vectors of *PCA*.

This approach is often used as a stand-alone classifier, which is referred to as *linear discriminant analysis*. A perpendicular hyperplane $\overline{W^*} \cdot \overline{X} + b = 0$ to the most discriminating direction is used as a binary class separator. The optimal value of b is selected based on the accuracy with respect to the training data. This approach can also be viewed as projecting the training points along the most discriminating vector $\overline{W^*}$, and then selecting the value of b to decide the point on the line that best separates the two classes. The Fisher's discriminant for binary classes can be shown to be a special case of *least-squares regression* for numeric classes, in which the response variables are set to $-1/p_0$ and $+1/p_1$, respectively, for the two classes (cf. Sect. 11.5.1.1 of Chap. 11).

[1]The unscaled versions of the two scatter matrices are $np_0p_1S_b$ and nS_w, respectively. The sum of these two matrices is the total scatter matrix, which is n times the covariance matrix (see Exercise 21).

[2]Maximizing $FS(\overline{W}) = \frac{\overline{W}S_b\overline{W}^T}{\overline{W}S_w\overline{W}^T}$ is the same as maximizing $\overline{W}S_b\overline{W}^T$ subject to $\overline{W}S_w\overline{W}^T = 1$. Setting the gradient of the Lagrangian relaxation $\overline{W}S_b\overline{W}^T - \lambda(\overline{W}S_w\overline{W}^T - 1)$ to 0 yields the generalized eigenvector condition $S_b\overline{W}^T = \lambda S_w\overline{W}^T$. Because $S_b\overline{W}^T = (\overline{\mu_1}^T - \overline{\mu_0}^T)\left[(\overline{\mu_1} - \overline{\mu_0})\overline{W}^T\right]$ always points in the direction of $(\overline{\mu_1}^T - \overline{\mu_0}^T)$, it follows that $S_w\overline{W}^T \propto \overline{\mu_1}^T - \overline{\mu_0}^T$. Therefore, we have $\overline{W} \propto (\overline{\mu_1} - \overline{\mu_0})S_w^{-1}$.

10.2.2 Wrapper Models

Different classification models are more accurate with different sets of features. Filter models are agnostic to the particular classification algorithm being used. In some cases, it may be useful to leverage the characteristics of the specific classification algorithm to select features. As you will learn later in this chapter, a linear classifier may work more effectively with a set of features where the classes are best modeled with linear separators, whereas a distance-based classifier works well with features in which distances reflect class distributions.

Therefore, one of the inputs to wrapper-based feature selection is a specific classification induction algorithm, denoted by $\mathcal{A}$. Wrapper models can optimize the feature selection process to the classification algorithm at hand. The basic strategy in wrapper models is to iteratively refine a current set of features F by successively adding features to it. The algorithm starts by initializing the current feature set F to $\{\}$. The strategy may be summarized by the following two steps that are executed iteratively:

1. Create an augmented set of features F by adding one or more features to the current feature set.

2. Use a classification algorithm $\mathcal{A}$ to evaluate the accuracy of the set of features F. Use the accuracy to either accept or reject the augmentation of F.

The augmentation of F can be performed in many different ways. For example, a greedy strategy may be used where the set of features in the previous iteration is augmented with an additional feature with the greatest discriminative power with respect to a filter criterion. Alternatively, features may be selected for addition via random sampling. The accuracy of the classification algorithm $\mathcal{A}$ in the second step may be used to determine whether the newly augmented set of features should be accepted, or one should revert to the set of features in the previous iteration. This approach is continued until there is no improvement in the current feature set for a minimum number of iterations. Because the classification algorithm $\mathcal{A}$ is used in the second step for evaluation, the final set of identified features will be sensitive to the choice of the algorithm $\mathcal{A}$.

10.2.3 Embedded Models

The core idea in embedded models is that the solutions to many classification formulations provide important hints about the most relevant features to be used. In other words, knowledge about the features is *embedded* within the solution to the classification problem. For example, consider a linear classifier that maps a training instance $\overline{X}$ to a class label y_i in $\{-1, 1\}$ using the following linear relationship:

$$y_i = \text{sign}\{\overline{W} \cdot \overline{X} + b\}. \tag{10.7}$$

Here, $\overline{W} = (w_1, \ldots w_d)$ is a d-dimensional vector of coefficients, and b is a scalar that is learned from the training data. The function "sign" maps to either -1 or $+1$, depending on the sign of its argument. As we will see later, many linear models such as Fisher's discriminant, support vector machine (SVM) classifiers, logistic regression methods, and neural networks use this model.

Assume that all features have been normalized to unit variance. If the value of $|w_i|$ is relatively[3] small, the ith feature is used very weakly by the model and is more likely to be noninformative. Therefore, such dimensions may be removed. It is then possible to train the

[3]Certain variations of linear models, such as L_1-regularized SVMs or *Lasso* (cf. Sect. 11.5.1 of Chap. 11), are particularly effective in this context. Such methods are also referred to as *sparse learning* methods.

Table 10.1: Training data snapshot relating the salary and age features to charitable donation propensity

Name	Age	Salary	Donor?
Nancy	21	37,000	N
Jim	27	41,000	N
Allen	43	61,000	Y
Jane	38	55,000	N
Steve	44	30,000	N
Peter	51	56,000	Y
Sayani	53	70,000	Y
Lata	56	74,000	Y
Mary	59	25,000	N
Victor	61	68,000	Y
Dale	63	51,000	Y

same (or a different) classifier on the data with the pruned feature set. If desired, statistical tests may be used to decide when the value of $|w_i|$ should be considered sufficiently small. Many decision tree classifiers, such as *ID3*, also have feature selection methods embedded in them.

In *recursive* feature elimination, an iterative approach is used. A small number of features are removed in each iteration. Then, the classifier is retrained on the pruned set of features to re-estimate the weights. The re-estimated weights are used to again prune the features with the least absolute weight. This procedure is repeated until all remaining features are deemed to be sufficiently relevant. Embedded models are generally designed in an ad hoc way, depending on the classifier at hand.

10.3 Decision Trees

Decision trees are a classification methodology, wherein the classification process is modeled with the use of a set of *hierarchical* decisions on the feature variables, arranged in a tree-like structure. The decision at a particular node of the tree, which is referred to as the *split criterion*, is typically a condition on one or more feature variables in the training data. The split criterion divides the training data into two or more parts. For example, consider the case where *Age* is an attribute, and the split criterion is $Age \leq 30$. In this case, the left branch of the decision tree contains all training examples with age at most 30, whereas the right branch contains all examples with age greater than 30. The goal is to identify a split criterion so that the level of "mixing" of the class variables in each branch of the tree is reduced as much as possible. Each node in the decision tree logically represents a subset of the data space defined by the combination of split criteria in the nodes above it. The decision tree is typically constructed as a hierarchical partitioning of the training examples, just as a top-down clustering algorithm partitions the data hierarchically. The main difference from clustering is that the partitioning criterion in the decision tree is *supervised* with the class label in the training instances. Some classical decision tree algorithms include *C4.5*, *ID3*, and *CART*. To illustrate the basic idea of decision tree construction, an illustrative example will be used.

In Table 10.1, a snapshot of a hypothetical charitable donation data set has been illustrated. The two feature variables represent the age and salary attributes. Both attributes are related to the donation propensity, which is also the class label. Specifically, the likelihood of an individual to donate is positively correlated with his or her age and salary. However, the best separation of the classes may be achieved only by combining the two attributes. The goal in the decision tree construction process is to perform a sequence of splits in top-down fashion to create nodes at the leaf level in which the donors and nondonors are separated well. One way of achieving this goal is depicted in Fig. 10.3a. The figure illustrates a hierarchical arrangement of the training examples in a treelike structure. The first-level split uses the age attribute, whereas the second-level split for both branches uses the salary attribute. Note that different splits at the same decision tree level need not be on the same attribute. Furthermore, the decision tree of Fig. 10.3a has two branches at each node, but this need not always be the case. In this case, the training examples in all leaf nodes belong to the same class, and, therefore, there is no point in growing the decision tree beyond the leaf nodes. The splits shown in Fig. 10.3a are referred to as *univariate* splits because they use a single attribute. To classify a test instance, a single relevant path in the tree is traversed in top-down fashion by using the split criteria to decide which branch to follow at each node of the tree. The dominant class label in the leaf node is reported as the relevant class. For example, a test instance with age less than 50 and salary less than 60,000 will traverse the leftmost path of the tree in Fig. 10.3a. Because the leaf node of this path contains only nondonor training examples, the test instance will also be classified as a nondonor.

Multivariate splits use more than one attribute in the split criteria. An example is illustrated in Fig. 10.3b. In this particular case, a single split leads to full separation of the classes. This suggests that multivariate criteria are more powerful because they lead to shallower trees. For the same level of class separation in the training data, shallower trees are generally more desirable because the leaf nodes contain more examples and, therefore, are statistically less likely to overfit the noise in the training data.

A decision tree induction algorithm has two types of nodes, referred to as the *internal nodes* and *leaf nodes*. Each leaf node is labeled with the dominant class at that node. A special internal node is the root node that corresponds to the entire feature space. The generic decision tree induction algorithm starts with the full training data set at the root node and recursively partitions the data into lower level nodes based on the split criterion. Only nodes that contain a mixture of different classes need to be split further. Eventually, the decision tree algorithm stops the growth of the tree based on a *stopping criterion*. The simplest stopping criterion is one where all training examples in the leaf belong to the same class. One problem is that the construction of the decision tree to this level may lead to overfitting, in which the model fits the noisy nuances of the training data. Such a tree will not generalize to *unseen* test instances very well. To avoid the degradation in accuracy associated with overfitting, the classifier uses a *postpruning* mechanism for removing overfitting nodes. The generic decision tree training algorithm is illustrated in Fig. 10.4.

After a decision tree has been constructed, it is used for classification of unseen test instances with the use of top-down traversal from the root to a unique leaf. The split condition at each internal node is used to select the correct branch of the decision tree for further traversal. The label of the leaf node that is reached is reported for the test instance.

10.3.1 Split Criteria

The goal of the split criterion is to maximize the separation of the different classes among the children nodes. In the following, only univariate criteria will be discussed. Assume that

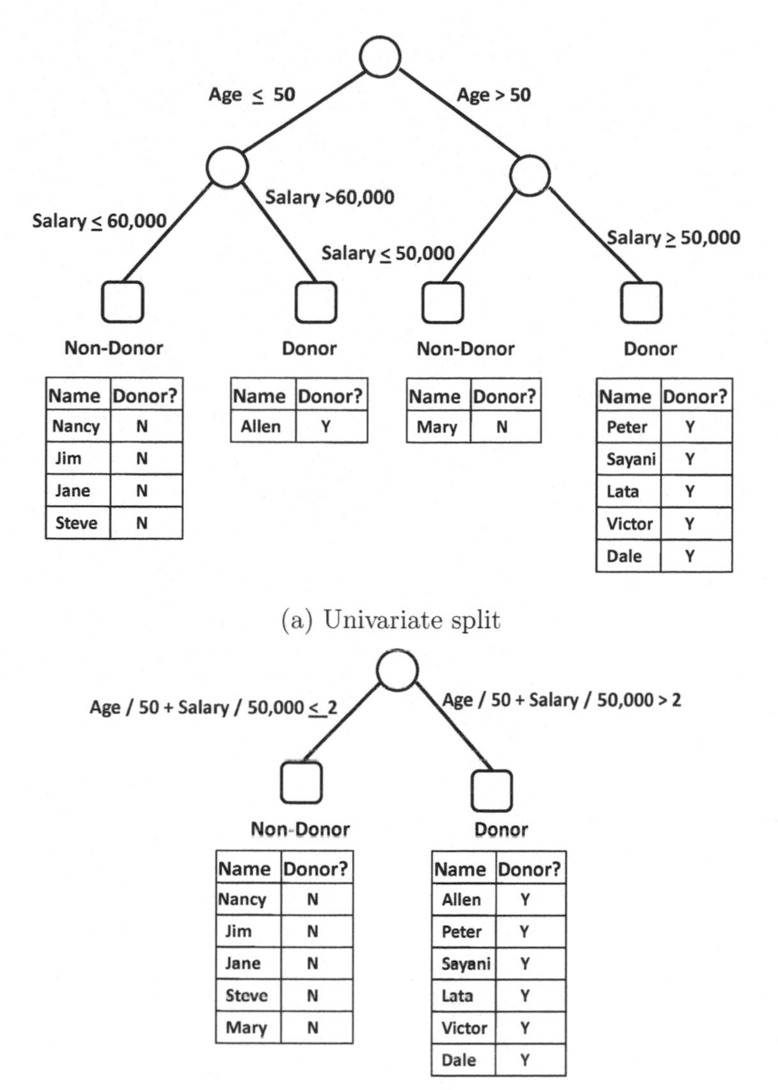

Figure 10.3: Illustration of univariate and multivariate splits for decision tree construction

a quality criterion for evaluating a split is available. The design of the split criterion depends on the nature of the underlying attribute:

1. *Binary attribute:* Only one type of split is possible, and the tree is always binary. Each branch corresponds to one of the binary values.

2. *Categorical attribute:* If a categorical attribute has r different values, there are multiple ways to split it. One possibility is to use an r-way split, in which each branch of the split corresponds to a particular attribute value. The other possibility is to use a binary split by testing each of the $2^r - 1$ combinations (or groupings) of categorical attributes, and selecting the best one. This is obviously not a feasible option when the value of r is large. A simple approach that is sometimes used is to convert categorical

```
Algorithm GenericDecisionTree(Data Set: D)
begin
  Create root node containing D;
  repeat
    Select an eligible node in the tree;
    Split the selected node into two or more nodes
       based on a pre-defined split criterion;
  until no more eligible nodes for split;
  Prune overfitting nodes from tree;
  Label each leaf node with its dominant class;
end
```

Figure 10.4: Generic decision tree training algorithm

data to binary data with the use of the binarization approach discussed in Chap. 2. In this case, the approach for binary attributes may be used.

3. *Numeric attribute:* If the numeric attribute contains a small number r of ordered values (e.g., integers in a small range $[1, r]$), it is possible to create an r-way split for each distinct value. However, for continuous numeric attributes, the split is typically performed by using a binary condition, such as $x \leq a$, for attribute value x and constant a.

   Consider the case where a node contains m data points. Therefore, there are m possible split points for the attribute, and the corresponding values of a may be determined by sorting the data in the node along this attribute. One possibility is to test all the possible values of a for a split and select the best one. A faster alternative is to test only a smaller set of possibilities for a, based on equi-depth division of the range.

Many of the aforementioned methods requires the determination of the "best" split from a set of choices. Specifically, it is needed to choose from multiple attributes and from the various alternatives available for splitting each attribute. Therefore, quantifications of split quality are required. These quantifications are based on the same principles as the feature selection criteria discussed in Sect. 10.2.

1. *Error rate:* Let p be the fraction of the instances in a set of data points S belonging to the dominant class. Then, the error rate is simply $1 - p$. For an r-way split of set S into sets $S_1 \dots S_r$, the overall error rate of the split may be quantified as the weighted average of the error rates of the individual sets S_i, where the weight of S_i is $|S_i|$. The split with the lowest error rate is selected from the alternatives.

2. *Gini index:* The Gini index $G(S)$ for a set S of data points may be computed according to Eq. 10.1 on the class distribution $p_1 \dots p_k$ of the training data points in S.

$$G(S) = 1 - \sum_{j=1}^{k} p_j^2 \tag{10.8}$$

   The overall Gini index for an r-way split of set S into sets $S_1 \dots S_r$ may be quantified as the weighted average of the Gini index values $G(S_i)$ of each S_i, where the weight of S_i is $|S_i|$.

$$\text{Gini-Split}(S \Rightarrow S_1 \ldots S_r) = \sum_{i=1}^{r} \frac{|S_i|}{|S|} G(S_i) \tag{10.9}$$

The split with the lowest Gini index is selected from the alternatives. The *CART* algorithm uses the Gini index as the split criterion.

3. *Entropy:* The entropy measure is used in one of the earliest classification algorithms, referred to as *ID3*. The entropy $E(S)$ for a set S may be computed according to Eq. 10.3 on the class distribution $p_1 \ldots p_k$ of the training data points in the node.

$$E(S) = -\sum_{j=1}^{k} p_j \log_2(p_j) \tag{10.10}$$

As in the case of the Gini index, the overall entropy for an r-way split of set S into sets $S_1 \ldots S_r$ may be computed as the weighted average of the Gini index values $G(S_i)$ of each S_i, where the weight of S_i is $|S_i|$.

$$\text{Entropy-Split}(S \Rightarrow S_1 \ldots S_r) = \sum_{i=1}^{r} \frac{|S_i|}{|S|} E(S_i) \tag{10.11}$$

Lower values of the entropy are more desirable. The entropy measure is used by the *ID3* and *C4.5* algorithms.

The information gain is closely related to entropy, and is equal to the *reduction* in the entropy $E(S)$ – Entropy-Split$(S \Rightarrow S_1 \ldots S_r)$ as a result of the split. Large values of the reduction are desirable. At a conceptual level, there is no difference between using either of the two for a split although a normalization for the degree of the split is possible in the case of information gain. Note that the entropy and information gain measures should be used only to compare two splits of the same degree because both measures are naturally biased in favor of splits with larger degree. For example, if a categorical attribute has many values, attributes with many values will be preferred. It has been shown by the *C4.5* algorithm that dividing the overall information gain with the normalization factor of $-\sum_{i=1}^{r} \frac{|S_i|}{|S|} \log_2(\frac{|S_i|}{|S|})$ helps in adjusting for the varying number of categorical values.

The aforementioned criteria are used to select the choice of the split attribute and the precise criterion on the attribute. For example, in the case of a numeric database, different split points are tested for each numeric attribute, and the best split is selected.

10.3.2 Stopping Criterion and Pruning

The stopping criterion for the growth of the decision tree is intimately related to the underlying pruning strategy. When the decision tree is grown to the very end until every leaf node contains only instances belonging to a particular class, the resulting decision tree exhibits 100 % accuracy on instances belonging to the training data. However, it often generalizes poorly to unseen test instances because the decision tree has now *overfit* even to the random characteristics in the training instances. Most of this noise is contributed by the lower level nodes, which contain a smaller number of data points. In general, simpler models (shallow decision trees) are preferable to more complex models (deep decision trees) if they produce the same error on the training data.

To reduce the level of overfitting, one possibility is to stop the growth of the tree early. Unfortunately, there is no way of knowing the correct point at which to stop the growth of the tree. Therefore, a natural strategy is to prune overfitting portions of the decision tree and convert internal nodes to leaf nodes. Many different criteria are available to decide whether a node should be pruned. One strategy is to explicitly penalize model complexity with the use of the *minimum description length principle (MDL)*. In this approach, the cost of a tree is defined by a weighted sum of its (training data) error and its complexity (e.g., the number of nodes). Information-theoretic principles are used to measure tree complexity. Therefore, the tree is constructed to optimize the cost rather than only the error. The main problem with this approach is that the cost function is itself a heuristic that does not work consistently well across different data sets. A simpler and more intuitive strategy is to a hold out a small fraction (say 20 %) of the training data and build the decision tree on the remaining data. The impact of pruning a node on the classification accuracy is tested on the holdout set. If the pruning improves the classification accuracy, then the node is pruned. Leaf nodes are iteratively pruned until it is no longer possible to improve the accuracy with pruning. Although such an approach reduces the amount of training data for building the tree, the impact of pruning generally outweighs the impact of training-data loss in the tree-building phase.

10.3.3 Practical Issues

Decision trees are simple to implement and highly interpretable. They can model arbitrarily complex decision boundaries, *given sufficient training data*. Even a univariate decision tree can model a complex decision boundary with piecewise approximations by building a sufficiently deep tree. The main problem is that the amount of training data required to properly approximate a complex boundary with a treelike model is very large, and it increases with data dimensionality. With limited training data, the resulting decision boundary is usually a rather coarse approximation of the true boundary. Overfitting is common in such cases. This problem is exacerbated by the sensitivity of the decision tree to the split criteria at the higher levels of the tree. A closely related family of classifiers, referred to as *rule-based classifiers*, is able to alleviate these effects by moving away from the strictly hierarchical structure of a decision tree.

10.4 Rule-Based Classifiers

Rule-based classifiers use a set of "if–then" rules $\mathcal{R} = \{R_1 \dots R_m\}$ to match *antecedents* to *consequents*. A rule is typically expressed in the following form:

IF *Condition* THEN *Conclusion*.

The condition on the left-hand side of the rule, also referred to as the antecedent, may contain a variety of logical operators, such as $<$, $\leq$, $>$, $=$, $\subseteq$, or $\in$, which are applied to the feature variables. The right-hand side of the rule is referred to as the consequent, and it contains the class variable. Therefore, a rule R_i is of the form $Q_i \Rightarrow c$ where Q_i is the antecedent, and c is the class variable. The "$\Rightarrow$" symbol denotes the "THEN" condition. The rules are generated from the training data during the training phase. The notation Q_i represents a precondition on the feature set. In some classifiers, such as association pattern classifiers, the precondition may correspond to a pattern in the feature space, though this may not always be the case. In general, the precondition may be any arbitrary condition on the feature variables. These rules are then used to classify a test instance. A rule is said to *cover* a training instance when the condition in its antecedent matches the training instance.

A decision tree may be viewed as a special case of a rule-based classifier, in which each path of the decision tree corresponds to a rule. For example, the decision tree in Fig. 10.3a corresponds to the following set of rules:

$$Age \leq 50 \text{ AND } Salary \leq 60,000 \Rightarrow \neg Donor$$
$$Age \leq 50 \text{ AND } Salary > 60,000 \Rightarrow Donor$$
$$Age > 50 \text{ AND } Salary \leq 50,000 \Rightarrow \neg Donor$$
$$Age > 50 \text{ AND } Salary > 50,000 \Rightarrow Donor$$

Note that each of the four aforementioned rules corresponds to a path in the decision tree of Fig. 10.3a. The logical expression on the left is expressed in conjunctive form, with a set of "AND" logical operators. Each of the primitive conditions in the antecedent, (such as $Age \leq 50$) is referred to as a *conjunct*. The rule set from a training data set is not unique and depends on the specific algorithm at hand. For example, only two rules are generated from the decision tree in Fig. 10.3b.

$$Age/50 + Salary/50,000 \leq 2 \Rightarrow \neg Donor$$
$$Age/50 + Salary/50,000 > 2 \Rightarrow Donor$$

As in decision trees, succinct rules, both in terms of the cardinality of the rule set and the number of conjuncts in each rule, are generally more desirable. This is because such rules are less likely to overfit the data, and will generalize well to unseen test instances. Note that the antecedents on the left-hand side always correspond to a rule *condition*. In many rule-based classifiers, such as association-pattern classifiers, the logical operators such as "$\subseteq$" are implicit and are omitted from the rule antecedent description. For example, consider the case where the age and salary are discretized into categorical attribute values.

$$Age\ [50:60],\quad Salary\ [50,000:60,000] \Rightarrow Donor$$

In such a case, the discretized attributes for age and salary will be represented as "items," and an association pattern-mining algorithm can discover the itemset on the left-hand side. The operator "$\subseteq$" is implicit in the rule antecedent. Associative classifiers are discussed in detail later in this section.

The training phase of a rule-based algorithm creates a set of rules. The classification phase for a test instance discovers all rules that are *triggered* by the test instance. A rule is said to be triggered by the test instance when the logical condition in the antecedent is satisfied by the test instance. In some cases, rules with conflicting consequent values are triggered by the test instance. In such cases, methods are required to resolve the conflicts in class label prediction. Rule sets may satisfy one or more of the following properties:

1. *Mutually exclusive rules:* Each rule covers a disjoint partition of the data. Therefore, at most one rule can be triggered by a test instance. The rules generated from a decision tree satisfy this property. However, if the extracted rules are subsequently modified to reduce overfitting (as in some classifiers such as *C4.5rules*), the resulting rules may no longer remain mutually exclusive.

2. *Exhaustive rules:* The entire data space is covered by at least one rule. Therefore, every test instance triggers at least one rule. The rules generated from a decision tree

also satisfy this property. It is usually easy to construct an exhaustive rule set by creating a single catch-all rule whose consequent contains the dominant class in the portion of the training data not covered by other rules.

It is relatively easy to perform the classification when a rule set satisfies both the aforementioned properties. The reason for this is that each test instance maps to exactly one rule, and there are no conflicts in class predictions by multiple rules. In cases where rule sets are not mutually exclusive, conflicts in the rules triggered by a test instance can be resolved in one of two ways:

1. *Rule ordering:* The rules are ordered by priority, which may be defined in a variety of ways. One possibility is to use a quality measure of the rule for ordering. Some popular classification algorithms, such as *C4.5rules* and *RIPPER*, use class-based ordering, where rules with a particular class are prioritized over the other. The resulting set of ordered rules is also referred to as a *decision list*. For an arbitrary test instance, the class label in the consequent of the top triggered rule is reported as the relevant one for the test instance. Any other triggered rule is ignored. If no rule is triggered then a default catch-all class is reported as the relevant one.

2. *Unordered rules:* No priority is imposed on the rule ordering. The dominant class label among *all* the triggered rules may be reported. Such an approach can be more robust because it is not sensitive to the choice of the *single* rule selected by a rule-ordering scheme. The training phase is generally more efficient because all rules can be extracted simultaneously with pattern-mining techniques without worrying about relative ordering. Ordered rule-mining algorithms generally have to integrate the rule ordering into the rule generation process with methods such as *sequential covering*, which are computationally expensive. On the other hand, the testing phase of an unordered approach can be more expensive because of the need to compare a test instance against all the rules.

How should the different rules be ordered for test instance classification? The first possibility is to order the rules on the basis of a quality criterion, such as the confidence of the rule, or a weighted measure of the support and confidence. However, this approach is rarely used. In most cases, the rules are ordered by class. In some rare class applications, it makes sense to order all rules belonging to the rare class first. Such an approach is used by *RIPPER*. In other classifiers, such as *C4.5rules*, various accuracy and information-theoretic measures are used to prioritize classes.

10.4.1 Rule Generation from Decision Trees

As discussed earlier in this section, rules can be extracted from the different paths in a decision tree. For example, *C4.5rules* extracts the rules from the *C4.5* decision tree. The sequence of split criteria on each path of the decision tree corresponds to the antecedent of a corresponding rule. Therefore, it would seem at first sight that rule ordering is not needed because the generated rules are exhaustive and mutually exclusive. However, the rule-extraction process is followed by a rule-pruning phase in which many conjuncts are pruned from the rules to reduce overfitting. Rules are processed one by one, and conjuncts are pruned from them in greedy fashion to improve the accuracy as much as possible on the covered examples in a separate holdout validation set. This approach is similar to decision tree pruning except that one is no longer restricted to pruning the conjuncts at the lower levels of the decision tree. Therefore, the pruning process is more flexible than that of a decision tree, because it is not restricted by an underlying tree structure. Duplicate rules may result from pruning of conjuncts. These rules are removed. The rule-pruning phase increases the coverage of the individual rules and, therefore, the mutually exclusive nature of the rules is lost. As a result, it again becomes necessary to order the rules.

In *C4.5rules*, all rules that belong to the class whose rule set has the smallest description length are prioritized over other rules. The total description length of a rule set is a weighted sum of the number of bits required to encode the size of the model (rule set) and the number of examples covered by the class-specific rule set in the training data, which belong to a different class. Typically, classes with a smaller number of training examples are favored by this approach. A second approach is to order the class first whose rule set has the least number of false-positive errors on a separate holdout set. A rule-based version of a decision tree generally allows the construction of a more flexible decision boundary with limited training data than the base tree from which the rules are generated. This is primarily because of the greater flexibility in the model which is no longer restrained by the straitjacket of an exhaustive and mutually exclusive rule set. As a result, the approach generalizes better to unseen test instances.

10.4.2 Sequential Covering Algorithms

Sequential covering methods are used frequently for creating ordered rule lists. Thus, in this case, the classification process uses the top triggered rule to classify unseen test instances. Examples of sequential covering algorithms include *AQ*, *CN2*, and *RIPPER*. The sequential covering approach iteratively applies the following two steps to grow the rules from the training data set $\mathcal{D}$ until a stopping criterion is met:

1. (*Learn-One-Rule*) Select a particular class label and determine the "best" rule from the current training instances $\mathcal{D}$ with this class label as the consequent. Add this rule to the bottom of the ordered rule list.

2. (*Prune training data*) Remove the training instances in $\mathcal{D}$ that are covered by the rule learned in the previous step. All training instances matching the *antecedent* of the rule must be removed, whether or not the class label of the training instance matches the consequent.

The aforementioned generic description applies to all sequential covering algorithms. The various sequential covering algorithms mainly differ in the details of how the rules are ordered with respect to each other.

1. *Class-based ordering:* In most sequential covering algorithms such as *RIPPER*, all rules corresponding to a particular class are generated and placed contiguously on the ordered list. Typically, rare classes are ordered first. Therefore, classes that are placed earlier on the list may be favored more than others. This can sometimes cause artificially lower accuracy for test instances belonging to the less favored class.

   When class-based ordering is used, the rules for a particular class are generated contiguously. The addition of rules for each class has a stopping criterion that is algorithm dependent. For example, *RIPPER* uses an MDL criterion that stops adding rules when further addition increases the description length of the model by at least a predefined number of units. Another simpler stopping criterion is when the error rate of the next generated rule on a separate validation set exceeds a predefined threshold. Finally, one might simply use a threshold on the number of uncovered training instances remaining for a class as the class-specific stopping criterion. When the number of uncovered training instances remaining for a class falls below a threshold, rules for that class consequent are no longer grown. At this point, rules corresponding to the next class are grown. For a k-class problem, this approach is repeated $(k-1)$ times. Rules for the kth class are not grown. The least prioritized rule is a single catch-all rule with its consequent as the kth class. When the test instance does not fire rules belonging to the other classes, this class is assumed as the relevant label.

2. *Quality-based ordering:* In some covering algorithms, class-based ordering is not used. A quality measure is used to select the next rule. For example, one might generate the rule with the highest confidence in the remaining training data. The catch-all rule corresponds to the dominant class among remaining test instances. Quality-based ordering is rarely used in practice because of the difficulty in interpreting a quality criterion which is defined only over the *remaining* test instances.

Because class-based ordering is more common, the Learn-One-Rule procedure will be described under this assumption.

10.4.2.1 Learn-One-Rule

The Learn-One-Rule procedure grows rules from the general to the specific, in much the same way a decision tree grows a tree hierarchically from general nodes to specific nodes. Note that a path in a decision tree is a rule in which the antecedent corresponds to the conjunction of the split criteria at the different nodes, and the consequent corresponds to the label of the leaf nodes. While a decision tree grows many different disjoint paths at one time, the Learn-One-Rule procedure grows a single "best" path. This is yet another example of the close relationship between decision trees and rule-based methods.

The idea of Learn-One-Rule is to successively add conjuncts to the left-hand side of the rule to grow a single decision *path* (rather than a decision tree) based on a quality criterion. The root of the tree corresponds to the rule $\{\} \Rightarrow c$. The class c represents the consequent of the rule being grown. In the simplest version of the procedure, a single path is grown at one time by successively adding conjuncts to the antecedent. In other words, conjuncts are added to increase the *quality* as much as possible. The simplest quality criterion is the accuracy of the rule. The problem with this criterion is that rules with high accuracy but very low coverage are generally not desirable because of overfitting. The precise choice of the quality criterion that regulates the trade-off between accuracy and coverage will be discussed in detail later. As in the case of a decision tree, various logical conditions (or split choices) must be tested to determine the best conjunct to be added. The process of enumeration of the various split choices is similar to a decision tree. The rule is grown until a particular stopping criterion is met. A natural stopping criterion is one where the quality of the rule does not improve by further growth.

One challenge with the use of this procedure is that if a mistake is made early on during tree growth, it will lead to suboptimal rules. One way of reducing the likelihood of suboptimal rules is to always maintain the m best paths during rule-growth rather than a single one. An example of rule growth with the use of a single decision path, for the donor example of Table 10.1, is illustrated in Fig. 10.5. In this case, the rule is grown for the donor class. The first conjunct added is $Age > 50$, and the second conjunct added is $Salary > 50,000$. Note the intuitive similarity between the decision tree of Figs. 10.3a and 10.5.

It remains to describe the quality criterion for the growth of the paths during the Learn-One-Rule procedure. On what basis is a particular path selected over the others? The

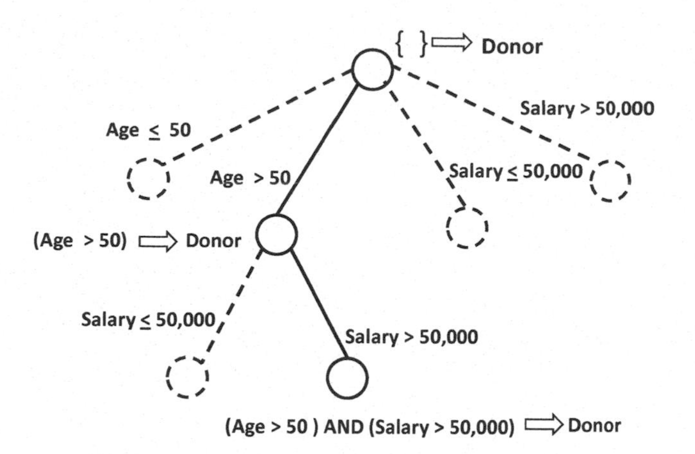

Figure 10.5: Rule growth is analogous to decision tree construction

similarity between rule growth and decision trees suggests the use of analogous measures such as the accuracy, entropy, or the Gini index, as used for split criteria in decision trees.

The criteria do need to be modified because a rule is relevant only to the training examples covered by the antecedent and the single class in the consequent, whereas decision tree splits are evaluated with respect to all training examples at a given node and all classes. Furthermore, decision tree split measures do not need to account for issues such as the coverage of the rule. One would like to determine rules with high coverage in order to avoid overfitting. For example, a rule that covers only a single training instance will always have 100 % accuracy, but it does not usually generalize well to unseen test instances. Therefore, one strategy is to combine the accuracy and coverage criteria into a single integrated measure.

The simplest combination approach is to use Laplacian smoothing with a parameter β that regulates the level of smoothing in a training data set with k classes:

$$\text{Laplace}(\beta) = \frac{n^+ + \beta}{n^+ + n^- + k\beta}. \tag{10.12}$$

The parameter $\beta > 0$ controls the level of smoothing, n^+ represents the number of correctly classified (positive) examples covered by the rule and n^- represents the number of incorrectly classified (negative) examples covered by the rule. Therefore, the total number of covered examples is $n^+ + n^-$. For cases where the absolute number of covered examples $n^+ + n^-$ is very small, Laplacian smoothing penalizes the accuracy to account for the unreliability of low coverage. Therefore, the measure favors greater coverage.

A second possibility is the *likelihood ratio statistic*. Let n_j be the observed number of training data points covered by the rule that belong to class j, and let n_j^e be the expected number of covered examples that would belong to class j, if the class distribution of the covered examples is the same as the full training data. In other words, if $p_1 \dots p_k$ be the fraction of examples belonging to each class in the full training data, then we have:

$$n_i^e = p_i \sum_{i=1}^{k} n_i. \tag{10.13}$$

Then, for a k-class problem, the likelihood ratio statistic R may be computed as follows:

$$R = 2\sum_{j=1}^{k} n_j \log(n_j/n_j^e). \tag{10.14}$$

When the distribution of classes in the covered examples is significantly different than that in the original training data, the value of R increases. Therefore, the statistic tends to favor covered examples whose distributions are very different from the original training data. Furthermore, the presence of raw frequencies $n_1 \ldots n_k$ as multiplicative factors of the individual terms in the right-hand side of Eq. 10.14 ensures that larger rule coverage is rewarded. This measure is used by the *CN2* algorithm.

Another criterion is *FOIL's information gain*. The term "FOIL" stands for *first order inductive learner*. Consider the case where a rule covers n_1^+ positive examples and n_1^- negative examples, where positive examples are defined as training examples matching the class in the consequent. Consider the case where the addition of a conjunct to the antecedent changes the number of positive examples and negative examples to n_2^+ and n_2^-, respectively. Then, FOIL's information gain FG is defined as follows:

$$FG = n_2^+ \left(\log_2 \frac{n_2^+}{n_2^+ + n_2^-} - \log_2 \frac{n_1^+}{n_1^+ + n_1^-} \right). \tag{10.15}$$

This measure tends to select rules with high coverage because n_2^+ is a multiplicative factor in FG. At the same time, the information gain increases with higher accuracy because of the term inside the parentheses. This particular measure is used by the *RIPPER* algorithm.

As in the case of decision trees, it is possible to grow the rules until 100 % accuracy is achieved on the training data, or when the added conjunct does not improve the accuracy of the rule. Another criterion used by *RIPPER* is that the minimum description length of the rule must not increase by more than a certain threshold because of the addition of a conjunct. The description length of a rule is defined by a weighted function of the size of the conjuncts and the misclassified examples.

10.4.3 Rule Pruning

Rule-pruning is relevant not only for rules generated by the Learn-One-Rule method, but also for methods such as *C4.5rules* that extract the rules from a decision tree. Irrespective of the approach used to extract the rules, overfitting may result from the presence of too many conjuncts. As in decision tree pruning, the MDL principle can be used for pruning. For example, for each conjunct in the rule, one can add a penalty term δ to the quality criterion in the rule-growth phase. This will result in a *pessimistic error rate*. Rules with many conjuncts will therefore have larger aggregate penalties to account for their greater model complexity. A simpler approach for computing pessimistic error rates is to use a separate holdout validation set that is used for computing the error rate (without a penalty) but is not used by Learn-One-Rule during rule generation.

The conjuncts successively added during rule growth (in sequential covering) are then tested for pruning in reverse order. If pruning reduces the pessimistic error rate on the training examples covered by the rule, then the generalized rule is used. While some algorithms such as *RIPPER* test the most recently added conjunct first for rule pruning, it is not a strict requirement to do so. It is possible to test the conjuncts for removal in any order, or in greedy fashion, to reduce the pessimistic error rate as much as possible. Rule pruning may result in some of the rules becoming identical. Duplicate rules are removed from the rule set before classification.

10.4.4 Associative Classifiers

Associative classifiers are a popular strategy because they rely on association pattern mining, for which many efficient algorithmic alternatives exist. The reader is referred to Chap. 4 for algorithms on association pattern mining. The discussion below assumes binary attributes, though any data type can be converted to binary attributes with the process of discretization and binarization, as discussed in Chap. 2. Furthermore, unlike sequential covering algorithms in which rules are always ordered, the rules created by associative classifiers may be either ordered or unordered, depending upon application-specific criteria. The main characteristic of class-based association rules is that they are mined in the same way as regular association rules, except that they have a single class variable in the consequent. The basic strategy for an associative classifier is as follows:

1. Mine all class-based association rules at a given level of minimum support and confidence.

2. For a given test instance, use the mined rules for classification.

A variety of choices exist for the implementation of both steps. A naive way of implementing the first step would be to mine all association rules and then filter out only the rules in which the consequent corresponds to an individual class. However, such an approach is rather wasteful because it generates many rules with nonclass consequents. Furthermore, there is significant redundancy in the rule set because many rules that have 100 % confidence are special cases of other rules with 100 % confidence. Therefore, pruning methods are required during the rule-generation process.

The *classification based on associations (CBA)* approach uses a modification of the *Apriori* method to generate associations that satisfy the corresponding constraints. The first step is to generate *1-rule-items*. These are newly created items corresponding to combinations of items and class attributes. These rule items are then extended using traditional *Apriori*-style processing. Another modification is that, when patterns are generated corresponding to rules with 100 % confidence, those rules are not extended in order to retain greater generality in the rule set. This broader approach can be used in conjunction with almost any tree enumeration algorithm. The bibliographic notes contain pointers to several recent algorithms that use other frequent pattern mining methods for rule generation.

The second step of associative classification uses the generated rule set to make predictions for unseen test instances. Both ordered or unordered strategies may be used. The ordered strategy prioritizes the rules on the basis of the support (analogous to coverage), and the confidence (analogous to accuracy). A variety of heuristics may be used to create an integrated measure for ordering, such as using a weighted combination of support and confidence. The reader is referred to Chap. 17 for discussion of a representative rule-based classifier, *XRules*, which uses different types of measures. After the rules have been ordered, the top m matching rules to the test instance are determined. The dominant class label from the matching rules is reported as the relevant one for the test instance. A second strategy does not order the rules but determines the dominant class label from all the triggered rules. Other heuristic strategies may weight the rules differently, depending on their support and confidence, for the prediction process. Furthermore, many variations of associative classifiers do not use the support or confidence for mining the rules, but directly use class-based discriminative methods for pattern mining. The bibliographic notes contain pointers to these methods.

10.5 Probabilistic Classifiers

Probabilistic classifiers construct a model that quantifies the relationship between the feature variables and the target (class) variable as a probability. There are many ways in which such a modeling can be performed. Two of the most popular models are as follows:

1. *Bayes classifier:* The Bayes rule is used to model the probability of each value of the target variable for a given set of feature variables. Similar to mixture modeling in clustering (cf. Sect. 6.5 in Chap. 6), it is assumed that the data points within a class are generated from a specific probability distribution such as the Bernoulli distribution or the multinomial distribution. A *naive Bayes assumption* of class-conditioned feature independence is often (but not always) used to simplify the modeling.

2. *Logistic regression:* The target variable is assumed to be drawn from a Bernoulli distribution whose mean is defined by a parameterized logit function on the feature variables. Thus, the probability distribution of the *class* variable is a parameterized function of the feature variables. This is in contrast to the Bayes model that assumes a specific generative model of the *feature* distribution of each class.

The first type of classifier is referred to as a *generative classifier*, whereas the second is referred to as a *discriminative classifier*. In the following, both classifiers will be studied in detail.

10.5.1 Naive Bayes Classifier

The Bayes classifier is based on the Bayes theorem for conditional probabilities. This theorem quantifies the conditional probability of a random variable (class variable), given known observations about the value of another set of random variables (feature variables). The Bayes theorem is used widely in probability and statistics. To understand the Bayes theorem, consider the following example, based on Table 10.1:

Example 10.5.1 *A charitable organization solicits donations from individuals in the population of which* $6/11$ *have age greater than 50. The company has a success rate of* $6/11$ *in soliciting donations, and among the individuals who donate, the probability that the age is greater than 50 is* $5/6$. *Given an individual with age greater than 50, what is the probability that he or she will donate?*

Consider the case where the event E corresponds to $(Age > 50)$, and event D corresponds to an individual being a donor. The goal is to determine the *posterior* probability $P(D|E)$. This quantity is referred to as the "posterior" probability because it is conditioned on the observation of the event E that the individual has age greater than 50. The "prior" probability $P(D)$, before observing the age, is $6/11$. Clearly, knowledge of an individual's age influences posterior probabilities because of the obvious correlations between age and donor behavior.

Bayes theorem is useful for estimating $P(D|E)$ when it is hard to estimate $P(D|E)$ directly from the training data, but other conditional and prior probabilities such as $P(E|D)$, $P(D)$, and $P(E)$ can be estimated more easily. Specifically, Bayes theorem states the following:

$$P(D|E) = \frac{P(E|D)P(D)}{P(E)}. \tag{10.16}$$

Each of the expressions on the right-hand side is already known. The value of $P(E)$ is 6/11, and the value of $P(E|D)$ is 5/6. Furthermore, the prior probability $P(D)$ before knowing the age is 6/11. Consequently, the posterior probability may be estimated as follows:

$$P(D|E) = \frac{(5/6)(6/11)}{6/11} = 5/6. \tag{10.17}$$

Therefore, if we had 1-dimensional training data containing only the *Age*, along with the class variable, the probabilities could be estimated using this approach. Table 10.1 contains an example with training instances satisfying the aforementioned conditions. It is also easy to verify from Table 10.1 that the fraction of individuals above age 50 who are donors is 5/6, which is in agreement with Bayes theorem. In this particular case, the Bayes theorem is not really essential because the classes can be predicted directly from a *single* attribute of the training data. A question arises, as to why the indirect route of using the Bayes theorem is useful, if the posterior probability $P(D|E)$ could be estimated directly from the training data (Table 10.1) in the first place. The reason is that the conditional event E usually corresponds to a *combination of* constraints on d different feature variables, rather than a single one. This makes the direct estimation of $P(D|E)$ much more difficult. For example, the probability $P(Donor|Age > 50, Salary > 50,000)$ is harder to robustly estimate from the training data because there are fewer instances in Table 10.1 that satisfy *both* the conditions on age and salary. This problem increases with increasing dimensionality. In general, for a d-dimensional test instance, with d conditions, it may be the case that not even a single tuple in the training data satisfies all these conditions. Bayes rule provides a way of expressing $P(Donor|Age > 50, Salary > 50,000)$ in terms of $P(Age > 50, Salary > 50,000|Donor)$. The latter is much easier to estimate with the use of a product-wise approximation known as the *naive Bayes approximation*, whereas the former is not.

For ease in discussion, it will be assumed that all feature variables are categorical. The numeric case is discussed later. Let C be the random variable representing the class variable of an unseen test instance with d-dimensional feature values $\overline{X} = (a_1 \ldots a_d)$. The goal is to estimate $P(C = c|\overline{X} = (a_1 \ldots a_d))$. Let the random variables for the individual dimensions of $\overline{X}$ be denoted by $\overline{X} = (x_1 \ldots x_d)$. Then, it is desired to estimate the conditional probability $P(C = c|x_1 = a_1, \ldots x_d = a_d)$. This is difficult to estimate directly from the training data because the training data may not contain even a single record with attribute values $(a_1 \ldots a_d)$. Then, by using Bayes theorem, the following equivalence can be inferred:

$$P(C = c|x_1 = a_1, \ldots x_d = a_d) = \frac{P(C = c)P(x_1 = a_1, \ldots x_d = a_d|C = c)}{P(x_1 = a_1, \ldots x_d = a_d)} \tag{10.18}$$

$$\propto P(C = c)P(x_1 = a_1, \ldots x_d = a_d|C = c). \tag{10.19}$$

The second relationship above is based on the fact that the term $P(x_1 = a_1, \ldots x_d = a_d)$ in the denominator of the first relationship is independent of the class. Therefore, it suffices to only compute the numerator to determine the class with the maximum conditional probability. The value of $P(C = c)$ is the prior probability of the class identifier c and can be *estimated* as the fraction of the training data points belonging to class c. The key usefulness of the Bayes rule is that the terms on the right-hand side can now be effectively approximated from the training data with the use of a naive Bayes approximation. The naive Bayes approximation assumes that the values on the different attributes $x_1 \ldots x_d$ are independent of one another conditional on the class. When two random events A and B are independent of one another conditional on a third event F, it follows that $P(A \cap B|F) = P(A|F)P(B|F)$. In the case of the naive Bayes approximation, it is assumed that the feature values are independent of one another conditional on a fixed value of the class variable. This implies the following for the conditional term on the right-hand side of Eq. 10.19.

$$P(x_1 = a_1, \ldots x_d = a_d | C = c) = \prod_{j=1}^{d} P(x_j = a_j | C = c) \tag{10.20}$$

Therefore, by substituting Eq. 10.20 in Eq. 10.19, the Bayes probability can be estimated within a constant of proportionality as follows:

$$P(C = c | x_1 = a_1, \ldots x_d = a_d) \propto P(C = c) \prod_{j=1}^{d} P(x_j = a_j | C = c). \tag{10.21}$$

Note that each term $P(x_j = a_j | C = c)$ is much easier to estimate from the training data than $P(x_1 = a_1, \ldots x_d = a_d | C = c)$ because enough training examples will exist in the former case to provide a robust estimate. Specifically, the *maximum likelihood estimate* for the value of $P(x_j = a_j | C = c)$ is the fraction of training examples taking on value a_j, conditional on the fact, that they belong to class c. In other words, if $q(a_j, c)$ is the number of training examples corresponding to feature variable $x_j = a_j$ and class c, and $r(c)$ is the number of training examples belonging to class c, then the estimation is performed as follows:

$$P(x_j = a_j | C = c) = \frac{q(a_j, c)}{r(c)}. \tag{10.22}$$

In some cases, enough training examples may still not be available to estimate these values robustly. For example, consider a rare class c with a *single training example* satisfying $r(c) = 1$, and $q(a_j, c) = 0$. In such a case, the conditional probability is estimated to 0. Because of the productwise form of the Bayes expression, the entire probability will be estimated to 0. Clearly, the use of a small number of training examples belonging to the rare class cannot provide robust estimates. To avoid this kind of overfitting, Laplacian smoothing is used. A small value of α is added to the numerator, and a value of $\alpha \cdot m_j$ is added to the denominator, where m_j is the number of distinct values of the jth attribute:

$$P(x_j = a_j | C = c) = \frac{q(a_j, c) + \alpha}{r(c) + \alpha \cdot m_j}. \tag{10.23}$$

Here, α is the Laplacian smoothing parameter. For the case where $r(c) = 0$, this has the effect of estimating the probability to an unbiased value of $1/m_j$ for all m_j distinct attribute values. This is a reasonable estimate in the absence of any training data about class c. Thus, the training phase only requires the estimation of these conditional probabilities $P(x_j = a_j | C = c)$ of each class–attribute–value combination, and the estimation of the prior probabilities $P(C = c)$ of each class.

This model is referred to as the *binary* or *Bernoulli* model for Bayes classification when it is applied to categorical data with only two outcomes of each feature attribute. For example, in text data, the two outcomes could correspond to the presence or absence of a word. In cases where more than two outcomes are possible for a feature variable, the model is referred to as the *generalized* Bernoulli model. The implicit generative assumption of this model is similar to that of mixture modeling algorithms in clustering (cf. Sect. 6.5 of Chap. 6). The features within each class (mixture component) are independently generated from a distribution whose probabilities are the productwise approximations of Bernoulli distributions. The estimation of model parameters in the training phase is analogous to the M-step in expectation–maximization (EM) clustering algorithms. Note that, unlike EM clustering algorithms, the labels on only the *training* data are used to compute the maximum likelihood estimates of parameters in the training phase. Furthermore, the E-step (or the iterative approach) is not required because the (deterministic) assignment "probabilities" of labeled data are already known. In Sect. 13.5.2.1 of Chap. 13, a more sophisticated model, referred to as the *multinomial model*, will be discussed. This model can address sparse frequencies associated with attributes, as in text data. In general, the Bayes model can assume any parametric form of the conditional feature distribution $P(x_1 = a_1, \ldots x_d = a_d|C = c)$ of each class (mixture component), such as a Bernoulli model, a multinomial model, or even a Gaussian model for numeric data. The parameters of the distribution of each class are estimated in a data-driven manner. The approach discussed in this section, therefore, represents only a single instantiation from a wider array of possibilities.

The aforementioned description is based on categorical data. It can also be generalized to numeric data sets by using the process of discretization. Each discretized range becomes one of the possible categorical values of an attribute. Such an approach can, however, be sensitive to the granularity of the discretization. A second approach is to assume a specific form of the probability distribution of each mixture component (class), such as a Gaussian distribution. The mean and variance parameters of the Gaussian distribution of each class are estimated in a data-driven manner, just as the class conditioned feature probabilities are estimated in the Bernoulli model. Specifically, the mean and variance of each Gaussian can be estimated directly as the mean and variance of the training data for the corresponding class. This is similar to the M-step in EM clustering algorithms with Gaussian mixtures. The conditional class probabilities in Eq. 10.21 for a test instance are replaced with the class-specific Gaussian densities of the test instance.

10.5.1.1 The Ranking Model for Classification

The aforementioned algorithms predict the labels of *individual* test instances. In some scenarios, a *set* of test instances is provided to the learner, and it is desired to *rank* these test instances by their propensity to belong to a particularly important class c. This is a common scenario in rare-class learning, which will be discussed in Sect. 11.3 of Chap. 11.

As discussed in Eq. 10.21, the probability of a test instance $(a_1 \ldots a_d)$ belonging to a particular class can be estimated within a constant of proportionality as follows:

$$P(C = c|x_1 = a_1, \ldots x_d = a_d) \propto P(C = c) \prod_{j=1}^{d} P(x_j = a_j|C = c). \tag{10.24}$$

The constant of proportionality is irrelevant while comparing the scores across different *classes* but is not irrelevant while comparing the scores across different *test instances*. This is because the constant of proportionality is the inverse of the generative probability of the specific test instance. An easy way to estimate the proportionality constant is to use normalization so that the sum of probabilities across different classes is 1. Therefore, if the class label c is assumed to be an integer drawn from the range $\{1 \ldots k\}$ for a k-class problem, then the Bayes probability can be estimated as follows:

$$P(C = c|x_1 = a_1, \ldots x_d = a_d) = \frac{P(C = c) \prod_{j=1}^{d} P(x_j = a_j|C = c)}{\sum_{c=1}^{k} P(C = c) \prod_{j=1}^{d} P(x_j = a_j|C = c)}. \tag{10.25}$$

These normalized values can then be used to rank different test instances. It should be pointed out that most classification algorithms return a numerical score for each class, and therefore an analogous normalization can be performed for virtually any classification algorithm. However, in the Bayes method, it is more natural to intuitively interpret the normalized values as probabilities.

10.5.1.2 Discussion of the Naive Assumption

The Bayes model is referred to as "naive" because of the assumption of conditional independence. This assumption is obviously not true in practice because the features in real data sets are almost always correlated even when they are conditioned on a specific class. Nevertheless, in spite of this approximation, the naive Bayes classifier seems to perform quite well in practice in many domains. Although it is possible to implement the Bayes model using more general multivariate estimation methods, such methods can be computationally more expensive. Furthermore, the estimation of multivariate probabilities becomes inaccurate with increasing dimensionality, especially with limited training data. Therefore, significant practical accuracy is often not gained with the use of theoretically more accurate assumptions. The bibliographic notes contain pointers to theoretical results on the effectiveness of the naive assumption.

10.5.2 Logistic Regression

While the Bayes classifier assumes a specific form of the feature probability distribution for each class, logistic regression directly models the class-membership probabilities in terms of the feature variables with a discriminative function. Thus, the nature of the modeling assumption is different in the two cases. Both are, however, probabilistic classifiers because they use a specific modeling assumption to map the feature variables to a class-membership probability. In both cases, the parameters of the underlying probabilistic model need to be estimated in a data-driven manner.

In the simplest form of logistic regression, it is assumed that the class variable is binary, and is drawn from $\{-1, +1\}$, although it is also possible to model nonbinary class variables. Let $\overline{\Theta} = (\theta_0, \theta_1 \dots \theta_d)$ be a vector of $d+1$ different parameters. The ith parameter θ_i is a coefficient related to the ith dimension in the underlying data, and θ_0 is an offset parameter. Then, for a record $\overline{X} = (x_1 \dots x_d)$, the probability that the class variable C takes on the values of $+1$ or -1, is modeled with the use of a logistic function.

$$P(C = +1|\overline{X}) = \frac{1}{1 + e^{-(\theta_0 + \sum_{i=1}^{d} \theta_i x_i)}} \tag{10.26}$$

$$P(C = -1|\overline{X}) = \frac{1}{1 + e^{(\theta_0 + \sum_{i=1}^{d} \theta_i x_i)}} \tag{10.27}$$

It is easy to verify that the sum of the two aforementioned probability values is 1. Logistic regression can be viewed as either a probabilistic classifier or a *linear classifier.* In linear classifiers, such as Fisher's discriminant, a linear hyperplane is used to separate the two classes. Other linear classifiers such as SVMs and neural networks will be discussed in Sects. 10.6 and 10.7 of this chapter. In logistic regression, the parameters $\overline{\Theta} = (\theta_0 \dots \theta_d)$ can be viewed as the coefficients of a separating hyperplane $\theta_0 + \sum_{i=1}^{d} \theta_i x_i = 0$ between the two classes. The term θ_i is the linear coefficient of dimension i, and the term θ_0 is the constant term. The value of $\theta_0 + \sum_{i=1}^{d} \theta_i x_i$ will be either positive or negative, depending on the side of the separating hyperplane on which $\overline{X}$ is located. A positive value is predictive of the class $+1$, whereas a negative value is predictive of the class -1. In many other linear classifiers, the sign of this expression yields the class label of $\overline{X}$ from $\{-1, +1\}$. Logistic

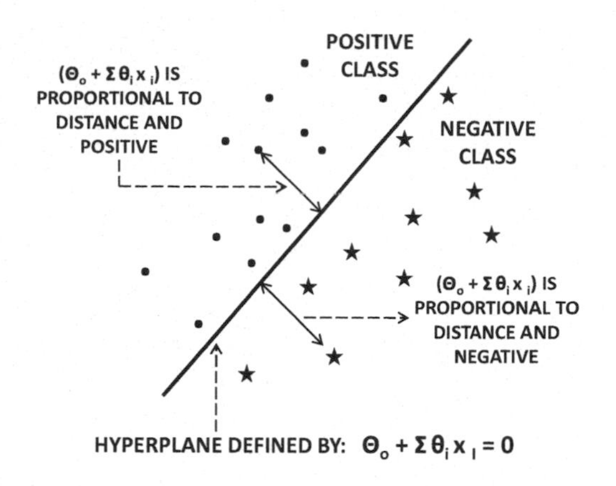

Figure 10.6: Illustration of logistic regression in terms of linear separators

regression achieves the same result in the form of probabilities defined by the aforementioned discriminative function.

The term $\theta_0 + \sum_{i=1}^{d} \theta_i x_i$, within the exponent of the logistic function is proportional to the distance of the data point from the separating hyperplane. When the data point lies exactly on this hyperplane, both classes are assigned the probability of 0.5 according to the logistic function. Positive values of the distance will assign probability values greater than 0.5 to the positive class. Negative values of the distance will assign (symmetrically equal) probability values greater than 0.5 to the negative class. This scenario is illustrated in Fig. 10.6. Therefore, the logistic function neatly exponentiates the distances, shown in Fig. 10.6, to convert them to intuitively interpretable probabilities in $(0, 1)$. The setup of logistic regression is similar to classical least squares linear regression, with the difference that the logit function is used to estimate probabilities of class membership instead of constructing a squared error objective. Consequently, instead of the least-squares optimization in linear regression, a maximum likelihood optimization model is used for logistic regression.

10.5.2.1 Training a Logistic Regression Classifier

The maximum likelihood approach is used to estimate the best fitting parameters of the logistic regression model. Let $\mathcal{D}_+$ and $\mathcal{D}_-$ be the segments of the training data belonging to the positive and negative classes, respectively. Let the kth data point be denoted by $\overline{X_k} = (x_k^1 \dots x_k^d)$. Then, the likelihood function $\mathcal{L}(\overline{\Theta})$ for the entire data set is defined as follows:

$$\mathcal{L}(\overline{\Theta}) = \prod_{\overline{X_k} \in \mathcal{D}_+} \frac{1}{1 + e^{-(\theta_0 + \sum_{i=1}^{d} \theta_i x_k^i)}} \prod_{\overline{X_k} \in \mathcal{D}_-} \frac{1}{1 + e^{(\theta_0 + \sum_{i=1}^{d} \theta_i x_k^i)}}. \tag{10.28}$$

This likelihood function is the product of the probabilities of all the training examples taking on their assigned labels according to the logistic model. The goal is to maximize this function to determine the optimal value of the parameter vector $\overline{\Theta}$. For numerical convenience, the log likelihood is used to yield the following:

$$\mathcal{LL}(\overline{\Theta}) = \log(\mathcal{L}(\overline{\Theta})) = - \sum_{\overline{X_k} \in \mathcal{D}_+} \log(1 + e^{-(\theta_0 + \sum_{i=1}^{d} \theta_i x_k^i)}) - \sum_{\overline{X_k} \in \mathcal{D}_-} \log(1 + e^{(\theta_0 + \sum_{i=1}^{d} \theta_i x_k^i)}). \tag{10.29}$$

There is no closed-form solution for optimizing the aforementioned expression with respect to the vector $\overline{\Theta}$. Therefore, a natural approach is to use a gradient ascent method to determine the optimal value of the parameter vector $\overline{\Theta}$ iteratively. The gradient vector is obtained by differentiating the log-likelihood function with respect to each of the parameters:

$$\nabla \mathcal{LL}(\overline{\Theta}) = \left(\frac{\partial \mathcal{LL}(\overline{\Theta}}{\partial \theta_0} \ldots \frac{\partial \mathcal{LL}(\overline{\Theta}}{\partial \theta_d} \right). \tag{10.30}$$

It is instructive to examine the ith component[4] of the aforementioned gradient, for $i > 0$. By computing the partial derivative of both sides of Eq. 10.29 with respect to θ_i, the following can be obtained:

$$\frac{\partial \mathcal{LL}(\overline{\Theta})}{\partial \theta_i} = \sum_{\overline{X_k} \in \mathcal{D}_+} \frac{x_k^i}{1 + e^{(\theta_0 + \sum_{i=1}^{d} \theta_i x_i)}} - \sum_{\overline{X_k} \in \mathcal{D}_-} \frac{x_k^i}{1 + e^{-(\theta_0 + \sum_{i=1}^{d} \theta_i x_i)}} \tag{10.31}$$

$$= \sum_{\overline{X_k} \in \mathcal{D}_+} P(\overline{X_k} \in \mathcal{D}_-) x_k^i - \sum_{\overline{X_k} \in \mathcal{D}_-} P(\overline{X_k} \in \mathcal{D}_+) x_k^i \tag{10.32}$$

$$= \sum_{\overline{X_k} \in \mathcal{D}_+} P(\text{Mistake on } \overline{X_k}) x_k^i - \sum_{\overline{X_k} \in \mathcal{D}_-} P(\text{Mistake on } \overline{X_k}) x_k^i. \tag{10.33}$$

It is interesting to note that the terms $P(\overline{X_k} \in \mathcal{D}_-)$ and $P(\overline{X_k} \in \mathcal{D}_+)$ represent the probability of an *incorrect prediction* of $\overline{X_k}$ in the positive and negative classes, respectively. Thus, the mistakes of the current model are used to identify the steepest ascent directions. This approach is generally true of many linear models, such as neural networks, which are also referred to as *mistake-driven methods.* In addition, the multiplicative factor x_k^i impacts the magnitude of the ith component of the gradient direction contributed by $\overline{X_k}$. Therefore, the update condition for θ_i is as follows:

$$\theta_i \leftarrow \theta_i + \alpha \left(\sum_{\overline{X_k} \in \mathcal{D}_+} P(\overline{X_k} \in \mathcal{D}_-) x_k^i - \sum_{\overline{X_k} \in \mathcal{D}_-} P(\overline{X_k} \in \mathcal{D}_+) x_k^i \right). \tag{10.34}$$

The value of α is the step size, which can be determined by using binary search to maximize the improvement in the objective function value. The aforementioned equation uses a batch ascent method, wherein all the training data points contribute to the gradient in a single update step. In practice, it is possible to cycle through the data points one by one for the update process. It can be shown that the likelihood function is concave. Therefore, a global optimum will be found by the gradient ascent method. A number of regularization methods are also used to reduce overfitting. A typical example of a regularization term, which is added to the log-likelihood function $\mathcal{LL}(\overline{\Theta})$ is $-\lambda \sum_{i=1}^{d} \theta_i^2/2$, where λ is the balancing parameter. The only difference to the gradient update is that the term $-\lambda\theta_i$ needs to be added to the ith gradient component for $i \geq 1$.

10.5.2.2 Relationship with Other Linear Models

Although the logistic regression method is a probabilistic method, it is also a special case of a broader class of *generalized linear models* (cf. Sect. 11.5.3 of Chap. 11). There are many ways of formulating a linear model. For example, instead of using a logistic function to set

[4]For the case where $i = 0$, the value of x_k^i is replaced by 1.

up a likelihood criterion, one might directly optimize the squared error of the prediction. In other words, if the class label for $\overline{X_k}$ is $y_k \in \{-1, +1\}$, one might simply attempt to optimize the squared error $\sum_{\overline{X_k} \in \mathcal{D}} (y_k - \text{sign}(\theta_0 + \sum_{i=1}^{d} \theta_i x_i^k))^2$ over all test instances. Here, the function "sign" evaluates to $+1$ or -1, depending on whether its argument is positive or negative. As will be evident in Sect. 10.7, such a model is (approximately) used by neural networks. Similarly, Fisher's linear discriminant, which was discussed at the beginning of this chapter, is also a linear least-squares model (cf. Sect. 11.5.1.1 of Chap. 11) but with a different coding of the class variable. In the next section, a linear model that uses the *maximum margin principle* to separate the two classes, will be discussed.

10.6 Support Vector Machines

Support vector machines (SVMs) are naturally defined for binary classification of numeric data. The binary-class problem can be generalized to the multiclass case by using a variety of tricks discussed in Sect. 11.2 of Chap. 11. Categorical feature variables can also be addressed by transforming categorical attributes to binary data with the binarization approach discussed in Chap. 2.

It is assumed that the class labels are drawn from $\{-1, 1\}$. As with all linear models, SVMs use separating hyperplanes as the decision boundary between the two classes. In the case of SVMs, the optimization problem of determining these hyperplanes is set up with the notion of *margin*. Intuitively, a *maximum margin hyperplane* is one that cleanly separates the two classes, and for which a large region (or *margin*) exists on each side of the boundary with no training data points in it. To understand this concept, the very special case where the data is *linearly separable* will be discussed first. In linearly separable data, it is possible to construct a linear hyperplane which cleanly separates data points belonging to the two classes. Of course, this special case is relatively unusual because real data is rarely fully separable, and at least a few data points, such as mislabeled data points or outliers, will violate linear separability. Nevertheless, the linearly separable formulation is crucial in understanding the important principle of maximum margin. After discussing the linear separable case, the modifications to the formulation required to enable more general (and realistic) scenarios will be addressed.

10.6.1 Support Vector Machines for Linearly Separable Data

This section will introduce the use of the maximum margin principle in linearly separable data. When the data is linearly separable, there are an infinite number of possible ways of constructing a linear separating hyperplane between the classes. Two examples of such hyperplanes are illustrated in Fig. 10.7a as hyperplane 1 and hyperplane 2. Which of these hyperplanes is better? To understand this, consider the test instance (marked by a square), which is very obviously much closer to class A than class B. The hyperplane 1 will correctly classify it to class A, whereas the hyperplane 2 will incorrectly classify it to class B.

The reason for the varying performance of the two classifiers is that the test instance is placed in a noisy and uncertain boundary region between the two classes, which is not easily *generalizable* from the available training data. In other words, there are few training data points in this uncertain region that are quite like the test instance. In such cases, a separating hyperplane like hyperplane 1, whose minimum perpendicular distance to training points from both classes is as large as possible, is the most robust one for correct classification. This distance can be quantified using the *margin* of the hyperplane.

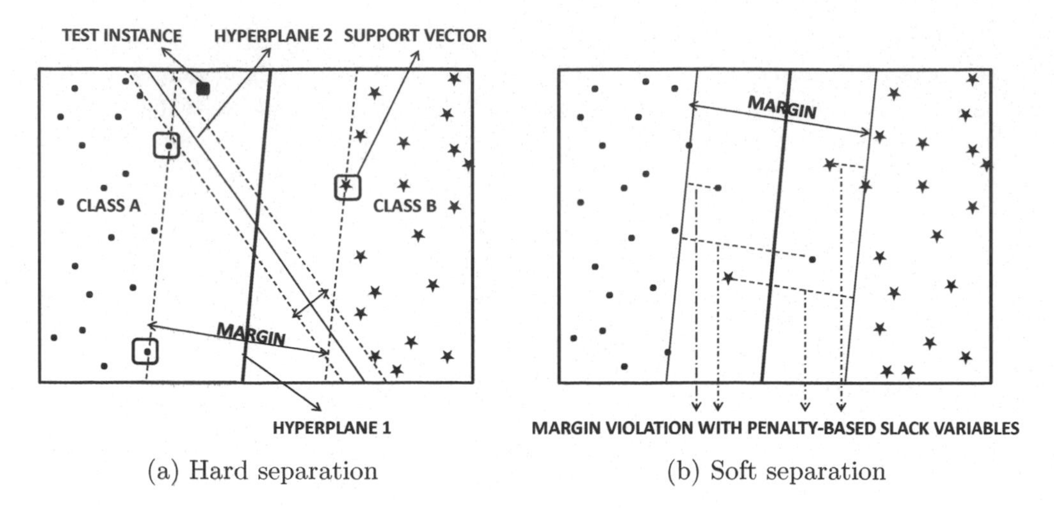

(a) Hard separation (b) Soft separation

Figure 10.7: Hard and soft SVMs

Consider a hyperplane that cleanly separates two linearly separable classes. The margin of the hyperplane is defined as the sum of its distances to the closest training points belonging to each of the two classes on the opposite side of the hyperplane. A further assumption is that the distance of the separating hyperplane to its closest training point of either class is the same. With respect to the separating hyperplane, it is possible to construct parallel hyperplanes that touch the training data of opposite classes on either side, and have no data point between them. The training data points on these hyperplanes are referred to as the *support vectors*, and the distance between the two hyperplanes is the *margin*. The separating hyperplane, or decision boundary, is precisely in the middle of these two hyperplanes in order to achieve the most accurate classification. The margins for hyperplane 1 and hyperplane 2 are illustrated in Fig. 10.7a by dashed lines. It is evident that the margin for hyperplane 1 is larger than that for hyperplane 2. Therefore, the former hyperplane provides better generalization power for unseen test instances in the "difficult" uncertain region separating the two classes where classification errors are most likely. This is also consistent with our earlier example-based observation about the more accurate classification with hyperplane 1.

How do we determine the maximum margin hyperplane? The way to do this is to set up a nonlinear programming optimization formulation that maximizes the margin by expressing it as a function of the coefficients of the separating hyperplane. The optimal coefficients can be determined by solving this optimization problem. Let the n data points in the training set $\mathcal{D}$ be denoted by $(\overline{X_1}, y_1) \dots (\overline{X_n}, y_n)$, where $\overline{X_i}$ is a d-dimensional row vector corresponding to the ith data point, and $y_i \in \{-1, +1\}$ is the binary class variable of the ith data point. Then, the separating hyperplane is of the following form:

$$\overline{W} \cdot \overline{X} + b = 0. \tag{10.35}$$

Here, $\overline{W} = (w_1 \dots w_d)$ is the d-dimensional row vector representing the normal direction to the hyperplane, and b is a scalar, also known as the *bias*. The vector $\overline{W}$ regulates the orientation of the hyperplane and the bias b regulates the distance of the hyperplane from the origin. The $(d + 1)$ coefficients corresponding to $\overline{W}$ and b need to be learned from the training data to maximize the margin of separation between the two classes. Because it is assumed that the classes are linearly separable, such a hyperplane can also be assumed to exist. All data points $\overline{X_i}$ with $y_i = +1$ will lie on one side of the hyperplane satisfying $\overline{W} \cdot \overline{X_i} + b \geq 0$. Similarly, all points with $y_i = -1$ will lie on the other side of the hyperplane satisfying $\overline{W} \cdot \overline{X_i} + b \leq 0$.

$$\overline{W} \cdot \overline{X_i} + b \geq 0 \quad \forall i : y_i = +1 \tag{10.36}$$

$$\overline{W} \cdot \overline{X_i} + b \leq 0 \quad \forall i : y_i = -1 \tag{10.37}$$

These constraints do not yet incorporate the margin requirements on the data points. A stronger set of constraints are defined using these margin requirements. It may be assumed that the separating hyperplane $\overline{W} \cdot \overline{X} + b = 0$ is located in the center of the two margin-defining hyperplanes. Therefore, the two symmetric hyperplanes touching the support vectors can be expressed by introducing another parameter c that regulates the distance between them.

$$\overline{W} \cdot \overline{X} + b = +c \tag{10.38}$$

$$\overline{W} \cdot \overline{X} + b = -c \tag{10.39}$$

It is possible to assume, without loss of generality, that the variables $\overline{W}$ and b are appropriately scaled, so that the value of c can be set to 1. Therefore, the two separating hyperplanes can be expressed in the following form:

$$\overline{W} \cdot \overline{X} + b = +1 \tag{10.40}$$

$$\overline{W} \cdot \overline{X} + b = -1. \tag{10.41}$$

These constraints are referred to as *margin constraints*. The two hyperplanes segment the data space into three regions. It is assumed that no training data points lie in the uncertain decision boundary region between these two hyperplanes, and all training data points for each class are mapped to one of the two remaining (extreme) regions. This can be expressed as pointwise constraints on the training data points as follows:

$$\overline{W} \cdot \overline{X_i} + b \geq +1 \quad \forall i : y_i = +1 \tag{10.42}$$

$$\overline{W} \cdot \overline{X_i} + b \leq -1 \quad \forall i : y_i = -1. \tag{10.43}$$

Note that the constraints for both the positive and negative classes can be written in the following succinct and algebraically convenient, but rather cryptic, form:

$$y_i(\overline{W} \cdot \overline{X_i} + b) \geq +1 \quad \forall i. \tag{10.44}$$

The distance between the two hyperplanes for the positive and negative instances is also referred to as the margin. As discussed earlier, the goal is to maximize this margin. What is the distance (or margin) between these two parallel hyperplanes? One can use linear algebra to show that the distance between two parallel hyperplanes is the normalized difference between their constant terms, where the normalization factor is the L_2-norm $||\overline{W}|| = \sqrt{\sum_{i=1}^{d} w_i^2}$ of the coefficients. Because the difference between the constant terms of the two aforementioned hyperplanes is 2, it follows that the distance between them is $2/||\overline{W}||$. This is the margin that needs to be maximized with respect to the aforementioned constraints. This form of the objective function is inconvenient because it incorporates a square root in the denominator of the objective function. However, maximizing $2/||\overline{W}||$ is the same as minimizing $||\overline{W}||^2/2$. This is a convex quadratic programming problem, because the quadratic objective function $||\overline{W}||^2/2$ needs to be minimized subject to a set of linear constraints (Eqs. 10.42–10.43) on the training points. Note that each training data point leads to a constraint, which tends to make the optimization problem rather large, and explains the high computational complexity of SVMs.

Such constrained nonlinear programming problems are solved using a method known as *Lagrangian relaxation.* The broad idea is to associate a nonnegative n-dimensional set of Lagrangian multipliers $\overline{\lambda} = (\lambda_1 \ldots \lambda_n) \geq 0$ for the different constraints. The multiplier λ_i corresponds to the margin constraint of the ith training data point. The constraints are then relaxed, and the objective function is augmented by incorporating a Lagrangian penalty for constraint violation:

$$L_P = \frac{||\overline{W}||^2}{2} - \sum_{i=1}^{n} \lambda_i \left[y_i(\overline{W} \cdot \overline{X_i} + b) - 1 \right]. \tag{10.45}$$

For *fixed* nonnegative values of λ_i, margin constraint violations increase L_p. Therefore, the penalty term pushes the optimized values of $\overline{W}$ and b towards constraint nonviolation for minimization of L_P with respect to $\overline{W}$ and b. Values of $\overline{W}$ and b that satisfy the margin constraints will always result in a nonpositive penalty. Therefore, for any fixed nonnegative value of $\overline{\lambda}$, the minimum value of L_P will always be at most equal to that of the original optimal objective function value $||\overline{W^*}||^2/2$ because of the impact of the non-positive penalty term for any feasible $(\overline{W^*}, b^*)$.

Therefore, if L_P is minimized with respect to $\overline{W}$ and b for any particular $\overline{\lambda}$, and then maximized with respect to nonnegative Lagrangian multipliers $\overline{\lambda}$, the resulting *dual* solution L_D^* will be a lower bound on the optimal objective function $O^* = ||\overline{W^*}||^2/2$ of the SVM formulation. Mathematically, this *weak duality* condition can be expressed as follows:

$$O^* \geq L_D^* = \max_{\overline{\lambda} \geq 0} \min_{\overline{W}, b} L_P. \tag{10.46}$$

Optimization formulations such as SVM are special because the objective function is convex, and the constraints are linear. Such formulations satisfy a property known as *strong duality.* According to this property, the minimax relationship of Eq. 10.46 yields an optimal and feasible solution to the original problem (i.e., $O^* = L_D^*$) in which the Lagrangian penalty term has zero contribution. Such a solution $(\overline{W^*}, b^*, \overline{\lambda^*})$ is referred to as the *saddle point* of the Lagrangian formulation. Note that zero Lagrangian penalty is achieved by a feasible solution only when each training data point $\overline{X_i}$ satisfies $\lambda_i \left[y_i(\overline{W} \cdot \overline{X_i} + b) - 1 \right] = 0$. These conditions are equivalent to the *Kuhn–Tucker optimality conditions*, and they imply that data points $\overline{X_i}$ with $\lambda_i > 0$ are support vectors. The Lagrangian formulation is solved using the following steps:

1. The Lagrangian objective L_P can be expressed more conveniently as a pure maximization problem by eliminating the minimization part from the awkward minimax formulation. This is achieved by eliminating the minimization variables $\overline{W}$ and b with gradient-based optimization conditions on these variables. By setting the gradient of L_P with respect to $\overline{W}$ to 0, we obtain the following:

$$\nabla L_P = \nabla \frac{||\overline{W}||^2}{2} - \nabla \sum_{i=1}^{n} \lambda_i \left[y_i(\overline{W} \cdot \overline{X_i} + b) - 1 \right] = 0 \tag{10.47}$$

$$\overline{W} - \sum_{i=1}^{n} \lambda_i y_i \overline{X_i} = 0. \tag{10.48}$$

Therefore, one can now derive an expression for $\overline{W}$ in terms of the Lagrangian multipliers and the training data points:

$$\overline{W} = \sum_{i=1}^{n} \lambda_i y_i \overline{X_i}. \tag{10.49}$$

Furthermore, by setting the partial derivative of L_P with respect to b to 0, we obtain $\sum_{i=1}^{n} \lambda_i y_i = 0$.

2. The optimization condition $\sum_{i=1}^{n} \lambda_i y_i = 0$ can be used to eliminate the term $-b \sum_{i=1}^{n} \lambda_i y_i$ from L_P. The expression $\overline{W} = \sum_{i=1}^{n} \lambda_i y_i \overline{X_i}$ from Eq. 10.49 can then be substituted in L_P to create a dual problem L_D in terms of only the *maximization* variables $\overline{\lambda}$. Specifically, the maximization objective function L_D for the Lagrangian dual is as follows:

$$L_D = \sum_{i=1}^{n} \lambda_i - \frac{1}{2} \sum_{i=1}^{n} \sum_{j=1}^{n} \lambda_i \lambda_j y_i y_j \overline{X_i} \cdot \overline{X_j}. \tag{10.50}$$

The dual problem maximizes L_D subject to the constraints $\lambda_i \geq 0$ and $\sum_{i=1}^{n} \lambda_i y_i = 0$. Note that L_D is expressed only in terms of λ_i, the class labels, and the pairwise dot products $\overline{X_i} \cdot \overline{X_j}$ between training data points. Therefore, solving for the Lagrangian multipliers requires knowledge of only the class variables and dot products between training instances but it does not require *direct* knowledge of the feature values $\overline{X_i}$. The dot products between training data points can be viewed as a kind of similarity between the points, which can easily be defined for data types beyond numeric domains. This observation is important for generalizing linear SVMs to nonlinear decision boundaries and arbitrary data types with the kernel trick.

3. The value of b can be derived from the constraints in the original SVM formulation, for which the Lagrangian multipliers λ_r are *strictly* positive. For these training points, the margin constraint $y_r(\overline{W} \cdot \overline{X_r} + b) = +1$ is satisfied exactly according to the Kuhn–Tucker conditions. The value of b can be derived from *any* such training point $(\overline{X_r}, y_r)$ as follows:

$$y_r \left[\overline{W} \cdot \overline{X_r} + b\right] = +1 \qquad \forall r : \lambda_r > 0 \tag{10.51}$$

$$y_r \left[\left(\sum_{i=1}^{n} \lambda_i y_i \overline{X_i} \cdot \overline{X_r}\right) + b \right] = +1 \qquad \forall r : \lambda_r > 0. \tag{10.52}$$

The second relationship is derived by substituting the expression for $\overline{W}$ in terms of the Lagrangian multipliers according to Eq. 10.49. Note that this relationship is expressed only in terms of Lagrangian multipliers, class labels, and dot products between training instances. The value of b can be solved from this equation. To reduce numerical error, the value of b may be averaged over all the support vectors with $\lambda_r > 0$.

4. For a test instance $\overline{Z}$, its class label $F(\overline{Z})$ is defined by the decision boundary obtained by substituting for $\overline{W}$ in terms of the Lagrangian multipliers (Eq. 10.49):

$$F(\overline{Z}) = \text{sign}\{\overline{W} \cdot \overline{Z} + b\} = \text{sign}\{(\sum_{i=1}^{n} \lambda_i y_i \overline{X_i} \cdot \overline{Z}) + b\}. \tag{10.53}$$

It is interesting to note that $F(\overline{Z})$ can be fully expressed in terms of the dot product between training instances and test instances, class labels, Lagrangian multipliers, and bias b. Because the Lagrangian multipliers λ_i and b can also be expressed in terms of the dot products between training instances, it follows that the classification can be fully performed using knowledge of only the dot product between different instances (training and test), without knowing the exact feature values of either the training or the test instances.

The observations about dot products are crucial in generalizing SVM methods to nonlinear decision boundaries and arbitrary data types with the use of a technique known as the *kernel trick*. This technique simply substitutes dot products with kernel similarities (cf. Sect. 10.6.4).

It is noteworthy from the derivation of $\overline{W}$ (see Eq. 10.49) and the aforementioned derivation of b, that only training data points that are support vectors (with $\lambda_r > 0$) are used to define the solution $\overline{W}$ and b in SVM optimization. As discussed in Chap. 11, this observation is leveraged by scalable SVM classifiers, such as *SVMLight*. Such classifiers shrink the size of the problem by discarding irrelevant training data points that are easily identified to be far away from the separating hyperplanes.

10.6.1.1 Solving the Lagrangian Dual

The Lagrangian dual L_D may be optimized by using the gradient ascent technique in terms of the n-dimensional parameter vector $\overline{\lambda}$.

$$\frac{\partial L_D}{\partial \lambda_i} = 1 - y_i \sum_{j=1}^{n} y_j \lambda_j \overline{X_i} \cdot \overline{X_j} \tag{10.54}$$

Therefore, as in logistic regression, the corresponding gradient-based update equation is as follows:

$$(\lambda_1 \ldots \lambda_n) \Leftarrow (\lambda_1 \ldots \lambda_n) + \alpha \left(\frac{\partial L_D}{\partial \lambda_1} \ldots \frac{\partial L_D}{\partial \lambda_n} \right). \tag{10.55}$$

The step size α may be chosen to maximize the improvement in objective function. The initial solution can be chosen to be the vector of zeros, which is also a feasible solution for $\overline{\lambda}$.

One problem with this update is that the constraints $\lambda_i \geq 0$ and $\sum_{i=1}^{n} \lambda_i y_i = 0$ may be violated after an update. Therefore, the gradient vector is projected along the hyperplane $\sum_{i=1}^{n} \lambda_i y_i = 0$ before the update to create a modified gradient vector. Note that the projection of the gradient ∇L_D along the normal to this hyperplane is simply $\overline{H} = (\overline{y} \cdot \nabla L_D)\,\overline{y}$, where $\overline{y}$ is the unit vector $\frac{1}{\sqrt{n}}(y_1 \ldots y_n)$. This component is subtracted from ∇L_D to create a modified gradient vector $\overline{G} = \nabla L_D - \overline{H}$. Because of the projection, updating along the modified gradient vector $\overline{G}$ will not violate the constraint $\sum_{i=1}^{n} \lambda_i y_i = 0$. In addition, any negative values of λ_i after an update are reset to 0.

Note that the constraint $\sum_{i=1}^{n} \lambda_i y_i = 0$ is derived by setting the gradient of L_P with respect to b to 0. In some alternative formulations of SVMs, the bias vector b can be included within $\overline{W}$ by adding a synthetic dimension to the data with a constant value of 1. In such cases, the gradient vector update is simplified to Eq. 10.55 because one no longer needs to worry about the constraint $\sum_{i=1}^{n} \lambda_i y_i = 0$. This alternative formulation of SVMs is discussed in Chap. 13.

10.6.2 Support Vector Machines with Soft Margin for Nonseparable Data

The previous section discussed the scenario where the data points of the two classes are linearly separable. However, perfect linear separability is a rather contrived scenario, and real data sets usually will not satisfy this property. An example of such a data set is illustrated in Fig. 10.7b, where no linear separator may be found. Many real data sets may, however, be approximately separable, where *most* of the data points lie on correct sides of well-chosen separating hyperplanes. In this case, the notion of margin becomes a softer one because training data points are allowed to violate the margin constraints *at the expense of a penalty*. The two margin hyperplanes separate out "most" of the training data points but not all of them. An example is illustrated in Fig. 10.7b.

The level of violation of each margin constraint by training data point $\overline{X_i}$ is denoted by a slack variable $\xi_i \geq 0$. Therefore, the new set of soft constraints on the separating hyperplanes may be expressed as follows:

$$\begin{aligned}
\overline{W} \cdot \overline{X_i} + b &\geq +1 - \xi_i \quad \forall i : y_i = +1 \\
\overline{W} \cdot \overline{X_i} + b &\leq -1 + \xi_i \quad \forall i : y_i = -1 \\
\xi_i &\geq 0 \quad \forall i.
\end{aligned}$$

These slack variables ξ_i may be interpreted as the distances of the training data points from the separating hyperplanes, as illustrated in Fig. 10.7b, when they lie on the "wrong" side of the separating hyperplanes. The values of the slack variables are 0 when they lie on the correct side of the separating hyperplanes. It is not desirable for too many training data points to have positive values of ξ_i, and therefore such violations are penalized by $C \cdot \xi_i^r$, where C and r are user-defined parameters regulating the level of softness in the model. Small values of C would result in relaxed margins, whereas large values of C would minimize training data errors and result in narrow margins. Setting C to be sufficiently large would disallow any training data error in separable classes, which is the same as setting all slack variables to 0 and defaulting to the hard version of the problem. A popular choice of r is 1, which is also referred to as *hinge loss*. Therefore, the objective function for soft-margin SVMs, with hinge loss, is defined as follows:

$$O = \frac{||\overline{W}||^2}{2} + C \sum_{i=1}^{n} \xi_i. \tag{10.56}$$

As before, this is a convex quadratic optimization problem that can be solved using Lagrangian methods. A similar approach is used to set up the Lagrangian relaxation of the problem with penalty terms and additional multipliers $\beta_i \geq 0$ for the slack constraints $\xi_i \geq 0$:

$$L_P = \frac{||\overline{W}||^2}{2} + C \sum_{i=1}^{n} \xi_i - \sum_{i=1}^{n} \lambda_i \left[y_i(\overline{W} \cdot \overline{X_i} + b) - 1 + \xi_i \right] - \sum_{i=1}^{n} \beta_i \xi_i. \tag{10.57}$$

A similar approach to the hard SVM case can be used to eliminate the minimization variables $\overline{W}$, ξ_i, and b from the optimization formulation and create a purely maximization dual formulation. This is achieved by setting the gradient of L_P with respect to these variables to 0. By setting the gradients of L_P with respect to $\overline{W}$ and b to 0, it can be respectively shown that the value of $\overline{W}$ is identical to the hard-margin case (Eq. 10.49), and the same multiplier constraint $\sum_{i=1}^{n} \lambda_i y_i = 0$ is satisfied. This is because the additional slack terms in L_P involving ξ_i do not affect the respective gradients with respect to $\overline{W}$ and b. Furthermore, it can be shown that the objective function L_D of the Lagrangian dual in the soft-margin case is identical to that of the hard-margin case, according to Eq. 10.50, because the linear terms involving each ξ_i evaluate[5] to 0. The *only* change to the dual optimization problem is that the nonnegative Lagrangian multipliers satisfy additional constraints of the form $C - \lambda_i = \beta_i \geq 0$. This constraint is derived by setting the partial derivative of L_P with respect to ξ_i to 0. One way of viewing this additional constraint $\lambda_i \leq C$ is that the influence of any training data point $\overline{X_i}$ on the weight vector $\overline{W} = \sum_{i=1}^{n} \lambda_i y_i \overline{X_i}$ is capped by C because of the softness of the margin. *The dual problem in soft SVMs maximizes L_D (Eq. 10.50) subject to the constraints $0 \leq \lambda_i \leq C$ and $\sum_{i=1}^{n} \lambda_i y_i = 0$.*

The Kuhn–Tucker optimality conditions for the slack nonnegativity constraints are $\beta_i \xi_i = 0$. Because we have already derived $\beta_i = C - \lambda_i$, we obtain $(C - \lambda_i)\xi_i = 0$. In other words, training points $\overline{X_i}$ with $\lambda_i < C$ correspond to zero slack ξ_i and they might either lie on the margin or on the correct side of the margin. However, in this case, the support vectors are defined as data points that satisfy the *soft* SVM constraints exactly and some of them might have nonzero slack. Such points might lie on the margin, between the margin, or on the wrong side of the decision boundary. Points that satisfy $\lambda_i > 0$ are always support vectors. The support vectors that lie on the margin will therefore satisfy $0 < \lambda_i < C$. These points are very useful in solving for b. Consider any such support vector $\overline{X_r}$ with zero slack, which satisfies $0 < \lambda_r < C$. The value of b may be obtained as before:

$$y_r \left[(\sum_{i=1}^{n} \lambda_i y_i \overline{X_i} \cdot \overline{X_r}) + b \right] = +1. \tag{10.58}$$

Note that this expression is the same as for the case of hard SVMs, except that the relevant training points are identified by using the condition $0 < \lambda_r < C$. The gradient-ascent update is also identical to the separable case (cf. Sect. 10.6.1.1), except that any multiplier λ_i exceeding C because of an update needs to be reset to C. The classification of a test instance also uses Eq. 10.53 in terms of Lagrangian multipliers because the relationship between the weight vector and the Lagrangian multipliers is the same in this case. Thus, the soft SVM formulation with hinge loss is strikingly similar to the hard SVM formulation. This similarity is less pronounced for other slack penalty functions such as quadratic loss.

The soft version of SVMs also allows an *unconstrained* primal formulation by eliminating the margin constraints and slack variables simultaneously. This is achieved by substituting $\xi_i = \max\{0, 1 - y_i[\overline{W} \cdot \overline{X_i} + b]\}$ in the primal objective function of Eq. 10.56. This results in an unconstrained optimization (minimization) problem purely in terms of $\overline{W}$ and b:

$$O = \frac{||\overline{W}||^2}{2} + C \sum_{i=1}^{n} \max\{0, 1 - y_i[\overline{W} \cdot \overline{X_i} + b]\}. \tag{10.59}$$

One can use a gradient descent approach, which is analogous to the gradient ascent method used in logistic regression. The partial derivatives of nondifferentiable function O with respect to $w_1, \ldots w_d$ and b are approximated on a casewise basis, depending on whether or not the term inside the maximum function evaluates to a positive quantity. The precise derivation of the gradient descent steps is left as an exercise for the reader. While the dual

[5]The additional term in L_P involving ξ_i is $(C - \beta_i - \lambda_i)\xi_i$. This term evaluates to 0 because the partial derivative of L_P with respect to ξ_i is $(C - \beta_i - \lambda_i)$. This partial derivative must evaluate to 0 for optimality of L_P.

approach is more popular, the primal approach is intuitively simpler, and it is often more efficient when an approximate solution is desired.

10.6.2.1 Comparison with Other Linear Models

The normal vector to a linear separating hyperplane can be viewed as a direction along which the data points of the two classes are best separated. Fisher's linear discriminant also achieves this goal by maximizing the ratio of the between-class scatter to the within-class scatter along an optimally chosen vector. However, an important distinguishing feature of SVMs is that they focus extensively on the *decision boundary* region between the two classes because this is the most uncertain region, which is prone to classification error. Fisher's discriminant focuses on the global separation between the two classes and may not necessarily provide the best separation in the uncertain boundary region. This is the reason that SVMs often have better generalization behavior for noisy data sets that are prone to overfitting.

It is instructive to express logistic regression as a minimization problem by using the negative of the log-likelihood function and then comparing it with SVMs. The coefficients $(\theta_0, \ldots \theta_d)$ in logistic regression are analogous to the coefficients $(b, \overline{W})$ in SVMs. SVMs have a margin component to increase the generalization power of the classifier, just as logistic regression uses regularization. Interestingly, the margin component $||\overline{W}||^2/2$ in SVMs has an identical form to the regularization term $\sum_{i=1}^{d} \theta_i^2/2$ in logistic regression. SVMs have slack penalties just as logistic regression implicitly penalizes the *probability of* mistakes in the log-likelihood function. However, the slack is computed using *margin violations* in SVMs, whereas the penalties in logistic regression are computed as a smooth function of the distances from the *decision boundary.* Specifically, the log-likelihood function in logistic regression creates a smooth loss function of the form $log(1 + e^{-y_i[\theta_0 + \overline{\theta} \cdot \overline{X_i}]})$, whereas the hinge loss $max\{0, 1 - y_i[\overline{W} \cdot \overline{X_i} + b]\}$ in SVMs is not a smooth function. The nature of the misclassification penalty is the only difference between the two models. Therefore, there are several conceptual similarities among these models, but they emphasize different aspects of optimization. SVMs and regularized logistic regression show similar performance in many practical settings with poorly separable classes. However, SVMs and Fisher's discriminant generally perform better than logistic regression for the special case of well-separated classes. All these methods can also be extended to nonlinear decision boundaries in similar ways.

10.6.3 Nonlinear Support Vector Machines

In many cases, linear solvers are not appropriate for problems in which the decision boundary is not linear. To understand this point, consider the data distribution illustrated in Fig. 10.8. It is evident that no linear separating hyperplanes can delineate the two classes. This is because the two classes are separated by the following decision boundary:

$$8(x_1 - 1)^2 + 50(x_2 - 2)^2 = 1. \qquad (10.60)$$

Now, if one already had some insight about the nature of the decision boundary, one might transform the training data into the new 4-dimensional space as follows:

$$z_1 = x_1^2$$
$$z_2 = x_1$$
$$z_3 = x_2^2$$
$$z_4 = x_2.$$

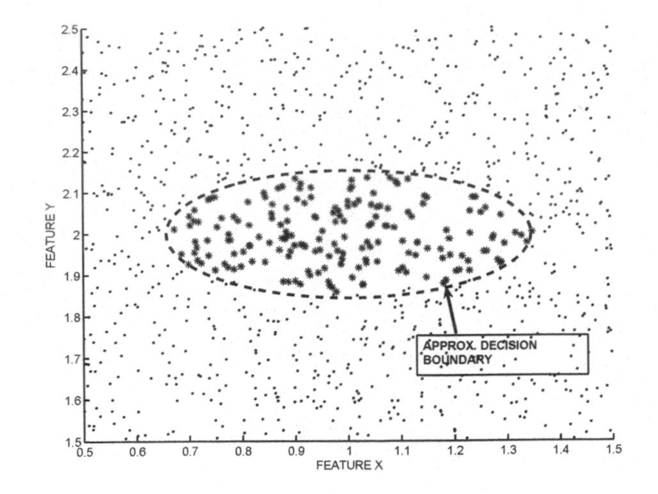

Figure 10.8: Nonlinear decision surface

The decision boundary of Eq. 10.60 can be expressed linearly in terms of the variables $z_1 \ldots z_4$, by expanding Eq. 10.60 in terms of x_1, x_1^2, x_2, and x_2^2:

$$8x_1^2 - 16x_1 + 50x_2^2 - 200x_2 + 207 = 0$$
$$8z_1 - 16z_2 + 50z_3 - 200z_4 + 207 = 0.$$

Thus, each training data point is now expressed in terms of these four newly transformed dimensions, and the classes will be linearly separable in this space. The SVM optimization formulation can then be solved in the transformed space as a linear model, and used to classify test instances that are also transformed to 4-dimensional space. It is important to note that the complexity of the problem effectively increased because of the increase in the size of the hyperplane coefficient vector $\overline{W}$.

In general, it is possible to approximate any polynomial decision boundary by adding an additional set of dimensions for each exponent of the polynomial. High-degree polynomials have significant expressive power in approximating many nonlinear functions well. This kind of transformation can be very effective in cases where one does not know whether the decision boundary is linear or nonlinear. This is because the additional degrees of freedom in the model, in terms of the greater number of coefficients to be learned, can determine the linearity or nonlinearity of the decision boundary in a data-driven way. In our previous example, if the decision boundary had been linear, the coefficients for z_1 and z_3 would automatically have been learned to be almost 0, given enough training data. The price for this additional flexibility is the increased computational complexity of the training problem, and the larger number of coefficients that need to be learned. Furthermore, if enough training data is not available, then this may result in overfitting where even a simple linear decision boundary is incorrectly approximated as a nonlinear one. A different approach, which is sometimes used to learn nonlinear decision boundaries, is known as the "kernel trick." This approach is able to learn arbitrary decision boundaries without performing the transformation explicitly.

10.6.4 The Kernel Trick

The kernel trick leverages the important observation that the SVM formulation can be fully solved in terms of dot products (or similarities) between pairs of data points. One does not need to know the feature values. Therefore, the key is to define the pairwise dot product (or similarity function) directly in the d'-dimensional transformed representation $\Phi(\overline{X})$, with the use of a kernel function $K(\overline{X_i}, \overline{X_j})$.

$$K(\overline{X_i}, \overline{X_j}) = \Phi(\overline{X_i}) \cdot \Phi(\overline{X_j}) \tag{10.61}$$

To effectively solve the SVM, recall that the transformed feature values $\Phi(\overline{X})$ need not be explicitly computed, as long as the dot product (or kernel similarity) $K(\overline{X_i}, \overline{X_j})$ is known. This implies that the term $\overline{X_i} \cdot \overline{X_j}$ may be replaced by the *transformed-space* dot product $K(\overline{X_i}, \overline{X_j})$ in Eq. 10.50, and the term $\overline{X_i} \cdot \overline{Z}$ in Eq. 10.53 can be replaced by $K(\overline{X_i}, \overline{Z})$ to perform SVM classification.

$$L_D = \sum_{i=1}^{n} \lambda_i - \frac{1}{2} \cdot \sum_{i=1}^{n} \sum_{j=1}^{n} \lambda_i \lambda_j y_i y_j K(\overline{X_i}, \overline{X_j}) \tag{10.62}$$

$$F(\overline{Z}) = \text{sign}\{(\sum_{i=1}^{n} \lambda_i y_i K(\overline{X_i}, \overline{Z})) + b\} \tag{10.63}$$

Note that the bias b is also expressed in terms of dot products according to Eq. 10.58. These modifications are carried over to the update equations discussed in Sect. 10.6.1.1, all of which are expressed in terms of dot products.

Thus, all computations are performed in the *original* space, and the actual transformation $\Phi(\cdot)$ does not need to be known as long as the kernel similarity function $K(\cdot, \cdot)$ is known. By using kernel-based similarity with carefully chosen kernels, arbitrary nonlinear decision boundaries can be approximated. There are different ways of modeling similarity between $\overline{X_i}$ and $\overline{X_j}$. Some common choices of the kernel function are as follows:

Function	Form
Gaussian radial basis kernel	$K(\overline{X_i}, \overline{X_j}) = e^{-\|\overline{X_i} - \overline{X_j}\|^2/2\sigma^2}$
Polynomial kernel	$K(\overline{X_i}, \overline{X_j}) = (\overline{X_i} \cdot \overline{X_j} + c)^h$
Sigmoid kernel	$K(\overline{X_i}, \overline{X_j}) = \tanh(\kappa \overline{X_i} \cdot \overline{X_j} - \delta)$

Many of these kernel functions have parameters associated with them. In general, these parameters may need to be tuned by holding out a portion of the training data, and using it to test the accuracy of different choices of parameters. Many other kernels are possible beyond the ones listed in the table above. Kernels need to satisfy a property known as *Mercer's theorem* to be considered valid. This condition ensures that the $n \times n$ kernel similarity matrix $S = [K(\overline{X_i}, \overline{X_j})]$ is positive semidefinite, and similarities can be expressed as dot products in some transformed space. Why must the kernel similarity matrix always be positive semidefinite for similarities to be expressed as dot products? Note that if the $n \times n$ kernel similarity matrix S can be expressed as the $n \times n$ dot-product matrix AA^T of some $n \times r$ transformed representation A of the points, then for any n-dimensional column vector $\overline{V}$, we have $\overline{V}^T S \overline{V} = (A\overline{V})^T (A\overline{V}) \geq 0$. In other words, S is positive semidefinite. Conversely, if the kernel matrix S is positive semi-definite then it can be expressed as a dot product with the eigen decomposition $S = Q\Sigma^2 Q^T = (Q\Sigma)(Q\Sigma)^T$, where Σ^2 is an $n \times n$ diagonal matrix of nonnegative eigenvalues and Q is an $n \times n$ matrix containing the eigenvectors of S in columns. The matrix $Q\Sigma$ is the n-dimensional transformed representation of the points, and it also sometimes referred to as the *data-specific Mercer kernel map*. This map is data set-specific, and it is used in many nonlinear dimensionality reduction methods such as kernel *PCA*.

What kind of kernel function works best for the example of Fig. 10.8? In general, there are no predefined rules for selecting kernels. Ideally, if the similarity values $K(\overline{X_i}, \overline{X_j})$ were defined so that a space exists, in which points with this similarity structure are linearly separable, then a linear SVM in the transformed space $\Phi(\cdot)$ will work well.

To explain this point, we will revisit the example of Fig. 10.8. Let $\overline{X2_i}$ and $\overline{X2_j}$ be the d-dimensional vectors derived by squaring each coordinate of $\overline{X_i}$ and $\overline{X_j}$, respectively. In the case of Fig. 10.8, consider the transformation (z_1, z_2, z_3, z_4) in the previous section. It can be shown that the dot product between two transformed data points can be captured by the following kernel function:

$$\textit{Transformed-Dot-Product}(\overline{X_i}, \overline{X_j}) = \overline{X_i} \cdot \overline{X_j} + \overline{X2_i} \cdot \overline{X2_j}. \tag{10.64}$$

This is easy to verify by expanding the aforementioned expression in terms of the transformed variables $z_1 \ldots z_4$ of the two data points. The kernel function *Transformed-Dot-Product*$(\overline{X_i}, \overline{X_j})$ would obtain the same Lagrangian multipliers and decision boundary as obtained with the explicit transformation $z_1 \ldots z_4$. Interestingly, this kernel is closely related to the second-order polynomial kernel.

$$K(\overline{X_i}, \overline{X_j}) = (0.5 + \overline{X_i} \cdot \overline{X_j})^2 \tag{10.65}$$

Expanding the second-order polynomial kernel results in a superset of the additive terms in *Transformed-Dot-Product*$(\overline{X_i}, \overline{X_j})$. The additional terms include a constant term of 0.25 and some inter-dimensional products. These terms provide further modeling flexibility. In the case of the 2-dimensional example of Fig. 10.8, the use of the second-order polynomial kernel is equivalent to using an extra transformed variable $z_5 = \sqrt{2}x_1x_2$ representing the product of the values on the two dimensions and a constant dimension $z_6 = 0.5$. These variables are in addition to the original four variables (z_1, z_2, z_3, z_4). Since these additional variables are redundant in this case, they will not affect the ability to discover the correct decision boundary, although they might cause some overfitting. On the other hand, a variable such as $z_5 = \sqrt{2}x_1x_2$ would have come in handy, if the ellipse of Fig. 10.8 had been arbitrarily oriented with respect to the axis system. A full separation of the classes would not have been possible with a linear classifier on the original four variables (z_1, z_2, z_3, z_4). Therefore, the second-order polynomial kernel can discover more general decision boundaries than the transformation of the previous section. Using even higher-order polynomial kernels can model increasingly complex boundaries but at a greater risk of overfitting.

In general, different kernels have different levels of flexibility. For example, a transformed feature space that is implied by the Gaussian kernel of width σ can be shown to have an infinite number of dimensions by using the polynomial expansion of the exponential term. The parameter σ controls the relative scaling of various dimensions. A smaller value of σ results in a greater ability to model complex boundaries, but it may also cause overfitting. Smaller data sets are more prone to overfitting. Therefore, the optimal values of kernel parameters depend not only on the shape of the decision boundary but also on the size of the training data set. Parameter tuning is important in kernel methods. With proper tuning, many kernel functions can model complex decision boundaries. Furthermore, kernels provide a natural route for using SVMs in complex data types. This is because kernel methods only need the pairwise similarity between objects, and are agnostic to the feature values of the data points. Kernel functions have been defined for text, images, sequences, and graphs.

10.6.4.1 Other Applications of Kernel Methods

The use of kernel methods is not restricted to SVM methods. These methods can be extended to any technique in which the solutions are directly or indirectly expressed in terms of dot products. Examples include the Fisher's discriminant, logistic regression, linear regression (cf. Sect. 11.5.4 of Chap. 11), dimensionality reduction, and k-means clustering.

1. *Kernel k-means:* The key idea is that the Euclidean distance between a data point $\overline{X}$ and the cluster centroid $\overline{\mu}$ of cluster $\mathcal{C}$ can be computed as a function of the dot product between $\overline{X}$ and the data points in $\mathcal{C}$:

$$||\overline{X}-\overline{\mu}||^2 = ||\overline{X}-\frac{\sum_{\overline{X_i}\in\mathcal{C}}\overline{X_i}}{|\mathcal{C}|}||^2 = \overline{X}\cdot\overline{X}-2\frac{\sum_{\overline{X_i}\in\mathcal{C}}\overline{X}\cdot\overline{X_i}}{|\mathcal{C}|}+\frac{\sum_{\overline{X_i},\overline{X_j}\in\mathcal{C}}\overline{X_i}\cdot\overline{X_j}}{|\mathcal{C}|^2}. \quad (10.66)$$

   In kernel k-means, the dot products $\overline{X_i}\cdot\overline{X_j}$ are replaced with kernel similarity values $K(\overline{X_i},\overline{X_j})$. For the data point $\overline{X}$, the index of its assigned cluster is obtained by selecting the minimum value of the (kernel-based) distance in Eq. 10.66 over all clusters. Note that the cluster centroids in the transformed space do not need to be explicitly maintained over the different k-means iterations, although the cluster assignment indices for each data point need to be maintained for computation of Eq. 10.66. Because of its implicit nonlinear transformation approach, kernel k-means is able to discover arbitrarily shaped clusters like spectral clustering in spite of its use of the spherically biased Euclidean distance.

2. *Kernel PCA:* In conventional *SVD* and *PCA* of an $n \times d$ mean-centered data matrix D, the basis vectors are given by the eigenvectors of D^TD (columnwise dot product matrix), and the coordinates of the transformed points are extracted from the scaled eigenvectors of DD^T (rowwise dot product matrix). While the basis vectors can no longer be derived in kernel *PCA*, the coordinates of the transformed data can be extracted. The rowwise dot product matrix DD^T can be replaced with the kernel similarity matrix $S = [K(\overline{X_i},\overline{X_j})]_{n\times n}$. The similarity matrix is then adjusted for mean-centering of the data in the transformed space as $S \Leftarrow (I-\frac{U}{n})S(I-\frac{U}{n})$, where U is an $n\times n$ matrix containing all 1s (see Exercise 17). The assumption is that the matrix S can be approximately expressed as a dot product of the reduced data points in some k-dimensional transformed space. Therefore, one needs to approximately factorize S into the form AA^T to extract its reduced $n\times k$ embedding A in the transformed space. This is achieved by eigen-decomposition. Let Q_k be the $n \times k$ matrix containing the largest k eigenvectors of S, and Σ_k be the $k \times k$ diagonal matrix containing the square root of the corresponding eigenvalues. Then, it is evident that $S \approx Q_k\Sigma_k^2Q_k^T = (Q_k\Sigma_k)(Q_k\Sigma_k)^T$, and the k-dimensional embeddings of the data points are given[6] by the rows of the $n \times k$ matrix $A = Q_k\Sigma_k$. Note that this is a truncated version of the data-specific Mercer kernel map. This nonlinear embedding is similar to that obtained

[6] The original result [450] uses a more general argument to derive $S'Q_k\Sigma_k^{-1}$ as the $m \times k$ matrix of k-dimensional embedded coordinates of any *out-of-sample* $m \times d$ matrix D'. Here, $S' = D'D^T$ is the $m \times n$ matrix of kernel similarities between out-of-sample points in D' and in-sample points in D. However, when $D' = D$, this expression is (more simply) equivalent to $Q_k\Sigma_k$ by expanding $S' = S \approx Q_k\Sigma_k^2Q_k^T$.

by *ISOMAP*; however, unlike *ISOMAP*, out-of-sample points can also be transformed to the new space. It is noteworthy that the embedding of spectral clustering is also expressed in terms of the large eigenvectors[7] of a *sparsified* similarity matrix, which is better suited to preserving *local* similarities for clustering. In fact, most forms of nonlinear embeddings can be shown to be large eigenvectors of similarity matrices (cf. Table 2.3 of Chap. 2), and are therefore special cases of kernel *PCA*.

10.7 Neural Networks

Neural networks are a model of simulation of the human nervous system. The human nervous system is composed of cells, referred to as neurons. Biological neurons are connected to one another at contact points, which are referred to as synapses. Learning is performed in living organisms by changing the strength of synaptic connections between neurons. Typically, the strength of these connections change in response to external stimuli. Neural networks can be considered a simulation of this biological process.

As in the case of biological networks, the individual nodes in artificial neural networks are referred to as *neurons*. These neurons are units of computation that receive input from some other neurons, make computations on these inputs, and feed them into yet other neurons. The computation function at a neuron is defined by the weights on the input connections to that neuron. This weight can be viewed as analogous to the strength of a synaptic connection. By changing these weights appropriately, the computation function can be learned, which is analogous to the learning of the synaptic strength in biological neural networks. The "external stimulus" in artificial neural networks for learning these weights is provided by the training data. The idea is to incrementally modify the weights whenever incorrect predictions are made by the current set of weights.

The key to the effectiveness of the neural network is the *architecture* used to arrange the connections among nodes. A wide variety of architectures exist, starting from a simple single-layer *perceptron* to complex multilayer networks.

10.7.1 Single-Layer Neural Network: The Perceptron

The most basic architecture of a neural network is referred to as the *perceptron*. An example of the perceptron architecture is illustrated in Fig. 10.10a. The perceptron contains two layers of nodes, which correspond to the input nodes, and a single output node. The number of input nodes is exactly equal to the dimensionality d of the underlying data. Each input node receives and transmits a single numerical attribute to the output node. Therefore, the input nodes only *transmit* input values and do not perform any *computation* on these values. In the basic perceptron model, the output node is the only node that performs a mathematical function on its inputs. The individual features in the training data are assumed to be numerical. Categorical attributes are handled by creating a separate binary input for each value of the categorical attribute. This is logically equivalent to binarizing the categorical attribute into multiple attributes. For simplicity of further discussion, it will be assumed that all input variables are numerical. Furthermore, it will be assumed that the classification problem contains two possible values for the class label, drawn from $\{-1, +1\}$.

[7]Refer to Sect. 19.3.4 of Chap. 19. The small eigenvectors of the symmetric Laplacian are the same as the large eigenvectors of $S = \Lambda^{-1/2} W \Lambda^{-1/2}$. Here, W is often defined by the *sparsified* heat-kernel similarity between data points, and the factors involving $\Lambda^{-1/2}$ provide local normalization of the similarity values to handle clusters of varying density.

As discussed earlier, each input node is connected by a weighted connection to the output node. These weights define a function from the values transmitted by the input nodes to a binary value drawn from $\{-1, +1\}$. This value can be interpreted as the perceptron's prediction of the class variable of the test instance fed to the input nodes, for a binary-class value drawn from $\{-1, +1\}$. Just as learning is performed in biological systems by modifying synaptic strengths, the learning in a perceptron is performed by modifying the weights of the links connecting the input nodes to the output node whenever the predicted label does not match the true label.

The function learned by the perceptron is referred to as the *activation function*, which is a signed linear function. This function is very similar to that learned in SVMs for mapping training instances to binary class labels. Let $\overline{W} = (w_1 \dots w_d)$ be the weights for the connections of d different inputs to the output neuron for a data record of dimensionality d. In addition, a bias b is associated with the activation function. The output $z_i \in \{-1, +1\}$ for the feature set $(x_i^1 \dots x_i^d)$ of the ith data record $\overline{X_i}$, is as follows:

$$z_i = \text{sign}\{\sum_{j=1}^{d} w_j x_i^j + b\} \tag{10.67}$$

$$= \text{sign}\{\overline{W} \cdot \overline{X_i} + b\} \tag{10.68}$$

The value z_i represents the *prediction* of the perceptron for the class variable of $\overline{X_i}$. It is, therefore, desired to learn the weights, so that the value of z_i is equal to y_i for as many training instances as possible. The error in prediction $(z_i - y_i)$ may take on any of the values of -2, 0, or $+2$. A value of 0 is attained when the predicted class is correct. The goal in neural network algorithms is to learn the vector of weights $\overline{W}$ and bias b, so that z_i approximates the true class variable y_i as closely as possible.

The basic perceptron algorithm starts with a random vector of weights. The algorithm then feeds the input data items X_i into the neural network one by one to create the prediction z_i. The weights are then updated, based on the error value $(z_i - y_i)$. Specifically, when the data point $\overline{X_i}$ is fed into it in the tth iteration, the weight vector $\overline{W}^t$ is updated as follows:

$$\overline{W}^{t+1} = \overline{W}^t + \eta(y_i - z_i)\overline{X_i}. \tag{10.69}$$

The parameter η regulates the learning rate of the neural network. The perceptron algorithm repeatedly cycles through all the training examples in the data and iteratively adjusts the weights until convergence is reached. The basic perceptron algorithm is illustrated in Fig. 10.9. Note that a single training data point may be cycled through many times. Each such cycle is referred to as an *epoch*.

Let us examine the incremental term $(y_i - z_i)\overline{X_i}$ in the update of Eq. 10.69, without the multiplicative factor η. It can be shown that this term is a heuristic approximation[8] of the negative of the gradient of the least-squares prediction error $(y_i - z_i)^2 = (y_i - \text{sign}(\overline{W} \cdot \overline{X_i} - b))^2$ of the class variable, with respect to the vector of weights $\overline{W}$. The update in this case is performed on a tuple-by-tuple basis, rather than globally, over the entire data set, as one would expect in a global least-squares optimization. Nevertheless, the basic perceptron algorithm can be considered a modified version of the gradient descent method, which implicitly minimizes the squared error of prediction. It is easy to see that nonzero updates are made to the weights only when errors are made in categorization. This is because the incremental term in Eq. 10.69 will be 0 whenever the predicted value z_i is the same as the class label y_i.

[8]The derivative of the sign function is replaced by only the derivative of its argument. The derivative of the sign function is zero everywhere, except at zero, where it is indeterminate.

Algorithm *Perceptron*(Training Data: $\mathcal{D}$)
begin
  Initialize weight vector $\overline{W}$ to random values;
  repeat
    Receive next training tuple $(\overline{X_i}, y_i)$;
    $z_i = \overline{W} \cdot \overline{X_i} + b$;
    $\overline{W} = \overline{W} + \eta(y_i - z_i)\overline{X_i}$;
  until convergence;
end

Figure 10.9: The perceptron algorithm

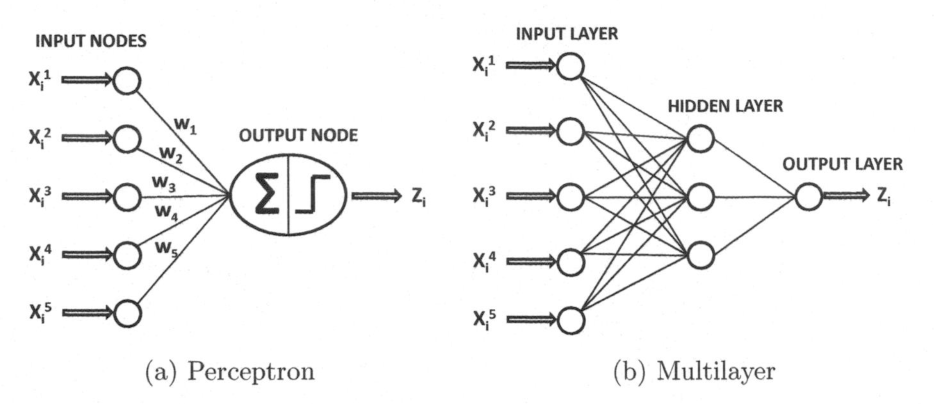

(a) Perceptron (b) Multilayer

Figure 10.10: Single and multilayer neural networks

A question arises as to how the learning rate η may be chosen. A high value of η will result in fast learning rates, but may sometimes result in suboptimal solutions. Smaller values of η will result in a convergence to higher-quality solutions, but the convergence will be slow. In practice, the value of η is initially chosen to be large and gradually reduced, as the weights become closer to their optimal values. The idea is that large steps are likely to be helpful early on, but may result in oscillation between suboptimal solutions at later stages. For example, the value of η is sometimes selected to be proportional to the inverse of the number of cycles through the training data (or epochs) so far.

10.7.2 Multilayer Neural Networks

The perceptron model is the most basic form of a neural network, containing only a single input layer and an output layer. Because the input layers only transmit the attribute values without actually applying any mathematical function on the inputs, the function learned by the perceptron model is only a simple linear model based on a single output node. In practice, more complex models may need to be learned with multilayer neural networks.

Multilayer neural networks have a *hidden layer*, in addition to the input and output layers. The nodes in the hidden layer can, in principle, be connected with different types of topologies. For example, the hidden layer can itself consist of multiple layers, and nodes in one layer might feed into nodes of the next layer. This is referred to as the *multilayer feed-forward network*. The nodes in one layer are also assumed to be fully connected to the nodes in the next layer. Therefore, the topology of the multilayer feed-forward network is automatically determined, after the number of layers, and the number of nodes in each layer, have been specified by the analyst. The basic perceptron may be viewed as a single-layer feed-forward network. A popularly used model is one in which a multilayer feed-forward network contains only a single hidden layer. Such a network may be considered a two-layer feed-forward network. An example of a two-layer feed-forward network is illustrated in Fig. 10.10b. Another aspect of the multilayer feed-forward network is that it is not restricted to the use of linear signed functions of the inputs. Arbitrary functions such as the logistic, sigmoid, or hyperbolic tangents may be used in different nodes of the hidden layer and output layer. An example of such a function, when applied to the training tuple $\overline{X_i} = (x_i^1 \dots x_i^d)$, to yield an output value of z_i, is as follows:

$$z_i = \sum_{j=1}^{d} w_j \frac{1}{1 + e^{-x_i^j}} + b. \tag{10.70}$$

The value of z_i is no longer a predicted output of the final class label in $\{-1, +1\}$, if it refers to a function computed at the hidden layer nodes. This output is then propagated forward to the next layer.

In the single-layer neural network, the training process was relatively straightforward because the *expected* output of the output node was known to be equal to the training label value. The known *ground truth* was used to create an optimization problem in least squares form, and update the weights with a gradient-descent method. Because the output node is the only neuron with weights in a single-layer network, the update process is easy to implement. In the case of multilayer networks, the problem is that the ground-truth output of the hidden layer nodes are not known because there are no training labels associated with the outputs of these nodes. Therefore, a question arises as to how the weights of these nodes should be updated when a training example is classified incorrectly. Clearly, when a classification error is made, some kind of "feedback" is required from the nodes in the forward layers to the nodes in earlier layers about the *expected* outputs (and corresponding errors). This is achieved with the use of the *backpropagation* algorithm. Although this algorithm is not discussed in detail in this chapter, a brief summary is provided here. The backpropagation algorithm contains two main phases, which are applied in the weight update process for each training instance:

1. *Forward phase:* In this phase, the inputs for a training instance are fed into the neural network. This results in a forward cascade of computations across the layers, using the current set of weights. The final predicted output can be compared to the class label of the training instance, to check whether or not the predicted label is an error.

2. *Backward phase:* The main goal of the backward phase is to learn weights in the backward direction by providing an error estimate of the output of a node in the earlier layers from the errors in later layers. The error estimate of a node in the hidden layer is computed as a function of the error estimates and weights of the nodes in the layer ahead of it. This is then used to compute an error gradient with respect to the weights in the node and to update the weights of this node. The actual update equation is not very different from the basic perceptron at a conceptual level. The only differences that arise are due to the nonlinear functions commonly used in hidden layer nodes, and the fact that errors at hidden layer nodes are estimated via backpropagation, rather than directly computed by comparison of the output to a training label. This entire process is propagated backwards to update the weights of all the nodes in the network.

The basic framework of the multilayer update algorithm is the same as that for the single-layer algorithm illustrated in Fig. 10.9. The major difference is that it is no longer possible to use Eq. 10.69 for the hidden layer nodes. Instead, the update procedure is substituted with the forward–backward approach discussed above. As in the case of the single-layer network, the process of updating the nodes is repeated to convergence by repeatedly cycling through the training data in epochs. A neural network may sometimes require thousands of epochs through the training data to learn the weights at the different nodes.

A multilayer neural network is more powerful than a kernel SVM in its ability to capture arbitrary functions. A multilayer neural network has the ability to not only capture decision boundaries of arbitrary shapes, but also capture noncontiguous class distributions with different decision boundaries in different regions of the data. Logically, the different nodes in the hidden layer can capture the different decision boundaries in different regions of the data, and the node in the output layer can combine the results from these different decision boundaries. For example, the three different nodes in the hidden layer of Fig. 10.10b could conceivably capture three different nonlinear decision boundaries of different shapes in different localities of the data. With more nodes and layers, virtually any function can be approximated. This is more general than what can be captured by a kernel-based SVM that learns a single nonlinear decision boundary. In this sense, neural networks are viewed as *universal function approximators.* The price of this generality is that there are several implementation challenges in neural network design:

1. The initial design of the topology of the network presents many trade-off challenges for the analyst. A larger number of nodes and hidden layers provides greater generality, but a corresponding risk of overfitting. Little guidance is available about the design of the topology of the neural network because of poor interpretability associated with the multilayer neural network classification process. While some hill climbing methods can be used to provide a limited level of learning of the correct neural network topology, the issue of good neural network design still remains somewhat of an open question.

2. Neural networks are slow to train and sometimes sensitive to noise. As discussed earlier, thousands of epochs may be required to train a multilayer neural network. A larger network is likely to have a very slow learning process. While the training process of a neural network is slow, it is relatively efficient to classify test instances.

The previous discussion addresses only binary class labels. To generalize the approach to multiclass problems, a multiclass meta-algorithm discussed in the next chapter may be used. Alternatively, it is possible to modify both the basic perceptron model and the general neural network model to allow multiple output nodes. Each output node corresponds to the predicted value of a specific class label. The overall training process is exactly identical to the previous case, except that the weights of each output node now need to be trained.

10.7.3 Comparing Various Linear Models

Like neural networks, logistic regression also updates model parameters based on mistakes in categorization. This is not particularly surprising because both classifiers are linear classifiers but with different forms of the objective function for optimization. In fact, the use of some forms of logistic activation functions in the perceptron algorithm can be shown to be approximately equivalent to logistic regression. It is also instructive to examine the relationship of neural networks with SVM methods. In SVMs, the optimization function is based on the principle of maximum margin separation. This is different from neural networks, where the errors of predictions are directly penalized and then optimized with the use of a hill-climbing approach. In this sense, the SVM model has greater sophistication than the *basic* perceptron model by using the maximum margin principle to better focus on the more important decision boundary region. Furthermore, the generalization power of neural networks can be improved by using a (weighted) regularization penalty term $\lambda||\overline{W}||^2/2$ in the objective function. Note that this regularization term is similar to the maximum margin term in SVMs. The maximum margin term is, in fact, also referred to as the regularization term for SVMs. Variations of SVMs exist, in which the maximum margin term is replaced with an L_1 penalty $\sum_{i=1}^{d} |w_i|$. In such cases, the regularization interpretation is more natural than a margin-based interpretation. Furthermore, certain forms of the slack term in SVMs (e.g., quadratic slack) are similar to the main objective function in other linear models (e.g., least-squares models). The main difference is that the slack term is computed from the margin separators in SVMs rather than the decision boundary. This is consistent with the philosophy of SVMs that discourages training data points from not only being on the wrong side of the decision boundary, but also from being close to the decision boundary. Therefore, various linear models share a number of conceptual similarities, but they emphasize different aspects of optimization. This is the reason that maximum margin models are generally more robust to noise than linear models that use only distance-based penalties to reduce the number of data points on the wrong side of the separating hyperplanes. It has experimentally been observed that neural networks are sensitive to noise. On the other hand, multilayer neural networks can approximate virtually any complex function in principle.

10.8 Instance-Based Learning

Most of the classifiers discussed in the previous sections are *eager* learners in which the classification model is constructed *up front* and then used to classify a specific test instance. In instance-based learning, the training is delayed until the last step of classification. Such classifiers are also referred to as *lazy learners*. The simplest principle to describe instance-based learning is as follows:

Similar instances have similar class labels.

A natural approach for leveraging this general principle is to use nearest-neighbor classifiers. For a given test instance, the closest m training examples are determined. The dominant label among these m training examples is reported as the relevant class. In some variations of the model, an inverse distance-weighted scheme is used, to account for the varying importance of the m training instances that are closest to the test instance. An example of such an inverse weight function of the distance δ is $f(\delta) = e^{-\delta^2/t^2}$, where t is a user-defined parameter. Here, δ is the distance of the training point to the test instance. This weight is used as a vote, and the class with the largest vote is reported as the relevant label.

If desired, a nearest-neighbor index may be constructed up front, to enable more efficient retrieval of instances. The major challenge with the use of the nearest-neighbor classifier is the choice of the parameter m. In general, a very small value of m will not lead to robust classification results because of noisy variations within the data. On the other hand, large values of m will lose sensitivity to the underlying data locality. In practice, the appropriate value of m is chosen in a heuristic way. A common approach is to test different values of m for accuracy over the training data. While computing the m-nearest neighbors of a training instance $\overline{X}$, the data point $\overline{X}$ is not included[9] among the nearest neighbors. A similar approach can be used to learn the value of t in the distance-weighted scheme.

10.8.1 Design Variations of Nearest Neighbor Classifiers

A number of design variations of nearest-neighbor classifiers are able to achieve more effective classification results. This is because the Euclidean function is usually not the most effective distance metric in terms of its sensitivity to feature and class distribution. The reader is advised to review Chap. 3 on distance function design. Both unsupervised and supervised distance design methods can typically provide more effective classification results. Instead of using the Euclidean distance metric, the distance between two d-dimensional points $\overline{X}$ and $\overline{Y}$ is defined with respect to a $d \times d$ matrix A.

$$Dist(\overline{X}, \overline{Y}) = \sqrt{(\overline{X} - \overline{Y})A(\overline{X} - \overline{Y})^T} \tag{10.71}$$

This distance function is the same as the Euclidean metric when A is the identity matrix. Different choices of A can lead to better sensitivity of the distance function to the local and global data distributions. These different choices will be discussed in the following subsections.

10.8.1.1 Unsupervised Mahalanobis Metric

The Mahalanobis metric is introduced in Chap. 3. In this case, the value of A is chosen to be the inverse of the $d \times d$ covariance matrix Σ of the data set. The (i, j)th entry of the matrix Σ is the covariance between the dimensions i and j. Therefore, the Mahalanobis distance is defined as follows:

$$Dist(\overline{X}, \overline{Y}) = \sqrt{(\overline{X} - \overline{Y})\Sigma^{-1}(\overline{X} - \overline{Y})^T}. \tag{10.72}$$

The Mahalanobis metric adjusts well to the different scaling of the dimensions and the redundancies across different features. Even when the data is uncorrelated, the Mahalanobis metric is useful because it auto-scales for the naturally different ranges of attributes describing different physical quantities, such as age and salary. Such a scaling ensures that no single attribute dominates the distance function. In cases where the attributes are correlated, the Mahalanobis metric accounts well for the varying redundancies in different features. However, its major weakness is that it does not account for the varying shapes of the class distributions in the underlying data.

10.8.1.2 Nearest Neighbors with Linear Discriminant Analysis

To obtain the best results with a nearest-neighbor classifier, the distance function needs to account for the varying distribution of the different classes. For example, in the case of Fig. 10.11, there are two classes A and B, which are represented by "." and "*," respectively. The test instance denoted by X lies on the side of the boundary related to class A. However, the Euclidean metric does not adjust well to the arrangement of the class distribution, and a circle drawn around the test instance seems to include more points from class B than class A.

One way of resolving the challenges associated with this scenario, is to weight the most discriminating directions more in the distance function with an appropriate choice of the

[9]This approach is also referred to as leave-one-out cross-validation, and is described in detail in Sect. 10.9 on classifier evaluation.

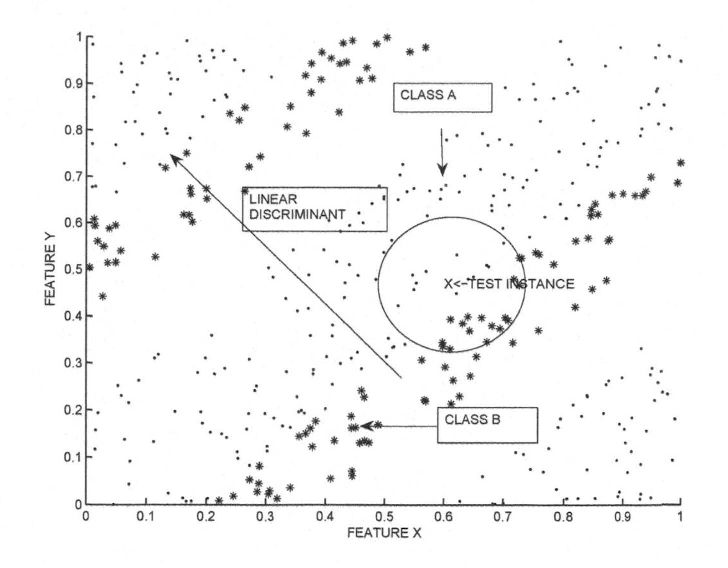

Figure 10.11: Importance of class sensitivity in distance function design

matrix A in Eq. 10.71. In the case of Fig. 10.11, the best discriminating direction is illustrated pictorially. Fisher's linear discriminant, discussed in Sect. 10.2.1.4, can be used to determine this direction, and map the data into a 1-dimensional space. In this 1-dimensional space, the different classes are separated out perfectly. The nearest-neighbor classifier will work well in this newly projected space. This is a very special example where only a 1-dimensional projection works well. However, it may not be generalizable to an arbitrary data set.

A more general way of computing the distances in a class-sensitive way, is to use a soft weighting of different directions, rather than selecting specific dimensions in a hard way. This can be achieved with the use of an appropriate choice of matrix A in Eq. 10.71. The choice of matrix A defines the shape of the neighborhood of a test instance. A distortion of this neighborhood from the circular Euclidean contour corresponds to a soft weighting, as opposed to a hard selection of specific directions. A soft weighting is also more robust in the context of smaller training data sets where the optimal linear discriminant cannot be found without overfitting. Thus, the core idea is to "elongate" the neighborhoods along the less discriminative directions and "shrink" the neighborhoods along the more discriminative directions with the use of matrix A. Note that the elongation of a neighborhood in a direction by a particular factor $\alpha > 1$, is equivalent to de-emphasizing that direction by that factor because distance components in that direction need to be divided by α. This is also done in the case of the Mahalanobis metric, except that the Mahalanobis metric is an unsupervised approach that is agnostic to the class distribution. In the case of the unsupervised Mahalanobis metric, the level of elongation achieved by the matrix A is inversely dependent on the variance along the different directions. In the supervised scenario, the goal is to elongate the directions, so that the level of elongation is inversely dependent on the ratio of the interclass variance to intraclass variance along the different directions.

Let $\mathcal{D}$ be the full database, and $\mathcal{D}_i$ be the portion of the data set belonging to class i. Let $\overline{\mu}$ represent the mean of the entire data set. Let $p_i = |\mathcal{D}_i|/|\mathcal{D}|$ be the fraction of data points belonging to class i, $\overline{\mu_i}$ be the d-dimensional row vector of means of $\mathcal{D}_i$, and Σ_i be the

$d \times d$ covariance matrix of $\mathcal{D}_i$. Then, the scaled[10] within-class scatter matrix S_w is defined as follows:

$$S_w = \sum_{i=1}^{k} p_i \Sigma_i. \tag{10.73}$$

The between-class scatter matrix S_b may be computed as follows:

$$S_b = \sum_{i=1}^{k} p_i (\overline{\mu_i} - \overline{\mu})^T (\overline{\mu_i} - \overline{\mu}). \tag{10.74}$$

Note that the matrix S_b is a $d \times d$ matrix because it results from the product of a $d \times 1$ matrix with a $1 \times d$ matrix. Then, the matrix A (of Eq. 10.71), which provides the desired distortion of the distances on the basis of class distribution, can be shown to be the following:

$$A = S_w^{-1} S_b S_w^{-1}. \tag{10.75}$$

It can be shown that this choice of the matrix A provides an excellent discrimination between the different classes, where the elongation in each direction depends inversely on ratio of the between-class variance to within-class variance along the different directions. The reader is referred to the bibliographic notes for pointers to the derivation of the aforementioned steps.

10.9 Classifier Evaluation

Given a classification model, how do we quantify its accuracy on a given data set? Such a quantification has several applications, such as evaluation of classifier effectiveness, comparing different models, selecting the best one for a particular data set, parameter tuning, and several meta-algorithms such as *ensemble analysis.* The last of these applications will be discussed in the next chapter. This leads to several challenges, both in terms of *methodology* used for the evaluation, and the specific approach used for quantification. These two challenges are stated as follows:

1. *Methodological issues:* The methodological issues are associated with dividing the labeled data appropriately into training and test segments for evaluation. As will become apparent later, the choice of methodology has a direct impact on the evaluation process, such as the underestimation or overestimation of classifier accuracy. Several approaches are possible, such as *holdout, bootstrap*, and *cross-validation.*

2. *Quantification issues:* The quantification issues are associated with providing a numerical measure for the quality of the method after a specific methodology (e.g., cross-validation) for evaluation has been selected. Examples of such measures could include the accuracy, the cost-sensitive accuracy, or a receiver operating characteristic curve quantifying the trade-off between true positives and false positives. Other types of numerical measures are specifically designed to compare the relative performance of classifiers.

In the following, these different aspects of classifier evaluation will be studied in detail.

[10]The unscaled version may be obtained by multiplying S_w with the number of data points. There is no difference to the final result whether the scaled or unscaled version is used, within a constant of proportionality.

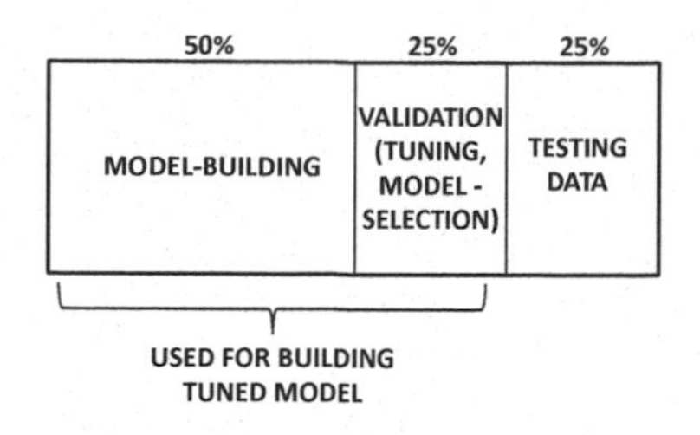

Figure 10.12: Segmenting the labeled data for parameter tuning and evaluation

10.9.1 Methodological Issues

While the problem of classification is defined for unlabeled test examples, the evaluation process does need labels to be associated with the test examples as well. These labels correspond to the *ground truth* that is required in the evaluation process, but not used in the training. The classifier cannot use the same examples for both training and testing because such an approach will overestimate the accuracy of the classifier due to overfitting. It is desirable to construct models with high generalizability to *unseen* test instances.

A common mistake in the process of bench-marking classification models is that analysts often use the test set to tune the parameters of the classification algorithm or make other choices about classifier design. Such an approach might overestimate the true accuracy because *knowledge of the test set has been implicitly used in the training process.* In practice, the labeled data should be divided into three parts, which correspond to (a) the model-building part of the labeled data, (b) the validation part of the labeled data, and (c) the testing data. This division is illustrated in Fig. 10.12. The validation part of the data should be used for parameter tuning or *model selection*. Model selection (cf. Sect. 11.8.3.4 of Chap. 11) refers to the process of deciding which classification algorithm is best suited to a particular data set. The testing data should not even be looked at during this phase. After tuning the parameters, the classification model is sometimes reconstructed on the entire training data (including the validation but not test portion). Only at this point, the testing data can be used for evaluating the classification algorithm *at the very end.* Note that if an analyst uses insights gained from the resulting performance on the test data to again adjust the algorithm in some way, then the results will be contaminated with knowledge from the test set.

This section discusses how the labeled data may be divided into the data used for constructing the tuned model (i.e., first two portions) and testing data (i.e., third portion) to accurately estimate the classification accuracy. The methodologies discussed in this section are also used for dividing the first two portions into the first and second portions (e.g., for parameter tuning), although we consistently use the terminologies "training data" and "testing data" to describe the two portions of the division. One problem with segmenting the labeled data is that it affects the measured accuracy depending on how the segmentation is done. This is especially the case when the amount of labeled data is small because one might accidently sample a small test data set which is not an accurate representative of the training data. For cases in which the labeled data is small, careful methodological variations are required to prevent erroneous evaluations.

10.9.1.1 Holdout

In the holdout method, the labeled data is randomly divided into two disjoint sets, corresponding to the training and test data. Typically a majority (e.g., two-thirds or three-fourths) is used as the training data, and the remaining is used as the test data. The approach can be repeated several times with multiple samples to provide a final estimate. The problem with this approach is that classes that are overrepresented in the training data are also underrepresented in the test data. These random variations can have a significant impact when the original class distribution is imbalanced to begin with. Furthermore, because only a subset of the available labeled data is used for training, the full power of the training data is not reflected in the error estimate. Therefore, the error estimates obtained are pessimistic. By repeating the process over b different holdout samples, the mean and variance of the error estimates can be determined. The variance can be helpful in creating statistical confidence intervals on the error.

One of the challenges with using the holdout method robustly is the case when the classes are imbalanced. Consider a data set containing 1000 data points, with 990 data points belonging to one class and 10 data points belonging to the other class. In such cases, it is possible for a test sample of 200 data points to contain not even one data point belonging to the rare class. Clearly, in such cases, it will be difficult to estimate the classification accuracy, especially when cost-sensitive accuracy measures are used that weigh the various classes differently. Therefore, a reasonable alternative is to implement the holdout method by independently sampling the two classes at the same level. Therefore, exactly 198 data points will be sampled from the first class, and 2 data points will be sampled from the rare class to create the test data set. Such an approach ensures that the classes are represented to a similar degree in both the training and test sets.

10.9.1.2 Cross-Validation

In cross-validation, the labeled data is divided into m disjoint subsets of equal size n/m. A typical choice of m is around 10. One of the m segments is used for testing, and the other $(m-1)$ segments are used for training. This approach is repeated by selecting each of the m different segments in the data as a test set. The average accuracy over the different test sets is then reported. The size of the training set is $(m-1)n/m$. When m is chosen to be large, this is almost equal to the labeled data size, and therefore the estimation error is close to what would be obtained with the original training data, but only for a small set of test examples of size n/m. However, because every labeled instance is represented exactly once in the testing over the m different test segments, the overall accuracy of the cross-validation procedure tends to be a highly representative, but pessimistic estimate, of model accuracy. A special case is one where m is chosen to be n. Therefore, $(n-1)$ examples are used for training, and one example is used for testing. This is averaged over the n different ways of picking the test example. This is also referred to as *leave-one-out* cross-validation. This special case is rather expensive for large data sets because it requires the application of the training procedure n times. Nevertheless, such an approach is particularly natural for lazy learning methods, such as the nearest-neighbor classifier, where a training model does not need to be constructed up front. By repeating the process over b different random m-way partitions of the data, the mean and variance of the error estimates may be determined. The variance can be helpful in determining statistical confidence intervals on the error. *Stratified cross-validation* uses proportional representation of each class in the different folds and usually provides less pessimistic results.

10.9.1.3 Bootstrap

In the bootstrap method, the labeled data is sampled uniformly *with replacement*, to create a training data set, which might possibly contain duplicates. The labeled data of size n is sampled n times with replacement. This results in a training data with the same size as the original labeled data. However, the training typically contains duplicates and also misses some points in the original labeled data.

The probability that a particular data point is not included in a sample is given by $(1-1/n)$. Therefore, the probability that the data point is not included in n samples is given by $(1-1/n)^n$. For large values of n, this expression evaluates to approximately $1/e$, where e is the base of the natural logarithm. The fraction of the labeled data points included at least once in the training data is therefore $1 - 1/e \approx 0.632$. The training model $\mathcal{M}$ is constructed on the bootstrapped sample containing duplicates. The overall accuracy is computed using the original set of full labeled data as the test examples. The estimate is highly optimistic of the true classifier accuracy because of the large overlap between training and test examples. In fact, a 1-nearest neighbor classifier will always yield 100 % accuracy for the portion of test points included in the bootstrap sample and the estimates are therefore not realistic in many scenarios. By repeating the process over b different bootstrap samples, the mean and the variance of the error estimates may be determined.

A better alternative is to use *leave-one-out bootstrap*. In this approach, the accuracy $A(\overline{X})$ of each labeled instance $\overline{X}$ is computed using the classifier performance on only the subset of the b bootstrapped samples in which $\overline{X}$ is not a part of the bootstrapped sample of training data. The overall accuracy A_l of the leave-one-out bootstrap is the mean value of $A(\overline{X})$ over all labeled instances $\overline{X}$. This approach provides a pessimistic accuracy estimate. The 0.632-bootstrap further improves this accuracy with a "compromise" approach. The average *training data* accuracy A_t over the b bootstrapped samples is computed. This is a highly optimistic estimate. For example, A_t will always be 100 % for a 1-nearest neighbor classifier. The overall accuracy A is a weighted average of the leave-one-out accuracy and the training-data accuracy.

$$A = (0.632) \cdot A_l + (0.368) \cdot A_t \tag{10.76}$$

In spite of the compromise approach, the estimates of 0.632 bootstrap are usually optimistic. The bootstrap method is more appropriate when the size of the labeled data is small.

10.9.2 Quantification Issues

This section will discuss how the quantification of the accuracy of a classifier is performed after the training and test set for a classifier are fixed. Several measures of accuracy are used depending on the nature of the classifier output:

1. In most classifiers, the output is predicted in the form of a label associated with the test instance. In such cases, the ground-truth label of the test instance is compared with the predicted label to generate an overall value of the classifier accuracy.

2. In many cases, the output is presented as a numerical score *for each labeling possibility* for the test instance. An example is the Bayes classifier where a probability is reported for a test instance. As a convention, it will be assumed that higher values of the score imply a greater likelihood to belong to a particular class.

The following sections will discuss methods for quantifying accuracy in both scenarios.

10.9.2.1 Output as Class Labels

When the output is presented in the form of class labels, the ground-truth labels are compared to the predicted labels to yield the following measures:

1. *Accuracy:* The accuracy is the fraction of test instances in which the predicted value matches the ground-truth value.

2. *Cost-sensitive accuracy:* Not all classes are equally important in all scenarios while comparing the accuracy. This is particularly important in imbalanced class problems, which will be discussed in more detail in the next chapter. For example, consider an application in which it is desirable to classify tumors as *malignant* or *nonmalignant* where the former is much rarer than the latter. In such cases, the misclassification of the former is often much less desirable than misclassification of the latter. This is frequently quantified by imposing differential costs $c_1 \dots c_k$ on the misclassification of the different classes. Let $n_1 \dots n_k$ be the number of test instances belonging to each class. Furthermore, let $a_1 \dots a_k$ be the accuracies (expressed as a fraction) on the subset of test instances belonging to each class. Then, the overall accuracy A can be computed as a weighted combination of the accuracies over the individual labels.

$$A = \frac{\sum_{i=1}^{k} c_i n_i a_i}{\sum_{i=1}^{k} c_i n_i} \tag{10.77}$$

   The cost sensitive accuracy is the same as the unweighted accuracy when all costs $c_1 \dots c_k$ are the same.

Aside from the accuracy, the statistical robustness of a model is also an important issue. For example, if two classifiers are trained over a small number of test instances and compared, the difference in accuracy may be a result of random variations, rather than a truly *statistically significant* difference between the two classifiers. Therefore, it is important to design statistical measures to quantify the specific advantage of one classifier over the other.

Most statistical methodologies such as holdout, bootstrap, and cross-validation use $b > 1$ different randomly sampled rounds to obtain multiple estimates of the accuracy. For the purpose of discussion, let us assume that b different rounds (i.e., b different m-way partitions) of cross-validation are used. Let $\mathcal{M}_1$ and $\mathcal{M}_2$ be two models. Let $A_{i,1}$ and $A_{i,2}$ be the respective accuracies of the models $\mathcal{M}_1$ and $\mathcal{M}_2$ on the partitioning created by the ith round of cross-validation. The corresponding difference in accuracy is $\delta a_i = A_{i,1} - A_{i,2}$. This results in b estimates $\delta a_1 \dots \delta a_b$. Note that δa_i might be either positive or negative, depending on which classifier provides superior performance on a particular round of cross-validation. Let the average difference in accuracy between the two classifiers be ΔA.

$$\Delta A = \frac{\sum_{i=1}^{b} \delta a_i}{b} \tag{10.78}$$

The standard deviation σ of the difference in accuracy may be estimated as follows:

$$\sigma = \sqrt{\frac{\sum_{i=1}^{b} (\delta a_i - \Delta A)^2}{b-1}}. \tag{10.79}$$

Note that the sign of ΔA tells us which classifier is better than the other. For example, if $\Delta A > 0$ then model $\mathcal{M}_1$ has higher average accuracy than $\mathcal{M}_2$. In such a case, it is desired to determine a statistical measure of the confidence (or, a probability value) that $\mathcal{M}_1$ is truly better than $\mathcal{M}_2$.

The idea here is to assume that the different samples $\delta a_1 \ldots \delta a_b$ are sampled from a normal distribution. Therefore, the estimated mean and standard deviations of this distribution are given by ΔA and σ, respectively. The standard deviation of the estimated mean ΔA of b samples is therefore $\sigma/\sqrt{b}$ according to the central-limit theorem. Then, the number of standard deviations s by which ΔA is different from the break-even accuracy difference of 0 is as follows:

$$s = \frac{\sqrt{b}|\Delta A - 0|}{\sigma}. \tag{10.80}$$

When b is large, the standard normal distribution with zero mean and unit variance can be used to quantify the probability that one classifier is truly better than the other. The probability in any one of the symmetric tails of the standard normal distribution, more than s standard deviations away from the mean, provides the probability that this variation is not significant, and it might be a result of chance. This probability is subtracted from 1 to determine the confidence that one classifier is truly better than the other.

It is often computationally expensive to use large values of b. In such cases, it is no longer possible to estimate the standard deviation σ robustly with the use of a small number b of samples. To adjust for this, the Student's t-distribution with $(b - 1)$ degrees of freedom is used instead of the normal distribution. This distribution is very similar to the normal distribution, except that it has a heavier tail to account for the greater estimation uncertainty. In fact, for large values of b, the t-distribution with $(b - 1)$ degrees of freedom converges to the normal distribution.

10.9.2.2 Output as Numerical Score

In many scenarios, the output of the classification algorithm is reported as a numerical score associated with each test instance and label value. In cases where the numerical score can be reasonably compared across test instances (e.g., the probability values returned by a Bayes classifier), it is possible to compare the different test instances in terms of their relative propensity to belong to a specific class. Such scenarios are more common when one of the classes of interest is rare. Therefore, for this scenario, it is more meaningful to use the binary class scenario where one of the classes is the positive class, and the other class is the negative class. The discussion below is similar to the discussion in Sect. 8.8.2 of Chap. 8 on external validity measures for outlier analysis. This similarity arises from the fact that outlier validation with class labels is identical to classifier evaluation.

The advantage of a numerical score is that it provides more flexibility in evaluating the overall trade-off between labeling a varying number of data points as positives. This is achieved by using a threshold on the numerical score for the positive class to define the binary label. If the threshold is selected too aggressively to minimize the number of declared positive class instances, then the algorithm will miss true-positive class instances (false negatives). On the other hand, if the threshold is chosen in a more relaxed way, this will lead to too many false positives. This leads to a trade-off between the false positives and false negatives. The problem is that the "correct" threshold to use is never known exactly in a real scenario. However, the entire trade-off curve can be quantified using a variety of measures, and two algorithms can be compared over the entire trade-off curve. Two examples of such curves are the *precision–recall* curve, and the *receiver operating characteristic (ROC)* curve.

For any given threshold t on the predicted positive-class score, the declared positive class set is denoted by $\mathcal{S}(t)$. As t changes, the size of $\mathcal{S}(t)$ changes as well. Let $\mathcal{G}$ represent the true set (ground-truth set) of positive instances in the data set. Then, for any given threshold t, the *precision* is defined as the percentage of *reported* positives that truly turn out to be positive.

$$Precision(t) = 100 * \frac{|\mathcal{S}(t) \cap \mathcal{G}|}{|\mathcal{S}(t)|}$$

The value of $Precision(t)$ is *not* necessarily monotonic in t because both the numerator and denominator may change with t differently. The *recall* is correspondingly defined as the percentage of *ground-truth* positives that have been reported as positives at threshold t.

$$Recall(t) = 100 * \frac{|\mathcal{S}(t) \cap \mathcal{G}|}{|\mathcal{G}|}$$

While a natural trade-off exists between precision and recall, this trade-off is not necessarily monotonic. One way of creating a single measure that summarizes both precision and recall is the F_1-measure, which is the harmonic mean between the precision and the recall.

$$F_1(t) = \frac{2 \cdot Precision(t) \cdot Recall(t)}{Precision(t) + Recall(t)} \tag{10.81}$$

While the $F_1(t)$ measure provides a better quantification than either precision or recall, it is still dependent on the threshold t, and is therefore still not a complete representation of the trade-off between precision and recall. It is possible to visually examine the entire trade-off between precision and recall by varying the value of t, and examining the trade-off between the two quantities, by plotting the precision versus the recall. As shown later with an example, the lack of monotonicity of the precision makes the results harder to intuitively interpret.

A second way of generating the trade-off in a more intuitive way is through the use of the ROC curve. The *true-positive rate*, which is the same as the recall, is defined as the percentage of ground-truth positives that have been predicted as positive instances at threshold t.

$$TPR(t) = Recall(t) = 100 * \frac{|\mathcal{S}(t) \cap \mathcal{G}|}{|\mathcal{G}|}$$

The false-positive rate $FPR(t)$ is the percentage of the falsely reported positives out of the ground-truth negatives. Therefore, for a data set $\mathcal{D}$ with ground-truth positives $\mathcal{G}$, this measure is defined as follows:

$$FPR(t) = 100 * \frac{|\mathcal{S}(t) - \mathcal{G}|}{|\mathcal{D} - \mathcal{G}|}. \tag{10.82}$$

The ROC curve is defined by plotting the $FPR(t)$ on the X-axis, and $TPR(t)$ on the Y-axis for varying values of t. Note that the end points of the ROC curve are always at $(0, 0)$ and $(100, 100)$, and a random method is expected to exhibit performance along the diagonal line connecting these points. The *lift* obtained above this diagonal line provides an idea of the accuracy of the approach. The area under the ROC curve provides a concrete quantitative evaluation of the effectiveness of a particular method.

To illustrate the insights gained from these different graphical representations, consider an example of a data set with 100 points from which 5 points belong to the positive class. Two algorithms A and B are applied to this data set that rank all data points from 1 to 100, with lower rank representing greater propensity to belong to the positive class. Thus, the true-positive rate and false-positive rate values can be generated by determining the ranks of the five ground-truth positive label points. In Table 10.2, some hypothetical ranks for the

Table 10.2: Rank of ground-truth positive instances

Algorithm	Rank of positive class instances
Algorithm A	1, 5, 8, 15, 20
Algorithm B	3, 7, 11, 13, 15
Random Algorithm	17, 36, 45, 59, 66
Perfect Oracle	1, 2, 3, 4, 5

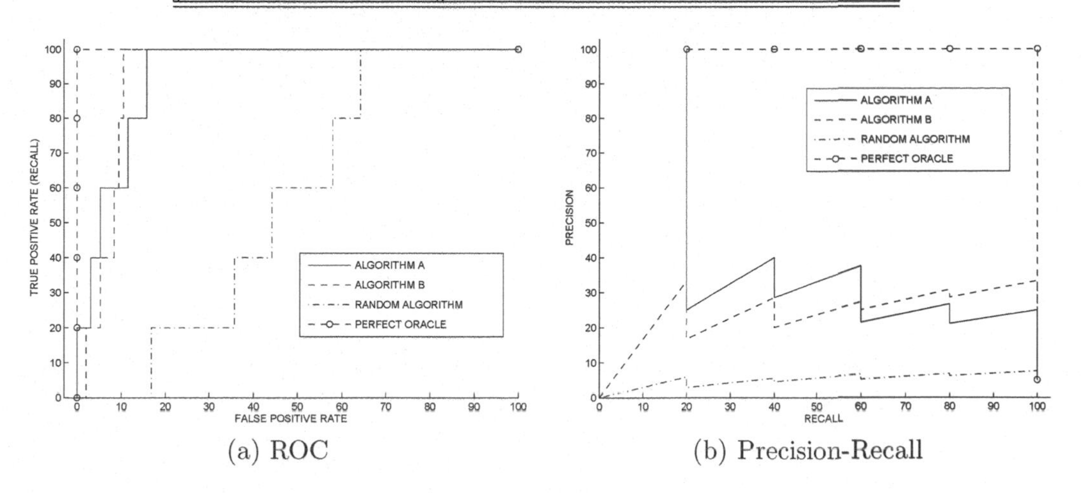

(a) ROC

(b) Precision-Recall

Figure 10.13: ROC curve and precision–recall curves

five ground-truth positive label instances have been illustrated for the different algorithms. In addition, ranks of the ground-truth positives for a random algorithm have been indicated. The random algorithm outputs a random score for each data point. Similarly, the ranks for a "perfect oracle" algorithm that ranks the correct top five points to belong to the positive class have also been illustrated in the table. The resulting ROC curves are illustrated in Fig. 10.13a. The corresponding precision–recall curves are illustrated in Fig. 10.13b. While the precision–recall curves are not quite as nicely interpretable as the ROC curves, it is easy to see that the *relative trends* between different algorithms, are the same in both cases. In general, ROC curves are used more frequently because of the ease in interpreting the quality of the algorithm with respect to a random classifier.

What do these curves really tell us? For cases in which one curve strictly dominates another, it is clear that the algorithm for the former curve is superior. For example, it is immediately evident that the oracle algorithm is superior to all algorithms, and the random algorithm is inferior to all the other algorithms. On the other hand, algorithms A and B show domination at different parts of the ROC curve. In such cases, it is hard to say that one algorithm is strictly superior. From Table 10.2, it is clear that Algorithm A ranks three of the correct positive instances very highly, but the remaining two positive instances are ranked poorly. In the case of Algorithm B, the highest ranked positive instances are not as well ranked as Algorithm A, though all five positive instances are determined much earlier in terms of rank threshold. Correspondingly, Algorithm A dominates on the earlier part of the ROC curve, whereas Algorithm B dominates on the later part. It is possible to use the area under the ROC curve as a proxy for the overall effectiveness of the algorithm.

10.10 Summary

The problem of data classification can be considered a supervised version of data clustering, in which a predefined set of groups is provided to a learner. This predefined set of groups is used for training the classifier to categorize unseen test examples into groups. A wide variety of models have been proposed for data classification.

Decision trees create a hierarchical model of the training data. For each test instance, the optimal path in the tree is used to classify unseen test instances. Each path in the tree can be viewed as a rule that is used to classify unseen test instances. Rule-based classifiers can be viewed as a generalization of decision trees, in which the classifier is not necessarily restricted to characterizing the data in a hierarchical way. Therefore, multiple conflicting rules can be used to cover the same training or test instance. Probabilistic classifiers map feature values to unseen test instances with probabilities. The naive Bayes rule or a logistic function may be used for effective estimation of probabilities. SVMs and neural networks are two forms of linear classifiers. The objective functions that are optimized are different. In the case of SVMs, the maximum margin principle is used, whereas for neural networks, the least squares error of prediction is approximately optimized. Instance-based learning methods are classifiers that delay learning to classification time as opposed to eager learners that construct the classification models up front. The simplest form of instance-based learning is the nearest-neighbor classifier. Many complex variations are possible by using different types of distance functions and locality-centric models.

Classifier evaluation is important for testing the relative effectiveness of different training models. Numerous models such as holdout, stratified sampling, bootstrap, and cross-validation have been proposed in the literature. Classifier evaluation can be performed either in the context of label assignment or numerical scoring. For label assignment, either the accuracy or the cost-sensitive accuracy may be used. For numerical scoring, the ROC curve is used to quantify the trade-off between the true-positive and false-positive rates.

10.11 Bibliographic Notes

The problem of data classification has been studied extensively by the data mining, machine learning, and pattern recognition communities. A number of books on these topics are available from these different communities [33, 95, 189, 256, 389]. Two surveys on the topic of data classification may be found in [286, 330]. A recent book [33] contains surveys on various aspects of data classification.

Feature selection is an important problem in data classification, to ensure the modeling algorithm is not confused by noise in the training data. Two books on feature selection may be found in [360, 366]. Fisher's discriminant analysis was first proposed in [207], although a slightly different variant with the assumption of normally distributed data used in linear discriminant analysis [379]. The most well-known decision tree algorithms include *ID3* [431], *C4.5* [430], and *CART* [110]. Decision tree methods are also used in the context of multivariate splits [116], though these methods are computationally more challenging. Surveys on decision tree algorithms may be found in [121, 393, 398]. Decision trees can be converted into rule-based classifiers where the rules are mutually exclusive. For example, the *C4.5* method has also been extended to the *C4.5rules* algorithm [430]. Other popular rule-based systems include *AQ* [386], *CN2* [177], and *RIPPER* [178]. Much of the discussion in this chapter was based on these algorithms. Popular associative classification algorithms include *CBA* [358], *CPAR* [529], and *CMAR* [349]. Methods for classification with discriminative patterns are discussed in [149]. A recent overview discussion of pattern-based classification algorithms may be found in [115]. The naive Bayes classifier has been discussed in detail in [187, 333, 344]. The work in [344] is particularly notable, in that it provides an understanding and justification of the naive Bayes assumption. A brief discussion of logistic regression models may be found in Chap. 3 of [33]. A more detailed discussion may be found in [275].

Numerous books are available on the topic of SVMs [155, 449, 478, 494]. An excellent tutorial on SVMs may be found in [124]. A detailed discussion of the Lagrangian relaxation technique for solving the resulting quadratic optimization problem may be found in [485]. It has been pointed out [133] that the advantages of the primal approach in SVMs seem to have been largely overlooked in the literature. It is sometimes mistakenly understood that the kernel trick can only be applied to the dual; the trick can be applied to the primal formulation as well [133]. A discussion of kernel methods for SVMs may be found in [451]. Other applications of kernels, such as nonlinear k-means and nonlinear PCA, are discussed in [173, 450]. The perceptron algorithm was due to Rosenblatt [439]. Neural networks are discussed in detail in several books [96, 260]. The back-propagation algorithm is described in detail in these books. The earliest work on instance-based classification was discussed in [167]. The method was subsequently extended to symbolic attributes [166]. Two surveys on instance-based classification may be found in [14, 183]. Local methods for nearest-neighbor classification are discussed in [216, 255]. Generalized instance-based learning methods have been studied in the context of decision trees [217], rule-based methods [347], Bayes methods [214], SVMs [105, 544], and neural networks [97, 209, 281]. Methods for classifier evaluation are discussed in [256].

10.12 Exercises

1. Compute the Gini index for the entire data set of Table 10.1, with respect to the two classes. Compute the Gini index for the portion of the data set with age at least 50.

2. Repeat the computation of the previous exercise with the use of the entropy criterion.

3. Show how to construct a (possibly overfitting) rule-based classifier that always exhibits 100 % accuracy on the training data. Assume that the feature variables of no two training instances are identical.

4. Design a univariate decision tree with a soft maximum-margin split criterion borrowed from SVMs. Suppose that this decision tree is generalized to the multivariate case. How does the resulting decision boundary compare with SVMs? Which classifier can handle a larger variety of data sets more accurately?

5. Discuss the advantages of a rule-based classifier over a decision tree.

6. Show that an SVM is a special case of a rule-based classifier. Design a rule-based classifier that uses SVMs to create an ordered list of rules.

7. Implement an associative classifier in which only maximal patterns are used for classification, and the majority consequent label of rules fired, is reported as the label of the test instance.

8. Suppose that you had d-dimensional numeric training data, in which it was known that the probability density of d-dimensional data instance $\overline{X}$ in each class i is proportional

to $e^{-||\overline{X}-\overline{\mu_i}||_1}$, where $||\cdot||_1$ is the Manhattan distance, and $\overline{\mu_i}$ is known for each class. How would you implement the Bayes classifier in this case? How would your answer change if $\overline{\mu_i}$ is unknown?

9. Explain the relationship of mutual exclusiveness and exhaustiveness of a rule set, to the need to order the rule set, or the need to set a class as the default class.

10. Consider the rules $Age > 40 \Rightarrow Donor$ and $Age \leq 50 \Rightarrow \neg Donor$. Are these two rules mutually exclusive? Are these two rules exhaustive?

11. For the example of Table 10.1, determine the prior probability of each class. Determine the conditional probability of each class for cases where the *Age* is at least 50.

12. Implement the naive Bayes classifier.

13. For the example of Table 10.1, provide a single linear separating hyperplane. Is this separating hyperplane unique?

14. Consider a data set containing four points located at the corners of the square. The two points on one diagonal belong to one class, and the two points on the other diagonal belong to the other class. Is this data set linearly separable? Provide a proof.

15. Provide a systematic way to determine whether two classes in a labeled data set are linearly separable.

16. For the soft SVM formulation with hinge loss, show that:

(a) The weight vector is given by the same relationship $\overline{W} = \sum_{i=1}^{n} \lambda_i y_i \overline{X_i}$, as for hard SVMs.

(b) The condition $\sum_{i=1}^{n} \lambda_i y_i = 0$ holds as in hard SVMs.

(c) The Lagrangian multipliers satisfy $\lambda_i \leq C$.

(d) The Lagrangian dual is identical to that of hard SVMs.

17. Show that it is possible to omit the bias parameter b from the decision boundary of SVMs by suitably preprocessing the data set. In other words, the decision boundary is now $\overline{W} \cdot \overline{X} = 0$. What is the impact of eliminating the bias parameter on the gradient ascent approach for Lagrangian dual optimization in SVMs?

18. Show that an $n \times d$ data set can be mean-centered by premultiplying it with the $n \times n$ matrix $(I - U/n)$, where U is a unit matrix of all ones. Show that an $n \times n$ kernel matrix K can be adjusted for mean centering of the data in the transformed space by adjusting it to $K' = (I - U/n)K(I - U/n)$.

19. Consider two classifiers A and B. On one data set, a 10-fold cross validation shows that classifier A is better than B by 3 %, with a standard deviation of 7 % over 100 different folds. On the other data set, classifier B is better than classifier A by 1 %, with a standard deviation of 0.1 % over 100 different folds. Which classifier would you prefer on the basis of this evidence, and why?

20. Provide a nonlinear transformation which would make the data set of Exercise 14 linearly separable.

21. Let S_w and S_b be defined according to Sect. 10.2.1.3 for the binary class problem. Let the fractional presence of the two classes be p_0 and p_1, respectively. Show that $S_w + p_0 p_1 S_b$ is equal to the covariance matrix of the data set.

Chapter 11

Data Classification: Advanced Concepts

*"Labels are for filing. Labels are for clothing.
Labels are not for people."*—Martina Navratilova

11.1 Introduction

In this chapter, a number of advanced scenarios related to the classification problem will be addressed. These include more difficult special cases of the classification problem and various ways of enhancing classification algorithms with the use of additional inputs or a combination of classifiers. The enhancements discussed in this chapter belong to one of the following two categories:

1. *Difficult classification scenarios:* Many scenarios of the classification problem are much more challenging. These include multiclass scenarios, rare-class scenarios, and cases where the size of the training data is large.

2. *Enhancing classification:* Classification methods can be enhanced with additional data-centric input, user-centric input, or multiple models.

The difficult classification scenarios that are addressed in this chapter are as follows:

1. *Multiclass learning:* Although many classifiers such as decision trees, Bayesian methods, and rule-based classifiers, can be directly used for multiclass learning, some of the models, such as support-vector machines, are naturally designed for binary classification. Therefore, numerous meta-algorithms have been designed for adapting binary classifiers to multiclass learning.

2. *Rare class learning:* The positive and negative examples may be imbalanced. In other words, the data set contains only a small number of positive examples. A direct use of traditional learning models may often result in the classifier assigning all examples to the negative class. Such a classification is not very informative for imbalanced scenarios in which misclassification of the rare class incurs much higher cost than misclassification of the normal class.

C. C. Aggarwal, *Data Mining: The Textbook*, DOI 10.1007/978-3-319-14142-8_11

3. *Scalable learning:* The sizes of typical training data sets have increased significantly in recent years. Therefore, it is important to design models that can perform the learning in a scalable way. In cases where the data is not memory resident, it is important to design algorithms that can minimize disk accesses.

4. *Numeric class variables:* Most of the discussion in this book assumes that the class variables are categorical. Suitable modifications are required to classification algorithms, when the class variables are numeric. This problem is also referred to as *regression modeling.*

The addition of more training data or the simultaneous use of a larger number of classification models can improve the learning accuracy. A number of methods have been proposed to enhance classification methods. Examples include the following:

1. *Semisupervised learning:* In these cases, unlabeled examples are used to improve the effectiveness of classifiers. Although unlabeled data does not contain any information about the label distribution, it does contain a significant amount of information about the manifold and clustering structure of the underlying data. Because the classification problem is a supervised version of the clustering problem, this connection can be leveraged to improve the classification accuracy. The core idea is that in most real data sets, labels vary in a smooth way over dense regions of the data. The determination of dense regions in the data only requires unlabeled information.

2. *Active learning:* In real life, it is often expensive to acquire labels. In active learning, the user (or an oracle) is actively involved in determining the most *informative* examples for which the labels need to be acquired. Typically, these are examples that provide the user the more accurate knowledge about the uncertain regions in the data, where the distribution of the class label is unknown.

3. *Ensemble learning:* Similar to the clustering and the outlier detection problems, ensemble learning uses the power of multiple models to provide more robust results for the classification process. The motivation is similar to that for the clustering and outlier detection problems.

This chapter is organized as follows. Multiclass learning is addressed in Sect. 11.2. Rare class learning methods are introduced in Sect. 11.3. Scalable classification methods are introduced in Sect. 11.4. Classification with numeric class variables is discussed in Sect. 11.5. Semisupervised learning methods are introduced in Sect. 11.6. Active learning methods are discussed in Sect. 11.7. Ensemble methods are proposed in Sect. 11.8. Finally, a summary of the chapter is given in Sect. 11.9.

11.2 Multiclass Learning

Some models such as support vector machines (*SVMs*), neural networks, and logistic regression are naturally designed for the binary class scenario. While multiclass generalizations of these methods are available, it is helpful to design generic meta-frameworks that can directly use the binary methods for multiclass classification. These frameworks are designed as meta-algorithms that can take a binary classification algorithm $\mathcal{A}$ as input and use it to make multilabel predictions. Several strategies are possible to convert binary classifiers into multilabel classifiers. In the following discussion, it will be assumed that the number of classes is denoted by k.

The first strategy is the *one-against-rest* approach. In this approach, k different binary classification problems are created, such that one problem corresponds to each class. In the ith problem, the ith class is considered the set of positive examples, whereas all the remaining examples are considered negative examples. The binary classifier $\mathcal{A}$ is applied to each of these training data sets. This creates a total of k models. If the positive class is predicted in the ith problem, then the ith class is rewarded with a vote. Otherwise, each of the remaining classes is rewarded with a vote. The class with the largest number of votes is predicted as the relevant one. In practice, more than one model may predict an example to belong to a positive class. This may result in ties. To avoid ties, one may also use the numeric output of a classifier (e.g., Bayes posterior probability) to weight the corresponding vote. The highest numeric score for a particular class is selected to predict the label. Note that the choice of the numeric score for weighting the votes depends on the classifier at hand. Intuitively, the score represents the "confidence" of that classifier in a particular label.

The second strategy is the *one-against-one* approach. In this strategy, a training data set is constructed for each of the $\binom{k}{2}$ pairs of classes. The algorithm $\mathcal{A}$ is applied to each training data set. This results in a total of $k(k-1)/2$ models. For each model, the prediction provides a vote to the winner. The class label with the most votes is declared as the winner in the end. At first sight, it seems that this approach is computationally more expensive, because it requires us to train $k(k-1)/2$ classifiers, rather than training k classifiers, as in the one-against-rest approach. However, the computational cost is ameliorated by the smaller size of the training data in the one-against-one approach. Specifically, the training data size in the latter case is approximately $2/k$ of the training data size used in the one-against-rest approach on the average. If the running time of each individual classifier scales super-linearly with the number of training points, then the overall running time of this approach may actually be lower than the first approach that requires us to train only k classifiers. This is usually the case for kernel SVM classifiers, in which the running times scale up more than linearly with the number of data points. Note that the size of the kernel matrix scales up quadratically with the number of data points. The one-against-one approach may also result in ties between different classes that receive the same number of votes. In such cases, the numeric scores output by the classifier may be used to weight the votes for the different classes. As in the previous case, the choice of the numeric score depends on the choice of the base classifier model.

11.3 Rare Class Learning

The class distribution in many applications is not balanced. Consider a scenario in which data points representing credit card activity are labeled as either "normal" or "fraudulent." In such cases, the class distribution is typically very imbalanced. For example, 99 % of the data points may be normal, whereas only 1% of the data points may be fraudulent. The straightforward application of classification algorithms may lead to misleading results because of the preponderance of the normal class.

Consider a test instance $\overline{X}$ whose nearest 100 neighbors contain 49 rare class instances and 51 normal class instances. In such a case, it is evident that the test instance is surrounded by large fraction of rare instances *relative to expectation*. Yet, a k-nearest neighbor classifier with $k = 100$ will categorize instance $\overline{X}$ into the normal class. Such a classifier does not provide informative results, because its behavior approximately mimics a trivial classifier that classifies every instance as normal.

This behavior is not restricted to nearest-neighbor classifiers. A Bayesian classifier will have biased priors that favor the normal class. A decision-tree will find it difficult to separate out instances belonging to the rare class. As a result, most of these classifiers, if not modified appropriately, will classify many rare instances to the majority class. Interestingly, even a trivial classifier that labels all instances as normal might provide a high *absolute* accuracy. However, achieving a high classification accuracy on the rare class is more important in such application domains. This is because the applications associated with rare class detection are typically such that the *consequences* of misclassifying a rare class are much higher than those of misclassifying the normal class. For example, in the credit card scenario, it is much costlier to the credit card company to accept fraudulent activity as normal, rather than warning a customer incorrectly about suspicious activity on their card.

These observations suggest that rare-class learning algorithms need to have an explicit mechanism for emphasizing the greater importance of the rare class. This mechanism is provided by a *cost-matrix* $C(i, j)$ that quantifies the cost of misclassifying the class i to class j where $i \neq j$. In practice, for multiclass problems, it is often difficult to populate the full $k \times k$ matrix of misclassification possibilities. Therefore, a simplification is to associate the misclassification costs with the source class, rather than a source-destination pair. In other words, the cost of misclassifying class i is denoted by $C(i)$, irrespective of the incorrect destination class j to which it is predicted. Typically, the cost of misclassifying a rare class is much larger than that of misclassifying a normal class. Therefore, the goal is to maximize the *cost-weighted accuracy*, rather than the absolute accuracy.

Fortunately, these goals can be achieved by making modest changes to existing classification algorithms. Some examples of these modifications are as follows:

1. *Example reweighting:* The training examples from various classes are reweighted according to their misclassification costs. This approach naturally leads to a bias in classifying rare class examples more accurately than normal class examples. Therefore, classification algorithms need to be modified to work with weighted examples.

2. *Example resampling:* The examples from different classes are resampled to undersample normal classes and/or oversample rare classes. In such cases, unweighted classifiers can be directly used.

Each of these different methods will be discussed in the following sections.

11.3.1 Example Reweighting

In this case, the examples are weighted in proportion to their costs. Because the original classification problem is designed to maximize accuracy, the analogous solution to the weighted problem maximizes cost-weighted accuracy. Therefore, all instances belonging to the ith class are weighted by $C(i)$. Therefore, the existing classification algorithms need to be modified to work with these additional weights. In most cases, the required changes are relatively minor. The following contains a brief description of the required changes to various classification algorithms:

1. *Decision trees:* Weights can be incorporated in decision-tree algorithms easily. The split criterion requires the computation of the Gini index and entropy, all of which can be computed using weights on the examples. Both the Gini index and the entropy are computed as a function of the proportionate class distribution of the training examples. This proportionate class distribution can be computed with the use of

weights on the examples. Tree-pruning can also be modified to measure the impact of removing nodes on the weighted accuracy.

2. *Rule-based classifiers:* Sequential covering algorithms are similar to decision-tree construction. The main difference is in terms of the criteria used to grow rules. Measures such as the Laplace measure and FOIL's information gain use the raw number of positive and negative examples covered by the rule. In this case, the weighted number of examples are used as substitute for the raw number of examples. Rule-pruning uses weighted accuracy to measure the impact of conjunct pruning. For associative classifiers, the weights on the instances need to be used in computation of support and confidence.

3. *Bayes classifiers:* The implementation of Bayes classifiers remains virtually the same as the unweighted case except for one crucial difference in the probability estimation process. The class priors and conditional feature probabilities are now estimated using weights on the instances.

4. *Support vector machines:* Interestingly, the hard-margin support vector machines are not affected by reweighting of examples because the support vectors do not depend on example weights. However, in practice, soft margin is used. In such cases, the slack penalty terms in the objective function are appropriately weighted, and it results in modifications to both the primal and dual methods for soft SVMs (see Exercises 3 and 4). This typically leads to a movement of the boundary of the support-vector machine toward the normal class side of the separation. This ensures that fewer rare class examples are penalized for (the more costly) margin violation, and more normal class examples are penalized. The result is a lower likelihood of incorrectly misclassifying rare class examples but a greater likelihood of misclassifying normal class examples.

5. *Instance-based methods:* Weighted votes are used for the different classes, after determining the m nearest neighbors to a given test instance.

Thus, most classifiers can be made to work with the weighted case with relatively small changes. The advantage of weighting techniques is that they work with the original training data, and are therefore less prone to overfitting than sampling methods that manipulate the training data.

11.3.2 Sampling Methods

In adaptive resampling, the different classes are differentially sampled to enhance the impact of the rare class on the classification model. Sampling can be performed either with or without replacement. The rare class can be oversampled, or the normal class can be undersampled, or both can occur. The classification model is learned on the resampled data. The sampling probabilities are typically chosen in proportion to their misclassification costs. This enhances the proportion of the rare costs in the sample used for learning, and the approach is generally applicable to multiclass scenarios as well. It has generally been observed that undersampling the normal class has a number of advantages over oversampling the rare class. When undersampling is used, the sampled training data is much smaller than the original data set, which leads to better training efficiency.

In some variations, all instances of the rare class are used in combination with a small sample of the normal class. This is also referred to as *one-sided selection*. The logic of this approach is that rare class instances are too valuable as training data to modify any type of sampling. Undersampling has several advantages with respect to oversampling because of the following reasons:

1. The model construction phase for a smaller training data set requires much less time.
2. The normal class is less important for modeling purposes, and all instances from the more valuable rare class are included for modeling. Therefore, the discarded instances do not impact the modeling effectiveness in a significant way.

11.3.2.1 Relationship Between Weighting and Sampling

Resampling methods can be understood as methods that sample the data in proportion to their weights, and then treat all examples equally. Therefore, the two methods are almost equivalent although sampling methods have greater randomness associated with them. A direct weight-based technique is generally more reliable because of the absence of this randomness. On the other hand, sampling can be more naturally combined with ensemble methods (cf. Sect. 11.8) such as bagging to improve accuracy. Furthermore, sampling has distinct *efficiency* advantages because it works with a much smaller data set. For example, for a data set containing a rare to normal ratio of 1:99, it is possible for a resampling technique to work effectively with 2 % of the original data when the data is resampled into an equal mixture of the normal and anomalous classes. This kind of resampling translates to a performance improvement of a factor of 50.

11.3.2.2 Synthetic Oversampling: SMOTE

One of the problems with oversampling the minority class is that a larger number of samples with replacement leads to repeated samples of the same data point. Repeated samples cause overfitting and reduce classification accuracy. In order to address this issue, a recent approach is to use synthetic oversampling that creates synthetic examples without repetition.

The *SMOTE* approach works as follows. For each minority instance, its k nearest neighbors belonging to the same class are determined. Then, depending on the level of oversampling required, a fraction of them are chosen randomly. For each sampled example-neighbor pair, a synthetic data example is generated on the line segment connecting that minority example to its nearest neighbor. The exact position of the example is chosen uniformly at random along the line segment. These new minority examples are added to the training data, and the classifier is trained with the augmented data. The *SMOTE* algorithm is generally more accurate than a vanilla oversampling approach. This approach forces the decision region of the resampled data to become more general than one in which only members from the rare classes in the *original* training data are oversampled.

11.4 Scalable Classification

In many applications, the training data sizes are rather large. This leads to numerous scalability challenges in building classification models. In such cases, the data will typically not fit in main memory, and therefore the algorithms need to be designed to optimize the accesses to disk. Although the traditional decision-tree algorithms, such as *C4.5*, work well for smaller data sets, they are not optimized to disk-resident data. One solution is to sample the training data, but this has the disadvantage of losing the learning knowledge in the discarded training instances. Some classifiers, such as associative classifiers and nearest-neighbor methods, can be made faster by using more efficient subroutines for frequent pattern mining and nearest-neighbor indexing, respectively. Other classifiers, such as decision trees and support vector machines, require more careful redesign because they do not rely on any specific computationally intensive subroutines. These two classifiers are also particularly popular, and each of them is used widely in various data domains. Therefore, this chapter will specifically focus on these two classifiers in the context of scalability. An additional scalability challenge is created by *streaming* data, although such algorithms are not discussed in this chapter. The discussion of streaming data is deferred to Chap. 12.

11.4.1 Scalable Decision Trees

The construction of a decision tree can be computationally expensive because the evaluation of a split criterion at a node can sometimes be very slow. In the following, we will discuss two well-known methods for scalable decision tree construction.

11.4.1.1 RainForest

The *RainForest* approach is based on the insight that the evaluation of the split criteria in univariate decision trees do not need access to the data in its multidimensional form. Because each attribute value is analyzed independently in a univariate split, only the *count statistics* of distinct attributes values need to be maintained over different classes. For numeric data, it is assumed that they are discretized into categorical attribute values. The count statistics are collectively referred to as the *AVC-set*. The AVC-set is specific to a decision-tree node, and provides the counts of the distinct values of the attribute in the data records relevant to that node for different classes. Therefore, the size of the AVC-set depends *only* on the number of distinct attribute values and the number of classes. This size is often extremely small in comparison to the number of data records. Therefore, the memory requirement is dependent on the dimensionality of the data, the number of distinct values per dimension, and the number of classes. The larger the base training data set, the greater the proportional savings.

These AVC-sets are stored in main memory and used for efficiently evaluating the split criteria at the nodes. The splits are performed at nodes, until the AVC-sets no longer fit in main memory. The data does need to be scanned when the AVC-sets are constructed for newly created nodes. By carefully interleaving the splits and the AVC-set construction, significant computational and disk-access savings can be achieved.

11.4.1.2 BOAT

The *Bootstrapped Optimistic Algorithm for Tree construction (BOAT)* algorithm uses *bootstrapped samples* for decision-tree construction. In bootstrapping, the data is sampled with replacement to create b different bootstrapped samples. These are used to create b different trees denoted by $T_1 \dots T_b$. Then, it is checked whether the choice of the split attributes and the splitting subsets are identical, at a particular node in the different bootstrapped trees. For nodes where this is not the case, they are deleted along with the corresponding subtrees. The bootstrapping is used to create an *information-coarse splitting criterion* where a confidence interval is imposed on the numeric attribute at each node. The width of this confidence interval can be controlled with the number of bootstrapped samples. At a later stage of the algorithm, the coarse splitting criterion is converted to an exact one by integrating the various confidence intervals of the splits into a crisp criterion. In effect, *BOAT* uses the trees $T_1 \ldots T_b$ to create a new tree that is very close to one that would have been constructed, even if all the data had been available. The *BOAT* algorithm is faster than *RainForest*, and it requires only two scans over the database. Furthermore, *BOAT* also has the capability of performing incremental decision tree induction and can also handle tuple deletions.

11.4.2 Scalable Support Vector Machines

A major problem with support vector machines is that the size of the optimization problem scales with the number of training data points, and that the memory requirements may scale with the *square* of the number of data points in the case of kernel-based support vector machines. For example, consider the optimization problem for SVM discussed in Sect. 10.6 of Chap. 10. The kernel-based Lagrangian dual of the problem, as adapted from Eq. 10.62 in Chap. 10, may be written as follows:

$$L_D = \sum_{i=1}^{n} \lambda_i - \frac{1}{2} \sum_{i=1}^{n} \sum_{j=1}^{n} \lambda_i \lambda_j y_i y_j K(\overline{X_i}, \overline{X_j}). \tag{11.1}$$

The number of Lagrangian parameters λ_i (or optimization variables) is equal to the number of training data points n, and the size of the kernel matrix $K(\overline{X_i}, \overline{X_j})$ is $O(n^2)$. As a result, the coefficients of the entire optimization problem cannot even be loaded in main memory for large values of n. The *SVMLight* approach is designed to address this issue. This approach is mainly based on the following two observations:

1. It is not necessary to solve the entire problem at one time. A subset (or *working set*) of the variables $\lambda_1 \ldots \lambda_n$ may be selected for optimization at a given time. Different working sets are selected and optimized iteratively to arrive at the global optimal solution.

2. The support vectors for the SVMs correspond to only a small number of training data points. Even if most of the other training data points were removed, it would have no impact on the decision boundary of the SVM. Therefore, the *early* identification of such data points during the computationally intensive training process is crucial for efficiency maximization.

The following observations discuss how each of the aforementioned observations may be leveraged. In the case of the first observation, an iterative approach is used, in which the set of variables of the optimization problem are improved iteratively by fixing the majority of the variables to their current value, and improving only a small working set of the variables. Note that the size of the *relevant* kernel matrix within each local optimization scales with the square of the size q of the working set S_q, rather than the number n of training points. The *SVMLight* algorithm repeatedly executes the following two iterative steps until global optimality conditions are satisfied:

1. Select q variables as the active working set S_q, and fix the remaining $n - q$ variables to their current value.

2. Solve $L_D(S_q)$, a smaller optimization subproblem, with only q variables.

A key issue is how the working set of size q may be identified in each iteration. Ideally, it is desired to select a working set for which the maximum improvement in the objective function is achieved. Let $\overline{V}$ be a vector with length equal to the number of Lagrangian variables and at most q nonzero elements. The goal is to determine the optimal choice for the q nonzero elements to determine the working set. An optimization problem is set up for determining $\overline{V}$ in which the dot product of $\overline{V}$ with the gradient of L_D (with respect to the Lagrangian variables) is optimized. This is a separate optimization problem that needs to be solved in each iteration to determine the optimal working set.

The second idea for speeding up support vector machines is that of *shrinking* the training data. In the support vector machine formulation, the focus is primarily on the decision boundary. Training examples that are on the correct size of the margin, and far away from it, have no impact on the solution to the optimization problem, even if they are removed. The early identification of these training examples is required during the optimization process to benefit as much as possible from their removal. A heuristic approach, based on the Lagrangian multiplier estimates, is used in the *SVMLight* approach. The specific details of determining these training examples are beyond the scope of this book but pointers are provided in the bibliographic notes. Another later approach, known as *SVMPerf*, shows how to achieve linear scale-up, but for the case of the linear model only. For some domains, such as text, the linear model works quite well in practice. Furthermore, the *SVMPerf* method has $O(s \cdot n)$ complexity where s is the number of *nonzero features*, and n is the number of training examples. In cases where $s \ll d$, such a classifier is very effective. This is the case for sparse high-dimensional domains such as text and market basket data. Therefore, this approach will be described in Sect. 13.5.3 of Chap. 13 on text data.

11.5 Regression Modeling with Numeric Classes

In many applications, the class variables are numerical. In this case, the goal is to minimize the squared error of prediction of the numeric class variable. This variable is also referred to as the *response variable*, *dependent variable*, or *regressand*. The feature variables are referred to as *explanatory variables*, *input variables*, *predictor variables*, *independent variables*, or *regressors*. The prediction process is referred to as *regression modeling*. This section will discuss a number of such regression modeling algorithms.

11.5.1 Linear Regression

Let D be an $n \times d$ data matrix whose ith data point (row) is the d-dimensional input feature vector $\overline{X_i}$, and the corresponding response variable is y_i. Let the n-dimensional column-vector of response variables be denoted by $\overline{y} = (y_1, \ldots y_n)^T$. In linear regression, the dependence of each response variable y_i on the corresponding independent variables $\overline{X_i}$ is modeled in the form of a linear relationship:

$$y_i \approx \overline{W} \cdot \overline{X_i} \quad \forall i \in \{1 \ldots n\}. \tag{11.2}$$

Here, $\overline{W} = (w_1 \ldots w_d)$ is a d-dimensional row vector of coefficients that needs to be learned from the training data so as to minimize the unexplained error $\sum_{i=1}^{n}(\overline{W} \cdot \overline{X_i} - y_i)^2$ of modeling. The response values of test instances can be predicted with this linear relationship. Note that a constant (bias) term is not needed on the right-hand side, because we can append an artificial dimension[1] with a value of 1 to each data point to include the constant term within $\overline{W}$. Alternatively, instead of using an artificial dimension, one can mean-center the

[1]Here, we assume that the total number of dimensions is d, including the artificial column.

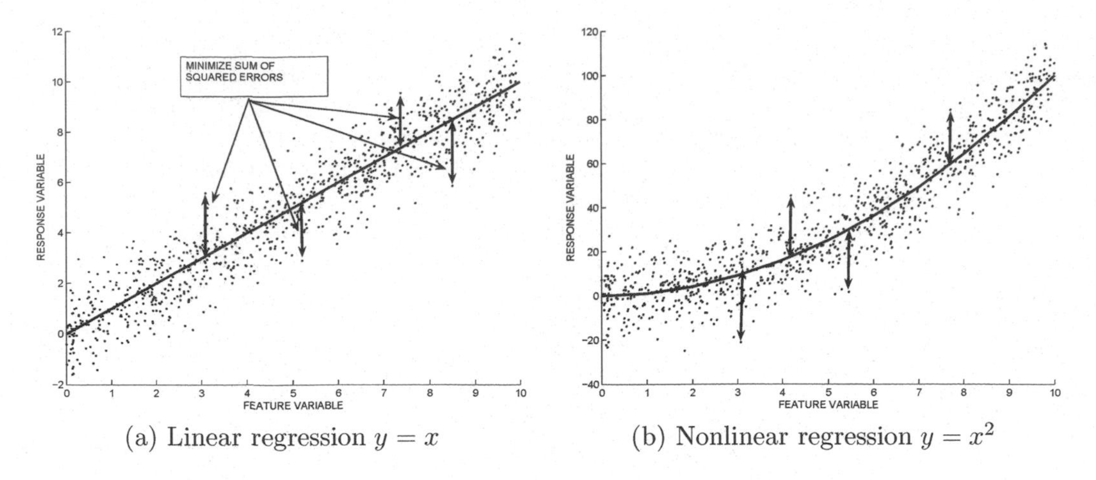

(a) Linear regression $y = x$

(b) Nonlinear regression $y = x^2$

Figure 11.1: Examples of linear and nonlinear regression

data matrix and the response variable. In such a case, it can be shown that the bias term is not necessary (see Exercise 8). Furthermore, the standard deviations of all columns of the data matrix, except for the artificial column, are assumed to have been scaled to 1. In general, it is common to *standardize* the data in this way to ensure similar scaling and weighting for all attributes. An example of a linear relationship for a 1-dimensional feature variable is illustrated in Fig. 11.1a.

To minimize the squared-error of prediction on the training data, one must determine $\overline{W}$ that minimizes the following objective function O:

$$O = \sum_{i=1}^{n} (\overline{W} \cdot \overline{X_i} - y_i)^2 = ||D\overline{W}^T - \overline{y}||^2. \tag{11.3}$$

Using[2] matrix calculus, the gradient of O with respect to $\overline{W}$ can be shown to be the d-dimensional vector $2D^T(D\overline{W}^T - \overline{y})$. Setting the gradient to 0 yields the following d-dimensional vector of optimization conditions:

$$D^T D\overline{W}^T = D^T\overline{y}. \tag{11.4}$$

If the symmetric matrix $D^T D$ is invertible, then the solution for $\overline{W}$ can be derived from the aforementioned condition as $\overline{W}^T = (D^T D)^{-1} D^T \overline{y}$. The numerical class value of a previously unseen test instance $\overline{T}$ can then be predicted as the dot product between $\overline{W}$ and $\overline{T}$.

It is noteworthy that the matrix $(D^T D)^{-1} D^T$ is also referred to as the *Moore–Penrose pseudoinverse* D^+ of the matrix D. Therefore, the solution to linear regression can also be expressed as $D^+\overline{y}$. The pseudoinverse is more generally defined even for the case where $D^T D$ is not invertible:

$$D^+ = \lim_{\delta \to 0}(D^T D + \delta^2 I)^{-1} D^T. \tag{11.5}$$

[2] Excluding constant terms, the objective function $O = (D\overline{W}^T - \overline{y})^T(D\overline{W}^T - \overline{y})$ can be expanded to the two additive terms $\overline{W}D^T D\overline{W}^T$ and $-(\overline{W}D^T\overline{y} + \overline{y}^T D\overline{W}^T) = -2\overline{W}D^T\overline{y}$. The gradients of these terms are $2D^T D\overline{W}^T$ and $-2D^T\overline{y}$, respectively. In the event that the Tikhonov regularization term $\lambda||\overline{W}||^2$ is added to the objective function, an additional term of $2\lambda\overline{W}^T$ will appear in the gradient.

Here, I is a $d \times d$ identity matrix. When the number of training data points is small, all the training examples might lie on a hyperplane with dimensionality less than d. As a result, the $d \times d$ matrix $D^T D$ is not of full rank and therefore not invertible. In other words, the system of equations $D^T D\overline{W}^T = D^T \overline{y}$ is *underdetermined* and has infinitely many solutions. In this case, the general definition of the Moore–Penrose pseudoinverse in Eq. 11.5 is useful. Even though the inverse of $D^T D$ does not exist, the limit of Eq. 11.5 can still be computed. It is possible to compute D^+ using the *SVD* of D (cf. Sect. 2.4.3.4 of Chap. 2). More efficient computation methods are possible with the use of the following matrix identity (see Exercise 15):

$$D^+ = (D^T D)^+ D^T = D^T (DD^T)^+. \tag{11.6}$$

This identity is useful when $d \ll n$ or $n \ll d$. Here, we will show only the case where $d \ll n$ because the other case is very similar. The first step to diagonalize the $d \times d$ symmetric matrix $D^T D$:

$$D^T D = P \Lambda P^T. \tag{11.7}$$

The columns of P are the orthonormal eigenvectors of $D^T D$ and Λ is a diagonal matrix containing the eigenvalues. When the matrix $D^T D$ is of rank $k < d$, the pseudoinverse $(D^T D)^+$ of $D^T D$ is computed as follows:

$$(D^T D)^+ = P \Lambda^+ P^T. \tag{11.8}$$

Λ^+_{ii} is derived from Λ by setting it to $1/\Lambda_{ii}$ for the k nonzero entries, and 0, otherwise. Then, the solution for $\overline{W}$ is defined as follows:

$$\overline{W}^T = (D^T D)^+ D^T \overline{y}. \tag{11.9}$$

Even though the underdetermined system of equations $D^T D\overline{W}^T = D^T \overline{y}$ has infinitely many solutions, the pseudoinverse always provides a solution $\overline{W}$ with the smallest L_2-norm $||\overline{W}||$ among the alternatives. Smaller coefficients are desirable because they reduce overfitting. Overfitting is a significant problem in general, and especially so when the matrix $D^T D$ is not of full rank. A more effective approach is to use *Tikhonov regularization* or *Lasso*. In Tikhonov regularization, also known as *ridge regression*, a penalty term $\lambda ||\overline{W}||^2$ is added to the objective function O of Eq. 11.3, where $\lambda > 0$ is a regularization parameter. In that case, the solution for $\overline{W}^T$ becomes $(D^T D + \lambda I)^{-1} D^T \overline{y}$, where I is a $d \times d$ identity matrix. The matrix $(D^T D + \lambda I)$ can be shown to be always positive-definite and therefore invertible. The compact solution provided by the Moore–Penrose pseudoinverse is a special case of Tikhonov regularization in which λ is infinitesimally small (i.e., $\lambda \rightarrow 0$). In general, the value of λ should be selected adaptively by cross-validation. In *Lasso*, an L_1-penalty, $\lambda \sum_{i=1}^{d} |w_i|$, is used instead of the L_2-penalty term. The resulting problem does not have a closed form solution, and it is solved using iterative techniques such as *proximal gradient methods* and *coordinate descent* [256]. *Lasso* has the tendency to select sparse solutions (i.e., few nonzero components) for $\overline{W}$, and it is particularly effective for high-dimensional data with many irrelevant features. *Lasso* can also be viewed as an embedded model (cf. Sect. 10.2 of Chap. 10) for feature selection because features with zero coefficients are effectively discarded. The main advantage of *Lasso* over ridge regression is not necessarily one of performance, but that of its highly interpretable feature selection.

Although the use of a penalty term for regularization might seem arbitrary, it creates stability by discouraging very large coefficients. By penalizing all regression coefficients, noisy features are often deemphasized to a greater degree. A common manifestation of overfitting in linear regression is that the additive contribution of a large coefficient to $\overline{W} \cdot \overline{X}$ may be frequently canceled by another large coefficient in a small training data set. Such features might be noisy. This situation can lead to inaccurate predictions over unseen test instances because the response predictions are very sensitive to small perturbations in feature values. Regularization prevents this situation by penalizing large coefficients. Bayesian interpretations also exist for these regularization methods. For example, Tikhonov regularization assumes Gaussian priors on the parameters $\overline{W}$ and class variable. Such assumptions are helpful in obtaining a unique and probabilistically interpretable solution when the available training data is limited.

11.5.1.1 Relationship with Fisher's Linear Discriminant

Fisher's linear discriminant for binary classes (cf. Sect. 10.2.1.4 of Chap. 10) can be shown to be a special case of least-squares regression. Consider a problem with two classes, in which the two classes 0 and 1 contain a fraction p_0 and p_1, respectively, of the n data points. Assume that the d-dimensional mean vectors of the two classes are $\overline{\mu_0}$ and $\overline{\mu_1}$, and the covariance matrices are Σ_0 and Σ_1, respectively. Furthermore, it is assumed that the data matrix D is mean-centered. The response variables $\overline{y}$ are set to $-1/p_0$ for class 0 and $+1/p_1$ for class 1. Note that the response variables are also mean-centered as a result. Let us now examine the solution for $\overline{W}$ obtained by least-squares regression. The term $D^T\overline{y}$ is proportional to $\overline{\mu_1}^T - \overline{\mu_0}^T$, because the value of $\overline{y}$ is $-1/p_0$ for a fraction p_0 of the data records belonging to class 0, and it is equal to $1/p_1$ for a fraction p_1 of the data records belonging to class 1. In other words, we have:

$$\begin{aligned}(D^T D)\overline{W}^T &= D^T\overline{y} \\ &\propto \overline{\mu_1}^T - \overline{\mu_0}^T.\end{aligned}$$

For mean-centered data, $\frac{D^T D}{n}$ is equal to the covariance matrix. It can be shown using some simple algebra (see Exercise 21 of Chap. 10) that the covariance matrix is equal to $S_w + p_0 p_1 S_b$, where $S_w = (p_0\Sigma_0 + p_1\Sigma_1)$ and $S_b = (\overline{\mu_1} - \overline{\mu_0})^T(\overline{\mu_1} - \overline{\mu_0})$ are the (scaled) $d \times d$ within-class and between-class scatter matrices, respectively. Therefore, we have:

$$(S_w + p_0 p_1 S_b)\overline{W}^T \propto \overline{\mu_1}^T - \overline{\mu_0}^T. \quad (11.10)$$

Furthermore, the vector $S_b\overline{W}^T$ always points in the direction $\overline{\mu_1}^T - \overline{\mu_0}^T$ because $S_b\overline{W}^T = (\overline{\mu_1}^T - \overline{\mu_0}^T)\left[(\overline{\mu_1} - \overline{\mu_0})\overline{W}^T\right]$. This implies that we can drop the term involving S_b from Eq. 11.10 without affecting the constant of proportionality:

$$\begin{aligned}S_w\overline{W}^T &\propto (\overline{\mu_1}^T - \overline{\mu_0}^T) \\ (p_0\Sigma_0 + p_1\Sigma_1)\overline{W}^T &\propto (\overline{\mu_1}^T - \overline{\mu_0}^T) \\ \overline{W}^T &\propto (p_0\Sigma_0 + p_1\Sigma_1)^{-1}(\overline{\mu_1}^T - \overline{\mu_0}^T).\end{aligned}$$

It is easy to see that the vector $\overline{W}$ is the same as the Fisher's linear discriminant of Sect. 10.2.1.4 in Chap. 10.

11.5.2 Principal Component Regression

Because overfitting is caused by the large number of parameters in $\overline{W}$, a natural approach is to work with a reduced dimensionality data matrix. In principal component regression, the largest $k \ll d$ principal components of the input data matrix D (cf. Sect. 2.4.3.1 of Chap. 2) with nonzero eigenvalues are determined. These principal components are the top-k eigenvectors of the $d \times d$ covariance matrix of D. Let the top-k eigenvectors be arranged in matrix form as the orthonormal columns of the $d \times k$ matrix P_k. The original $n \times d$ data matrix D is transformed to a new $n \times k$ data matrix $R = DP_k$. The new derived set of k-dimensional input variables $\overline{Z_1} \ldots \overline{Z_n}$, which are rows of R, are used as training data to learn a reduced k-dimensional set of coefficients $\overline{W}$:

$$y_i \approx \overline{W} \cdot \overline{Z_i}. \tag{11.11}$$

In this case, the k-dimensional vector of regression coefficients $\overline{W}$ can be expressed in terms of R as $(R^T R)^{-1} R^T \overline{y}$. This solution is identical to the previous case, except that a smaller and full-rank $k \times k$ matrix $R^T R$ is inverted. Prediction on a test instance $\overline{T}$ is performed after transforming it to this new k-dimensional space as $\overline{T}P_k$. The dot product between $\overline{T}P_k$ and $\overline{W}$ provides the numerical prediction of the test instance. The effectiveness of principal component regression is because of the discarding of the low-variance dimensions, which are either redundant directions (zero eigenvalues) or noisy directions (very small eigenvalues). If all directions are included after *PCA*-based axis rotation (i.e., $k = d$), then the approach will yield the same results as linear regression on the original data. It is common to standardize the data matrix D to zero mean and unit variance before performing *PCA*. In such cases, the test instances also need to be scaled and translated in an identical way.

11.5.3 Generalized Linear Models

The implicit assumption in linear models is that a constant change in the ith feature variable leads to a constant change in the response variable, which is proportional to w_i. However, such assumptions are inappropriate in many settings. For example, if the response variable is the height of a person, and the feature variable is the age, the height is not expected to vary linearly with age. Furthermore, the model needs to account for the fact that such variables can never be negative. In other cases, such as customer ratings, the response variables might take on integer values from a bounded range. Nevertheless, the elegant simplicity of linear models can still be leveraged in these settings. In generalized linear models (*GLM*), each response variable y_i is modeled as an *outcome* of a (typically exponential) probability distribution with mean $f(\overline{W} \cdot \overline{X_i})$ as follows:

$$y_i \sim \text{Probability distribution with mean } f(\overline{W} \cdot \overline{X_i}) \quad \forall i \in \{1 \ldots n\}. \tag{11.12}$$

This function $f(\cdot)$ is referred to as the *mean function*, and its inverse $f^{-1}(\cdot)$ is referred to as the *link function*. Although the same mean/link function can be used with different probability distributions, the selected mean/link functions and probability distributions are usually paired carefully to maximize effectiveness and interpretability of the model. If the observed responses are discrete (e.g., binary), it is possible to use a discrete probability distribution for y_i (e.g., Bernoulli), as long as its mean is $f(\overline{W} \cdot \overline{X_i})$. An example of this scenario is logistic regression. Some common examples of mean functions with their associated probability distribution assumptions are illustrated in the table below:

Link function	Mean function	Distribution assumption
Identity	$\overline{W} \cdot \overline{X}$	Normal
Inverse	$-1/(\overline{W} \cdot \overline{X})$	Exponential, Gamma
Log	$\exp(\overline{W} \cdot \overline{X})$	Poisson
Logit	$1/[1 + \exp(-\overline{W} \cdot \overline{X})]$	Bernoulli, Categorical
Probit	$\Phi(\overline{W} \cdot \overline{X})$	Bernoulli, Categorical

The link function regulates the nature of the response variable and its usability in a specific application. For example, the log, logit, and probit link functions are typically used to model the relative frequency of a discrete or categorical outcome. Because of the probabilistic modeling of the response variable, a maximum likelihood approach is used to determine the optimal parameter set $\overline{W}$, where the product of the probabilities (or probability densities) of the response variable outcomes is maximized. After estimating the parameters in $\overline{W}$, the expected response value of a test instance $\overline{T}$ is estimated as $f(\overline{W} \cdot \overline{T})$. Furthermore, the probability distribution of the response variable (with mean $f(\overline{W} \cdot \overline{T})$) may be used for detailed analysis.

An important special case of *GLM* is least-squares regression. In this case, the probability distribution of the response y_i is the normal distribution with mean $f(\overline{W} \cdot \overline{X_i}) = \overline{W} \cdot \overline{X_i}$ and constant variance σ^2. The relationship $f(\overline{W} \cdot \overline{X_i}) = \overline{W} \cdot \overline{X_i}$ follows from the fact that the link function is the identity function. The likelihood of the training data is as follows:

$$\begin{aligned}\text{Likelihood}(\{y_1 \dots y_n\}) = \prod_{i=1}^{n} \text{Probability}(y_i) &= \prod_{i=1}^{n} \frac{1}{\sqrt{2\pi}\sigma} \exp\left(-\frac{(y_i - f(\overline{W} \cdot \overline{X_i}))^2}{2\sigma^2}\right) \\ &= \prod_{i=1}^{n} \frac{1}{\sqrt{2\pi}\sigma} \exp\left(-\frac{(y_i - \overline{W} \cdot \overline{X_i})^2}{2\sigma^2}\right) \\ &\propto \exp\left(-\frac{\sum_{i=1}^{n}(y_i - \overline{W} \cdot \overline{X_i})^2}{2\sigma^2}\right).\end{aligned}$$

In this special case, the maximum likelihood approach can be shown to be equivalent to the least-squares approach because the logarithm of the likelihood yields the scaled objective function of linear regression. Another specific example of the process of maximum likelihood estimation with the logit function and Bernoulli distribution is discussed in detail in Sect. 10.6 of Chap. 10. In this case, the *discrete* binary variable y_i is modeled from a Bernoulli distribution with mean function $f(\overline{W} \cdot \overline{X_i}) = 1/[1 + \exp(-\overline{W} \cdot \overline{X_i})]$:

$$y_i = \begin{cases} 1 & \text{with probability } 1/[1 + \exp(-\overline{W} \cdot \overline{X_i})] \\ 0 & \text{with probability } 1/[1 + \exp(\overline{W} \cdot \overline{X_i})]. \end{cases} \tag{11.13}$$

Note that[3] the mean of y_i still satisfies the mean function according to the table above. This special case of GLMs is referred to as *logistic regression.* Logistic regression can also be used for k-way categorical response values. In that case, a k-way categorical distribution is used, and its mean function maps to a k-dimensional vector to represent each outcome of the categorical variable. An added restriction is that the components of the k-dimensional vector must add to 1. Probit regression is a sister family of models to logit regression, in which the cumulative density function (CDF) $\Phi(\cdot)$ of a standard normal distribution

[3]A slightly different convention of $y_i \in \{-1, +1\}$ is used in Chap. 10 for notational convenience. In that case, the mean function would need to be adjusted to $\frac{1-\exp(-\overline{W} \cdot \overline{X})}{1+\exp(-\overline{W} \cdot \overline{X})}$.

is used instead of the logit function. *Ordered* probit regression can model ordered integer values within a range (e.g., ratings) for the response variable by using the quantiles of a standard normal distribution. The key insight of *GLM* is to choose the link function and distribution assumption judiciously depending on the nature of the observed response in a specific application. Generalized linear models can be viewed as a unification of large classes of regression models, such as linear regression, logistic regression, probit regression, and Poisson regression.

11.5.4 Nonlinear and Polynomial Regression

Linear regression cannot capture nonlinear relationships such as those in Fig. 11.1b. The basic linear regression approach can be used for nonlinear regression by using *derived input features*. For example, consider a new set of m features denoted by $h_1(\overline{X_j}) \ldots h_m(\overline{X_j})$ for the jth data point. Here, $h_i(\cdot)$ represents a nonlinear transformation function from the d-dimensional input feature space to 1-dimensional space. This results in a new $n \times m$ input data matrix. By applying linear regression on this derived data matrix, one is able to model relationships of the following form:

$$y = \sum_{i=1}^{m} w_i h_i(\overline{X}). \tag{11.14}$$

For example, in *polynomial regression*, the higher powers of each dimension up to order r are used as a new set of derived features. This approach expands the number of dimensions by a factor of r, but it allows greater expressiveness in terms of nonlinear relationships. The main disadvantage of the approach is that it expands the dimensionality of the parameter set $\overline{W}$, and can therefore result in overfitting. Therefore, it is important to use regularization.

Arbitrary nonlinear relationships can also be captured by methods such as *kernel ridge regression*. In order to use kernels, the main goal is to show that the closed-form solution to linear ridge regression can be expressed in terms of dot products between training and test instances. One way of achieving this goal is by formulating the dual of the linear ridge regression problem [448], and then using the kernel trick as in SVMs. A simpler approach is to make use of a specialized variant of the Sherman–Morrison–Woodbury identity in matrix algebra (see Exercise 14), which is true for any $n \times d$ data matrix D and scalar λ:

$$(D^T D + \lambda I_d)^{-1} D^T = D^T (DD^T + \lambda I_n)^{-1}. \tag{11.15}$$

Note that I_d is a $d \times d$ identity matrix, whereas I_n is an $n \times n$ identity matrix. For an unseen test instance $\overline{Z}$, which is expressed as a row vector, the prediction $F(\overline{Z})$ of linear regression is given by $\overline{Z}\,\overline{W}^T$. By substituting the closed-form solution of ridge regression for $\overline{W}^T$ and then making use of the aforementioned identity, we obtain:

$$F(\overline{Z}) = \overline{Z}(D^T D + \lambda I_d)^{-1} D^T \overline{y} = \overline{Z} D^T (DD^T + \lambda I_n)^{-1} \overline{y}. \tag{11.16}$$

Note that $\overline{Z} D^T$ is an n-dimensional row vector of dot products between the test instance $\overline{Z}$ and the n training instances. According to the kernel trick, we can replace this row vector with a vector $\overline{\kappa}$ containing the n kernel similarities between the test and training instances. Furthermore, the matrix DD^T contains the $n \times n$ dot products between the training instances. We can replace this matrix with the $n \times n$ kernel matrix K constructed on the training instances. Then, the prediction for test instance $\overline{Z}$ is as follows:

$$F(\overline{Z}) = \overline{\kappa}(K + \lambda I_n)^{-1} \overline{y}. \tag{11.17}$$

The kernel trick can also be applied to other variants of linear regression, such as Fisher's discriminant and logistic regression. The extension to Fisher's discriminant is straightforward because it is a special case of linear regression, whereas the derivation for kernel logistic regression uses the dual optimization formulation like SVMs.

11.5.5 From Decision Trees to Regression Trees

Regression trees are designed to model nonlinear relationships between the features and the response variable. If the regression model is constructed at each leaf node in a hierarchical partitioning of the data, locally optimized linear regression models can be obtained within each partition. Even when the relationship between the class variable and feature variables is nonlinear, a *local* linear approximation is quite effective. Each test instance can then be classified with its locally optimized linear regression model by determining its appropriate partition. This hierarchical partitioning is essentially a decision tree because the assigned partition of a test instance is determined by the split criteria at the internal nodes. The overall strategy of constructing a decision tree remains the same as in the case of categorical class variables. Similarly, the splits can use univariate (axis-parallel) splits on the feature variables, as in a traditional decision tree. However, changes need to be made to the splitting and pruning criteria because of the numeric class variable:

1. *Splitting criterion:* In the case of categorical classes, the splitting criterion uses the Gini index or entropy of the class variable as a qualitative measure to decide the splitting attribute. However, in the case of numeric classes, an error-based measure is used. The regression modeling approach of the previous section is applied to each child resulting from a potential split. The *aggregate* squared error of prediction of all the training data points in the different child nodes is computed. The split with the minimum aggregate squared error is selected among all possible splits at a particular node.

   The main *computational* problem with this approach is that a linear regression model needs to be constructed for *each possible* split. An alternative is to not use linear regression in the tree construction phase. The average variance of the numeric class variable in the children nodes resulting from a possible split is used as the quality criterion for split evaluation. In other words, the Gini index splitting criterion for the categorical class variable in traditional decision-tree construction is replaced with the variance of the numeric class variable. The linear regression models are constructed at the leaf nodes for prediction only *after* the entire tree has already been constructed. While this approach will result in larger trees, it is more practical from a computational point of view.

2. *Pruning criterion:* To minimize overfitting, a portion of the training data is not used for constructing the decision tree. This training data is then used for evaluating the squared error of prediction of the decision tree. A similar post-pruning strategy is used as the case of categorical class variables. Leaf nodes are iteratively removed if their removal improves accuracy on the validation set, until no more nodes can be removed.

The main drawback of this approach is that overfitting of the linear regression model is a real possibility when leaf nodes do not contain enough data. Therefore, a sufficient amount of training data is required to begin with. In such cases, regression trees can be very powerful because they can model complex nonlinear relationships.

11.5.6 Assessing Model Effectiveness

The effectiveness of linear regression models can be evaluated with a measure known as the R^2*-statistic*, or the *coefficient of determination.* The term $SSE = \sum_{i=1}^{n}(y_i - g(\overline{X_i}))^2$ yields the sum-of-squared error of prediction of regression. Here, $g(\overline{X})$ represents the linear model used for regression. The squared error of the response variable about its mean (or *total sum of squares*) is $SST = \sum_{i=1}^{n}\left(y_i - \sum_{j=1}^{n}\frac{y_j}{n}\right)^2$. Then the fraction of unexplained variance is given by SSE/SST, and the R^2-statistic is as follows:

$$R^2 = 1 - \frac{SSE}{SST}. \tag{11.18}$$

This statistic always ranges between 0 and 1 for the case of linear models. Higher values are desirable. When the dimensionality is large, the adjusted R^2-statistic provides a more accurate measure:

$$R^2 = 1 - \frac{(n-d)}{(n-1)}\frac{SSE}{SST}. \tag{11.19}$$

The R^2-statistic is appropriate only for the case of linear models. For nonlinear models, it is possible for the R^2-statistic to be highly misleading or even negative. In such cases, one might directly use the SSE as a measure of the error.

11.6 Semisupervised Learning

In many applications, labeled data is expensive and hard to acquire. On the other hand, unlabeled data is often copiously available. It turns out that unlabeled data can be used to significantly improve the accuracy of many mining algorithms. Unlabeled data is useful because of the following two reasons:

1. Unlabeled data can be used to estimate the low-dimensional manifold structure of the data. The available variation in label distribution can then be extrapolated on this manifold structure.

2. Unlabeled data can be used to estimate the joint probability distribution of features. The joint probability distributions of features are useful for indirectly relating feature values to labels.

The two aforementioned points are closely related. To explain these points, we will use two examples. In Fig. 11.2, an example has been illustrated where only *two* labeled examples are available. Based only on this training data, a reasonable decision boundary is illustrated in Fig. 11.2a. Note that this is the best decision boundary that one can hope to find with the use of this limited training data. Portions of this decision boundary are in regions of the space where almost no feature values are available. Therefore, the decision boundaries in these regions may not reflect the class behavior of *unseen* test instances.

Now, suppose that a large number of unlabeled examples are added to the training data, as illustrated in Fig. 11.2b. Because of the addition of these unlabeled examples, it becomes immediately evident that the data is distributed along two manifolds, each of which contains one of the training examples. A key assumption here is that the class variables are likely to vary *smoothly* over dense regions of the space, but it may vary significantly over sparse regions of the space. This leads to a new decision boundary that takes the underlying feature correlations into account in addition to the labeled instances. In the particular example of

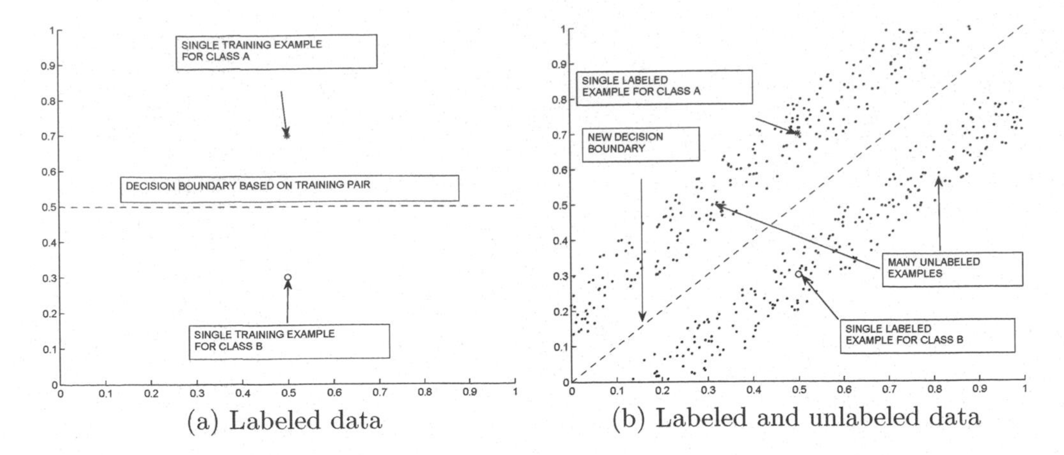

(a) Labeled data (b) Labeled and unlabeled data

Figure 11.2: Impact of unlabeled data on classification

Fig. 11.2a, if a test instance were provided near the coordinates $(1, 0.7)$ with only the original training data, then almost any classifier, such as the nearest-neighbor classifier, will assign the data points to class A. However, this prediction is not reliable because of few previously seen labeled examples in the locality of the test instance. However, the unlabeled examples could be used to *expand* the labeled examples appropriately, by incrementally labeling the unlabeled examples in each hyperplane of Fig. 11.2b with the appropriate class. At this point, it becomes evident that test instances near the coordinates $(1, 0.7)$ really belong to class B.

A different way of understanding the impact of feature correlation estimation is by examining the intuitively interpretable text domain. Consider a scenario where one were trying to determine whether documents belong to the "Science" category. It is possible, that not enough *labeled* documents may contain the word "Einstein" in the documents. However, the word "Einstein" may often co-occur with other (more common) words such as "Physics" in *unlabeled* documents. At the same time, these more common words may already have been associated with the "Science" category because of their presence in labeled documents. Thus, the unlabeled documents provide the insight that the word "Einstein" is also relevant to the "Science" category. This example shows that unlabeled data can be used to learn *joint* feature distributions that are very relevant to the classification process.

Many of the semisupervised methods are often termed as *transductive* because they cannot handle out-of-sample test instances. In other words, all test instances need to be specified at the time of constructing the training model. New out-of-sample instances cannot be classified after the model has been constructed. This is different from most of the *inductive classifiers* discussed in the previous chapter in which training and testing phases are cleanly separated.

There are two primary types of techniques that are used for semisupervised learning. Some of these methods are *meta-algorithms* that can use any existing classification algorithm as a subroutine, and leverage it to incorporate the impact of unlabeled data. The second type of methods are those in which a number of modifications are incorporated in specific classifiers to account for the impact of unlabeled data. Two examples of the second type of methods are semisupervised Bayes classifiers, and transductive support vector machines. This section will discuss both these classes of techniques.

11.6.1 Generic Meta-algorithms

The goal of generic meta-algorithms is to use existing classification algorithms to enhance the classification process with unlabeled data. The simplest method is *self-training*, in which the smoothness assumption is used to incrementally expand the labeled portions of the training data. The major drawback of this approach is that it might lead to overfitting. One way of avoiding overfitting is by using *co-training*. Co-training partitions the feature space and independently labels instances using classifiers trained on each of these feature spaces. The labeled instances from one classifier are used as feedback to the other, and vice versa.

11.6.1.1 Self-Training

The self-training procedure can use any existing classification algorithm $\mathcal{A}$ as input. The classifier $\mathcal{A}$ is used to incrementally assign labels to unlabeled examples for which it has the most confident prediction. As input, the self-training procedure uses the initial labeled set L, the unlabeled set U, and a user-defined parameter k that may sometimes be set to 1. The self-training procedure iteratively uses the following steps:

1. Use algorithm $\mathcal{A}$ on the current labeled set L to identify the k instances in the unlabeled data U for which the classifier $\mathcal{A}$ is the most confident.

2. Assign labels to the k most confidently predicted instances and add them to L. Remove these instances from U.

It is easy to see that the self-training procedure will work very well for the simple example of Fig. 11.2. However, in practice, the different classes may not be quite as cleanly separated. The major drawback of self-training is that the addition of predicted labels to the training data can lead to propagation of errors in the presence of noise. Another procedure, known as *co-training*, is able to avoid such overfitting more effectively.

11.6.1.2 Co-training

In co-training, it is assumed that the feature set can be partitioned into two *disjoint* groups F_1 and F_2, such that each of them is sufficient to learn the target classification function. It is important to select the two feature subsets so that they are as independent from one another as possible. Two classifiers are constructed, such that one classifier is constructed on each of these groups. These classifiers are not allowed to interact with one another directly for prediction of unlabeled examples though they are used to build up training sets for each other. This is the reason that the approach is referred to as *co-training*.

Let L be the labeled training data and U be the unlabeled data. Let L_1 and L_2 be the labeled sets for each of these classifiers. The sets L_1 and L_2 are initialized to the available labeled data L, except that they are represented in terms of disjoint feature sets F_1 and F_2, respectively. Over the course of the co-training process, as different examples from the initially unlabeled set U are added to L_1 and L_2, respectively, the training instances in L_1 and L_2 may vary from one another. Two classifier models $\mathcal{A}_1$ and $\mathcal{A}_2$ are constructed using the training sets L_1 and L_2, respectively. The following steps are then iteratively applied:

1. Train classifier $\mathcal{A}_1$ using labeled set L_1, and add k most confidently predicted instances from unlabeled set $U - L_2$ to training data set L_2 for classifier $\mathcal{A}_2$.

2. Train classifier $\mathcal{A}_2$ using labeled set L_2, and add k most confidently predicted instances from unlabeled set $U - L_1$ to training data set L_1 for classifier $\mathcal{A}_1$.

In many implementations of the method, the most confidently labeled examples *for each class* are added to the training sets of the other classifier. This procedure is repeated until all instances are labeled. The two classifiers are then retrained with the expanded training data sets. This approach can be used to label not only the unlabeled data set U, but also unseen test instances. At the end of the procedure, two classifiers are returned. For an unseen test instance, each classifier may be used to determine the class label scores. The score for a test instance is determined by combining the scores of the two classifiers. For example, if the Bayes method is used as the base classifier, then the product of the posterior probabilities returned by the two classifiers may be used.

The co-training approach is more robust to noise because of the disjoint feature sets used by the two algorithms. An important assumption is that of *conditional independence* of the features in the two sets with respect to a particular class. In other words, after the class label is fixed, the features in one subset are conditionally independent of the other. The intuition for this is that instances generated by one classifier appear to be randomly distributed to the other, and vice versa. As a result, the approach will generally be more robust to noise than the self-training method.

11.6.2 Specific Variations of Classification Algorithms

The algorithms in the previous section were designed as generic meta-algorithms that can use virtually any known classification algorithm $\mathcal{A}$ for semisupervised learning. A few methods have also been designed that rely on variations of other classification algorithms, such as variations of the Bayes classifier and support vector machines.

11.6.2.1 Semisupervised Bayes Classification with EM

An important observation is that both the EM-clustering algorithm (cf. Sect. 6.5 of Chap. 6) and the naive Bayes classifier (cf. Sect. 10.5.1 of Chap. 10) use the same generative mixture model, wherein examples from each cluster (class) are generated from a predefined distribution, such as the Bernoulli or the Gaussian. In the case of the naive Bayes classifier, the iterative approach of the EM-algorithm is not required because the class memberships of the training data are already fixed, which makes the E-step unnecessary. In the case of semisupervised classification, however, the unlabeled examples need to be assigned to classes in order to expand the training data. Therefore, the iterative approach of the EM-algorithm again becomes essential. Semisupervised Bayes classification can be viewed as a combination of EM clustering and the naive Bayes classifier.

This method was originally proposed in the context of text data, although this discussion will assume categorical data for simplicity. Note that the binary representation of text data may also be considered categorical data. The naive Bayes algorithm requires the estimation of the conditional probabilities of the feature values for each class. Specifically, Eq. 10.22 in Sect. 10.5.1 of Chap. 10 requires the estimation of $P(x_j = a_j|C = c)$. This expression represents the conditional probability of the feature value, given the class and is estimated from the training data. The estimation cannot be performed accurately, if the number of training examples is small. Consider the case of the text domain. If only five to ten labeled documents are available for a particular class, and x_j is a binary variable corresponding to the presence or absence of a particular word j, then this estimation cannot be performed robustly. As discussed earlier, the joint distribution of features with labeled and unlabeled data can be very helpful in this respect.

Intuitively, the idea is to use the EM clustering algorithm to determine the clusters of documents most similar to the labeled classes. A partially supervised EM clustering method associates each cluster with a particular class. The conditional feature distributions in these clusters are used as a more robust proxy for the feature distributions of the corresponding classes.

The basic idea is to use a generative model to create semisupervised clusters from the data. A one-to-one correspondence between the mixture components and the classes is retained in this case. The use of EM algorithms for clustering categorical data and its semisupervised variant are discussed in Sects. 7.2.3 and 7.5.1, respectively, of Chap. 7. The reader is advised to revisit these sections for the relevant background before reading further.

For initialization, the labeled examples are used as the seeds for the EM algorithm, and the number of mixture components is set to the number of classes. A Bayes classifier is used to assign documents to clusters (classes) in the E-step. In the first iteration, the Bayes classifier uses only the labeled data to determine the initial set of posterior cluster (class) membership probabilities, as in a standard Bayes classifier. This results in a set of "soft" clusters, in which the (unlabeled) data point $\overline{X}$ has a weight $w(\overline{X}, c)$ in the range $(0, 1)$ associated with each class c, corresponding to its posterior Bayes membership probability. Only labeled documents have binary weights that are either 0 or 1 for each class, depending on their fixed assignments. The value of $P(x_j = a_j|C = c)$ is now estimated using a weighted variant of Eq. 10.22 in Chap. 10 that leverages *both* the labeled *and* the unlabeled documents.

$$P(x_j = a_j|C = c) = \frac{\sum_{\overline{X} \in \mathcal{L} \cup \mathcal{U}} w(\overline{X}, c) I(x_j, a_j)}{\sum_{\overline{X} \in \mathcal{L} \cup \mathcal{U}} w(\overline{X}, c)} \tag{11.20}$$

Here, $I(x_j, a_j)$ is an indicator variable that takes on the value of 1, if the jth feature x_j of $\overline{X}$ is a_j, and 0 otherwise. The major difference from Eq. 10.22 is that the posterior Bayes estimates of unlabeled documents are *also* used to estimate class-conditional feature distributions. As in the standard Bayes method, the same Laplacian smoothing approach may be incorporated to reduce overfitting. The prior probabilities $P(C = c)$ for each cluster may also be estimated by computing the average assignment probability of the data points to the corresponding class. This is the M-step of the EM algorithm. The next E-step uses these modified values of $P(x_j = a_j|C = c)$ and the prior probability to derive the posterior Bayes probability with a standard Bayes classifier. Therefore, the Bayes classifier implicitly incorporates the impact of unlabeled data. The algorithm may be summarized by the following two iterative steps that are continually repeated to convergence:

1. (E-step) Estimate posterior probability of membership of data points to clusters (classes) using Bayes rule.

$$P(C = c|\overline{X}) \propto P(C = c) \prod_{j=1}^{d} P(x_j = a_j|C = c) \tag{11.21}$$

2. (M-step) Estimate conditional distribution of features for different clusters (classes), using the current estimated posterior probabilities (unlabeled data) and known memberships (labeled data) of data points to clusters (classes).

One challenge with the use of the approach is that the clustering structure may sometimes not correspond to the class distribution very well. In such cases, the use of unlabeled data can harm the classification accuracy, as the clusters found by the EM algorithm drift away from the true class structure. After all, unlabeled data are plentiful compared to labeled data, and therefore the estimation of $P(x_j = a_j|C = c)$ in Eq. 11.20 will be dominated by the unlabeled data. To ameliorate this effect, the labeled and unlabeled data are weighted differently during the estimation of $P(x_j = a_j|C = c)$. The unlabeled data are weighted down by a predefined discount factor $\mu < 1$ to ensure better correspondence between the clustering structure and the class distribution. In other words, the value of $w(\overline{X}, c)$ is multiplied with μ for only the unlabeled examples before estimating $P(x_j = a_j|C = c)$ in Eq. 11.20. The EM-approach for semisupervised classification is particularly remarkable because it demonstrates the link between semisupervised clustering and semisupervised classification, even though these two kinds of semisupervision are motivated by different application scenarios.

11.6.2.2 Transductive Support Vector Machines

The general assumption for most of the semisupervised methods is that the label values of unsupervised examples do not vary abruptly at densely populated regions of the data. In transductive support vector machines, this assumption is implicitly encoded by assigning labels to unsupervised examples that maximize the margin of the support vector machine. To understand this point, consider the example of Fig 11.2b. In this case, the margin of the SVM will be optimized only when the labels of the examples in the cluster containing the single example for class A, are also set to the same value A. The same is true for the unlabeled examples in the cluster containing the single label for class B. Therefore, the SVM formulation now needs to be modified to incorporate additional margin constraints, and binary decision variables for each unlabeled example. Recall from the discussion in Sect. 10.6 of Chap. 10 that the original SVM formulation was to minimize the objective function $\frac{||W||^2}{2} + C\sum_{i=1}^{n} \xi_i$, subject to the following constraints:

$$y_i(\overline{W} \cdot \overline{X_i} + b) \geq 1 - \xi_i \quad \forall i. \tag{11.22}$$

In addition, the nonnegativity constraint $\xi_i \geq 0$ on the slack variables is observed. Note that the value of y_i is *known*, because the training examples are labeled. For the case of unlabeled examples, binary decision *variables* $z_i \in \{-1, +1\}$ (with corresponding slack penalties) are incorporated for each *unlabeled* training example $\overline{X_i} \in \mathcal{U}$. These decision variables correspond to the assignment of the unlabeled examples to a particular class. The following constraint is added to the optimization problem:

$$z_i(\overline{W} \cdot \overline{X_i} + b) \geq 1 - \xi_i \quad \forall i : \overline{X_i} \in \mathcal{U}. \tag{11.23}$$

The slack penalties for the unlabeled examples can also be included in the optimization objective function. Note that, unlike y_i, the value of z_i is not known, and it is a binary *integer* variable that becomes a part of the optimization problem. Furthermore, the modified optimization formulation is an integer program, which is far more difficult than the original convex optimization problem for support vector machines.

A number of techniques have, therefore, been designed to approximately solve this problem with iterative mechanisms. One of these methods starts by labeling the most confidently predicted examples and iteratively expanding them. The number of positive examples initially labeled from the unlabeled instances, is based on the required trade-off between precision and recall. This ratio of positive to negative examples is maintained throughout the iterative algorithm. In each iteration, one positive example is changed to negative, and one negative example to positive to improve the soft margin of the classifier as much as possible. The bibliographic notes contain a discussion of the methods commonly used in this context.

11.6.3 Graph-Based Semisupervised Learning

The conversion of arbitrary data types to graphs is discussed in Sect. 2.2.2.9 of Chap. 2. Therefore, one advantage of this approach is that it can be used for semisupervised classification of arbitrary data types, as long as a distance function is available for quantifying proximity between data objects. This is a property that graph-based methods inherit from their origins in nearest-neighbor classification. The steps in graph-based semisupervised learning are as follows:

1. Construct a similarity graph on both the labeled and the unlabeled data records. Each data object O_i is associated with a node in the similarity graph. Each object is connected to its k-nearest neighbors.

2. The weight w_{ij} of the edge (i, j) is equal to a kernelized function of the distance $d(O_i, O_j)$ between the objects O_i and O_j, so that larger weights indicate greater similarity. A typical example of the weight is based on the *heat kernel* [90]:

$$w_{ij} = e^{-d(O_i,O_j)^2/t^2}. \tag{11.24}$$

   Here, t is a user-defined parameter.

This problem is one where we have a graph containing both labeled and unlabeled nodes. It is now desired to infer the labels of the unlabeled nodes with the use of these proximity relationships. This problem is exactly identical to the *collective classification problem* introduced in Sect. 19.4 of Chap. 19. Readers are advised to refer to the methods discussed in that section.

Graph-based semisupervised learning may be viewed as a semisupervised extension of nearest-neighbor classifiers. The only difference of graph-based semisupervised methods from nearest-neighbor classifiers is the way in which similarity graphs are constructed. Nearest-neighbor methods can be conceptually viewed as collective classification methods on similarity graphs in which edges are added only between pairs of labeled and unlabeled instances. Nearest-neighbor classification simply selects the dominant label from the labeled nodes incident on an unlabeled node. In the semisupervised case, edges can be added between any pair of nodes, whether they are labeled or unlabeled. The addition of these extra edges is necessary in semisupervised learning because of the scarcity of the labeled nodes in the similarity graph. Such edges are able to associate unlabeled clusters of arbitrary shape to their closest labeled instances more effectively. The reader is referred to Sect. 19.4 of Chap. 19 for discussions on collective classification.

11.6.4 Discussion of Semisupervised Learning

An important question in semisupervised learning is whether unlabeled data always helps in improving classification accuracy. Semisupervised learning depends on the inherent class structure of the underlying data. For semisupervised learning to be effective, the class structure of the data should approximately match its clustering structure. This assumption is obvious in the case of the semisupervised EM algorithm. The assumption is, however, implicitly used by other methods as well.

In practice, semisupervised learning is most effective when the number of labeled examples is extremely small, and there is no realistic way of making confident predictions about scarcely populated regions of the space. In some domains, such as node classification of graphs, this is almost always true. Therefore, in such domains, the transductive setting is the only way in which classification can be performed. These methods will be discussed in detail in Sect. 19.4 of Chap. 19. On the other hand, when a lot of labeled data is already available, then the unlabeled examples do not provide much advantage to the learner, and can, in fact, be harmful in some cases.

11.7 Active Learning

From the discussion in the previous section on semisupervised classification, it is evident that labeled data are often scarce in real applications. While labeled data are often expensive to obtain, the cost of procuring labeled data can often be quantified. Some examples of costly labeling mechanisms are as follows:

- *Document collections:* Large amounts of document data, which are usually unlabeled, are available on the Web. A common approach is to manually label the documents, which is a slow, painstaking, and laborious process. Alternatively, crowdsourcing mechanisms, such as Amazon Mechanical Turk, may be used. However, such mechanisms typically incur a dollar-cost on a per-instance basis.

- *Privacy-constrained data sets:* In many scenarios, the labels on records may be sensitive information that can be acquired at a significant query cost (e.g., obtaining permission from the relevant entity). In such cases, costs are harder to quantify explicitly, but can nevertheless be estimated through modeling.

- *Social networks:* In social networks, it may be desirable to identify nodes with specific properties. For example, an advertising company may desire to identify social network nodes that are interested in "cosmetics." However, it is rare that labels will be explicitly associated with the nodes. Identification of relevant nodes may require either manual examination of social network posts or user surveys. Both processes are time-consuming and costly.

It is clear from the aforementioned examples that the acquisition of labels should be viewed as a *cost-centric process* that helps improve modeling accuracy. The goal in active learning is to maximize the accuracy of classification at a specific cost of label acquisition. Therefore, active learning *integrates label acquisition and model construction.* This is different from all the other algorithms discussed in this book, where it is assumed that training data labels are already available.

Not all training examples are equally *informative.* To illustrate this point, consider the two-class problem illustrated in Fig. 11.3. The two classes, labeled by A and B, respectively, have a vertical decision boundary separating them. Suppose that the acquisition of labels is so costly that one is only allowed to acquire the labels of four examples from the entire data set and use this set of four examples to train a model. Clearly, this is a very small number of training examples, and the wrong choice of training examples may lead to significant overfitting. For example, in the case of Fig. 11.3a, the four examples have been randomly sampled from the data set. A typical linear classifier, such as logistic regression, may determine a decision boundary, corresponding to the dashed line in Fig. 11.3a. It is evident that this decision boundary is a poor representation of the true (vertical) decision boundary. On the other hand, in the case of Fig. 11.3b, the sampled examples are chosen more carefully to align along the true decision boundary. This set of labeled examples will result in a much better classification model for the entire data set. The goal in active learning is to integrate the labeling and classification process in a single framework to create

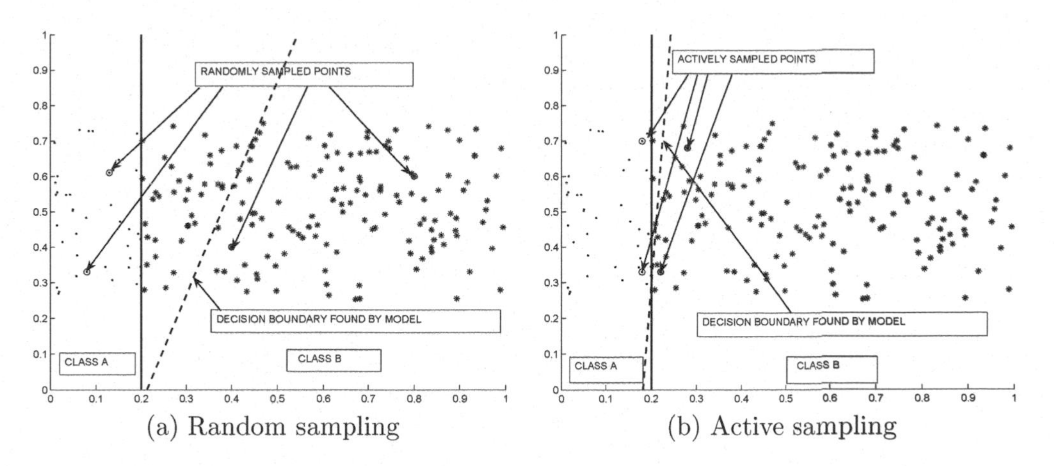

(a) Random sampling (b) Active sampling

Figure 11.3: Impact of active sampling on decision boundary

robust models. In practice, the determination of the correct choice of query instances is a very challenging problem. The key is to use the knowledge gained from the labels already acquired to "guess" the most informative regions in which to query the labels. Such an approach can help discover the true shape of the decision boundary as quickly as possible. Therefore, the key question in active learning is as follows:

How do we select instances to label to create the most accurate model at a given cost?

In some scenarios, the labeling cost may be instance-specific cost, although most models use the simplifying assumption of equal costs over all instances. Every active learning system has two primary components, one of which is already given:

1. *Oracle:* The oracle provides the responses to the underlying query in the form of labels of specified test instances. The oracle may be a human labeler, or a cost-driven data-acquisition system, such as Amazon Mechanical Turk. In general, for modeling purposes, the oracle is viewed as a black-box that is part of the input to the process.

2. *Query system:* The job of the query system is to pose queries to the oracle for labels of specific records. The querying strategy typically uses the distribution of currently known set of training instance labels to determine the most informative regions for querying.

The design of the query system may depend on the application at hand. For example, some query systems use *selective sampling*, in which a sequence of examples are presented to the user who makes a decision about whether or not to query them. The pool-based sampling approach assumes the availability of a base "pool" of instances from which to query the labels of data points. The task of the learner is to, therefore, determine (informative) instances one by one from this pool for querying.

The pool-based approach is the most common scenario for active learning, and will therefore be discussed in this chapter. The overall approach in the procedure is an iterative one. In each iteration, a number of interesting instances are identified, for which the addition of labels would be most informative for further classification. These are considered the "important" instances. The identification of the important instances is the job of the query-system, whereas the determination of the labels of queried instances is the job of the oracle, which, in some cases, might be a human expert. The iterative process is repeated until either the cost budget is exhausted or the classification accuracy no longer improves with further addition of labels.

It is evident that the crucial part of active learning is the choice of the querying strategy. How should this querying be performed? From the example of Fig. 11.3, it is evident that the most effective querying strategies can map out the *boundaries* of separation most clearly. Because the boundary regions often contain instances of multiple classes, they are characterized by class label uncertainty or disagreements between different learners about the class label. This is, of course, not always true because uncertain regions may sometimes contain unrepresentative outliers. Therefore, the various models work with different assumptions about the most appropriate methodology for identifying the most informative query points.

1. *Heterogeneity-based models:* These models attempt to sample regions of the space that are uncertain, heterogeneous, or dissimilar to what has already been seen so far. Examples of such models include *uncertainty sampling*, *query-by-committee*, and *expected model change*. These models are based on the assumption that regions near the decision boundary are more likely to be heterogeneous and instances in these regions are more valuable for learning the decision boundary.

2. *Performance-based models:* These models directly use performance measures of classifiers such as expected error or variance reduction. Therefore, these models quantify the impact of adding the queried instance to the classifier performance on remaining unlabeled instances.

3. *Representativeness-based models:* These models attempt to create data, that is as representative as possible, of the underlying population of training instances. For example, it may be desired that the density distribution of the queried instances matches that of the training data. However, a heterogeneity criterion is often retained within the query model.

In the following, a brief discussion of each of these different types of models is provided.

11.7.1 Heterogeneity-Based Models

The goal in these models is to determine regions of greatest heterogeneity. The typical approach is to use the current set of training labels to examine the classification uncertainty of unseen instances with respect to available labels. This heterogeneity may be quantified in various ways, such as by measuring the uncertainty of classification, dissimilarity with the current model, or disagreement between a committee of classifiers.

11.7.1.1 Uncertainty Sampling

In uncertainty sampling, the learner attempts to label those instances for which the value of the label is the least certain. For example, the posterior probability of a Bayes classifier may be used to quantify its uncertainty. The Bayes classifier is trained on instances whose labels are already available. A binary-label instance is deemed as uncertain when its posterior class probabilities are as close to 0.5 as possible. The corresponding criterion may be formalized as follows:

$$\text{Certain}(\overline{X}) = \sum_{i=1}^{k} ||p_i - 0.5||. \tag{11.25}$$

The value lies in the range $(0, 1)$, and lower values are indicative of greater uncertainty. In the multiclass scenario, a formal entropy measure may be used to quantify uncertainty. If the Bayes posterior probabilities of the k classes are $p_1 \ldots p_k$, respectively, based on the current set of labeled instances, then the entropy measure Entropy($\overline{X}$) is defined as follows:

$$\text{Entropy}(\overline{X}) = -\sum_{i=1}^{k} p_i \text{log}(p_i). \tag{11.26}$$

In this case, larger values of the entropy indicate greater uncertainty and are more desirable for label acquisition.

11.7.1.2 Query-by-Committee

In this case, the heterogeneity is measured in terms of the disagreement of different classifiers rather than the posterior probabilities of a single classifier over different labels. This criterion, however, tries to achieve the same intuitive goal, but in a different way. Intuitively, when the posterior probability of a Bayes classifier is the same across different classes, a significant disagreement may exist between different classification models about the predicted label. Therefore, this approach uses a committee of different classifiers that are trained on the current set of labeled instances. These classifiers are then used to predict the class label of each unlabeled instance. The instance for which the classifiers disagree the most is selected as the relevant one in this scenario.

At an intuitive level, the query-by-committee method achieves similar heterogeneity goals as the uncertainty sampling method. Different classifiers are more likely to disagree on the class label for instances near the true decision boundary. The mathematical formula for quantifying the disagreement is also the same as uncertainty sampling. In particular, the posterior probability p_i of each class i in Eq. 11.26 is replaced with the fraction of votes received by each class i. It is particularly beneficial to use diverse classifiers that use fundamentally different modeling methodologies.

11.7.1.3 Expected Model Change

In this approach, the instance with the greatest expected change from the current classification model by adding a particular instance to the training data is selected. In many optimization-based classification models, such as discriminative probabilistic models, the gradient of the model objective function with respect to the model parameters can be quantified. By adding a queried instance to the training data, the gradient will change as well. The instance with the greatest change in the gradient when the queried instance is added to set of labeled instances. The intuition is that such an instance is likely to be very different from the model constructed using already labeled instances. Let $\delta g_i(\overline{X})$ be the change in the gradient with respect to the model parameters, conditional on the fact that the correct training label of the candidate instance $\overline{X}$ is the ith class. In other words, if the current labeled training set is L and $\overline{\nabla G(L)}$ is the gradient of the objective function with respect to model parameters, we have:

$$\delta g_i(\overline{X}) = ||\overline{\nabla G(L \cup (\overline{X}, i))} - \overline{\nabla G(L)}||. \tag{11.27}$$

Of course, we do not yet know the training label of $\overline{X}$, but we can only estimate the posterior probability of each label with a Bayes classifier. Let p_i be the posterior probability of the class i with respect to the current label set of known labels in the training data. Then, the expected model change $C(\overline{X})$ with respect to the instance $\overline{X}$ is defined as follows:

$$C(\overline{X}) = \sum_{i=1}^{k} p_i \cdot \delta g_i(\overline{X}).$$

The instance $\overline{X}$ with the largest value of $C(\overline{X})$ is queried for the label.

11.7.2 Performance-Based Models

Although the motivation of heterogeneity-based models is that uncertain regions are the most informative by virtue of their proximity to decision boundaries, they have a drawback as well. Querying uncertain regions can inadvertently lead to the addition of unrepresentative outliers to the training data. Performance-based classifiers are therefore focused directly on the classification objective function. Therefore, these methods evaluate the accuracy of classification on the *remaining unlabeled instances.*

11.7.2.1 Expected Error Reduction

For the purpose of discussion, the remaining set of instances that has not yet been labeled is denoted by V. This set is used as the validation set on which the expected error reduction is computed. This approach is related to uncertainty sampling in a complementary way. Whereas uncertainty sampling *maximizes* the label uncertainty of the *queried* instance, the expected error reduction *minimizes* the expected label uncertainty of the *remaining* instances V when the queried instance is added to the training data. Thus, in the case of a binary-classification problem, the predicted posterior probabilities of the instances in V should be as far away from 0.5 as possible *after adding the queried instance.* The idea here is that greater certainty in prediction of class labels *of the remaining unlabeled instances*, will eventually result in a lower error rate on an unseen test set as well. Thus, error-reduction models can also be considered as *greatest certainty* models, except that the certainty criterion is applied to the instances in V rather than the query instance itself. Let $p_i(\overline{X})$ denote the posterior probability of the label i for the query candidate instance $\overline{X}$ with a Bayes model trained on the current set of labeled instances. Let $P_j^{(\overline{X},i)}(\overline{Z})$ be the posterior probability of class label j, when the instance-label combination $(\overline{X}, i)$ is added to the current set of labeled instances. Then, the error objective function $E(\overline{X}, V)$ for the *binary* class problem (i.e., $k = 2$) is defined as follows:

$$E(\overline{X}, V) = \sum_{i=1}^{k} p_i(\overline{X}) \left(\sum_{j=1}^{k} \sum_{\overline{Z} \in V} ||P_j^{(\overline{X},i)}(\overline{Z}) - 0.5|| \right). \quad (11.28)$$

The objective function can be interpreted as the expected label certainty of *remaining test instances.* Therefore, the objective function is maximized rather than minimized, as in the case of uncertainty-based models.

This result can easily be extended to the case of k-way models by using the same entropy criterion that was discussed for uncertainty-based models. In that case, the aforementioned expression is modified to replace $||P_j^{(\overline{X},i)}(\overline{Z}) - 0.5||$ with the class-specific entropy term $-P_j^{(\overline{X},i)}(\overline{Z})\log(P_j^{(\overline{X},i)}(\overline{Z}))$. Furthermore, this criterion needs to be minimized.

11.7.2.2 Expected Variance Reduction

One observation about the aforementioned error-reduction method of Eq. 11.28 is that it needs to be computed in terms of the entire set of unlabeled instances in V, and a new model needs to be trained incrementally to test the effect of adding a new instance. This can be computationally expensive. It should be pointed out that when the error of an instance set reduces, the corresponding variance also typically reduces. The overall generalization error can be expressed[4] as a sum of the true label noise, model bias, and variance. Of these, only the last term is highly dependent on the choice of instances selected. Therefore, it is possible to reduce the variance instead of the error, and the main advantage of doing so is the reduction in computational requirements. The main advantage of these techniques is the ability to express the variance in *closed form*, and therefore achieve greater computational efficiency. A detailed description of this class of methods is beyond the scope of this book. Refer to the bibliographic notes.

11.7.3 Representativeness-Based Models

The main advantage of performance-based models over heterogeneity-based models is that they intend to improve the error behavior on the *aggregate* set of unlabeled instances, rather than evaluating the uncertainty behavior of the *queried* instance. Therefore, unrepresentative or outlier-like queries are avoided. In some models, the representativeness itself becomes a part of the criterion for querying. One way of measuring representativeness is with the use of a density-based criterion, in which the density of a region in the space is used to weight the querying criterion. This weight is combined with a heterogeneity-based query criterion. Therefore, such methods can be considered a variation of the heterogeneity-based model, but with a representativeness weighting to ensure that outliers are not selected.

Therefore, these methods *combine* the heterogeneity behavior of the queried instance with a representativeness function from the unlabeled set V to decide on the queried instance. The representativeness function weights dense regions of the input space. The objective function $O(\overline{X}, V)$ of such a model is expressed as the product of a heterogeneity component $H(\overline{X})$ and a representativeness component $R(\overline{X}, V)$.

$$O(\overline{X}, V) = H(\overline{X})R(\overline{X}, V)$$

The value of $H(\overline{X})$ (assumed to be a maximization function) can be any of the heterogeneity criteria (transformed appropriately for maximization), such as the entropy criterion from uncertainty sampling, or the expected model change criterion. The representativeness criterion $R(\overline{X}, V)$ is simply a measure of the density of $\overline{X}$ with respect to the instances in V. A simple version of this density is the average similarity of $\overline{X}$ to the instances in V. Many other sophisticated variations of this simple measure are used. The reader is referred to the bibliographic notes for a discussion of the available measures.

11.8 Ensemble Methods

Ensemble methods are motivated by the fact that different classifiers may make different predictions on test instances due to the specific characteristics of the classifier, or their sensitivity to the random artifacts in the training data. An ensemble method is an approach to increase the prediction accuracy by combining the results from multiple classifiers. The

[4]This theoretical concept is discussed in detail in the next section.

Algorithm *EnsembleClassify*(Training Data Set: $\mathcal{D}$
  Base Algorithms: $\mathcal{A}_1 \ldots \mathcal{A}_r$, Test Instances: $\mathcal{T}$)
begin
  $j = 1$;
  repeat
    Select an algorithm $\mathcal{Q}_j$ from $\mathcal{A}_1 \ldots \mathcal{A}_r$;
    Create a new training data set $f_j(\mathcal{D})$ from $\mathcal{D}$;
    Apply $\mathcal{Q}_j$ to $f_j(\mathcal{D})$ to learn model $\mathcal{M}_j$;
    $j = j + 1$;
  until(termination);
  report labels of each $T \in \mathcal{T}$ based on combination of
    predictions from all learned models $\mathcal{M}_j$;
end

Figure 11.4: The generic ensemble framework

basic approach of ensemble analysis is to apply the *base ensemble learners* multiple times by using either different models, or by using the same model on different subsets of the training data. The results from different classifiers are then combined into a single robust prediction.

Although there are significant differences in how the individual learners are constructed and combined by various ensemble models, we start with a very generic description of ensemble algorithms. Later in this section, we will discuss specific *instantiations* of this broad framework, such as bagging, boosting, and random decision trees. The ensemble approach uses a set of base classification algorithms $\mathcal{A}_1 \ldots \mathcal{A}_r$. Note that these learners might be completely different algorithms, such as decision trees, SVMs, or the Bayes classifier. In some types of ensembles, such as boosting and bagging, a single learning algorithm is used but with different choices of training data. Different learners are used to leverage the greater robustness of different algorithms in different regions of the data. Let the learning algorithm selected in the jth iteration be denoted by $\mathcal{Q}_j$. It is assumed that $\mathcal{Q}_j$ is selected from the base learners. At this point, a *derivative training data set* $f_j(\mathcal{D})$ from the base training data is selected. This may be a random sample of the training data, as in bagging, or it may be based on the results of the past execution of ensemble components, as in boosting. A model $\mathcal{M}_j$ is learned in the jth iteration by applying the selected learning algorithm $\mathcal{Q}_j$ to $f_j(\mathcal{D})$. For each test instance T, a prediction is made by combining the results of different models $\mathcal{M}_j$ on T. This combination may be performed in various ways. Examples include the use of simple averaging, the use of a weighted vote, or the treatment of the model combination process as a learning problem. The overall ensemble framework is illustrated in Fig. 11.4.

The description of Fig. 11.4 is very generic, and allows significant flexibility in terms of how the ensemble components may be learned and the combination may be performed. The two primary types of ensembles are special cases of the description of Fig. 11.4:

1. *Data-centered ensembles:* A single base learning algorithm (e.g., an SVM or a decision tree) is used, and the primary variation is in terms of how the derivative data set $f_j(\mathcal{D})$ for the jth ensemble component is constructed. In this case, the input to the algorithm contains only a single learning algorithm $\mathcal{A}_1$. The data set $f_j(\mathcal{D})$ for the jth component of the ensemble may be constructed by sampling the data, focusing on incorrectly classified portions of the training data in previously executed ensemble components, manipulating the features of the data, or manipulating the class labels in the data.

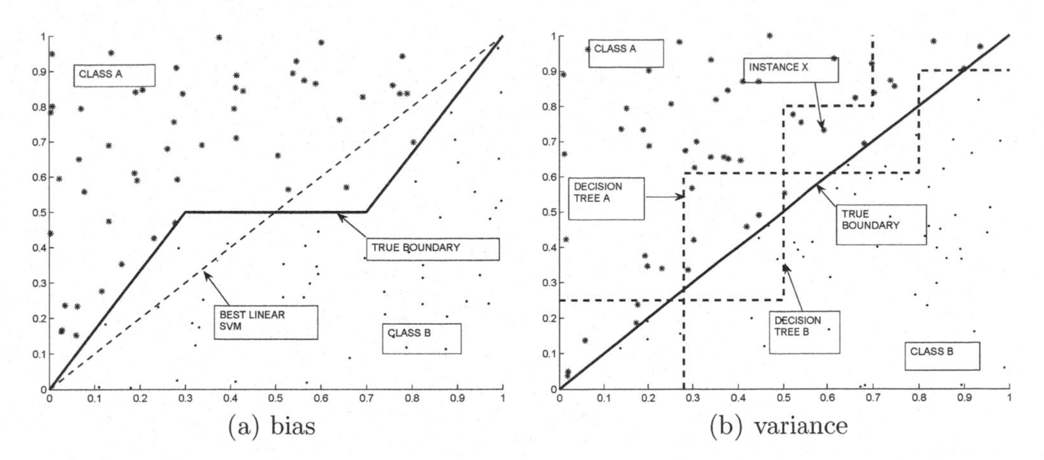

Figure 11.5: Impact of bias and variance on classification accuracy

2. *Model-centered ensembles:* Different algorithms $\mathcal{Q}_j$ are used in each ensemble iteration. In these cases, the data set $f_j(\mathcal{D})$ for each ensemble component is the same as the original data set $\mathcal{D}$. The rationale for these methods is that different models may work better in different regions of the data, and therefore the combination of the models may be more effective for any given test instance, as long as the specific errors of a classification algorithm are not reflected by the majority of the ensemble components on any particular test instance.

The following discussion introduces the rationale for ensemble analysis before presenting specific instantiations.

11.8.1 Why Does Ensemble Analysis Work?

The rationale for ensemble analysis can be best understood by examining the different components of the error of a classifier, as discussed in statistical learning theory. There are three primary components to the error of a classifier:

1. *Bias:* Every classifier makes its own modeling assumptions about the nature of the decision boundary between classes. For example, a linear SVM classifier assumes that the two classes may be separated by a linear decision boundary. This is, of course, not true in practice. For example, in Fig. 11.5a, the decision boundary between the different classes is clearly not linear. The correct decision boundary is shown by the solid line. Therefore, no (linear) SVM classifier can classify all the possible test instances correctly even if the best possible SVM model is constructed with a very large training data set. Although the SVM classifier in Fig. 11.5a seems to be the best possible approximation, it obviously cannot match the correct decision boundary and therefore has an inherent error. In other words, any given linear SVM model will have an inherent *bias.* When a classifier has high bias, it will make *consistently incorrect* predictions over particular choices of test instances near the incorrectly modeled decision-boundary, even when different samples of the training data are used for the learning process.

2. *Variance:* Random variations in the choices of the training data will lead to different models. Consider the example illustrated in Fig. 11.5b. In this case, the true decision

boundary is linear. A *sufficiently deep* univariate decision tree can approximate a linear boundary quite well with axis-parallel piecewise approximations. However, *with limited training data*, even when the trees are grown to full depth without pruning, the piecewise approximations will be coarse like the boundaries illustrated for hypothetical decision trees A and B in Fig. 11.5b. Different choices of training data might lead to different split choices, as a result of which the decision boundaries of trees A and B are very different. Therefore, (test) instances such as X are *inconsistently classified* by decision trees which were created by different choices of training data sets. This is a manifestation of model *variance*. Model variance is closely related to overfitting. When a classifier has an overfitting tendency, it will make *inconsistent* predictions for the same test instance over different training data sets.

3. *Noise:* The noise refers to the intrinsic errors in the target class labeling. Because this is an intrinsic aspect of data quality, there is little that one can do to correct it. Therefore, the focus of ensemble analysis is generally on reducing bias and variance.

Note that the design choices of a classifier often reflect a trade-off between the bias and the variance. For example, pruning a decision tree results in a more stable classifier and therefore reduces the variance. On the other hand, because the pruned decision tree makes stronger assumptions about the simplicity of the decision boundary than the unpruned tree, the former leads to greater bias. Similarly, using a larger number of neighbors for a nearest-neighbor classifier will lead to larger bias but lower variance. In general, simplified assumptions about the decision boundary lead to greater bias but lower variance. On the other hand, complex assumptions reduce bias but are harder to robustly estimate with limited data. The bias and variance are affected by virtually every design choice of the model, such as the choice of the base algorithm or the choice of model parameters.

Ensemble analysis can often be used to reduce both the bias and variance of the classification process. For example, consider the case of the example illustrated in Fig. 11.5a, in which the decision boundary is not linear, and therefore any linear SVM classifier will not find the correct decision boundary. However, by using different choices of model parameters, or data subset selection, it is possible to create three different linear SVM hyperplanes A, B, and C, as illustrated in Fig. 11.6a. Note that these different classifiers tend to work well in different parts of the data and have different *directions* of bias in any particular part of the data. This kind of differential performance on different parts of the data is sometimes artificially induced in ensemble components in some methods, such as boosting. In other cases, it may be a natural result of using ensemble model components that are very different from one another (e.g., decision trees and Bayes classifiers). Now consider a new ensemble classifier that is created using the majority vote of the three aforementioned classifiers corresponding to hyperplanes A, B, and C. The decision boundary of this ensemble classifier is illustrated in Fig. 11.6a as well. *This decision boundary is not linear and has lower bias with respect to the true decision boundary.* The reason for this is that different classifiers have different levels and directions of bias in different parts of the training data, and the majority vote across the different classifiers is able to obtain results that are generally less biased in any specific region than each of the component classifiers.

A similar argument applies to the variance example illustrated in Fig. 11.5b. Although instances such as X are inconsistently classified because of model variance, they will often be classified correctly when the model bias is low. As a result, by using the aggregation over sufficiently independent classifiers, it becomes increasingly likely that instances close to the decision boundary, such as X, will be correctly classified. For example, a majority vote of just three *independent* trees, each of which classifies X correctly with 80 % probability,

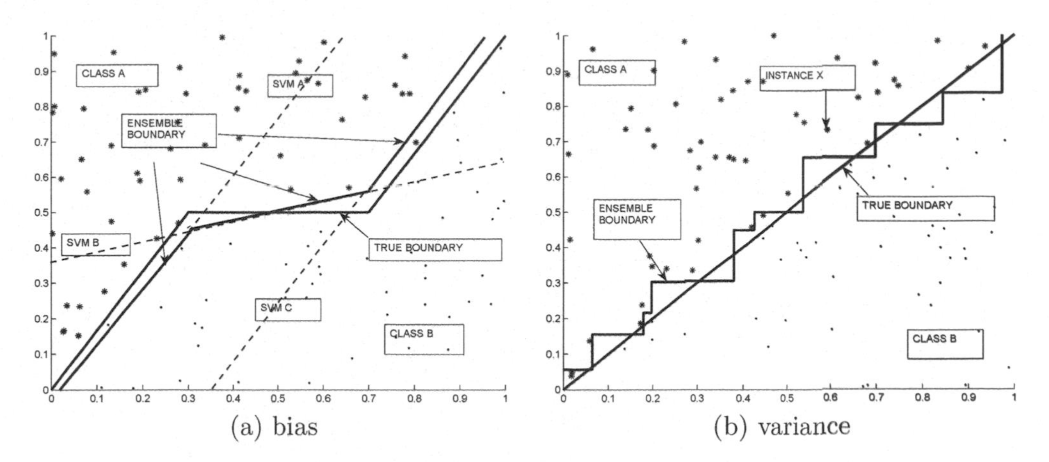

(a) bias (b) variance

Figure 11.6: Ensemble decision boundaries are more refined than those of component classifiers

will be correct $(0.8^3 + \binom{3}{2} \times 0.8^2 \times 0.2) \times 100 \approx 90\%$ of the time. In other words, the ensemble decision boundary of the majority classifier will be much closer to the true decision boundary than that of *any* of its component classifiers. In fact, a realistic example of how an ensemble boundary might look like after combining a set of relatively coarse decision trees, is illustrated in Fig. 11.6b. Note that the ensemble boundary is much closer to the true boundary because it is not regulated by the unpredictable variations in decision-tree behavior for a training data set of limited size. Such an ensemble is, therefore, better able to make use of the knowledge in the training data.

In general, different classification models have different sources of bias and variance. Models that are too simple (such as a linear SVM or shallow decision tree) make too many assumptions about the shape of the decision boundary, and will therefore have high bias. Models that are too complex (such as a deep decision tree) will overfit the data, and will therefore have high variance. Sometimes a different parameter setting in the same classifier will favor different parts of the bias-variance trade-off curve. For example, a small value of k in a nearest-neighbor classifier will result in lower bias but higher variance. Because different kinds of ensembles learners have different impacts on bias and variance, it is important to choose the component classifiers, so as to optimize the impact on the bias-variance trade-off. An overview of the impact of different models on bias and variance is provided in Table 11.1.

11.8.2 Formal Statement of Bias-Variance Trade-off

In the following, a formal statement of the bias-variance trade-off will be provided. Consider a classification problem with a training data set $\mathcal{D}$. The classification problem can be viewed as that of learning the function $f(\overline{X})$ between the feature variables $\overline{X}$ and the binary class variable y:

$$y = f(\overline{X}) + \epsilon. \tag{11.29}$$

Here, $f(\overline{X})$ is the function representing the true (but unknown) relationship between the feature variables and the class variable, and ϵ is the intrinsic error in the data that cannot be modeled. Therefore, over the n test instances $(\overline{X_1}, y_1) \ldots (\overline{X_n}, y_n)$, the intrinsic noise ϵ_a^2

Table 11.1: Impact of different techniques on bias-variance trade-off

Technique	Source/level of bias	Source/level of variance
Simple models	Oversimplification increases bias in decision boundary	Low variance. Simple models do not overfit
Complex models	Generally lower than simple models. Complex boundary can be modeled	High variance. Complex assumptions will be overly sensitive to data variation
Shallow decision trees	High bias. Shallow tree will ignore many relevant split predicates	Low variance. The top split levels do not depend on minor data variations
Deep decision trees	Lower bias than shallow decision tree. Deep levels model complex boundary	High variance because of overfitting at lower levels
Rules	Bias increases with fewer antecedents per rule	Variance increases with more antecedents per rule
Naive Bayes	High bias from simplified model (e.g., Bernoulli) and naive assumption	Variance in estimation of model parameters. More parameters increase variance
Linear models	High bias. Correct boundary may not be linear	Low variance. Linear separator can be modeled robustly
Kernel SVM	Bias lower than linear SVM. Choice of kernel function	Variance higher than linear SVM
k-NN model	Simplified distance function such as Euclidean causes bias. Increases with k	Complex distance function such as local discriminant causes variance. Decreases with k
Regularization	Increases bias	Reduces variance

may be estimated as follows:

$$\epsilon_a^2 = \frac{1}{n}\sum_{i=1}^{n}(y_i - f(\overline{X_i}))^2. \tag{11.30}$$

Because the precise form of the function $f(\overline{X})$ is unknown, most classification algorithms construct a model $g(\overline{X}, \mathcal{D})$ based on certain *modeling assumptions*. An example of such a modeling assumption is the linear decision boundary in SVM. This function $g(\overline{X}, \mathcal{D})$ may be defined either algorithmically (e.g., decision tree), or it may be defined in closed form, such as the SVM classifier, discussed in Sect. 10.6 of Chap. 10. In the latter case, the function $g(\overline{X}, \mathcal{D})$ is defined as follows:

$$g(\overline{X}, \mathcal{D}) = \text{sign}\{\overline{W} \cdot \overline{X} + b\}. \tag{11.31}$$

Note that the coefficients $\overline{W}$ and b can only be *estimated* from the training data set $\mathcal{D}$. The notation $\mathcal{D}$ occurs as an argument of $g(\overline{X}, \mathcal{D})$ because the training data set $\mathcal{D}$ is required to estimate model coefficients such as $\overline{W}$ and b. For a training data set $\mathcal{D}$ of limited size, it is often not possible to estimate $g(\overline{X}, \mathcal{D})$ very accurately, as in the case of the coarse decision boundaries of Fig. 11.5b. Therefore, the specific impact of the training data set on the estimated result $g(\overline{X}, \mathcal{D})$ can be quantified by comparing it with its *expected value* $E_{\mathcal{D}}[g(\overline{X}, \mathcal{D})]$ over all possible outcomes of training data sets $\mathcal{D}$.

Aside from the intrinsic error ϵ_a^2, which is *data-set specific*, there are two primary sources of error in the modeling and estimation process:

1. The *modeling assumptions* about $g(\overline{X}, \mathcal{D})$ may not reflect the true model. Consider the case where the linear SVM modeling assumption is used for $g(\overline{X}, \mathcal{D})$, and the true boundary between the classes is not linear. This would result in a *bias* in the model. In

practice, the bias usually arises from oversimplification in the modeling assumptions. Even if we had a very large training data set and could somehow estimate the expected value of $g(\overline{X}, \mathcal{D})$, the value of $(f(\overline{X}) - E_{\mathcal{D}}[g(\overline{X}, \mathcal{D})])^2$ would be nonzero because of the *bias* arising from the difference between the true model and assumed model. This is referred to as the (squared) *bias*. Note that it is often possible to reduce the bias by assuming a more complex form for $g(\overline{X}, \mathcal{D})$, such as using a kernel SVM instead of a linear SVM in Fig. 11.5a.

2. Even if our assumptions about $g(\overline{X}, \mathcal{D})$ were correct, it is not possible to *exactly* estimate $E_{\mathcal{D}}[g(\overline{X}, \mathcal{D})]$ with any given training data set $\mathcal{D}$. Over different instantiations of a training data set $\mathcal{D}$ and fixed test instance $\overline{X}$, the predicted class label $g(\overline{X}, \mathcal{D})$ would be different. This is the model *variance*, which corresponds to $E_{\mathcal{D}}[(g(\overline{X}, \mathcal{D}) - E_{\mathcal{D}}[g(\overline{X}, \mathcal{D})])^2]$. Note that the expectation function $E_{\mathcal{D}}[g(\overline{X}, \mathcal{D})]$ defines a decision boundary which is usually much closer to the true decision boundary (e.g., ensemble boundary estimate in Fig. 11.6b) as compared to that defined by a specific instantiation $\mathcal{D}$ of the training data (e.g., boundaries A and B in Fig. 11.5b).

It can be shown that the expected mean squared error of the prediction $E_{\mathcal{D}}[MSE] = \frac{1}{n}\sum_{i=1}^{n} E_{\mathcal{D}}[(y_i - g(\overline{X_i}, \mathcal{D}))^2]$ of test data points $(\overline{X_1}, y_1) \ldots (\overline{X_n}, y_n)$ over different choices of training data set $\mathcal{D}$ can be decomposed into the bias, variance, and the intrinsic error as follows:

$$E_{\mathcal{D}}[MSE] = \frac{1}{n}\sum_{i=1}^{n}\left(\underbrace{(f(\overline{X_i}) - E_{\mathcal{D}}[g(\overline{X_i}, \mathcal{D})])^2}_{\text{Bias}^2} + \underbrace{E_{\mathcal{D}}[(g(\overline{X_i}, \mathcal{D}) - E_{\mathcal{D}}[g(\overline{X_i}, \mathcal{D})])^2]}_{\text{Variance}}\right) + \underbrace{\epsilon_a^2}_{\text{Error}}.$$

The ensemble method reduces the classification error by reducing the bias and variance of the combined model. By carefully choosing the component models with specific bias-variance properties and combining them appropriately, it is often possible to reduce both the bias and the variance over an individual model.

11.8.3 Specific Instantiations of Ensemble Learning

A number of ensemble approaches have been designed to increase accuracy by reducing bias, variance, or a combination of both. In the following, a discussion of selected models is provided.

11.8.3.1 Bagging

Bagging, which is also referred to as bootstrapped aggregating, is an approach that attempts to reduce the variance of the prediction. It is based on the idea that if the variance of a prediction is σ^2, then the variance of the average of k independent and identically distributed (i.i.d.) predictions is reduced to $\frac{\sigma^2}{k}$. Given sufficiently independent predictors, such an approach of averaging will reduce the variance significantly.

So how does one approximate i.i.d. predictors? In bagging, data points are sampled uniformly from the original data with replacement. This sampling approach is referred to as *bootstrapping* and is also used for model evaluation. A sample of approximately the same size as the original data is drawn. This sample may contain duplicates, and typically contains approximately $1 - (1 - 1/n)^n \approx 1 - 1/e$ fraction of the *distinct* data points in the original data. Here, e represents the base of the natural logarithm. This result is easy to show because the probability of a data point *not* being included in the sample is given by $(1 - 1/n)^n$. A total of k different bootstrapped samples, each of size n, are drawn independently, and the classifier is trained on each of them. For a given test instance, the predicted class label is reported by the ensemble as the dominant vote of the different classifiers.

The primary advantage of bagging is that it reduces the model variance. However, bagging does not reduce the model bias. Therefore, this approach will not be effective for the example of Fig. 11.5a, but it will be effective for the example of Fig. 11.5b. In Fig. 11.5b, the different decision tree boundaries are created by the random variations in the bootstrapped samples. The *majority* vote of these bootstrapped samples will, however, perform better than a model constructed on the full data set because of a reduction in the variance component. For cases in which the bias is the primary component of error, the bootstrapping approach may show a slight degradation in accuracy. Therefore, while designing a bootstrapping approach, it is advisable to design the individual components to reduce the bias at the possible expense of variance, because bagging will take care of the latter. Such a choice optimizes the bias-variance trade-off. For example, one might want to grow a deep decision tree with 100 % class purity at the leaves. In fact, decision trees are an ideal choice for bagging, because they have low bias and high variance when they are grown sufficiently deep. The main problem with bagging is that the i.i.d. assumption is usually not satisfied because of correlations between ensemble components.

11.8.3.2 Random Forests

Bagging works best when the individual predictions from various ensemble components satisfy the i.i.d. property. If the k different predictors, each with variance σ^2, have a positive pairwise correlation of ρ between them, then the variance of the averaged prediction can be shown to be $\rho \cdot \sigma^2 + \frac{(1-\rho)\cdot\sigma^2}{k}$. The term $\rho \cdot \sigma^2$ is invariant to the number of components k in the ensemble. This term limits the performance gains from bagging. As we will discuss below, the predictions from bootstrapped decision trees are usually positively correlated.

Random forests can be viewed as a generalization of the basic bagging method, as applied to decision trees. Random forests are defined as an ensemble of decision trees, in which randomness has explicitly been inserted into the model building process of each decision tree. While the bootstrapped sampling approach of bagging is also an indirect way of adding randomness to model-building, there are some disadvantages of doing so. The main drawback of using decision-trees directly with bagging is that the split choices at the top levels of the tree are statistically likely to remain approximately invariant to bootstrapped sampling. Therefore, the trees are more correlated, which limits the amount of error reduction obtained from bagging. In such cases, it makes sense to directly increase the diversity of the component decision-tree models. The idea is to use a randomized decision tree model with less correlation between the different ensemble components. The underlying variability can then be more effectively reduced by an averaging approach. The final results are often more accurate than a direct application of bagging on decision trees. Figure 11.6b provides a typical example of the effect of the approach on the decision boundary, which is similar to that of bagging, but it is usually more pronounced.

The *random-split selection* approach directly introduces randomness into the split criterion. An integer parameter $q \leq d$ is used to regulate the amount of randomness introduced in split selection. The split selection at each node is preceded by the randomized selection of a subset S of attributes of size q. The splits at that node are then executed using only this subset. Larger values of q will result in correlated trees that are similar to a tree without any injected randomness. By selecting small values of q relative to the full dimensionality

d, the correlations across different components of the random forest can be reduced. The resulting trees can also be grown more efficiently, because a smaller number of attributes need to be considered at each node. On the other hand, when a larger size q of the selected feature set S is used, then each individual ensemble component will have better accuracy, which is also desirable. Therefore, the goal is to select a good trade-off point. It has been shown that, when the number of input attributes is d, a total of $q = \log_2(d) + 1$ attributes should be selected to achieve the best trade-off. Interestingly, even a choice of $q = 1$ seems to work well in terms of accuracy, though it requires the use of a larger number of ensemble components. After the ensemble has been constructed, the dominant label from the various predictions of a test instance is reported. This approach is referred to as *Forest-RI* because it is based on *random input selection.*

This approach does not work well when the overall dimensionality d is small, and therefore it is no longer possible to use values of q much smaller than d. In such cases, a value $L \leq d$ is specified, which corresponds to the number of input features that are *combined* together. At each node, L features are randomly selected, and combined linearly with coefficients generated uniformly at random from the range $[-1, 1]$. A total of q such combinations are generated in order to create a new subset S of multivariate attributes. As in the previous case, the splits are executed using only the attribute set S. This results in multivariate random splits. This approach is referred to as *Forest-RC* because it uses *random linear combinations.*

In the random forest approach, deep trees are grown without pruning. Each tree is grown on a bootstrapped sample of the training data to increase the variance reduction. As in bagging, the variance is reduced significantly by the random forest approach. However, the component classifiers may have higher bias because of the restricted split selection of each component classifier. This can sometimes cause a problem when the fraction of informative features in the training data is small. Therefore, the gains in random forests are a result of variance reduction. In practice, the random forests approach usually performs better than bagging and comparable to boosting. It is also resistant to noise and outliers.

11.8.3.3 Boosting

In boosting, a weight is associated with each training instance, and the different classifiers are trained with the use of these weights. The weights are modified iteratively based on classifier performance. In other words, the future models constructed are dependent on the results from previous models. Thus, each classifier in this model is constructed using a the same algorithm $\mathcal{A}$ on a weighted training data set. The basic idea is to focus on the incorrectly classified instances in future iterations by increasing the relative weight of these instances. The hypothesis is that the errors in these misclassified instances are caused by classifier bias. Therefore, increasing the instance weight of misclassified instances will result in a new classifier that corrects for the bias on these *particular* instances. By iteratively using this approach and creating a weighted combination of the various classifiers, it is possible to create a classifier with lower *overall* bias. For example, in Fig. 11.6a, each individual SVM is not globally optimized and is accurate only near specific portions of the decision boundary, but the combination ensemble provides a very accurate decision boundary.

The most well-known approach to boosting is the *AdaBoost* algorithm. For simplicity, the following discussion will assume the binary class scenario. It is assumed that the class labels are drawn from $\{-1, +1\}$. This algorithm works by associating each training example with a *weight* that is updated in each iteration, depending on the results of the classification in the last iteration. The base classifiers therefore need to be able to work with weighted

Algorithm *AdaBoost*(Data Set: $\mathcal{D}$, Base Classifier: $\mathcal{A}$, Maximum Rounds: T)
begin
  $t = 0$;
  for each i initialize $W_1(i) = 1/n$;
  repeat
    $t = t + 1$;
    Determine weighted error rate ϵ_t on $\mathcal{D}$ when base algorithm $\mathcal{A}$
        is applied to weighted data set with weights $W_t(\cdot)$;
    $\alpha_t = \frac{1}{2}\log_e((1 - \epsilon_t)/\epsilon_t)$;
    for each misclassified $\overline{X_i} \in \mathcal{D}$ **do** $W_{t+1}(i) = W_t(i)e^{\alpha_t}$;
      else (correctly classified instance) **do** $W_{t+1}(i) = W_t(i)e^{-\alpha_t}$;
    for each instance $\overline{X_i}$ **do** normalize $W_{t+1}(i) = W_{t+1}(i)/[\sum_{j=1}^{n} W_{t+1}(j)]$;
  until $((t \geq T)$ OR $(\epsilon_t = 0)$ OR $(\epsilon_t \geq 0.5))$;
  Use ensemble components with weights α_t for test instance classification;
end

Figure 11.7: The *AdaBoost* algorithm

instances. Weights can be incorporated either by direct modification of training models, or by (biased) bootstrap sampling of the training data. The reader should revisit the section on rare class learning for a discussion on this topic. Instances that are misclassified are given higher weights in successive iterations. Note that this corresponds to intentionally biasing the classifier in later iterations with respect to the *global* training data, but reducing the bias in certain *local* regions that are deemed "difficult" to classify by the specific model $\mathcal{A}$.

In the tth round, the weight of the ith instance is $W_t(i)$. The algorithm starts with equal weight of $1/n$ for each of the n instances, and updates them in each iteration. In the event that the ith instance is misclassified, then its (relative) weight is increased to $W_{t+1}(i) = W_t(i)e^{\alpha_t}$, whereas in the case of a correct classification, the weight is decreased to $W_{t+1}(i) = W_t(i)e^{-\alpha_t}$. Here α_t is chosen as the function $\frac{1}{2}\log_e((1-\epsilon_t)/\epsilon_t)$, where ϵ_t is the fraction of incorrectly predicted training instances (computed after weighting with $W_t(i)$) by the model in the tth iteration. The approach terminates when the classifier achieves 100 % accuracy on the training data ($\epsilon_t = 0$), or it performs worse than a random (binary) classifier ($\epsilon_t \geq 0.5$). An additional termination criterion is that the number of boosting rounds is bounded above by a user-defined parameter T. The overall training portion of the algorithm is illustrated in Fig. 11.7.

It remains to be explained how a particular test instance is classified with the ensemble learner. Each of the models induced in the different rounds of boosting is applied to the test instance. The prediction $p_t \in \{-1, +1\}$ of the test instance for the tth round is weighted with α_t and these weighted predictions are aggregated. The sign of this aggregation $\sum_t p_t \alpha_t$ provides the class label prediction of the test instance. Note that less accurate components are weighted less by this approach.

An error rate of $\epsilon_t \geq 0.5$ is as bad or worse than the expected error rate of a random (binary) classifier. This is the reason that this case is also used as a termination criterion. In some implementations of boosting, the weights $W_t(i)$ are reset to $1/n$ whenever $\epsilon_t \geq 0.5$, and the boosting process is continued with the reset weights. In other implementations, ϵ_t is allowed to increase beyond 0.5, and therefore some of the prediction results p_t for a test instance are effectively inverted with negative values of the weight $\alpha_t = \log_e((1 - \epsilon_t)/\epsilon_t)$.

Boosting primarily focuses on reducing the bias. The bias component of the error is reduced because of the greater focus on misclassified instances. The combination decision boundary is a complex combination of the simpler decision boundaries, which are each optimized to specific parts of the training data. An example of how simpler decision boundaries combine to provide a complex decision boundary was already illustrated in Fig. 11.6a. Because of its focus on the bias of classifier models, such an approach is capable of combining many weak (high bias) learners to create a strong learner. Therefore, the approach should generally be used with simpler (high bias) learners with low variance in the individual ensemble components. In spite of its focus on bias, boosting can also reduce the variance slightly when reweighting is implemented with sampling. This reduction is because of the repeated construction of models on randomly sampled, albeit reweighted, instances. The amount of variance reduction depends on the reweighting scheme used. Modifying the weights less aggressively between rounds will lead to better variance reduction. For example, if the weights are not modified at all between boosting rounds, then the boosting approach defaults to bagging, which only reduces variance. Therefore, it is possible to leverage variants of boosting to explore the bias-variance trade-off in various ways.

Boosting is vulnerable to data sets with significant noise in them. This is because boosting assumes that misclassification is caused by the bias component of instances near the *incorrectly modeled decision boundary*, whereas it might simply be a result of the mislabeling of the *data*. This is the noise component that is intrinsic to the *data*, rather than the *model*. In such cases, boosting inappropriately overtrains the classifier to low-quality portions of the data. Indeed, there are many noisy real-world data sets where boosting does not perform well. Its accuracy is typically superior to bagging in scenarios where the data sets are not excessively noisy.

11.8.3.4 Bucket of Models

The bucket of models is based on the intuition that it is often difficult to know *a priori*, which classifier will work well for a particular data set. For example, a particular data set may be suited to decision trees, whereas another data set may be well suited to support vector machines. Therefore, *model selection* methods are required. Given a data set, how does one determine, which algorithm to use? The idea is to first divide the data set into two subsets A and B. Each algorithm is trained on subset A. The set B is then used to evaluate the performance of each of these models. The winner in this "bake-off" contest is selected. Then, the winner is retrained using the full data set. If desired, cross-validation can be used for the contest, instead of the "hold-out" approach of dividing the training data into two subsets.

Note that the accuracy of the bucket of models is no better than the accuracy of the best classifier for a *particular* data set. However, over *many* data sets, the approach has the advantage of being able to use the best model that is suited to *each* data set, because different classifiers may work differently on different data sets. The bucket of models is used commonly for model selection and parameter tuning in classification algorithms. Each individual model is the same classifier over a different choice of the parameters. The winner therefore provides the optimal parameter choice across all models.

The bucket of models approach is based on the idea that different classifiers will have different kinds of bias on different data sets. This is because the "correct" decision boundary varies with the data set. By using a "winner-takes-all" contest, the classifier with the most accurate decision boundary will be selected for each data set. Because the bucket of models evaluates the classifier accuracy based on *overall* accuracy, it will also tend to select a model with lower variance. Therefore, the approach can reduce both the bias and the variance.

11.8.3.5 Stacking

The stacking approach is a very general one, in which two levels of classification are used. As in the case of the bucket of models approach, the training data is divided into two subsets A and B. The subset A is used for the first-level classifiers that are the ensemble components. The subset B is used for the second-level classifier that combines the results from different ensemble components in the previous phase. These two steps are described as follows:

1. Train a set of k classifiers (ensemble components) on the training data subset A. These k ensemble components can be generated in various ways, such as drawing k bootstrapped samples (bagging) on data subset A, k rounds of boosting on data subset A, k different random decision trees on data subset A, or simply training k heterogeneous classifiers on data subset A.

2. Determine the k outputs of each of the classifiers on the training data subset B. Create a new set of k features, in which each feature value is the output of one of these k classifiers. Thus, each point in training data subset B is transformed to this k-dimensional space based on the prediction of the k first-level classifiers. Its class label is its (known) ground-truth value. The second-level classifier is trained on this new representation of subset B.

The result is a set of k first-level models used to transform the feature space, and a combiner classifier at the second-level. For a test instance, the first-level models are used to create a new k-dimensional representation. The second-level classifier is then used to predict the test instance. In many implementations of stacking, the original features of data subset B are retained along with the k new features for learning the second-level classifier. It is also possible to use class probabilities as features rather than the class label predictions. To prevent loss of training data in the first-level and second-level models, this method can be combined with m-way cross-validation. In this approach, a new feature set is derived for *each* training data point by iteratively using $(m - 1)$ segments for training the first-level classifier, and using it to derive the features of the remainder. The second-level classifier is trained on the newly created data set, which represents *all* the training data points. Furthermore, the first-level classifiers are re-trained on the full training data in order to enable more robust feature transformations of test instances during classification.

The stacking approach is able to reduce both bias and variance, because its combiner learns from the errors of different ensemble components. Many other ensemble methods can be viewed as special cases of stacking in which a data-independent model combination algorithm, such as a majority vote, is used. The main advantage of stacking is the flexible learning approach of its combiner, which makes it potentially more powerful than other ensemble methods.

11.9 Summary

In this chapter, we studied several advanced topics in data classification, such as multiclass learning, scalable learning, and rare class learning. These are more challenging scenarios for data classification that require dedicated methods. Classification can often be enhanced with additional unlabeled data in semisupervised learning, or by selective acquisition of the user, as in active learning. Ensemble methods can also be used to significantly improve classification accuracy.

In multiclass learning methods, binary classifiers are combined in a meta-algorithm framework. Typically, either a one-against-rest, or a one-against-one approach is used. The voting from different classifiers is used to provide a final result. In many scenarios, the one-against-one approach is more efficient than the one-against-rest approach. Many scalable methods have been designed for data classification. For decision trees, two scalable methods include *RainForest* and *BOAT*. Numerous fast variations of SVM classifiers have also been designed.

The rare class learning problem is very common, especially because class distributions are very imbalanced in many real scenarios. Typically, the objective function for the classification problem is changed with the use of cost-weighting. Cost-weighting can be achieved either with example weighting, or with example resampling. Typically, the normal class is undersampled in example resampling, which results in better training efficiency.

The paucity of training data is common in real domains. Semisupervised learning is one way of addressing the paucity of training data. In these methods, the copiously available unlabeled data is used to make estimations about class distributions in regions where little labeled data is available. A second way of leveraging the copious unlabeled data, is by actively assigning labels so that the most informative labels are determined for classification.

Ensemble methods improve classifier accuracy by reducing their bias and variance. Some ensemble methods, such as bagging and random forests, are designed only to reduce the variance, whereas other ensemble methods, such as boosting and stacking, can help reduce both the bias and variance. In some cases, such as boosting, it is possible for the ensemble to overfit the noise in the data and thereby lead to lower accuracy.

11.10 Bibliographic Notes

Multiclass strategies are used for those classifiers that are designed to be binary. A typical example of such a classifier is the support vector machine. The one-against-rest strategy is introduced in [106]. The one-against-one strategy is discussed in [318].

Some examples of scalable decision tree methods include *SLIQ* [381], *BOAT* [227], and *RainForest* [228]. Some early parallel implementations of decision trees include the *SPRINT* method [458]. An example of a scalable SVM method is *SVMLight* [291]. Other methods, such as *SVMPerf* [292], reformulate the SVM optimization to reduce the number of slack variables, and increase the number of constraints. A cutting plane approach that works with a small subset of constraints at a time is used to make the SVM classifier scalable. This approach is discussed in detail in Chap. 13 on text mining.

Detailed discussions on imbalance and cost-sensitive learning may be found in [136, 139, 193]. A variety of general methods have been proposed for cost-sensitive learning, such as *MetaCost* [174], weighting [531], and sampling [136, 531]. The *SMOTE* method is discussed in [137]. Boosting algorithms have also been studied for the problem of cost-sensitive learning. The *AdaCost* algorithm was proposed in [203]. Boosting techniques can also be combined with sampling methods, as in the case of the *SMOTEBoost* algorithm [138]. An evaluation of boosting algorithms for rare class detection is provided in [296]. Discussions of linear regression models and regression trees may be found in [110, 256, 391].

Recently, the semisupervised and active learning problems have been studied to use external information for better supervision. The co-training method was discussed in [100]. The EM algorithm for combining labeled and unlabeled data was proposed in [410]. Transductive SVM methods are proposed in [293, 496]. The method in [293] is a scalable SVM method that uses an iterative approach. Graph-based methods for semisupervised learning are discussed in [101, 294]. Surveys on semisupervised classification may be found in [33, 555].

A detailed survey on active learning may be found in [13, 454]. Methods for uncertainty sampling [345], query-by-committee [457], greatest model change [157], greatest error reduction [158], and greatest variance reduction [158] have been proposed. Representativeness-based models have been discussed in [455]. Another form of active learning queries the data *vertically*. In other words, instead of examples, it is learned which *attributes* to collect to minimize the error at a given cost level [382].

The problem of meta-algorithm analysis has recently become very important because of its significant impact on improving the accuracy of classification algorithms. The bagging and random forests methods were proposed in [111, 112]. The boosting method was proposed in [215]. Bayesian model averaging and combination methods are proposed in [175]. The stacking method is discussed in [491, 513], and the bucket-of-models approach is explained in [541].

11.11 Exercises

1. Suppose that a classification training algorithm requires $O(n^r)$ time for training on a data set of size n. Here r is assumed to be larger than 1. Consider a data set $\mathcal{D}$ with an exactly even distribution across k different classes. Compare the running time of the one-against-rest approach with that of the one-against-one approach.

2. Discuss some general meta-strategies for speeding up classifiers. Discuss some strategies that you might use to scale up (a) nearest-neighbor classifiers, and (b) associative classifiers.

3. Describe the changes required to the dual formulation of the soft SVM classifier with hinge loss to make it a weighted classifier for rare-class learning.

4. Describe the changes required to the primal formulation of the soft SVM classifier with hinge loss to make it a weighted classifier for rare-class learning.

5. Implement the one-against-rest and one-against-one multiclass approach. Use the nearest-neighbor algorithm as the base classifier.

6. Design a semisupervised classification algorithm with the use of a supervised modification of the k-means algorithm. How is this algorithm related to the EM-based semisupervised approach?

7. Suppose that your data was distributed into two thin and separated concentric rings, each of which belonged to one of the two classes. Suppose that you had only a small number of labeled examples from each of the rings but you had a lot of unlabeled examples. Would your rather use the (a) EM-algorithm, (b) transductive SVM algorithm, or the (c) graph-based approach for semisupervised classification? Why?

8. Write the optimization formulation for least-squares regression of the form $y = \overline{W} \cdot \overline{X} + b$ with a bias term b. Provide a closed-form solution for the optimal values of $\overline{W}$ and b in terms of the data matrix D and response variable vector $\overline{y}$. Show that the optimal value of the bias term b always evaluates to 0 when the data matrix D and response variable vector $\overline{y}$ are both mean-centered.

9. Design a modification of the uncertainty sampling approach in which the dollar-costs of querying various instances are known to be different. Assume that the cost of querying instance i is known to be c_i.

10. Consider a situation where a classifier gives very consistent class-label predictions when trained on samples of the (training) data. Which ensemble method should you not use? Why?

11. Design a heuristic variant of the *AdaBoost* algorithm, which will perform better than *AdaBoost* in terms of reducing the variance component of the error. Does this mean that the overall error of this ensemble variant will be lower than that of *AdaBoost*?

12. Would you rather use a linear SVM to create the ensemble component in bagging or a kernel SVM? What would you do in the case of boosting?

13. Consider a d-dimensional data set. Suppose that you used the 1-nearest neighbor class label in a randomly chosen subspace with dimensionality $d/2$ as a classification model. This classifier is repeatedly used on a test instance to create a majority-vote prediction. Discuss the bias-variance mechanism with which such a classifier will reduce error.

14. For any $d \times n$ matrix A and scalar λ, use its singular value decomposition to show that the following is always true:

$$(AA^T + \lambda I_d)^{-1}A = A(A^TA + \lambda I_n)^{-1}.$$

Here, I_d and I_n are $d \times d$ and $n \times n$ identity matrices, respectively.

15. Let the singular value decomposition of an $n \times d$ matrix D be $Q\Sigma P^T$. According to Chap. 2, its pseudoinverse is $P\Sigma^+Q^T$. Here, Σ^+ is obtained by inverting the nonzero diagonal entries of the $n \times d$ matrix Σ and then transposing the resulting matrix.

(a) Use this result to show that:

$$D^+ = (D^TD)^+D^T.$$

(b) Show that an alternative way of computing the pseudoinverse is as follows:

$$D^+ = D^T(DD^T)^+.$$

(c) Discuss the efficiency of various methods of computing the pseudoinverse of D with varying values of n and d.

(d) Discuss the usefulness of any of the aforementioned methods for computing the pseudoinverse in the context of incorporating the kernel trick in linear regression.

Chapter 12

Mining Data Streams

"*You never step into the same stream twice.*"—Heraclitus

12.1 Introduction

Advances in hardware technology have led to new ways of collecting data at a more rapid rate than before. For example, many transactions of everyday life, such as using a credit card or a phone, lead to automated data collection. Similarly, new ways of collecting data, such as wearable sensors and mobile devices, have added to the deluge of dynamically available data. An important assumption in these forms of data collection is that the data *continuously accumulate over time at a rapid rate.* These dynamic data sets are referred to as *data streams.*

A key assumption in the streaming paradigm is that it is no longer possible to store all the data because of resource constraints. While it is possible to archive such data using distributed "big data" frameworks, this approach comes at the expense of enormous storage costs and the loss of real-time processing capabilities. In many cases, such frameworks are not practical because of high costs and other analytical considerations. The streaming framework provides an alternative approach, where real-time analysis can often be performed with carefully designed algorithms, without a significant investment in specialized infrastructure. Some examples of application domains relevant to streaming data are as follows:

1. *Transaction streams:* Transaction streams are typically created by customer buying activity. An example is the data created by using a credit card, point-of-sale transaction at a supermarket, or the online purchase of an item.

2. *Web click-streams:* The activity of users at a popular Web site creates a Web click stream. If the site is sufficiently popular, the rate of generation of the data may be large enough to necessitate the need for a streaming approach.

C. C. Aggarwal, *Data Mining: The Textbook*, DOI 10.1007/978-3-319-14142-8_12

3. *Social streams:* Online social networks such as Twitter continuously generate massive text streams because of user activity. The speed and volume of the stream typically scale superlinearly with the number of actors in the social network.

4. *Network streams:* Communication networks contain large volumes of traffic streams. Such streams are often mined for intrusions, outliers, or other unusual activity.

Data streams present a number of unique challenges because of the processing constraints associated with the large volumes of continuously arriving data. In particular, data streaming algorithms typically need to operate under the following constraints, at least a few of which are always present, whereas others are occasionally present:

1. *One-pass constraint:* Because volumes of data are generated continuously and rapidly, it is assumed that the data can be processed *only once*. This is a hard constraint in all streaming models. The data are almost never assumed to be archived for future processing. This has significant consequences for algorithmic development in streaming applications. In particular, many data mining algorithms are inherently iterative and require multiple passes over the data. Such algorithms need to be appropriately modified to be usable in the context of the streaming model.

2. *Concept drift:* In most applications, the data may evolve over time. This means that various statistical properties, such as correlations between attributes, correlations between attributes and class labels, and cluster distributions may change over time. This aspect of data streams is almost always present in practical applications, but is not necessarily a universal assumption for all algorithms.

3. *Resource constraints:* The data stream is typically generated by an external process, over which a user may have very little control. Therefore, the user also has little control over the arrival rate of the stream. In cases, where the arrival rates vary with time, it may be difficult to execute online processing continuously during peak periods. In these cases, it may be necessary to drop tuples that cannot be processed in a timely fashion. This is referred to as *loadshedding*. Even though resource constraints are almost universal to the streaming paradigm, surprisingly few algorithms incorporate them.

4. *Massive-domain constraints:* In some cases, when the attribute values are discrete, they may have a large number of distinct values. For example, consider a scenario, where analysis of pairwise communications in an e-mail network is desired. The number of distinct pairs of e-mail addresses in an e-mail network with 10^8 participants is of the order of 10^{16}. When expressed in terms of required storage, the number of possibilities easily exceeds the *petabyte* order. In such cases, storing even simple statistics such as the counts or the number of distinct stream elements becomes very challenging. Therefore, a number of specialized data structures for synopsis construction of massive-domain data streams have been designed.

Because of the large volume of data streams, virtually all streaming methods use an online synopsis construction approach in the mining process. The basic idea is to create an online synopsis that is then leveraged for mining. Many different kinds of synopsis can be constructed depending upon the application at hand. The nature of a synopsis highly influences the type of insights that can be mined from it. Some examples of synopsis structures include random samples, bloom filters, sketches, and distinct element-counting data structures. In addition, some traditional data mining applications, such as clustering, can be leveraged to create effective synopses from the data.

This chapter is organized as follows. Section 12.2 introduces various types of synopsis construction methods for data streams. Section 12.3 discusses frequent pattern mining methods for data streams. Clustering methods are discussed in Sect. 12.4, and outlier analysis methods are discussed in Sect. 12.5. Classification methods are introduced in Sect. 12.6. Section 12.7 gives a summary.

12.2 Synopsis Data Structures for Streams

A wide variety of synopsis data structures have been designed for different applications. Synopsis data structures are of two types:

1. *Generic:* In this case, the synopsis can be used for most applications directly. The only such synopsis is a random sample of the data points, although it cannot be used for some applications such as distinct element counting. In the context of data streams, the process of maintaining a random sample from the data is also referred to as *reservoir sampling.*

2. *Specific:* In this case, the synopsis is designed for a specific task, such as frequent element counting or distinct element counting. Examples of such data structures include the Flajolet–Martin data structure for distinct element counting, and sketches for frequent element counting or moment computation.

In the following, different types of synopsis structures will be discussed.

12.2.1 Reservoir Sampling

Sampling is one of the most flexible methods for stream summarization. The main advantage of sampling over other synopsis data structures is that it can be used for an arbitrary application. After a sample of points has been drawn from the data, virtually any offline algorithm can be applied to the sample. By default, sampling should be considered the method of choice in streaming scenarios, although it does have limitations for a small number of applications such as distinct-element counting. In the context of data streams, the methodology used to maintain a dynamic sample from the data is referred to as *reservoir sampling.* The resulting sample is referred to as a *reservoir sample.* The method of reservoir sampling is introduced briefly in Sect. 2.4.1.2 of Chap. 2.

The streaming scenario creates some interesting challenges for a simple problem such as sampling. The challenge arises from the fact that one cannot store the entire stream on disk for sampling. In reservoir sampling, the goal is to *continuously* maintain a *dynamically updated* sample of k points from a data stream without explicitly storing the stream on disk at any given point in time. Therefore, for each incoming data point in the stream, one must use a set of efficiently implementable operations to *maintain* the sample. In the static case, the probability of including a data point in the sample is k/n, where k is the sample size, and n is the number of points in the data set. In the streaming scenario, the "data set" is not static, and the value of n continually increases with time. Furthermore, previously arrived data points, which are not included in the sample, have been irrevocably lost. Thus, the sampling approach works with *incomplete knowledge* about the previous history of the stream at any given moment in time. In other words, for each incoming data point in the stream, one needs to make two simple *admission control* decisions dynamically:

1. What sampling rule should be used to decide whether to include the incoming data point in the sample?

2. What rule should be used to decide how to eject a data point from the sample to "make room" for the newly inserted data point?

The reservoir sampling algorithm proceeds as follows. For a reservoir of size k, the first k data points in the stream are always included in the reservoir. Subsequently, for the nth incoming stream data point, the following two admission control decisions are applied.

1. Insert the nth incoming stream data point in the reservoir with probability k/n.

2. If the newly incoming data point was inserted, then eject one of the old k data points in the reservoir at random to make room for the newly arriving point.

It can be shown that the aforementioned rule maintains an unbiased reservoir sample from the data stream.

Lemma 12.2.1 *After n stream points have arrived, the probability of any stream point being included in the reservoir is the same, and is equal to k/n.*

Proof: This result is easy to show by induction. At initialization of the first k data points, the theorem is trivially true. Let us (inductively) assume that it is also true after $(n-1)$ data points have been received. Therefore, the probability of each point being included in the reservoir is $k/(n-1)$. The lemma is trivially true for the arriving data point because the probability of its being included in the stream is k/n. It remains to prove the result for the remaining points in the data stream. There are two *disjoint* case events that can arise for an incoming data point, and the final probability of a point being included in the reservoir is the sum of these two cases:

I: The incoming data point is not inserted into the reservoir. The probability of this is $(n-k)/n$. Because the original probability of any point being included in the reservoir by the inductive assumption, is $k/(n-1)$, the overall probability of a point being included in the reservoir *and* the occurrence of the Case I event, is the multiplicative value of $p_1 = \frac{k(n-k)}{n(n-1)}$.

II: The incoming data point is inserted into the reservoir. The probability of Case II is equal to insertion probability k/n of incoming data points. Subsequently, existing reservoir points are retained with probability $(k-1)/k$ because exactly one of them is ejected. Because the inductive assumption implies that any of the earlier points in the stream was originally present in the reservoir with probability $k/(n-1)$, it implies that the probability of a point being included in the reservoir *and* Case II event is given by the product p_2 of the three aforementioned probabilities:

$$p_2 = \left(\frac{k}{n}\right)\left(\frac{k-1}{k}\right)\left(\frac{k}{n-1}\right) = \frac{k(k-1)}{n(n-1)} \quad (12.1)$$

Therefore, the total probability of a stream point being retained in the reservoir after the nth data point arrival is given by the sum of p_1 and p_2. It can be shown that this is equal to k/n. ■

The major problem with this approach is that it cannot handle concept drift because the data is uniformly sampled without decay.

12.2.1.1 Handling Concept Drift

In streaming scenarios, recent data are generally considered more important than older data. This is because the data generating process may change over time, and the older data are often considered "stale" from the perspective of analytical insights. A uniform random sample from the reservoir will contain data points that are distributed uniformly over time. Typically, most streaming applications use a decay-based framework to regulate the relative importance of data points, so that more recent data points have a higher probability to be included in the sample. This is achieved with the use of a *bias function*.

The bias function associated with the rth data point, at the time of arrival of the nth data point, is given by $f(r, n)$. This function is related to the probability $p(r, n)$ of the rth data point belonging to the reservoir at the time of arrival of the nth data point. In other words, the value of $p(r, n)$ is proportional to $f(r, n)$. It is reasonable to assume that the function $f(r, n)$ decreases monotonically with n (for fixed r), and increases monotonically with r (for fixed n). In other words, recent data points have a higher probability of belonging to the reservoir. This kind of sampling will result in a *bias-sensitive sample* $\mathcal{S}(n)$ of data points.

Definition 12.2.1 *Let $f(r, n)$ be the bias function for the rth point at the time of arrival of the nth point. A* biased *sample $\mathcal{S}(n)$ at the time of arrival of the nth point in the stream is defined as a sample such that the relative probability $p(r, n)$ of the rth point belonging to the sample $S(n)$ (of size n) is proportional to $f(r, n)$.*

In general, it is an open problem to perform reservoir sampling with arbitrary bias functions. However, methods exist for the case of the commonly used *exponential bias* function:

$$f(r, n) = e^{-\lambda(n-r)} \tag{12.2}$$

The parameter λ defines the bias rate and typically lies in the range $[0, 1]$. In general, this parameter λ is chosen in an application-specific way. A choice of $\lambda = 0$ represents the unbiased case. The exponential bias function defines the class of *memoryless functions* in which the future probability of retaining a current point in the reservoir is independent of its past history or arrival time. It can be shown that this problem is interesting only in space constrained scenarios, where the size of the reservoir k is strictly less than $1/\lambda$. This is because it can be shown [35] that an exponentially biased sample from a stream of infinite length, will not exceed $1/\lambda$ in expected size. This is also referred to as the *maximum reservoir requirement*. The following discussion is based on the assumption that $k < 1/\lambda$.

The algorithm starts with an empty reservoir. The following replacement policy is used to fill up the reservoir. Assume that at the time of (just before) the arrival of the nth point, the fraction of the reservoir filled is $F(n) \in [0, 1]$. When the $(n+1)$th point arrives, it is inserted into the reservoir with insertion probability[1] $\lambda \cdot k$. However, one of the old points in the reservoir is not necessarily deleted because the reservoir is only partially full. A coin is flipped with the success probability $F(n)$. In the event of a success, one of the points in the reservoir is randomly selected and replaced with the incoming $(n+1)$th point. In the event of a failure, there is no deletion, and the $(n+1)$th point is added to the reservoir. In the latter case, the number of points in the reservoir (the current sample size) increases by 1. In this approach, the reservoir fills up fast early in the process, but then levels off, as it reaches near its capacity. The reader is referred to the bibliographic notes for the proof of correctness of this approach. A variant of this approach that fills up the reservoir even faster is also discussed in the same work.

[1]This value is always at most 1, because $k < 1/\lambda$.

12.2.1.2 Useful Theoretical Bounds for Sampling

While the reservoir method provides data samples, it is often desirable to obtain quality bounds on the results obtained with the use of the sampling process. A common application of sampling is the estimation of statistical aggregates with the reservoir sample. The accuracy of these aggregates is often quantified with the use of *tail inequalities.*

The simplest tail inequality is the *Markov inequality.* This inequality is defined for probability distributions that take on only nonnegative values. Let X be a random variable with the probability distribution $f_X(x)$, a mean of $E[X]$, and a variance of $Var[X]$.

Theorem 12.2.1 (Markov Inequality) *Let X be a random variable that takes on only nonnegative random values. Then, for any constant α satisfying $E[X] < \alpha$, the following is true:*

$$P(X > \alpha) \leq E[X|/\alpha \qquad (12.3)$$

Proof: Let $f_X(x)$ represent the density function for the random variable X. Then, we have:

$$\begin{aligned} E[X] &= \int_x x f_X(x)\mathrm{d}x \\ &= \int_{0 \leq x \leq \alpha} x f_X(x)\mathrm{d}x + \int_{x > \alpha} x f_X(x)\mathrm{d}x \\ &\geq \int_{x > \alpha} x f_X(x)\mathrm{d}x \\ &\geq \int_{x > \alpha} \alpha f_X(x)\mathrm{d}x. \end{aligned}$$

The first inequality follows from the nonnegativity of x, and the second follows from the fact that the integral is computed only over cases where $x > \alpha$. The term on the right-hand side of the last line is exactly equal to $\alpha P(X > \alpha)$. Therefore, the following is true:

$$E[X] \geq \alpha P(X > \alpha) \qquad (12.4)$$

The above inequality can be rearranged to obtain the final result. ■

The Markov inequality is defined only for probability distributions of nonnegative values and provides a bound only on the upper tail. In practice, it is often desired to bound both tails of probability distributions over both positive and negative values.

Consider the case where X is a random variable that is not necessarily nonnegative. The *Chebychev inequality* is a useful approach to derive symmetric tail bounds on X. The Chebychev inequality is a direct application of the Markov inequality to a nonnegative (square deviation-based) derivative of X.

Theorem 12.2.2 (Chebychev Inequality) *Let X be an arbitrary random variable. Then, for any constant α, the following is true:*

$$P(|X - E[X]| > \alpha) \leq Var[X|/\alpha^2 \qquad (12.5)$$

Proof: The inequality $|X - E[X]| > \alpha$ is true if and only if $(X - E[X])^2 > \alpha^2$. By defining $Y = (X - E[X])^2$ as a (nonnegative) derivative random variable from X, it is easy to see that $E[Y] = Var[X]$. Then, the expression on the left-hand side of the theorem statement is the same as determining the probability $P(Y > \alpha^2)$. By applying the Markov inequality to the random variable Y, one can obtain the desired result. ■

The main trick used in the aforementioned proof was to apply the Markov inequality to a nonnegative function of the random variable. This technique can generally be very useful for proving other kinds of bounds, when the distribution of X has a specific form (such as the sum of Bernoulli random variables). In such cases, a parameterized function of the random variable can be used to obtain a parameterized bound. The underlying parameter can then be optimized for the tightest possible bound. Several well-known bounds such as the Chernoff bound and the Hoeffding inequality are derived with the use of this approach. Such bounds are significantly tighter than the (much weaker) Markov and Chebychev inequalities. This is because the parameter optimization process implicitly creates a bound that is optimized for the special form of the corresponding probability distribution of the random variable X.

Many practical scenarios can be captured with the use of special families of random variables. A particular case is one in which a random variable X may be expressed as a sum of other independent bounded random variables. For example, consider the case where the data points have binary class labels associated with them and one wishes to use a stream sample to estimate the fraction of examples belonging to each class. While the fraction of points in the sample belonging to a class provides an estimate, how can one bound the probabilistic accuracy of this bound? Note that the estimated fraction can be expressed as a (scaled) sum of independent and identically distributed (i.i.d.) binary random variables, depending on the binary class associated with each sample instance. The *Chernoff bound* provides an excellent bound on the accuracy of the estimate.

A second example is one where the underlying random variables are not necessarily binary, but are bounded. For example, consider the case where the stream data points correspond to individuals of a particular age. The average age of the individuals is estimated with the use of an average of the points in the reservoir sample. Note that the age can be (realistically) assumed to be a *bounded* random variable from the range $(0, 125)$. In such cases, the *Hoeffding* bound can be used to determine a tight bound on the estimate.

First, the Chernoff bound will be introduced. Because the expressions for the lower tail and upper tail are slightly different, they will be addressed separately. The lower-tail Chernoff bound is introduced below.

Theorem 12.2.3 (Lower-Tail Chernoff Bound) *Let X be a random variable that can be expressed as the sum of n independent binary (Bernoulli) random variables, each of which takes on the value of 1 with probability p_i.*

$$X = \sum_{i=1}^{n} X_i$$

Then, for any $\delta \in (0, 1)$, we can show the following:

$$P(X < (1 - \delta)E[X]) < e^{-E[X]\delta^2/2}, \tag{12.6}$$

where e is the base of the natural logarithm.

Proof: The first step is to show the following inequality:

$$P(X < (1 - \delta)E[X]) < \left(\frac{e^{-\delta}}{(1 - \delta)^{(1-\delta)}} \right)^{E[X]} \tag{12.7}$$

The *unknown* parameter $t > 0$ is introduced to create a parameterized bound. The lower-tail inequality of X is converted into an upper-tail inequality on e^{-tX}. This can be bounded by the Markov inequality, and it provides a bound that is a function of t. This function of

t can be optimized, to obtain the tightest possible bound. By using the Markov inequality on the exponentiated form, the following can be derived:

$$P(X < (1-\delta)E[X]) \leq \frac{E[e^{-tX}]}{e^{-t(1-\delta)E[X]}}.$$

By expanding $X = \sum_{i=1}^{n} X_i$ in the exponent, the following can be obtained:

$$P(X < (1-\delta)E[X]) \leq \frac{\prod_i E[e^{-tX_i}]}{e^{-t(1-\delta)E[X]}}. \tag{12.8}$$

The aforementioned simplification uses the fact that the expectation of the product of independent variables is equal to the product of the expectations. Because each X_i is Bernoulli, the following can be shown by summing up the probabilities over the cases where $X_i = 0$ and 1, respectively:

$$E[e^{-tX_i}] = 1 + E[X_i](e^{-t} - 1) < e^{E[X_i](e^{-t}-1)}.$$

The second inequality follows from the polynomial expansion of $e^{E[X_i](e^{-t}-1)}$. By substituting this inequality back into Eq. 12.8, and using $E[X] = \sum_i E[X_i]$, the following may be obtained:

$$P(X < (1-\delta)E[X]) \leq \frac{e^{E[X](e^{-t}-1)}}{e^{-t(1-\delta)E[X]}}.$$

The expression on the right is true for any value of $t > 0$. It is desired to pick a value t that provides the *tightest possible* bound. Such a value of t may be obtained by using the standard optimization process of using the derivative of the expression with respect to t. It can be shown by working out the details of this optimization process that the optimum value of $t = t^*$ is as follows:

$$t^* = \ln(1/(1-\delta)). \tag{12.9}$$

By using this value of t^* in the inequality above, it can be shown to be equivalent to Eq. 12.7. This completes the first part of the proof.

The first two terms of the Taylor expansion of the logarithmic term in $(1-\delta)\ln(1-\delta)$ can be expanded to show that $(1-\delta)^{(1-\delta)} > e^{-\delta+\delta^2/2}$. By substituting this inequality in the denominator of Eq. 12.7, the desired result is obtained. ∎

A similar result for the upper-tail Chernoff bound may be obtained that has a slightly different form.

Theorem 12.2.4 (Upper-Tail Chernoff Bound) *Let X be a random variable that can be expressed as the sum of n independent binary (Bernoulli) random variables, each of which takes on the value of 1 with probability p_i.*

$$X = \sum_{i=1}^{n} X_i.$$

Then, for any $\delta \in (0, 2e-1)$, the following is true:

$$P(X > (1+\delta)E[X]) < e^{-E[X]\delta^2/4}, \tag{12.10}$$

where e is the base of the natural logarithm.

Proof: The first step is to show the following inequality:

$$P(X > (1+\delta)E[X]) < \left(\frac{e^{\delta}}{(1+\delta)^{(1+\delta)}}\right)^{E[X]}. \tag{12.11}$$

As before, this can be done by introducing the unknown parameter $t > 0$, and converting the upper-tail inequality on X into that on e^{tX}. This can be bounded by the Markov inequality, and it provides a bound that is a function of t. This function of t can be optimized to obtain the tightest possible bound.

It can be further shown by algebraic simplification that the inequality in Eq. 12.11 provides the desired result, when $\delta \in (0, 2e-1)$. ■

Next, the Hoeffding inequality will be introduced. The Hoeffding inequality is a more general tail inequality than the Chernoff bound because it does not require the underlying data values to be Bernoulli. In this case, the ith data value needs to be drawn from *the bounded interval* $[l_i, u_i]$. The corresponding probability bound is expressed in terms of the parameters l_i and u_i. Thus, the scenario for the Chernoff bound is a special case of that for the Hoeffding's inequality. We state the Hoeffding inequality below, for which both the upper- and lower-tail inequalities are identical.

Theorem 12.2.5 (Hoeffding Inequality) *Let X be a random variable that can be expressed as the sum of n independent random variables, each of which is bounded in the range $[l_i, u_i]$.*

$$X = \sum_{i=1}^{n} X_i.$$

Then, for any $\theta > 0$, the following can be shown:

$$P(X - E[X] > \theta) \leq e^{-\frac{2\theta^2}{\sum_{i=1}^{n}(u_i - l_i)^2}} \tag{12.12}$$

$$P(E[X] - X > \theta) \leq e^{-\frac{2\theta^2}{\sum_{i=1}^{n}(u_i - l_i)^2}} \tag{12.13}$$

Proof: The proof for the upper tail will be briefly described here. The proof of the lower-tail inequality is identical. For an unknown parameter t, the following is true:

$$P(X - E[X] > \theta) = P(e^{t(X - E[X])} > e^{t\theta}) \tag{12.14}$$

The Markov inequality can be used to show that the right-hand probability is at most $E[e^{(X-E[X])}]e^{-t\theta}$. The expression within $E[e^{(X-E[X])}]$ can be expanded in terms of the individual components X_i. Because the expectation of the product is equal to the product of the expectations of independent random variables, the following can be shown:

$$P(X - E[X] > \theta) \leq e^{-t\theta} \prod_i E[e^{t(X_i - E[X_i])}]. \tag{12.15}$$

The key is to show that the value of $E[e^{t(X_i - E[X_i])}]$ is at most equal to $e^{t^2(u_i - l_i)^2/8}$. This can be shown with the use of an argument that uses the convexity of the exponential function $e^{t(X_i - E[X_i])}$ in conjunction with Taylor's theorem (see Exercise 12).

Therefore, the following is true:

$$P(X - E[X] > \theta) \leq e^{-t\theta} \prod_i e^{t^2(u_i - l_i)^2/8}. \tag{12.16}$$

Table 12.1: Comparison of different methods used to bound tail probabilities

Result	Scenario	Strength
Chebychev	Any random variable	Weak
Markov	Nonnegative random variable	Weak
Hoeffding	Sum of independent bounded random variables	Strong
Chernoff	Sum of independent Bernoulli variables	Strong

This inequality holds for any nonnegative value of t. Therefore, to find the tightest bound, the value of t that minimizes the right-hand side of the above equation needs to be determined. The optimal value of $t = t^*$ can be shown to be:

$$t^* = \frac{4\theta}{\sum_{i=1}^{n}(u_i - l_i)^2}. \tag{12.17}$$

By substituting the value of $t = t^*$ in Eq. 12.16, the desired result may be obtained. The lower-tail bound may be derived by applying the aforementioned steps to $P(E[X] - X > \theta)$ rather than $P(X - E[X] > \theta)$. ■

Thus, the different inequalities may apply to scenarios of different generality, and they may also have different levels of strength. These different scenarios are illustrated in Table 12.1.

12.2.2 Synopsis Structures for the Massive-Domain Scenario

As discussed in the introduction, many streaming applications contain discrete attributes, whose domain is drawn on a large number of distinct values. A classical example would be the value of the IP address in a network stream, or the e-mail address in an e-mail stream. Such scenarios are more common in data streams, simply because the massive number of data items in the stream are often associated with discrete identifiers of different types. E-mail addresses and IP addresses are examples of such identifiers. The streaming objects are often associated with *pairs* of identifiers. For example, each element of an e-mail stream may have both a sender and recipient. In some applications, it may be desirable to store statistics using pairwise identifiers, and therefore the pairwise combination is treated as a single integrated attribute. The *domain of possible values* can be rather large. For example, for an e-mail application with over a hundred million different senders and receivers, the number of possible pairwise combinations is 10^{16}. In such cases, *even storing simple summary statistics such as set membership, frequent item counts, and distinct element counts becomes more challenging from the perspective of space constraints.*

If the number of distinct elements were small, one might simply use an array, and update the counts in these arrays in order to create an effective summary. Such a summary could address all the aforementioned queries. However, such an approach would not be practical in the massive-domain scenario because an array with 10^{16} elements would require more than 10 petabytes. Furthermore, for many queries, such as set membership and distinct element counting, a reservoir sample would not work well. This is because the vast majority of the stream may contain infrequent elements, and the reservoir will disproportionately overrepresent the frequent elements for queries that are agnostic to the absolute frequency. Set membership and distinct-element counting are examples of such queries.

It is often difficult to create a single synopsis structure that can address all queries. Therefore, different synopsis structures are designed for different classes of queries. In the following, a number of different synopsis structures will be described. Each of these synopsis structures is optimized to different kinds of queries. For each of the synopsis structures discussed in this chapter, the relevant queries and query-processing approach will also be described.

12.2.2.1 Bloom Filter

Bloom filters are designed for set-membership queries of discrete elements. The set-membership query is as follows:

Given a particular element, has it ever occurred in the data stream?

Bloom filters provide a way to maintain a synopsis of the stream, so that this query can be resolved with a probabilistic bound on the accuracy. One property of this data structure is that false positives are possible, but false negatives are not. In other words, if the bloom filter reports that an element does not belong to the stream, then this will always be the case. Bloom filters are referred to as "filters" because they can be used for making important selection decisions in a stream in real time. This is because the knowledge of membership of an item in a set of stream elements plays an important role in filtering decisions, such as the removal of duplicates. This will be discussed in more detail later. First, the simple case of stream membership queries will be discussed.

A bloom filter is a binary bit array of length m. Thus, the space requirement of the bloom filter is $m/8$ bytes. The elements of the bit array are indexed starting with 0 and ending at $(m-1)$. Therefore, the index range is $\{0, 1, 2, \ldots m-1\}$. The bloom filter is associated with a set of w *independent* hash functions denoted by $h_1(\cdot) \ldots h_w(\cdot)$. The argument of each of these hash functions is an element of the data stream. The hash function maps uniformly at random to an integer in the range $\{0 \ldots m-1\}$.

Consider a stream of discrete elements. These discrete elements could be e-mail addresses (either individually or sender–receiver pairs), IP addresses, or another set of discrete values drawn on a massive domain of possible values. The bits on the bloom filter are used to keep track of the distinct values encountered. The hash functions are used to map the stream elements to the bits in the bloom filter. For the following discussion, it will be assumed that the bloom filter data structure is denoted by $\mathcal{B}$.

The bloom filter is constructed from a stream $\mathcal{S}$ of values as follows. All bits in the bloom filter are initialized to 0. For each incoming stream element x, the functions $h_1(x) \ldots h_w(x)$ are applied to it. For each $i \in \{1 \ldots w\}$, the element $h_i(x)$ in the bloom filter is set to 1. In many cases, the value of some of these bits might already be 1. In such cases, the value does not need to be changed. A pictorial representation of the bloom filter and the update process is illustrated in Fig. 12.1. The pseudocode for the overall update process is illustrated in Fig. 12.2. In the pseudocode, the stream is denoted by $\mathcal{S}$, and the bloom filter data structure is denoted by $\mathcal{B}$. The input parameters include the size of the bloom filter m, and the number of hash functions w. It is important to note that multiple elements can map onto the same bit in the bloom filter. This is referred to as a *collision*. As discussed later, collisions may lead to false positives in set-membership checking.

The bloom filter can be used to check for membership of an item y in the data stream. The first step is to compute the hash functions $h_1(y) \ldots h_w(y)$. Then, it is checked whether the $h_i(y)$th element is 1. If at least one of the these values is 0, we are *guaranteed* that the element has not occurred in the data stream. This is because, if that element had occurred in the stream, the entry would have already been set to 1. Thus, false negatives

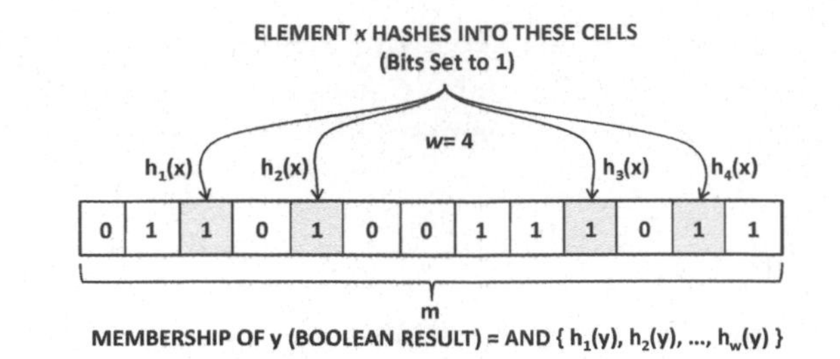

Figure 12.1: The bloom filter

Algorithm *BloomConstruct*(Stream: $\mathcal{S}$, Size: m, Num. Hash Functions: w)
begin
  Initialize all elements in a bit array $\mathcal{B}$ of size m to 0;
  repeat
    Receive next stream element $x \in \mathcal{S}$;
    for $i = 1$ to w **do**
      Update $h_i(x)$th element in bit array $\mathcal{B}$ to 1;
  until end of stream $\mathcal{S}$;
  return $\mathcal{B}$;
end

Figure 12.2: Update of bloom filter

are not possible with the bloom filter. On the other hand, if all the entries $h_1(y) \ldots h_w(y)$ in the bit array have a value of 1, then it is reported that y has occurred in the data stream. This can be checked efficiently by applying an "AND" logical operator to all the bit entries corresponding to the indices $h_1(y) \ldots h_w(y)$ in the bit array. The overall procedure of membership checks is illustrated in Fig. 12.3. The binary result for the decision problem for checking membership is tracked by the variable *BooleanFlag*. This binary flag is reported at the end of the procedure.

The bloom filter approach can lead to false positives, but not false negatives. A false positive occurs, if all the w different bloom filter array elements $h_i(y)$ for $i \in \{1 \ldots w\}$ have been set to 1 by some spurious element other than y. This is a direct result of collisions. As the number of elements in the data stream increases, all elements in the bloom filter are eventually set to 1. In such a case, all set-membership queries will yield a positive response. This is, of course, not a useful application of the bloom filter. Therefore, it is instructive to bound the false positive probability in terms of the size of the filter and the number of distinct elements in the data stream.

Lemma 12.2.2 *Consider a bloom filter $\mathcal{B}$ with m elements, and w different hash functions. Let n be the number of distinct elements in the stream $\mathcal{S}$ so far. Consider an element y, which has not appeared in the stream so far. Then, the probability F that an element y is reported as a false positive is given by the following:*

$$F = \left[1 - \left(1 - \frac{1}{m}\right)^{w \cdot n}\right]^w \tag{12.18}$$

Algorithm *BloomQuery*(Element: y, Bloom Filter: $\mathcal{B}$)
begin
  Initialize $BooleanFlag = 1$;
  for $i = 1$ to w **do**
    $BooleanFlag = BooleanFlag$ AND $h_i(y)$;
  return $BooleanFlag$;
end

Figure 12.3: Membership check using bloom filter

Proof: Consider a particular bit corresponding to the bit array element $h_r(y)$ for some fixed value of the index $r \in \{1 \dots w\}$. Each element $x \in \mathcal{S}$ sets w different bits $h_1(x) \dots h_w(x)$ to 1. The probability that *none* of these bits is the same as $h_r(y)$ is given by $(1 - 1/m)^w$. Over n distinct stream elements, this probability is $(1 - 1/m)^{w \cdot n}$. Therefore, the probability that the bit array index $h_r(y)$ is set to 1, by at least one of the n spurious elements in $\mathcal{S}$ is given by $Q = 1 - (1 - 1/m)^{w \cdot n}$. A false positive occurs, when *all* bit array indices $h_r(y)$ (over varying values of $r \in \{1 \dots w\}$) have been set to 1. The probability of this is $F = Q^w$. The result follows. ∎

While the false-positive probability is expressed above in terms of the number of *distinct* stream elements, it is trivially true for the total number of stream elements (including repetitions), as an upper bound.

The expression in the aforementioned lemma can be simplified by observing that $(1 - 1/m)^m \approx e^{-1}$, where e is the base of the natural logarithm. Correspondingly, the expression can be rewritten as follows:

$$F = (1 - e^{-n \cdot w/m})^w. \tag{12.19}$$

Values of w that are too small or too large lead to poor performance. The value of w needs be selected optimally in terms of m and n to minimize the number of false positives. The number of false positives is minimized, when $w = m \cdot \ln(2)/n$. Substituting this value in Eq. 12.19, it can be shown that the probability of false positives for optimal number of hash functions is:

$$F = 2^{-m \cdot \ln(2)/n}. \tag{12.20}$$

The expression above can be written purely as an expression of m/n. Therefore, *for a fixed value of the false-positive probability F, the length of the bloom filter m needs to be proportional to the number of distinct elements n in the data stream.* Furthermore, the constant of proportionality for a particular false-positive probability F can be shown to be $\frac{m}{n} = \frac{\ln(1/F)}{(\ln(2))^2}$. While this may not seem like a significant compression, it needs to be pointed out that bloom filters use elementary *bits* to track the membership of arbitrary elements, such as strings. Furthermore, because of bitwise operations, which can be implemented very efficiently with low-level implementations, the overall approach is generally very efficient.

It does need to be kept in mind that the value of n is not known in advance for many applications. Therefore, one strategy is to use a cascade of bloom filters for geometrically increasing values of w, and to use a logical AND of the membership query result over different bloom filters. This is a practical approach that provides more stable performance over the life of the data stream.

The bloom filter is referred to as a "filter" because it is often used to make decisions on which elements to exclude from a data stream, when they meet the membership condition.

For example, if one wanted to filter all duplicates from the data stream, the bloom filter is an effective strategy. Another strategy is to filter forbidden elements from a universe of values, such as a set of spammer e-mail addresses in an e-mail stream. In such a case, the bloom filter needs to be constructed up front with the spam e-mail addresses.

Many variations of the basic bloom filter strategy provide different capabilities and trade-offs:

1. The bloom filter can be used to approximate the number of distinct elements in a data stream. If $m_0 < m$ is the number of bits with a value of 0 in the bloom filter, then the number of distinct elements n can be estimated as follows (see Exercise 13):

$$n \approx \frac{m \cdot \ln(m/m_0)}{w} \tag{12.21}$$

   The accuracy of this estimate reduces drastically, as the bloom filter fills up. When $m_0 = 0$, the value of n is estimated to be ∞, and therefore the estimate is practically useless.

2. The bloom filter can be used to estimate the size of the intersection and union of sets corresponding to different streams, by creating one bloom filter for each stream. To determine the size of the union, the bitwise OR of the two filters is determined. The bitwise OR of the filter can be shown to be *exactly* the same as the bloom filter representation of the union of the two sets. Then, the formula of Eq. 12.21 is used. However, such an approach cannot be used for determining the size of the intersection. While the intersection of two sets can be approximated by using a bitwise AND operation on the two filters, the resulting bit positions in the filter will not be the same as that obtained by constructing the filter directly on the intersection. The resulting filter might contain false negatives, and, therefore, such an approach is *lossy*. To estimate the size of the intersection, one can first estimate the size of the union and then use the following simple setwise relationship:

$$|\mathcal{S}_1 \cap \mathcal{S}_2| = |\mathcal{S}_1| + |\mathcal{S}_2| - |\mathcal{S}_1 \cup \mathcal{S}_2| \tag{12.22}$$

3. The bloom filter is primarily designed for membership queries, and is not the most space-efficient data structure, when used purely for distinct element counting. In a later section, a space-efficient technique, referred to as the Flajolet–Martin algorithm, will be discussed.

4. The bloom filter can allow a limited (one-time) tracking of deletions by setting the corresponding bit elements to zero, when an element is deleted. In such a case, false negatives are also possible.

5. Variants of the bloom filter can be designed in which the w hash functions can map onto separate bit arrays. A further generalization of this principle is to track counts of elements rather than simply binary bit values to enable richer queries. This generalization, discussed in the next section, is also referred to as the *count-min* sketch.

Bloom filters are commonly used in many streaming settings in the text domain.

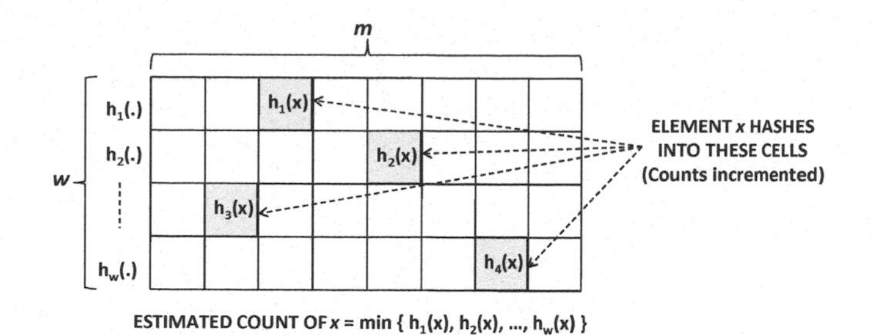

Figure 12.4: The count-min sketch

12.2.2.2 Count-Min Sketch

While the bloom filter is effective for set-membership queries, it is not designed for methods that require *count-based* summaries. This is because the bloom filter tracks only binary values. The count-min sketch is designed for resolving such queries and is intuitively related to the bloom filter. A count-min sketch consists of a set of w different *numeric* arrays, each of which has a length m. Thus, the space requirement of the count-min sketch is equal to $m \cdot w$ cells containing *numeric values*. The elements of each of the w numeric arrays are indexed starting with 0, corresponding to an index range of $\{0 \dots m-1\}$. The count-min sketch can also be viewed as a $w \times m$ 2-dimensional array of cells.

Each of the w numeric arrays corresponds to a hash function. The ith numeric array corresponds to the ith hash function $h_i(\cdot)$. The hash functions have the following properties:

1. The ith hash function $h_i(\cdot)$ maps a stream element to an integer in the range $[0 \dots m-1]$. This mapping can also be viewed as one of the index values in the ith numeric array.

2. The w hash functions $h_1(\cdot) \dots h_w(\cdot)$ are fully independent of one another, but *pairwise* independent over different arguments. In other words, for any two values x_1 and x_2, $h_i(x_1)$ and $h_i(x_2)$ are independent.

The *pairwise* independence requirement is a weaker one than the full independence requirement. This is a convenient property of the count-min sketch because it is usually easier to construct pairwise independent hash functions rather than fully independent ones.

The procedure for updating the sketch is as follows. All $m \cdot w$ entries in the count-min sketch are initialized to 0. For each incoming stream element x, the functions $h_1(x) \dots h_w(x)$ are applied to it. For the ith array, the element $h_i(x)$ is incremented by 1. Thus, if the count-min sketch $\mathcal{CM}$ is visualized as a 2-dimensional $w \times m$ numeric array, then the element $(i, h_i(x))$ is incremented[2] by 1. Note that the value of $h_i(x)$ maps to an integer in the range $[0, m-1]$. This is also the range of the indices of each numeric array. A pictorial illustration of the count-min sketch and the corresponding update process is provided in Fig. 12.4. The pseudocode for the overall update process is illustrated in Fig. 12.5. In the pseudocode, the stream is denoted by $\mathcal{S}$, and the count-min sketch data structure is denoted by $\mathcal{CM}$. The inputs to the algorithm are the stream $\mathcal{S}$ and two parameters (w, m) specifying the size of the 2-dimensional array for the count-min sketch. A 2-dimensional $w \times m$ array $\mathcal{CM}$ is initialized with all values set to 0. For each incoming stream element, the counts of all the

[2]In the event that each distinct element is associated with a nonnegative frequency, the count-min sketch can be updated with the frequency value. Only the simple case of unit updates is discussed here.

```
Algorithm CountMinConstruct(Stream: S, Width: w, Height: m)
begin
  Initialize all entries of w × m array CM to 0;
  repeat
    Receive next stream element x ∈ S;
    for i = 1 to w do
         Increment (i, h_i(x))th element in CM by 1;
  until end of stream S;
  return CM;
end
```

Figure 12.5: Update of count-min sketch

```
Algorithm CountMinQuery(Element: y, Count-min Sketch: CM)
begin
  Initialize Estimate = ∞;
  for i = 1 to w do
       Estimate = min{Estimate, V_i(y)};
  { V_i(y) is the count of the (i, h_i(y))th element in CM }
  return Estimate;
end
```

Figure 12.6: Frequency queries for count-min sketch

cells $(i, h_i(x))$ are updated for $i \in \{1 \dots w\}$. In the pseudocode description, the resulting sketch $\mathcal{CM}$ is returned after processing all the stream elements. In practice, the count-min sketch can be used at any time during the progression of the stream $\mathcal{S}$. As in the case of the bloom filter, it is possible for multiple stream elements to map to the same cell in the count-min sketch. Therefore, different stream elements will increment the same cell, and the resulting cell counts are always overestimates of one or more stream element counts.

The count-min sketch can be used for many different queries. The simplest query is to determine the frequency of an element y. The first step is to compute the hash functions $h_1(y) \dots h_w(y)$. For the ith numeric array in $\mathcal{CM}$, the value $V_i(y)$ of the $(i, h_i(y))$th array element is retrieved. Each value $V_i(y)$ is an overestimate of the true frequency of y because of potential collisions. Therefore, the tightest possible estimate may be obtained by using the minimum value $\min_i\{V_i(y)\}$ over the different hash functions. The overall procedure for frequency estimation is illustrated in Fig. 12.6.

The count-min sketch causes an overestimation of frequency values because of collisions of nonnegative frequency counts of distinct stream items. It is therefore helpful to determine an upper bound on the estimation quality.

Lemma 12.2.3 *Let $E(y)$ be the estimate of the frequency of the item y, using a count-min sketch of size $w \times m$. Let n_f be the total frequencies of all items received so far, and $G(y)$ be true frequency of item y. Then, with probability at least $1 - e^{-w}$, the upper bound on the estimate $E(y)$ is as follows:*

$$E(y) \leq G(y) + \frac{n_f \cdot e}{m}. \tag{12.23}$$

Here, e represents the base of the natural logarithm.

Proof: The expected number of spurious items hashed to the cells belonging to item y is about[3] n_f/m, if all spurious items are hashed uniformly at random to the different cells. This result uses pairwise independence of hash functions because it relies on the fact that the mapping of y to a cell does not affect the distribution of another spurious item in its cells. The probability of the number of spurious items exceeding $n_f \cdot e/m$ in any of the w cells belonging to y is given by at most e^{-1} by the Markov inequality. For $E(y)$ to exceed the upper bound of Eq. 12.23, this violation needs to be repeated for all the w cells to which y is mapped. The probability of a violation of Eq. 12.23 is therefore e^{-w}. The result follows. ∎

In many cases, it is desirable to directly control the error level ϵ and the error probability δ. By setting $m = e/\epsilon$ and $w = \ln(1/\delta)$, it is possible to bound the error with a user-defined tolerance $n_f \cdot \epsilon$ and probability at least $1 - \delta$. Two natural generalizations of the point query can be implemented as follows:

1. If the stream elements have arbitrary positive frequencies associated with them, the only change required is to the update operation, where the counts are incremented by the relevant frequency. The frequency bound is identical to Eq. 12.23, with n_f representing the sum of the frequencies of the stream items.

2. If the stream elements have either positive or negative frequencies associated with them, then a further change is required to the query procedure. In this case, the *median* of the counts is reported. The corresponding error bound of Eq. 12.23 now needs to be modified. With a probability of at least $1 - e^{-w/4}$, the estimated frequency $E(y)$ of item y lies in the following ranges:

$$G(y) - \frac{3n_f \cdot e}{m} \leq E(y) \leq G(y) + \frac{3n_f \cdot e}{m}. \tag{12.24}$$

   In this case, n_f represents the sum of the *absolute* frequencies of the incoming items in the data stream. The bounds in this case are much weaker than those for nonnegative elements.

A useful application is to determine the dot product of the frequency counts of the discrete attribute values in two data streams. This has a useful application in estimating the join size on the massive-domain attribute in two data streams. The dot product between the frequency counts of the items in a pair of nonnegative data streams can be estimated by first constructing a count-min sketch representation for each of the two data streams in a separate $w \times m$ count-min data structure. The same hash functions are used for both sketches. The dot product of their corresponding count-min arrays for each hash function is computed. The minimum value of the dot product over the w different arrays is reported as the estimation. As in the previous case, this is an overestimate, and an upper bound on the estimate may be obtained with a probability of at least $1 - e^{-w}$. The corresponding error tolerance for the upper bound is $n_f^1 \cdot n_f^2 \cdot e/m$, where n_f^1 and n_f^2 are the aggregate frequencies of the items in each of the two streams. Other useful queries with the use of the count-min sketch include the determination of quantiles and frequent elements. Frequent elements are also referred to as *heavy hitters*. The bibliographic notes contain pointers to various queries and applications that can be addressed with the use of the count-min sketch.

[3]It is exactly equal to n_s/m, where n_s is the frequency of all items other than y. However, n_s is less than n_f by the frequency of y.

12.2.2.3 AMS Sketch

As discussed at the beginning of this section, different synopsis structures are designed for different kinds of queries. While the bloom filter and count-min sketch provide good estimations of various queries, some queries, such as second moments, can be better addressed with the *Alon–Matias–Szegedy (AMS)* sketch. In the AMS sketch, a random binary value is generated from $\{-1, 1\}$ for each stream element by applying a hash function to the stream element. These binary values are assumed to be 4-wise independent. This means that, if at most four values generated from the same hash function are sampled, they will be statistically independent of one another. It is easier to design a 4-wise independent hash function than a fully independent hash function. The details of 4-wise independent hash functions may be found in the bibliographic notes.

Consider a stream in which the ith stream element is associated with the *aggregate* frequency f_i. The second-order moment F_2 of the data stream, for a stream with n *distinct* elements, is defined as follows:

$$F_2 = \sum_{i=1}^{n} f_i^2 \tag{12.25}$$

In the massive-domain scenario, where the number of distinct elements is large, this quantity is hard to estimate because running counts of the frequencies f_i cannot be maintained with an array. However, it can be estimated effectively using the AMS sketch. As a practical application, the second-order moment yields an estimate of the self-join size of a data stream with respect to the massive-domain attribute. The second-order moment can also be viewed as a variant of the Gini index, which measures the level of frequency skew over different items in the data stream. When the skew is large, the value of F_2 is large, and very close to its upper bound $(\sum_{i=1}^{n} f_i)^2$.

The AMS sketch contains m different sketch components, each of which is associated with an independent hash function. Each hash function generates its corresponding sketch component as follows. A random binary value, with equal probability for both realizations, is generated for the incoming stream element. This binary value is denoted by $r \in \{-1, 1\}$, and is generated using the hash function for that component. The frequency of each incoming stream element is multiplied by r, and added to the corresponding component of the sketch. Let $r_i \in \{-1, 1\}$ be the random value generated by a particular hash function for the ith distinct element. Then, the corresponding component Q of the sketch, for a stream of n distinct elements with aggregate frequencies $f_1 \ldots f_n$, can be shown to be equal to the following:

$$Q = \sum_{i=1}^{n} f_i \cdot r_i. \tag{12.26}$$

This relationship is because of the incremental way in which Q is updated, each time a stream item is received. Note that the value of Q is a random variable, dependent on how the binary random values $r_1 \ldots r_n$ are generated by the hash function. The value of Q is useful in estimating the second moment.

Lemma 12.2.4 *The second moment of the data stream can be estimated by the square of the AMS sketch component Q:*

$$F_2 = E[Q^2]. \tag{12.27}$$

Proof: It is easy to see that $Q^2 = \sum_{i=1}^{n} f_i^2 r_i^2 + 2 \sum_{i=1}^{n} \sum_{j=1}^{n} f_i \cdot f_j \cdot r_i \cdot r_j$. For any pair of hash values r_i, r_j, we have $r_i^2 = r_j^2 = 1$ and $E[r_i \cdot r_j] = E[r_i] \cdot E[r_j] = 0$. The last of these results uses 2-wise independence, which is implied by 4-wise independence. Therefore, $E[Q^2] = \sum_{i=1}^{n} f_i^2 = F_2$. ■

The 4-wise independence can also be used to bound the variance of the estimate (see Exercise 16).

Lemma 12.2.5 *The variance of the square of a component Q of the AMS sketch is bounded above by twice the frequency moment.*

$$Var[Q^2] \leq 2 \cdot F_2^2 \tag{12.28}$$

The bound on the variance can be reduced further by averaging over the m different sketch components $Q_1 \ldots Q_m$. The reduced variance can be used to create a (weak) probabilistic estimate on the quality of the second moment estimate with the Chebychev inequality. This can be tightened further by using a "mean–median combination trick" that is commonly used in such a probabilistic analysis. This trick can be used to robustly estimate a random variable, whenever its variance is no larger than a modest factor of the square of its expected value. This is the case for the random variable Q^2.

The mean–median combination trick works as follows. It is desired to establish a bound with probability at least $(1 - \delta)$ that the second moment can be estimated to within a multiplicative factor of $1 \pm \epsilon$. Let $Q_1 \ldots Q_m$ be m different sketch components, each of which is generated using a different hash function. The value of m is chosen to be $O(\ln(1/\delta)/\epsilon^2)$. These m sketch components are further partitioned into $O(\ln(1/\delta))$ different groups of size $O(1/\epsilon^2)$ each. The sketch values in each group are averaged. The median of these $O(\ln(1/\delta))$ averages is reported. A combination of the Chebychev inequality and the Chernoff bounds can be used to show the following result:

Lemma 12.2.6 *By selecting the median of $O(ln(1/\delta))$ averages of $O(1/\epsilon^2)$ copies of Q_i^2, it is possible to guarantee the accuracy of the sketch-based second-moment approximation to within $1 \pm \epsilon$ with a probability of at least $1 - \delta$.*

Proof: According to Lemma 12.2.5, the variance of each sketch component is at most $2 \cdot F_2^2$. By using the average of $16/\epsilon^2$ independent sketch components, the variance of the averaged estimate can be reduced to $F_2^2 \cdot \epsilon^2/8$. In this case, the Chebychev inequality shows that the ϵ-bound is violated by the averaged estimate with probability at most $1/8$. Assume that a total of $4 \cdot \ln(1/\delta)$ such averaged and independent estimates are available. The random variable Y is defined as the sum of the Bernoulli indicator variables of ϵ-bound violations over these $q = 4 \cdot \ln(1/\delta)$ averages. The expected value of Y is $q/8 = \ln(1/\delta)/2$. The Chernoff bound is used to show the following:

$$P(Y > q/2) = P(Y > (1+3) \cdot q/8) = P(Y > (1+3)E[Y]) \leq e^{-3^2 \cdot \ln(1/\delta)/8} = \delta^{9/8} \leq \delta.$$

The median can violate the ϵ-bound only when more than half the averages violate the bound. The probability of this event is exactly $P(Y > q/2)$. Therefore, the median violates the ϵ-bound with probability at most δ. ■

The AMS sketch can be used to estimate many other values in a similar way, with corresponding quality bounds. For example, consider two streams with the sketch components Q_i and R_i.

1. The dot product between the frequency counts of the items in a pair of streams is estimated as the product of the corresponding sketch components Q_i and R_i. By using

the median of $O(\ln(1/\delta))$ averages of different sets of $O(1/\epsilon^2)$ values of $Q_i \cdot R_i$, it is possible to bound the approximation within $1 \pm \epsilon$ with probability at least $1 - \delta$. This estimation can be performed using the count-min sketch as well, though with a different bound.

2. The Euclidean distance between the frequency counts of a pair of streams can be estimated as $Q_i^2 + R_i^2 - 2Q_i \cdot R_i$. The Euclidean distance can be viewed as a linear combination of three different dot products (including self-products) between the frequency counts of the two streams. Because each dot product is itself bounded using the "mean–median trick" discussed above, the approach can be used to determine similar quality bounds in this case as well.

3. Like the count-min sketch, the AMS sketch can be used to estimate frequency values. For the jth distinct stream element with frequency f_j, the product of the random variable r_j and Q_i provides an estimate of the frequency.

$$E[f_j] = r_j \cdot Q_i. \tag{12.29}$$

The mean, median, or mean–median combination of these values over different sketch components Q_i can be reported as a robust estimate. The AMS sketch can also be used to identify heavy hitters from the data stream.

Some of the queries resolved by the AMS and count-min sketch are similar, although others are different. The bounds provided by the two techniques are also different, although none of them is strictly better than the other in all scenarios. The count-min sketch does have the advantage of being intuitively easy to interpret because of its natural hash-table data structure. As a result, it can be more naturally integrated in data mining applications such as clustering and classification in a seamless way.

12.2.2.4 Flajolet–Martin Algorithm for Distinct Element Counting

Sketches are designed for determining stream statistics that are dominated by *large aggregate signals* of frequent items. However, they are not optimized for estimating stream statistics that are dominated by infrequently occurring items. Problems such as distinct element counting are more directly influenced by the much larger number of infrequent items in a data stream. Distinct element counting can be performed efficiently with the Flajolet–Martin algorithm.

The Flajolet–Martin algorithm uses a hash function $h(\cdot)$ to render a mapping from a given element x in the data stream to an integer in the range $[0, 2^L - 1]$. The value of L is selected to be large enough, so that 2^L is an upper bound on the number of distinct elements. Usually, the value L is selected to be 64 for implementation convenience, and because the value of 2^{64} is large enough for most practical applications. Therefore, the binary representation of the integer $h(x)$ will have length L. The position[4] R of the rightmost 1 bit of the binary representation of the integer $h(x)$ is determined. Thus, the value of R represents the number of trailing zeros in this binary representation. Let R_{max} be the maximum value of R over all stream elements. The value of R_{max} can be maintained incrementally in the streaming scenario by applying the hash function to each incoming stream element, determining its rightmost bit, and then updating R_{max}. The key idea in the Flajolet–Martin algorithm is that the dynamically maintained value of R_{max} is logarithmically related to the number of distinct elements encountered so far in the stream.

[4]The position of the least significant bit is 0, the next most significant bit is 1, and so on.

The intuition behind this result is quite simple. For a uniformly distributed hash function, the probability of R trailing zeros in the binary representation of a stream element is equal to 2^{-R-1}. Therefore, for n distinct elements and a fixed value of R, the expected number of times that exactly R trailing zeros are achieved is equal to $2^{-R-1} \cdot n$. Therefore, for values of R larger than $\log(n)$, the expected number of such bitstrings falls off exponentially less than 1. Of course, in our application, the value of R is not fixed, but it is a random variable that is generated by the outcome of the hash function. It has been rigorously shown that the expected value $E[R_{max}]$ of the maximum value of R over all stream elements is logarithmically related to the number of distinct elements as follows:

$$E[R_{max}] = \log_2(\phi n), \quad \phi = 0.77351. \tag{12.30}$$

The standard deviation is $\sigma(R_{max}) = 1.12$. Therefore, the value of $2^{R_{max}}/\phi$ provides an estimate for the number of distinct elements n. To further improve the estimate of R_{max}, the following techniques can be used:

1. Multiple hash functions can be used, and the average value of R_{max} over the different hash functions is used.

2. The averages are still somewhat susceptible to large variations. Therefore, the "mean–median trick" may be used. The medians of a set of averages are reported. Note that this is similar to the trick used in the AMS sketch. As in that case, a combination of the Chebychev inequality and Chernoff bounds can be used to establish qualitative guarantees.

It should be pointed out that the bloom filter can also be used to estimate the number of distinct elements. However, the bloom filter is not a space-efficient way to count the number of distinct elements when set-membership queries are not required.

12.3 Frequent Pattern Mining in Data Streams

The frequent pattern mining problem in data streams is studied in the context of two different scenarios. The first scenario is the massive-domain scenario, in which the number of possible items is very large. In such cases, even the problem of finding frequent items becomes difficult. Frequent items are also referred to as *heavy hitters*. The second case is the conventional scenario of a large (but manageable) number of items that fit in main memory. In such cases, the frequent item problem is no longer quite as interesting, because the frequent counts can be directly maintained in an array. In such cases, one is more interested in determining frequent *patterns*. This is a difficult problem, because most frequent pattern mining algorithms require multiple passes over the entire data set. The one-pass constraint of the streaming scenario makes this difficult. In the following, two different approaches will be described. The first of these approaches leverages generic synopsis structures in conjunction with traditional frequent pattern mining algorithms and the second designs streaming versions of frequent pattern mining algorithms.

12.3.1 Leveraging Synopsis Structures

Synopsis structures can be used effectively in most streaming data mining problems, including frequent pattern mining. In the context of frequent pattern mining methods, synopsis structures are particularly attractive because of the ability to use a wider array of algorithms, or for incorporating temporal decay into the frequent pattern mining process.

12.3.1.1 Reservoir Sampling

Reservoir sampling is the most flexible approach for frequent pattern mining in data streams. It can be used either for frequent *item* mining (in the massive-domain scenario) or for frequent pattern mining. The basic idea in using reservoir sampling is simple:

1. Maintain a reservoir sample S from the data stream.
2. Apply a frequent pattern mining algorithm to the reservoir sample S and report the patterns.

It is possible to derive qualitative guarantees on the frequent patterns mined as a function of the sample size S. The probability of a pattern being a false positive can be determined by using the Chernoff bound. By using modestly lower support thresholds, it is also possible to obtain a guaranteed reduction in the number of false negatives. The bibliographic notes contain pointers to such guarantees. Reservoir sampling has several flexibility advantages because of its clean separation of the sampling and the mining process. Virtually, any efficient frequent pattern mining algorithm can be used on the memory-resident reservoir sample. Furthermore, different variations of pattern mining algorithms, such as constrained pattern mining or interesting pattern mining, can be applied as well. Concept drift is also relatively easy to address. The use of a decay-biased reservoir sample with off-the-shelf frequent pattern mining methods translates to a decay-weighted definition of the support.

12.3.1.2 Sketches

Sketches can be used for determining frequent *items*, though they cannot be used for determining frequent *itemsets* quite as easily. The core idea is that sketches are generally much better at estimating the counts of more frequent items accurately on a relative basis. This is because the bound on the frequency estimation of any item is an absolute one, in which the error depends on the aggregate frequency of the stream items rather than that of the item itself. This is evident from Lemma 12.2.3. As a result, the frequencies of heavy hitters can generally be estimated more accurately on a relative basis. Both the AMS sketch and the count-min sketch can be used to determine the heavy hitters. The bibliographic notes contain pointers to some of these algorithms.

12.3.2 Lossy Counting Algorithm

The lossy counting algorithm can be used either for frequent item, or frequent itemset counting. The approach divides the stream into segments $S_1 \dots S_i \dots$ such that each segment S_i has a size $w = \lfloor 1/\epsilon \rfloor$. The parameter ϵ is a user-defined tolerance on the required accuracy.

First, the easier problem of frequent item mining will be described. The algorithm maintains the frequencies of all the items in an array and increments them as new items arrive. If the number of distinct items is not very large, then one can maintain all the counts and report the frequent ones. The problem arises when the total available space is less than that required to maintain the counts of the distinct items. In such cases, whenever the boundary of a segment S_i is reached, infrequent items are dropped. This results in the removal of many items because the vast majority of the items in the stream are infrequent in practice. How does one decide which items should be dropped, to retain a quality bound on the approximation? For this purpose, a *decremental* trick is used.

Whenever the boundary of a segment S_i is reached, the frequency count of *every* item in the array is decreased by 1. After the decrease, items with zero frequencies are pruned from the array. Consider the situation where n items have already been processed. Because each segment contains w items, a total of $r = O(n/w) = O(n \cdot \epsilon)$ segments have been processed. This implies that any particular item has been decremented at most $r = O(n \cdot \epsilon)$ times. Therefore, if $\lfloor n \cdot \epsilon \rfloor$ were to be added to the counts of the items after processing n items, then no count will be underestimated. Furthermore, this is a good overestimate on the frequency that is proportional to the user-defined tolerance ϵ. If the frequent items are reported with the use of this overestimate, it may result in some false positives, but no false negatives. Under some uniformity assumptions, it has been shown that the lossy counting algorithm requires $O(1/\epsilon)$ space.

The approach can be generalized to the case of frequent patterns by *batching* multiple segments, each of size $w = \lfloor 1/\epsilon \rfloor$. In this case, arrays containing counts of patterns (rather than items) are maintained. However, patterns can obviously not be generated efficiently from individual transactions. The idea here is to batch η segments that are read into main memory. The value of η is decided on the basis of memory availability. When the η segments have been read in, the frequent patterns with (absolute) support of at least η are determined using any memory-based frequent pattern mining algorithm. First, all the old counts in the array are decremented by η, and then the counts of the corresponding patterns from the current segment are added to the array. Those itemsets with zero or negative supports are removed from the array. Over the entire processing of the stream of length n, the count of any itemset is decreased by at most $\lfloor \epsilon \cdot n \rfloor$. Therefore, by adding $\lfloor \epsilon \cdot n \rfloor$ to all array counts at the end of the process, no counts would be underestimated. The overestimate is the same as in the previous case. Thus, it is possible to report the frequent patterns with no false negatives, and false positives that are regulated by user-defined tolerance ϵ. Conceptually, the main difference of this algorithm for frequent *itemset* counting from the aforementioned algorithm for frequent *item* counting is that batching is used. The main goal of batching is to reduce the number of frequent patterns generated at support level of η during the application of the frequent pattern mining algorithm. If batching is not used, then a large number of irrelevant frequent patterns will be generated at an absolute support level of 1. The main shortcoming of lossy counting is that it cannot adjust to concept drift. In this sense, reservoir sampling has a number of advantages over the lossy counting algorithm.

12.4 Clustering Data Streams

The problem of clustering is especially significant in the data stream scenario because of its ability to provide a compact synopsis of the data stream. A clustering of the data stream can often be used as a heuristic substitute for reservoir sampling, especially if a fine-grained clustering is used. For these reasons, stream clustering is often used as a precursor to other applications such as streaming classification. In the following, a few representative stream clustering algorithms will be discussed.

12.4.1 STREAM Algorithm

The *STREAM* algorithm is based on the k-medians clustering methodology. The core idea is to break the stream into smaller memory-resident segments. Thus, the original data stream $\mathcal{S}$ is divided into segments $S_1 \ldots S_r$. Each segment contains at most m data points. The value of m is fixed on the basis of a predefined memory budget.

Because each segment S_i fits in main memory, a more complex clustering algorithm can be applied to it, without worrying about the one-pass constraint. One can use a variety of different k-medians[5] style algorithms for this purpose. In k-medians algorithms, a set $\mathcal{Y}$ of k representatives from each chunk S_i is selected, and each point in S_i is assigned to its closest representative. The goal is to select the representatives to minimize the *sum of squared distances (SSQ)* of the assigned data points to these representatives. For a set of m data points $\overline{X_1} \ldots \overline{X_m}$ in segment S, and a set of k representatives $\mathcal{Y} = \overline{Y_1} \ldots \overline{Y_k}$, the objective function is defined as follows:

$$Objective(S, \mathcal{Y}) = \sum_{\overline{X_i} \in S, \overline{X_i} \Leftarrow \overline{Y_{j_i}}} dist(\overline{X_i}, \overline{Y_{j_i}}). \tag{12.31}$$

The assignment operator is denoted by "$\Leftarrow$" above. The squared distance between a data point and its assigned cluster center is denoted by $dist(\overline{X_i}, \overline{Y_{j_i}})$, where the data record $\overline{X_i}$ is assigned to the representative $\overline{Y_{j_i}}$. In principle, any partitioning algorithm, such as k-means or k-medoids, can be applied to the segment S_i in order to determine the representatives $\overline{Y_1} \ldots \overline{Y_k}$. For the purpose of discussion, this algorithm will be treated as a black box.

After the first segment S_1 has been processed, we now have a set of k medians that are stored away. The number of points assigned to each representative is stored as a "weight" for that representative. Such representatives are considered *level-1* representatives. The next segment S_2 is independently processed to find its k optimal median representatives. Thus, at the end of processing the second segment, one will have $2 \cdot k$ such representatives. Thus, the memory requirement for storing the representatives also increases with time, and after processing r segments, one will have a total of $r \cdot k$ representatives. When the number of representatives exceeds m, a second level of clustering is applied to these set of $r \cdot k$ points, except that the stored weights on the representatives are also used in the clustering process. The resulting representatives are stored as level-2 representatives. In general, when the number of representatives of level-p reaches m, they are converted to k level-$(p+1)$ representatives. Thus, the process will result in increasing the number of representatives of all levels, though the number of representatives at higher levels will increase exponentially slower than those at the lower levels. At the end of processing the entire data stream (or when a specific need for the clustering result arises), all remaining representatives of different levels are clustered together in one final application of the k-medians subroutine.

The specific choice of the algorithm used for the k-medians problem is critical in ensuring a high-quality clustering. The other factor that affects the quality of the final output is the effect of the problem decomposition into chunks followed by hierarchical clustering. How does such a problem decomposition affect the final quality of the output? It has been shown in the *STREAM* paper [240], that the final quality of the output cannot be arbitrarily worse than the particular subroutine that is used at the intermediate stage for k-medians clustering.

Lemma 12.4.1 *Let the subroutine used for k-medians clustering in the STREAM algorithm have an approximation factor of c. Then, the STREAM algorithm will have an approximation factor of no worse than* $5 \cdot c$.

A variety of solutions are possible for the k-medians problem. In principle, virtually any approximation algorithm can be used as a black box. A particularly effective solution is based on the problem of facility location. The reader is referred to the bibliographic notes for pointers to the relevant approach.

[5] This terminology is different from the k-medians approach introduced in Chap. 6. The relevant subroutines in the *STREAM* algorithm are more similar to a k-medoids algorithm. Nevertheless, the "k-medians" terminology is used here to ensure consistency with the original research paper describing *STREAM* [240].

A major limitation of the *STREAM* algorithm is that it is not particularly sensitive to evolution in the underlying data stream. In many cases, the patterns in the underlying stream may evolve and change significantly. Therefore, it is critical for the clustering process to be able to adapt to such changes and provide insights over different time horizons. In this sense, the *CluStream* algorithm is able to provide significantly better insights at different levels of temporal granularity.

12.4.2 CluStream Algorithm

The concept drift in an evolving data stream changes the clusters significantly over time. The clusters over the past day are very different from the clusters over the past month. In many data mining applications, analysts may wish to have the flexibility to determine the clusters based on one or more time horizons, which are unknown at the beginning of the stream clustering process. Because stream data naturally imposes a one-pass constraint on the design of the algorithms, it is difficult to compute clusters over different time horizons using conventional algorithms. A direct extension of the *STREAM* algorithm to such a case would require the simultaneous maintenance of the intermediate results of clustering algorithms over all possible time horizons. The computational burden of such an approach increases with progression of the data stream and can rapidly become a bottleneck for online implementation.

A natural approach to address this issue is to apply the clustering process with a two-stage methodology, including an online microclustering stage, and an offline macroclustering stage. The online microclustering stage processes the stream in real time to continuously maintain summarized but detailed cluster statistics of the stream. These are referred to as *microclusters*. The offline macroclustering stage further summarizes these detailed clusters to provide the user with a more concise understanding of the clusters over different time horizons and levels of temporal granularity. This is achieved by retaining sufficiently detailed statistics in the microclusters, so that it is possible to re-cluster these detailed representations over user-specified time horizons.

12.4.2.1 Microcluster Definition

It is assumed that the multidimensional records in the data stream are denoted by $\overline{X_1} \dots \overline{X_k} \dots$, arriving at time stamps $T_1 \dots T_k \dots$. Each $\overline{X_i}$ is a multidimensional record containing d dimensions that are denoted by $\overline{X_i} = (x_i^1 \dots x_i^d)$. The microclusters capture summary statistics of the data stream to facilitate clustering and analysis over different time horizons. These summary statistics are defined by the following structures:

1. *Microclusters:* The microclusters are defined as a temporal extension of the *cluster feature vector* used in the *BIRCH* algorithm of Chap. 7. This concept can be viewed as a temporally optimized representation of the CF-vector specifically designed for the streaming scenario. To achieve this goal, the microclusters contain temporal statistics in addition to the feature statistics.

2. *Pyramidal Time Frame:* The microclusters are stored at snapshots in time that follow a pyramidal pattern. This pattern provides an effective trade-off between the storage requirements and the ability to recall summary statistics from different time horizons. This is important for enabling the ability to re-cluster the data over different time horizons.

Microclusters are defined as follows.

Definition 12.4.1 *A microcluster for a set of d-dimensional points $X_{i_1} \ldots X_{i_n}$ with time stamps $T_{i_1} \ldots T_{i_n}$ is the $(2 \cdot d + 3)$ tuple $(\overline{CF2^x}, \overline{CF1^x}, CF2^t, CF1^t, n)$, wherein $\overline{CF2^x}$ and $\overline{CF1^x}$ each correspond to a vector of d entries. The definition of each of these entries is as follows:*

1. *For each dimension, the sum of the squares of the data values is maintained in $\overline{CF2^x}$. Thus, $\overline{CF2^x}$ contains d values. The p-th entry of $\overline{CF2^x}$ is equal to $\sum_{j=1}^{n} (x_{i_j}^p)^2$.*

2. *For each dimension, the sum of the data values is maintained in $\overline{CF1^x}$. Thus, $\overline{CF1^x}$ contains d values. The p-th entry of $\overline{CF1^x}$ is equal to $\sum_{j=1}^{n} x_{i_j}^p$.*

3. *The sum of the squares of the time stamps $T_{i_1} \ldots T_{i_n}$ is maintained in $CF2^t$.*

4. *The sum of the time stamps $T_{i_1} \ldots T_{i_n}$ is maintained in $CF1^t$.*

5. *The number of data points is maintained in n.*

An important property of microclusters is that they are *additive*. In other words, the microclusters can be updated by purely additive operations. Note that each of the $2 \cdot d + 3$ components of the microcluster can be expressed as a linearly separable sum over the constituent data points in the microcluster. This is an important property for enabling the efficient maintenance of the microclusters in the online streaming scenario. When a data point $\overline{X_i}$ is added to a microcluster, the corresponding statistics of the data point $\overline{X_i}$ need to be added to each of the $(2 \cdot d + 3)$ components. Similarly, the microclusters for the stream period (t_1, t_2) can be obtained by subtracting the microclusters at time t_1 from those at time t_2. This property is important for enabling the computation of the higher-level macroclusters over an arbitrary time horizon (t_1, t_2) from the microclusters stored at different times.

12.4.2.2 Microclustering Algorithm

The data stream clustering algorithm can generate approximate clusters in any user-specified length of history from the current instant. This is achieved by storing the microclusters at particular moments in the stream that are referred to as *snapshots*. At the same time, the current snapshot of microclusters is always maintained by the algorithm. The additive property can be used to extract microclusters from any time horizon. The macroclustering phase is applied to this representation.

The input to the algorithm is the number of microclusters, denoted by k. The online phase of the algorithm works in an iterative fashion, by always maintaining a current set of microclusters. Whenever a new data point $\overline{X_i}$ arrives, the microclusters are updated to reflect the changes. Each data point either needs to be absorbed by a microcluster, or it needs to be put in a cluster of its own. The first preference is to absorb the data point into a currently existing microcluster. The distance of the data point to the current microcluster centroids $\mathcal{M}_1 \ldots \mathcal{M}_k$ is determined. The distance value of the data point $\overline{X_i}$ to the centroid of the microcluster $\mathcal{M}_j$ is denoted by $dist(\mathcal{M}_j, \overline{X_i})$. Because the centroid of the microcluster can be derived from the cluster feature vector, this distance value can be computed easily. The closest centroid $\mathcal{M}_p$ is determined. The data point $\overline{X_i}$ is assigned to its closest cluster $\mathcal{M}_p$, unless it is deemed that the data point does not "naturally" belong to that (or any other) microcluster. In such cases, the data point $\overline{X_i}$ needs to be assigned a (new) microcluster of its own. Therefore, before assigning a data point to a microcluster, it first needs to be decided whether it naturally belongs to its closest microcluster centroid $\mathcal{M}_p$.

To make this decision, the cluster feature vector of $\mathcal{M}_p$ is used to decide if this data point falls within the *maximum boundary* of the microcluster $\mathcal{M}_p$. If so, then the data point $\overline{X_i}$ is added to the microcluster $\mathcal{M}_p$ by using the additivity property of microclusters. The maximum boundary of the microcluster $\mathcal{M}_p$ is defined as a factor t of the root-mean-square deviation of the data points in $\mathcal{M}_p$ from the centroid. The value of t is a user-defined parameter, and it is typically set to 3.

If the data point does not lie within the maximum boundary of the nearest microcluster, then a new microcluster must be created containing the data point $\overline{X_i}$. However, to create this new microcluster, the number of other microclusters must be reduced by 1 to free memory availability. This can be achieved by either deleting an old microcluster or merging two of the older clusters. This decision is made by examining the staleness of the different clusters, and the number of points in them. The time-stamp statistics of the microclusters are examined to determine whether one of them is "sufficiently" stale to merit removal. If this is not the case, then a merging of the two microclusters is initiated.

How is staleness of a microcluster determined? The microclusters are used to approximate the average time-stamp of the last m data points of the cluster $\mathcal{M}$. This value is not known explicitly because the last m data points are not explicitly retained in order to minimize memory requirements. The mean μ and variance σ^2 of the time-stamps in the microcluster can be used together with a normal distribution assumption of the distribution of time stamps to *estimate* this value. Thus, if the cluster contains $m_0 > m$ data points, then the $m/(2 \cdot m_0)$th percentile of the normal distribution with mean μ and variance σ^2 may be used as the estimate. This value is referred to as the *relevance stamp* of cluster $\mathcal{M}$. Note that μ and σ^2 can be computed from the temporal components of the cluster feature vectors. When the smallest such relevance stamp of any microcluster is below a user-defined threshold δ, it can be eliminated. In cases where no microclusters can be safely deleted, the closest microclusters are merged. The merging operation can be effectively performed because of the existence of the cluster feature vector. Distances between microclusters can be easily computed using the cluster-feature vector. When two microclusters are merged, their statistics are added together, because of the additivity property of microclusters.

12.4.2.3 Pyramidal Time Frame

The microclusters statistics are stored periodically to enable horizon-specific analysis of the clusters. This maintenance is performed during the microclustering phase. In this approach, the microcluster snapshots are stored at varying levels of granularity depending on the recency of the snapshot. Snapshots are classified into different *orders* that can vary from 1 to $\log(T)$, where T is the clock time elapsed since the beginning of the stream. The order of a snapshot regulates the level of temporal granularity at which it is stored, according to the following rules:

- Snapshots of the ith order are stored at time intervals of α^i, where α is an integer and $\alpha \geq 1$. Specifically, each snapshot of the ith order is stored when the clock value is exactly divisible by α^i.
- At any given time, only the last $\alpha^l + 1$ snapshots of order i are stored.

The aforementioned definition allows for considerable redundancy in storage of snapshots. For example, the clock time of 8 is divisible by 2^0, 2^1, 2^2, and 2^3 (where $\alpha = 2$). Therefore, the state of the microclusters at a clock time of 8 simultaneously corresponds to order 0, order 1, order 2, and order 3 snapshots. From an implementation point of view, a snapshot needs to be maintained only once.

Table 12.2: An example [39] of snapshots stored for $\alpha = 2$ and $l = 2$

Order of Snapshots	Clock times (last five snapshots)
0	55 54 53 52 51
1	54 52 50 48 46
2	52 48 44 40 36
3	48 40 32 24 16
4	48 32 16
5	32

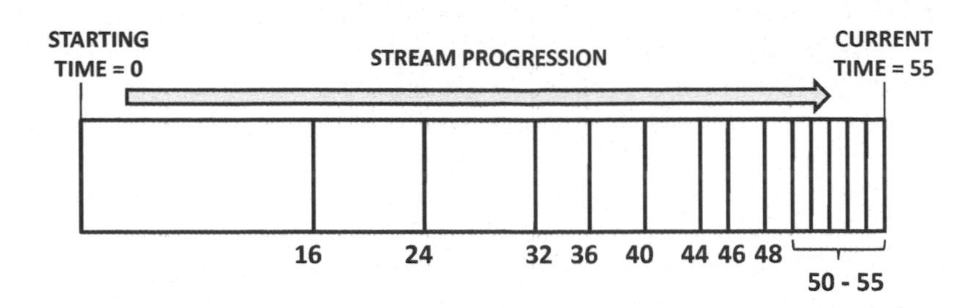

Figure 12.7: Recent snapshots are stored more frequently by pyramidal time frame

To illustrate the snapshots, an example will be used. Consider the case when the stream has been running starting at a clock time of 1, and a use of $\alpha = 2$ and $l = 2$. Therefore, $2^2 + 1 = 5$ snapshots of each order are stored. Then, at a clock time of 55, snapshots at the clock times illustrated in Table 12.2 are stored. While some snapshots are redundant in this case, they are not stored in a redundant way. The corresponding pattern of storage is illustrated in Fig. 12.7. It is evident that recent snapshots are stored more frequently in the pyramidal pattern of storage.

The following observations are true at any moment in time over the course of the data stream:

- The maximum order of any snapshot stored at T time units since the beginning of the stream mining process is $\log_\alpha(T)$.
- The maximum number of snapshots maintained at T time units since the beginning of the stream mining process is $(\alpha^l + 1) \cdot \log_\alpha(T)$.
- For any user-specified time horizon h, at least one stored snapshot can be found, which corresponds to a horizon of length within a factor $(1 + 1/\alpha^{l-1})$ units of the desired value h. This property is important because the microcluster statistics of time horizon $(t_c - h, t_c)$ can be constructed by subtracting the statistics at time $(t_c - h)$ from those at time t_c. Therefore, the microcluster within the approximate temporal locality of $(t_c - h)$ can be used instead. This enables the approximate clustering of data stream points within an arbitrary time horizon $(t_c - h, t_c)$ from the stored pyramidal pattern of microcluster statistics.

For larger values of l, the time horizon can be approximated as closely as desired. It is instructive to use an example to illustrate the combination of the effectiveness and compactness achieved by the pyramidal pattern of snapshot storage. For example, by choosing

$l = 10$, it is possible to approximate any time horizon within 0.2 %. At the same time, a total of only $(2^{10}+1) \cdot \log_2(100 * 365 * 24 * 60 * 60) \approx 32,343$ snapshots are required for a stream with a clock granularity of 1 s and running over 100 years. If each snapshot of size $k \cdot (2 \cdot d+3)$ requires storage of less than a megabyte, the overall storage required is of order of a few gigabytes. Because historical snapshots can be stored on disk and only the current snapshot needs to be maintained in main memory, this requirement is modest from a practical point of view. As the clustering algorithm progresses, only the relevant snapshots according to the pyramidal time frame are maintained. The remaining snapshots are discarded. This enables the computation of horizon-specific clusters at a modest storage cost.

12.4.3 Massive-Domain Stream Clustering

As discussed earlier, the massive-domain scenario is ubiquitous in the stream context. In many cases, one may need to work with a multidimensional data stream, in which the individual attributes are drawn on a massive domain of possible values. In such cases, stream analysis becomes much more difficult because "concise" summaries of the clusters become much more space-intensive. This is also the motivation for many synopsis structures, such as the bloom filter, the count-min sketch, the AMS sketch, and the Flajolet–Martin algorithm.

The data clustering problem also becomes more challenging in the massive-domain scenario, because of the difficulty in maintaining concise statistics of the clusters. A recent method *CSketch* has been designed for clustering massive-domain data streams. The idea in this method is to use a count-min sketch to store the frequencies of attribute–value combinations in each cluster. Thus, the number of count-min sketches used is equal to the number of clusters. An online k-means style clustering is applied, in which the sketch is used as the representative for the (discrete) attributes in the cluster. For any incoming data point, a dot product is computed with respect to each cluster.

The computation is performed as follows. For each attribute–value combination in the d-dimensions, the hash function $h_r(\cdot)$ is applied to it for a particular value of r. The frequency of the corresponding sketch cell is determined. The frequencies of all the relevant sketch cells for the d different dimensions are added together. This provides an estimate of the dot product. To obtain a tighter estimate, the minimum value over different hash functions (different values of r) is used. The dot product is divided by the total frequency of items in the cluster, to avoid bias towards clusters with many data items.

This computation can be performed accurately because the count-min sketch can compute the dot product accurately in a small space. The data point is assigned to the cluster with which it has the largest dot product. The statistics in the sketch representing that particular cluster are then updated. Thus, this approach shares a common characteristic with microclustering in terms of how data points are incrementally assigned to clusters. However, it does not implement the merging and removal steps. Furthermore, the sketch representation is used instead of the microcluster representation for cluster statistics maintenance. Theoretical guarantees can be shown on clustering quality, with respect to a clustering that has infinite space availability. The bibliographic notes contain pointers to these results.

12.5 Streaming Outlier Detection

The problem of streaming outlier detection typically arises either in the context of multidimensional data or time-series data streams. Outlier detection in multidimensional data streams is generally quite different from time series outlier detection. In the latter case, each time series is treated as a unit, whereas temporal correlations are much weaker for multidimensional data. This chapter will address only multidimensional streaming outlier detection, whereas time-series methods will be addressed in Chap. 14.

The multidimensional stream scenario is similar to static multidimensional outlier analysis. The only difference is the addition of a temporal component to the analysis, though this temporal component is much weaker than in the case of time series data. In the context of multidimensional data streams, efficiency is an important concern because the outliers need to be discovered quickly. There are two kinds of outliers that may arise in the context of multidimensional data streams.

1. One is based on the outlier detection of individual records. For example, a first news story on a specific topic represents an outlier of this type. Such an outlier is also referred to as a *novelty.*

2. The second is based on changes in the *aggregate trends* of the multidimensional data. For example, an unusual event such as a terrorist attack may lead to a burst of news stories on a specific topic. This represents an aggregated outlier based on a specific time window. The second kind of change point almost always begins with an individual outlier of the first type. However, an individual outlier of the first type may not always develop into an aggregate change point. This is closely related to the concept of concept drift. While concept drift is generally gentle, an abrupt change may be viewed as an outlier *instant in time* rather than an outlier *data point.*

Both kinds of outliers (or change points) will be discussed in this section.

12.5.1 Individual Data Points as Outliers

The problem of detecting individual data points as outliers is closely related to the problem of *unsupervised novelty detection*, especially when the entire history of the data stream is used. This problem is studied extensively in the text domain in the context of the problem of *first story detection.* Such novelties are often trend setters and may eventually become a part of the normal data. However, when an individual record is declared an outlier in the context of a *window* of data points, it may not necessarily be a novelty. In this context, proximity-based algorithms are particularly easy to generalize to the incremental scenario by almost direct applications of the corresponding algorithms to the window of data points.

Distance-based algorithms can be easily generalized to the streaming scenario. The original distance-based definition of outliers is modified in the following way:

The outlier score of a data point is defined in terms of its k-nearest neighbor distance to data points in a time window of length W.

Note that this is a relatively straightforward modification of the original distance-based definition. When the entire window of data points can be maintained in main memory, it is fairly easy to determine the outliers by computing the score of every data point in the window. However, incremental maintenance of the scores of data points is more challenging because of the addition and removal of data points from the window. Furthermore, some algorithms such as *LOF* require the re-computation of statistics such as reachability distances. The *LOF* algorithm has been extended to the incremental scenario. Two steps are performed in the process:

1. The statistics of the newly inserted data points are computed such as its reachability distance and *LOF* score.

2. The *LOF* scores of the existing points in the window are updated along with their densities and reachability distances. In other words, the scores of many of the existing data points need to be updated because they are affected by the addition of a new data point. However, not all scores need to be updated because only the locality of the new data point is affected. Similarly, when data points are deleted, only the *LOF* scores in the locality of the deleted point are affected.

Because distance-based methods are well-known to be computationally expensive, many of the aforementioned methods are still quite expensive in the context of the data stream. Therefore, the complexity of the outlier detection process can be greatly improved by using an online clustering-based approach. The microclustering approach discussed earlier in this chapter automatically discovers outliers, together with clusters.

While clustering-based methods are generally not advisable when the number of data points are limited, this is not the case in streaming analysis. In the context of a data stream, a sufficient number of data points are typically available to maintain the clusters at a very high level of granularity. *In the context of a streaming clustering algorithm, the formation of new clusters is often associated with unsupervised novelties.* For example, the *CluStream* algorithm explicitly regulates the creation of new clusters in the data stream when an incoming data point does not lie within a specified statistical radius of the existing clusters in the data. Such data points may be considered outliers. In many cases, this is the beginning of a new trend, as more data points are added to the cluster at later stages of the algorithm. In some cases, such data points may correspond to novelties, and in other cases, they may correspond to trends that were seen a long time ago, but are no longer reflected in the current clusters. In either case, such data points are interesting outliers. However, it is not possible to distinguish between these different kinds of outliers unless one is willing to allow the number of clusters in the stream to increase over time.

12.5.2 Aggregate Change Points as Outliers

The sudden changes in aggregate local and global trends in the underlying data are often indicative of anomalous events in the data. Many methods also provide statistical ways of quantifying the level of the changes in the underlying data stream. One way of measuring concept drift is to use the concept of velocity density. The idea in velocity density estimation is to construct a density-based velocity profile of the data. This is analogous to the concept of kernel density estimation in static data sets. The kernel density estimation $\overline{f}(\overline{X})$ for n data points and kernel function $K'_h(\cdot)$ is defined as follows:

$$f(\overline{X}) = \frac{1}{n}\sum_{i=1}^{n} K'_h(\overline{X} - \overline{X_i})$$

The kernel function used is a Gaussian kernel with width h.

$$K'_h(\overline{X} - \overline{X_i}) \propto e^{-||\overline{X}-\overline{X_i}||^2/(2h^2)}$$

The estimation error is defined by the kernel width h that is chosen in a data-driven manner based on Silverman's approximation rule [471].

The velocity density computations are performed over a temporal window of size h_t. Intuitively, the value of h_t defines the time horizon over which the evolution is measured.

Thus, if h_t is chosen to be large, then the velocity density estimation technique provides long term trends, whereas if h_t is chosen to be small then the trends are relatively short term. This provides the user flexibility in analyzing the changes in the data over different time horizons. In addition, a spatial smoothing parameter h_s is used that is analogous to the kernel width h in conventional kernel density estimation.

Let t be the current instant and S be the set of data points that have arrived in the time window $(t - h_t, t)$. The rate of increase in density at spatial location $\overline{X}$ and time t is estimated with two measures the *forward time-slice density estimate* and the *reverse time-slice density estimate*. Intuitively, the forward time-slice estimate measures the density function for all spatial locations at a given time t based on the set of data points that have arrived in the *past* time window $(t-h_t, t)$. Similarly, the reverse time-slice estimate measures the density function at a given time t based on the set of data points that will arrive in the *future* time window $(t, t + h_t)$. Obviously, this value cannot be computed until these points have actually arrived.

It is assumed that the ith data point in S is denoted by $(\overline{X_i}, t_i)$, where i varies from 1 to $|S|$. Then, the forward time-slice estimate $F_{(h_s,h_t)}(X, t)$ of the set S at the spatial location $\overline{X}$ and time t is given by:

$$F_{(h_s,h_t)}(\overline{X}, t) = C_f \cdot \sum_{i=1}^{|S|} K_{(h_s,h_t)}(\overline{X} - \overline{X_i}, t - t_i).$$

Here $K_{(h_s,h_t)}(\cdot, \cdot)$ is a spatiotemporal kernel smoothing function, h_s is the spatial kernel vector, and h_t is temporal kernel width. The kernel function $K_{(h_s,h_t)}(\overline{X} - \overline{X_i}, t - t_i)$ is a smooth distribution that decreases with increasing value of $t - t_i$. The value of C_f is a suitably chosen normalization constant, so that the entire density over the spatial plane is one unit. Thus, C_f is defined as follows:

$$\int_{\text{All } X} F_{(h_s,h_t)}(\overline{X}, t)\delta X = 1.$$

The reverse time-slice density estimate is calculated differently from the forward time-slice density estimate. Assume that the set of points in the time interval $(t, t + h_t)$ is denoted by U. As before, the value of C_r is chosen as a normalization constant. Correspondingly, the reverse time-slice density estimate $R_{(h_s,h_t)}(\overline{X}, t)$ is defined as follows:

$$R_{(h_s,h_t)}(\overline{X}, t) = C_r \cdot \sum_{i=1}^{|U|} K_{(h_s,h_t)}(\overline{X} - \overline{X_i}, t_i - t).$$

In this case, $t_i - t$ is used in the argument instead of $t - t_i$. Thus, the reverse time-slice density in the interval $(t, t+h_t)$ would be exactly the same as the forward time-slice density, if time were reversed, and the data stream arrived in reverse order, starting at $t + h_t$ and ending at t.

The velocity density $V_{(h_s,h_t)}(\overline{X}, T)$ at spatial location $\overline{X}$ and time T is defined as follows:

$$V_{(h_s,h_t)}(\overline{X}, T) = \frac{F_{(h_s,h_t)}(X, T) - R_{(h_s,h_t)}(\overline{X}, T - h_t)}{h_t}.$$

Note that the reverse time-slice density estimate is defined with a temporal argument of $(T - h_t)$, and therefore the future points *with respect to* $(T - h_t)$ are known at time T. A positive value of the velocity density corresponds to an increase in the data density at a given point. A negative value of the velocity density corresponds to a reduction in the data density at a given point. In general, it has been shown that when the spatiotemporal kernel function is defined as below, then the velocity density is directly proportional to a rate of change of the data density at a given point.

$$K_{(h_s,h_t)}(X,t) = (1 - t/h_t) \cdot K'_{h_s}(X).$$

This kernel function is defined only for values of t in the range $(0, h_t)$. The Gaussian spatial kernel function $K'_{h_s}(\cdot)$ was used because of its well-known effectiveness. Specifically, $K'_{h_s}(\cdot)$ is the product of d identical gaussian kernel functions, and $h_s = (h_s^1, \ldots h_s^d)$, where h_s^i is the smoothing parameter for dimension i.

The velocity density is associated with a data point as well as a time instant, and therefore this definition allows the labeling of both data points and time instants as outliers. However, the interpretation of a data point as an outlier in the context of aggregate change analysis is slightly different from the previous definitions in this section. An outlier is defined on an aggregate basis, rather than in a specific way for that point. Because outliers are data points in regions where abrupt change has occurred, *outliers are defined as data points* $\overline{X}$ *at time instants* t *with unusually large absolute values of the* **local** *velocity density.* If desired, a normal distribution could be used to determine the extreme values among the absolute velocity density values. Thus, the velocity density approach is able to convert the multidimensional data distributions into a quantification that can be used in conjunction with extreme-value analysis.

It is important to note that the data point $\overline{X}$ is an outlier only in the context of *aggregate* changes occurring in its locality, rather than its own properties as an outlier. In the context of the news-story example, this corresponds to a news story belonging to a particular burst of related articles. Thus, such an approach could detect the sudden emergence of local clusters in the data, and report the corresponding data points in a timely fashion. Furthermore, it is also possible to compute the aggregate absolute level of change (over all regions) occurring in the underlying data stream. This is achieved by computing the average *absolute* velocity density over the entire data space by summing the changes at sample points in the space. Time instants with large values of the aggregate velocity density may be declared as outliers.

12.6 Streaming Classification

The problem of streaming classification is especially challenging because of the impact of concept drift. One simple approach is to use a reservoir sample to create a concise representation of the training data. This concise representation can be used to create an offline model. If desired, a decay-based reservoir sample can be used to handle concept drift. Such an approach has the advantage that any conventional classification algorithm can be used since the challenges associated with the streaming paradigm have already been addressed at the sampling stage. A number of dedicated methods have also been proposed for streaming classification.

12.6.1 VFDT Family

Very fast decision trees (VFDT) are based on the principle of *Hoeffding trees.* The basic idea is that a decision tree can be constructed on a sample of a very large data set, using a carefully designed approach, so that the resulting tree is the same as what would have been achieved with the original data set with high probability. The Hoeffding bound is used to estimate this probability, and therefore the intermediate steps of the approach are designed with this bound in mind. This is the reason that such trees are referred to as *Hoeffding trees.*

The Hoeffding tree can be constructed incrementally by growing the tree simultaneously with stream arrival. An important assumption is that the stream does not evolve, and therefore the currently arrived set of points can be viewed as a sample of the full stream. The higher levels of the tree are constructed at earlier stages of the stream, when enough tuples have been collected to quantify the accuracy of the corresponding split criteria. The lower level nodes are constructed later because statistics about lower level nodes can be collected only after the higher level nodes have been constructed. Thus, successive levels of the tree are constructed, as more examples stream in and the tree continues to grow. The key in the Hoeffding tree algorithm is to quantify the point at which statistically sufficient tuples have been collected in order to perform a split, so that the split is approximately the same as what would have been performed with knowledge of the full stream.

The same decision tree will be constructed on the current stream sample and the full stream, as long as the same splits are used at each stage. Therefore, the goal of the approach is to ensure that the splits on the sample are identical to the splits on the full stream. For ease in discussion, consider the case where each attribute[6] is binary. In this case, two algorithms will produce exactly the same tree, as long as the same split attribute is selected at each point. The split attribute is selected using a measure such as the Gini index. Consider a particular node in the tree constructed on the original data, and the same node constructed on the sampled data. What is the probability that the same attribute will be selected for the stream sample as for the full stream?

Consider the best and second-best attributes for a split, indexed by i and j, respectively, in the sampled data. Let G_i an G'_i be the Gini index values of the split attribute i, as computed on the full stream, and the sampled data, respectively. Because the attribute i was selected for a split in the sampled data, it is evident that $G'_i < G'_j$. The problem is that the sampling might cause an error. In other words, for the *original data*, it might be the case that $G_j < G_i$. Let the difference $G'_j - G'_i$ between G'_j and G'_i be $\epsilon > 0$. If the number of samples n for evaluating the split is large enough, then it can be shown with the use of the Hoeffding bound that the undesirable case where $G_j < G_i$ will not occur with at least a user-defined probability $1-\delta$. The required value of n would be a function of ϵ and δ. In the context of data streams with continuously accumulating samples, the key is to *wait* for a large enough sample size n before performing the split. In the Hoeffding tree, the Hoeffding bound is used to determine the value of n in terms of ϵ and δ as follows:

$$n = \frac{R^2 \cdot \ln(1/\delta)}{2\epsilon^2}. \tag{12.32}$$

The value of R denotes the range of the split criterion. For the Gini index, the value of R is 1, and for the entropy criterion, the value is $\log(k)$, where k is the number of classes. Near ties in the split criterion correspond to small values of ϵ. According to Eq. 12.32, such ties will lead to large sample size requirements, and therefore a larger waiting time until one can be sufficiently confident of performing a split with the available stream sample.

The Hoeffding tree approach determines whether the current difference in the Gini index between the best and second-best split attributes is at least $\sqrt{\frac{R^2 \cdot \ln(1/\delta)}{2n}}$ in order to initiate a split. This provides a guarantee on the quality of a split at a particular node. In cases,

[6]The argument also applies to general attributes by first transforming them to binary data with discretization and binarization.

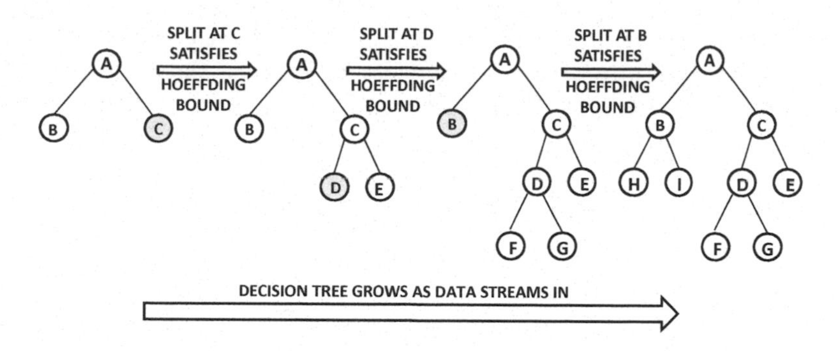

Figure 12.8: Incremental process of Hoeffding tree construction

where there are near ties in split quality (very small values of ϵ), the algorithm will need to wait for a larger value of n until the aforementioned split condition is satisfied. It can be shown that the probability that the Hoeffding tree makes the same classification on the instance as a tree constructed with infinite data is given by at least $1 - \delta/p$, where p is the probability that the instance will be assigned to a particular leaf. The memory requirements are modest because only the counts of the different discrete values of the attributes (over different classes) need to be maintained at various nodes to make split decisions.

The major theoretical implication of the Hoeffding tree algorithm is that one does not need all the data to grow exactly the same tree as what would be constructed by a potentially infinite data stream. Rather, the total number of required tuples is limited once the probabilistic certainty level δ is fixed. The major bottleneck of the approach is that the construction of some of nodes is delayed because of near ties during tree construction. Most of the time is spent in breaking near ties. In the Hoeffding tree algorithm, once a decision is made about a split (and it is a poor one), it cannot be reversed. The incremental process of Hoeffding tree construction is illustrated in Fig. 12.8. It is noteworthy that test instance classification can be performed at any point during stream progression, but the size of the tree increases over time together with classification accuracy.

The *VFDT* approach improves over the Hoeffding tree algorithm by breaking ties more aggressively and through the deactivation of less promising leaf nodes. It also has a number of optimizations to improve accuracy, such as the dropping of poor splitting attributes, and batching intermediate computations over multiple data points. However, it is not designed to handle concept drift. The *CVFDT* approach was subsequently designed to address concept drift. *CVFDT* incorporates two main ideas to address the additional challenges of drift:

1. A sliding window of training items is used to limit the impact of historical behavior.

2. Alternate subtrees at each internal node i are constructed to account for the fact that the best split attribute may no longer remain the top choice because of stream evolution.

Because of the sliding window approach, a difference from the previous method is in the update of the attribute frequency statistics at the nodes, as the sliding window moves forward. For the incoming items, their statistics are added to the attribute value frequencies in the current window, and the expiring items at the other end of the window are removed from the statistics as well. Therefore, when these statistics are updated, some nodes may no longer meet the Hoeffding bound. Such nodes are replaced. *CVFDT* associates each internal node i with a list of alternate subtrees corresponding to splits on different attributes. These alternate subtrees are grown along with the main tree used for classification. These alternate trees are used periodically to perform the replacement once the best split attribute has changed. Experimental results show that the *CVFDT* approach generally achieves higher accuracy in concept-drifting data streams.

12.6.2 Supervised Microcluster Approach

The supervised microcluster is essentially an instance-based classification approach. In this model, it is assumed that a training and a test stream are simultaneously received over time. Because of concept drift, it is important to adjust the model dynamically over time.

In the nearest-neighbor classification approach, the dominant class label among the top-k nearest neighbors is reported as the relevant result. In the streaming scenario, it is difficult to efficiently compute the k nearest neighbors for a particular test instance because of the increasing size of the stream. However, fine-grained microclustering can be used to create a fixed-size summary of the data stream that does not increase with stream progression. A supervised variant of microclustering is used in which data points of different classes are not allowed to mix within clusters. It is relatively easy to maintain such microclusters with minor changes to the *CluStream* algorithm. The main difference is that data points are assigned to microclusters belonging to the same class during the cluster update process. Thus, labels are associated with microclusters rather than individual data points. The dominant label of the top-k nearest microclusters is reported as the relevant label.

This does not, however, account for the changes that need to be made to the algorithm as a result of concept drift. Because of concept drift, the trends in the stream will change. Therefore, it is more relevant to use microclusters from specific time horizons to increase accuracy. While the most recent horizon may often be relevant, this may sometimes not be the case when the trends in the stream revert back suddenly to older trends. Therefore, a part of the training stream is separated out as the validation stream. Recent parts of the validation stream are utilized as test cases to evaluate the accuracy over different time horizons. The optimal horizon is selected. The k-nearest neighbor approach is applied to test instances over this optimally selected horizon.

12.6.3 Ensemble Method

A robust ensemble method was also proposed for the classification of data streams. The method is also designed to handle concept drift because it can effectively account for evolution in the underlying data. The data stream is partitioned into chunks, and multiple classifiers are trained on each of these chunks. The final classification score is computed as a function of the score on each of these chunks. In particular, ensembles of classification models are scored, such as C4.5, RIPPER, naive Bayesian, from sequential chunks of the data stream. The classifiers in the ensemble are weighted based on their expected classification accuracy under the time-evolving environment. This ensures that the approach is able to achieve a higher degree of accuracy because the classifiers are dynamically tuned to optimize the accuracy for that part of the data stream. It was shown that an ensemble classifier produces a smaller error than a single classifier if the weights of all classifiers are assigned based on their expected classification accuracy.

12.6.4 Massive-Domain Streaming Classification

Many streaming applications contain multidimensional discrete attributes with very high cardinality. In such cases, it becomes difficult to use conventional classifiers because of memory limitations. The count-min sketch can be used to address these challenges. Each class is associated with a sketch that is used to track frequent r-combinations of items in the training data, where r is bounded above by a small number k. For each incoming training data point, all possible r-combinations (for $r \leq k$) are treated as pseudo-items that are added to the sketch of the relevant class. Different classes will have different relevant pseudo-items that will show up in the varying frequencies of the cells belonging to sketches of different classes. These differences can be used to determine the most discriminative cells in the different sketches. The frequent discriminative pseudo-items are determined to create *implicit* rules relating the pseudo-items to the different classes. These rules are implicit because they are not actually materialized, but implicitly stored in the sketches. They are retrieved only at the time of the classification of a test instance. For a given test instance, it is determined, which pseudo-items correspond to the combination of items inside them. The discriminative ones among them are determined by retrieving their statistics from the class-specific sketches. These are then used to perform the classification of the test instance, using the same general approach as a rule-based classifier. The bibliographic notes contain pointers to details of the massive-domain classification work.

12.7 Summary

In this chapter, algorithms for stream mining were presented. Streams present several challenges related to high volume, concept drift, the massive-domain nature of data items, and resource constraints. In this context, synopsis construction is one of the most fundamental problems in the streaming scenario. As long as a high-quality stream synopsis can be constructed, it can be leveraged for stream mining algorithms. The major issue with the use of synopsis methods is that different synopsis structures are suited to different applications. The most common synopsis structures used with data streams are reservoir samples and sketches. Reservoir samples provide the greatest flexibility and should be used where possible.

The core problems of frequent pattern mining, clustering, outlier detection, and classification have also been addressed in the streaming scenario. Most of these problems can be addressed with reservoir sampling effectively, where approximate solutions are desired. In the particular case of outlier detection, numerous variations of the problem definition are possible in the streaming scenario.

12.8 Bibliographic Notes

A detailed discussion of streaming algorithms may be found in [40]. The reservoir-sampling method was originally proposed in [498]. The biased reservoir sampling approach with decay was proposed in [35]. The count-min sketch was described in [165]. Numerous other applications of the count-min sketch are discussed in the same work. The AMS sketch was proposed in [72]. The Flajolet–Martin data structure for distinct element counting was proposed in [208]. A survey of synopsis construction algorithms in data streams is provided in [40]. A detailed discussion of the capabilities of some of these data structures may also be found in the same work.

The lossy frequent itemset counting algorithm was proposed in [376]. Surveys on streaming frequent pattern mining may be found in [34, 40]. The *STREAM* algorithm was proposed in [240]. The massive-domain scenario for stream clustering was addressed in [36]. A survey on stream clustering algorithms may be found in [32]. The *STORM* algorithm for point outlier detection was discussed in [67], and the extension of the *LOF* algorithm to data streams was proposed in [426]. The aggregate change detection algorithm was proposed in [21]. Methods for outlier detection in data streams are discussed in [5]. The *VFDT* and *CVFDT* algorithms were proposed in [176, 279]. The microcluster-based classification method was discussed in [20], and the ensemble method was discussed in [503]. The massive-domain scenario for streaming classification was discussed in [47]. A survey on stream classification methods may be found in [33].

12.9 Exercises

1. Let X be a random variable in $[0, 1]$ with mean of 0.5. Show that $P(X > 0.9) \leq 5/9$.

2. Suppose the standard deviation of a random variable X is r times its mean. Here, r can be any constant. Show how to combine the Chebychev inequality and Chernoff bound to show that repeated i.i.d. samples can be used to create a well-bounded estimate of X. In other words, we would like to create another random variable Z (using the multiple i.i.d. samples) with the same expected value of X, such that for small δ, we would like to show that:

$$P(|Z - E[Z]) > \alpha \cdot E[Z]) \leq \delta$$

(Hint: This is the "mean–median trick" discussed in the chapter.)

3. Discuss scenarios in which both the Hoeffding inequality and the Chernoff bound can be used. Which one applies more generally?

4. Suppose that you have a reservoir of size $k = 1000$, and you have a sample of a stream containing an exactly equal distribution of two classes. Use the upper-tail Chernoff bound to determine the probability that the reservoir contains more than 600 samples of one of the two classes. Can the lower tail be used?

5. (Difficult) Work out the full proof of the biased reservoir sampling algorithm.

6. (Difficult) Work out the proof of correctness of the dot-product estimate obtained with the use of the count-min sketch.

7. Discuss the generality of different synopsis construction methods to various stream mining problems. Why is it difficult to apply these methods to outlier analysis?

8. Implement the *CluStream* algorithm.

9. Extend the implementation of the previous exercise to the problem of classification with the microclustering method.

10. Implement the Flajolet–Martin algorithm for distinct element counting.

11. Suppose that X is a random variable, which always lies in the range $[1, 64]$. Suppose that Y is the geometric mean of a large number n of independent and identical realizations of X. Establish a bound on $\log_2(Y)$. Assume that you know the expected value of $\log_2(X)$.

12. Let Z be a random variable satisfying $E[Z] = 0$, and $Z \in [a, b]$.

(a) Show that $E[e^{t \cdot Z}] \leq e^{t^2 \cdot (b-a)^2/8}$.

(b) Use the aforementioned result to complete the proof of the Hoeffding inequality.

13. Suppose that n distinct items are loaded into a bloom filter of length m with w hash functions.

(a) Show that the probability of a bit taking on the value of 0 is equal to $(1-1/m)^{nw}$.

(b) Show that the probability in (a) is approximately equal to $e^{-nw/m}$.

(c) Show that the expected number of 0-bits m_0 in the bloom filter is related to n, m, and w as follows:

$$n \approx \frac{m \cdot \ln(m/m_0)}{w}$$

14. Show the proof of the bound discussed in the chapter for the count-min sketch when items with negative counts are included in the sketch.

15. Let a single component of an AMS sketch be constructed for each of two streams with the same hash-function. Show that the expected value of the product of these components is equal to the dot product of the frequency vector of distinct items in the two streams.

16. Show that the variance of the square of an AMS sketch component is bounded above by twice the square of the second-order moment of the items in the data stream.

17. Show the correctness of AMS point query frequency estimation methodology discussed in the chapter. In other words, the expected value of the $r_i \cdot Q$ should be equal to the point query result.

Chapter 13

Mining Text Data

"The first forty years of life give us the text; the next thirty supply the commentary on it."—Arthur Schopenhauer

13.1 Introduction

Text data are copiously found in many domains, such as the Web, social networks, newswire services, and libraries. With the increasing ease in archival of human speech and expression, the volume of text data will only increase over time. This trend is reinforced by the increasing digitization of libraries and the ubiquity of the Web and social networks. Some examples of relevant domains are as follows:

1. *Digital libraries:* A recent trend in article and book production is to rely on digitized versions, rather than hard copies. This has led to the proliferation of digital libraries in which effective document management becomes crucial. Furthermore mining tools are also used in some domains, such as biomedical literature, to glean useful insights.

2. *Web and Web-enabled applications:* The Web is a vast repository of documents that is further enriched with links and other types of side information. Web documents are also referred to as *hypertext*. The additional side information available with hypertext can be useful in the knowledge discovery process. In addition, many web-enabled applications, such as social networks, chat boards, and bulletin boards, are a significant source of text for analysis.

3. *Newswire services:* An increasing trend in recent years has been the de-emphasis of printed newspapers and a move toward electronic news dissemination. This trend creates a massive stream of news documents that can be analyzed for important events and insights.

The set of features (or dimensions) of text is also referred to as its *lexicon*. A collection of documents is referred to as a *corpus*. A document can be viewed as either a sequence, or a multidimensional record. A text document is, after all, a discrete sequence of words, also

C. C. Aggarwal, *Data Mining: The Textbook*, DOI 10.1007/978-3-319-14142-8_13

referred to as a *string*. Therefore, many sequence-mining methods discussed in Chap. 15 are theoretically applicable to text. However, such sequence mining methods are rarely used in the text domain. This is partially because sequence mining methods are most effective when the length of the sequences and the number of possible tokens are both relatively modest. On the other hand, documents can often be long sequences drawn on a lexicon of several hundred thousand words.

In practice, text is usually represented as multidimensional data in the form of frequency-annotated bag-of-words. Words are also referred to as *terms*. Although such a representation loses the ordering information among the words, it also enables the use of much larger classes of multidimensional techniques. Typically, a preprocessing approach is applied in which the very common words are removed, and the variations of the same word are consolidated. The processed documents are then represented as an *unordered* set of words, where normalized frequencies are associated with the individual words. The resulting representation is also referred to as the *vector space representation* of text. The vector space representation of a document is a multidimensional vector that contains a frequency associated with each word (dimension) in the document. The overall dimensionality of this data set is equal to the number of distinct words in the lexicon. The words from the lexicon that are not present in the document are assigned a frequency of 0. Therefore, text is not very different from the multidimensional data type that has been studied in the preceding chapters.

Due to the multidimensional nature of the text, the techniques studied in the aforementioned chapters can also be applied to the text domain with a modest number of modifications. What are these modifications, and why are they needed? To understand these modifications, one needs to understand a number of specific characteristics that are unique to text data:

1. *Number of "zero" attributes:* Although the base dimensionality of text data may be of the order of several hundred thousand words, a single document may contain only a few hundred words. If each word in the lexicon is viewed as an attribute, and the document word frequency is viewed as the attribute value, most attribute values are 0. This phenomenon is referred to as high-dimensional *sparsity*. There may also be a wide variation in the number of nonzero values across different documents. This has numerous implications for many fundamental aspects of text mining, such as distance computation. For example, while it is possible, in theory, to use the Euclidean function for measuring distances, the results are usually not very effective from a practical perspective. This is because Euclidean distances are extremely sensitive to the varying document lengths (the number of nonzero attributes). The Euclidean distance function cannot compute the distance between two short documents in a comparable way to that between two long documents because the latter will usually be larger.

2. *Nonnegativity:* The frequencies of words take on nonnegative values. When combined with high-dimensional sparsity, the nonnegativity property enables the use of specialized methods for document analysis. In general, all data mining algorithms must be cognizant of the fact that the presence of a word in a document is statistically more significant than its absence. Unlike traditional multidimensional techniques, incorporating the *global* statistical characteristics of the data set in pairwise distance computation is crucial for good distance function design.

3. *Side information:* In some domains, such as the Web, additional side information is available. Examples include hyperlinks or other metadata associated with the document. These additional attributes can be leveraged to enhance the mining process further.

This chapter will discuss the adaptation of many conventional data mining techniques to the text domain. Issues related to document preprocessing will also be discussed.

This chapter is organized as follows. Section 13.2 discusses the problem of document preparation and similarity computation. Clustering methods are discussed in Sect. 13.3. Topic modeling algorithms are addressed in Sect. 13.4. Classification methods are discussed in Sect. 13.5. The first story detection problem is discussed in Sect. 13.6. The summary is presented in Sect. 13.7.

13.2 Document Preparation and Similarity Computation

As the text is not directly available in a multidimensional representation, the first step is to convert raw text documents to the multidimensional format. In cases where the documents are retrieved from the Web, additional steps are needed. This section will discuss these different steps.

1. *Stop word removal:* Stop words are frequently occurring words in a language that are not very discriminative for mining applications. For example, the words "*a*," "*an*," and "*the*" are commonly occurring words that provide very little information about the actual content of the document. Typically, articles, prepositions, and conjunctions are stop words. Pronouns are also sometimes considered stop words. Standardized stop word lists are available in different languages for text mining. The key is to understand that almost all documents will contain these words, and they are usually not indicative of topical or semantic content. Therefore, such words add to the noise in the analysis, and it is prudent to remove them.

2. *Stemming:* Variations of the same word need to be consolidated. For example, singular and plural representations of the same word, and different tenses of the same word are consolidated. In many cases, stemming refers to common root extraction from words, and the extracted root may not even be a word in of itself. For example, the common root of *hoping* and *hope* is *hop*. Of course, the drawback is that the word *hop* has a different meaning and usage of its own. Therefore, while stemming usually improves *recall* in document retrieval, it can sometimes worsen *precision* slightly. Nevertheless, stemming usually enables higher quality results in mining applications.

3. *Punctuation marks:* After stemming has been performed, punctuation marks, such as commas and semicolons, are removed. Furthermore, numeric digits are removed. Hyphens are removed, if the removal results in distinct and meaningful words. Typically, a base dictionary may be available for these operations. Furthermore, the distinct parts of the hyphenated word can either be treated as separate words, or they may be merged into a single word.

After the aforementioned steps, the resulting document may contain only semantically relevant words. This document is treated as a bag-of-words, in which relative ordering is irrelevant. In spite of the obvious loss of ordering information in this representation, the bag-of-words representation is surprisingly effective.

13.2.1 Document Normalization and Similarity Computation

The problem of document normalization is closely related to that of similarity computation. While the issue of text similarity is discussed in Chap. 3, it is also discussed here for completeness. Two primary types of normalization are applied to documents:

1. *Inverse document frequency:* Higher frequency words tend to contribute noise to data mining operations such as similarity computation. The removal of stop words is motivated by this aspect. The concept of inverse document frequency generalizes this principle in a softer way, where words with higher frequency are weighted less.

2. *Frequency damping:* The repeated presence of a word in a document will typically bias the similarity computation significantly. To provide greater stability to the similarity computation, a damping function is applied to word frequencies so that the frequencies of different words become more similar to one another. It should be pointed out that frequency damping is optional, and the effects vary with the application at hand. Some applications, such as clustering, have shown comparable or better performance without damping. This is particularly true if the underlying data sets are relatively clean and have few spam documents.

In the following, these different types of normalization will be discussed. The inverse document frequency id_i of the ith term is a decreasing function of the number of documents n_i in which it occurs:

$$id_i = \log(n/n_i). \tag{13.1}$$

Here, the number of documents in the collection is denoted by n. Other ways of computing the inverse document frequency are possible, though the impact on the similarity function is usually limited.

Next, the concept of frequency damping is discussed. This normalization ensures that the excessive presence of a single word does not throw off the similarity computation. Consider a document with word-frequency vector $\overline{X} = (x_1 \dots x_d)$, where d is the size of the lexicon. A damping function $f(\cdot)$, such as the square root or the logarithm, is optionally applied to the frequencies before similarity computation:

$$f(x_i) = \sqrt{x_i}$$
$$f(x_i) = \log(x_i).$$

Frequency damping is optional and is often omitted. This is equivalent to setting $f(x_i)$ to x_i. The *normalized* frequency $h(x_i)$ for the ith word may be defined as follows:

$$h(x_i) = f(x_i) id_i. \tag{13.2}$$

This model is popularly referred to as the tf-idf model, where tf represents the term frequency and idf represents the inverse document frequency.

The normalized representation of the document is used for data mining algorithms. A popularly used measure is the cosine measure. The cosine measure between two documents with raw frequencies $\overline{X} = (x_1 \dots x_d)$ and $\overline{Y} = (y_1 \dots y_d)$ is defined using their normalized representations:

$$\cos(\overline{X}, \overline{Y}) = \frac{\sum_{i=1}^{d} h(x_i) h(y_i)}{\sqrt{\sum_{i=1}^{d} h(x_i)^2} \sqrt{\sum_{i=1}^{d} h(y_i)^2}} \tag{13.3}$$

Another measure that is less commonly used for text is the *Jaccard coefficient* $J(\overline{X}, \overline{Y})$:

$$J(\overline{X}, \overline{Y}) = \frac{\sum_{i=1}^{d} h(x_i) h(y_i)}{\sum_{i=1}^{d} h(x_i)^2 + \sum_{i=1}^{d} h(y_i)^2 - \sum_{i=1}^{d} h(x_i) h(y_i)}. \tag{13.4}$$

The Jaccard coefficient is rarely used for the text domain, but it is used very commonly for sparse binary data as well as sets. Many forms of transaction and market basket data use the Jaccard coefficient. It needs to be pointed out that the transaction and market basket data share many similarities with text because of their sparse and nonnegative characteristics. Most text mining techniques discussed in this chapter can also be applied to these domains with minor modifications.

13.2.2 Specialized Preprocessing for Web Documents

Web documents require specialized preprocessing techniques because of some common properties of their structure, and the richness of the links inside them. Two major aspects of Web document preprocessing include the removal of specific parts of the documents (e.g., tags) that are not useful, and the leveraging of the actual structure of the document. HTML tags are generally removed by most preprocessing techniques.

HTML documents have numerous fields in them, such as the title, the metadata, and the body of the document. Typically, analytical algorithms treat these fields with different levels of importance, and therefore weigh them differently. For example, the title of a document is considered more important than the body and is weighted more heavily. Another example is the anchor text in Web documents. Anchor text contains a description of the Web page pointed to by a link. Due to its descriptive nature, it is considered important, but it is sometimes not relevant to the topic of the page itself. Therefore, it is often removed from the text of the document. In some cases, where possible, anchor text could even be added to the text of the document *to which it points*. This is because anchor text is often a summary description of the document to which it points.

A Web page may often be organized into content blocks that are not related to the primary subject matter of the page. A typical Web page will have many irrelevant blocks, such as advertisements, disclaimers, or notices, that are not very helpful for mining. It has been shown that the quality of mining results improve when only the text in the main block is used. However, the (automated) determination of main blocks from web-scale collections is itself a data mining problem of interest. While it is relatively easy to decompose the Web page into blocks, it is sometimes difficult to identify the main block. Most automated methods for determining main blocks rely on the fact that a *particular* site will typically utilize a similar layout for the documents on the site. Therefore, if a collection of documents is available from the site, two types of automated methods can be used:

1. *Block labeling as a classification problem:* The idea in this case is to create a new training data set that extracts visual rendering features for each block in the training data, using Web browsers such as Internet Explorer. Many browsers provide an API that can be used to extract the coordinates for each block. The main block is then manually labeled for some examples. This results in a training data set. The resulting training data set is used to build a classification model. This model is used to identify the main block in the remaining (unlabeled) documents of the site.

2. *Tree matching approach:* Most Web sites generate the documents using a fixed template. Therefore, if the template can be extracted, then the main block can be identified relatively easily. The first step is to extract *tag trees* from the HTML pages. These represent the frequent tree patterns in the Web site. The tree matching algorithm, discussed in the bibliographic section, can be used to determine such templates from these tag trees. After the templates have been found, it is determined, which block is the main one in the extracted template. Many of the peripheral blocks often have similar content in different pages and can therefore be eliminated.

13.3 Specialized Clustering Methods for Text

Most of the algorithms discussed in Chap. 6 can be extended to text data. This is because the vector space representation of text is also a multidimensional data point. The discussion in this chapter will first focus on generic modifications to multidimensional clustering algorithms, and then present specific algorithms in these contexts. Some of the clustering methods discussed in Chap. 6 are used more commonly than others in the text domain. Algorithms that leverage the nonnegative, sparse, and high-dimensional features of the text domain are usually preferable to those that do not. Many clustering algorithms require significant adjustments to address the special structure of text data. In the following, the required modifications to some of the important algorithms will be discussed in detail.

13.3.1 Representative-Based Algorithms

These correspond to the family of algorithms such as k-means, k-modes, and k-median algorithms. Among these, the k-means algorithms are the most popularly used for text data. Two major modifications are required for effectively applying these algorithms to text data.

1. The first modification is the choice of the similarity function. Instead of the Euclidean distance, the cosine similarity function is used.

2. Modifications are made to the computation of the cluster centroid. All words in the centroid are not retained. The low-frequency words in the cluster are projected out. Typically, a maximum of 200 to 400 words in each centroid are retained. This is also referred to as a *cluster digest*, and it provides a representative set of *topical words* for the cluster. Projection-based document clustering has been shown to have significant effectiveness advantages. A smaller number of words in the centroid speeds up the similarity computations as well.

A specialized variation of the k-means for text, which uses concepts from hierarchical clustering, will be discussed in this section. Hierarchical methods can be generalized easily to text because they are based on generic notions of similarity and distances. Furthermore, combining them with the k-means algorithm results in both stability and efficiency.

13.3.1.1 Scatter/Gather Approach

Strictly speaking, the *scatter/gather* terminology does not refer to the clustering algorithm itself but the browsing ability enabled by the clustering. This section will, however, focus on the clustering algorithm. This algorithm uses a combination of k-means clustering and hierarchical partitioning. While hierarchical partitioning algorithms are very robust, they typically scale worse than $\Omega(n^2)$, where n is the number of documents in the collection. On the other hand, the k-means algorithm scales as $O(k \cdot n)$, where k is the number of clusters. While the k-means algorithm is more efficient, it can sometimes be sensitive to the choice of seeds. This is particularly true for text data in which each document contains only a small part of the lexicon. For example, consider the case where the document set is to be partitioned into five clusters. A vanilla k-means algorithm will select five documents from the original data as the initial seeds. The number of distinct words in these five documents will typically be a *very small* subset of the entire lexicon. Therefore, the first few iterations of k-means may not be able to assign many documents meaningfully to clusters when they do not contain a significant number of words from this small lexicon subset. This initial incoherence can sometimes be inherited by later iterations, as a result of which the quality of the final results will be poor.

To address this issue, the *scatter/gather* approach uses a combination of hierarchical partitioning and k-means clustering in a two-phase approach. An efficient and simplified form of hierarchical clustering is applied to a sample of the corpus, to yield a robust set of seeds in the first phase. This is achieved by using either of two possible procedures that are referred to as *buckshot* and *fractionation*, respectively. Both these procedures are different types of hierarchical procedures. In the second phase, the robust seeds generated in the first phase are used as the starting point of a k-means algorithm, as adapted to text data. The size of the sample in the first phase is carefully selected to balance the time required by the first phase and the second phase. Thus, the overall approach may be described as follows:

1. Apply either the *buckshot* or *fractionation* procedures to create a robust set of initial seeds.

2. Apply a k-means approach on the resulting set of seeds to generate the final clusters. Additional refinements may be used to improve the clustering quality.

Next, the *buckshot* and *fractionation* procedures will be described. These are two alternatives for the first phase with a similar running time. The fractionation method is the more robust one, but the buckshot method is faster in many practical settings.

- *Buckshot:* Let k be the number of clusters to be found and n be the number of documents in the corpus. The buckshot method selects a seed superset of size $\sqrt{k \cdot n}$ and then agglomerates them to k seeds. Straightforward agglomerative hierarchical clustering algorithms (requiring[1] quadratic time) are applied to this initial sample of $\sqrt{k \cdot n}$ seeds. As we use quadratically scalable algorithms in this phase, this approach requires $O(k \cdot n)$ time. This seed set is more robust than a naive data sample of k seeds because it represents the summarization of a larger sample of the corpus.

- *Fractionation:* Unlike the buckshot method, which uses a sample of $\sqrt{k \cdot n}$ documents, the fractionation method works with all the documents in the corpus. The fractionation algorithm initially breaks up the corpus into n/m buckets, each of size $m > k$ documents. An agglomerative algorithm is applied to each of these buckets to reduce them by a factor $\nu \in (0, 1)$. This step creates $\nu \cdot m$ agglomerated documents in each bucket, and therefore $\nu \cdot n$ agglomerated documents over all buckets. An "agglomerated document" is defined as the concatenation of the documents in a cluster. The process is repeated by treating each of these agglomerated documents as a single document. The approach terminates when a total of k seeds remains.

  It remains to be explained how the documents are partitioned into buckets. One possibility is to use a random partitioning of the documents. However, a more carefully designed procedure can achieve more effective results. One such procedure is to sort the documents by the index of the jth most common word in the document. Here, j is chosen to be a small number, such as 3, that corresponds to medium frequency words in the documents. Contiguous groups of m documents in this sort order are mapped to clusters. This approach ensures that the resulting groups have at least a few common words in them and are therefore not completely random. This can sometimes help in improving the quality of the centers.

[1] As discussed in Chap. 6, standard agglomerative algorithms require more than quadratic time, though some simpler variants of single-linkage clustering [469] can be implemented in approximately quadratic time.

The agglomerative clustering of m documents in the first iteration of the fractionation algorithm requires $O(m^2)$ time for each group, and sums to $O(n \cdot m)$ over the n/m different groups. As the number of individuals reduces geometrically by a factor of ν in each iteration, the total running time over all iterations is $O(n \cdot m \cdot (1 + \nu + \nu^2 + \ldots))$. For $\nu < 1$, the running time over all iterations is still $O(n \cdot m)$. By selecting $m = O(k)$, one still ensure a running time of $O(n \cdot k)$ for the initialization procedure.

The *buckshot* and *fractionation* procedures require $O(k \cdot n)$ time. This is equivalent to the running time of a single iteration of the k-means algorithm. As discussed below, this is important in (asymptotically) balancing the running time of the two phases of the algorithm.

When the initial cluster centers have been determined with the use of the *buckshot* or *fractionation* algorithms, one can apply the k-means algorithm with the seeds obtained in the first step. Each document is assigned to the nearest of the k cluster centers. The centroid of each such cluster is determined as the concatenation of the documents in that cluster. Furthermore, the less frequent words of each centroid are removed. These centroids replace the seeds from the previous iteration. This process can be iteratively repeated to refine the cluster centers. Typically, only a small constant number of iterations is required because the greatest improvements occur only in the first few iterations. This ensures that the overall running time of each of the first and second phases is $O(k \cdot n)$.

It is also possible to use a number of enhancements after the second clustering phase. These enhancements are as follows:

- *Split operation:* The process of splitting can be used to further refine the clusters into groups of better granularity. This can be achieved by applying the buckshot procedure on the individual documents in a cluster by using $k = 2$ and then reclustering around these centers. This entire procedure requires $O(k \cdot n_i)$ time for a cluster containing n_i documents, and therefore splitting all the groups requires $O(k \cdot n)$ time. However, it is not necessary to split *all* the groups. Instead, only a subset of the groups can be split. These are the groups that are not very coherent and contain documents of a disparate nature. To measure the coherence of a group, the self-similarity of the documents in the cluster is computed. This self-similarity provides an understanding of the underlying coherence. This quantity can be computed either in terms of the average similarity of the documents in a cluster to its centroid or in terms of the average similarity of the cluster documents to each other. The split criterion can then be applied selectively only to those clusters that have low self-similarity. This helps in creating more coherent clusters.

- *Join operation:* The join operation merges similar clusters into a single cluster. To perform the merging, the *topical* words of each cluster are computed, as the most frequent words in the centroid of the cluster. Two clusters are considered similar if there is significant overlap between the topical words of the two clusters.

The *scatter/gather* approach is effective because of its ability to combine hierarchical and k-means algorithms.

13.3.2 Probabilistic Algorithms

Probabilistic text clustering can be considered an unsupervised version of the naive Bayes classification method discussed in Sect. 10.5.1 of Chap. 10. It is assumed that the documents need to be assigned to one of k clusters $\mathcal{G}_1 \ldots \mathcal{G}_k$. The basic idea is to use the following generative process:

1. Select a cluster $\mathcal{G}_m$, where $m \in \{1 \ldots k\}$.
2. Generate the term distribution of $\mathcal{G}_m$ based on a generative model. Examples of such models for text include the Bernoulli model or the multinomial model.

The observed data are then used to estimate the parameters of the Bernoulli or multinomial distributions in the generative process. This section will discuss the Bernoulli model.

The clustering is done in an iterative way with the EM algorithm, where cluster assignments of documents are determined from conditional word distributions in the E-step with the Bayes rule, and the conditional word distributions are inferred from cluster assignments in the M-step. For initialization, the documents are randomly assigned to clusters. The initial prior probabilities $P(\mathcal{G}_m)$ and conditional feature distributions $P(w_j|\mathcal{G}_m)$ are estimated from the statistical distribution of this random assignment. A Bayes classifier is used to estimate the posterior probability $P(\mathcal{G}_m|\overline{X})$ in the E-step. The Bayes classifier commonly uses either a Bernoulli model or the multinomial model discussed later in this chapter. The posterior probability $P(\mathcal{G}_m|\overline{X})$ of the Bayes classifier can be viewed as a soft assignment probability of document $\overline{X}$ to the mth mixture component $\mathcal{G}_m$. The conditional feature distribution $P(w_j|\mathcal{G}_m)$ for word w_j is estimated from these posterior probabilities in the M-step as follows:

$$P(w_j|\mathcal{G}_m) = \frac{\sum_{\overline{X}} P(\mathcal{G}_m|\overline{X}) \cdot I(\overline{X}, w_j)}{\sum_{\overline{X}} P(\mathcal{G}_m|\overline{X})} \tag{13.5}$$

Here, $I(\overline{X}, w_j)$ is an indicator variable that takes on the value of 1, if the word w_j is present in $\overline{X}$, and 0, otherwise. As in the Bayes classification method, the same Laplacian smoothing approach may be incorporated to reduce overfitting. The prior probabilities $P(\mathcal{G}_m)$ for each cluster may also be estimated by computing the average assignment probability of all documents to $\mathcal{G}_m$. This completes the description of the M-step of the EM algorithm. The next E-step uses these modified values of $P(w_j|\mathcal{G}_m)$ and the prior probability to derive the posterior Bayes probability with a standard Bayes classifier. Therefore, the following two iterative steps are repeated to convergence:

1. (E-step) Estimate posterior probability of membership of documents to clusters using Bayes rule:

$$P(\mathcal{G}_m|\overline{X}) \propto P(\mathcal{G}_m) \prod_{w_j \in \overline{X}} P(w_j|\mathcal{G}_m) \prod_{w_j \notin \overline{X}} (1 - P(w_j|\mathcal{G}_m)). \tag{13.6}$$

   The aforementioned Bayes rule assumes a Bernoulli generative model. Note that Eq. 13.6 is identical to naive Bayes posterior probability estimation for classification. The multinomial model, which is discussed later in this chapter, may also be used. In such a case, the aforementioned posterior probability definition of Eq. 13.6 is replaced by the multinomial Bayes classifier.

2. (M-step) Estimate conditional distribution $P(w_j|\mathcal{G}_m)$ of words (Eq. 13.5) and prior probabilities $P(\mathcal{G}_m)$ of different clusters using the estimated probabilities in the E-step.

At the end of the process, the estimated value of $P(\mathcal{G}_m|\overline{X})$ provides a cluster assignment probability and the estimated value of $P(w_j|\mathcal{G}_m)$ provides the term distribution of each cluster. This can be viewed as a probabilistic variant of the notion of cluster digest discussed earlier. Therefore, the probabilistic method provides dual insights about cluster membership and the words relevant to each cluster.

13.3.3 Simultaneous Document and Word Cluster Discovery

The probabilistic algorithm discussed in the previous section can simultaneously discover document and word clusters. As discussed in Sect. 7.4 of Chap. 7 on high-dimensional clustering methods, this is important in the high-dimensional case because clusters are best characterized in terms of both rows and columns *simultaneously*. In the text domain, the additional advantage of such methods is that the topical words of a cluster provide *semantic* insights about that cluster. Another example is the nonnegative matrix factorization method, discussed in Sect. 6.8 of Chap. 6. This approach is very popular in the text domain because the factorized matrices have a natural interpretation for text data. This approach can simultaneously discover word clusters and document clusters that are represented by the columns of the two factorized matrices. This is also closely related to the concept of *co-clustering*.

13.3.3.1 Co-clustering

Co-clustering is most effective for nonnegative matrices in which many entries have zero values. In other words, the matrix is sparsely populated. This is the case for text data. Co-clustering methods can also be generalized to dense matrices, although these techniques are not relevant to the text domain. Co-clustering is also sometimes referred to as *bi-clustering* or *two-mode clustering* because of its exploitation of both "modes" (words and documents). While the co-clustering method is presented here in the context of text data, the broader approach is also used in the biological domain with some modifications.

The idea in co-clustering is to rearrange the rows and columns in the data matrix so that most of the nonzero entries become arranged into blocks. In the context of text data, this matrix is the $n \times d$ document term matrix D, where rows correspond to documents and columns correspond to words. Thus, the ith cluster is associated with a set of rows $\mathcal{R}_i$ (documents), and a set of columns $\mathcal{V}_i$ (words). The rows $\mathcal{R}_i$ are disjoint from one another over different values of i, and the columns $\mathcal{V}_i$ are also disjoint from one another over different values of i. Thus, the co-clustering method simultaneously leads to document clusters and word clusters. From an intuitive perspective, the words representing the columns of $\mathcal{V}_i$ are the most relevant (or topical) words for cluster $\mathcal{R}_i$. The set $\mathcal{V}_i$ therefore defines a cluster digest of $\mathcal{R}_i$.

In the context of text data, word clusters are just as important as document clusters because they provide insights about the topics of the underlying collection. Most of the methods discussed in this book for document clustering, such as the *scatter/gather* method, probabilistic methods, and nonnegative matrix factorization (see Sect. 6.8 of Chap. 6, produce word clusters (or cluster digests) in addition to document clusters. However, the words in the different clusters are *overlapping* in these algorithms, whereas document clusters are non overlapping in all algorithms except for the probabilistic (soft) EM method. In co-clustering, the word clusters and document clusters are *both* non overlapping. Each document and word is strictly associated with a particular cluster. One nice characteristic of co-clustering is that it explicitly explores the duality between word clusters and document clusters. Coherent word clusters can be shown to induce coherent document clusters and vice versa. For example, if meaningful word clusters were already available, then one might be able to cluster documents by assigning each document to the word cluster with which it has the most words in common. In co-clustering, the goal is to do this *simultaneously* so that word clusters and document clusters depend on each other in an optimal way.

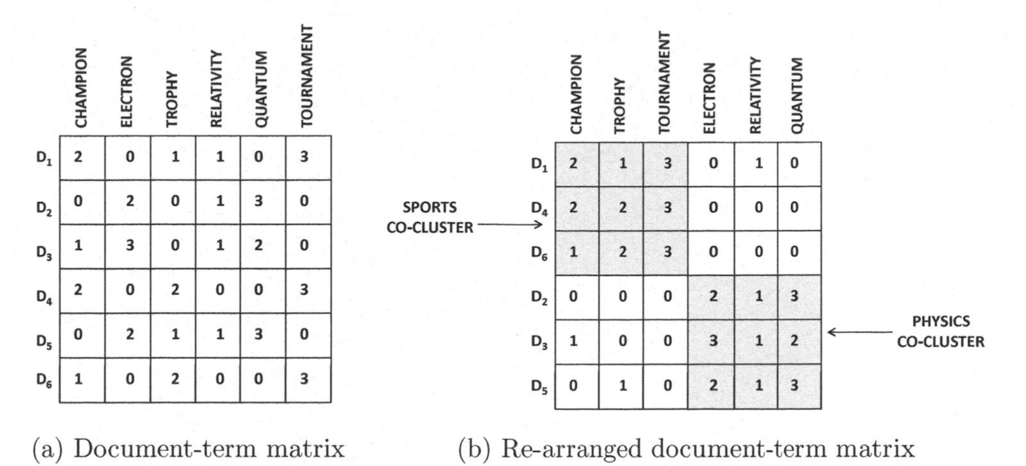

	CHAMPION	ELECTRON	TROPHY	RELATIVITY	QUANTUM	TOURNAMENT
D_1	2	0	1	1	0	3
D_2	0	2	0	1	3	0
D_3	1	3	0	1	2	0
D_4	2	0	2	0	0	3
D_5	0	2	1	1	3	0
D_6	1	0	2	0	0	3

(a) Document-term matrix

	CHAMPION	TROPHY	TOURNAMENT	ELECTRON	RELATIVITY	QUANTUM
D_1	2	1	3	0	1	0
D_4	2	2	3	0	0	0
D_6	1	2	3	0	0	0
D_2	0	0	0	2	1	3
D_3	1	0	0	3	1	2
D_5	0	1	0	2	1	3

(b) Re-arranged document-term matrix

Figure 13.1: Illustrating row and column reordering in co-clustering

To illustrate this point, a toy example[2] of a 6×6 document-word matrix has been illustrated in Fig. 13.1a. The entries in the matrix correspond to the word frequencies in six documents denoted by $D_1 \dots D_6$. The six words in this case are *champion*, *electron*, *trophy*, *relativity*, *quantum*, and *tournament*. It is easy to see that some of the words are from sports-related topics, whereas other words are from science-related topics. Note that the nonzero entries in the matrix of Fig. 13.1a seem to be arranged randomly. It should be noted that the documents $\{D_1, D_4, D_6\}$ seem to contain words relating to sports topics, whereas the documents $\{D_2, D_3, D_5\}$ seem to contain words relating to scientific topics. However, this is not evident from the random distribution of the entries in Fig. 13.1a. On the other hand, if the rows and columns were permuted, so that all the sports-related rows/columns occur earlier than all the science-related rows/columns, then the resulting matrix is shown in Fig. 13.1b. In this case, there is a clear block structure to the entries, in which disjoint rectangular blocks contain most of the nonzero entries. These rectangular blocks are shaded in Fig. 13.1b. The goal is to minimize the weights of the nonzero entries outside these shaded blocks.

How, then, can the co-clustering problem be solved? The simplest solution is to convert the problem to a bipartite graph partitioning problem, so that the aggregate weight of the nonzero entries in the nonshaded regions is equal to the aggregate weight of the edges across the partitions. A node set N_d is created, in which each node represents a document in the collection. A node set N_w is created, in which each node represents a word in the collection. An undirected bipartite graph $G = (N_d \cup N_w, A)$ is created, such that an edge (i, j) in A corresponds to a nonzero entry in the matrix, where $i \in N_d$ and $j \in N_w$. The weight of an edge is equal to the frequency of the term in the document. The bipartite graph for the co-cluster of Fig. 13.1 is illustrated in Fig. 13.2. A partitioning of this graph represents a simultaneous partitioning of the rows and columns. In this case, a 2-way partitioning has been illustrated for simplicity, although a k-way partitioning could be constructed in general. Note that each partition contains a set of documents and a corresponding set of words. It is easy to see that the corresponding documents and words in each graph partition of Fig. 13.2 represent the shaded areas in Fig. 13.1b. It is also easy to see that the weight

[2]While the document-term matrix is square in this specific toy example, this might not be the case in general because the corpus size n, and the lexicon size d are generally different.

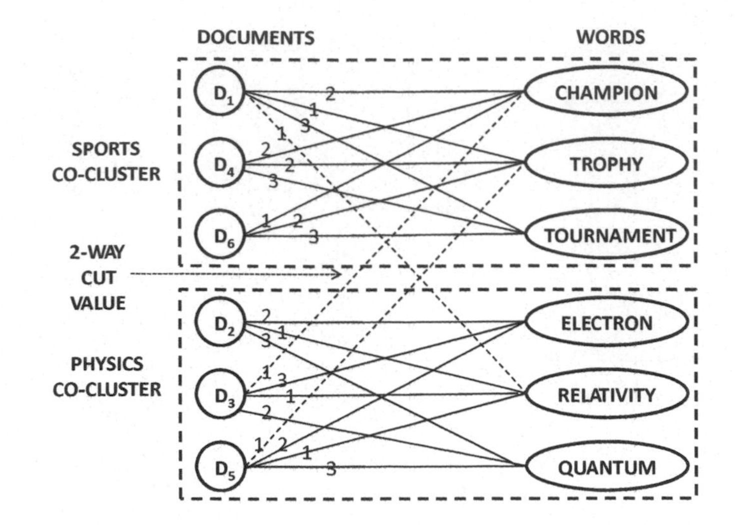

Figure 13.2: Graph partitioning for co-clustering

of edges across the partition represents the weight of the nonzero entries in Fig. 13.1b. Therefore, a k-way co-clustering problem can be converted to a k-way graph partitioning problem. The overall co-clustering approach may be described as follows:

1. Create a graph $G = (N_d \cup N_w, A)$ with nodes in N_d representing documents, nodes in N_w representing words, and edges in A with weights representing nonzero entries in matrix D.

2. Use a k-way graph partitioning algorithm to partition the nodes in $N_d \cup N_w$ into k groups.

3. Report row–column pairs $(\mathcal{R}_i \mathcal{V}_i)$ for $i \in \{1 \dots k\}$. Here, $\mathcal{R}_i$ represents the rows corresponding to nodes in N_d for the ith cluster, and $\mathcal{V}_i$ represents the columns corresponding to the nodes in N_w for the ith cluster.

It remains to be explained, how the k-way graph partitioning may be performed. The problem of graph partitioning is addressed in Sect. 19.3 of Chap. 19. Any of these algorithms may be leveraged to determine the required graph partitions. Specialized methods for bipartite graph partitioning are also discussed in the bibliographic notes.

13.4 Topic Modeling

Topic modeling can be viewed as a probabilistic version of the latent semantic analysis (LSA) method, and the most basic version of the approach is referred to as *Probabilistic Latent Semantic Analysis (PLSA)*. It provides an alternative method for performing dimensionality reduction and has several advantages over traditional LSA.

Probabilistic latent semantic analysis is an expectation maximization-based mixture modeling algorithm. However, the way in which the EM algorithm is used is different than the other examples of the EM algorithm in this book. This is because the underlying generative process is different, and is optimized to discovering the *correlation structure* of the words rather than the clustering structure of the documents. This is because the approach

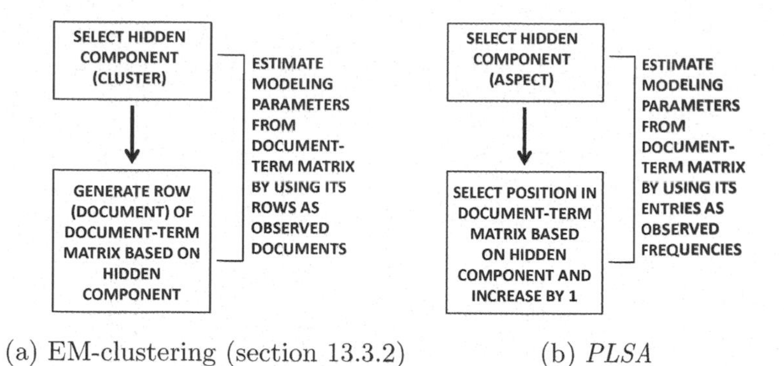

(a) EM-clustering (section 13.3.2) (b) *PLSA*

Figure 13.3: Varying generative process of EM-clustering and *PLSA*

can be viewed as a probabilistic variant of *SVD* and *LSA*, rather than a probabilistic variant of clustering. Nevertheless, soft clusters can also be generated with the use of this method. There are many other dimensionality reduction methods, such as nonnegative matrix factorization, which are intimately related to clustering. *PLSA* is, in fact, a nonnegative matrix factorization method with a maximum-likelihood objective function.

In most of the EM clustering algorithms of this book, a mixture component (cluster) is selected, and then the data record is generated based on a particular form of the distribution of that component. An example is the Bernoulli clustering model, which is discussed in Sect. 13.3.2. In *PLSA*, the generative process[3] is inherently designed for dimensionality reduction rather than clustering, and different parts of the same document can be generated by different mixture components. It is assumed that there are k *aspects* (or latent topics) denoted by $\mathcal{G}_1 \dots \mathcal{G}_k$. The generative process builds the document-term matrix as follows:

1. Select a latent component (aspect) $\mathcal{G}_m$ with probability $P(\mathcal{G}_m)$.

2. Generate the indices (i, j) of a document–word pair $(\overline{X_i}, w_j)$ with probabilities $P(\overline{X_i}|\mathcal{G}_m)$ and $P(w_j|\mathcal{G}_m)$, respectively. Increment the frequency of entry (i, j) in the document-term matrix by 1. The document and word indices are generated in a probabilistically independent way.

All the parameters of this generative process, such as $P(\mathcal{G}_m)$, $P(\overline{X_i}|\mathcal{G}_m)$, and $P(w_j|\mathcal{G}_m)$, need to be estimated from the observed frequencies in the $n \times d$ document-term matrix.

Although the aspects $\mathcal{G}_1 \dots \mathcal{G}_k$ are analogous to the clusters of Sect. 13.3.2, they are not the same. Note that each iteration of the generative process of Sect. 13.3.2 creates the final frequency vector of an *entire row* of the document-term matrix. In *PLSA*, even a single matrix entry may have frequency contributions from various mixture components. Indeed, even in deterministic latent semantic analysis, a document is expressed as a linear combination of different latent directions. Therefore, the interpretation of each mixture component as a cluster is more direct in the method of Sect. 13.3.2. The generative differences between these models are illustrated in Fig. 13.3. Nevertheless, *PLSA* can also be used for clustering because of the highly interpretable and nonnegative nature of the underlying latent factorization. The relationship with and applicability to clustering will be discussed later.

[3]The original work [271] uses an asymmetric generative process, which is equivalent to the (simpler) symmetric generative process discussed here.

An important assumption in *PLSA* is that the selected documents and words are *conditionally* independent after the latent topical component $\mathcal{G}_m$ has been fixed. In other words, the following is assumed:

$$P(\overline{X_i}, w_j|\mathcal{G}_m) = P(\overline{X_i}|\mathcal{G}_m) \cdot P(w_j|\mathcal{G}_m) \tag{13.7}$$

This implies that the joint probability $P(\overline{X_i}, w_j)$ for selecting a document–word pair can be expressed in the following way:

$$P(\overline{X_i}, w_j) = \sum_{m=1}^{k} P(\mathcal{G}_m) \cdot P(\overline{X_i}, w_j|\mathcal{G}_m) = \sum_{m=1}^{k} P(\mathcal{G}_m) \cdot P(\overline{X_i}|\mathcal{G}_m) \cdot P(w_j|\mathcal{G}_m). \tag{13.8}$$

It is important to note that *local* independence between documents and words within a latent component does not imply *global* independence between the same pair over the entire corpus. The local independence assumption is useful in the derivation of EM algorithm.

In *PLSA*, the posterior probability $P(\mathcal{G}_m|\overline{X_i}, w_j)$ of the latent component associated with a particular *document–word pair* is estimated. The EM algorithm starts by initializing $P(\mathcal{G}_m)$, $P(\overline{X_i}|\mathcal{G}_m)$, and $P(w_j|\mathcal{G}_m)$ to $1/k$, $1/n$, and $1/d$, respectively. Here, k, n, and d denote the number of clusters, number of documents, and number of words, respectively. The algorithm iteratively executes the following E- and M-steps to convergence:

1. (E-step) Estimate posterior probability $P(\mathcal{G}_m|\overline{X_i}, w_j)$ in terms of $P(\mathcal{G}_m)$, $P(\overline{X_i}|\mathcal{G}_m)$, and $P(w_j|\mathcal{G}_m)$.

2. (M-step) Estimate $P(\mathcal{G}_m)$, $P(\overline{X_i}|\mathcal{G}_m)$ and $P(w_j|\mathcal{G}_m)$ in terms of the posterior probability $P(\mathcal{G}_m|\overline{X_i}, w_j)$, and observed data about word-document co-occurrence using log-likelihood maximization.

These steps are iteratively repeated to convergence. It now remains to discuss the details of the E-step and the M-step. First, the E-step is discussed. The posterior probability estimated in the E-step can be expanded using the Bayes rule:

$$P(\mathcal{G}_m|\overline{X_i}, w_j) = \frac{P(\mathcal{G}_m) \cdot P(\overline{X_i}, w_j|\mathcal{G}_m)}{P(\overline{X_i}, w_j)}. \tag{13.9}$$

The numerator of the right-hand side of the aforementioned equation can be expanded using Eq. 13.7, and the denominator can be expanded using Eq. 13.8:

$$P(\mathcal{G}_m|\overline{X_i}, w_j) = \frac{P(\mathcal{G}_m) \cdot P(\overline{X_i}|\mathcal{G}_m) \cdot P(w_j|\mathcal{G}_m)}{\sum_{r=1}^{k} P(\mathcal{G}_r) \cdot P(\overline{X_i}|\mathcal{G}_r) \cdot P(w_j|\mathcal{G}_r)}. \tag{13.10}$$

This shows that the E-step can be implemented in terms of the estimated values $P(\mathcal{G}_m)$, $P(\overline{X_i}|\mathcal{G}_m)$, and $P(w_j|\mathcal{G}_m)$.

It remains to show how these values can be estimated using the *observed* word-document co-occurrences in the M-step. The posterior probabilities $P(\mathcal{G}_m|\overline{X_i}, w_j)$ may be viewed as weights attached with word-document co-occurrence pairs for each aspect $\mathcal{G}_m$. These weights can be leveraged to estimate the values $P(\mathcal{G}_m)$, $P(\overline{X_i}|\mathcal{G}_m)$, and $P(w_j|\mathcal{G}_m)$ for each aspect using maximization of the log-likelihood function. The details of the log-likelihood function, and the differential calculus associated with the maximization process will not be discussed here. Rather, the final estimated values will be presented directly. Let $f(\overline{X_i}, w_j)$ represent the observed frequency of the occurrence of word w_j in document $\overline{X_i}$ in the corpus. Then, the estimations in the M-step are as follows:

$$P(\overline{X_i}|\mathcal{G}_m) \propto \sum_{w_j} f(\overline{X_i}, w_j) \cdot P(\mathcal{G}_m|\overline{X_i}, w_j) \;\; \forall i \in \{1 \dots n\}, m \in \{1 \dots k\} \tag{13.11}$$

$$P(w_j|\mathcal{G}_m) \propto \sum_{\overline{X_i}} f(\overline{X_i}, w_j) \cdot P(\mathcal{G}_m|\overline{X_i}, w_j) \;\; \forall j \in \{1 \dots d\}, m \in \{1 \dots k\} \tag{13.12}$$

$$P(\mathcal{G}_m) \propto \sum_{\overline{X_i}} \sum_{w_j} f(\overline{X_i}, w_j) \cdot P(\mathcal{G}_m|\overline{X_i}, w_j) \;\; \forall m \in \{1 \dots k\}. \tag{13.13}$$

Each of these estimations may be scaled to a probability by ensuring that they sum to 1 over all the outcomes for that random variable. This scaling corresponds to the constant of proportionality associated with the "$\propto$" notation in the aforementioned equations. Furthermore, these estimations can be used to decompose the original document-term matrix into a product of three matrices, which is very similar to *SVD/LSA*. This relationship will be explored in the next section.

13.4.1 Use in Dimensionality Reduction and Comparison with Latent Semantic Analysis

The three key sets of parameters estimated in the M-step are $P(\overline{X_i}|\mathcal{G}_m)$, $P(w_j|\mathcal{G}_m)$, and $P(\mathcal{G}_m)$, respectively. These sets of parameters provide an *SVD*-like matrix factorization of the $n \times d$ document-term matrix D. Assume that the document-term matrix D is scaled by a constant to sum to an aggregate probability value of 1. Therefore, the (i, j)th entry of D can be viewed as an *observed instantiation* of the probabilistic quantity $P(\overline{X_i}, w_j)$. Let Q_k be the $n \times k$ matrix, for which the (i, m)th entry is $P(\overline{X_i}|\mathcal{G}_m)$, let Σ_k be the $k \times k$ diagonal matrix for which the mth diagonal entry is $P(\mathcal{G}_m)$, and let P_k be the $d \times k$ matrix for which the (j, m)th entry is $P(w_j|\mathcal{G}_m)$. Then, the (i, j)th entry $P(\overline{X_i}, w_j)$ of the matrix D can be expressed in terms of the entries of the aforementioned matrices according to Eq. 13.8, which is replicated here:

$$P(\overline{X_i}, w_j) = \sum_{m=1}^{k} P(\mathcal{G}_m) \cdot P(\overline{X_i}|\mathcal{G}_m) \cdot P(w_j|\mathcal{G}_m). \tag{13.14}$$

This LHS of the equation is equal to the (i, j)th entry of D, whereas the RHS of the equation is the (i, j)th entry of the matrix product $Q_k \Sigma_k P_k^T$. Depending on the number of components k, the LHS can only approximate the matrix D, which is denoted by D_k. By stacking up the $n \times d$ conditions of Eq. 13.14, the following matrix condition is obtained:

$$D_k = Q_k \Sigma_k P_k^T. \tag{13.15}$$

It is instructive to note that the matrix decomposition in Eq. 13.15 is similar to that in *SVD/LSA* (cf. Eq. 2.12 of Chap. 2). Therefore, as in *LSA*, D_k is an approximation of the document-term matrix D, and the transformed representation in k-dimensional space is given by $Q_k \Sigma_k$. However, the transformed representations will be different in *PLSA* and *LSA*. This is because different objective functions are optimized in the two cases. *LSA* minimizes the mean-squared error of the approximation, whereas *PLSA* maximizes the log-likelihood fit to a probabilistic generative model. One advantage of *PLSA* is that the entries of Q_k and P_k and the transformed coordinate values are nonnegative and have clear

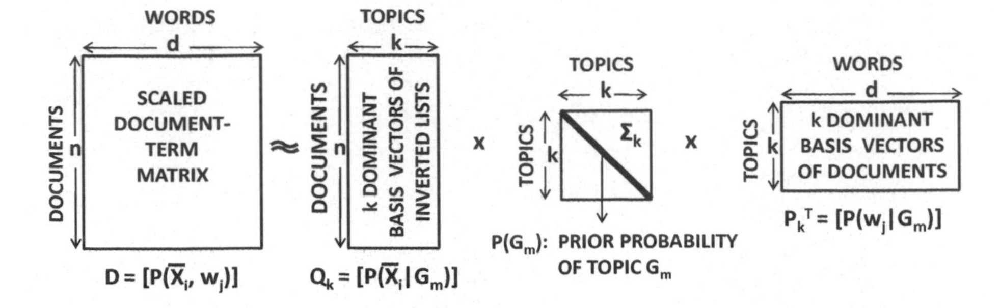

Figure 13.4: Matrix factorization of *PLSA*

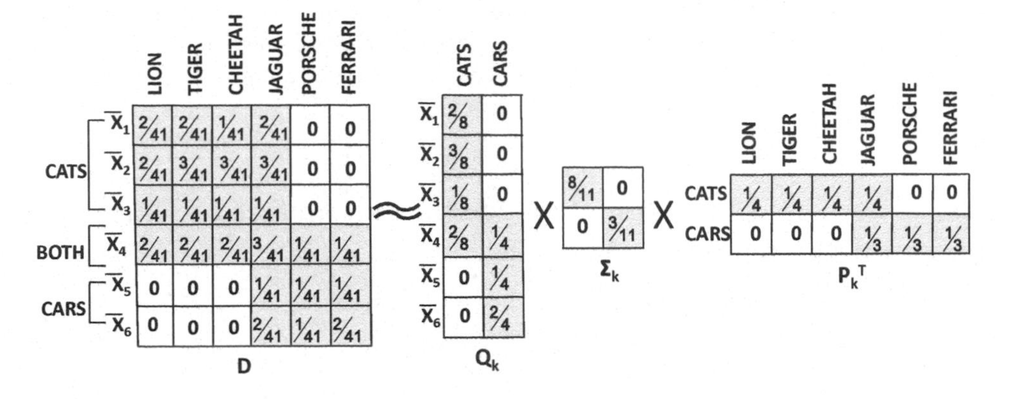

Figure 13.5: An example of *PLSA* (Revisiting Fig. 6.22 of Chap. 6)

probabilistic interpretability. By examining the probability values in each column of P_k, one can immediately infer the topical words of the corresponding aspect. This is not possible in *LSA*, where the entries in the corresponding matrix P_k do not have clear probabilistic significance and may even be negative. One advantage of *LSA* is that the transformation can be interpreted in terms of the rotation of an orthonormal axis system. In *LSA*, the columns of P_k are a set of orthonormal vectors representing this rotated basis. This is not the case in *PLSA*. Orthogonality of the basis system in *LSA* enables straightforward projection of *out-of-sample* documents (i.e., documents not included in D) onto the new rotated axis system.

Interestingly, as in *SVD/LSA*, the latent properties of the transpose of the document matrix are revealed by *PLSA*. Each row of $P_k \Sigma_k$ can be viewed as the transformed coordinates of the *vertical or inverted list representation* (rows of the transpose) of the document matrix D in the basis space defined by columns of Q_k. These complementary properties are illustrated in Fig. 13.4. *PLSA* can also be viewed as a kind of nonnegative matrix factorization method (cf. Sect. 6.8 of Chap. 6) in which matrix elements are interpreted as probabilities and the maximum-likelihood estimate of a generative model is maximized rather than minimizing the Frobenius norm of the error matrix.

An example of an approximately optimal *PLSA* matrix factorization of a toy 6×6 example, with 6 documents and 6 words, is illustrated in Fig. 13.5. This example is the same (see Fig. 6.22) as the one used for nonnegative matrix factorization (*NMF*) in Chap. 6. Note that the factorizations in the two cases are very similar except that all basis vectors are normalized to sum to 1 in *PLSA*, and the dominance of the basis vectors is reflected in a separate diagonal matrix containing the prior probabilities. Although the factorization presented here for *PLSA* is identical to that of *NMF* for intuitive understanding, the factorizations will usually be slightly different[4] because of the difference in objective functions in the two cases. Also, most of the entries in the factorized matrices will not be exactly 0 in a real example, but many of them might be quite small.

As in *LSA*, the problems of synonymy and polysemy are addressed by *PLSA*. For example, if an aspect $\mathcal{G}_1$ explains the topic of cats, then two documents $\overline{X}$ and $\overline{Y}$ containing the words "cat" and "kitten," respectively, will have positive values of the transformed coordinate for aspect $\mathcal{G}_1$. Therefore, similarity computations between these documents will be improved in the transformed space. A word with multiple meanings (polysemous word) may have positive components in different aspects. For example, a word such as "jaguar" can either be a cat or a car. If $\mathcal{G}_1$ be an aspect that explains the topic of cats, and $\mathcal{G}_2$ is an aspect that explains the topic of cars, then both $P(\text{"jaguar"}|\mathcal{G}_1)$ and $P(\text{"jaguar"}|\mathcal{G}_2)$ may be highly positive. However, the other words in the document will provide the context necessary to reinforce one of these two aspects. A document $\overline{X}$ that is mostly about cats will have a high value of $P(\overline{X}|\mathcal{G}_1)$, whereas a document $\overline{Y}$ that is mostly about cars will have a high value of $P(\overline{Y}|\mathcal{G}_2)$. This will be reflected in the matrix $Q_k = [P(\overline{X_i}|\mathcal{G}_m)]_{n \times k}$ and the new transformed coordinate representation $Q_k\Sigma_k$. Therefore, the computations will also be robust in terms of adjusting for polysemy effects. In general, semantic concepts will be amplified in the transformed representation $Q_k\Sigma_k$. Therefore, many data mining applications will perform more robustly in terms of the $n \times k$ transformed representation $Q_k\Sigma_k$ rather than the original $n \times d$ document-term matrix.

13.4.2 Use in Clustering and Comparison with Probabilistic Clustering

The estimated parameters have intuitive interpretations in terms of clustering. In the Bayes model for clustering (Fig. 13.3a), the generative process is optimized to clustering documents, whereas the generative process in topic modeling (Fig. 13.3b) is optimized to discovering the latent semantic components. The latter can be shown to cluster document–word *pairs*, which is different from clustering documents. Therefore, although the same parameter set $P(w_j|\mathcal{G}_m)$ and $P(\overline{X}|\mathcal{G}_m)$ is estimated in the two cases, qualitatively different results will be obtained. The model of Fig. 13.3a generates a document from a unique hidden component (cluster), and the final soft clustering is a result of uncertainty in *estimation* from observed data. On the other hand, in the probabilistic latent semantic model, different parts of the same document may be generated by different aspects, even at the generative *modeling* level. Thus, documents are not generated by individual mixture components, but by a combination of mixture components. In this sense, *PLSA* provides a more realistic model because the diverse words of an unusual document discussing both cats and cars (see Fig. 13.5) can be generated by distinct aspects. In Bayes clustering, even though such a document is generated *in entirety* by one of the mixture components, it may have similar assignment (posterior) probabilities with respect to two or more clusters because of *estimation uncertainty*. This difference is because *PLSA* was originally intended as a data transformation and dimensionality reduction method, rather than as a clustering method. Nevertheless, good document clusters can usually be derived from *PLSA* as well. The value $P(\mathcal{G}_m|\overline{X_i})$ provides an assignment probability of the document $\overline{X_i}$ to aspect (or "cluster")

[4]The presented factorization for *PLSA* is approximately optimal, but not exactly optimal.

$\mathcal{G}_m$ and can be derived from the parameters estimated in the M-step using the Bayes rule as follows:

$$P(\mathcal{G}_m|\overline{X_i}) = \frac{P(\mathcal{G}_m) \cdot P(\overline{X_i}|\mathcal{G}_m)}{\sum_{r=1}^{k} P(\mathcal{G}_r) \cdot P(\overline{X_i}|\mathcal{G}_r)}. \quad (13.16)$$

Thus, the *PLSA* approach can also be viewed a *soft* clustering method that provides assignment probabilities of documents to clusters. In addition, the quantity $P(w_j|\mathcal{G}_m)$, which is estimated in the M-step, provides probabilistic information about the probabilistic affinity of different words to aspects (or *topics*). The terms with the highest probability values for a specific aspect $\mathcal{G}_m$ can be viewed as a *cluster digest* for that topic.

As the *PLSA* approach also provides a multidimensional $n \times k$ coordinate representation $Q_k \Sigma_k$ of the documents, a different way of performing the clustering would be to represent the documents in this new space and use a k-means algorithm on the transformed corpus. Because the noise impact of synonymy and polysemy has been removed by *PLSA*, the k-means approach will generally be more effective on the reduced representation than on the original corpus.

13.4.3 Limitations of PLSA

Although the *PLSA* method is an intuitively sound model for probabilistic modeling, it does have a number of practical drawbacks. The number of parameters grows linearly with the number of documents. Therefore, such an approach can be slow and may overfit the training data because of the large number of estimated parameters. Furthermore, while *PLSA* provides a generative model of document–word pairs in the training data, it cannot easily assign probabilities to previously unseen documents. Most of the other EM mixture models discussed in this book, such as the probabilistic Bayes model, are much better at assigning probabilities to previously unseen documents. To address these issues, *Latent Dirichlet Allocation (LDA)* was defined. This model uses Dirichlet priors on the topics, and generalizes relatively easily to new documents. In this sense, *LDA* is a fully generative model. The bibliographic notes contain pointers to this model.

13.5 Specialized Classification Methods for Text

As in clustering, classification algorithms are affected by the nonnegative, sparse and high-dimensional nature of text data. An important effect of sparsity is that the presence of a word in a document is more informative than the absence of the word. This observation has implications for classification methods such as the Bernoulli model used for Bayes classification that treat the presence and absence of a word in a symmetric way.

Popular techniques in the text domain include instance-based methods, the Bayes classifier, and the SVM classifier. The Bayes classifier is very popular because Web text is often combined with other types of features such as URLs or side information. It is relatively easy to incorporate these features into the Bayes classifier. The sparse high-dimensional nature of text also necessitates the design of more refined multinomial Bayes models for the text domain. SVM classifiers are also extremely popular for text data because of their high accuracy. The major issue with the use of the SVM classifier is that the high-dimensional nature of text necessitates performance enhancements to such classifiers. In the following, some of these algorithms will be discussed.

13.5.1 Instance-Based Classifiers

Instance-based classifiers work surprisingly well for text, especially when a preprocessing phase of clustering or dimensionality reduction is performed. The simplest form of the nearest neighbor classifier returns the dominant class label of the top-k nearest neighbors with the cosine similarity. Weighting the vote with the cosine similarity value often provides more robust results. However because of the sparse and high-dimensional nature of text collections, this basic procedure can be modified in two ways to improve both the efficiency and the effectiveness. The first method uses dimensionality reduction in the form of latent semantic indexing. The second method uses fine-grained clustering to perform centroid-based classification.

13.5.1.1 Leveraging Latent Semantic Analysis

A major source of error in instance-based classification is the noise inherent in text collections. This noise is often a result of *synonymy* and *polysemy*. For example, the words *comical* and *hilarious* mean approximately the same thing. Polysemy refers to the fact that the same word may mean two different things. For example, the word *jaguar* could refer to a car or a cat. Typically, the significance of a word can be understood only in the context of other words in the document. These characteristics of text create challenges for classification algorithms because the computation of similarity with the use of word frequencies may not be completely accurate. For example, two documents containing the words *comical* and *hilarious*, respectively, may not be deemed sufficiently similar because of synonymy effects. In latent semantic indexing, dimensionality reduction is applied to the collection to reduce these effects.

Latent semantic analysis (*LSA*) is an approach that relies on singular value decomposition (*SVD*) to create a reduced representation for the text collection. The reader is advised to refer to Sect. 2.4.3.3 of Chap. 2 for details of *SVD* and *LSA*. The latent semantic analysis (*LSA*) method is an application of the *SVD* method to the $n \times d$ document-term matrix D, where d is the size of the lexicon, and n is the number of documents. The eigenvectors with the largest eigenvalues of the square $d \times d$ matrix $D^T D$ are used for data representation. The sparsity of the data set results in a low intrinsic dimensionality. Therefore, in the text domain, the reduction in dimensionality resulting from *LSA* is rather drastic. For example, it is not uncommon to be able to represent a corpus drawn on a lexicon of size 100,000 in less than 300 dimensions. The removal of the dimensions with small eigenvalues typically leads to a reduction in the noise effects of synonymy and polysemy. This data representation is no longer sparse and resembles multidimensional numeric data. A conventional k-nearest neighbor classifier with cosine similarity can be used on this transformed corpus. The *LSA* method does require an additional effort up front to create the eigenvectors.

13.5.1.2 Centroid-Based Classification

Centroid-based classification is a fast alternative to k-nearest neighbor classifiers. The basic idea is to use an off-the-shelf clustering algorithm to partition the documents of each class into clusters. The number of clusters derived from the documents of each class is proportional to the number of documents in that class. This ensures that the clusters in each class are of approximately the same granularity. Class labels are associated with individual clusters rather than the actual documents.

The cluster digests from the centroids are extracted by retaining only the most frequent words in that centroid. Typically, about 200 to 400 words are retained in each centroid. The lexicon in each of these centroids provides a stable and topical representation of the subjects in each class. An example of the (weighted) word vectors for two classes corresponding to the labels "*Business schools*" and "*Law schools*" could be as follows:

1. *Business schools:* business (35), management (31), school (22), university (11), campus (15), presentation (12), student (17), market (11), ...
2. *Law schools:* law (22), university (11), school (13), examination (15), justice (17), campus (10), courts (15), prosecutor (22), student (15), ...

Typically, most of the noisy words have been truncated from the cluster digest. Similar words are represented in the same centroid, and words with multiple meanings can be represented in contextually different centroids. Therefore, this approach also indirectly addresses the issues of synonymy and polysemy, with the *additional* advantage that the k-nearest neighbor classification can be performed more efficiently with a smaller number of centroids. The dominant label from the top-k matching centroids, based on cosine similarity, is reported. Such an approach can provide comparable or better accuracy than the vanilla k-nearest neighbor classifier in many cases.

13.5.1.3 Rocchio Classification

The Rocchio method can be viewed as a special case of the aforementioned description of the centroid-based classifier. In this case, all documents belonging to the same class are aggregated into a *single* centroid. For a given document, the class label of the closest centroid is reported. This approach is obviously extremely fast because it requires a small constant number of similarity computations that is dependent on the number of classes in the data. On the other hand, the drawback is that the accuracy depends on the assumption of *class contiguity*. The class-contiguity assumption, as stated in [377], is as follows:

"*Documents in the same class form a contiguous region, and regions of different classes do not overlap.*"

Thus, Rocchio's method would not work very well if documents of the same class were separated into distinct clusters. In such cases, the centroid of a class of documents may not even lie in one of the clusters of that class. A bad case for Rocchio's method is illustrated in Fig. 13.6, in which two classes and four clusters are depicted. Each class is associated with two distinct clusters. In this case, the centroids for each of the classes are approximately the same. Therefore, the Rocchio method would have difficulty in distinguishing between the classes. On the other hand, a k-nearest neighbor classifier for small values of k, or a centroid-based classifier would perform quite well in this case. As discussed in Chap. 11, an increase in the value of k for a k-nearest neighbor classifier increases its bias. The Rocchio classifier can be viewed as a k-nearest neighbor classifier with a high value of k.

13.5.2 Bayes Classifiers

The Bayes classifier is described in Sect. 10.5.1 of Chap. 10. The particular classifier described was a binary (or *Bernoulli*) model in which the posterior probability of a document belonging to a particular class was computed using only the presence or the absence of a word. This special case corresponds to the fact that each feature (word) takes on the value of either 0 or 1 depending on whether or not it is present in the document. However, such an approach does not account for the frequencies of the words in the documents.

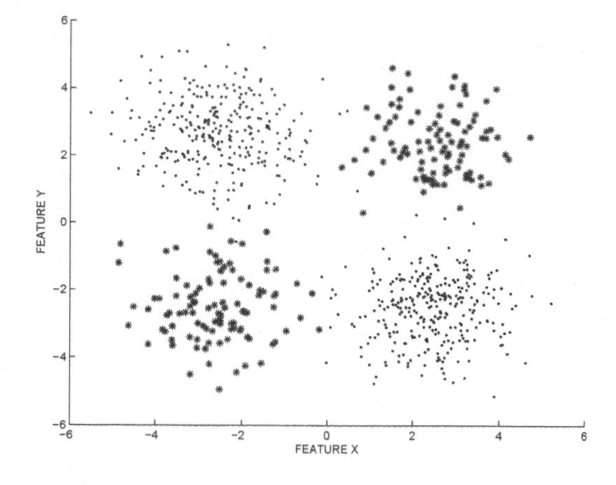

Figure 13.6: A bad case for the Rocchio method

13.5.2.1 Multinomial Bayes Model

A more general approach is to use a multinomial Bayes model, in which the frequencies of the words are used explicitly. The Bernoulli model is helpful mostly for cases where the documents are short, and drawn over a lexicon of small size. In the general case of documents of longer sizes over a large lexicon, the multinomial model is more effective. Before discussing the multinomial model, the Bernoulli model (cf. Sect. 10.5.1 of Chap. 10) will be revisited in the context of text classification.

Let C be the random variable representing the class variable of an unseen test instance, with d-dimensional feature values $\overline{X} = (a_1 \dots a_d)$. For the Bernoulli model on text data, each value of a_i is 1 or 0, depending on whether or not the ith word of the lexicon is present in the document $\overline{X}$. The goal is to estimate the posterior probability $P(C = c|\overline{X} = (a_1 \dots a_d))$. Let the random variables for the individual dimensions of $\overline{X}$ be denoted by $\overline{X} = (x_1 \dots x_d)$. Then, it is desired to estimate the conditional probability $P(C = c|x_1 = a_1, \dots x_d = a_d)$. Then, by using Bayes' theorem, the following equivalence can be inferred.

$$P(C = c|x_1 = a_1, \dots x_d = a_d) = \frac{P(C = c)P(x_1 = a_1, \dots x_d = a_d|C = c)}{P(x_1 = a_1, \dots x_d = a_d)} \tag{13.17}$$

$$\propto P(C = c)P(x_1 = a_1, \dots x_d = a_d|C = c) \tag{13.18}$$

$$\approx P(C = c) \prod_{i=1}^{d} P(x_i = a_i|C = c). \tag{13.19}$$

The last of the aforementioned relationships is based on the *naive* assumption of conditional independence. In the binary model discussed in Chap. 10, each attribute value a_i takes on the value of 1 or 0 depending on the presence or the absence of a word. Thus, if the fraction of the documents in class c containing word i is denoted by $p(i, c)$, then the value of $P(x_i = a_i|C = c)$ is estimated[5] as either $p(i, c)$ or $1 - p(i, c)$ depending upon whether a_i is 1 or 0, respectively. Note that this approach explicitly penalizes nonoccurrence of words in documents. Larger lexicon sizes will result in many words that are absent in a document. Therefore, the Bernoulli model may be dominated by word absence rather than

[5] The exact value will be slightly different because of Laplacian smoothing. Readers are advised to refer to Sect. 10.5.1 of Chap. 10.

word presence. Word absence is usually weakly related to class labels. This leads to greater noise in the evaluation. Furthermore, differential frequencies of words are ignored by this approach. Longer documents are more likely to have repeated words. The multinomial model is designed to address these issues.

In the multinomial model, the L terms in a document are treated as samples from a multinomial distribution. The total number of terms in the document (or document length) is denoted by $L = \sum_{j=1}^{d} a_i$. In this case, the value of a_i is assumed to be the raw frequency of the term in the document. The posterior class probabilities of a test document with the frequency vector $(a_1 \ldots a_d)$ are defined and estimated using the following generative approach:

1. Sample a class c with a class-specific prior probability.
2. Sample L terms *with replacement* from the term distribution of the chosen class c. The term distribution is defined using a multinomial model. The sampling process generates the frequency vector $(a_1 \ldots a_d)$. All training and test documents are assumed to be observed samples of this generative process. Therefore, all model parameters of the generative process are estimated from the training data.
3. *Test instance classification:* What is the posterior probability that the class c is selected in the first generative step, conditional on the *observed* word frequency $(a_1 \ldots a_d)$ in the test document?

When the sequential ordering of the L different samples are considered, the number of possible ways to sample the different terms to result in the representation $(a_1 \ldots a_d)$ is given by $\frac{L!}{\prod_{i:a_i>0} a_i!}$. The probability of *each* of these sequences is given by $\prod_{i:a_i>0} p(i,c)^{a_i}$, by using the naive independence assumption. In this case, $p(i,c)$ is estimated as the fractional number of occurrences of word i in class c *including repetitions.* Therefore, unlike the Bernoulli model, repeated presence of word i in a document belonging to class c will increase $p(i,c)$. If $n(i,c)$ is the number of occurrences of word i in all documents belonging to class c, then $p(i,c) = \frac{n(i,c)}{\sum_i n(i,c)}$. Then, the class conditional feature distribution is estimated as follows:

$$P(x_1 = a_1, \ldots x_d = a_d | C = c) \approx \frac{L!}{\prod_{i:a_i>0} a_i!} \prod_{i:a_i>0} p(i,c)^{a_i}. \tag{13.20}$$

Using the Bayes rule, the multinomial Bayes model computes the posterior probability for a test document as follows:

$$P(C = c | x_1 = a_1, \ldots x_d = a_d) \propto P(C = c) \cdot P(x_1 = a_1, \ldots x_d = a_d | C = c) \tag{13.21}$$

$$\approx P(C = c) \cdot \frac{L!}{\prod_{i:a_i>0} a_i!} \prod_{i:a_i>0} p(i,c)^{a_i} \tag{13.22}$$

$$\propto P(C = c) \cdot \prod_{i:a_i>0} p(i,c)^{a_i}. \tag{13.23}$$

The constant factor $\frac{L!}{\prod_{i:a_i>0} a_i!}$ has been removed from the last condition because it is the same across all classes. Note that in this case, the product on the right-hand side only uses those words i, for which a_i is strictly larger than 0. Therefore, nonoccurrence of words is ignored. In this case, we have assumed that each a_i is the raw frequency of a word, which is an integer. It is also possible to use the multinomial Bayes model with the tf-idf frequency of a word, in which the frequency a_i might be fractional. However, the generative explanation becomes less intuitive in such a case.

13.5.3 SVM Classifiers for High-Dimensional and Sparse Data

The number of terms in the Lagrangian dual of the SVM formulation scales with the square of the number of dimensions. The reader is advised to refer to Sect. 10.6 of Chap. 10, and 11.4.2 of Chap. 11 for the relevant discussions. While the *SVMLight* method in Sect. 11.4.2 of Chap. 11 addresses this issue by making changes to the algorithmic approach, it does not make modifications to the SVM formulation itself. Most importantly, the approach does not make any modifications to address the high dimensional and sparse nature of text data.

The text domain is high dimensional and sparse. Only a small subset of the dimensions take on nonzero values for a given text document. Furthermore, linear classifiers tend to work rather well for the text domain, and it is often not necessary to use the kernelized version of the classifier. Therefore, it is natural to focus on linear classifiers, and ask whether it is possible to improve the complexity of SVM classification further by using the special domain-specific characteristics of text. *SVMPerf* is a linear-time algorithm designed for text classification. Its training complexity is $O(n \cdot s)$, where s is the average number of nonzero attributes per training document in the collection.

To explain the approach, we first briefly recap the soft penalty-based SVM formulation introduced in Sect. 10.6 of Chap. 10. The problem definition, referred to as the optimization formulation (OP1), is as follows:

$$\text{(OP1): Minimize } \frac{||\overline{W}||^2}{2} + C\frac{\sum_{i=1}^{n} \xi_i}{n}$$

$$\text{subject to:}$$

$$y_i \overline{W} \cdot \overline{X_i} \geq 1 - \xi_i \quad \forall i$$
$$\xi_i \geq 0 \quad \forall i.$$

One difference from the conventional SVM formulation of Chap. 10 is that the constant term b is missing. The conventional SVM formulation uses the constraint $y_i(\overline{W} \cdot \overline{X_i} + b) \geq 1 - \xi_i$. The two formulations are, however, equivalent because it can be shown that adding a dummy feature with a constant value of 1 to each training instance has the same effect. The coefficient in $\overline{W}$ of this feature will be equal to b. Another minor difference from the conventional formulation is that the slack component in the objective function is scaled by a factor of n. This is not a significant difference either because the constant C can be adjusted accordingly. These minor variations in the notation are performed without loss of generality for algebraic simplicity.

The *SVMPerf* method reformulates this problem with a *single* slack variable ξ, and 2^n constraints that are generated by summing a random subset of the n constraints in (OP1). Let $\overline{U} = (u_1 \ldots u_n) \in \{0,1\}^n$ represent the indicator vector for the constraints that are summed up to create this new synthetic constraint. An alternative formulation of the SVM model is as follows:

$$\text{(OP2): Minimize } \frac{||\overline{W}||^2}{2} + C\xi$$

$$\text{subject to:}$$

$$\frac{1}{n}\sum_{i=1}^{n} u_i y_i \overline{W} \cdot \overline{X_i} \geq \frac{\sum_{i=1}^{n} u_i}{n} - \xi \quad \forall\, \overline{U} \in \{0,1\}^n$$
$$\xi \geq 0.$$

The optimization formulation (OP2) is different from (OP1) in that it has only one slack variable ξ but 2^n constraints that represent the sum of every subset of constraints in (OP1). It can be shown that a one-to-one correspondence exists between the solutions of (OP1) and (OP2).

Lemma 13.5.1 *A one-to-one correspondence exists between solutions of (OP1) and (OP2), with equal values of* $\overline{W} = \overline{W^*}$ *in both models, and* $\xi^* = \frac{\sum_{i=1}^n \xi_i^*}{n}$.

Proof: We will show that if the same value of $\overline{W}$ is fixed for (OP1), and (OP2), then it will lead to the same objective function value. The first step is to derive the slack variables in terms of this value of $\overline{W}$ for (OP1) and (OP2). For problem (OP1), it can be derived from the slack constraints that the optimal value of ξ_i is achieved for $\xi_i = \max\{0, 1 - y_i\overline{W} \cdot \overline{X_i}\}$ in order to minimize the slack penalty. For the problem OP2, a similar result for ξ can be obtained:

$$\xi = \max_{u_1 \dots u_n} \left\{ \frac{\sum_{i=1}^n u_i}{n} - \frac{1}{n}\sum_{i=1}^n u_i y_i \overline{W} \cdot \overline{X_i} \right\}. \tag{13.24}$$

Because this function is linearly separable in u_i, one can push the maximum inside the summation, and independently optimize for each u_i:

$$\xi = \sum_{i=1}^n \max_{u_i} u_i \left\{ \frac{1}{n} - \frac{1}{n} y_i \overline{W} \cdot \overline{X_i} \right\}. \tag{13.25}$$

For optimality, the value of u_i should be picked as 1 for only the positive values of $\{\frac{1}{n} - \frac{1}{n} y_i \overline{W} \cdot \overline{X_i}\}$ and 0, otherwise. Therefore, one can show the following:

$$\xi = \sum_{i=1}^n \max \left\{ 0, \frac{1}{n} - \frac{1}{n} y_i \overline{W} \cdot \overline{X_i} \right\} \tag{13.26}$$

$$= \frac{1}{n} \sum_{i=1}^n \max\{0, 1 - y_i \overline{W} \cdot \overline{X_i}\} = \frac{\sum_{i=1}^n \xi_i}{n}. \tag{13.27}$$

This one-to-one correspondence between optimal values of $\overline{W}$ in (OP1) and (OP2) implies that the two optimization problems are equivalent. ■

Thus, by determining the optimal solution to problem (OP2), it is possible to determine the optimal solution to (OP1) as well. Of course, it is not yet clear, why (OP2) is a better formulation than (OP1). After all, problem (OP2) contains an exponential number of constraints, and it seems to be intractable to even *enumerate* the constraints, let alone solve them.

Even so, the optimization formulation (OP2) does have some advantages over (OP1). First, a single slack variable measures the feasibility of all the constraints. This implies that all constraints can be expressed in terms of $(\overline{W}, \xi)$. Therefore, if one were to solve the optimization problem with only a subset of the 2^n constraints and the remaining were satisfied to a precision of ϵ by $(\overline{W}, \xi)$, then it is guaranteed that $(\overline{W}, \xi + \epsilon)$ is feasible for the *full* set of constraints.

The key is to never use all the constraints *explicitly*. Rather, a small subset $\mathcal{WS}$ of the 2^n constraints is used as the working set. We start with an empty working set $\mathcal{WS}$. The corresponding optimization problem is solved, and the most violated constraint among the constraints not in $\mathcal{WS}$ is added to the working set. The vector $\overline{U}$ for the most violated constraint is relatively easy to find. This is done by setting u_i to 1, if $y_i \overline{W} \cdot \overline{X_i} < 1$, and 0 otherwise. Therefore, the iterative steps for adding to the working set $\mathcal{WS}$ are as follows:

1. Determine optimal solution $(\overline{W}, \xi)$ for objective function of (OP2) using only constraints in the working set $\mathcal{WS}$.

2. Determine most violated constraint among the 2^n constraints of (OP2) by setting u_1 to 1 if $y_i \overline{W} \cdot \overline{X_i} < 1$, and 0 otherwise.

3. Add the most violated constraint to $\mathcal{WS}$.

The termination criterion is the case when the most violated constraint is violated by no more than ϵ. This provides an approximate solution to the problem, depending on the desired precision level ϵ.

This algorithm has several desirable properties. It can be shown that the time required to solve the problem for a constant size working set $\mathcal{WS}$ is $O(n \cdot s)$, where n is the number of training examples, and s is the number of *nonzero attributes* per example. This is important for the text domain, where the number of non-zero attributes is small. Furthermore, the algorithm usually terminates in a small constant number of iterations. Therefore, the working set $\mathcal{WS}$ never exceeds a constant size, and the entire algorithm terminates in $O(n \cdot s)$ time.

13.6 Novelty and First Story Detection

The problem of first story detection is a popular one in the context of temporal text stream mining applications. The goal is to determine novelties from the underlying text stream based on the history of previous text documents in the stream. This problem is particularly important in the context of streams of news documents, where a first story on a new topic needs to be reported as soon as possible.

A simple approach is to compute the maximum similarity of the current document with all previous documents, and report the documents with very low maximum similarity values as novelties. Alternatively, the inverse of the maximum similarity value could be continuously reported as a streaming novelty score or alarm level. The major problem with this approach is that the stream size continuously increases with time, and one has to compute similarity with all previous documents. One possibility is to use reservoir sampling to maintain a constant sample of documents. The inverse of the maximum similarity of the document to any incoming document is reported as the novelty score. The major drawback of this approach is that similarity between individual pairs of documents is often not a stable representation of the aggregate trends. Text documents are sparse, and pairwise similarity often does not capture the impact of synonymy and polysemy.

13.6.1 Micro-clustering Method

The micro-clustering method can be used to maintain online clusters of the text documents. The idea is that micro-clustering simultaneously determines the clusters and novelties from the underlying text stream. The basic micro-clustering method is described in Sect. 12.4 of Chap. 12. The approach maintains k different cluster centroids, or cluster digests. For an incoming document, its similarity to all the centroids is computed. If this similarity is larger than a user-defined threshold, then the document is added to the cluster. The frequencies of the words in the corresponding centroid are updated, by adding the frequency of the word in the document to it. For each document, only the r most frequent words in the centroid are retained. The typical value of r varies between 200 and 400. On the other hand, when the incoming document is not sufficiently similar to one of the centroids, then it is reported as a novelty, or as a first story. A new cluster is created containing the singleton document. To make room for the new centroid, one of the old centroids needs to be removed. This is achieved by maintaining the last update time of each cluster. The most stale cluster is removed. This algorithm provides the online ability to report the novelties in the text stream. The bibliographic notes contain pointers to more detailed versions of this method.

13.7 Summary

The text domain is sometimes challenging for mining purposes because of its sparse and high-dimensional nature. Therefore, specialized algorithms need to be designed.

The first step is the construction of a bag-of-words representation for text data. Several preprocessing steps need to be applied, such as stop-word removal, stemming, and the removal of digits from the representation. For Web documents, preprocessing techniques are also required to remove the anchor text and to extract text from the main block of the page.

Algorithms for problems such as clustering and classification need to be modified as well. For example, density-based methods are rarely used for clustering text. The k-means methods, hierarchical methods, and probabilistic methods can be suitably modified to work for text data. Two popular methods include the scatter/gather approach, and the probabilistic EM-algorithm. The co-clustering method is also commonly used for text data. Topic modeling can be viewed as a probabilistic modeling approach that shares characteristics of both dimensionality reduction and clustering. The problem of novelty detection is closely related to text clustering. Streaming text clustering algorithms can be used for novelty detection. Data points that do not fit in any cluster are reported as novelties.

Among the classification methods, decision trees are not particularly popular for text data. On the other hand, instance-based methods, Bayes methods, and SVM methods are used more commonly. Instance-based methods need to be modified to account for the noise effects of synonymy and polysemy. The multinomial Bayes model is particularly popular for text classification of long documents. Finally, the *SVMPerf* method is commonly used for efficient text classification with support vector machines.

13.8 Bibliographic Notes

An excellent book on text mining may be found in [377]. This book covers both information retrieval and mining problems. Therefore, issues such as preprocessing and similarity computation are covered well by this book. Detailed surveys on text mining may be found in [31]. Discussions of the tree matching algorithm may be found in [357, 542].

The scatter/gather approach discussed in this chapter was proposed in [168]. The importance of projecting out infrequent words for efficient document clustering was discussed in [452]. The *PLSA* discussion is adopted from the paper by Hofmann [271]. The *LDA* method is a further generalization, proposed in [98]. A survey on topic modeling may be found in [99]. Co-clustering methods for text were discussed in [171, 172, 437]. The co-clustering problem is also studied more generally as biclustering in the context of biological data. A general survey on biclustering methods may be found in [374]. General surveys on text clustering may be found in [31, 32].

The text classification problem has been explored extensively in the literature. The *LSA* approach was discussed in [184]. Centroid-based text classification was discussed in [249]. A detailed description of different variations of the Bayes model in may be found in [31, 33].

The *SVMPerf* and *SVMLight* classifiers were described in [291] and [292], respectively. A survey on SVM classification may be found in [124]. General surveys on text classification may be found in [31, 33, 453].

The first-story detection problem was first proposed in the context of the *topic detection and tracking* effort [557]. The micro-cluster-based novelty detection method described in this chapter was adapted from [48]. Probabilistic models for novelty detection may be found in [545]. A general discussion on the topic of first-story detection may be found in [5].

13.9 Exercises

1. Implement a computer program that parses a set of text, and converts it to the vector space representation. Use tf-idf normalization. Download a list of stop words from `http://www.ranks.nl/resources/stopwords.html` and remove them from the document, before creating the vector space representation.

2. Discuss the weaknesses of the k-medoids algorithm when applied to text data.

3. Suppose you paired the shared nearest neighbor similarity function (see Chap. 2) with cosine similarity to implement the k-means clustering algorithm for text. What is its advantage over the direct use of cosine similarity?

4. Design a combination of hierarchical and k-means algorithms in which merging operations are interleaved with the assignment operations. Discuss its advantages and disadvantages with respect to the *scatter/gather* clustering algorithm in which merging strictly precedes assignment.

5. Suppose that you have a large collection of short tweets from Twitter. Design a Bayes classifier which uses the identity as well as the exact position of each of the first ten words in the tweet to perform classification. How would you handle tweets containing less than ten words?

6. Design a modification of single-linkage text clustering algorithms, which is able to avoid excessive chaining.

7. Discuss why the multinomial Bayes classification model works better on longer documents with large lexicons than the Bernoulli Bayes model.

8. Suppose that you have class labels associated with documents. Describe a simple supervised dimensionality reduction approach that uses *PLSA* on a derivative of the document-term matrix to yield basis vectors which are each biased towards one or more of the classes. You should be able to control the level of supervision with a parameter λ.

9. Design an EM algorithm for clustering text data, in which the documents are generated from the multinomial distribution instead of the Bernoulli distribution. Under what scenarios would you prefer this clustering algorithm over the Bernoulli model?

10. For the case of binary classes, show that the Rocchio method defines a linear decision boundary. How would you characterize the decision boundary in the multiclass case?

11. Design a method which uses the EM algorithm to discover outlier documents.

Chapter 14

Mining Time Series Data

"The only reason for time is so that everything doesn't happen at once.—Albert Einstein

14.1 Introduction

Temporal data is common in data mining applications. Typically, this is a result of continuously occurring processes in which the data is collected by hardware or software monitoring devices. The diversity of domains is quite significant and extends from the medical to the financial domain. Some examples of such data are as follows:

- *Sensor data:* Sensor data is often collected by a wide variety of hardware and other monitoring devices. Typically, this data contains continuous readings about the underlying data objects. For example, environmental data is commonly collected with different kinds of sensors that measure temperature, pressure, humidity, and so on. Sensor data is the most common form of time series data.

- *Medical devices:* Many medical devices such as electrocardiogram (ECG) and electroencephalogram (EEG) produce continuous streams of time series data. These represent measurements of the functioning of the human body, such as the heart beat, pulse rate, blood pressure, etc. Real-time data is also collected from patients in intensive care units (ICU) to monitor their condition.

- *Financial market data:* Financial data, such as stock prices, is often temporal. Other forms of temporal data include commodity prices, industrial trends, and economic indicators.

In general, temporal data may be either *discrete* or *continuous*. For example, Web log data contains a series of *discrete* events corresponding to user clicks, whereas environmental data may contain a series of continuous *values* such as temperature. Continuous temporal data sets are referred to as *time series*, whereas discrete temporal data sets are referred to as *sequences*. This chapter focuses on continuous time series data. The next chapter

C. C. Aggarwal, *Data Mining: The Textbook*, DOI 10.1007/978-3-319-14142-8_14

studies data mining methods for discrete sequence data. While time series and discrete sequence data are conceptually similar, there are significant differences in the algorithmic methodologies used in each domain. However, in many cases, time series data is converted to discrete sequence data through discretization to facilitate the application of rich classes of sequence mining techniques. This chapter also discusses such cases.

Unlike multidimensional data, in which all attributes are treated equally, time series data are viewed as *contextual data* representations. In contextual data representations, the attributes are of two types:

- *Contextual attribute(s):* These represent the attributes that provide the *context* in which the measurements are made. In other words, the contextual attributes provide the reference points at which the behavioral values are measured. For the case of time series data, the *single* contextual attribute corresponds to the time dimension. Some data types, such as spatial data, may contain multiple contextual attributes corresponding to spatial coordinates. The time stamps could correspond to actual time values at which the data points are measured, or they could correspond to *consecutive indices* (or *ticks*) at which these values are measured.

- *Behavioral attribute(s):* These represent the behavioral values at the reference points. For example, in an environmental sensor, this could correspond to the temperature attribute. In general, each contextual attribute value (e.g., time stamp) has a corresponding behavioral attribute value (e.g., temperature). The behavioral attributes are usually the interesting ones from an application-specific perspective, but they cannot be properly interpreted without the knowledge of the contextual attributes. When more than one behavioral attribute is associated with each series, the corresponding series is referred to as a *multivariate* time series.

The analysis of contextual data types is more difficult because behavioral attribute values cannot be interpreted effectively without using the contextual attribute. For example, a sudden change of the behavioral attribute between successive time stamps (contextual attribute) is often indicative of outlier behavior. Thus, unlike multidimensional data, problem definitions are dependent on a combination of the interrelationships between contextual and behavioral attributes. Thus, problems such as clustering, classification, and outlier detection need to be significantly modified to account for the impact of the contextual attribute. Several data types discussed in subsequent chapters fall within this class. Other examples include sequence data and spatial data.

The greater complexity of time series data enables a larger number of problem definitions. Most of the models can be categorized into one of two types:

1. *Real-time analysis:* In real-time analysis, the data points in one or more series are analyzed in real time, to make predictions. Typically, a small window of recent history is used over the different data streams for the analysis. Examples of such analysis include forecasting, deviation detection, or event detection. When multiple series are available, they are typically analyzed in a temporally synchronized way. Even in cases where data mining applications such as clustering are applied to these problems, the analysis is typically performed in real time.

2. *Retrospective analysis:* In retrospective analysis, the time series data is already available, and subsequently analyzed. The analysis of different time series within a database is sometimes not synchronized over time. For example, in a time series database of ECG readings, the data may have been recorded over different periods.

Both these forms of analysis are useful in different kinds of applications. Furthermore, these two scenarios have different interpretations for the same applications such as clustering or outlier detection. These issues are discussed in more detail in later sections.

This chapter is organized as follows. The next section presents methods for time series preparation and similarity. Because the methods for time series similarity have already been discussed in detail in Chap. 3, they are summarized only briefly in this chapter. The reader is referred to the relevant sections of Chap. 3 for the different time series similarity measures. The problem of time series forecasting is discussed in Sect. 14.3. Time series motif discovery is discussed in Sect. 14.4. Section 14.5 addresses the problem of clustering time series. Outlier detection is discussed in Sect. 14.6. Time series classification is discussed in Sect. 14.7. The summary of the chapter is presented in Sect. 14.8.

14.2 Time Series Preparation and Similarity

Time series data may be either univariate or multivariate. In univariate time series data, a single behavioral attribute is associated with each time instant. In multivariate time series data, multiple behavioral attributes are associated with each time instant. The dimensionality of the time series, therefore, refers to the number of behavioral attributes being tracked.

Definition 14.2.1 (Multivariate Time Series Data) *A time series of length n and dimensionality d contains d numeric features at each of n timestamps $t_1 \dots t_n$. Each timestamp contains a component for each of the d series. Therefore, the set of values received at timestamp t_i is $\overline{Y_i} = (y_i^1 \dots y_i^d)$. The value of the jth series at timestamp t_i is y_i^j*

In a univariate time series, the value of d is 1. In such cases, a series of length n is represented as a set of scalar behavioral values $y_1 \dots y_n$, associated with the timestamps $t_1 \dots t_n$.

14.2.1 Handling Missing Values

It is common for time series data to contain missing values. Furthermore, the values of the series may not be synchronized in time when they are collected by independent sensors. It is often convenient to have time series values that are equally spaced and synchronized across different behavioral attributes for data processing. The most common methodology used for handling missing, unequally spaced, or unsynchronized values is linear interpolation. The idea is to create *estimated* values at the desired time stamps. These can be used to generate multivariate time series that are synchronized, equally spaced, and have no missing values.

Consider the scenario where y_i and y_j are values of the time series at times t_i and t_j, respectively, where $i < j$. Let t be a time drawn from the interval (t_i, t_j). Then, the interpolated value of the series is given by:

$$y = y_i + \left(\frac{t - t_i}{t_j - t_i}\right) \cdot (y_j - y_i) \tag{14.1}$$

This is simple linear interpolation, although other more complex methods, such as polynomial interpolation or spline interpolation, are possible. However, such methods require a larger number of data points in a time window for the estimation. In many cases, such methods do not provide significantly superior results over the straightforward linear interpolation method.

14.2.2 Noise Removal

Noise-prone hardware, such as sensors, are often used for time series data collection. The approach used by most of the noise removal methods is to remove *short-term fluctuations*. It should be pointed out that the distinction between noise and *interesting* outliers is often a difficult one to make. Interesting outliers are fluctuations, caused by specific aspects of the data *generation* process, rather than artifacts of the data *collection* process. Therefore, such cleaning and smoothing methods are sometimes not appropriate for problems such as outlier detection. Two methods, referred to as *binning* and *smoothing*, are often used for noise removal.

Binning

The method of binning divides the data into time intervals of size k denoted by $[t_1, t_k], [t_{k+1}, t_{2k}]$, etc. It is assumed that the timestamps are equally spaced apart. Therefore, each bin is of the same size, and it contains an equal number of points. The average value of the data points in each interval are reported as the smoothed values. Let $y_{i \cdot k+1} \ldots y_{i \cdot k+k}$ be the values at timestamps $t_{i \cdot k+1} \ldots t_{i \cdot k+k}$. Then, the new binned value will be y'_{i+1}, where

$$y'_{i+1} = \frac{\sum_{r=1}^{k} y_{i \cdot k+r}}{k}$$

Therefore, this approach uses the mean of the values in the bins. It is also possible to use the median of the behavioral attribute values. Typically, the median provides more robust estimates than the mean because the outlier points do not affect the median in a disproportionate way. The main problem with binning is that it reduces the number of available data points by a factor of k. Binning is also referred to as piecewise aggregate approximation (PAA). Such an approach can be rather lossy for large values of k, although it can also be advantageous for fast distance computations [309] because it provides a compressed representation.

Moving-Average Smoothing

Moving-average methods reduce the loss in binning by using overlapping bins, over which the averages are computed. As in the case of binning, averages are computed over windows of the time series. The main difference is that a bin is constructed starting at *each* timestamp in the series rather than only the timestamps at the boundaries of the bins. Therefore, the bin intervals are chosen to be $[t_1, t_k], [t_2, t_{k+1}]$, etc. This results in a set of overlapping intervals. The time series values are averaged over each of these intervals. Moving averages are also referred to as *rolling averages* and they reduce the noise in the time series because of the smoothing effect of averages.

In a real-time application, the moving average becomes available only after the *last* timestamp of the window. Therefore, moving averages introduce lags into the analysis and also lose some points at the beginning of the series because of boundary effects. Furthermore, short-term trends are sometimes lost because of smoothing. Larger bin sizes result in greater smoothing and lag. Because of the impact of lag, it is possible for the moving average to contain troughs (or downtrends) where there are peaks (or uptrends) in the original series, and vice versa. This can sometimes lead to a misleading understanding of recent trends.

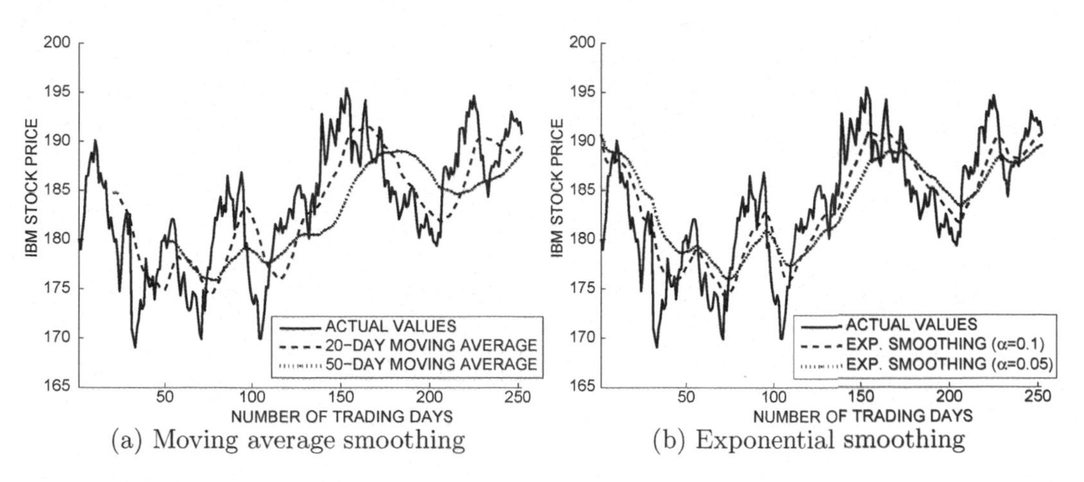

(a) Moving average smoothing (b) Exponential smoothing

Figure 14.1: Various smoothing methods applied to IBM stock price from September 5, 2013 to September 4, 2014

Exponential Smoothing

In exponential smoothing, the smoothed value y'_i is defined as a linear combination of the current value y_i, and the previously smoothed value y'_{i-1}. The smoothing parameter $\alpha \in (0, 1)$ is used for this purpose.

$$y'_i = \alpha \cdot y_i + (1 - \alpha) \cdot y'_{i-1} \tag{14.2}$$

The value of y'_0 is typically set to the first point in the series. When the value of α is 1, there are no smoothing effects, and the smoothed series is the same as the original series. When the value of α is 0, the entire series becomes smoothed to the constant value of y'_0. The approach is referred to as exponential smoothing because the value of y'_i can be expressed as an exponentially decayed sum of the series values. By recursively substituting the aforementioned equation into itself, the following can be shown:

$$y'_i = (1 - \alpha)^i \cdot y'_0 + \alpha \cdot \sum_{j=1}^{i} y_j \cdot (1 - \alpha)^{i-j}. \tag{14.3}$$

The choice of α regulates the decay factor. Unlike moving averages, exponential smoothing provides more importance to recent data points. Data points are not lost at the beginning of the series, and the impact of the lag is reduced for the same level of smoothing. Examples of moving average and exponential smoothing are illustrated in Fig. 14.1a, b, respectively. It is evident that exponential smoothing does not lose any points at the beginning of the series and generally provides slightly better smoothing for lower lag.

14.2.3 Normalization

Time series typically need to be normalized, especially when multiple series are analyzed simultaneously. For example, one series might measure temperature, whereas another might measure pressure. Because these values are measured on different scales, they cannot be compared meaningfully. Therefore, two normalization methods are commonly used to adjust for such variations.

1. *Range-based normalization:* In range-based normalization, the minimum and maximum value of the time series are determined. Let these values be denoted by min and max, respectively. Then, the time series value y_i is mapped to the new value y'_i in the range (0, 1) as follows:
$$y'_i = \frac{y_i - min}{max - min}. \tag{14.4}$$

2. *Standardization:* In standardization, the mean and standard deviation of the series are used for normalization. This is essentially the Z-value of the time series. Let μ and σ represent the mean and standard deviation of the values in the time series. Then, the time series value y_i is mapped to a new value z_i as follows:
$$z_i = \frac{y_i - \mu}{\sigma}. \tag{14.5}$$

Standardization is generally the preferred method. However, it does not guarantee a specific range of the time series values.

14.2.4 Data Transformation and Reduction

A variety of preprocessing methods exist for transforming and reducing the time series data into a reduced representation. Some of these methods transform the data into a smaller number of numeric coefficients, whereas other methods transform the data into discrete values.

14.2.4.1 Discrete Wavelet Transform

The discrete wavelet transform (DWT) converts a time series to multidimensional data. While time series can also be considered as multidimensional data by viewing[1] the values at the different timestamps as dimensions, the values in successive timestamps are highly related to one another. A direct application of multidimensional methods ignores the temporal continuity in data values. In wavelets, the coefficients describe properties of different *contiguous temporal regions* of the series. Each coefficient is equal to half the difference in the average value of the behavioral attribute between a pair of carefully chosen contiguous segments of the series. The resulting representation can be more easily analyzed like multidimensional data because temporal locality is already incorporated within the coefficients. By using only the largest coefficients for representation, it is possible to reconstruct the entire time series accurately. Typically, the number of retained coefficients is much smaller than the length of the original time series. Thus, the approach is a dimensionality reduction method as well. DWT is described in detail in Sect. 2.4.4.1 of Chap. 2.

14.2.4.2 Discrete Fourier Transform

Wavelets are most effective when most of the variations in the series can be captured in specific local regions of the series. In cases where the series contain global periodicity, the discrete Fourier transform (DFT) is more effective. Examples of scenarios in which either of these methods would perform well are provided in Fig. 14.2. The basic idea is that any series

[1]The concept of "dimension" can be defined in two ways for time series data. Each behavioral attribute in a multivariate series can be viewed as a dimension. Alternatively, the different values in a univariate time series can be viewed as dimensions. The usage is often dependent on the semantics of the application at hand.

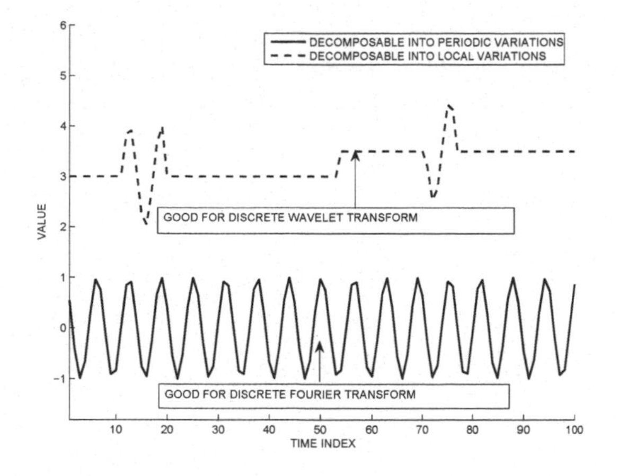

Figure 14.2: Preferred scenarios for *DFT* and *DWT*

of length n can be expressed as a linear combination of smooth periodic sinusoidal series. Along with a single constant term, the $n-1$ sinusoidal series have periodicity drawn from $n, n/2, n/3, \ldots n/(n-1)$. The data can be reduced using this decomposition because only a small number of these constituent series have large enough contributions to be included. Consider a time series $x_0 \ldots x_{n-1}$. Each coefficient X_k of the Fourier transform is a *complex* value which is defined as follows:

$$X_k = \sum_{r=0}^{n-1} x_r \cdot e^{-ir\omega k} = \sum_{r=0}^{n-1} x_r \cdot \cos(r\omega k) - i \sum_{r=0}^{n-1} x_r \cdot \sin(r\omega k) \quad \forall k \in \{0 \ldots n-1\}. \quad (14.6)$$

Here, ω is set to $\frac{2\pi}{n}$ radians, and the notation i denotes the imaginary number $\sqrt{-1}$. Therefore, X_k is a complex value. One property of the Fourier coefficients is that X_{n-k} can be derived from X_k by flipping the sign of the imaginary part for $k \geq 1$ (see Exercise 7). Therefore, only the first $n/2$ complex coefficients need to be retained. Furthermore, only the coefficients $X_k = a_k + ib_k$ with large energy $a_k^2 + b_k^2$ need to be retained. The top-m retained coefficients (together with their index k) can be used to approximate the time series in a compact way. Both the real and imaginary parts of the coefficients can be stored in a real-valued vector data structure. This vector provides the reduced representation of the series. The original series can be reconstructed from the coefficients as follows:

$$x_r = \frac{1}{n} \sum_{k=0}^{n-1} X_k \cdot e^{ir\omega k} = \frac{1}{n} \left(\sum_{k=0}^{n-1} X_k \cdot \cos(r\omega k) + i \sum_{k=0}^{n-1} X_k \cdot \sin(r\omega k) \right) \quad \forall r \in \{0 \ldots n-1\}.$$

Note that each X_k is a complex value. However, the imaginary part of the right-hand side of this equation will always evaluate to zero to yield the real series value x_r.

DFT has several properties, which make it useful for data mining applications. It satisfies the *additivity property*; the Fourier coefficients of the sum (or difference) of two series can be obtained as the sum (or difference) of their Fourier coefficients. It also satisfies Parseval's theorem, which states that if $X_k = a_k + ib_k$ is the kth Fourier coefficient, then we have $\sum_{r=0}^{n-1} x_r^2 = \frac{1}{n} \sum_{k=0}^{n-1} (a_k^2 + b_k^2)$. Because of these properties, one can compute the (scaled) Euclidean distance between two time series by computing the Euclidean distance between their Fourier coefficients. Like *DWT*, *DFT* can also be viewed as the transformation of the time series to a new (rotated) orthogonal basis system, except that each basis vector $\overline{B_k} =$

$[1, e^{i\omega k}, e^{2i\omega k}, \ldots, e^{(n-1)i\omega k}]$ of the Fourier coefficient X_k is a complex vector. Therefore, the time series may be decomposed in terms of the mutually orthogonal basis vectors $\overline{B_0} \ldots \overline{B_{n-1}}$ as follows:

$$(x_0 \ldots x_{n-1}) = \frac{1}{n} \sum_{k=0}^{n-1} X_k \overline{B_k} \tag{14.7}$$

Typically, off-the-shelf mathematical packages are available to compute the coefficients with the use of the fast Fourier transform (FFT). A closely related transform, known as the discrete cosine transform (DCT), provides even better compression.

14.2.4.3 Symbolic Aggregate Approximation (SAX)

This approach converts a time series to discrete sequence data. The basic idea is to determine piecewise aggregate approximates by averaging behavioral attribute values over successive and equally-spaced windows of the time series. The resulting continuous values are then discretized into a small number of discrete values. Depending on the application, the number of breakpoints may vary between 3 and 10. The approach selects the break points of the discretization, so that each of the symbolic values has an approximately equal frequency of representation. One possibility is to use equi-depth discretization of the continuous values, though this can be impractical or infeasible for long series or streaming series. For long series or streaming series, a Gaussian distribution assumption of the resulting averages is used to determine the discretization breakpoints. The idea is to select points on the Gaussian curve, so that the area between successive breakpoints is equal, and therefore the different symbols have approximately the same frequency.

14.2.5 Time Series Similarity Measures

Time series similarity measures are typically designed with application-specific goals in mind. The most common methods for time series similarity computation are Euclidean distance and dynamic time warping (DTW). The Euclidean distance is defined in an identical way to multidimensional data where the behavioral attribute values at the different timestamps are interpreted as dimensions. The Euclidean distance can be used only when the two series have the same length, and a one-to-one correspondence exists between the data points. This is not appropriate in unsynchronized time series where the data may be generated at different rates over different portions of the time series. The DTW method stretches and shrinks the time dimension differently in different portions of one of the series to create an optimal matching. As discussed in Sect. 16.3.4.1 of Chap. 16, DTW can also be extended to multivariate time series such as trajectory data. Two other similarity/distance functions include the Edit Distance and the Longest Common Subsequence. These measures are used more commonly for discrete sequences, rather than continuous time series. All these measures are described in detail in Sect. 3.4.1 of Chap. 3.

14.3 Time Series Forecasting

Forecasting is one of the most common applications of time series analysis. The prediction of future trends has applications in retail sales, economic indicators, weather forecasting, stock markets, and many other application scenarios. In this case, we have one or more series of data values, and it is desirable to predict the future values of the series using the history of previous values.

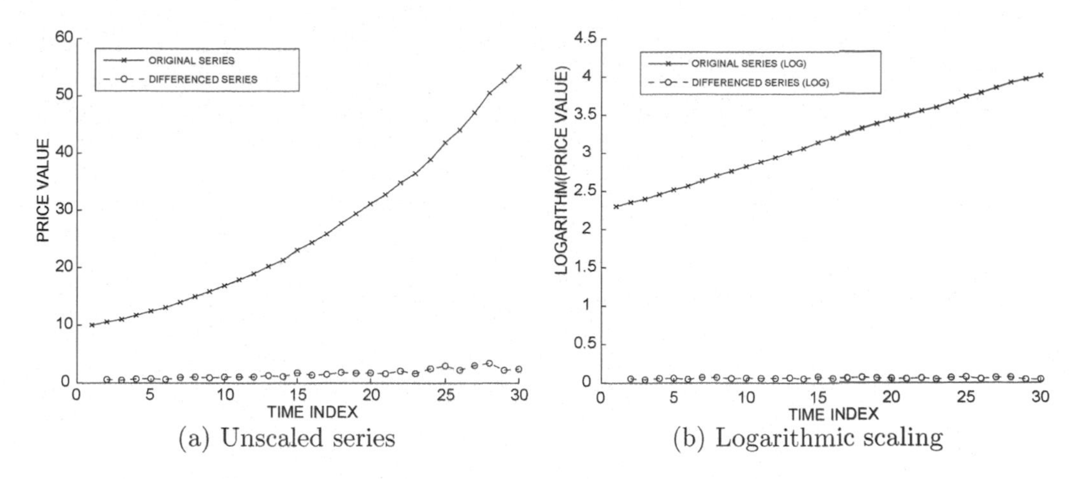

(a) Unscaled series (b) Logarithmic scaling

Figure 14.3: Impact of different operations on stationary and non-stationary series

Time series can be either *stationary* or *nonstationary*. A stationary stochastic process is one whose parameters, such as the mean and variance, do not change with time. A nonstationary process is one whose parameters change with time. Some kinds of time series such as white noise are stationary. White noise is the strongest form of stationarity with zero mean, constant variance, and zero covariance between series values separated by a fixed lag. On the other hand, consider the case, where the behavioral attribute corresponds to the price level of an industrial commodity such as crude oil. This is typically nonstationary because the average price level may increase over time as a result of inflation. In fact, most time series in real applications are nonstationary. A stationary series will usually be characterized as a noisy series with a level trend, constant variance, and zero covariance between different series values. For example, in Fig. 14.3a, both the series are nonstationary because the average values increase with time. On the other hand, in Fig. 14.3b, the dashed curve is stationary because the trends do not change significantly with time. A *strictly stationary* time series is defined as follows:

Definition 14.3.1 (Strictly Stationary Time Series) *A strictly stationary time series is one in which the probabilistic distribution of the values in any time interval* $[a, b]$ *is identical to that in the shifted interval* $[a + h, b + h]$ *for any value of the time shift* h.

In other words, all multivariate distributions of subsets of variables must match with their shifted counterparts. The window-based statistical parameters of a stationary time series can be estimated in a meaningful way because the parameters do not vary over different time windows. In such cases, the estimated statistical parameters are good predictors of future behavior. On the other hand, the *current* mean, variances, and statistical correlations of the series are not necessarily good predictors of *future* behavior in regression-based forecasting models for nonstationary series. Therefore, it is often advantageous to convert nonstationary series to stationary ones before forecasting analysis. After the forecasting has been performed on the stationary series, the predicted values are transformed back to the original representation, using the inverse transformation. The strict stationarity concept of Definition 14.3.1 is, however, too restrictive to be meaningfully used in real applications. For example, it is difficult even to determine whether or not a time series is strictly stationary from a single instance because one must comprehensively characterize all multivariate distributions of subsets of variables.

A key observation is that it is much easier to either obtain or convert to series that exhibit *weak* stationarity properties. In such cases, unlike white noise, the mean of the series, and the covariance between approximately adjacent time series values may be non-zero but constant over time. This is referred to as *covariance stationarity*. This kind of weak stationarity can be assessed relatively easily and is also useful for forecasting models that are dependent on specific parameters such as the mean and covariance. In other nonstationary series, the average value of the series can be described by a *trend-line* that is not necessarily horizontal, as required by a stationary series. Periodically, the series will deviate from the trend line, possibly because of some changes in the generative process, and then return to the trend line. This is referred to as a *trend stationary* series. Such weak forms of stationarity are also very useful for creating effective forecasting models. In the following, some practical methods that are commonly used to convert nonstationary series to stationary series will be discussed.

Differencing

A common approach used for converting time series to stationary forms is differencing. In differencing, the time series value y_i is replaced by the difference between it and the previous value. Therefore, the new value y_i' is as follows:

$$y_i' = y_i - y_{i-1}. \tag{14.8}$$

If the series is stationary after differencing, then an appropriate model for the data is:

$$y_{i+1} = y_i + e_{i+1}. \tag{14.9}$$

Here, e_{i+1} corresponds to white noise with zero mean. A differenced time series would have $t-1$ values for a series of length t because it is not possible for the first value to be reflected in the transformed series.

Higher order differencing can be used to achieve stationarity in second order changes. Therefore, the higher order differenced value y_i'' is defined as follows:

$$\begin{aligned} y_i'' &= y_i' - y_{i-1}' && (14.10)\\ &= y_i - 2 \cdot y_{i-1} + y_{i-2} && (14.11) \end{aligned}$$

This model allows the series to drift over time, since the noise has non-zero mean. The corresponding model is as follows:

$$y_{i+1} = y_i + c + e_{i+1}. \tag{14.12}$$

Here, c is a non-zero constant that accounts for the drift. Generally, it is rare to use differences beyond the second order.

A different approach is to use seasonal differences when it is known that the series is stationary after seasonal differencing. The seasonal differences are defined as follows:

$$y_i' = y_i - y_{i-m} \tag{14.13}$$

Here m is an integer greater than 1.

In some cases, such as geometrically increasing series, the logarithm function is applied to the values in the series, before the differencing operation. For example, consider a time series of prices that increase at an approximately constant inflation factor. In such cases, it may be useful to apply the logarithm function to the time series values, before the differencing operation. An example is provided in Fig. 14.3a, where the variation in inflation is illustrated with time. It is evident that the differencing operation does not help in making the series stationary. In Fig. 14.3b, the logarithm function is applied to the series before the differencing operation. In this case, the series becomes stationary after the differencing operation.

In the following, a number of univariate time series forecasting models will be discussed. These models work effectively under different assumptions on the time series patterns. Some of these models assume a stationary time series, whereas others do not.

14.3.1 Autoregressive Models

Univariate time series contain a single variable that is predicted using *autocorrelations.* Autocorrelations represent the correlations between adjacently located timestamps in a series. Typically, the behavioral attribute values at adjacently located timestamps are positively correlated. The autocorrelations in a time series are defined with respect to a particular value of the lag L. Thus, for a time series $y_1, \ldots y_n$, the autocorrelation at lag L is defined as the Pearson coefficient of correlation between y_t and y_{t+L}.

$$\text{Autocorrelation}(L) = \frac{\text{Covariance}_t(y_t, y_{t+L})}{\text{Variance}_t(y_t)}. \tag{14.14}$$

The autocorrelation always lies in the range $[-1, 1]$, although the value is almost always positive for very small values of L, and gradually drops off with increasing lag L. The positive correlation is a result of the fact that adjacent values of most time series are very similar, though the similarity drops off with increasing distance. High (absolute) values of the autocorrelation imply that the value at a given position in the series can be predicted as a function of the values in the immediately preceding window. This is, in fact, the key property that enables the use of the autoregressive model. For example, the variation in autocorrelation with lag for the IBM stock example (Fig. 14.1) is illustrated in Fig. 14.4a. Such a figure is referred to as the *autocorrelation plot* and is used commonly in AR models. While the autocorrelation is usually positive and falls off with lag, the precise behavior is highly application-specific. For periodic series, the autocorrelation may be periodic and negative at certain lag intervals. An example of the autocorrelations for a periodic sine wave is illustrated in Fig. 14.4b.

In the autoregressive model, the value of y_t at time t is defined as a linear combination of the values in the immediately preceding window of length p.

$$y_t = \sum_{i=1}^{p} a_i \cdot y_{t-i} + c + \epsilon_t \tag{14.15}$$

A model that uses the preceding window of length p is referred to as an $AR(p)$ model. The values of the regression coefficients $a_1 \ldots a_p, c$ need to be learned from the training data. The larger the value of p, the greater the lag that one is willing to incorporate in the autocorrelations. The choice of p should be guided by the level of autocorrelation of Eq. 14.14. Because the autocorrelation often reduces with increasing values of the lag L, a value of p should be selected, so that the autocorrelation at lag $L = p$ is small. In such cases, increasing the window of regression further may not help the accuracy of the modeling process, and may sometimes result in overfitting. Typically, the autocorrelation plot (Fig. 14.4) is used to identify the window. Instead of using a window of coefficients in

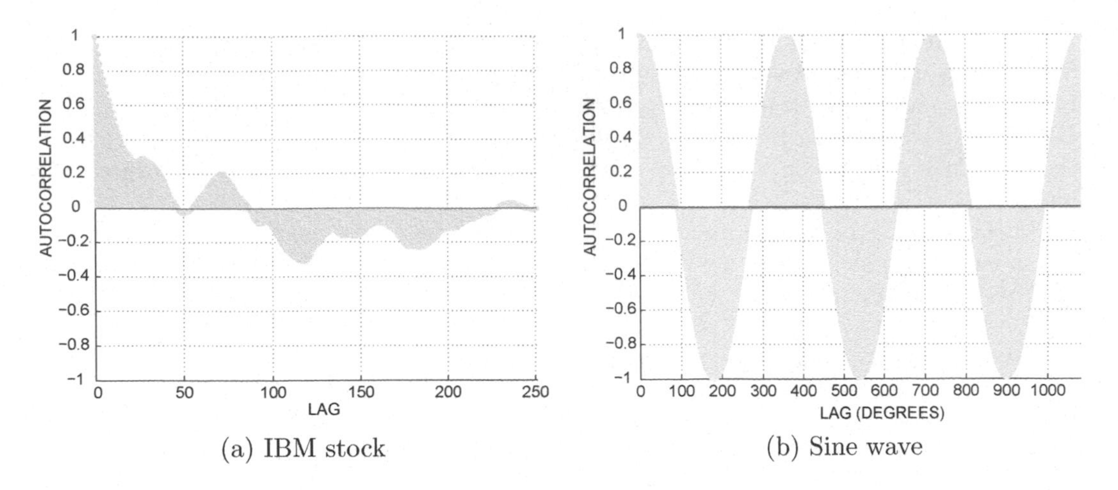

(a) IBM stock (b) Sine wave

Figure 14.4: Autocorrelation plots for various series

Eq. 14.15, it is also possible to select coefficients with specific lag values. In particular, lag values with high absolute autocorrelation in the autocorrelation plot may be selected. Such an approach is also helpful for forecasting periodic series.

Each timestamp in the past history of the time series data creates a linear equation between the time series variables. A set of linear equations between the coefficients can be created by using the value at each timestamp in the training data, along with its immediately preceding window of length p. When the number of timestamps available is much larger than p, this is an over-determined system of equations, which is infeasible. Therefore, any (infeasible) solution will have an error associated with it. The coefficients $a_1, \ldots a_p, c$ can be approximated with *least-squares regression*, to minimize the square-error of the over-determined system (cf. Sect. 11.5 of Chap. 11). Note that the model can be used effectively for forecasting *future* values, only if the key properties of the time series, such as the mean, variance, and autocorrelation do not change significantly with time. Many off-the-shelf commercial solvers are available for these models. The effectiveness of the forecasting model may be quantified by using the noise level in the estimated coefficients. Specifically, the R^2-value, which is also referred to as the *coefficient of determination*, measures the ratio of the white noise to the series variance:

$$R^2 = 1 - \frac{\text{Mean}_t(\epsilon_t^2)}{\text{Variance}_t(y_t)} \tag{14.16}$$

The coefficient of determination quantifies the fraction of variability in the series that is explained by the regression, as opposed to random noise. It is therefore desirable for this coefficient to be as close to 1 as possible.

14.3.2 Autoregressive Moving Average Models

While autocorrelation is a useful predictive property of time series, it does not always explain all the variations. In fact, the unexpected component of the variations (shocks), does impact future values of the time series. This component can be captured with the use of a moving average model (MA). The autoregressive model can therefore be made more robust by combining it with an MA. Before discussing the autoregressive moving average model (ARMA), the MA will be introduced.

The moving average model predicts subsequent series values on the basis of the past history of deviations from predicted values. A deviation from a predicted value can be viewed as white noise, or a shock. This model is best used in scenarios where the behavioral attribute value at a timestamp is dependent on the history of shocks in the time series, rather than the actual series values. The moving average model is defined as follows:

$$y_t = \sum_{i=1}^{q} b_i \cdot \epsilon_{t-i} + c + \epsilon_t$$

The aforementioned model is also referred to as $MA(q)$. The parameter c is the mean of the time series. The values of $b_1 \dots b_q$ are the coefficients that need to be learned from the data. The moving average model is quite different from the autoregressive model, in that it relates the current value to the mean of the series and the previous history of *deviations from forecasts*, rather than the actual values. Here the values of ϵ_t are assumed to be white noise error terms that are uncorrelated with one another. A problem here is that the error terms ϵ_t are not part of observed data, but also need to be derived from the forecasting model. This circularity implies that the system of equations is inherently nonlinear when expressed purely in terms of the coefficients and the observed values y_i. Typically, iterative nonlinear fitting procedures are used instead of the linear least-squares approach to determine a solution to the moving average model. It is rare that the series values can be predicted in terms of *only* the shocks, and not the autocorrelations. Autocorrelations are extremely important in time series analysis because of the inherent temporal continuity of time series data. At the same time, the history of shocks do impact the future values of the series. Therefore, neither the autoregressive nor the moving average model can fully capture all the correlations needed for forecasting in isolation.

A more general model may be obtained by combining the power of both the autoregressive model and the moving average model. The idea is to learn the appropriate impact of *both* the autocorrelations and the shocks in predicting time series values. The two models can be combined with p autoregressive terms and q moving average terms. This model is referred to as the *ARMA* model. In this case, the relationships between the different terms may be expressed as follows:

$$y_t = \sum_{i=1}^{p} a_i \cdot y_{t-i} + \sum_{i=1}^{q} b_i \cdot \epsilon_{t-i} + c + \epsilon_t$$

The aforementioned model is the $ARMA(p, q)$ model. A key question here is about the choice of the parameters p and q in these models. If the values of p and q are set to be too small, then the model will not fit the data well. On the other hand if the values of p and q are set to be too large, then the model is likely to overfit the data. In general, it is advisable to select the values of p and q as small as possible, so that the model fits the data well. As in the previous case, autoregressive moving average models are best used with stationary data.

In many cases, nonstationary data can be addressed by combining differencing with the autoregressive moving average model. This results in the *autoregressive integrated moving average model (ARIMA)*. In principle, differences of any order may be used, although first- and second-order differences are most commonly used. Consider the case where the first order differenced value y'_t is used. Then, the *ARIMA* model can be expressed as follows:

$$y'_t = \sum_{i=1}^{p} a_i \cdot y'_{t-i} + \sum_{i=1}^{q} b_i \cdot \epsilon_{t-i} + c + \epsilon_t$$

Thus, this model is virtually identical to the $ARMA(p, q)$ model, except that differencing is used within the model. If the order of the differencing is d, then this model is referred to as the $ARIMA(p, d, q)$ model.

14.3.3 Multivariate Forecasting with Hidden Variables

All the aforementioned models are designed for a single time series. In practice, a given application may have thousands of time series, and there may be significant correlations both across different series and across time. Therefore, models are required that can combine the autoregressive correlations with the cross-series correlations for making forecasts.

While there are many different ways of multivariate forecasting, hidden variables are often used to achieve this goal. This is because the hidden variable approach is able to cleanly separate out the cross-series correlations from the autoregressive correlations in the modeling process. The idea in hidden variable modeling is to transform the large number of cross-correlated time series into a small number of uncorrelated time series. Typically, principal component analysis (PCA) is used for this transformation. Because these different series are uncorrelated with one another, it is possible to use any of the AR, $ARMA$ or $ARIMA$ models individually on the series to predict the hidden values. Then, the predicted values are mapped back to their original representation. This provides the forecasted values for all the different series with the use of a small number of hidden variable predictions. Readers are advised to revisit Sect. 2.4.3.1 of Chap. 2 for the discussion on PCA before reading further.

It is assumed that there are d synchronized time series of length n. The d different time series values received at the ith timestamp are denoted by $\overline{Y_i} = (y_i^1 \dots y_i^d)$. The goal is to predict $\overline{Y_{n+1}}$ from $\overline{Y_1} \dots \overline{Y_n}$. The steps of the multivariate forecasting approach are as follows:

1. Construct the $d \times d$ covariance matrix of the multidimensional time series. Let the $d \times d$ covariance matrix be denoted by C. The (i, j)th entry of C is the covariance between the ith and jth series. This step is identical to the case of multidimensional data, and the temporal ordering among the different values of $\overline{Y_i}$ is not used at this stage. Thus, the covariance matrix only captures information about correlations across series, rather than correlations across time. Note that covariance matrices can also be maintained incrementally in the streaming setting, using an approach discussed in Sect. 20.3.1.4 of Chap. 20.

2. Determine the eigenvectors of the covariance matrix C as follows:

$$C = P \Lambda P^T \tag{14.17}$$

   Here, P is a $d \times d$ matrix, whose d columns contain the orthonormal eigenvectors. The matrix Λ is a diagonal matrix containing the eigenvalues. Let $P_{truncated}$ be a $d \times p$ matrix obtained by selecting the $p \ll d$ columns of P with the largest eigenvalues. Typically, the value of p is much smaller than d. This represents a basis for the hidden series with the greatest variability.

3. A new multivariate time series with p hidden time series variables is created. Each d-dimensional time series data point $\overline{Y_i}$ at the ith timestamp is expressed in terms of a p-dimensional hidden series data point. This is achieved by using the p basis vectors

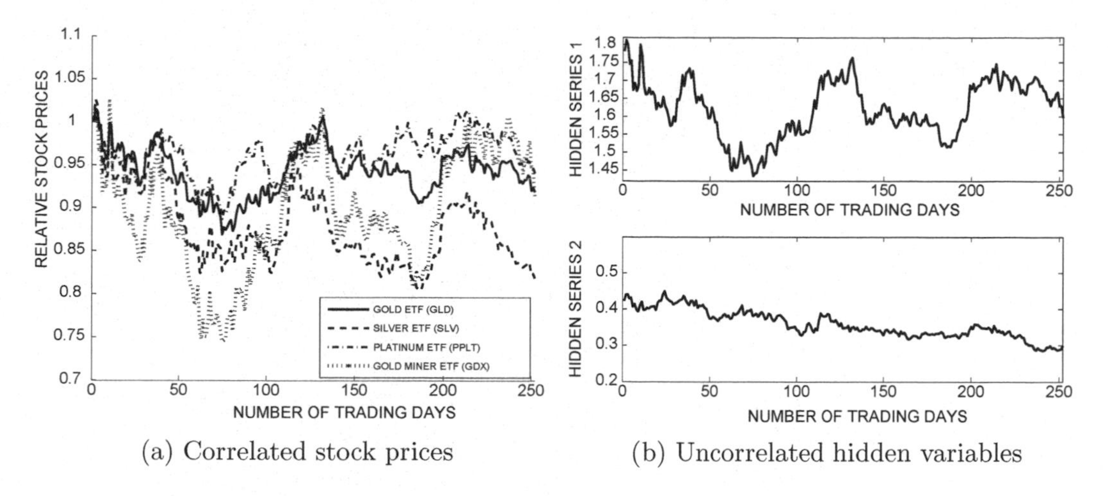

(a) Correlated stock prices

(b) Uncorrelated hidden variables

Figure 14.5: Normalized prices of four precious metal exchange traded funds (*ETFs*) from September 5, 2013 to September 4, 2014 and corresponding uncorrelated hidden variables

derived in the previous step. Therefore, the p-dimensional hidden value $\overline{Z_i} = (z_i^1 \dots z_i^p)$ is derived as follows:

$$\overline{Z_i} = \overline{Y_i} P_{truncated} \tag{14.18}$$

The value of $\overline{Z_i}$ represents the p different values for the hidden series variables at the ith timestamp. Thus, this step creates p different hidden variable time series that are approximately independent of one another. Note that the other $(d-p)$ hidden variables in $\overline{Y_i}P$ are approximately constant over time because of their small eigenvalues (variance). The means of these $(d-p)$ approximately constant values are noted as well. No predictive modeling is required for the vast majority of these hidden variables with constant values. In Fig. 14.5a, the stock prices of four precious metal-related exchange traded funds (ETFs) are illustrated for a period of 1 year. Each series was multiplicatively scaled to a relative value starting at 1. The top two hidden variable series are illustrated in Fig. 14.5b. Note that these derived series are uncorrelated and the first hidden variable has much higher variance than the second. The remaining two hidden variables are not shown because their variance is even smaller. In fact, each of the four correlated series in Fig. 14.5a can be approximately expressed as a different linear combination of the two hidden-variable series in Fig. 14.5b. Therefore, forecasting the hidden variables yields approximate forecasts of the original series.

4. For each of the p uncorrelated and high-variance series, use any univariate forecasting model to predict the values of the p hidden variables at the $(n+1)$th timestamp. A univariate approach can be used effectively because the different hidden variables are uncorrelated by design. This provides a set of values $\overline{Z_{n+1}} = (z_{n+1}^1 \dots z_{n+1}^p)$. Append the means of the approximately constant values of the remaining $(d-p)$ hidden series to $\overline{Z_{n+1}}$ to create a new d-dimensional hidden variable vector $\overline{W_{n+1}}$.

5. Transform back the predicted hidden variables $\overline{W_{n+1}}$ to the original d-dimensional representation by using the reverse transformation. This provides the forecasted values of the original series:

$$\overline{Y_{n+1}} = \overline{W_{n+1}} P^T \tag{14.19}$$

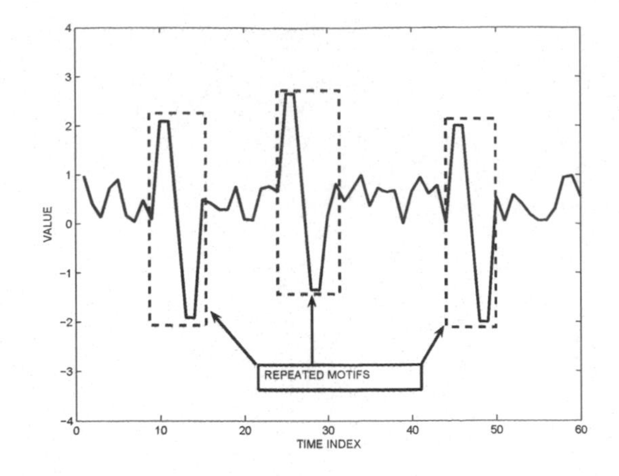

Figure 14.6: Repeated motif in a single time series

The aforementioned description is a simplified version of the *SPIRIT* framework. It reduces the computational effort of prediction because simplified univariate modeling is performed only on a small number $p \ll d$ of independent time series. On the other hand, it does incur the overhead of computing eigenvectors. The hidden-variable series is a linear combination of many different series. Therefore, the noise effects of individual series are often smoothed out within the hidden variables, which increases the robustness of the forecasting process.

14.4 Time Series Motifs

A *motif* is a frequently occurring pattern or shape in the time series. Motif discovery can be formulated in a wide variety of ways, depending on application-specific requirements. These different formulations vary in terms of the input data and the nature of the motifs discovered. These variations are as follows:

1. *Single series versus database of many series:* In the first case, a single series is available, and the frequently occurring shapes in specific windows of the series are determined. For example, in Fig. 14.6, the highlighted shape appears three times in the same series and therefore has a count of 3. A different formulation is one in which we have N different series, and the occurrence of a shape *at least once* in a particular series is given a credit of exactly 1. The frequency is therefore computed in terms of the number of series in which the pattern occurs. The second formulation is much closer to sequential pattern mining in discrete data. Different applications may require different definitions of motif discovery.

2. *Contiguous versus noncontiguous motifs:* Contiguous motifs require that the shapes are discovered over a contiguous window of the time series. Noncontiguous motifs may allow gaps between different elements of the motif. Much of the work in time series analysis assumes that the motifs are defined over contiguous windows. Non-contiguous motifs are more common in discrete sequence analysis. Nevertheless, noncontiguous motifs may have utility in some applications.

3. *Multigranularity motifs:* Many formulations fix the window size in which the motifs are discovered. However, in practice, the frequent motifs may occur over windows of

different sizes. Such motifs are very useful in many application-specific scenarios. For example, in the financial-market series of Fig. 14.11a, an important motif is caused by a "flash crash" event over the course of a day. On the other hand, in Fig. 14.11b, the (recessionary) trend occurs over several months. In the second case, it may be needed to smooth out the local variations to discover motifs. Thus, different techniques are required to discover different types of motifs.

When does a motif belong to a time series? Two methods are typically used by different applications.

1. *Distance-based support:* A particular segment of a sequence is said to support a motif when the distance between the segment and the motif is less than a particular threshold.

2. *Transformation to sequential pattern mining:* A variety of discretizations can be used to convert time series into discrete sequences. After the conversion, motifs can be defined as discrete subsequences of the sequences.

The latter method lends itself to richer classes of algorithms from sequential pattern mining. Furthermore, the approach used for discretization can be varied to discover motifs of different kinds. It also allows the discovery of noncontiguous patterns because sequential pattern mining algorithms do not assume contiguity by default. This section will discuss both kinds of methods. In addition, the notion of periodic patterns will be introduced.

14.4.1 Distance-Based Motifs

Distance-based motifs are always defined on *contiguous* segments of the time series. First, the concept of approximate distance match between a motif and a contiguous segment in a time series needs to be defined.

Definition 14.4.1 (Approximate Distance Match) *A sequence (or motif) $S = s_1 \dots s_w$ of real values is said to approximately match a contiguous subsequence of length w in the time series $(y_1, \dots y_n)$ $(w \leq n)$ starting at position i, if the distance between $(s_1, \dots s_w)$ and $(y_i, \dots y_{i+w-1})$ is at most ϵ.*

A wide variety of distance functions may be used, and the Euclidean distance function is a common choice. The aforementioned definition assumes that the two subsequences being matched have the same length. This is a conservative assumption that allows the use of distance functions such as the Euclidean function. However, if other distance functions, such as dynamic time warping, are used, it may not be necessary for both the matched motifs to have the same length.

The number of occurrences of the motif in a single long series is used to quantify the frequency of the motif. In addition to the series itself, the window length w and the approximation threshold ϵ are the two main inputs to the algorithm.

Definition 14.4.2 (Motif Count) *The number of matches of a time series window $S = s_1 \dots s_w$ to the time series $(y_1 \dots y_n)$ at threshold level ϵ, is equal to the number of windows of length w in $(y_1 \dots y_n)$, for which the distance between the corresponding subsequences is at most ϵ.*

The goal is to discover the top k motifs for a user-defined parameter k. Furthermore, to ensure that the k motifs discovered are very different from one another, a constraint is

Algorithm *FindBestMotif*(Time Series: $y_1 \ldots y_n$, Window: w
Distance Threshold: ϵ)
begin
  for $i = 1$ to $n - w + 1$ **do begin**
    Candidate-Motif= $(y_i, \ldots y_{i+w-1})$;
    for $j = 1$ to $n - w + 1$ **do begin**
      Comparison-Motif= $(y_j \ldots y_{j+w-1})$;
      $D = ComputeDistance$(Candidate-Motif, Comparison-Motif);
      if $(D \leq \epsilon)$ and (non-trivial match)
          then increment count of Candidate-Motif by 1;
    endfor
    if Candidate-Motif has highest count found so far
      then update Best-Candidate to Candidate-Motif;
  endfor
  return Best-Candidate;
end

Figure 14.7: Determining the most frequent motif

imposed; the distances between any pairs of motifs discovered among the top-k motifs must be at least $2 \cdot \epsilon$. In the following, the discovery of the most frequent occurring single motif will be described. The generalization to the case of top k motifs is relatively straightforward. The overall approach [356] uses a nested-loop algorithm to discover the most frequent motif. The approach is described in Fig. 14.7.

The approach extracts all of the candidate motifs of length w from a time series and computes the distances to all of the windows of length w. The number of windows over which the match occurs is counted. Care is taken to exclude *trivial matches* in the count. Trivial matches are defined as those matches where approximately the same (overlapping) window is being matched. For example, the case when $i = j$ is a trivial match. Furthermore, in the case where $i < j$, if the window starting at i matches with all windows starting at $i+1, i+2 \ldots j$, then the match at j is trivial as well. In the case where $i > j$, if the window starting at i matches with all windows starting at $i-1, i-2 \ldots j$, then the match at j is trivial. Therefore, this condition is explicitly checked in the counting. The best candidate is tracked over the course of the algorithm, and reported at termination. As evident from Fig. 14.7, the approach requires a nested loop, and the number of iterations in each loop is almost equal to the size of the series n. Thus, the approach requires $O(n^2)$ distance computations. In principle, any time series distance function, such as DTW, can be used for the computation, although it is generally more expensive.

The majority of the time is spent in distance computations. In many cases, a fast computation of the lower bound on the distance can be used to speed up the approach. If the computed lower bound between a pair of windows is greater than ϵ, then the pair is guaranteed to be irrelevant for adding to the candidate motif count. Therefore, the distance computation does not need to be explicitly performed. The piecewise aggregate approximation (PAA) can be used to speed up the distance computations. Consider a scenario where the PAA has been performed over windows of length m. The resulting series has been compressed by a factor of m, and therefore the distance computations are much faster. If the series X' be the PAA of $X = (x_1 \ldots x_n)$, and Y' be the PAA of $Y = (y_1 \ldots y_n)$, then it can be shown that:

$$Dist(X, Y) \geq \sqrt{m} \cdot Dist(X', Y') \quad (14.20)$$

The proof of this result is as follows. Consider the time series $Z = X - Y$. Over any window of m data points, the second moment of elements of Z in that window, is at least[2] equal to m times the square of the mean of the same elements. Other faster methods for approximation exist, such as the use of the SAX representation. When the SAX representation is used, a table of precomputed distances can be maintained for all pairs of discrete values, and a simple table lookup is required for lower bounding. Furthermore, some other time series distance functions such as dynamic time warping can also be bounded from below. The bibliographic notes contain pointers to some of these bounds. Many variations of the basic approach are possible by adding another layer of nesting, which accounts for variations in the window size.

14.4.2 Transformation to Sequential Pattern Mining

A particularly convenient method for discovering motifs in time series is to transform the problem to the sequential pattern mining problem. The setting for this case is somewhat different, where a database of N series is available, and it is desired to determine all frequent motifs at a specified minimum support level. Since motif (pattern) mining is more naturally defined in the discrete case, this transformation facilitates the use of a wide variety of tools available for the discrete scenario. Furthermore, such an approach can also enable the discovery of noncontiguous patterns in the time series. This is because the subsequences in sequential pattern mining are allowed to be noncontiguous.

The first step is to convert the time series into discrete sequences, by discretizing the behavioral attribute value at each timestamp into categorical values. It is possible to combine discretization with binning to create a robust sequence representation. It should be pointed out that there are many different ways of converting a time series to discrete sequences, depending on application-specific goals. For example, the discretization of the difference of the behavioral attribute values between successive timestamps is equivalent to using discretized wavelet coefficients of the most detailed level of granularity. Lower order wavelet coefficients will provide insights into trends over larger segments of the time series. Thus, it is even possible to perform multiresolution motif analysis by using discretized wavelet coefficients of different orders, and creating separate base sequences for wavelets of each order. In general, the approach for converting time series to discrete sequences will heavily influence the nature of the motifs found.

For all these methods, the final result of the discretization is a sequence of discrete values for each of the N time series in the database. After this new database of sequences has been constructed, any sequential pattern mining algorithm can be applied. The *GSP* algorithm is described in Sect. 15.2 of Chap. 15. It is important to note that the algorithms in Chap. 15 allow gaps between successive elements of the sequence. However, these algorithms can be trivially generalized to the contiguous case, by adding a maximum gap constraint in the sequential pattern mining algorithm. Constrained sequential pattern mining algorithms are briefly discussed in Sect. 15.2.2 of Chap. 15. It should be pointed out that the different constraints discussed in Sect. 15.2.2 correspond to different kinds of motifs. Because of the wide variation in the kinds of motifs that can be found by varying either the discretization approach or the sequential pattern mining approach, this methodology is very flexible, and it can be tailored to many different application scenarios.

[2] The mean of the squares is always no less than the square of the mean for any set of numeric elements. The difference between the two is equal to the variance, which is always nonnegative.

14.4.3 Periodic Patterns

Just as *DWT* is used for discovering *local* patterns in a time series, *DFT* is often used for discovering *periodic* patterns. Recall from Sect. 14.2.4.2 that the rth component of a time series $x_0 \dots x_{n-1}$ can be expressed in terms of n complex Fourier coefficients $X_0 \dots X_{n-1}$ as follows:

$$x_r = \frac{1}{n}\left(\sum_{k=0}^{n-1} X_k \cdot \cos(r\omega k) + i\sum_{k=0}^{n-1} X_k \cdot \sin(r\omega k)\right) \quad \forall r \in \{0 \dots n-1\}$$

Here ω is set to $\frac{2\pi}{n}$ radians. Since the imaginary part of this summation is always 0 for real values of x_r, let us expand the real part by assuming $X_k = a_k + ib_k$:

$$x_r = \frac{1}{n}\left(\sum_{k=0}^{n-1} (a_k + ib_k) \cdot \cos(r\omega k) + i\sum_{k=0}^{n-1} (a_k + ib_k) \cdot \sin(r\omega k)\right) \quad \forall r \in \{0 \dots n-1\}$$

By ignoring the imaginary part, we obtain:

$$x_r = \frac{1}{n}\left(\sum_{k=0}^{n-1} a_k \cdot \cos(r\omega k) - \sum_{k=0}^{n-1} b_k \cdot \sin(r\omega k)\right) \quad \forall r \in \{0 \dots n-1\}$$

$$= \frac{1}{n}\sqrt{a_k^2 + b_k^2} \cdot \sum_{k=0}^{n-1} \cos(r\omega k + \theta_k) \quad \forall r \in \{0 \dots n-1\}$$

Here, we have $\theta_k = \cos^{-1}\left(\frac{a_k}{\sqrt{a_k^2+b_k^2}}\right)$. All terms with $k \geq 1$ are periodic. In other words, *the time series can be decomposed into $n-1$ periodic sinusoidal components, so that the kth component has a periodicity of $\frac{n}{k}$ and amplitude of $\sqrt{a_k^2 + b_k^2}$.* Therefore, if a periodic component has very high amplitude relative to other components, the entire series will be dominated by its periodic behavior. In order to detect such components, the mean and standard deviation of all the n amplitudes are determined. Any amplitude $\sqrt{a_k^2 + b_k^2}$, which is at least δ standard deviations greater than the mean is flagged. Such a component has periodicity $\frac{n}{k}$, and its periodicity will be apparent in the series because of its high amplitude. Note that the smaller Fourier coefficients are also discarded in the case of dimensionality reduction. However, when the threshold δ is chosen more aggressively (i.e., very large positive values such as 3), only 2 or 3 coefficients remain, and the periodicity of the residual series becomes apparent. Furthermore, only values of $k \in (\beta, \frac{n}{\alpha})$ are relevant for discovering patterns that have period at least $\alpha \geq 2$ and have appeared at least $\beta \geq 2$ times in the series. The bibliographic notes contain pointers to methods for discovering *partial* periodic patterns.

14.5 Time Series Clustering

Time series data clustering can be defined in two different ways, depending on the application-specific scenario.

1. In the first approach, real-time clustering is performed on time series that are received simultaneously in time. For example, in a financial market application, it may be desirable to segment the series into groups that coevolve over time. In this case, the

values in the different time series are compared to one another in an approximately synchronized way. Typically, the analysis is performed on a small window of the recent history. The time series are clustered into groups based on correlations between series in the window. Furthermore, the clustering is performed in online fashion, and the different series may move across different clusters. For example, a stock ticker for IBM may move along with Microsoft on one day, but not the next.

2. In the second approach, a database of time series is available. These different time series may or may not have been collected at the same instant. It is desirable to cluster these series, on the basis of their *shapes*. For example, in an application containing electrocardiogram (ECG) time series, the different patients may have contributed a time series to the database at different instants. Shape matching typically requires the use of time series similarity functions discussed in Sect. 3.4.1 of Chap. 3. Thus, both the contextual attribute and the behavioral attribute(s) may be warped or scaled, depending on the nature of the similarity function. In such cases, the different time series may not even be of equal length.

In this section, the different kinds of clustering methods will be discussed in detail. The problem becomes much more difficult when shape-based clustering is applied to multivariate time series. One solution is to generalize the similarity functions to the multivariate case. Time series similarity functions can be generalized to the multivariate case, though a full discussion of this topic is beyond the scope of this book. Relevant pointers may be found in the bibliographic notes.

For shape-based clustering, the special case of bivariate and trivariate series can also be addressed with the use of trajectory clustering. An example of how multivariate series may be converted to trajectory data is found in Sect. 1.3.2.3 of Chap. 1. Methods for trajectory clustering are discussed in Sect. 16.3.4 of Chap. 16.

14.5.1 Online Clustering of Coevolving Series

The problem of online clustering of coevolving series is based on determining correlations across the series, in online fashion. This is useful in many real-time applications such as financial markets because it provides an understanding of the aggregate trends in the series. In these cases, the time series are clustered based on their *correlations* in a window of length p. Because of the use of correlations to define similarity, the approach is referred to as *time series correlation clustering*. The *ORCLUS* correlation clustering algorithm for multidimensional data was discussed in Chap. 7. A similar principle applies to time series data, except that the correlation is measured between different components (behavioral dimensions) of the multivariate time series. The same temporal window is used for the different time series in order to compute the correlations. Therefore, the analysis of the different streams is temporally synchronized.

A natural approach is to use regression-based similarity functions to compute the similarities between different streams. It is not necessary for the two streams to be positively correlated. Rather, the streams may be highly negatively correlated. The key issue here is the *predictability* of the different time series with respect to each other. For example, in Fig. 14.8, the series A and B are very similar because they are perfectly negatively correlated with one another. This is because these two series can be predicted from one another. On the other hand, series C is very different, and has low predictability with respect to either stream, and it is useful in applications where it is desired to maximize the predictive power of cluster representatives. An example is sensor selection, where a subset of sensors

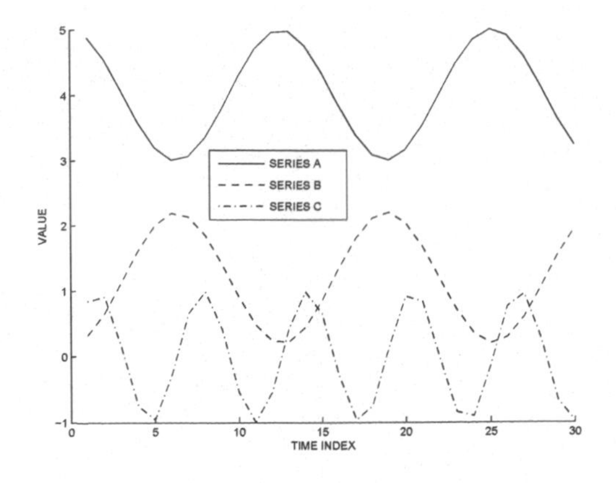

Figure 14.8: Time series correlation clustering

need to be selected which maximize the ability to predict the values of all other sensors. Because prediction is one of the most fundamental problems in real-time time series analysis, the use of regression-based similarity is natural in such scenarios. This is different from offline shape-based analysis where more conventional time series similarity functions, such as *DTW*, are used. A method that directly uses regression analysis for real-time time series clustering is referred to as *online time series correlation clustering.*

For ease in discussion, we will treat the d time series as a single multivariate series with d behavioral attributes. The multivariate time series of length t is denoted by $\overline{Y_1} \dots \overline{Y_t}$. The value $\overline{Y_t}$ of each of the d streams at the tth tick is $(y_t^1 \dots y_t^d)$. The goal is to therefore always to maintain a partition of the d series, so that highly correlated components are assigned to the same partition. A representative-based approach is used for clustering. The basic idea is to incrementally maintain a set of k representative time series from the d series in real-time. This representative set, denoted by J, is similar to the representative set of a k-medoids algorithm. After the representatives have been determined, all of the time series streams can be assigned to one of the representatives with the use of a time series similarity function. Each series can be assigned to its closest representative. This similarity function will be discussed later in more detail.

A natural approach is to incrementally maintain the representatives, and add or drop streams to the set J where necessary. The clustering is implicitly defined by assignment of the d time series to their closest representatives. Therefore, when a new time series data point arrives, the current set of representatives J need to be *updated.* Streams are iteratively exchanged between the current cluster representatives and the non-representatives to optimize a quality criterion based on minimizing the error. The similarity between a representative stream i and nonrepresentative stream j, is the regression error of predicting stream j from stream i. The idea is that true cluster representatives can be used to accurately predict the other streams. To predict the stream j from stream i, a similar model as the autoregressive model is used, except that the elements of stream i are used to predict stream j, instead of its own elements. Thus, the regression model is as follows:

$$y_t^j = \sum_{r=1}^{p} a_r \cdot y_{t-r}^i + c + \epsilon_t$$

This is similar to the $AR(p)$ model, except that the elements of stream i are being used to predict those of stream j. As in the case of the $AR(p)$ model, least-squares regression

Algorithm *UpdateClusters*(Multivariate Stream: $\overline{Y_1} \ldots \overline{Y_t} \ldots$
Current Set of Representatives: J)
begin
Receive next time-stamp $\overline{Y_t}$ of multivariate stream;
repeat
Add a stream to J that leads to the maximum
decrease in regression error of the clustering;
Drop the stream from J that leads to the least
increase in regression error of the clustering;
Assign each series to closest representative in J to
create the clustering $\mathcal{C}$;
until(J did not change in previous iteration);
return(J, $\mathcal{C}$);
end

Figure 14.9: Dynamically maintaining cluster representatives

can be used to learn p coefficients. Furthermore, the training data is restricted to a window of size $w > p$ to allow for stream evolution. The squared average of the white noise error terms, or the R^2-statistic over the window of size $w > p$, can be used as the distance (similarity) between the two streams. Note that the regression coefficients can also be maintained incrementally because they are already known at the previous timestamp, and the model simply needs to be updated with the addition of a single data point. Most iterative optimization methods for least-squares regression, such as gradient descent, converge very fast when starting with a near-optimal solution.

This regression-based similarity function is not symmetric because the error of predicting stream j from stream i is different from the error of predicting stream i from stream j. The representative set J can also be used to create a clustering $\mathcal{C}$ of the streams, by assigning each stream to the representative, that best predicts it. Thus, at each step, the set of representatives J and clusters $\mathcal{C}$ can be incrementally reported, after updating the model. The pseudocode for the online stream clustering approach is illustrated in Fig. 14.9. This approach can be useful for trend analysis in financial markets, where a representative set of stocks needs be tracked from the vast universe of stocks. Another relevant application is that of *sensor selection*, where a subset of representative sensors need to be determined to lower the operational costs of sensor networks. The description in this section is based on a simplification of a cost-based dynamic sensor selection algorithm [50].

14.5.2 Shape-Based Clustering

The second type of clustering is derived from shape-based clustering. In this case, the different time series may not be synchronized in time. The time series are clustered on the basis of the similarity of the shape of the overall series. A first step is the design of a shape-based similarity function. A major challenge in this case is that the different series may be scaled, translated, or stretched differently. This issue was discussed in Sect. 3.4.1 of Chap. 3. The illustration of Fig. 3.7 is replicated in Fig. 14.10. This figure illustrates different hypothetical stock tickers. In these cases, the three stocks show similar patterns, but with different scaling and random variations. Furthermore, in some cases, the time dimension may also be warped. For example, in Fig. 14.10, the entire set of values for

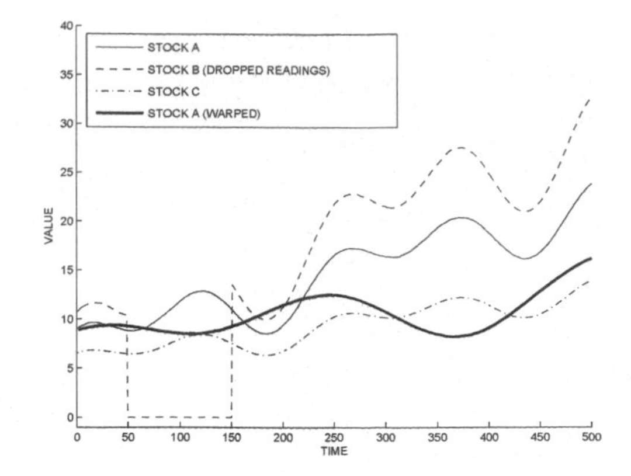

Figure 14.10: Impact of scaling, translation and noise on clustering (revisiting Fig. 3.7)

stock A was stretched because the time-granularity information was not available to the analyst. This is referred to as *time warping*. Fortunately, the *dynamic time warping (DTW)* similarity function, discussed in Sect. 3.4.1 of Chap. 3, can address these issues. The design of an effective similarity function is, therefore, one of the most crucial steps in time series clustering.

Many existing methods can be adapted to shape-based time series clustering with the use of different time series similarity functions. The k-medoids and graph-based methods can be used with almost any similarity function. Methods such as k-means can also be used, though in a more limited way. This is because the different time series need to be of the same length in order for the mean of a cluster to be defined meaningfully.

14.5.2.1 k-Means

The k-means method for multidimensional data is discussed in Sect. 6.3.1 of Chap. 6. This method can be adapted to time series data, by changing the similarity function and the computation of the means of the time series. The computation of the similarity function can be adapted from Sect. 3.4.1 of Chap. 3. The precise choice of the similarity function may depend on the application at hand, though the k-means approach is optimized for the Euclidean distance function. This is because the k-means approach can be viewed as an iterative solution to an optimization problem, in which the objective function is constructed with the Euclidean distance. This aspect is discussed in more detail in Chap. 6.

The Euclidean distance function on a time series is defined in the same way as multidimensional data. The means of the different time series are also defined in a similar way to multidimensional data. The k-means method is best used for databases of series of the same length with a one-to-one correspondence between the time points. Thus, the centroid for each time point can be defined using the correspondence. Time warping is typically difficult to use in k-means algorithms in a meaningful way, because of the assumption of one-to-one correspondence between the time series data points. For more generic distance functions such as DTW, other methods for time series clustering are more appropriate.

14.5.2.2 k-Medoids

The main problem with the k-means approach is the fact that it cannot incorporate arbitrary similarity (or distance) functions. The k-medoids approach can be used more effectively in this case because it does not make any assumptions on the relative lengths of the different time series. The approach is described in detail in Sect. 6.3.4 of Chap. 6. The main difference from the description provided in this section is that of the choice of the similarity function. Any of the similarity functions described in Sect. 3.4.1 of Chap. 3 may be used. The *CLARANS* method discussed in Sect. 7.3.1 of Chap. 7 can also be generalized to this case.

14.5.2.3 Hierarchical Methods

The hierarchical methods, discussed in Sect. 6.4 of Chap. 6, can also be generalized to any data type because they work with pairwise distances between the different data objects. In these methods, the main challenge is that distance computations between all pairs of time series are required. Many time series distance and similarity functions require expensive dynamic programming methods. This is a major disadvantage in the use of hierarchical methods. Nevertheless, the approach can still be used quite effectively in cases where the total number of time series is small.

14.5.2.4 Graph-Based Methods

Graph-based methods provide a transformational approach to time series data clustering. The idea is to transform the time series data set into a single large graph, on which community detection algorithms can be applied. As discussed in Sect. 2.2.2.9 of Chap. 2, any data type can be converted to a similarity graph, once a similarity function has been defined. Each node in this graph corresponds to a data object. Each node is connected to its k-nearest neighbors, and the weight of the edge is equal to the similarity between the corresponding pair of objects. Once a similarity graph has been defined, any of the graph clustering algorithms discussed in Sect. 19.3 of Chap. 19 can be used to determine node clusters. The spectral method of Sect. 19.3.4 is most commonly used. The clusters (communities) of nodes can then be mapped back to clusters of time series by using the correspondence between nodes and time series data objects.

14.6 Time Series Outlier Detection

As in the case of time series clustering, the problem of outlier detection in time series can be defined in two different ways.

1. *Point outliers:* A point outlier is a sudden change in a time series value at a given timestamp. This problem is closely related to forecasting, because an outlier is defined as a significant deviation from expected (or *forecasted*) values. Such outliers are referred to as *contextual* outliers because they are outliers in the *context* of their immediate history.

2. *Shape outliers:* In this case, a *consecutive pattern* of data points in a contiguous window may be defined as an anomaly. For example, in an ECG series, an irregular heart beat may be considered an anomaly *when considered together*, although no individual point in the series may be considered an anomaly. Such outliers are referred to as *collective outliers* because they are defined by combining the patterns from multiple data items.

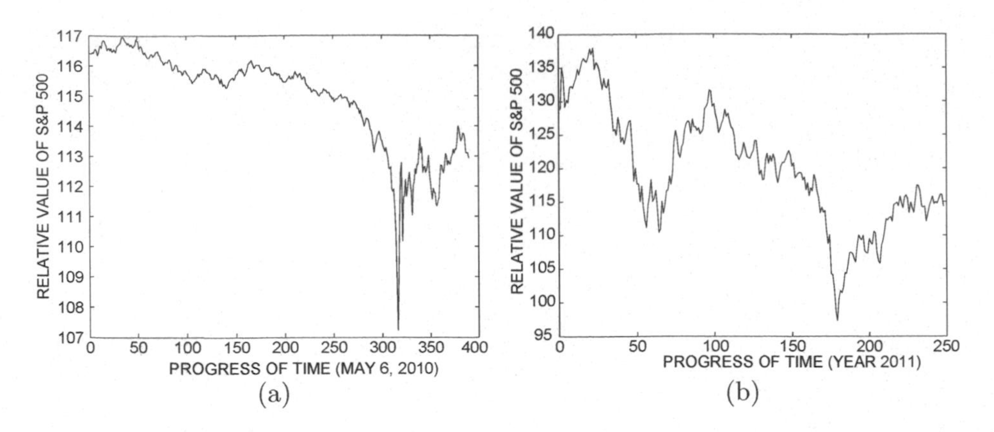

Figure 14.11: Behavior of the *S&P* 500 on the day of the flash crash (May 6, 2010) (**a**), and year 2001 (**b**)

To illustrate the distinction between these two kinds of anomalies, an example from financial markets will be used. The two cases illustrated in Fig. 14.11a, b show the behavior[3] of the *S&P* 500 over different periods in time. Figure 14.11a illustrates the movement of the *S&P* 500 on May 16, 2010. This was the date of the stock market flash crash. This is a very unusual event *both* from the perspective of the *point* deviation at the time of the drop, *and* from the perspective of the *shape* of the drop. A different scenario is illustrated in Fig. 14.11b. Here, the variation of the *S&P* 500 during the year 2001 is illustrated. There are two significant drops over the course of the year, both because of stock market weakness, and also because of the 9/11 terrorist attacks. While the specific timestamps of drop may be considered somewhat abnormal based on deviation analysis over specific windows, the actual shape of these time series is not unusual because it is frequently encountered during bear markets (periods of market weakness). Thus, these two kinds of outliers require dedicated methods for analysis. It should be pointed out that a similar distinction between the two kinds of outliers can be defined in many contextual data types such as discrete sequence data. These are referred to as point outliers and combination outliers, respectively, for the case of discrete sequence data. The combination outliers in discrete sequence data are analogous to shape outliers in continuous time series data. This is discussed in greater detail in Chap. 15.

14.6.1 Point Outliers

Point outliers are closely related to the problem of forecasting in time series data. A data point is considered an outlier if it deviates significantly from its expected (or *forecasted*) value. Such point outliers correspond to unsupervised *events* in the underlying data. Event detection is often considered a synonym for temporal outlier detection when it performed in real time.

Point outliers can be defined for either univariate or multivariate data. The case of univariate data and multivariate data is almost identical. Therefore, the more general case of multivariate data will be discussed. As in previous sections, assume that the multivariate series on which the outliers are to be detected is denoted by $\overline{Y_1} \dots \overline{Y_n}$. The overall approach comprises four steps:

[3]The tracking Exchange Traded Fund (ETF) SPY was used.

1. Determine the forecasted values of the time series at each timestamp. Depending on the nature of the underlying series, any of the univariate or multivariate methodologies discussed in Sect. 14.3 may be used. Let the forecasted value at the rth timestamp t_r be denoted by $\overline{W_r}$

2. Compute the (possibly multivariate) time series of deviations $\overline{\Delta_1} \ldots \overline{\Delta_r} \ldots$. In other words, for the rth timestamp t_r, the deviation is computed as follows:

$$\overline{\Delta_r} = \overline{W_r} - \overline{Y_r}. \tag{14.21}$$

3. Let the d different components of $\overline{\Delta_r}$ be denoted by $(\delta_r^1 \ldots \delta_r^d)$. These can be separated out into d different univariate series of deviations along each dimension. The values of the ith series are denoted by $\delta_1^i \ldots \delta_n^i$. Let the mean and standard deviation of the ith series of deviations be denoted by μ_i and σ_i.

4. Compute the normalized deviations δz_r^i as follows:

$$\delta z_r^i = \frac{\delta_r^i - \mu_i}{\sigma_i}. \tag{14.22}$$

   The resulting deviation is essentially equal to the Z-value of a normal distribution. This approach provides a continuous alarm level of outlier scores for each of the d dimensions of the multivariate time series. Unusual time instants can be detected by using thresholding on these scores. Because of the Z-value interpretation, an absolute threshold of 3 is usually considered sufficient.

In some cases, it is desirable to create a unified alarm level of deviation scores rather than creating a separate alarm level for each of the series. This problem is closely related to that of outlier ensemble analysis that is discussed in Sect. 9.4 of Chap. 9. The unified alarm level U_r at timestamp r can be reported as the maximum of the scores across the different components of the multivariate series:

$$U_r = \max_{i \in \{1 \ldots d\}} \delta z_r^i. \tag{14.23}$$

The score across different detectors can be combined in other ways, such as by using the average or squared aggregate over different series.

14.6.2 Shape Outliers

One of the earliest methods for finding shape-based outliers is the *Hotsax* approach. In this approach, outliers are defined on windows of the time series. A k-nearest neighbor method is used to determine the outlier scores. Specifically, the Euclidean distance of a data point to its kth-nearest neighbors is used to define the outlier score.

The outlier analysis is performed over windows of length W. Therefore, the approach reports windows of unusual shapes from a time series of data points. The first step is to extract all windows of length W from the time series by using a sliding-window approach. The analysis is then performed over these newly created data objects. For each extracted window, its Euclidean distance to the other *nonoverlapping* windows is computed. The windows with the highest k-nearest neighbor distance values are reported as outliers. The reason for using nonoverlapping windows is to minimize the impact of trivial matches to overlapping windows. While a brute-force k-nearest neighbor approach can determine the outliers, the complexity will scale with the square of the number of data points. Therefore, a pruning method is used for improving the efficiency. While this method optimizes the efficiency, and it does not affect the final result reported by the method.

The general principle of pruning for more efficient outlier detection in nearest neighbor methods was introduced in Sect. 8.5.1.2 of Chap. 8. The algorithm examines the candidate subsequences iteratively in an outer loop. For each such candidate subsequence, the k-nearest neighbors are computed progressively in an inner loop with distance computations to other subsequences. Each candidate subsequence is either included in the current set of best n outlier estimates at the end of an outer loop iteration, or discarded via *early abandonment* of the inner loop without computing the *exact* value of the k-nearest neighbor. This inner loop can be terminated early when the currently approximated k-nearest neighbor distance for that candidate subsequence is less than the score for the nth best outlier found so far. Clearly, such a subsequence cannot be an outlier. To obtain the best pruning results, the subsequences need to be heuristically ordered so that the earliest candidate subsequences examined in the outer loop have the greatest tendency to be outliers. Furthermore, the pruning performance is also most effective when the true outliers are found early. It remains to explain, how the heuristic orderings required for good pruning are achieved.

Pruning is facilitated by an approach that can measure the clustering behavior of the underlying subsequences. Clustering has a well known relationship of complementarity with outlier analysis. Therefore it is useful to examine those subsequences first in the outer loop that are members of clusters containing very few (or one) members. The SAX representation is used to create a simple mapping of the subsequences into clusters. Subsequences that map to the same SAX word, are assumed to belong to a single cluster. The piecewise aggregate approximations of SAX are performed over windows of length $w < W$. Therefore, the length of a SAX word is W/w, and the number of distinct possibilities for a SAX word is small if W/w is small. These distinct words correspond to the different clusters. Multiple subsequences map to the same cluster. Therefore, the ordering of the candidates is based on the number of data objects in the same cluster. Candidates in clusters with fewer objects are examined first because they are more likely to be outliers.

This cluster-based ordering is used to design an efficient pruning mechanism for outlier analysis. The candidates in the clusters are examined one by one in an outer loop. The k-nearest neighbor distances to these candidates are computed in an inner loop. For each candidate subsequence, those subsequences that map to the same word as the candidate may be considered first for computing the nearest neighbor distances in the inner loop. This provides quick and tight upper bounds on the nearest neighbor distances. As these distances are computed one by one, a tighter and tighter upper bound on the nearest neighbor distance is computed over the progression of the inner loop. *A candidate can be pruned when an upper bound on its nearest neighbor distance is guaranteed to be smaller (i.e., more similar) than the nth best outlier distance found so far.* Therefore, for any given candidate series, it is not necessary to determine its exact nearest neighbor by comparing to all subsequences. Rather, early termination of the inner loop is often possible during the computation of the nearest neighbor distance. This forms the core of the pruning methodology of *Hotsax*, and is similar in principle to the nested-loop pruning methodology discussed in Sect. 8.5.1.2 of Chap. 8 on multidimensional outlier analysis. The main difference is in terms of how the SAX representation is used both for ordering the candidates in the outer loop, and ordering the distance computations in the inner loop.

14.7 Time Series Classification

Time series classification can be defined in several ways, depending on the association of the underlying class labels to either individual timestamps, or the whole series.

1. *Point labels:* In this case, the class labels are associated with individual timestamps. In most cases, the class of interest is rare in nature and corresponds to unusual activity at that timestamp. This problem is also referred to as *event detection*. This version of the event detection problem can be distinguished from the unsupervised outlier detection problem discussed in Sect. 14.6, in that it is *supervised* with labels.

2. *Whole-series labels:* In this case, the class labels are associated with the full series. Therefore, the series needs to be classified on the basis of the shapes inside it.

Both these problems will be discussed in this chapter.

14.7.1 Supervised Event Detection

The problem of supervised event detection is one in which the class labels are associated with the timestamps rather than the full series. In most cases, one or more of the class labels are rare, and the remaining labels correspond to the "normal" periods. While it is possible in principle to define the problem with a balanced distribution of labels, this is rarely the case in application-specific settings. Therefore, the discussion in this subsection will focus only on the imbalanced label distribution scenario.

These rare class labels correspond to the events in the underlying data. For example, consider a scenario, in which the performance of a machine is tracked using sensors. In some cases, a rare event, such as the malfunctioning of the machine, may cause unusual sensor readings. Such unusual events need to be tracked *in a timely fashion*. Therefore, this problem is similar to point anomaly detection, except that it is done in a supervised way.

In many application-specific scenarios, the time series data collection is inherently designed in such a way that the unusual events are reflected in unexpected deviations of the time series. This is particularly true of many sensor-based collection mechanisms. While this can be captured by unsupervised methods, the addition of supervision helps in the removal of spurious events that may have different underlying causes. For example, consider the case of an environmental monitoring application. Many deviations may be the result of the failure of the sensor equipment, or another spurious event that causes deviations in sensor values. This may not necessarily reflect an anomaly of interest. While anomalous events often correspond to extreme deviations in sensor stream values, the precise causality of different kinds of deviations may be quite different. These other *noisy* or *spurious* abnormalities may not be of any interest to an analyst. For example, consider the case illustrated in Fig. 14.12, in which temperature and pressure values inside pressurized pipes containing heating fluids are illustrated. Figures 14.12 a and b illustrate values on two sensors in a pipe rupture scenario. Figures 14.12 c and d illustrate the values of the two sensors in a situation where the pressure sensor malfunctions, and this results in a value of 0 at each timestamp in the pressure sensor. In the first case, the readings of both pressure and temperature sensors are affected by the malfunction, though the final pressure values are not zero, but they reflect the pressure in the external surroundings. The readings on the temperature sensor are not affected at all in the second scenario, since the malfunction is specific to the pressure sensor.

Thus, the key is to *differentiate* among the deviations of different behavioral attributes in a multivariate scenario. The use of supervision is very helpful because it can be used

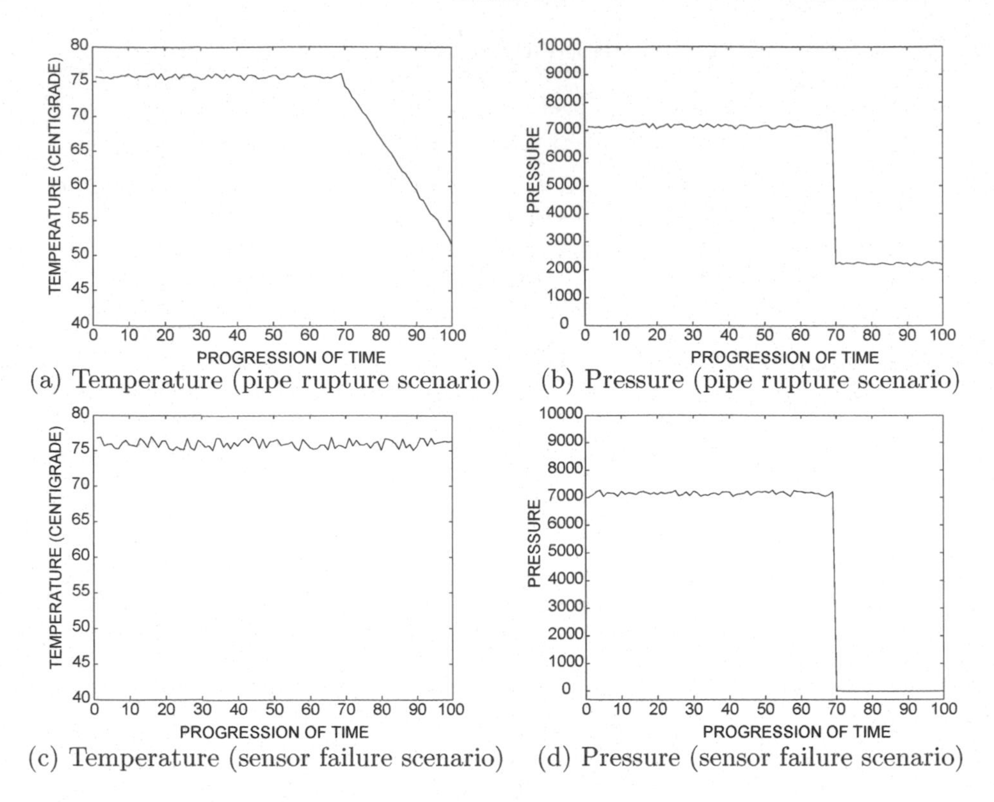

(a) Temperature (pipe rupture scenario) (b) Pressure (pipe rupture scenario)

(c) Temperature (sensor failure scenario) (d) Pressure (sensor failure scenario)

Figure 14.12: Behavior of temperature and pressure sensors due to pipe rupture (**a, b**), and failure of pressure sensor (**c, d**)

to determine the *differential* behavior of the deviations across different streams. In the aforementioned pipe rupture scenario, the relative deviations in the two events are quite different. In the labeling input, it is assumed that most of the timestamps are labeled "normal." A few *ground truth timestamps*, $T_1 \ldots T_r$, are labeled "rare." These are used for supervision. These are referred to as *primary abnormal events.* In addition, spurious events may also cause large deviations. These timestamps are referred to as *secondary abnormal events.* In some application-specific scenarios, the timestamps for the secondary abnormal events may be provided, though this is not assumed here. The bibliographic notes contain pointers to these enhanced methods.

It is assumed that a total of d different time series data streams are available, and the differential patterns in the d streams are used to detect the abnormal events. The overall process of event prediction is to create a composite alarm level from the error terms in the time series prediction. The first step is to use a univariate time series prediction model to determine the error terms at a given timestamp. Any of the models discussed in Sect. 14.3 may be used. These are then combined together to create a composite alarm level with the use of coefficients $\alpha_1 \ldots \alpha_d$ for the d different time series data streams. The values of $\alpha_1 \ldots \alpha_d$ are learned from the training data in an offline (or periodic batch) phase, so as to best discriminate the true event from the normal periods. The actual prediction can be performed using an online approach in real time. Therefore, the steps may be summarized as follows:

1. (Offline Batch) Learn the coefficients $\alpha_1 \ldots \alpha_d$ that best distinguish between the true and normal periods. The details of this step are discussed later in this section.

2. (Real Time) Determine the (absolute) deviation level for each timeseries data stream, with the use of any forecasting method discussed in Sect. 14.3. These correspond the absolute values of the white noise error terms. Let the absolute deviation level of stream j at timestamp n be denoted by z_n^j.

3. (Real Time) Combine the deviation levels for the different streams as follows, to create the composite alarm level:

$$Z_n = \sum_{i=1}^{d} \alpha_i z_n^i \tag{14.24}$$

   The value of Z_n is reported as the alarm level at timestamp n. Thresholding can be used on the alarm level to generate discrete labels.

The main step in the last section, which has not yet been discussed, is the determination of the *discrimination coefficients* $\alpha_1 \ldots \alpha_d$. These should be selected in the training phase, so as to maximize the differences in the alarm level between the primary events and the normal periods.

To learn the coefficients $\alpha_1 \ldots \alpha_d$ in the training phase, the composite alarm level is averaged at the timestamps $T_1 \ldots T_r$ for all primary events of interest. Note that the composite alarm level at each timestamp T_i is an algebraic expression, which is a linear function of the coefficients $\alpha_1 \ldots \alpha_d$ according to Eq. 14.24. These expressions are added up over the time stamps $T_1 \ldots T_r$ to create an alarm level $Q^p(\alpha_1 \ldots \alpha_d)$ which is a function of $(\alpha_1, \ldots \alpha_d)$.

$$Q^p(\alpha_1 \ldots \alpha_d) = \frac{\sum_{i=1}^{r} Z_{T_i}}{r}. \tag{14.25}$$

A similar algebraic expression for the normal alarm level $Q^n(\alpha_1 \ldots \alpha_d)$ is also computed by using all of the available timestamps, the majority of which are assumed to be normal.

$$Q^n(\alpha_1 \ldots \alpha_d) = \frac{\sum_{i=1}^{n} Z_i}{n} \tag{14.26}$$

As in the case of the event signature, the normal alarm level is also a linear function of $\alpha_1 \ldots \alpha_d$. Then, the optimization problem is that of determining the optimal values of α_i that increase the differential signature between the primary events and the normal alarm level. This optimization problem is as follows:

$$\text{Maximize } Q^p(\alpha_1 \ldots \alpha_d) - Q^n(\alpha_1 \ldots \alpha_d)$$

$$\text{subject to: } \sum_{i=1}^{d} \alpha_i^2 = 1$$

This optimization problem can be solved using any off-the-shelf iterative optimization solver. In practice, the online event detection and offline learning processes are executed simultaneously, as new events are encountered. In such cases, the values of α_i can be updated incrementally within the iterative optimization solver. The composite alarm level can be reported as an event score. Alternatively, thresholding on the alarm level can be used to generate discrete timestamps at which the events are predicted. The choice of threshold will regulate the trade-off between the precision and recall of the predicted events.

14.7.2 Whole Series Classification

In whole-series classification, the labels are associated with the entire series, rather than events associated with individual timestamps. It is assumed that a database of N different series is available, and each series has a length of n. Each of the series is associated with a class label drawn from $\{1 \ldots k\}$.

Many proximity-based classifiers are designed with the help of time series similarity functions. Thus, the effective design of similarity functions is crucial in classification, as is the case in many other time series data mining applications.

In the following, three classification methods will be discussed. Two of these methods are *inductive* methods, in which only the training instances are used to build a model. These are then used for classification. The third method is a *transductive semisupervised method*, in which the training and test instances are used together for classification. The semisupervised approach is a graph-based method in which the unlabeled test instances are leveraged for more effective classification.

14.7.2.1 Wavelet-Based Rules

A major challenge in time series classification is that much of the series may be noisy and irrelevant. The classification properties may be exhibited only in temporal segments of varying length in the series. For example, consider the scenario where the series in Fig. 14.11 are presented to a learner with labels. In the case where the label corresponds to a recession (Fig. 14.11a), it is important for a learner to analyze the trends for a period of a few weeks or months in order to determine the correct labels. On the other hand, where the label corresponds to the occurrence of a flash crash (Fig. 14.11b), it is important for a learner to be able to extract out the trends over the period of a day.

For a given learning problem, it may not be known *a priori* what level of granularity should be used for the learning process. The Haar wavelet method provides a multigranularity decomposition of the time series data to handle such scenarios. As discussed in Sect. 14.4 on time series motifs, wavelets are an effective way to determine frequent trends over varying levels of granularity. It is therefore natural to combine multigranular motif discovery with associative classifiers.

Readers are advised to refer to Sect. 2.4.4.1 of Chap. 2 for a discussion of wavelet decomposition methods. The Haar wavelet coefficient of order i analyzes trends over a time period, which is proportional to $2^{-i} \cdot n$, where n is the full length of the series. Specifically, the coefficient value is equal to half the difference between the average values of the first half and second half of the time period of length $2^{-i} \cdot n$. Because the Haar wavelet represents the coefficients of different orders in the transformation, it automatically accounts for trends of different granularity. In fact, an arbitrary shape in any particular window of the series can usually be well approximated by an appropriate subset of wavelet coefficients. These can be considered signatures that are specific to a particular class label. The goal of the rule-based method is to discover signatures that are specific to particular class labels. Therefore, the overall training approach in the rule-based method is as follows:

1. Generate wavelet representation of each of the N time series to create N numeric multidimensional representations.

2. Discretize wavelet representation to create categorical representations of the time series wavelet transformation. Thus, each categorical attribute value represents a range of numeric values of the wavelet coefficients.

3. Generate rule set using any rule-based classifier described in Sect. 10.4 of Chap. 10. The combination of wavelet coefficients in the rule antecedent correspond to the "signature" shapes in the time series, which are relevant to classification.

Once the rule set has been generated, it can be used to classify arbitrary time series. A given test series is converted into its wavelet representation. The rules that are fired by this series are determined. These are used to classify the test instance. The methods for using a rule set to classify a test instance are discussed in Sect. 10.4 of Chap. 10. When it is known that the class labels are sensitive to periodicity rather than local trends, this approach should be used with Fourier coefficients instead of wavelet coefficients.

14.7.2.2 Nearest Neighbor Classifier

Nearest neighbor classifiers are introduced in Sect. 10.8 of Chap. 10. The nearest neighbor classifier can be used with virtually any data type, as long as an appropriate distance function is available. Distance functions for time series data have already been introduced in Sect. 3.4.1 of Chap. 3. Any of these distance (similarity) functions may be used, depending on the domain-specific scenario. The basic approach is the same as in the case of multidimensional data. For any test instance, its k-nearest neighbors in the training data are determined. The dominant label from these k-nearest neighbors is reported as the relevant class label. The optimal value of k may be determined by using leave-one-out cross-validation.

14.7.2.3 Graph-Based Methods

Similarity graphs can be used for clustering and classification of virtually any data type. The use of similarity graphs for semisupervised classification was introduced in Sect. 11.6.3 of Chap. 11. The basic approach constructs a similarity graph from *both* the training and test instances. Thus, this approach is a transductive method because the test instances are used along with the training instances for classification. A graph $G = (N, A)$ is constructed, in which a node in N corresponds to each of the training and test instances. A subset of nodes in G is labeled. These correspond to instances in the training data, whereas the unlabeled nodes correspond to instances in the test data. Each node in N is connected to its k-nearest neighbors with an undirected edge in A. The similarity is computed using any of the distance functions discussed in Sect. 3.4.1 or 3.4.2 of Chap. 3. The specified labels of nodes in N are then used to derive labels for nodes where they are unknown. This problem is referred to as collective classification. Numerous methods for collective classification are discussed in Sect. 19.4 of Chap. 19.

14.8 Summary

Time series data is common in many domains, such as sensor networking, healthcare, and financial markets. Typically, time series data needs to be normalized, and missing values need to be imputed for effective processing. Numerous data reduction techniques such as Fourier and wavelet transforms are used in time series analysis. The choice of similarity function is the most crucial aspect of time series analysis, because many data mining applications such as clustering, classification, and outlier detection are dependent on this choice.

Forecasting is an important problem in time series analysis because it can be used to make predictions about data points in the future. Most time series applications use either point-wise or shape-wise analysis. For example, in the case of clustering, point-wise analysis results in temporal correlation clusters, where a cluster contains many different series that move together. On the other hand, shape-wise analysis is focused on determining groups of time series with approximately similar shapes.

The problem of point-wise outlier detection is closely related to forecasting. A time series data point is an outlier if it differs significantly from its expected (or *forecasted*) value. A shape outlier is defined in time series data with the use of similarity functions. When supervision is incorporated in point-wise outlier detection, the problem is referred to as *event detection*. Many existing classification techniques can be extended to shape-based classification.

14.9 Bibliographic Notes

The problem of time series analysis has been studied extensively by statisticians and computer scientists. Detailed books on temporal data mining and time series analysis may be found in [134, 467, 492]. Data preparation and normalization are important aspects of time series analysis. The binning approach is also referred to as piecewise aggregate approximation (PAA) [309]. The SAX approach is described in [355]. The *DWT*, *DFT*, and *DCT* transforms are discussed in [134, 467, 475, 492]. Time series similarity measures are discussed in detail in Chap. 3 of this book, and in an earlier tutorial by Gunopulos and Das [241].

The problem of time series motif discovery has been discussed in [151, 394, 395, 418, 524]. The distance-based motif discussion in this chapter is based on the description in [356]. A wavelet-based approach for multiresolution motif discovery is discussed in [51]. The discovered motifs are used for classification. Further discussions on periodic pattern mining may be found in [251, 411, 467]. The problem of time series forecasting is discussed in detail in [134]. The lower bounding of distance functions is useful for fast pruning and indexing. The lower bounding on PAA has been shown in [309]. It has been shown how to perform lower bounding on *DTW* in [308].

A recent survey on time series data clustering may be found in [324]. The problem of online clustering time series data streams is related to the problem of sensor selection. The Selective MUSCLES method was introduced in [527] that can potentially be used to select representatives from a set of time series. The online correlation method, discussed in this chapter, is based on the discussion in [50]. A survey of representative selection algorithms for sensor data may be found in [414]. Many of these algorithms may also be used for online correlation clustering.

A survey on outlier detection for temporal data may be found in [237]. A chapter on temporal outlier detection may also be found in a recent outlier detection book [5]. The online detection of *timestamps* is referred to as event detection. The supervised version of this problem is related to rare class detection. The supervised event detection method discussed in Sect. 14.7.1 was proposed in [52]. The *Hotsax* approach discussed in this book was proposed in [306]. A wavelet-based approach for classification of sequences is discussed in [51]. This approach has been adapted for time series data in this chapter. Surveys on temporal data classification may be found in [33, 516]. The latter survey is on sequence classification, although it also discusses many aspects of time series classification.

14.10 Exercises

1. For the time series $(2, 7, 5, 3, 3, 5, 5, 3)$, determine the binned time series where the bins are chosen to be of length 2.

2. For the time series of Exercise 1, construct the rolling average series for a window size of 2 units. Compare the results to those obtained in the previous exercise.

3. For the time series of Exercise 1, construct the exponentially smoothed series, with a smoothing parameter $\alpha = 0.5$. Set the initial smoothed value y_0 to the first point in the series.

4. Implement the binning, moving average, and exponential smoothing methods.

5. Consider a series, in which consecutive values are related as follows:

$$y_{i+1} = y_i \cdot (1 + R_i) \tag{14.27}$$

Here R_i is a random variable drawn from $[0.01, 0.05]$. What transformation would you apply to make this series stationary?

6. Consider the series in which y_i is defined as follows:

$$y_i = 1 + i + i^2 + R_i \tag{14.28}$$

Here R_i is a random variable drawn from $[0.01, 0.05]$. What transformation would you apply to make this series stationary?

7. For a real-valued time series $x_0 \ldots x_{n-1}$ with Fourier coefficients $X_0 \ldots X_{n-1}$, show that $X_k + X_{n-k}$ is real-valued for each $k \in \{1 \ldots n-1\}$.

8. Suppose that you wanted to implement the k-means algorithm for a set of time series, and you were given the same subset of complex Fourier coefficients for each dimensionality-reduced series. How would the implementation be different from that of using k-means on the original time series?

9. Use Parseval's theorem and additivity to show that the dot product of two series is proportional to the sum of the dot products of the real parts and the dot products of the imaginary parts of the Fourier coefficients of the two series. What is the proportionality factor?

10. Implement a shape-based k-nearest neighbor classifier for time series data.

11. Generalize the distance-based motif discovery algorithm, discussed in this chapter to the case where the motifs are allowed to be of any length $[a, b]$, and the Manhattan segmental distance is used for distance comparison. The Manhattan segmental distance between a pair of series is the same as the Manhattan distance, except that it divides the distance with the motif length for normalization.

12. Suppose you have a database of N series, and the frequency of motifs are counted, so that their occurrence once in any series is given a credit of one. Discuss the details of an algorithm that can use wavelets to determine motifs at different resolutions.

Chapter 15

Mining Discrete Sequences

"I am above the weakness of seeking to establish a sequence of cause and effect."—Edgar Allan Poe

15.1 Introduction

Discrete sequence data can be considered the categorical analog of timeseries data. As in the case of timeseries data, it contains a single *contextual* attribute that typically corresponds to time. However, the behavioral attribute is categorical. Some examples of relevant applications are as follows:

1. *System diagnosis:* Many automated systems generate discrete sequences containing information about the *system state*. Examples of system state are UNIX system calls, aircraft system states, mechanical system states, or network intrusion states.

2. *Biological data:* Biological data typically contains sequences of amino acids. Specific patterns in these sequences may define important properties of the data.

3. *User-action sequences:* Various sequences are generated by user actions in different domains.

   (a) Web logs contain long sequences of visits to Websites by different individuals.

   (b) Customer transactions may contain sequences of buying behavior. The individual elements of the sequence may correspond to the identifiers of the different items, or sets of identifiers of different items that are bought.

   (c) User actions on Websites, such as online banking sites, are frequently logged. This case is similar to that of Web logs, except that the logs of banking sites often contain more detailed information for security purposes.

Biological data is a special kind of sequence data, in which the contextual data is not temporal but relates to the *placement* of the different attributes. Methods for temporal

C. C. Aggarwal, *Data Mining: The Textbook*, DOI 10.1007/978-3-319-14142-8_15

sequence data can be leveraged for biological sequence data and vice versa. A discrete sequence is formally defined as follows:

Definition 15.1.1 (Discrete Sequence Data) *A discrete sequence $\overline{Y_1} \ldots \overline{Y_n}$ of length n and dimensionality d, contains d discrete feature values at each of n different timestamps $t_1 \ldots t_n$. Each of the n components $\overline{Y_i}$ contains d discrete behavioral attributes $(y_i^1 \ldots y_i^d)$, collected at the ith timestamp.*

In many practical scenarios, the timestamps $t_1 \ldots t_n$ may simply be tick values indexed from 1 through n. This is especially true in cases such as biological data, in which the contextual attribute represents placement. In general, the actual timestamps are rarely used in most sequence mining applications, and the discrete sequence values are assumed to be equally spaced in time. Furthermore, most of the analytical techniques are designed for the case where $d = 1$. Such discrete sequences are also referred to as *strings*. This chapter will, therefore, use these terms interchangeably. Most of the discussion in this chapter focuses on these more common and simpler cases.

In some applications, such as *sequential pattern mining*, each $\overline{Y_i}$ is not a vector but a set of unordered values. This is a variation from Definition 15.1.1. Therefore, the notation Y_i (without an overline) will be used to denote a set rather than a vector. For example, in a supermarket application, the set Y_i may represent a set of items bought by the customer at a given time. There is no temporal ordering among the items in Y_i. In a Web log analysis application, the set Y_i represents the Web pages browsed by a given user in a single session. Thus, discrete sequences can be defined in a wider variety of ways than timeseries data. This is because of the ability to define sets on discrete items. Each position in the sequence is also referred to as an *element* and is composed of individual *items* in the set. Throughout this chapter, the word "element" will refer to one of the sets of items within the sequence, including a 1-itemset.

These variations in definitions arise out of a natural variation in the different kinds of application scenarios. This chapter will study the different problem definitions relevant to discrete sequence mining. The four major problems of pattern mining, clustering, outlier analysis, and classification, are each defined differently for discrete sequence mining than for multidimensional data. These different definitions will be studied in detail. A few models, such as *Hidden Markov Models*, are used widely across many different application domains. These commonly used models will also be studied in this chapter.

This chapter is organized as follows. Section 15.2 introduces the problem of sequential pattern mining. Sequence clustering algorithms are discussed in Sect. 15.3. The problem of sequence outlier detection is discussed in Sect. 15.4. Section 15.5 introduces Hidden Markov Models (HMM) that can be used for either clustering, classification, or outlier detection. The problem of sequence classification is addressed in Sect. 15.6. Section 15.7 gives a summary.

15.2 Sequential Pattern Mining

The problem of sequential pattern mining can be considered the temporal analog of frequent pattern mining. In fact, most algorithms for frequent pattern mining can be directly adapted to sequential pattern mining with a systematic approach, although the latter problem is more complex. As in frequent pattern mining, the original motivating application for sequential pattern mining was market basket analysis, although the problem is now used in a wider variety of temporal application domains, such as computer systems, Web logs, and telecommunication applications.

The sequential pattern mining problem is defined on a set of N sequences. The ith sequence contains n_i *elements* in a specific temporal order. Each element contains a set of *items*. The complex element is, therefore, a set, such as a basket of items bought by a customer. For example, consider the sequence:

$\langle\{Bread, Butter\}, \{Butter, Milk\}, \{Bread, Butter, Cheese\}, \{Eggs\}\rangle$

Here, $\{Bread, Butter\}$ is an element, and $Bread$ is an item inside the element. A subsequence of this sequence is also a temporal ordering of sets, such that each element in the subsequence is a subset of an element in the base sequence *in the same temporal order*. For example, consider the following sequence:

$\langle\{Bread, Butter\}, \{Bread, Butter\}, \{Eggs\}\rangle$

The second sequence is a subsequence of the first because each element in the second sequence can be matched to a corresponding element in the first sequence *by a subset relationship*, so that the matching elements are in the same temporal order. Unlike transactions that are sets, note that sequences (and the mined subsequences) contain *ordered* (and possibly repeated) elements, each of which is itself like a transaction. For example, $\{Bread, Butter\}$ is a repeated element in one of the aforementioned sequences, and it may correspond to two separate visits of a customer to the supermarket at different times. Formally, a subsequence relationship is defined as follows:

Definition 15.2.1 (Subsequence) *Let $\mathcal{Y} = \langle Y_1 \ldots Y_n\rangle$ and $\mathcal{Z} = \langle Z_1 \ldots Z_k\rangle$ be two sequences, such that all the elements Y_i and Z_i in the sequences are sets. Then, the sequence $\mathcal{Z}$ is a subsequence of $\mathcal{Y}$, if k elements $Y_{i_1} \ldots Y_{i_k}$ can be found in $\mathcal{Y}$, such that $i_i < i_2 < \ldots < i_k$, and $Z_r \subseteq Y_{i_r}$ for each $r \in \{1 \ldots k\}$.*

Consider a sequence database $\mathcal{T}$ containing a set of N sequences $\mathcal{Y}_1 \ldots \mathcal{Y}_N$. The *support* of a subsequence $\mathcal{Z}$ with respect to database $\mathcal{T}$ is defined in an analogous way to frequent pattern mining.

Definition 15.2.2 (Support) *The support of a subsequence $\mathcal{Z}$ is defined as the fraction of sequences in the database $\mathcal{T} = \{\mathcal{Y}_1 \ldots \mathcal{Y}_N\}$, that contain $\mathcal{Z}$ as a subsequence.*

The sequential pattern mining problem is that of identifying all subsequences that satisfy the required level of minimum support *minsup*.

Definition 15.2.3 (Sequential Pattern Mining) *Given a sequence database $\mathcal{T} = \{\mathcal{Y}_1, \ldots \mathcal{Y}_N\}$, determine all subsequences whose support with respect to the database $\mathcal{T}$ is at least* minsup.

It is easy to see that this definition is very similar to that of the definition of association pattern mining in Chap. 4. The minimum support value *minsup* can be specified either as an absolute value, or as a relative support value. As in the case of frequent pattern mining, a relative value will be assumed, unless otherwise specified.

An *Apriori*-like algorithm, known as *Generalized Sequential Pattern Mining (GSP)*, was proposed as the first algorithm for sequential pattern mining. This algorithm is very similar to *Apriori*, in terms of how candidates are generated and counted. In fact, many frequent pattern mining algorithms, such as *TreeProjection* and *FP-growth*, have direct analogs in sequential pattern mining. This section describes only the *GSP* algorithm in detail. A later

Algorithm *GSP*(Sequence Database: $\mathcal{T}$, Minimum Support: *minsup*)
begin
  $k = 1$;
  $\mathcal{F}_k = \{$ All Frequent 1-item elements $\}$;
  while $\mathcal{F}_k$ is not empty **do begin**
    Generate $\mathcal{C}_{k+1}$ by joining pairs of sequences in $\mathcal{F}_k$, such that
      removing an item from the first element of one sequence matches the sequence
      obtained by removing an item from the last element of the other;
    Prune sequences from $\mathcal{C}_{k+1}$ that violate downward closure;
    Determine $\mathcal{F}_{k+1}$ by support counting on $(\mathcal{C}_{k+1}, \mathcal{T})$ and retaining
      sequences from $\mathcal{C}_{k+1}$ with support at least *minsup*;
    $k = k + 1$;
  end;
  return($\cup_{i=1}^{k} \mathcal{F}_i$);
end

Figure 15.1: The *GSP* algorithm is related to the *Apriori* algorithm. The reader is encouraged to compare this pseudocode with the *Apriori* algorithm described in Fig. 4.2 of Chap. 4

section provides a broad overview of how enumeration tree algorithms can be generalized to sequential pattern mining.

The *GSP* and *Apriori* algorithms are similar, except that the former needs to be designed for finding frequent sequences rather than sets. First, the notion of the *length* of candidates needs to be defined in sequential pattern mining. This notion needs to be defined more carefully, because the individual elements in the sequence are sets rather than items. The *length* of a candidate or a frequent sequence is equal to the number of items (not elements) in the candidate. In other words, a k-sequence) $\langle Y_1 \ldots Y_r \rangle$ is a sequence with length $\sum_{i=1}^{r} |Y_i| = k$. Thus, $\langle \{Bread, Butter, Cheese\}, \{Cheese, Eggs\} \rangle$ is a 5-candidate, even though it contains only 2 elements. This is because this sequence contains 5 items in total, including a repetition of "*Cheese*" in two distinct elements. A $(k-1)$-subsequence of a k-candidate can be generated by removing an item from any element in the k-sequence. The *Apriori* property continues to hold for sequences because any $(k-1)$-subsequence of a k-sequence will have support at least equal to that of the latter. This sets the stage for a candidate generate-and-test approach, together with downward closure pruning, which is analogous to *Apriori*.

The *GSP* algorithm starts by generating all frequent 1-item sequences by straightforward counting of individual items. This set of frequent 1-sequences is represented by $\mathcal{F}_1$. Subsequent iterations construct $\mathcal{C}_{k+1}$ by joining pairs of sequence patterns in $\mathcal{F}_k$. The join process is different from association pattern mining because of the greater complexity in the definition of sequences. Any pair of frequent k-sequences $\mathcal{S}_1$ and $\mathcal{S}_2$ can be joined, if removing an item from the first element in one frequent k-sequence $\mathcal{S}_1$ is identical to the sequence obtained by removing an item from the last element in the other frequent sequence $\mathcal{S}_2$. For example, the two 5-sequences $\mathcal{S}_1 = \langle \{Bread, Butter, Cheese\}, \{Cheese, Eggs\} \rangle$ and $\mathcal{S}_2 = \langle \{Bread, Butter\}, \{Milk, Cheese, Eggs\} \rangle$ can be joined because removing "*Cheese*" from the first element of $\mathcal{S}_1$ will result in an identical sequence to that obtained by removing "*Milk*" from the last element of $\mathcal{S}_2$. Note that if $\mathcal{S}_2$ were a 5-candidate with 3 elements corresponding to $\mathcal{S}_2 = \langle \{Bread, Butter\}, \{Cheese, Eggs\}, \{Milk\} \rangle$, then a join can also be performed. This is because removing the last item from $\mathcal{S}_2$ creates a sequence with 2 elements and 4 items, which is identical to $\mathcal{S}_1$. However, the nature of the join will be somewhat different in these cases. In general, cases where the last element of $\mathcal{S}_2$ is a 1-itemset need to be treated specially. The following rules can be used to execute the join:

1. If the last element of $\mathcal{S}_2$ is a 1-itemset, then the joined candidate may be obtained by *appending* the last element of $\mathcal{S}_2$ to $\mathcal{S}_1$ as a separate element. For example, consider the following two sequences:

   $\mathcal{S}_1 = \langle\{Bread, Butter, Cheese\}, \{Cheese, Eggs\}\rangle$
   $\mathcal{S}_2 = \langle\{Bread, Butter\}, \{Cheese, Eggs\}, \{Milk\}\rangle$

   The join of the two sequences is $\langle\{Bread, Butter, Cheese\}, \{Cheese, Eggs\}, \{Milk\}\rangle$.

2. If the last element of $\mathcal{S}_2$ is not a 1-itemset, but a superset of the last element of $\mathcal{S}_1$, then the joined candidate may be obtained by *replacing* the last element of $\mathcal{S}_1$ with the last element of $\mathcal{S}_2$. For example, consider the following two sequences:

   $\mathcal{S}_1 = \langle\{Bread, Butter, Cheese\}, \{Cheese, Eggs\}\rangle$
   $\mathcal{S}_2 = \langle\{Bread, Butter\}, \{Milk, Cheese, Eggs\}\rangle$

   The join of the two sequences is $\langle\{Bread, Butter, Cheese\}, \{Milk, Cheese, Eggs\}\rangle$.

These key differences from *Apriori* joins are a result of the temporal complexity and the set-based elements in sequential patterns. Alternative methods exist for performing the joins. For example, an alternative approach is to remove one item from the *last* elements of *both* $\mathcal{S}_1$ and $\mathcal{S}_2$ to check whether the resulting sequences are identical. However, in this case multiple candidates might be generated from the same pair. For example, $\langle a, b, c\rangle$ and $\langle a, b, d\rangle$ can join to any of $\langle a, b, c, d\rangle$, $\langle a, b, d, c\rangle$, and $\langle a, b, cd\rangle$. The first join rule of removing the first item from $\mathcal{S}_1$ and the last item from $\mathcal{S}_2$ has the merit of having a unique join result. For any specific join rule, it is important to ensure exhaustive and nonrepetitive generation of candidates. As we will see later, a similar notion to the frequent-pattern enumeration tree can be introduced in sequential pattern mining to ensure exhaustive and nonrepetitive candidate generation.

The *Apriori* trick is then used to prune sequences that violate downward closure. The idea is to check if each k-subsequence of a candidate in $\mathcal{C}_{k+1}$ is present in $\mathcal{F}_k$. The candidate set is pruned of those candidates that do not satisfy this closure property. The frequent $(k+1)$-candidate sequences $\mathcal{C}_{k+1}$ are then checked against the sequence database $\mathcal{T}$, and the support is counted. The counting of the support is executed according to the notion of a subsequence, as presented in Definition 15.2.1. All frequent candidates in $\mathcal{C}_{k+1}$ are retained in $\mathcal{F}_{k+1}$. The algorithm terminates, when no frequent sequences are generated in $\mathcal{F}_{k+1}$ in a particular iteration. All the frequent sequences generated during the different iterations of the levelwise approach are returned by the algorithm. The pseudocode of the *GSP* algorithm is illustrated in Fig. 15.1.

15.2.1 Frequent Patterns to Frequent Sequences

It is easy to see that the *Apriori* and *GSP* algorithms are structurally similar. This is not a coincidence. The basic structure of the frequent pattern and sequential pattern mining problems are similar. Aside from the differences in the support counting approach, the main difference between *GSP* and *Apriori* is in terms of how candidates are generated. The join

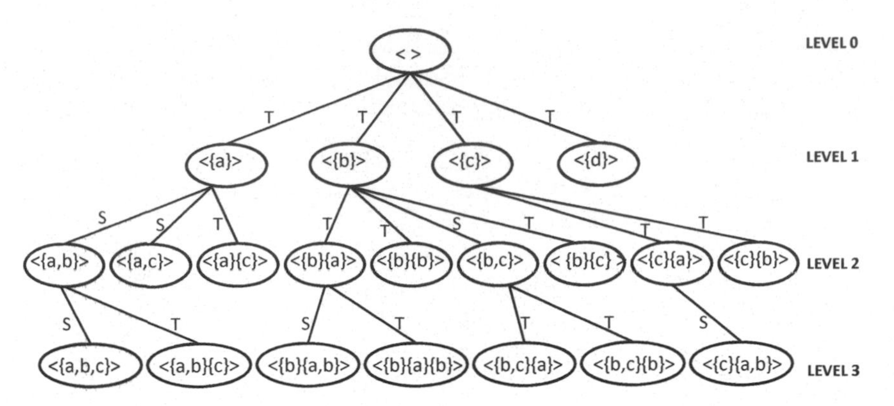

Figure 15.2: The equivalent of an enumeration tree for sequential pattern mining

generation in *GSP* is defined in terms of two separate cases. The two cases correspond to *temporal* extensions and *set-wise* extensions of candidates.

As discussed in Chap. 4, the *Apriori* algorithm can be viewed as an enumeration tree algorithm. It is also possible to define an analogous candidate tree in sequential pattern mining, albeit with a somewhat different structure than the enumeration tree in frequent pattern mining. The key differences in join-based candidate generation between *Apriori* and *GSP* algorithms translate to differences in the structure and growth of the candidate tree in sequential pattern mining. In general, candidate trees for sequential pattern mining are more complex because they need to accommodate both temporal and set-wise growth of sequences. Therefore, the definition of candidate extensions of a tree-node needs to be changed. A node for sequence $\mathcal{S}$ can be extended to a lower-level node in one of two ways:

1. *Set-wise extension:* In this case, an item is added to the last element in the sequence $\mathcal{S}$ to create a candidate pattern. Therefore, the number of elements does not increase. For an item to be added to the last element of $\mathcal{S}$, it must satisfy two properties; (a) the item successfully extends the parent sequence of $\mathcal{S}$ in the candidate tree with either a set-wise or temporal extension to another frequent sequence, and (b) the item must be *lexicographically* later than all items in the last element of $\mathcal{S}$. As in frequent pattern mining, a lexicographic ordering of items needs to be fixed up front.

2. *Temporal extension:* A new element with a single item is added to the end of the current sequence S. As in the previous case of set-wise extensions, any frequent item extension of the parent of $\mathcal{S}$ may be used to extend $\mathcal{S}$ (condition (a)). However, the added item need not be lexicographically later than the items in the last element of sequence $\mathcal{S}$.

These two kinds of extensions can be shown to be equivalent to the two kinds of joins in the *GSP* algorithm. As in frequent pattern mining, the candidate extensions of a node are a subset of the corresponding frequent extensions of its parent node. An example of the frequent portion of the candidate tree for sequential pattern mining is illustrated in Fig. 15.2. Note the greater complexity of the tree, because of set-wise and temporal addition of items at each level of the tree. The set-wise extensions are marked as "S", and the temporal extensions are marked as "T" on the corresponding tree edges. A particularly illuminating discussion on this topic may be found in [243], on which this example is based.

It is possible to convert any enumeration tree algorithm for frequent pattern mining to a sequential pattern mining algorithm by systematically making appropriate modifications. These changes account for the different structure of the candidate tree in sequential pattern mining compared to that in frequent pattern mining. This candidate tree is implicitly generated by all sequential pattern mining algorithms, such as *GSP* and *PrefixSpan*. Because the enumeration tree[1] is the generic framework describing all frequent pattern mining algorithms, it implies that virtually all frequent pattern mining algorithms can be modified to the sequential pattern mining scenario. For example, the work in [243] generalizes *TreeProjection*, and the *PrefixSpan* algorithm generalizes *FP-growth*. Savasere et al.'s vertical format [446] has also been generalized to sequential pattern mining algorithms. The main difference between these algorithms is the different efficiency of counting with the use of a variety of data structures, projection-based reuse tricks, and different candidate-tree exploration strategies such as breadth-first or depth-first, rather than a fundamental difference in search space size. The size of the candidate tree is fixed, although pruning strategies such as the *Apriori*-style pruning can reduce its size.

The notion of projection-based reuse can also be extended to sequential pattern mining. The projected representation $\mathcal{T}(\mathcal{P})$ of the sequence database $\mathcal{T}$ is associated with a sequential pattern $\mathcal{P}$ in the candidate enumeration tree (as modified for sequential pattern mining). Then, each sequence $\mathcal{Y} \in \mathcal{T}$ in the database is projected at $\mathcal{P}$ according to the following rules:

1. The sequential pattern $\mathcal{P}$ needs to be a subsequence of $\mathcal{Y}$ for the projection of $\mathcal{Y}$ to be included in the projected database $\mathcal{T}(\mathcal{P})$.

2. All items that are either not in the last element of $\mathcal{P}$, or are not successful frequent extensions (either temporal or set-wise) of the parent of $\mathcal{P}$ are not included in the projection of $\mathcal{Y}$ because they are irrelevant for counting frequent extensions of $\mathcal{P}$.

3. The *earliest* temporal occurrence of $\mathcal{P}$ in $\mathcal{Y}$, as a subsequence, is determined. Let the last element P_r in $\mathcal{P}$ be matched to the element Y_k in $\mathcal{Y}$ according to this subsequence matching. Then, the first element of the projected representation of $\mathcal{Y}$ is equal to the set of items in $Y_k - P_r$ that are lexicographically later than all items in P_r. If the resulting element Q is null, then it is not included in the projection of $\mathcal{Y}$. This first element, if non-null, is special because it may only be used for counting set-wise extensions of element P_r in the enumeration tree and is, therefore, denoted as $\_Q$ with an underscore in front of it.

4. The remaining elements in the projected sequence after $\_Q$ correspond to the elements in $\mathcal{Y}$ occurring temporally after the last matched element Y_k in $\mathcal{Y}$. All these elements are included in the projection of $\mathcal{Y}$ after removing the irrelevant items discussed in step 2. These remaining elements can be used for counting either set-wise extensions of the last element P_r in $\mathcal{P}$, or temporal extensions of $\mathcal{P}$. For any of these remaining elements (other than $\_Q$) to be used for counting the set-wise extensions of P_r, the element would already need to contain P_r.

The projected database $\mathcal{T}(\mathcal{P})$ can be used to count the frequent extensions of $\mathcal{P}$ more efficiently and determine the frequent ones. As in the frequent pattern mining, this projection can be performed successively in top-down fashion during the construction an enumeration-tree-like candidate structure. The projected database at a node can be generated by recursively projecting the database at its parent. The basic approach is exactly analogous to

[1]See discussion in Sect. 4.4.4.5 of Chap. 4. A similar argument applies to sequential pattern mining.

projection-based frequent pattern mining algorithms discussed in Chap. 4. The algorithm starts with a candidate tree $\mathcal{ET}$ which is the null node with the entire sequence database $\mathcal{T}$ at that node. This tree is extended repeatedly using the following step until no nodes remain in $\mathcal{ET}$ for further extension.

Select a node $(\mathcal{P}, \mathcal{T}(\mathcal{P}))$ in $\mathcal{ET}$ for extension;
Generate the candidate temporal and set-wise extensions of $\mathcal{P}$;
Determine the frequent extensions of $\mathcal{P}$ using support counting on $\mathcal{T}(P)$;
Extend $\mathcal{ET}$ with frequent extensions and their recursively projected databases;

The final candidate tree $\mathcal{ET}$ contains all the frequent sequential patterns. Different strategies for selecting the node $\mathcal{P}$ can lead to the generation of the candidate tree in a different order such as breadth-first or depth-first order. This simplified and generalized description is roughly based on the frameworks independently proposed in [243] and *PrefixSpan*, which are closely related. The reader is referred to the bibliographic notes for discussion of the specific algorithms.

15.2.2 Constrained Sequential Pattern Mining

In many cases, additional constraints are imposed on the sequential patterns, such as constraints on gaps between successive elements of the sequence. One solution is to use the unconstrained *GSP* algorithm, and then, as a postprocessing step, remove all subsequences not satisfying the constraint. However, this brute-force approach is a very inefficient solution because the number of constrained patterns may be orders of magnitude smaller than the unconstrained patterns. Therefore, the incorporation of the constraints *directly* into the *GSP* algorithm, during the candidate-generation or support-counting step, is significantly more efficient. Depending on the nature of the constraints, the changes required to the *GSP* algorithm may be minor, or significant. In all cases, the support-counting procedure for $\mathcal{F}_k$ needs to be modified. The constraints are explicitly checked during the support counting. This reduces the number of frequent patterns generated, and makes the process more efficient than the brute-force method. However, the incorporation of such constraints may sometimes result in invalidation of the downward closure property of the mined patterns. In such cases, appropriate changes may need to be made to the *GSP* algorithm. In cases where the downward closure property is not violated, the *GSP* algorithm can be used with very minor modifications for constraint checking during support counting.

An important constraint that does not violate the downward closure property is the *maxspan* constraint. This constraint specifies that the time difference between the first and last elements of a subsequence must be no larger than *maxspan*. Therefore, the *GSP* algorithm can be used directly, with the modification that the constraint is checked during support counting. Thus, the approach works with a much smaller set of subsequences in each step and is generally more efficient than the brute-force method.

Another common constraint in sequential pattern mining is the maximum gap constraint. This is also referred to as the *maxgap* constraint. Note that all $(k-1)$-subsequences of a particular frequent k-sequence may not be valid because of the maximum gap constraint. This is somewhat problematic because the *Apriori* principle cannot be used effectively. For example, the subsequence a_1a_5 is not supported by the transaction database sequence $a_1a_2a_3a_4a_5$ under the *maxgap* value of 1, because the gap value between a_1 and a_5 is 3. However, the subsequence $a_1a_3a_5$ is supported by the same transaction database sequence under this *maxgap* value. It is, therefore, possible for a_1a_5 to have lower support than $a_1a_3a_5$. Thus,

Apriori pruning cannot be applied. However, the sequence obtained by dropping items from the *first* or *last* elements of a frequent sequence will always be frequent. Therefore, the *specific* join-based approach discussed in this chapter can still be used to exhaustively generate all candidates without losing any frequent patterns. A modified pruning rule is used. While checking $(k-1)$-subsequences of a candidate for pruning, only subsequences containing the same number of *elements* as the candidate are checked. In other words, elements containing 1 item cannot be removed from the candidate for checking its subsequences. Such subsequences are referred to as *contiguous subsequences.* A noteworthy special case is one in which $maxgap = 0$. This case is often used for determining timeseries motifs after a time series has been converted to a discrete sequence using methods discussed in Sect. 14.4 of Chap. 14.

Another constraint of interest is the minimum gap constraint, or *mingap* constraint between successive elements. A minimum gap constraint between successive elements will always satisfy the downward closure property. Therefore, the *GSP* approach can be used with very minor modifications. The only modification is that this constraint is checked during support counting. This generates a smaller set of subsequences $\mathcal{F}_k$. The join and pruning steps remain unchanged. The bibliographic notes contain pointers to many other interesting constraints such as the window-size constraint.

15.3 Sequence Clustering

As in the case of timeseries data, the clustering of sequences is heavily dependent on the definition of similarity. When a similarity function has been defined, many of the traditional multidimensional methods such as k-medoids and graph-based methods can be easily adapted to sequence data. It should be pointed out that these two methods can be used for virtually any data type, and are dependent only on the choice of distance function.

Similarity measures for sequence data have been defined in Chap. 3. The most common similarity functions used for sequence data are as follows:

1. *Match-based measure:* This measure is equal to the number of matching positions between the two sequences. This can be meaningfully computed only when the two sequences are of equal length, and a one-to-one correspondence exists between the positions.

2. *Dynamic time warping (DTW):* In this case, the number of *nonmatches* between the two sequences can be used with dynamic time warping. The dynamic time warping (*DTW*) method is discussed in detail in Sect. 3.4.1.3 of Chap. 3. The idea is to stretch and shrink the time dimension dynamically to account for the varying speeds of data generation for different series.

3. *Longest common subsequence (LCSS):* As the name of this measure suggests, the longest matching subsequence between the two sequences is computed. This is then used to measure the similarity between the two sequences. The *LCSS* method is discussed in detail in Sect. 3.4.2.2 of Chap. 3.

4. *Edit distance:* This is defined as the cost of edit operations required to transform one sequence into another. The edit distance measure is described in Sect. 3.4.2.1 of Chap. 3. A number of alignment methods, such as BLAST, are specifically designed for biological sequences. Pointers to these methods may be found in the bibliographic notes.

5. *Keyword-based similarity:* In this case, a k-gram representation is used, in which each sequence is represented by a bag of segments of length k. These k-grams are extracted from the original data sequences by using a sliding window of length k on the sequences. Each such k-gram represents a new "keyword," and a *tf-idf* representation can be used in terms of these keywords. If desired, the infrequent k-grams can be dropped. Any text-based, vector-space similarity measure, discussed in Chap. 13, may be used. Since the ordering of the segments is no longer used after the transformation, such an approach allows the use of a wider range of data mining algorithms. In fact, any text mining algorithm can be used on this transformation.

6. *Kernel-based similarity:* Kernel-based similarity is particularly useful for SVM classification. Some examples of kernel-based similarity are discussed in detail in section 15.6.4.

The different measures are used in different application-specific scenarios. Many of these scenarios will be discussed in this chapter.

15.3.1 Distance-Based Methods

When a distance or similarity function has been defined, the k-medoids method can be generalized very simply to sequence data. The k-medoids method is agnostic as to the choice of data type and the similarity function because it is designed as a generic hill-climbing approach. In fact, the *CLARANS* algorithm, discussed in Sect. 7.3.1 of Chap. 7, can be easily generalized to work with any data type. The algorithm selects k representative sequences from the data, and assigns each data point to their closest sequence, using the selected distance (similarity) function. The quality of the representative sequences is improved iteratively with the use of a hill-climbing algorithm.

The hierarchical methods, discussed in Sect. 6.4 of Chap. 6, can also be generalized to any data type because they work with pairwise distances between the different data objects. The main challenge of using hierarchical methods is that they require $O(n^2)$ pairwise distance computations between n objects. Distance function computations on sequence data are generally expensive because they require expensive dynamic programming methods. Therefore, the applicability of hierarchical methods is restricted to smaller data sets.

15.3.2 Graph-Based Methods

Graph-based methods are a general approach to clustering that is agnostic to the underlying data type. The transformation of different data types to similarity graphs is described in Sect. 2.2.2.9 of Chap. 2. The broader approach in graph-based methods is as follows:

1. Construct a graph in which each node corresponds to a data object. Each node is connected to its k-nearest neighbors, with a weight equal to the similarity between the corresponding pairs of data objects. In cases where a distance function is used, it is converted to a similarity function as follows:

$$w_{ij} = e^{-d(O_i,O_j)^2/t^2} \tag{15.1}$$

Here, $d(O_i, O_j)$ represents the distance between the objects O_i and O_j and t is a parameter.

2. The edges are assumed to be undirected, and any parallel edges are removed by dropping one of the edges. Because the distance functions are assumed to be symmetric, the parallel edges will have the same weight.

3. Any of the clustering or community detection algorithms discussed in Sect. 19.3 of Chap. 19 may be used for clustering nodes of the newly created graph.

After the nodes have been clustered, these clusters can be mapped back to clusters of data objects by using the correspondence between nodes and data objects.

15.3.3 Subsequence-Based Clustering

The major problem with the aforementioned methods is that they are based on similarity functions that use *global* alignment between the sequences. For longer sequences, global alignment becomes increasingly ineffective because of the noise effects of computing similarity between pairs of long sequences. Many local portions of sequences are noisy and irrelevant to similarity computations even when large portions of two sequences are similar. One possibility is to design local alignment similarity functions or use the keyword-based similarity method discussed earlier.

A more direct approach is to use frequent subsequence-based clustering methods. Some related approaches also use k-grams extracted from the sequence instead of frequent subsequences. However, k-grams are generally more sensitive to noise than frequent subsequences. The idea is that the frequent subsequences represent the key structural characteristics that are common across different sequences. After the frequent subsequences have been determined, the original sequences can be transformed into this new feature space, and a "bag-of-words" representation can be created in terms of these new features. Then, the sequence objects can be clustered in the same way as any text-clustering algorithm. The overall approach can be described as follows:

1. Determine the frequent subsequences $\mathcal{F}$ from the sequence database $\mathcal{D}$ using any frequent sequential pattern mining algorithm. Different applications may vary in the specific constraints imposed on the sequences, such as a minimum or maximum length of the determined sequences.

2. Determine a subset $\mathcal{F}_S$ from the frequent subsequences $\mathcal{F}$ based on an appropriate selection criterion. Typically, a subset of frequent subsequences should be selected, so as to maximize coverage and minimize redundancy. The idea is to use only a modest number of relevant features for clustering. For example, the notion of *Frequent Summarized Subsequences (FSS)* is used to determine condensed groups of sequences [505]. The bibliographic notes contain specific pointers to these methods.

3. Represent each sequence in the database as a "bag of frequent subsequences" from $\mathcal{F}_S$. In other words, the transformed representation of a sequence contains all frequent subsequences from $\mathcal{F}_S$ that it contains.

4. Apply any text-clustering algorithm on this new representation of the database of sequences. Text-clustering algorithms are discussed in Chap. 13. The tf-idf weighting may be applied to the different features, as discussed in Chap. 13.

The aforementioned discussion is a broad overview of frequent subsequence-based clustering, although the individual steps are implemented in different ways by different methods. The key differences among the different algorithms are in terms of methodology for feature

Algorithm *CLUSEQ*(Sequence Database: $\mathcal{D}$, Similarity Threshold: t)
begin
  $k = f = 1$;
  Let $\mathcal{C}_1$ be a singleton cluster with randomly chosen sequence;
  repeat
    Add $k_a = k \cdot f$ new singleton clusters containing sequences
      that are as different as possible from existing clusters/each other;
    $k = k + k_a$;
    Assign (if possible) each sequence in $\mathcal{D}$ to each cluster in
      $\mathcal{C}_1 \ldots \mathcal{C}_k$ for which the similarity is at least t;
    Eliminate the k_r clusters containing less than *minthresh*
        sequences *uniquely* assigned to them;
    $k = k - k_r$;
    $f = \frac{\max\{k_a - k_r, 0\}}{k_a}$;
  until no change in clustering result;
  return clusters $\mathcal{C}_1 \ldots \mathcal{C}_k$;
end

Figure 15.3: The simplified *CLUSEQ* Algorithm

construction and the choice of the text-clustering algorithm. The *CONTOUR* method [505] uses a two-level hierarchical clustering, where fine-grained microclusters are generated in the first step. Then, these microclusters are agglomerated into higher-level clusters. The bibliographic notes contain pointers to specific instantiations of this framework.

15.3.4 Probabilistic Clustering

Probabilistic clustering methods are based on the generative principle, that a symbol in a given sequence is generated with a probability defined by statistical correlations with the symbols before it. This is based on the general principle of *Markovian Models*. Therefore, the similarity between a sequence and a cluster is computed using the generative probability of the symbols within that cluster. After a similarity function has been defined between a cluster and a sequence, it can be used to create a distance-based algorithm. The *CLUSEQ* algorithm is based on this principle.

15.3.4.1 Markovian Similarity-Based Algorithm: CLUSEQ

The *CLUstering SEQuences (CLUSEQ)* algorithm is based on the broader principle of Markovian Models. Markovian models are used to define a similarity function between a sequence and cluster. The *CLUSEQ* algorithm can otherwise be considered a similarity-based iterative partitioning algorithm. While traditional partitioning algorithms fix the number of clusters over multiple iterations, this is not the case in *CLUSEQ*. The *CLUSEQ* algorithm starts with only a single cluster. A carefully controlled number of new clusters containing individual sequences are added in each iteration, and older ones are removed when they are deemed to be too similar to existing clusters. The initial growth in the number of clusters is rapid but it slows down over the course of the algorithm. It is even possible for the number of clusters to shrink in later iterations. One advantage of this approach is that the algorithm can automatically determine the natural number of clusters.

Instead of using the number of clusters as an input parameter, the *CLUSEQ* algorithm works with a similarity threshold t. A sequence is assigned to a cluster, if its similarity to the cluster exceeds the threshold t. Sequences may be assigned to any number of clusters (or no cluster) as long as the similarity is greater than t. The *CLUSEQ* algorithm has three main steps corresponding to addition of new clusters, assignment of sequences to clusters, and elimination of clusters. These steps are repeated iteratively until there is no change in the clustering result. A simplified version[2] of the *CLUSEQ* algorithm is described in Fig. 15.3. A detailed description of the individual steps is provided below.

1. *Cluster addition:* The number of clusters added is $k \cdot f$, where k is the number of clusters at the end of the last iteration. The value of f is in the range $(0, 1)$, and is computed as follows. Let k_a be the number of clusters added in the previous iteration, and let k_r be the number of clusters removed because of elimination of overlapping clusters in the previous iteration. Then, the value of f is computed as follows:

$$f = \frac{\max\{k_a - k_r, 0\}}{k_a} \tag{15.2}$$

   The rationale for this is that when the algorithm reaches its "natural" number of clusters, eliminations will dominate. In such cases, f will be small or 0, and few new clusters need to be added. On the other hand, in cases where the current number of clusters is significantly lower than the "natural" number of clusters in the data, the value of f should be close to 1. In earlier iterations, the number of added clusters is much larger than the number of removed clusters, which results in rapid growth.

   The new clusters created are singleton clusters. The sequences that are as different as possible from both the existing clusters and each other are selected. Therefore pairwise similarity needs to be computed between each unclustered sequence and other clusters/unclustered sequences. Because it can be expensive to compute pairwise similarity between the clusters and all unclustered sequences, a sample of unclustered sequences is used to restrict the scope of new seed selection. The approach for computing similarity will be described later.

2. *Sequence assignment to clusters:* Sequences are assigned to clusters for which the similarity to the cluster is larger than a user-specified threshold t. The original *CLUSEQ* algorithm provides a way to adjust the threshold t as well, though the description in this chapter provides only a simplified version of the algorithm, where t is fixed and specified by the user. A given sequence may be assigned to either multiple clusters or may remain unassigned to any cluster. The actual similarity computation is performed using a *Markovian* similarity measure. This measure will be described later.

3. *Cluster elimination:* Many clusters are highly overlapping because of assignment of sequences to multiple clusters. It is desired to restrict this overlap to reduce redundancy in the clustering. If the number of sequences that are *unique* to a particular cluster is less than a predefined threshold, then such a cluster is eliminated.

The only step that remains to be described is the computation of the *Markovian similarity measure* between sequences and clusters. The idea is that if a sequence of symbols $S = s_1 s_2 \ldots s_n$ is similar to a cluster C_i, then it should be "easy" to generate S using the

[2]The original *CLUSEQ* algorithm also adjusts the similarity threshold t iteratively to optimize results.

conditional distribution of the symbols inside the cluster. Then, the probability $P(S|\mathcal{C}_i)$ is defined as follows:

$$P(S|\mathcal{C}_i) = P(s_1|\mathcal{C}_i) \cdot P(s_2|s_1, \mathcal{C}_i) \ldots P(s_n|s_1 \ldots s_{n-1}, \mathcal{C}_i) \tag{15.3}$$

This is the generative probability of the sequence S for cluster $\mathcal{C}_i$. Intuitively, the term $P(s_j|s_1 \ldots s_{j-1}, \mathcal{C}_i)$ represents the fraction of times that s_j follows $s_1 \ldots s_{j-1}$ in cluster $\mathcal{C}_i$. This term can be estimated in a data-driven manner from the sequences in $\mathcal{C}_i$. When a cluster is highly similar to a sequence, this value will be high. A *relative* similarity can be computed by comparing with a sequence generation model in which all symbols are generated randomly in proportion to their presence in the full data set. The probability of such a random generation is given by $\prod_{j=1}^{n} P(s_j)$, where $P(s_j)$ is estimated as the fraction of sequences containing symbol s_j. Then, the similarity of S to cluster $\mathcal{C}_i$ is defined as follows:

$$sim(S, \mathcal{C}_i) = \frac{P(S|\mathcal{C}_i)}{\prod_{j=1}^{n} P(s_j)} \tag{15.4}$$

One issue is that many parts of the sequence S may be noisy and not match the cluster well. Therefore, the similarity is computed as the maximum similarity of any contiguous segment of S to $\mathcal{C}_i$. In other words, if S_{kl} be the contiguous segment of S from positions k to l, then the final similarity $SIM(S, \mathcal{C}_i)$ is computed as follows:

$$SIM(S, \mathcal{C}_i) = \max_{1 \leq k \leq l \leq n} sim(S_{kl}, \mathcal{C}_i) \tag{15.5}$$

The maximum similarity value can be computed by computing $sim(S_{kl}, \mathcal{C}_i)$ over all pairs $[k, l]$. This is the similarity value used for assigning sequences to their relevant clusters.

One problematic issue is that the computation of each of the terms $P(s_j|s_1 \ldots s_{j-1}, \mathcal{C}_i)$ on the right-hand side of Eq. 15.3 may require the examination of all the sequences in the cluster $\mathcal{C}_i$ for probability estimation purposes. Fortunately, these terms can be estimated efficiently using a data structure, referred to as *Probabilistic Suffix Trees.* The *CLUSEQ* algorithm always dynamically maintains the *Probabilistic Suffix Trees (PST)* whenever new clusters are created or sequences are added to clusters. This data structure will be described in detail in Sect. 15.4.1.1.

15.3.4.2 Mixture of Hidden Markov Models

This approach can be considered the string analog of the probabilistic models discussed in Sect. 6.5 of Chap. 6 for clustering multidimensional data. Recall that a generative mixture model is used in that case, where each component of the mixture has a Gaussian distribution. A Gaussian distribution is, however, appropriate only for generating numerical data, and is not appropriate for generating sequences. A good generative model for sequences is referred to as *Hidden Markov Models (HMM).* The discussion of this section will assume the use of HMM as a black box. The actual details of HMM will be discussed in a later section. As we will see later in Sect. 15.5, the HMM can itself be considered a kind of mixture model, in which states represent dependent components of the mixture. Therefore, this approach can be considered a *two-level* mixture model. The discussion in this section should be combined with the description of HMMs in Sect. 15.5 to provide a complete picture of HMM-based clustering.

The broad principle of a mixture-based generative model is to assume that the data was generated from a mixture of k distributions with the probability distributions $\mathcal{G}_1 \ldots \mathcal{G}_k$, where each $\mathcal{G}_i$ is a Hidden Markov Model. As in Sect. 6.5 of Chap. 6, the approach assumes the use of prior probabilities $\alpha_1 \dots \alpha_k$ for the different components of the mixture. Therefore, the generative process is described as follows:

1. Select one of the k probability distributions with probability α_i where $i \in \{1 \dots k\}$. Let us assume that the rth one is selected.

2. Generate a sequence from $\mathcal{G}_r$, where $\mathcal{G}_r$ is a Hidden Markov Model.

One nice characteristic of mixture models is that the change in the data type and corresponding mixture distribution does not affect the broader framework of the algorithm. The analogous steps can be applied in the case of sequence data, as they are applied in multidimensional data. Let S_j represent the jth sequence and Θ be the entire set of parameters to be estimated for the different HMMs. Then, the E-step and M-step are exactly analogous to the case of the multidimensional mixture model.

1. (E-step) Given the current state of the trained HMM and priors α_i, determine the posterior probability $P(\mathcal{G}_i|S_j, \Theta)$ of each sequence S_j using the HMM generative probabilities $P(S_j|\mathcal{G}_i, \Theta)$ of S_j from the ith HMM, and priors $\alpha_1 \dots \alpha_k$ in conjunction with the Bayes rule. This is the posterior probability that the sequence S_j was generated by the ith HMM.

2. (M-step) Given the current probabilities of assignments of data points to clusters, use the *Baum–Welch algorithm* on each HMM to learn its parameters. The assignment probabilities are used as weights for averaging the estimated parameters. The Baum–Welch algorithm is described in Sect. 15.5.4 of this chapter. The value of each α_i is estimated to be proportional to average assignment probability of all sequences to cluster i. Thus, the M-step results in the estimation of the entire set of parameters Θ.

Note that there is an almost exact correspondence in the steps used here, and to those used for mixture modeling in Sect. 6.5 of Chap. 6. The major drawback of this approach is that it can be rather slow. This is because the process of training each HMM is computationally expensive.

15.4 Outlier Detection in Sequences

Outlier detection in sequence data shares a number of similarities with timeseries data. The main difference between sequence data and timeseries data is that sequence data is discrete, whereas timeseries data is continuous. The discussion in the previous chapter showed that time series outliers can be either *point outliers*, or *shape outliers.* Because sequence data is the discrete analog of timeseries data, an identical principle can be applied to sequence data. Sequence data outliers can be either *position outliers* or *combination outliers.*

1. *Position outliers:* In position-based outliers, the values at specific positions are predicted by a model. This is used to determine the *deviation* from the model and predict specific *positions* as outliers. Typically, Markovian methods are used for predictive outlier detection. This is analogous to deviation-based outliers discovered in timeseries data with the use of regression models. Unlike regression models, Markovian models are better suited to discrete data. Such outliers are referred to as *contextual* outliers because they are outliers in the *context* of their immediate temporal neighborhood.

2. *Combination outliers:* In combination outliers, an entire test sequence is deemed to be unusual because of the combination of symbols in it. This could be the case because this combination may rarely occur in a sequence database, or its distance (or similarity) to most other subsequences of similar size may be very large (or small). More complex models, such as Hidden Markov Models, can also be used to model the frequency of presence in terms of generative probabilities. For a longer test sequence, smaller subsequences are extracted from it for testing, and then the outlier score of the entire sequence is predicted as a combination of these values. This is analogous to the determination of unusual shapes in timeseries data. Such outliers are referred to as *collective outliers* because they are defined by combining the patterns from multiple data items.

The following section will discuss these different types of outliers.

15.4.1 Position Outliers

In the case of continuous timeseries data discussed in the previous chapter, an important class of outliers was designed by determining significant deviations from *expected* values at timestamps. Thus, these methods intimately combine the problems of *forecasting* and *deviation-detection*. A similar principle applies to discrete sequence data, in which the discrete positions at specific timestamps can be predicted with the use of different models. When a position has very low probability of matching its forecasted value, it is considered an outlier. For example, consider an RFID application, in which event sequences are associated with product items in a superstore with the use of semantic extraction from RFID tags. A typical example of a normal event sequence is as follows:

`PlacedOnShelf, RemovedFromShelf, CheckOut, ExitStore.`

On the other hand, in a shoplifting scenario, the event sequence may be *unusually* different. An example of an event sequence in the shoplifting scenario is as follows:

`PlacedOnShelf, RemovedFromShelf, ExitStore.`

Clearly, the sequence symbol *ExitStore* is anomalous in the second case but not in the first case. This is because it does not depict the *expected* or *forecasted* value for that position in the second case. It is desirable to detect such anomalous *positions* on the basis of expected values. Such anomalous positions may appear *anywhere* in the sequence and not necessarily in the last element, as in the aforementioned example. The basic problem definition for position outlier detection is as follows:

Definition 15.4.1 *Given a set of N training sequences* $\mathcal{D} = T_1 \dots T_N$*, and a test sequence* $V = a_1 \dots a_n$*, determine if the position* a_i *in the test sequence should be considered an anomaly based on its expected value.*

Some formulations do not explicitly distinguish between training and test sequences. This is because a sequence can be used for both model construction and outlier analysis when it is very long.

Typically, the position a_i can be predicted in temporal domains only from the positions before a_i, whereas in other domains, such as biological data, both directions may be relevant. The discussion below will assume the temporal scenario, though generalization to the placement scenario (as in biological data) is straightforward by examining windows on both sides of the position.

Just as regression modeling of continuous streams uses small windows of past history, discrete sequence prediction also uses small windows of the symbols. It is assumed that the prediction of the values at a position depends upon this short history. This is known as the *short memory property* of discrete sequences, and it generally holds true across a wide variety of temporal application domains.

Definition 15.4.2 (Short Memory Property) *For a sequence of symbols* $V = a_1 \ldots a_i \ldots$, *the value of the probability* $P(a_i|a_1 \ldots a_{i-1})$ *is well approximated by* $P(a_i|a_{i-k} \ldots a_{i-1})$ *for some small value of* k.

After the value of $P(a_i|a_{i-k} \ldots a_{i-1})$ is estimated, a position in a test sequence can be flagged as an outlier, if it has very low probability on the basis of the models derived from the training sequences. Alternatively, if a different symbol (than one present in the test sequence) is predicted with very high probability, then that position can be flagged as an outlier.

This section will discuss the use of *Markovian* models for the position outlier detection problem. This model exploits the *short memory* property of sequences to explicitly model the sequences as a set of states in a Markov Chain. These models represent the sequence-generation process with the use of transitions in a Markov Chain defined on the alphabet Σ. This is a special kind of *Finite State Automaton*, in which the states are defined by a short (immediately preceding) history of the sequences generated. Such models correspond to a set of states A that represent the different kinds of memory about the system events. For example, in *first-order* Markov Models, each state represents the last symbol from the alphabet Σ that was generated in the sequence. In kth order Markov Models, each state corresponds to the subsequence of the last k symbols $a_{n-k} \ldots a_{n-1}$ in the sequence. Each transition in this model represents an event a_n, the transition probability of which from the state $a_{n-k} \ldots a_{n-1}$ to the state $a_{n-k+1} \ldots a_n$ is given by the conditional probability $P(a_n|a_{n-k} \ldots a_{n-1})$. A Markov Model can be depicted as a set of nodes representing the states and a set of edges representing the events that cause movement from one state to another. The probability of an edge provides the conditional probability of the corresponding event. Clearly, the order of the model encodes the memory length of the string segment retained for the modeling process. First-order models correspond to the least amount of retained memory.

To understand how Markov Models work, the previous example of tracking items with RFID tags will be revisited. The actions performed an item can be viewed as a sequence drawn on the alphabet $\Sigma = \{P, R, C, E\}$. The semantic meaning of each of these symbols is illustrated in Fig. 15.4. A state of an order-k Markov model corresponds to the previous k (action) symbols of the sequence drawn on the alphabet $\Sigma = \{P, R, C, E\}$. Examples of different states along with transitions are illustrated in Fig. 15.4. Both a first-order and a second-order model have been illustrated in the figure. The edge transition probabilities are also illustrated in the figure. These are typically estimated from the training data. The transitions that correspond to the shoplifting anomaly are marked in both models. In each case, it is noteworthy that the corresponding transition probability for the actual shoplifting event is very low. This is a particularly simple example, in which a memory of one event is sufficient to completely represent the state of an item. This is not the case in general. For example, consider the case of a Web log in which the Markov Models correspond to sequences of Web pages visited by users. In such a case, the probability distribution of the next Web page visited depends not just on the last page visited, but also on the other preceding visits by the user.

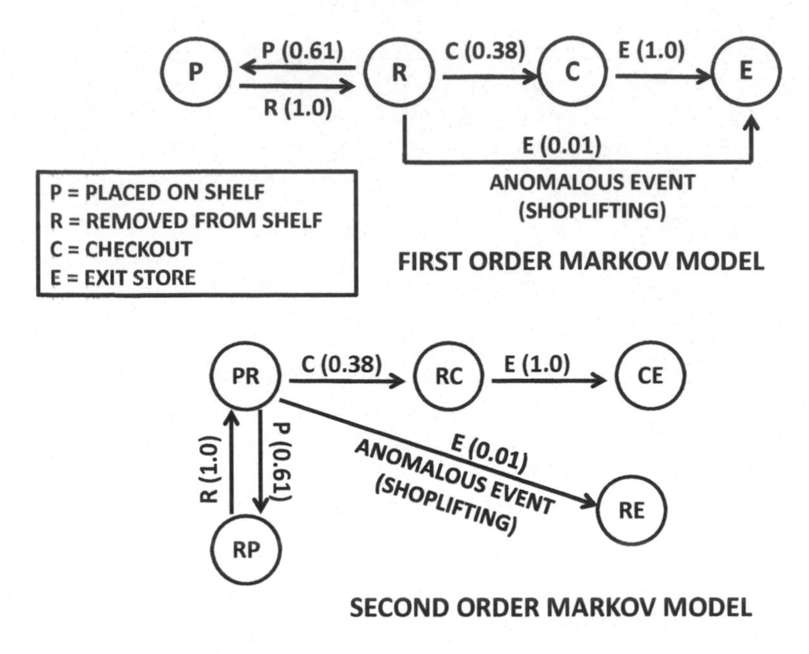

Figure 15.4: Markov model for the RFID-based shoplifting anomaly

An observation from Fig. 15.4 is that the number of states in the second-order model is larger than that in the first-order model. This is not a coincidence. As many as $|\Sigma|^k$ states may exist in an order-k model, though this provides only an upper bound. Many of the subsequences corresponding to these states may either not occur in the training data, or may be invalid in a particular application. For example, a PP state would be invalid in the example of Fig. 15.4 because the same item cannot be sequentially placed twice on the shelf without removing it at least once. Higher-order models represent complex systems more accurately at least at a theoretical level. However, choosing models of a higher order degrades the efficiency and may also result in overfitting.

15.4.1.1 Efficiency Issues: Probabilistic Suffix Trees

It is evident from the discussion in the previous sections that the Markovian and rule-based models are equivalent, with the latter being a simpler and easy-to-understand heuristic approximation of the former. Nevertheless, in both cases, the challenge is that the number of possible antecedents of length k can be as large as $|\Sigma|^k$. This can make the methods slow, when a lookup for a test subsequence $a_{i-k} \dots a_{i-1}$ is required to determine the probability of $P(a_i|a_{i-k} \dots a_{i-1})$. It is expensive to either compute these values on the fly, or even to retrieve their precomputed values, if they are not organized properly. The *Probabilistic Suffix Tree (PST)* provides an efficient approach for retrieval of such precomputed values. The utility of probabilistic suffix trees is not restricted to outlier detection, but is also applicable to clustering and classification. For example, the *CLUSEQ* algorithm, discussed in Sect. 15.3.4.1, uses PST to retrieve these prestored probability values.

Suffix trees are a classical data structure that store all subsequences in a given database. Probabilistic suffix trees represent a generalization of this structure that also stores the conditional probabilities of generation of the next symbol for a given sequence database. For order-k Markov Models, a suffix tree of depth at most k will store all the required conditional

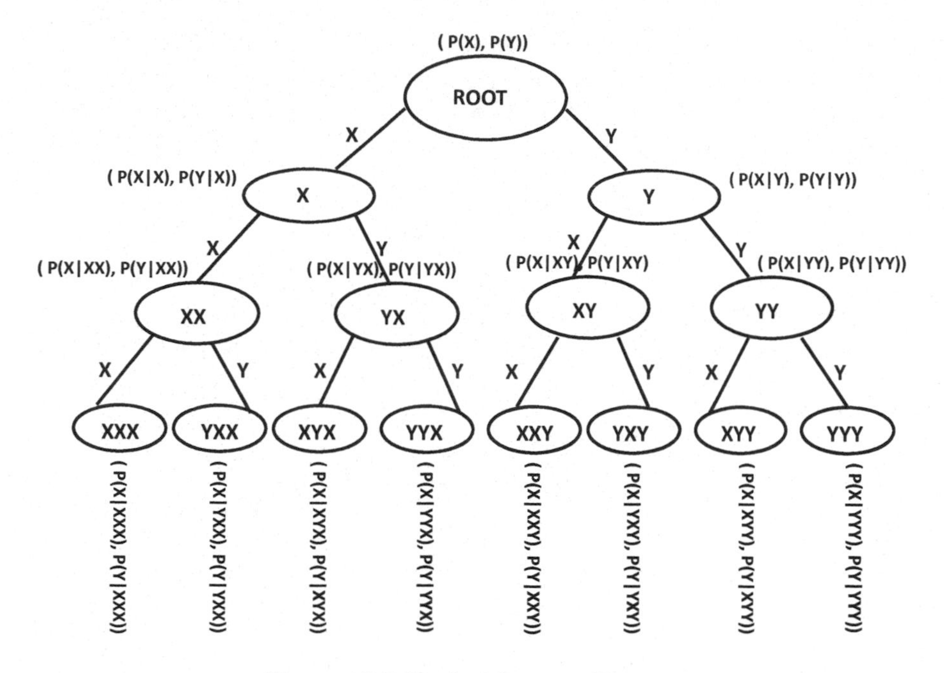

Figure 15.5: Probabilistic suffix tree

probability values for the kth order Markovian models, including the conditionals for all lower-order Markov Models. Therefore, such a structure encodes all the information required for variable-order Markov Models as well. A key challenge is that the number of nodes in such a suffix tree can be as large as $\sum_{i=0}^{k} |\Sigma|^i$, an issue that needs to be addressed with selective pruning.

A probabilistic suffix tree is a hierarchical data structure representing the different suffixes of a sequence. A node in the tree with depth k represents a suffix of length k and is, therefore, labeled with a sequence of length k. The parent of a node $a_{i-k} \ldots a_i$ corresponds to the sequence $a_{i-k+1} \ldots a_i$. The latter is obtained by removing the *first* symbol from the former. Each edge is labeled with the symbol that needs to be removed to derive the sequence at the parent node. Thus, a path in the tree corresponds to *suffixes* of the same sequence. Each node also maintains a vector Σ of probabilities that correspond to the conditional probability of the generation of any symbol from $\Sigma = \{\sigma_1 \ldots \sigma_{|\Sigma|}\}$ after that sequence. Therefore, for a node corresponding to the sequence $a_{i-k} \ldots a_i$, and for each $j \in \{1 \ldots |\Sigma|\}$, the values of $P(\sigma_j | a_{i-k} \ldots a_i)$ are maintained. As discussed earlier, this corresponds to the conditional probability that σ_j appears immediately *after* $a_{i-k} \ldots a_i$, once the latter sequence has already been observed. This provides the generative probability crucial to the determination of position outliers. Note that this generative probability is also useful for other algorithms such as the *CLUSEQ* algorithm discussed earlier in this chapter.

An example of a suffix tree with the symbol set $\Sigma = \{X, Y\}$ is illustrated in Fig. 15.5. The two possible symbol-generation probabilities at each node corresponding to either of the symbols X and Y are placed next to the corresponding nodes. It is also evident that a probabilistic suffix tree of depth k encodes *all* the transition probabilities for Markovian models up to order k. Therefore, such an approach can be used for higher-order Markovian models.

The probabilistic suffix true is pruned significantly to improve its compactness. For example, suffixes that correspond to very low counts in the original data can be pruned from consideration. Furthermore, nodes with low generative probabilities of their underlying sequences can be pruned from consideration. The generative probability of a sequence $a_1 \ldots a_n$ is approximated as follows:

$$P(a_1 \ldots a_n) = P(a_1) \cdot P(a_2|a_1) \ldots P(a_n|a_1 \ldots a_{n-1}) \quad (15.6)$$

For Markovian models of order $k < n$, the value of $P(a_r|a_1 \ldots a_{r-1})$ in the equation above is approximated by $P(a_r|a_{r-k} \ldots a_{r-1})$ for any value of k less than r. To create Markovian models of order k or less, it is not necessary to keep portions of the tree with depth greater than k.

Consider the sequence $a_1 \ldots a_i \ldots a_n$, in which it is desired to test whether position a_i is a position outlier. Then, it is desired to determine $P(a_i|a_1 \ldots a_{i-1})$. It is possible that the suffix $a_1 \ldots a_{i-1}$ may not be present in the suffix tree because it may have been pruned from consideration. In such cases, the *short memory* property is used to determine the longest suffix $a_j \ldots a_{i-1}$ present in the suffix tree, and the corresponding probability is estimated by $P(a_i|a_j \ldots a_{i-1})$. Thus, the probabilistic suffix tree provides an efficient way to store and retrieve the relevant probabilities. The length of the longest path that exists in the suffix tree containing a nonzero probability estimate of $P(a_i|a_j \ldots a_{i-1})$ also provides an idea of the level of rarity of this particular sequence of events. Positions that contain only short paths preceding them in the suffix tree are more likely to be outliers. Thus, outlier scores may be defined from the suffix tree in multiple ways:

1. If only short path lengths exist in the (pruned) suffix tree corresponding to a position a_i and its preceding history, then that position is more likely be an outlier.

2. For the paths of lengths $1 \ldots r$ that do exist in the suffix tree for position a_i, a combination score may be used based on the models of different orders. In some cases, only lower-order scores are combined. In general, the use of lower-order scores is preferable, since they are usually more robustly represented in the training data.

15.4.2 Combination Outliers

In combination outliers, the goal is to determine unusual combinations of symbols in the sequences. Consider a setting, where a set of training sequences is provided, together with a test sequence. It is desirable to determine whether a test sequence is an anomaly, based on the "normal" patterns in the training sequences. In many cases, the test sequences may be quite long. Therefore, the combination of symbols in the full sequence may be unique with respect to the training sequences. This means that it is hard to characterize "normal" sequences on the basis of the full sequence. Therefore, small windows are extracted from the training and test sequences for the purpose of comparison. Typically, all windows (including overlapping ones) are extracted from the sequences, though it is also possible to work with nonoverlapping windows. These are referred to as *comparison units*. The anomaly scores are defined with respect to these comparison units. Thus, unusual *windows* in the sequences are reported. The following discussion will focus exclusively on determining such unusual windows.

Some notations and definitions will be used to distinguish between the training database, test sequence, and the comparison units.

1. The training database is denoted by $\mathcal{D}$, and contains sequences denoted by $T_1 \ldots T_N$.

2. The test sequence is denoted by V.

3. The comparison units are denoted by $U_1 \ldots U_r$. Typically, each U_i is derived from small, contiguous windows of V. In domain-dependent cases, $U_1 \ldots U_r$ may be provided by the user.

The model may be a distance-based, or frequency-based or may be a Hidden Markov Model. Each of these will be discussed in subsequent sections. Because Hidden Markov Models are general constructs that are used for different problems such as clustering, classification, and outlier detection, they will be discussed in a section of their own, immediately following this section.

15.4.2.1 Distance-Based Models

In distance-based models, the absolute distance (or similarity) of the comparison unit is computed to equivalent windows of the training sequence. The distance of the k-th nearest neighbor window in the training sequence is used to determine the anomaly score. In the context of sequence data, many proximity-functions are *similarity* functions rather than *distance* functions. In the former case, higher values indicate greater proximity. Some common methods for computing the similarity between a pair of sequences are as follows:

1. *Simple matching coefficient:* This is the simplest possible function and determines the number of matching positions between two sequences of equal length. This is also equivalent to the Hamming distance between a pair of sequences.

2. *Normalized longest common subsequence:* The longest common subsequence can be considered the sequential analog of the cosine distance between two ordered sets. Let T_1 and T_2 be two sequences, and the length of (unnormalized) longest common subsequence between T_1 and T_2 be denoted by $L(T_1, T_2)$. The unnormalized longest common subsequence can be computed using methods discussed in Sect. 3.4.2 of Chap. 3. Then, the value $NL(T_1, T_2)$ of the normalized longest common subsequence is computed by normalizing $L(T_1, T_2)$ with the underlying sequence lengths in a way similar to the cosine computation between unordered sets:

$$NL(T_1, T_2) = \frac{L(T_1, T_2)}{\sqrt{|T_1|} \cdot \sqrt{|T_2|}} \tag{15.7}$$

The advantage of this approach is that it can match two sequences of unequal lengths. The drawback is that the computation process is relatively slow.

3. *Edit distance:* The edit distance is one of the most common similarity functions used for sequence matching. This similarity function is discussed in Chap. 3. This function measures the distance between two sequences by the minimum number of edits required to transform one sequence to the other. The computation of the edit distance can be computationally very expensive.

4. *Compression-based dissimilarity:* This measure is based on principles of information theory. Let W be a window of the training data, and $W \oplus U_i$ be the string representing the concatenation of W and U_i. Let $DL(S) < |S|$ be the description length of any string S after applying a standard compression algorithm to it. Then, the compression-based dissimilarity $CD(W, U_i)$ is defined as follows:

$$CD(W, U_i) = \frac{DL(W \oplus U_i)}{DL(W) + DL(U_i)} \tag{15.8}$$

This measure always lies in the range $(0, 1)$, and lower values indicate greater similarity. The intuition behind this approach is that when the two sequences are very similar, the description length of the combined sequence will be much smaller than that of the sum of the description lengths. On the other hand, when the sequences are very different, the description length of the combined string will be almost the same as the sum of the description lengths.

To compute the anomaly score for a comparison unit U_i with respect to the training sequences in $T_1 \ldots T_N$, the first step is to extract equivalent windows from $T_1 \ldots T_N$ as the size of the comparison unit. The k-th nearest neighbor distance is used as the anomaly score for that window. The unusual windows may be reported, or the scores from different windows may be consolidated into a single anomaly score.

15.4.2.2 Frequency-Based Models

Frequency-based models are typically used with domain-specific comparison units specified by the user. In this case, the relative frequency of the comparison unit needs to be measured in the training sequences and the test sequences, and the level of surprise is correspondingly determined.

When the comparison units are specified by the user, a natural way of determining the anomaly score is to test the frequency of the comparison unit U_j in the training and test patterns. For example, when a sequence contains a hacking attempt, such as a sequence of *Login* and *Password* events, this sequence will have much higher frequency in the test sequence, as compared to the training sequences. The specification of such relevant comparison units by a user provides very useful domain knowledge to an outlier analysis application.

Let $f(T, U_j)$ represent the number of times that the comparison unit U_j occurs in the sequence T. Since the frequency $f(T, U_j)$ depends on the length of T, the normalized frequency $\hat{f}(T, U_j)$ may be obtained by dividing the frequency by the length of the sequence:

$$\hat{f}(T, U_j) = \frac{f(T, U_j)}{|T|}$$

Then, the anomaly score of the training sequence T_i with respect to the test sequence V is defined by subtracting the relative frequency of the training sequence from the test sequence. Therefore, the anomaly score $A(T_i, V, U_j)$ is defined as follows:

$$A(T_i, V, U_j) = \hat{f}(V, U_j) - \hat{f}(T_i, U_j)$$

The absolute value of the average of these scores is computed over all the sequences in the database $\mathcal{D} = T_1 \ldots T_N$. This represents the final anomaly score.

A useful output of this approach is the specific subset of comparison units specified by the user that are the most anomalous. This provides intensional knowledge and feedback to the analyst about *why* a particular test sequence should be considered anomalous. A method called *TARZAN* uses suffix tree representations to efficiently determine all the anomalous subsequences in a comparative sense between a test sequence and a training sequence. Readers are referred to the bibliographic notes for pointers to this method.

15.5 Hidden Markov Models

Hidden Markov Models (HMM) are probabilistic models that generate sequences through a sequence of transitions between states in a Markov chain. Hidden Markov Models are used for clustering, classification, and outlier detection. Therefore, the applicability of these models is very broad in sequence analysis. For example, the clustering approach in Sect. 15.3.4.2 uses Hidden Markov Models as a subroutine. This section will use outlier detection as a specific application of HMM to facilitate understanding. In Sect. 15.6.5, it will also be shown how HMM may be used for classification.

So how are Hidden Markov Models different from the Markovian techniques introduced earlier in this chapter? Each state in the Markovian techniques introduced earlier in this chapter is well defined and is based on the last k positions of the sequence. This state is also directly visible to the user because it is defined by the latest sequence combination of length k. Thus, the generative behavior of the Markovian model is always known *deterministically*, in terms of the correspondence between states and sequence positions for a *particular* input string.

In a Hidden Markov Model, the states of the system are *hidden* and not directly visible to the user. Only a sequence of (typically) discrete observations is visible to the user that is generated by symbol emissions from the states after each transition. The generated sequence of symbols corresponds to the application-specific sequence data. In many cases, the states may be defined (during the modeling process) on the basis of an *understanding* of how the underlying system behaves, though the precise sequence of transitions may not be known to the analyst. This is why such models are referred to as "*hidden*."

Each state in an HMM is associated with a *set of emission probabilities* over the symbol Σ. In other words, a visit to the state j leads to an emission of one of the symbols $\sigma_i \in \Sigma$ with probability $\theta^j(\sigma_i)$. Correspondingly, a sequence of transitions in an HMM corresponds to an *observed data sequence*. Hidden Markov Models may be considered a kind of mixture model of the type discussed in Chap. 6, in which the different components of the mixture are not independent of one another, but are related through sequential transitions. Thus, each state is analogous to a component in the multidimensional mixture model discussed in Chap. 6. Each symbol generated by this model is analogous to a data point generated by the multidimensional mixture model. Furthermore, unlike multidimensional mixture models, the successive generation of individual data items (sequence symbols) are also not independent of one another. This is a natural consequence of the fact that the successive states emitting the data items are dependent on one another with the use of probabilistic transitions. Unlike multidimensional mixture models, Hidden Markov Models are designed for sequential data that exhibits temporal correlations.

To better explain Hidden Markov Models, an illustrative example will be used for the specific problem of using HMMs for anomaly detection. Consider the scenario where a set of students register for a course and generate a sequence corresponding to the grades received in each of their weekly assignments. This grade is drawn from the symbol set $\Sigma = \{A, B\}$. *The model created by the analyst is that* the class contains students who, at any given time, are either *doers* or *slackers* with different grade-generation probabilities. A student in a *doer* state may sometimes transition to a *slacker* state and vice versa. These represent the two states in the system. Weekly home assignments are handed out to each student and are graded with one of the symbols from Σ. This results in a *sequence* of grades for each student, and it represents the only *observable* output for the analyst. The state of a student represents only a *model* created by the analyst to *explain* the grade sequences and is, therefore, not observable in of itself. It is important to understand that if this model is a poor reflection of the true generative process, then it will impact the quality of the learning process.

Assume that a student in a *doer* state is likely to receive an A grade in a weekly assignment with 80% probability and a B with 20% probability. For *slackers*, these probability

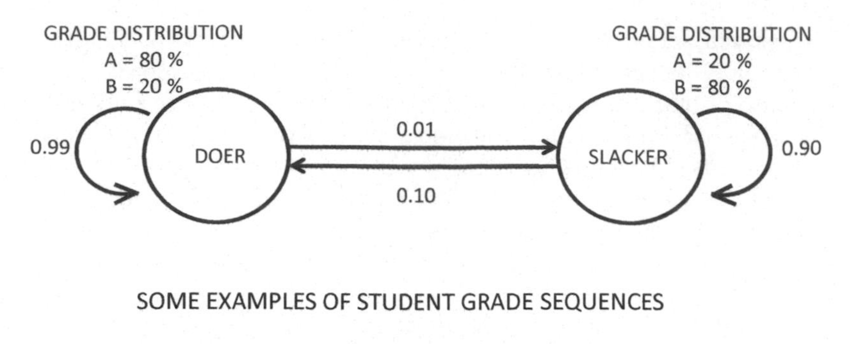

Figure 15.6: Generating grade sequences from a Hidden Markov Model

values are reversed. Although these probabilities are explicitly specified here for illustrative purposes, they need to be *learned* or *estimated* from the observed grade sequences for the different students and are not known *a priori*. The precise status (state) of any student in a given week is not known to the analyst at any given time. These grade sequences are, in fact, the only *observable* outputs for the analyst. Therefore, from the perspective of the analyst, this is a *Hidden* Markov Model, which generates the sequences of grades from an *unknown* sequence of states, representing the state transitions of the students. The precise sequence of transitions between the states can be only *estimated* for a particular observed sequence.

The two-state Hidden Markov Model for the aforementioned example is illustrated in Fig. 15.6. This model contains two states, denoted by *doer* and *slacker*, that represent the state of a student in a particular week. It is possible for a student to transition from one state to another each week, though the likelihood of this is rather low. It is assumed that set of initial state probabilities governs the *a priori* distribution of *doer*s and *slacker*s. This distribution represents the a priori understanding about the students when they join the course. Some examples of *typical sequences* generated from this model, along with their rarity level, are illustrated in Fig. 15.6. For example, the sequence AAABAAAABAAAA is most likely generated by a student who is consistently in a *doer* state, and the sequence BBBBABBBABBBB is most likely generated by a student who is consistently in *slacker* state. The second sequence is typically rarer than the first because the population mostly contains[3] *doers*. The sequence AAABAAABBABBB corresponds to a *doer* who eventually transitions into a *slacker*. This case is even rarer because it requires a transition from the *doer* state to a *slacker* state, which has a very low probability. The sequence ABABABABABABA is *extremely anomalous* because it does not represent temporally consistent *doer* or *slacker* behavior that is implied by the model. Correspondingly, such a sequence has very low probability of fitting the model.

A larger number of states in the Markov Model can be used to encode more complex scenarios. It is possible to encode domain knowledge with the use of states that describe different generating scenarios. In the example discussed earlier, consider the case that *doer*s sometimes slacks off for short periods and then return to their usual state. Alternatively,

[3]The assumption is that the initial set of state probabilities are approximately consistent with the steady state behavior of the model for the particular set of transition probabilities shown in Fig. 15.6.

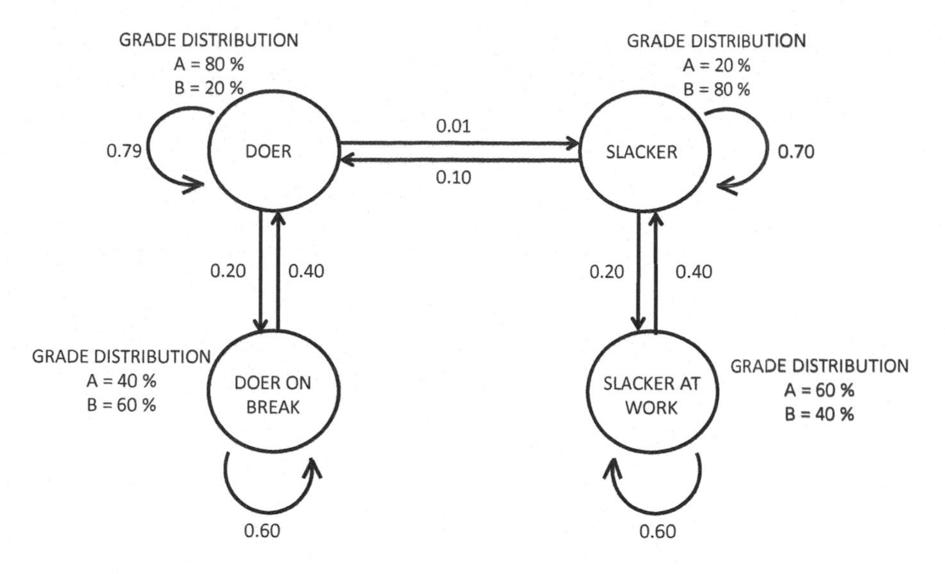

Figure 15.7: Extending the model in Fig. 15.6 with two more states provides greater expressive power for modeling sequences

slackers may sometimes become temporarily inspired to be *doers*, but may eventually return to what they are best at. Such episodes will result in local portions of the sequence that are distinctive from the remaining sequence. These scenarios can be captured with the four-state Markov Model illustrated in Fig. 15.7. The larger the number of states, the more complex the scenarios that can be captured. Of course, more training data is required to learn the (larger number of) parameters of such a model, or this may result in overfitting. For smaller data sets, the transition probabilities and symbol-generation probabilities are not estimated accurately.

15.5.1 Formal Definition and Techniques for HMMs

In this section, Hidden Markov Models will be formally introduced along with the associated training methods. It is assumed that a Hidden Markov Model contains n states denoted by $\{s_1 \dots s_n\}$. The symbol set from which the observations are generated is denoted by $\Sigma = \{\sigma_1 \dots \sigma_{|\Sigma|}\}$. The symbols are generated from the model by a sequence of transitions from one state to the other. Each visit to a state (including self-transitions) generates a symbol drawn from a categorical[4] probability distribution on Σ. The symbol emission distribution is specific to each state. The probability $P(\sigma_i|s_j)$ that the symbol σ_i is generated from state s_j is denoted by $\theta^j(\sigma_i)$. The probability of a transition from state s_i to s_j is denoted by p_{ij}. The initial state probabilities are denoted by $\pi_1 \dots \pi_n$ for the n different states. The topology of the model can be expressed as a network $G = (M, A)$, in which M is the set of states $\{s_1 \dots s_n\}$. The set A represents the possible transitions between the states. In the most common scenario, where the architecture of the model is constructed with a domain-specific understanding, the set A is not the complete network. In cases where domain-specific knowledge is not available, the set A may correspond to the complete

[4]HMMs can also generate continuous time series, though they are less commonly used in timeseries analysis.

network, including self-transitions. *The goal of training the HMM model is to learn the initial state probabilities, transition probabilities, and the symbol emission probabilities from the training database* $\{T_1 \dots T_N\}$. Three methodologies are commonly leveraged in creating and using a Hidden Markov Model:

- *Training:* Given a set of training sequences $T_1 \dots T_N$, estimate the model parameters, such as the initial probabilities, transition probabilities, and symbol emission probabilities with an Expectation-Maximization algorithm. The Baum–Welch algorithm is used for this purpose.

- *Evaluation:* Given a test sequence V (or comparison unit U_i), determine the probability that it fits the HMM. This is used to determine the anomaly scores. A recursive forward algorithm is used to compute this.

- *Explanation:* Given a test sequence V, determine the most likely sequence of states that generated this test sequence. This is helpful for providing an understanding of why a sequence should be considered an anomaly (in outlier detection) or belong to a specific class (in data classification). The idea is that the states correspond to an intuitive understanding of the underlying system. In the example of Fig. 15.6, it would be useful to know that an observed sequence is an anomaly because of the unusual oscillation of a student between *doer* and *slacker* states. This can provide the *intensional knowledge* for understanding the state of a system. This most likely sequence of states is computed with the Viterbi algorithm.

Since the description of the training procedure relies on technical ideas developed for the evaluation method, we will deviate from the natural order of presentation and present the training algorithms last. The evaluation and explanation techniques will assume that the model parameters, such as the transition probabilities, are already available from the training phase.

15.5.2 Evaluation: Computing the Fit Probability for Observed Sequence

One approach for determining the fit probability of a sequence $V = a_1 \dots a_m$ would be to compute all the n^m possible sequences of states (paths) in the HMM, and compute the probability of each, based on the observed sequence, symbol-generation probabilities, and transition probabilities. The sum of these values can be reported as the fit probability. Obviously, such an approach is not practical because it requires the enumeration of an exponential number of possibilities.

This computation can be greatly reduced by recognizing that the fit probability of the first r symbols (and a fixed value of the rth state) can be recursively computed in terms of the corresponding fit probability of first $(r-1)$ observable symbols (and a fixed $(r-1)$th state). Specifically, let $\alpha_r(V, s_j)$ be the probability that the first r symbols in V are generated by the model, and the last state in the sequence is s_j. Then, the recursive computation is as follows:

$$\alpha_r(V, s_j) = \sum_{i=1}^{n} \alpha_{r-1}(V, s_i) \cdot p_{ij} \cdot \theta^j(a_r)$$

This approach recursively sums up the probabilities for all the n different paths for different penultimate nodes. The aforementioned relationship is iteratively applied for $r = 1 \dots m$. The probability of the first symbol is computed as $\alpha_1(V, s_j) = \pi_j \cdot \theta^j(a_1)$ for initializing the recursion. This approach requires $O(n^2 \cdot m)$ time. Then, the overall probability is computed by summing up the values of $\alpha_m(V, s_j)$ over all possible states s_j. Therefore, the final fit $F(V)$ is computed as follows:

$$F(V) = \sum_{j=1}^{n} \alpha_m(V, s_j)$$

This algorithm is also known as the *Forward Algorithm.* Note that the fit probability has a direct application to many problems, such as classification and anomaly detection, depending upon whether the HMM is constructed in supervised or unsupervised fashion. By constructing separate HMMs for each class, it is possible to test the better-fitting class for a test sequence. The fit probability is useful in problems such as data clustering, classification and outlier detection. In data clustering and classification, the fit probability can be used to model the probability of a sequence belonging to a cluster or class, by creating a group-specific HMM. In outlier detection, it is possible to determine poorly fitting sequences with respect to a global HMM and report them as anomalies.

15.5.3 Explanation: Determining the Most Likely State Sequence for Observed Sequence

One of the goals in many data mining problems is to provide an explanation for why a sequence fits part (e.g. class or cluster) of the data, or does not fit the whole data set (e.g. outlier). Since the sequence of (hidden) generating states often provides an intuitive explanation for the observed sequence, it is sometimes desirable to determine the *most likely sequence of states* for the observed sequence. The Viterbi algorithm provides an efficient way to determine the most likely state sequence.

One approach for determining the most likely state path of the test sequence $V = a_1 \ldots a_m$ would be to compute all the n^m possible sequences of states (paths) in the HMM, and compute the probability of each of them, based on the observed sequence, symbol-generation probabilities, and transition probabilities. The maximum of these values can be reported as the most likely path. Note that this is a similar problem to the fit probability except that it is needed to determine the *maximum* fit probability, rather than the *sum* of fit probabilities, over all possible paths. Correspondingly, it is also possible to use a similar recursive approach as the previous case to determine the most likely state sequence.

Any subpath of an optimal state path must also be optimal for generating the corresponding subsequence of symbols. This property, in the context of an optimization problem of sequence selection, normally enables dynamic programming methods. The best possible state path for generating the first r symbols (with the rth state fixed to j) can be recursively computed in terms of the corresponding best paths for the first $(r-1)$ observable symbols and different penultimate states. Specifically, let $\delta_r(V, s_j)$ be the probability of the best state sequence for generating the first r symbols in V and also ending at state s_j. Then, the recursive computation is as follows:

$$\delta_r(V, s_j) = MAX_{i=1}^{n} \delta_{r-1}(V, s_i) \cdot p_{ij} \cdot \theta^j(a_r)$$

This approach recursively computes the maximum of the probabilities of all the n different paths for different penultimate nodes. The approach is iteratively applied for $r = 1 \ldots m$. The first probability is determined as $\delta_1(V, s_j) = \pi_j \cdot \theta^j(a_1)$ for initializing the recursion. This approach requires $O(n^2 \cdot m)$ time. Then, the final best path is computed by using the maximum value of $\delta_m(V, s_j)$ over all possible states s_j. This approach is, essentially, a dynamic programming algorithm. In the anomaly example of student grades, an oscillation between *doer* and *slacker* states will be discovered by the Viterbi algorithm as the causality for outlier behavior. In a clustering application, a consistent presence in the *doer* state will explain the cluster of diligent students.

15.5.4 Training: Baum–Welch Algorithm

The problem of learning the parameters of an HMM is a very difficult one, and no known algorithm is guaranteed to determine the global optimum. However, options are available to determine a reasonably effective solution in most scenarios. The Baum–Welch algorithm is one such method. It is also known as the *Forward-backward* algorithm, and it is an application of the EM approach to the generative Hidden Markov Model. First, a description of training with the use of a single sequence $T = a_1 \ldots a_m$ will be provided. Then, a straightforward generalization to N sequences $T_1 \ldots T_N$ will be discussed.

Let $\alpha_r(T, s_j)$ be the *forward* probability that the first r symbols in a sequence T of length m are generated by the model, and the last symbol in the sequence is s_j. Let $\beta_r(T, s_j)$ be the *backward* probability that the portion of the sequence after *and not including the* r*th position* is generated by the model, *conditional on the fact that* the state for the rth position is s_j. Thus, the forward and backward probability definitions are not symmetric. The forward and backward probabilities can be computed from model probabilities in a way similar to the evaluation procedure discussed above in Sect. 15.5.2. The major difference for the backward probabilities is that the computations start from the end of the sequence in the backward direction. Furthermore, the probability value $\beta_{|T|}(T, s_j)$ is initialized to 1 at the bottom of the recursion to account for the difference in the two definitions. Two additional probabilistic quantities need to be defined to describe the EM algorithm:

- $\psi_r(T, s_i, s_j)$: Probability that the rth position in sequence T corresponds to state s_i, the $(r+1)$th position corresponds to s_j.

- $\gamma_r(T, s_i)$: Probability that the rth position in sequence T corresponds to state s_i.

The EM procedure starts with a random initialization of the model parameters and then iteratively estimates $(\alpha(\cdot), \beta(\cdot), \psi(\cdot), \gamma(\cdot))$ from the model parameters, and vice versa. Specifically, the iteratively executed steps of the EM procedure are as follows:

- (E-step) Estimate $(\alpha(\cdot), \beta(\cdot), \psi(\cdot), \gamma(\cdot))$ from currently estimated values of the model parameters $(\pi(\cdot), \theta(\cdot), p_{\cdot\cdot})$.
- (M-step) Estimate model parameters $(\pi(\cdot), \theta(\cdot), p_{\cdot\cdot})$ from currently estimated values of $(\alpha(\cdot), \beta(\cdot), \psi(\cdot), \gamma(\cdot))$.

It now remains to explain how each of the above estimations is performed. The values of $\alpha(\cdot)$ and $\beta(\cdot)$ can be estimated using the forward and backward procedures, respectively. The forward procedure is already described in the evaluation section, and the backward procedure is analogous to the forward procedure, except that it works backward from the end of the sequence. The value of $\psi_r(T, s_i, s_j)$ is equal to $\alpha_r(T, s_i) \cdot p_{ij} \cdot \theta^j(a_{r+1}) \cdot \beta_{r+1}(T, s_j)$ because the sequence-generation procedure can be divided into three portions corresponding to that up to position r, the generation of the $(r+1)$th symbol, and the portion after the

$(r+1)$th symbol. The estimated values of $\psi_r(T, s_i, s_j)$ are normalized to a probability vector by ensuring that the sum over different pairs $[i, j]$ is 1. The value of $\gamma_r(T, s_i)$ is estimated by summing up the values of $\psi_r(T, s_i, s_j)$ over fixed i and varying j. This completes the description of the E-step.

The re-estimation formulas for the model parameters in the M-Step are relatively straightforward. Let $I(a_r, \sigma_k)$ be a binary indicator function, which takes on the value of 1 when the two symbols are the same, and 0 otherwise. Then the estimations can be performed as follows:

$$\pi(j) = \gamma_1(T, s_j), \; p_{ij} = \frac{\sum_{r=1}^{m-1} \psi_r(T, s_i, s_j)}{\sum_{r=1}^{m-1} \gamma_r(T, s_i)}$$

$$\theta^i(\sigma_k) = \frac{\sum_{r=1}^{m} I(a_r, \sigma_k) \cdot \gamma_r(T, s_i)}{\sum_{r=1}^{m} \gamma_r(T, s_i)}$$

The precise derivations of these estimations, on the basis of expectation-maximization principles, may be found in [327]. This completes the description of the M-step.

As in all EM methods, the procedure is applied iteratively to convergence. The approach can be generalized easily to N sequences by applying the steps to each of the sequences, and averaging the corresponding model parameters in each step.

15.5.5 Applications

Hidden Markov Models can be used for a wide variety of sequence mining problems, such as clustering, classification, and anomaly detection. The application of HMM to clustering has already been described in Sect. 15.3.4.2 of this chapter. The application to classification will be discussed in Sect. 15.6.5 of this chapter. Therefore, this section will focus on the problem of anomaly detection.

In theory, it is possible to compute anomaly scores directly for the test sequence V, once the training model has been constructed from the sequence database $\mathcal{D} = T_1 \ldots T_N$. However, as the length of the test sequence increases, the robustness of such a model diminishes because of the increasing noise resulting from the curse of dimensionality. Therefore, the comparison units (either extracted from the test sequence or specified by the domain expert), are used for computing the anomaly scores of windows of the sequence. The anomaly scores of the different windows can then be combined together by using a simple function such as determining the number of anomalous window units in a sequence.

Some methods also use the Viterbi algorithm on the test sequence to mine the most likely state sequence. In some domains, it is easier to determine anomalies in terms of the state sequence rather than the observable sequence. Furthermore, low transition probabilities on portions of the state sequence provide anomalous localities of the observable sequence. The downside is that the most likely state sequence may have a very low probability of matching the observed sequence. Therefore, the estimated anomalies may not reflect the true anomalies in the data when an *estimated* state sequence is used for anomaly detection. The real utility of the Viterbi algorithm is in providing an *explanation* of the anomalous behavior of sequences in terms of the intuitively understandable states, rather than anomaly score quantification.

15.6 Sequence Classification

It is assumed that a set of N sequences, denoted by $S_1 \ldots S_N$, is available for building the training model. Each of these sequences is annotated with a class label drawn from $\{1 \ldots k\}$. This training data is used to construct a model that can predict the label of unknown test sequences. Many modeling techniques, such as nearest neighbor classifiers, rule-based methods, and graph-based methods, are common to timeseries and discrete sequence classification because of the temporal nature of the two data types.

15.6.1 Nearest Neighbor Classifier

The nearest neighbor classifier is used frequently for different data types, including discrete sequence data. The basic description of the nearest neighbor classifier for multidimensional data may be found in Sect. 10.8 of Chap. 10. For discrete sequence data, the main difference is in the similarity function used for nearest neighbor classification. Similarity functions for discrete sequences are discussed in Sects. 3.4.1 and 3.4.2 of Chap. 3. The basic approach is the same as in multidimensional data. For any test instance, its k-nearest neighbors in the training data are determined. The dominant label from these k-nearest neighbors is reported as the relevant one for the test instance. The optimal value of k may be determined by using leave-one-out cross-validation. The effectiveness of the approach is highly sensitive to the choice of the distance function. The main problem is that the presence of noisy portions in the sequences throw off global similarity functions. A common approach is to use keyword-based similarity in which n-grams are extracted from the string to create a vector-space representation. The nearest-neighbor (or any other) classifier can be constructed with this representation.

15.6.2 Graph-Based Methods

This approach is a *semisupervised* algorithm because it combines the knowledge in the training and test instances for classification. Furthermore, the approach is transductive because out-of-sample classification of test instances is generally not possible. Training and testing instances must be specified at the same time. The use of similarity graphs for semisupervised classification was introduced in Sect. 11.6.3 of Chap. 11. Graph-based methods can be viewed as general semisupervised meta-algorithms that can be used for any data type. The basic approach constructs a similarity graph from *both* the training and test instances. A graph $G = (V, A)$ is constructed, in which a node in V corresponds to each of the training and test instances. A subset of nodes in G is labeled. These correspond to instances in the training data, whereas the unlabeled nodes correspond to instances in the test data. Each node in V is connected to its k-nearest neighbors with an undirected edge in A. The similarity is computed using any of the distance functions discussed in Sects. 3.4.1 and 3.4.2 of Chap. 3. In the resulting network, a subset of the nodes are labeled, and the remaining nodes are unlabeled. The specified labels of nodes in N are then used to predict labels for nodes where they are unknown. This problem is referred to as *collective classification*. Numerous methods for collective classification are discussed in Sect. 19.4 of Chap. 19. The derived labels on the nodes are then mapped back to the data objects. As in the case of nearest-neighbor classification, the effectiveness of the approach is sensitive to the choice of distance function used for constructing the graph.

15.6.3 Rule-Based Methods

A major challenge in sequence classification is that many parts of the sequence may be noisy and not very relevant to the class label. In some cases, a short pattern of two symbols may be relevant to classification, whereas in other cases, a longer pattern of many symbols may be discriminative for classification. In some cases, the discriminative patterns may not even occur contiguously. This issue was discussed in the context of timeseries classification in Sect. 14.7.2.1 of Chap. 14. However, discrete sequences can be converted into binary timeseries sequences, with the use of binarization. These binary timeseries can be converted to multidimensional wavelet representations. This is described in detail in Sect. 2.2.2.6, and the description is repeated here for completeness.

The first step is to convert the discrete sequence to a *set* of (binary) time series, where the number of time series in this set is equal to the number of distinct symbols. The second step is to map each of these time series into a multidimensional vector using the wavelet transform. Finally, the features from the different series are combined to create a single multidimensional record. A rule-based classifier is constructed on this multidimensional representation.

To convert a sequence to a binary time series, one can create a binary string, in which each position value denotes whether or not a particular symbol is present at a position. For example, consider the following nucleotide sequence drawn on four symbols:

```
ACACACTGTGACTG
```

This series can be converted into the following set of four binary time series corresponding to the symbols A, C, T, and G, respectively.

```
10101000001000
01010100000100
00000010100010
00000001010001
```

A wavelet transformation can be applied to each of these series to create a multidimensional set of features. The features from the four different series can be appended to create a single numeric multidimensional record. After a multidimensional representation has been obtained, any rule-based classifier can be utilized. Therefore, the overall approach for data classification is as follows:

1. Generate wavelet representation of each of the N sequences to create N numeric multidimensional representations, as discussed above.

2. Discretize wavelet representation to create categorical representations of the timeseries wavelet transformation. Thus, each categorical attribute value represents a range of numeric values of the wavelet coefficients.

3. Generate a set of rules using any rule-based classifier described in Sect. 10.4 of Chap. 10. The patterns on the left-hand represent the patterns of different granularities defined by the combination of wavelet coefficients on the left-hand side.

When the rule set has been generated, it can be used to classify arbitrary test sequences by first transforming the test sequence to the same wavelet-based numeric multidimensional representation. This representation is used with the fired rules to perform the classification.

Such methods are discussed in Sect. 10.4 of Chap. 10. It is not difficult to see that this approach is a discrete version of the rule-based classification of time series, as presented in Sect. 14.7.2.1 of Chap. 14.

15.6.4 Kernel Support Vector Machines

Kernel support vector machines can construct classifiers with the use of kernel similarity between training and test instances. As discussed in Sect. 10.6.4 of Chap. 10, kernel support vector machines do not need the feature values of the records, as long as the kernel-based similarity $K(Y_i, Y_j)$ between any pair of data objects are available. In this case, these data objects are strings. Different kinds of kernels are very popular for string classification.

15.6.4.1 Bag-of-Words Kernel

In the bag-of-words kernel, the string is treated as a bag of alphabets, with a frequency equal to the number of alphabets of each type in the string. This can be viewed as the vector-space representation of a string. Note that a text document is also a string, with an alphabet size equal to the lexicon. Therefore, the transformation $\Phi(\cdot)$ can be viewed as almost equivalent to the vector-space transformation for a text document. If $\overline{V(Y_i)}$ be the vector-space representation of a string, then the kernel similarity is equal to the dot product between the corresponding vector space representations.

$$\Phi(Y_i) = \overline{V(Y_i)}$$
$$K(Y_i, Y_j) = \Phi(Y_i) \cdot \Phi(Y_j) = \overline{V(Y_i)} \cdot \overline{V(Y_j)}$$

The main disadvantage of the kernel is that it loses all the positioning information between the alphabets. This can be an effective approach for cases where the alphabet size is large. An example is text, where the alphabet (lexicon) size is of a few hundred thousand words. However, for smaller alphabet sizes, the information loss can be too significant for the resulting classifier to be useful.

15.6.4.2 Spectrum Kernel

The bag-of-words kernel loses *all* the sequential information in the strings. The spectrum kernel addresses this issue by extracting k-mers from the strings and using them to construct the vector-space representation. The simplest spectrum kernel pulls out all k-mers from the strings and builds a vector space representation from them. For example, consider the string `ATGCGATGG` constructed on the alphabet $\Sigma = \{A, C, T, G\}$. Then, the corresponding spectrum representation for $k = 3$ is as follows:

```
ATG(2), TGC(1), GCG(1), CGA(1), GAT(1), TGG(1)
```

The values in the brackets correspond to the frequencies in the vector-space representation. This corresponds to the feature map $\Phi(\cdot)$ used to define the kernel similarity.

It is possible to enhance the spectrum kernel further by adding a *mismatch neighborhood* to the kernel. Thus, instead of adding only the extracted k-mers to the feature map, we add all the k-mers that are m mismatches away from the k-mer. For example, at a mismatch level of $m = 1$, the following k-mers are added to the feature map for each instance of ATG:

CTG, GTG, TTG, ACG, AAG, AGG, ATC, ATA, ATT

This procedure is repeated for each element in the k-mer, and each of the neighborhood elements are added to the feature map. is procedure is repeated for each element in the k-mer, and each of the neighborhood elements are added to the feature map. The dot product is performed on this expanded feature map $\Phi(\cdot)$. The rationale for adding mismatches is to allow for some noise in the similarity computation. The bag-of-words kernel can be viewed as a special case of the spectrum kernel with $k = 1$ and no mismatches. The spectrum kernel can be computed efficiently with the use of either the trie or the suffix tree data structure. Pointers to such efficient computational methods are provided in the bibliographic notes. One advantage of spectrum kernels is that they can compute the similarity between two strings in an intuitively appealing way, even when the lengths of the two strings are widely varying.

15.6.4.3 Weighted Degree Kernel

The previous two kernel methods directly define a feature map $\Phi(\cdot)$ explicitly that largely ignores the ordering between the different k-mers. The weighted degree kernel directly defines $K(Y_i, Y_j)$, without explicitly defining a feature map $\Phi(\cdot)$. This approach is in the spirit of exploiting the full power of kernel methods. Consider two strings Y_i and Y_j of the same length n. Let $KMER(Y_i, r, k)$ represent the k-mer extracted from Y_i starting from position r. Then, the weighted degree kernel computes the kernel similarity as the number of times the k-mers of a *maximum* specified length, in the two strings *at exactly corresponding positions*, match perfectly. Thus, unlike spectrum kernels, k-mers of varying lengths are used, and the contribution of a particular length s is weighted by coefficient β_s. In other words, weighted degree kernel of order k is defined as follows:

$$K(Y_i, Y_j) = \sum_{s=1}^{k} \beta_s \sum_{r=1}^{n-s+1} I(KMER(Y_i, r, s) = KMER(Y_j, r, s)) \tag{15.9}$$

Here, $I(\cdot)$ is an indicator function that takes on the value of 1 in case of a match, and 0 otherwise. One drawback of the weighted degree kernel over the spectrum kernel, is that it requires the two strings Y_i and Y_j to be of equal length. This can be partially addressed by allowing shifts in the matching process. Pointers to these enhancements may be found in the bibliographic notes.

15.6.5 Probabilistic Methods: Hidden Markov Models

Hidden Markov Models are an important tool that are utilized in a wide variety of tasks in sequence analysis. It has already been shown earlier in this chapter, in Sects. 15.3.4.2 and 15.5, how Hidden Markov Models can be utilized for both clustering and outlier detection. In this section, the use of Hidden Markov Models for sequence classification will be leveraged. In fact, the most common use of HMMs is for the problem of classification. HMMs are very popular in computational biology, where they are used for protein classification.

The basic approach for using HMMs for classification is to create a separate HMM for each of the classes in the data. Therefore, if there are a total of k classes, this will result in k different Hidden Markov Models. The Baum–Welch algorithm, described in Sect. 15.5.4, is used to train the HMMs for each class. For a given test sequence, the fit of each of the k models to the test sequence is determined using the approach described in Sect. 15.5.2. The best matching class is reported as the relevant one. The overall approach for training and testing with HMMs may be described as follows:

1. (Training) Use Baum–Welch algorithm of Sect. 15.5.4 to construct a separate HMM model for each of the k classes.

2. (Testing) For a given test sequence Y, determine the fit probability of the sequence to the k different Hidden Markov Models, using the evaluation procedure discussed in Sect. 15.5.2. Report the class, for which the corresponding HMM has the highest fit probability to the test sequence.

Many variations of this basic approach have been used to achieve different trade-offs between effectiveness and efficiency. The bibliographic notes contain pointers to some of these methods.

15.7 Summary

Discrete sequence mining is closely related to timeseries data mining, just as categorical data mining is closely related to numeric data mining. Therefore, many algorithms are very similar across the two domains. The work on discrete sequence mining originated in the field of computational biology, where DNA strands are encoded as strings.

The problem of sequential pattern mining discovers frequent sequences from a database of sequences. The *GSP* algorithm for frequent sequence mining is closely based on the *Apriori* method. Because of the close relationship between the two problems, most algorithms for frequent pattern mining can be generalized to discrete sequence mining in a relatively straightforward way.

Many of the methods for multidimensional clustering can be generalized to sequence clustering, as long as an effective similarity function can be defined between the sequences. Examples include the k-medoids method, hierarchical methods, and graph-based methods. An interesting line of work converts sequences to bags of k-grams. Text-clustering algorithms are applied to this representation. In addition, a number of specialized methods, such as *CLUSEQ*, have been developed. Probabilistic methods use a mixture of Hidden Markov Models for sequence clustering.

Outlier analysis for sequence data is similar to that for timeseries data. Position outliers are determined using Markovian models for probabilistic prediction. Combination outliers can be determined using distance-based, frequency-based, or Hidden Markov Models. Hidden Markov Models are a very general tool for sequence analysis and are used frequently for a wide variety of data mining tasks. HMMs can be viewed as mixture models, in which each state of the mixture is sequentially dependent on the previous states.

Numerous techniques from multidimensional classification can be adapted to discrete sequence classification. These include nearest neighbor methods, graph-based methods, rule-based methods, Hidden Markov Models, and Kernel Support Vector Machines. Numerous string kernels have been designed for more effective sequence classification.

15.8 Bibliographic Notes

The problem of sequence mining has been studied extensively in computational biology. The classical book by Gusfield [244] provides an excellent introduction of sequence mining algorithms from the perspective of computational biology. This book also contains an excellent survey on most of the other important similarity measures for strings, trees, and graphs. String indexing is discussed in detail in this work. The use of transformation rules for timeseries similarity has been studied in [283, 432]. Such rules can be used to create edit distance-like measures for continuous time series. Methods for the string edit distance are proposed in [438]. It has been shown in [141], how the L_p-norm may be combined with the edit distance. Algorithms for the longest common subsequence problem may be found in [77, 92, 270, 280]. A survey of these algorithms is available in [92]. Numerous other measures for timeseries and sequence similarity may be found in [32]. Timeseries and discrete sequence similarity measures are discussed in detail in Chap. 3 of this book, and in an earlier tutorial by Gunopulos and Das [241]. In the context of biological data, the BLAST system [73] is one of the most popular alignment tools.

The problem of mining sequential patterns was first proposed in [59]. The *GSP* algorithm was also proposed in the same work. The *GSP* algorithm is a straightforward modification of the *Apriori* algorithm. Most frequent pattern mining problems can be extended to sequential pattern mining because of the relationship between the two models. Subsequently, most of the algorithms discussed in Chap. 4 on frequent pattern mining have been generalized to sequential pattern mining. Savasere et al.'s vertical data structures [446] have been generalized to *SPADE* [535], and the *FP-growth* algorithm has been generalized to *PrefixSpan* [421]. The *TreeProjection* algorithm has also been generalized to sequential pattern mining [243]. Both *PrefixSpan* and the *TreeProjection*-based methods are based on combining database projection with exploration of the candidate search space with the use of an enumeration tree. The description of this chapter is a simplified and generalized description of these two related works [243, 421]. Methods for finding constraint-based sequences are discussed in [224, 346]. A recent survey on sequential pattern mining may be found in [392].

The problem of sequence data clustering has been studied extensively. A detailed survey on clustering sequence data, in the context of the biological domain, may be found in [32]. The *CLUSEQ* algorithm is described in detail in [523]. The Probabilistic Suffix Trees, used by *CLUSEQ*, are discussed in the same work. The earliest frequent sequence-based approach for clustering was proposed in [242]. The *CONTOUR* method for frequent sequence mining was proposed in [505]. This method uses a combination of frequent sequence mining and microclustering to create clusters from the sequences. The use of Hidden Markov Models for discrete sequence clustering is discussed in [474].

A significant amount of work has been done on the problem of temporal outlier detection in general, and discrete sequences in particular. A general survey on temporal outlier detection may be found in [237]. The book [5] contains chapters on temporal and discrete sequence outlier detection. A survey on anomaly detection in discrete sequences was presented in [132]. Two well-known techniques that use Markovian techniques for finding position outliers are discussed in [387, 525]. Combination outliers typically use windowing techniques in which comparison units are extracted from the sequence for the purposes of analysis [211, 274]. The information-theoretic measures for compression-based similarity were proposed in [311]. The frequency-based approach for determining the surprise level of comparison units is discussed in [310]. The *TARZAN* algorithm, proposed in this work, uses suffix trees for efficient computation. A general survey on Hidden Markov Models may be found in [327].

The problem of sequence classification is addressed in detail in the surveys [33, 516]. The use of wavelet methods for sequence classification was proposed in [51]. A description of a variety of string kernels for SVM classification is provided in [85]. The use of Hidden Markov Models for string classification is discussed in [327].

15.9 Exercises

1. Consider the sequence $ABCDDCBA$, defined on the alphabet $\Sigma = \{A, B, C, D\}$. Compute the vector-space representation for all k-mers of length 1, and the vector-space representation of all k-mers of length 2. Repeat the process for the sequence $CCCCDDDD$.

2. Implement the *GSP* algorithm for sequential pattern mining.

3. Consider a special case of the sequential pattern mining problem where elements are always singleton items. What difference would it make to (a) *GSP* algorithm, and (b) algorithms based on the candidate tree.

4. Discuss the generalizability of k-medoids and graph-based methods for clustering of arbitrary data types.

5. The chapter introduces a number of string kernels for classification with SVMs. Discuss some other data mining applications you can implement with string kernels.

6. Discuss the similarity and differences between Markovian models for discovering position outliers in sequential data, with autoregressive models for discovering point outliers in timeseries data.

7. Write a computer program to determine all maximal frequent subsequences from a collection using *GSP*. Implement a program to express the sequences in a database in terms of these subsequences, in vector space representation. Implement a k-means algorithm on this representation.

8. Write a computer program to determine position outliers using order-1 Markovian Models.

9. Consider the discrete sequence ACGTACGTACGTACGTATGT. Construct an order-1 Markovian model to determine the position outliers. Which positions are found as outliers?

10. For the discrete sequence of Exercise 9, determine all subsequences of length 2. Use a frequency-based approach to assign combination outlier scores to subsequences. Which subsequences should be considered combination outliers?

11. Write a computer program to learn the state transition and symbol emission probabilities for a Hidden Markov Model. Execute your program using the sequence of Exercise 9.

12. Compute the kernel similarity between the two sequences in Exercise 1 with a bag-of-words kernel and a spectrum kernel in which sequences of length 2 are used.

13. What is the maximum number of possible sequential patterns of length at most k, where the alphabet size is $|\Sigma|$. Compare this with frequent pattern mining. Which is larger?

14. Suppose that the speed of an athlete on a racetrack probabilistically depends upon whether the day is cold, moderate, or hot. Accordingly, the athlete runs a race that is graded either Fast (F), Slow (S), or Average (A). The weather on a particular day probabilistically depends on the weather on the previous day. Suppose that you have a sequence of performances of the athlete on successive days in the form of a string, such as $FSFAAF$. Construct a Hidden Markov Model that explains the athlete's performance, without any knowledge of the weather on those days.

Chapter 16

Mining Spatial Data

"Time and space are modes by which we think and not conditions in which we live."—Albert Einstein

16.1 Introduction

Spatial data arises commonly in geographical data mining applications. Numerous applications related to meteorological data, earth science, image analysis, and vehicle data are spatial in nature. In many cases, spatial data is integrated with temporal components. Such data is referred to as *spatiotemporal data*. Some examples of applications in which spatial data arise, are as follows:

1. *Meteorological data:* Quantifications of important weather characteristics, such as the temperature and pressure, are typically measured at different geographical locations. These can be analyzed to discover interesting events in the underlying data.

2. *Mobile objects:* Moving objects typically create trajectories. Such trajectories can be analyzed for a wide variety of insights, such as characteristic trends, or anomalous paths of objects.

3. *Earth science data:* The land cover types at different spatial locations may be represented as behavioral attributes. Anomalies in such patterns provide insights about anomalous trends in human activity, such as deforestation or other anomalous vegetation trends.

4. *Disease outbreak data:* Data about disease outbreaks are often aggregated by spatial locations such as ZIP code and county. The analysis of trends in such data can provide information about the causality of the outbreaks.

5. *Medical diagnostics:* Magnetic resonance imaging (MRI) and positron emission tomography (PET) scans are spatial data in 2 or 3 dimensions. The detection of unusual

C. C. Aggarwal, *Data Mining: The Textbook*, DOI 10.1007/978-3-319-14142-8_16

localized regions in such data can help in detecting diseases such as brain tumors, the onset of Alzheimer disease, and multiple sclerosis lesions. In general, any form of image data may be considered spatial data. The analysis of shapes in such data is of considerable importance in a variety of applications.

6. *Demographic data:* Demographic (behavioral) attributes such as age, sex, race, and salary can be combined with spatial (contextual) attributes to provide insights about the demographic patterns in distributions. Such information can be useful for target-marketing applications.

Most forms of spatial data may be classified as a contextual data type, in which the attributes are partitioned into contextual attributes and behavioral attributes. This partitioning is similar to that in time series and discrete sequence data:

- *Contextual attribute(s):* These represent the attributes that provide the *context* in which the measurements are made. In other words, the contextual attributes provide the reference points at which the behavioral values are measured. In most cases, the contextual attributes contain the spatial coordinates of a data point. In some cases, the contextual attribute might be a logical location, such as a building or a state. In the case of spatiotemporal data, the contextual attributes may include time. For example, in an application in which the sea surface temperatures are measured at sensors at specific locations, the contextual attributes may include both the position of the sensor, and the time at which the measurement is made.

- *Behavioral attribute(s):* These represent the behavioral values at the reference points. For example, in the aforementioned sea surface temperature application, these correspond to the temperature attribute values.

In most forms of spatial data, the spatial attributes are contextual, and may optionally include temporal attributes. An exception is the case of trajectory data, in which the spatial attributes are behavioral, and time is the only contextual attribute. In fact, trajectory data can be considered equivalent to multivariate time series data. This equivalence is discussed in greater detail in Sect. 16.3.

This chapter separately studies cases where the spatial attributes are contextual, and those in which the spatial attributes are behavioral. The latter case typically corresponds to trajectory data, in which the contextual attribute corresponds to time. Thus, trajectory data is a form of spatiotemporal data. In other forms of spatiotemporal data, both spatial and temporal attributes are contextual.

This chapter is organized as follows. Section 16.2 addresses data mining scenarios in which the spatial attributes are contextual. In this context, the chapter studies several important problems, such as pattern mining, clustering, outlier detection, and classification. Section 16.3 discusses algorithms for mining trajectory data. Section 16.4 discusses the summary.

16.2 Mining with Contextual Spatial Attributes

In many forms of meteorological data, a variety of (behavioral) attributes, such as temperature, pressure, and humidity, are measured at different spatial locations. In these cases, the spatial attributes are contextual. An example of sea surface temperature contour charts is illustrated in Fig. 16.1. The different shades in the chart represent the different sea surface temperatures. These correspond to the values of the behavioral attributes at different spatial locations.

Another example is the case of image data, where the intensity of an image is measured in pixels. Such data is often used to capture diagnostic images. Examples of PET scans for a cognitively healthy person and an Alzheimer's patient are illustrated in Fig. 16.2. In this case, the values of the pixels represent the behavioral attributes, and the spatial locations of these pixels represent the contextual attributes. The behavioral attributes in spatial data may present themselves in a variety of ways, depending on the application domain:

1. For some types of spatial data, such as images, the analysis may be performed on the contour of a specific shape extracted from the data. For example, in Fig. 16.3, the contour of the insect may be extracted and analyzed with respect to other images in the data.

2. For other types of spatial data, such as meteorological applications, the behavioral attributes may be abstract quantities such as temperature. Therefore the analysis can be performed in terms of the trends on these abstract quantities. In such cases, the spatial data needs to be treated as a contextual data type with multiple reference points corresponding to spatial coordinates. Such an analysis is generally more complex.

The specific choice of data mining methodology often depends on the application at hand. Both these forms of data are often transformed into other data types such as time series or multidimensional data before analysis.

16.2.1 Shape to Time Series Transformation

In many spatial data sets such as images, the data may be dominated by a particular shape. The analysis of such shapes is challenging because of the variations in sizes and orientations. One common technique for analyzing spatial data is to transform it into a different format that is much easier to analyze. In particular, the contours of a shape are often transformed to time series for further analysis. For example, the contours of the insect shapes in Fig. 16.3 are difficult to analyze directly because of their complexity. However, it is possible to create a representation that is friendly to data processing by transforming them into time series.

A common approach is to use the distance from the centroid to the boundary of the object, and compute a sequence of real numbers derived in a clockwise sweep of the boundary. This yields a time series of real numbers, and is referred to as the *centroid distance signature*. This transformation can be used to map the problem of mining shapes to that of mining time series. The latter domain is much easier to analyze. For example, consider the elliptical shape illustrated in Fig. 16.4a. Then, the time series representing the distance from the centroid, using 360 different equally spaced angular samples, is illustrated in Fig. 16.4b. Note that the contextual attribute here is the number of degrees, but one can "pretend" that this represents a timestamp. This facilitates the use of all the powerful data mining techniques available for time series analysis. In this case, the sample points are started at one of the major axes of the ellipse. If the sample point starts at a different position, or if the shape is rotated (with the same angular starting point), then this causes a cyclic translation of the time series. This is quite important because the precise orientation of a shape may not be known in advance. For example, the shapes in Figs. 16.3b and c are rotated from the

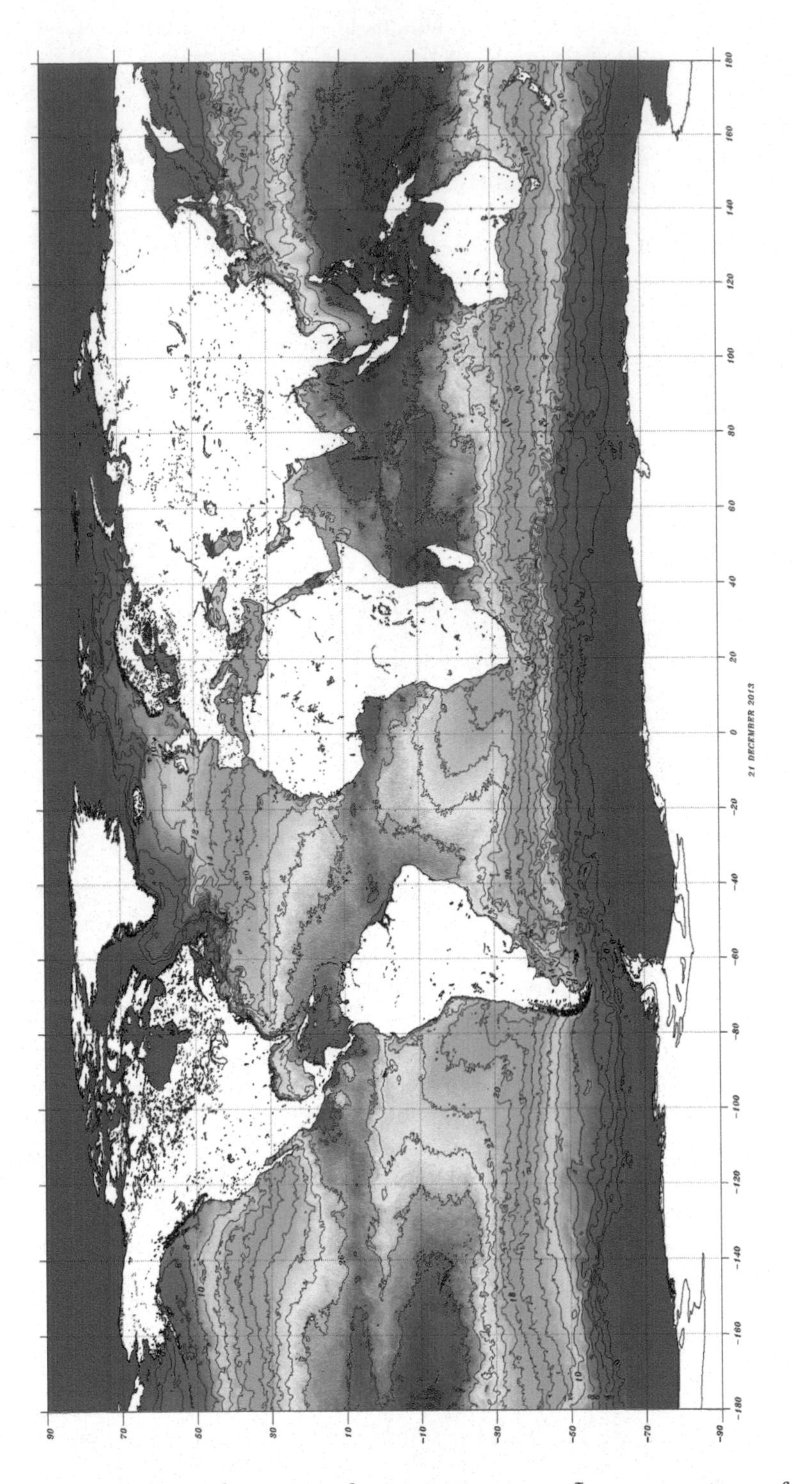

Figure 16.1: Contour charts for sea surface temperatures: Image courtesy of the NOAA Satellite and Information Service

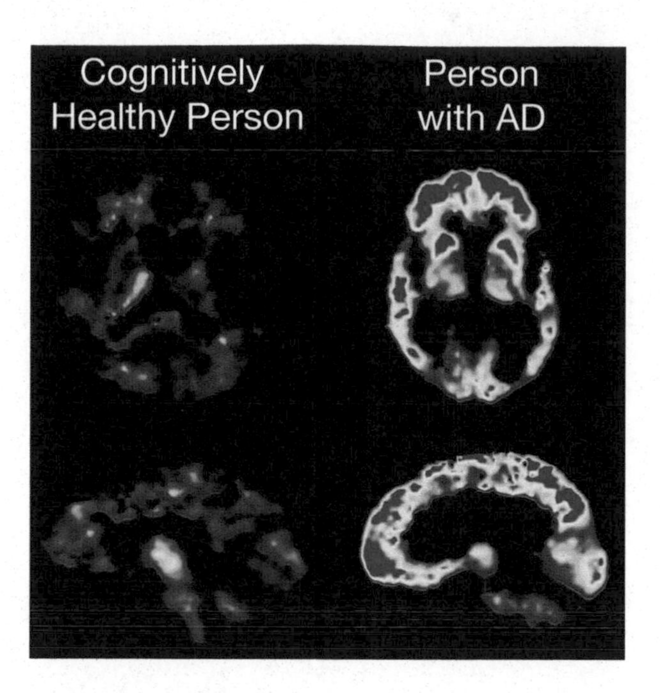

Figure 16.2: PET Scans of the brain of a cognitively healthy person versus an Alzheimer's patient. (Image courtesy of the National Institute on Aging/National Institutes of Health)

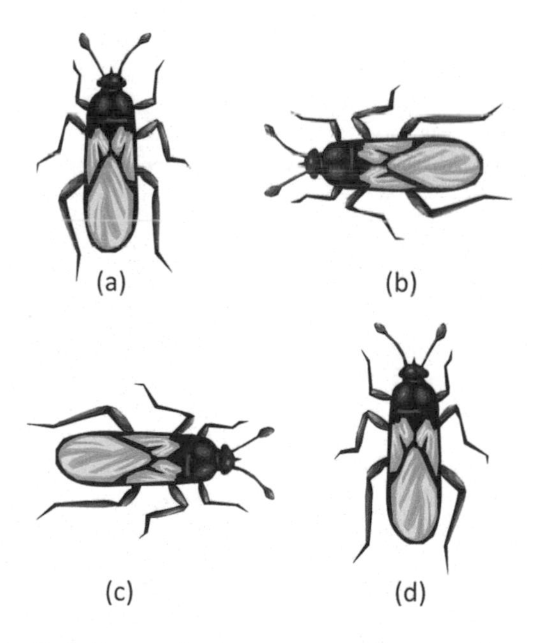

Figure 16.3: Rotation and mirror image effects on shape matching

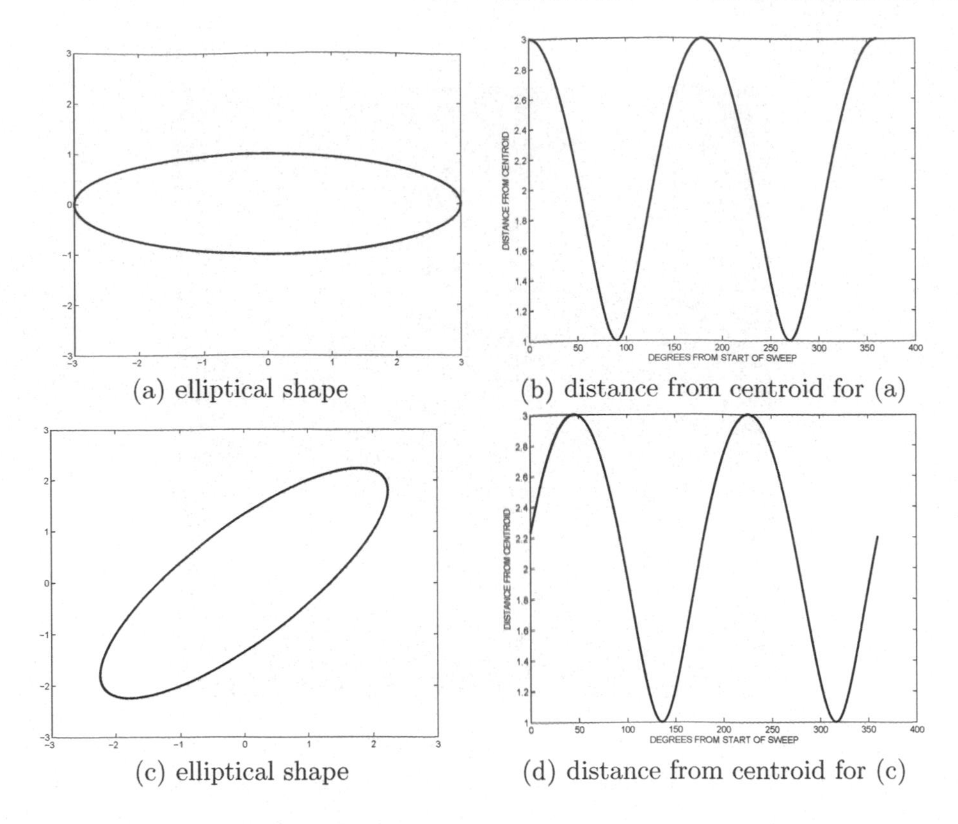

(a) elliptical shape (b) distance from centroid for (a)

(c) elliptical shape (d) distance from centroid for (c)

Figure 16.4: Conversion from shapes to time series

shape of Fig. 16.3a. The shape in Fig. 16.3d is a mirror image of the shape of Fig. 16.3a. While rotations result in cyclic translations, mirror images result in a reversal of the series.

Figure 16.4c represents a rotation of the shape of Fig. 16.4a by 45°. Correspondingly, the time series representation in Fig. 16.4d is a (cyclic) translation of time series representation in Fig. 16.4b. Similarly, the mirror image of a shape corresponds to a reversal of the time series. It will be evident later that the impact of rotation or mirror images needs to be explicitly incorporated into the distance or similarity function for the application at hand. After the time series has been extracted, it may be normalized in different ways, depending on the needs of the application:

- If no normalization is performed, then the data mining approach is sensitive to the absolute sizes of the underlying objects. This may be the case in many medical images such as MRI scans, in which all spatial objects are drawn to the same scale.

- If all time series values are multiplicatively scaled down by the same factor to unit mean, such an approach will allow the matching of shapes of different sizes, but discriminate between different levels of relative variations in the shapes. For example, two ellipses with very different ratios of the major and minor axes will be discriminated well.

- If all time series are standardized to zero mean and unit variance, then such an approach will match shapes where *relative* local variations in the shape are similar, but the overall shape may be quite different. For example, such an approach will not discriminate very well between two ellipses with very different ratios of the major and minor axes, but will discriminate between two such shapes with different relative local deviations in the boundaries. The only exception is a circular shape that appears as a straight line. Furthermore, noise effects in the contour will be differentially enhanced in shapes that are less elongated. For example, for two ellipses with similar noisy deviations at the boundaries, but different levels of elongation (major to minor axis ratio), the overall shape of the time series will be similar, but the local noisy deviations in the extracted time series will be *differentially* suppressed in the elongated shape. This can sometimes provide a distorted picture from the perspective of shape analysis. A perfectly circular shape may show unstable and large noisy deviations in the extracted time series because of trivial variations such as image rasterization effects. Thus, the usual mean and variance normalization of time series analysis often leads to unintended results.

In general, it is advisable to select the normalization method in an application-specific way. After the shapes have been converted to time series, they can be used in the context of a wide variety of applications. For example, motifs in the time series correspond to frequent contours in the spatial shapes. Similarly, clusters of similar shapes may be discovered by determining clusters in the time series. Similar observations apply to the problems of outlier detection and classification.

16.2.2 Spatial to Multidimensional Transformation with Wavelets

For data types such as meteorological data in which behavioral attribute values vary across the entire spatial domain, a contour-based shape may not be available for analysis. Therefore, the shape to time series transformation is not appropriate in these cases.

Wavelets are a popular method for the transformation of time series data to multidimensional data. Spatial data shares a number of similarities with time series data. Time series data has a single contextual attribute (time) along which a behaviorial attribute (e.g., temperature) may exhibit a smooth variation. Correspondingly, spatial data has two contextual attributes (spatial coordinates), along which a behavioral attribute (e.g., sea surface temperature) may exhibit a smooth variation. Because of this analogy, it is possible to generalize the wavelet-based approach to the case of multiple contextual attributes with appropriate modifications.

Assume that the spatial data is represented in the form of a 2-dimensional grid of size $q \times q$. Thus, each coordinate of the grid contains an instance of the behavioral attribute, such as the temperature. As discussed for the time series case in Sect. 2.4.4.1 of Chap. 2, differencing operations are applied over contiguous segments of the time series by successive division of the time series in hierarchical fashion. The corresponding basis vectors have $+1$ and -1 at the relevant positions. The 2-dimensional case is completely analogos, where contiguous *areas* of the spatial grid are used for successive divisions. These divisions are alternately performed along the different axes. The corresponding basis vectors are 2-dimensional *matrices* of size $q \times q$ that regulate how the differencing operations are performed. An example of how sea surface temperatures in a spatial data set may be converted to a multidimensional representation is provided in Fig. 16.5. This will result in a total of q^2 wavelet coefficients, though only the large coefficients need to be retained for analysis. A more detailed description of the generation of the spatial wavelet coefficients may be found in Sect. 2.4.4.1 of Chap. 2. The aforementioned description is for the case of a single behavioral attribute and multiple contextual attributes (spatial coordinates). Multiple behavioral attributes can also

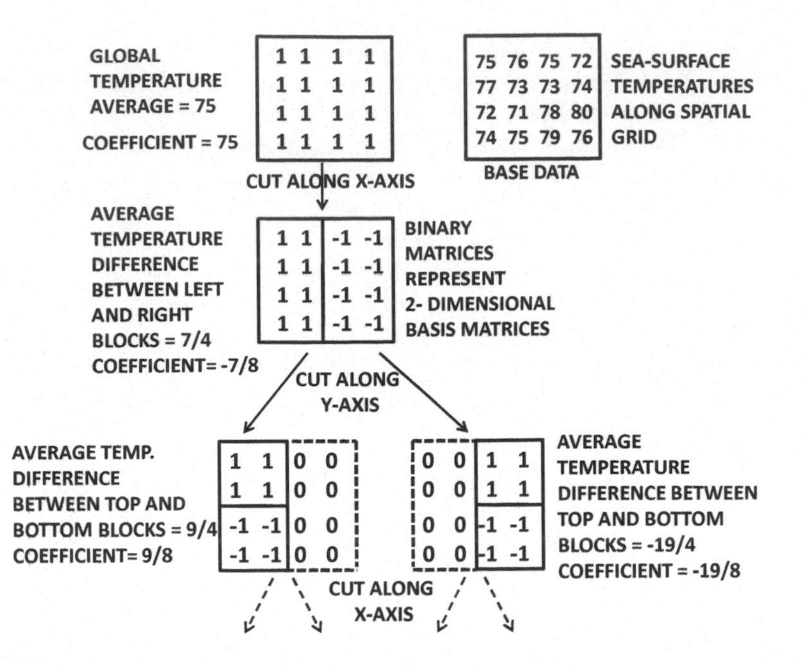

Figure 16.5: Illustration of the top levels of the wavelet decomposition for spatial data in a grid containing sea surface temperatures (Fig. 2.7 of Chap. 2 revisited)

be addressed by performing the decomposition separately for each behavioral attribute, and creating a separate set of dimensions for each behavioral attribute.

Like the time series wavelet, the spatial wavelet is a multiresolution representation. Trends at different levels of spatial granularity are represented in the coefficients. Higher-level coefficients represent trends in larger spatial areas, whereas lower-level coefficients represent trends in smaller spatial areas. Therefore, this approach is very powerful, and has broad usability for many spatial applications. Spatial wavelets can be used effectively for many image clustering and classification applications where (contextual) spatial data can be converted to (noncontextual) multidimensional data. Once the transformation has been performed, virtually all the multidimensional methods discussed in Chaps. 4 to 11 can be used on this representation. Such an approach opens the door to the use of a wide array of multidimensional data mining methods.

16.2.3 Spatial Colocation Patterns

In this problem, the contextual attributes are spatial and the behavioral attributes are typically boolean and nonspatial. Non-boolean behavioral attributes can be addressed with the use of type conversion via discretization or binarization. The goal of spatial colocation pattern mining is to discover combinations of features occurring at the same spatial location. Consider an ecology data set, where one has behavioral attributes such as fire ignition source, needle vegetation type, and a drought indicator. The spatial colocation of these features may often be a risk factor for forest fires. Therefore, the discovery of such patterns is useful in the context of data mining analysis. In many cases, a spatial event indicator of interest (e.g., disease outbreak, vegetation event, or climate event) is added to the other behavioral attributes. The discovery of useful patterns that include this indicator of interest can be used for discovering event causality. This problem is also closely related to rule-based spatial classification, where the likelihood of the event occurring in previously unseen test regions can be estimated from the resulting patterns.

One challenge in the mining process is that the different behavioral attributes may be derived from different data sources, and therefore may not have precisely the same value of the contextual (spatial) attribute in their measurements. Therefore, proper data preprocessing is crucial. The data can be homogenized by partitioning the spatial region into smaller regions. For each of these regions, each behavioral attribute's value is derived heuristically from the values in the original data source. For example, if the boolean attribute has a value of 1 more than predefined fraction of the time in a spatial region, then its value is set to 1. The contextual (spatial) attribute can be set to the centroid of that region. The mining can be performed on this preprocessed data. The overall approach is as follows:

1. Preprocess the data to create the behavioral attribute values at the same set of spatial locations.

2. For each spatial location, create a transaction containing the corresponding combination of boolean values.

3. Use any frequent pattern mining algorithm to discover the relevant patterns in these transactions.

4. For each discovered pattern, map it back to the spatial regions containing the pattern. Cluster the relevant spatial regions for each pattern, if necessary for summarization.

In cases where a particular behavioral attribute is an event of interest (e.g., disease outbreak), the transactions containing values of 0 and 1, respectively, for this attribute can be separately processed to discover two sets of patterns on the other behavioral attributes. The differences between these two sets of patterns can provide insights into discriminative factors for the event of interest at each spatial location. Such patterns are also useful for spatial classification of previously unseen test regions. This approach is identical to that of associative classifiers in Chap. 10.

This model can also address time-changing data in a seamless way. In such cases, the time becomes another contextual attribute in addition to the spatial attributes. Patterns can be discovered at different temporal snapshots using the aforementioned methodology. The key changes in these patterns over time can provide insights into the nature of the spatial evolution.

16.2.4 Clustering Shapes

In many applications, it may be desirable to cluster similar shapes prior to analysis. It is assumed that a database of N shapes is available and that a total of k groups of similar shapes need to be created. This can be a useful preprocessing task in many shape categorization applications. The conversion of a shape to a time series (Sect. 16.2.1) is the appropriate approach in this scenario. Many of the time series clustering algorithms discussed in Sect. 14.5 of Chap. 14 may be used effectively, once the shape has been converted to a time series. The k-medoids, hierarchical, and graph-based methods are particularly suitable because they require only the design of an appropriate similarity function for the corresponding time series. This is an issue that will be discussed in more detail later. The main steps of shape-based clustering are as follows:

1. Use the centroid-based sweep method discussed in Sect. 16.2.1 to convert each shape into a time series. This results in a database of N different time series.

2. Use any time series clustering algorithm, such as hierarchical, k-medoids or graph-based method on time series data as discussed in Sect. 14.5 of Chap. 14. This will cluster the N time series into k groups.

3. Map the k groups of time series clusters to k groups of shape clusters, by mapping each time series into its relevant shape.

The aforementioned clustering algorithm depends only on the choice of the distance function. Any of the time series measures discussed in Sect. 3.4.1 of Chap. 3 may be used, depending on the desired degree of error tolerance or distortion (warping) allowed in the matching. Another important issue is the adjustment of the distance function with the varying rotations of the different shapes. In the following, the Euclidean distance will be used as an example, although the general principle can be applied to any distance function.

It is evident from the example of Fig. 16.4 that *a rotation of the shape leads to a linear cyclic shifting of the time series generated by using the distances of the centroid of the shape to the contours of the shape.* For a time series of length n denoted by $a_1 a_2 \dots a_n$, a cyclic translation by i units leads to the time series $a_{i+1} a_{i+2} \dots a_n a_1 a_2 \dots a_i$. Then, the *rotation invariant Euclidean distance* $RIDist(T_1, T_2)$ between two time series $T_1 = a_1 \dots a_n$ and $T_2 = b_1 \dots b_n$ is given by the minimum distance between T_1 and all possible rotational translations of T_2 (or vice versa). Therefore, the rotation-invariant distance is expressed as follows:

$$RIDist(T_1, T_2) = \min_{i=1}^{n} \sum_{j=1}^{n} (a_j - b_{1+(j+i) \bmod n})^2.$$

In general, if a cyclic shift of the time series T_2 by i units is denoted by T_2^i, then the rotation invariant distance, using any distance function $Dist(T_1, T_2)$ may be expressed as follows:

$$RIDist(T_1, T_2) = \min_{i=1}^{n} Dist(T_1, T_2^i). \tag{16.1}$$

Note that the reversal of a time series corresponds to the mirror image of the underlying shape. Therefore, mirror images can also be addressed using this approach, by incorporating the reversals of the series (and its rotations) in the distance function. This will increase the computation by a factor of 2. The precise choice of distance function used is highly application-specific, depending on whether rotations or mirror image conversions are required.

16.2.5 Outlier Detection

In the context of spatial data, outliers can be either point outliers and shape outliers. These two kinds of outliers are also encountered in time series data, and in discrete sequences. In the case of spatial data, these two kinds of outliers are defined as follows:

1. *Point outliers:* These outliers are defined on a single spatial object with a variety of spatial and behavioral attributes. For example, a weather map is a spatial object that contains both spatial locations, and environmental measurements (behavioral values) at these locations. Abrupt changes in the behavioral attributes that violate spatial continuity provide useful information about the underlying contextual anomalies. For example, consider a meteorological application in which sea surface temperatures and

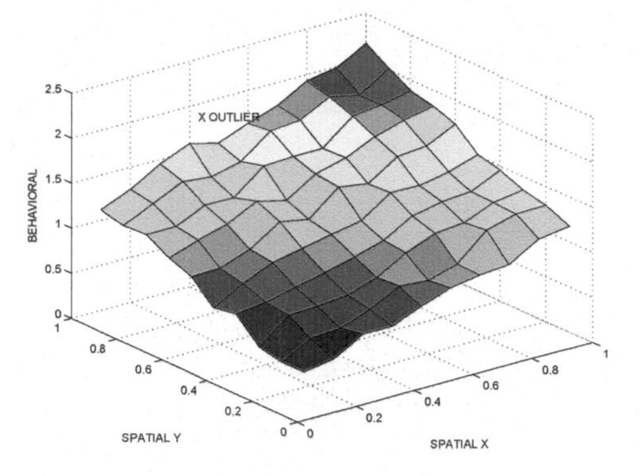

Figure 16.6: Example of point outlier for spatial data

pressure are measured. Unusually high sea surface temperature in a very small localized region is a hot-spot that may be the result of volcanic activity under the surface. Similarly, unusually low or high pressure in a small localized region may suggest the formation of hurricanes or cyclones. In all these cases, spatial continuity is violated by the attribute of interest. Such attributes are often tracked in meteorological applications on a daily basis. An example of a point outlier for spatial data is illustrated in Fig. 16.6.

2. *Shape outliers:* The application settings for these kinds of outliers are quite different. These outliers are defined in a database of multiple shapes. For example, the shapes may be extracted from the different images. In such cases, the unusual shapes in the different objects need to be reported as outliers.

This chapter studies both the aforementioned formulations.

16.2.5.1 Point Outliers

Neighborhood-based algorithms are generally used for discovering point outliers. In these algorithms, abrupt changes in the spatial neighborhood of a data point are used to diagnose outliers. Therefore, the first step is to define the concept of a spatial neighborhood. The behavioral values within the spatial neighborhood of a given data point are combined to create an *expected value* of the behavioral attribute. This expected value is then used to compute the deviation of the data point from the expected value. This provides an outlier score. This definition of point outliers in spatial data is similar to that in time series data.

Intuitively, it is unusual for the behavioral attribute value to vary abruptly within a small spatial locality. For example, a sudden variation of the temperature within a small spatial locality will be detected by this method. The neighborhood may be defined in many different ways:

- *Multidimensional neighborhoods:* In this case, the neighborhoods are defined with the use of multidimensional distances between data points. This approach is appropriate when the contextual attributes are defined as coordinates.

- *Graph-based neighborhoods:* In this case, the neighborhoods are defined by linkage relationships between spatial objects. Such neighborhoods may be more useful in cases where the location of the spatial objects may not correspond to exact coordinates (e.g., county or ZIP code). In such cases, graph-based representations provide a more general modeling tool.

Both multidimensional and graph-based methods will be discussed in the following sections.

Multidimensional Methods

While traditional multidimensional methods can also be used to detect outliers in spatial data, such methods do not distinguish between the contextual and the behavioral attributes. Therefore, such methods are not optimized for outlier detection in spatial data. This is because the (contextual) spatial attributes should be treated differently from the behavioral attributes. The basic idea is to adapt the k-nearest neighbor outlier detection methods to the case of spatial data.

The spatial neighborhood of the data is defined with the use of multidimensional distances on the spatial (contextual) attributes. Thus, the contextual attributes are used for determining the k nearest neighbors. The average of the behavioral attribute values provides an expected value for the behavioral attribute. The difference between the expected and true value is used to predict outliers. A variety of distance functions can be used on the multidimensional spatial data for the determination of proximity. The choice of the distance function is important because it defines the choice of the neighborhood that is used for computing the deviations in behavioral attributes. For a given spatial object o, with behavioral attribute value $f(o)$, let $o_1 \dots o_k$ be its k-nearest neighbors. Then, a variety of methods may be used to compute the predicted value $g(o)$ of the object o. The most straightforward method is the mean:

$$g(o) = \sum_{i=1}^{k} f(o_i)/k$$

Alternatively, $g(o)$ may be computed as the median of the surrounding values of $f(o_i)$, to reduce the impact of extreme values. Then, for each data object o, the value of $f(o) - g(o)$ represents a deviation from predicted values. The extreme values among these deviations may be computed using a variety of methods for univariate extreme value analysis. These are discussed in Chap. 8. The resulting extreme values are reported as outliers.

Graph-Based Methods

In graph-based methods, spatial proximity is modeled with the use of links between nodes in a graph representation of the spatial region. Thus, nodes are associated with behavioral attributes, and strong variations in the behavioral attribute across neighboring nodes are recognized as outliers. Graph-based methods are particularly useful when the individual nodes are not associated with point-specific coordinates, but they may correspond to regions of arbitrary shape. In such cases, the links between nodes correspond to neighborhood relationships between the different regions. Graph-based methods define spatial relationships in a more general way because semantic relationships can also be used to define neighborhoods. For example, two objects could be connected by an edge if they are in the same *semantic* location, such as a building, restaurant, or an office. In many applications, the links may be weighted on the basis of the strength of the proximity relationship. For example, consider a disease outbreak application in which the spatial objects correspond to county regions. In such a case, the strength of the links could correspond to the length of the boundary between two regions. Multidimensional data is a special case, where links correspond to distance-based proximity. Thus, graph representations allow more generic interpretations of the contextual attribute.

Let S be the set of neighbors of a given node i. Then, the concept of spatial continuity can be used to create a *predicted* value of the behavioral attribute based on those of its (spatial) neighbors. The strength of the links between i and its neighbors can also be used to compute the predicted values as either the weighted mean or median on the behavioral attribute of the k nearest spatial neighbors. For a given spatial object o with the behavioral attribute value $f(o)$, let $o_1 \ldots o_k$ be its k linked neighbors based on the relationship graph. Let the weight of the link (o, o_i) be $w(o, o_i)$. Then, the linkage-based weighted mean may be used to compute the predicted value $g(o)$ of the object o.

$$g(o) = \frac{\sum_{i=1}^{k} w(o, o_i) \cdot f(o_i)}{\sum_{i=1}^{k} w(o, o_i)}$$

Alternatively, the weighted median of the neighbor values may be used for predictive purposes. Since the true value of the behavioral attribute is known, this can be used to model the deviations of the behavioral attributes from their predicted values. As in the case of multidimensional methods, the value of $f(o) - g(o)$ represents a deviation from the predicted values. Extreme value analysis can be used on these deviations to determine the spatial outliers. This process is identical to that in the multidimensional case. The nodes with high values of the normalized deviation may be reported as outliers.

16.2.5.2 Shape Outliers

Shape-based outliers are relatively easy to determine in spatial data, with the use of the transformation from spatial data to time series described in Sect. 16.2.1. After the transformation has been performed, a k-nearest neighbor outlier detector can be applied to the resulting time series. The distance to the kth-nearest neighbor can be reported as the outlier score. A few key issues need to be kept in mind, while computing the outlier score.

1. The distance function needs to be modified to account for the rotation invariance of shape matching. This is achieved by comparing all cyclic shifts of one time series to the other. The rotation invariant distance can be captured by Eq. 16.1.

2. In some applications, mirror image invariance also needs to be accounted for. In such cases, all cyclic shifts and their reversals need to be included in the aforementioned comparison. The outliers are determined with respect to this enhanced database.

While a vanilla k-nearest neighbor detector can determine the outliers correctly, the approach can be made faster by pruning. The basic idea is similar to the *Hotsax* method discussed in Chap. 14, where a nested loop structure is used to maintain the top-n outliers. The outer loop corresponds to the selection of different candidates, and the inner loop corresponds to the computation of the k-nearest neighbors of each of these candidates. The inner loop can be terminated early, when the k-nearest neighbor value is less than the nth best outlier found so far. For optimal performance, the candidates in the outer loop and the computations in the inner loop need to be ordered appropriately.

This ordering is performed as follows. A combination of the SAX representation and LSH-hashing is used to create clusters on the candidates. Candidates which map to clusters with few members are examined first in the outer loop to discover high quality outliers early in the algorithm execution. Objects which appear in the same cluster as the outer loop candidate are examined first in the inner loop to ensure fast termination of the inner loop. This facilitates better pruning performance. The bibliographic notes contain pointers to specific details of the creation of SAX-based clusters in shape outlier detection.

16.2.6 Classification of Shapes

It is assumed that a set of N labeled shapes are used to conduct the training. This trained model is used to perform classification of test instances, for which the label is unknown. The transformation from spatial into time series data is a useful tool for distance-based classification algorithms. As in the case of clustering and outlier detection, the first step of the process is to transform the shapes into time series. This transforms the problem to the time series classification problem. A number of methods for the classification of time series are discussed in Sect. 14.7 of Chap. 14. The main difference is that the rotation invariance of the shapes needs to be accounted for. Any of the distance-based methods proposed in Sect. 14.7 of Chap. 14 for time series classification may be used after the shapes have been transformed into time series. This is because distance-based methods can be easily made rotation-invariant by using Eq. 16.1. The two main distance-based methods for time series classification include the nearest neighbor method and the graph-based collective classification approach. While the nearest-neighbor method is straightforward, the graph-based method is discussed in some detail below.

The graph-based method is transductive because it requires the test instances to be available at the time of training. When a larger number of test instances are available along with the training data, the latter method may be used. Therefore, the different methods may be more suitable in different scenarios. The overall approach for graph-based classification may be described as follows:

1. Transform both the training and test shapes into time series, by using the centroid sweep method described in Sect. 16.2.1.

2. Use any of the distance functions described in Sect. 3.4.1 of Chap. 3 to construct a neighborhood graph on the shapes. If needed, use a rotation-invariant version of the distance function, as discussed in Eq. 16.1. Each shape represents a node, which is connected to its k-nearest neighbors with edges. The labeled shapes correspond to labeled nodes. The collective classification method described in Sect. 19.4 of Chap. 19 is used to assign labels to the unlabeled nodes (i.e., the test shapes).

In some cases, rotation invariance may not be an application-specific need. In such cases, the efficiency of distance computation is improved.

16.3 Trajectory Mining

Trajectory data arises in a wide variety of spatial applications. The proliferation of GPS-enabled devices, such as mobile phones, has enabled the large-scale collection of trajectory data. Trajectory data can be analyzed for a very wide variety of insights, such as determining co-location patterns, clusters and outliers. Trajectory data is different from the other kinds of spatial data discussed in this chapter in the following respects:

1. In the spatial data applications addressed so far in this chapter, spatial attributes are contextual, whereas other types of attributes (e.g., temperature in a meteorological application) are behavioral. In the case of trajectory data, spatial attributes are behavioral.

2. The *only* contextual attribute in trajectory data is time. Therefore, trajectory data can be considered *spatiotemporal data*. While the scenarios discussed in previous sections may also be generalized further by *including* time among the contextual attributes, the spatial attributes are not behavioral in those cases. For example, when sea surface temperatures are tracked over time, both spatial and temporal attributes are contextual.

Trajectory analysis is typically performed in one of two different ways:

1. *Online analysis:* In online analysis, the trajectories are analyzed in *real time*, and the patterns in the trajectories at a given time are most relevant to the analysis.

2. *Shape-based analysis:* In shape-based analysis, the time variable has already been removed from the analysis. For example, two similar trajectories, formed at different periods, can be meaningfully compared to one another. For example, a cluster of trajectories is based on their shape, rather than the simultaneity in their movement.

The two kinds of analysis in trajectory data are similar to time series data. This is not particularly surprising because trajectory data is a form of time series data.

16.3.1 Equivalence of Trajectories and Multivariate Time Series

Trajectory data is a form of multivariate time series data. For a trajectory in two dimensions, the X-coordinate and Y-coordinate of the trajectory form two components of the multivariate series. A 3-dimensional trajectory will result in a trivariate series.

Because of the equivalence between multivariate time series and trajectory data, the transformation can be performed in either direction to facilitate the use of the methods designed for each domain. For example, trajectory mining methods can be utilized for applications that are nonspatial. In particular, any n-dimensional multivariate time series can be converted into trajectory data. In multivariate temporal data, the different behavioral attributes are typically measured with the use of multiple sensors simultaneously. Consider the example of the *Intel Research Berkeley Sensor data* [556] that measures different behavioral attributes, such as temperature, pressure, and voltage, in the Intel Berkeley laboratory over time. For example, the behavior of the temperature and voltage sensors in the same segment of time are illustrated in Figs. 16.7a, b, respectively.

It is possible to visualize the variation of the two behaviorial attributes by eliminating the common time attribute, or by creating a 3-dimensional trajectory containing the time and the other two behaviorial attributes. Examples of such trajectories are illustrated in Fig. 16.7c, d, respectively. The most generic of these trajectories is illustrated in Fig. 16.7d. This figure shows the simultaneous variation of all three attributes. In general, a multivariate time series with n behavioral attributes can be mapped to an $(n + 1)$-dimensional trajectory. Most of the trajectory analysis methods are designed under the assumption of 2 or 3 dimensions, though they can be generalized to n dimensions where needed.

16.3.2 Converting Trajectories to Multidimensional Data

Because of the equivalence between trajectories and multivariate time series, trajectories can also be converted to multidimensional data. This is achieved by using the wavelet transformation on the time series representation of the trajectory. The wavelet transformation for time series is described in detail in Sect. 2.4.4.1 of Chap. 2. In this case, the time series is multivariate, and therefore has two behavioral attributes. The wavelet representation for

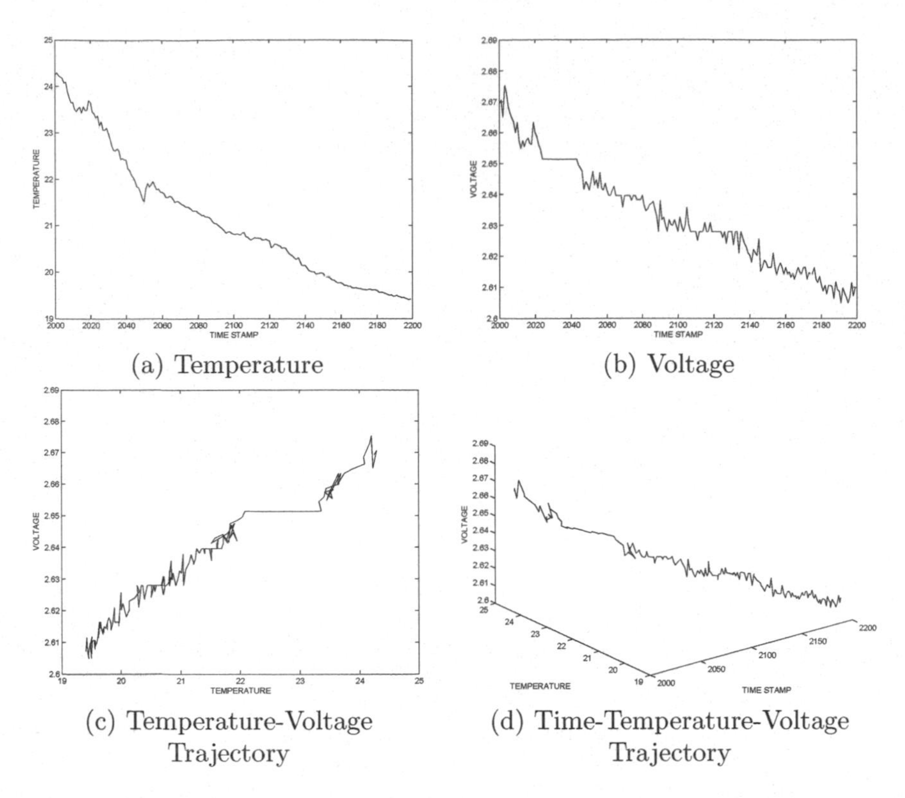

(a) Temperature

(b) Voltage

(c) Temperature-Voltage Trajectory

(d) Time-Temperature-Voltage Trajectory

Figure 16.7: Multivariate time series can be mapped to trajectory data

each of these behavioral attributes is extracted independently. In other words, the time series on the X-coordinate is converted into a wavelet representation, and so is the time series on the Y-coordinate. This yields two multidimensional representations, one of which is for the X-coordinate, and the other is for the Y-coordinate. The dimensions in these two representations are combined to create a single higher-dimensional representation for the trajectory. If desired, only the larger wavelet coefficients may be retained to reduce the dimensionality. The conversion of trajectory data to multidimensional data is an effective way to use the vast array of multidimensional methods for trajectory analysis.

16.3.3 Trajectory Pattern Mining

There are many different ways in which the problem of trajectory pattern mining may be formulated. This is because of the natural complexity of trajectory data that allows for multiple ways of defining patterns. In the following sections, some of the common definitions of trajectory pattern mining will be explored. These definitions are by no means exhaustive, although they do illustrate some of the most important scenarios in trajectory analysis.

16.3.3.1 Frequent Trajectory Paths

A key problem is that of determining frequent sequential paths in trajectory data. To determine the frequent sequential paths from a set of trajectories, the first step is to convert the multidimensional trajectory (with numeric coordinates) to a 1-dimensional discrete

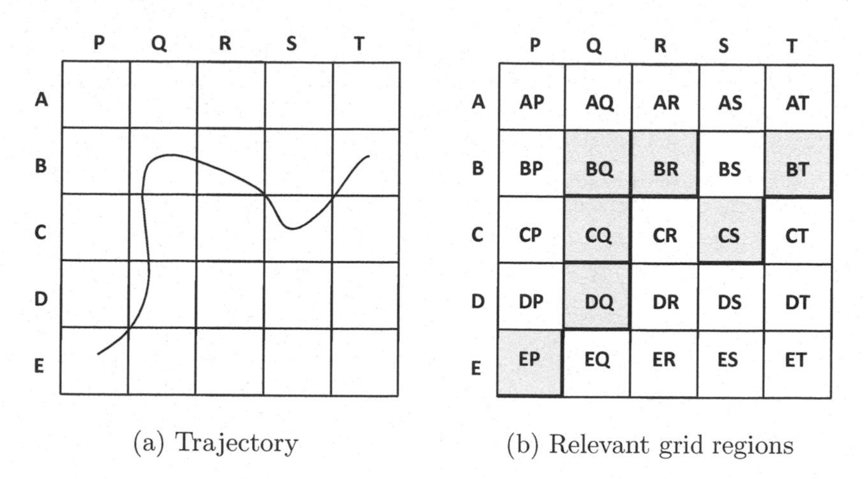

(a) Trajectory (b) Relevant grid regions

Figure 16.8: Grid-based discretization of trajectories

sequence. Once this conversion has been performed, any sequential pattern mining algorithm can be applied to the transformed data.

The most effective way to convert a multidimensional trajectory to a discrete sequence is to use grid-based discretization. In Fig. 16.8a, a trajectory has been illustrated, together with a 4×4 grid representation of the underlying data space. The grid ranges along one of the dimensions are denoted by A, B, C, D, and E. The grid ranges along the other dimension are denoted by P, Q, R, S, and T. The 2-dimensional tiles are denoted by a combination of the ranges along each of the dimensions. For example, the tile AP represents the intersection of the grid range A with the grid range P. Thus, each tile has a distinct (discrete) identifier, and a trajectory can be expressed in terms of the sequence of discrete identifiers through which it passes. The shaded tiles for the trajectory in Fig. 16.8a are illustrated in Fig. 16.8b. The corresponding 1-dimensional sequential pattern is as follows:

$$EP, DQ, CQ, BQ, BR, CS, BT$$

This transformation is referred to as the *spatial tile transformation*. In principle, it is possible to enhance the discretization further, by imposing a minimum time spent in each grid square, though this is not considered here. Consider a database containing N different trajectories. The frequent sequential paths can be determined from these trajectories by using a two-step approach:

1. Convert each of the N trajectories into a discrete sequence, using grid-based discretization.

2. Apply the sequential pattern mining algorithms discussed in Sect. 15.2 of Chap. 15 to discover frequent sequential patterns from the resulting data set.

By incorporating different types of constraints on the sequential pattern mining process, such as time-gap constraints, it is also possible to apply these constraints on the trajectories. One advantage of this transformation-based approach is that it can take advantage of the power of all the different variations of sequential pattern mining. Sequential pattern mining has a rich body of literature on constraint-based methods.

Another interesting aspect is that this formulation can be modified to address situations in which the patterns of movements occur at the *same period in time*. The time period over which the movement occurs is discretized into m periods denoted by $1 \ldots m$. For example, the intervals could be $[8AM, 9AM]$, $[9AM, 10AM]$, $[10AM, 11AM]$, and so on. Thus, for each time-interval, the grid identifier is tagged with the relevant time-period identifier. A time period identifier is tagged to a grid region if a minimum amount of time from that time period range was spent in that region by the trajectory. This results in a set of patterns defined on a new set of discrete symbols of the form $< GridId >:< TimeId >$. In the trajectory of Fig. 16.8a, a possible sequence that appends the time period identifier is as follows:

$$EP : 1, EP : 2, DQ : 2, DQ : 3, DQ : 4, CQ : 5, BQ : 5, BR : 5, CS : 6, CS : 7, BT : 7$$

This transformation is referred to as the *spatiotemporal tile transformation*. Note that this sequence is longer than that in Fig. 16.8b because the trajectory may spend more than one interval in the same grid region. A set of N different sequences are extracted, corresponding to the N different trajectories. The sequential pattern mining can be performed on this new representation. Because of the addition of the time identifiers, the resulting patterns will correspond to *simultaneous* movements in time. Thus, the sequential pattern mining approach has significant flexibility in terms of either detecting patterns of similar shapes, or patterns of simultaneous movements. Furthermore, because the sequential pattern mining formulation does not require successive symbols in a frequent sequential pattern to be contiguous in the original sequence, it can ignore noisy gaps in the underlying trajectories, while mining patterns. Furthermore, by using different constrained sequential pattern mining formulations, different kinds of constrained trajectory patterns can be discovered.

One drawback of the approach is that the granularity level of the discretization may affect the quality of the results. It is possible to address this issue by using a multigranularity discretization of the spatial regions. A different approach for conversion to symbolic representation is the use of spatial clustering on the different temporal snapshots of the object positions. The cluster identifiers of each object over different snapshots may be used to construct its sequence representation. The bibliographic notes contain pointers to several algorithms for transformation and pattern discovery from trajectories. The broader idea of many of these methods is to convert to a symbolic sequence representation for more effective pattern mining.

16.3.3.2 Colocation Patterns

Colocation patterns are designed to discover social connections between the trajectories of different individuals. The basic idea of colocation patterns is that individuals who frequently appear at the same point at the same time are likely to be related to one another. Colocation pattern mining attempts to discover patterns of individuals, rather than patterns of spatial trajectory paths. Because of the complementary nature of this analysis, a *vertical* representation of the sequence database is particularly convenient.

A similar grid discretization (as designed for the case of frequent trajectory patterns) can be used for preprocessing. However, in this case, a somewhat different (vertical) representation is used for the locations of the different individuals in the grid regions at different times. For each grid region and time-interval pair, a list of person identifiers (or trajectory identifiers) is determined. Thus, for the grid region EP and time interval 5, if the persons 3, 9, and 11 are present, then the corresponding set is constructed:

$EP : 5 \Rightarrow \{3, 9, 11\}$

Note that this is an *unordered* set, since it represents the individuals present in a particular $(space, time)$ pair. A similar set can be constructed over all the $(space, time)$ pairs that are populated with at least two individuals. This can be viewed as a *vertical representation* of the sequence database.

Any frequent pattern mining algorithm, discussed in Chap. 4, can be applied to the resulting database of sets. The frequent patterns correspond to the frequent sets of *colocated* individuals. These individuals are often likely to be socially related individuals.

16.3.4 Trajectory Clustering

In the following, a detailed discussion of the different kinds of trajectory clustering algorithms will be provided. Trajectory clustering algorithms are naturally related to trajectory pattern mining because of the close relationship between the two problems. Trajectory clustering methods are of two types.

1. The first type of methods use conventional clustering algorithms, with the use of a distance function between trajectories. Once a distance function has been designed, many different kinds of algorithms, such as k-medoids or graph-based methods, may be used.

2. The second type of methods use data transformation and discretization to convert trajectories into sequences of discrete symbols. Different types of transformations, such as segment extraction or grid-based discretization, may be applied to the trajectories. After the transformation, pattern mining algorithms are applied to the extracted sequence of symbols.

In addition, a number of other *ad hoc* methods have also been designed for trajectory clustering. This section will focus only on the systematic techniques. The bibliographic notes contain pointers to the *ad hoc* methods.

16.3.4.1 Computing Similarity Between Trajectories

A key aspect of trajectory clustering is the ability to compute similarity between different trajectories. At first sight, similarity function computation seems to be a daunting task because of the spatial and temporal aspects of trajectory analysis. However, in practice, similarity computation between trajectories is not very different from that of time series data. As discussed at the beginning of Sect. 16.3, trajectory data is equivalent to multivariate time series data. Several dynamic programming algorithms are discussed in Chap. 3 for similarity computation in univariate time series data. These algorithms can be generalized to the multivariate case. In the following, the extension of the *dynamic time warping* algorithm to the multivariate case will be discussed. A similar approach can be used for other dynamic programming methods. The reader is advised to revisit Sect. 3.4.1.3 of Chap. 3 on dynamic time warping before reading further.

First, the discussion on univariate time series distances will be revisited briefly. Let $DTW(i, j)$ be the optimal distance between the first i and first j elements of two *univariate* time series $\overline{X} = (x_1 \dots x_m)$ and $\overline{Y} = (y_1 \dots y_n)$ respectively. Then, the value of $DTW(i, j)$

is defined recursively as follows:

$$DTW(i,j) = distance(x_i, y_j) + \min \begin{cases} DTW(i, j-1) & \text{repeat } x_i \\ DTW(i-1, j) & \text{repeat } y_j \\ DTW(i-1, j-1) & \text{otherwise} \end{cases} \quad (16.2)$$

In the case of a 2-dimensional trajectory, we have a multivariate time series for each trajectory, corresponding to the two coordinates of each trajectory. Thus, the first trajectory has two coordinates corresponding to $\overline{X1} = (x1_1 \dots x1_m)$ and $\overline{X2} = (x2_1 \dots x2_m)$. The second trajectory has two coordinates, corresponding to $\overline{Y1} = (y1_1 \dots y1_n)$ and $\overline{Y2} = (y2_1 \dots y2_n)$. Let $\overline{X_i} = (x1_i, x2_i)$ represent the 2-dimensional position of the first trajectory at the ith timestamp, and let $\overline{Y_j} = (y1_j, y2_j)$ represent the 2-dimensional position of the second trajectory at the jth timestamp. *Then, the only difference from the case of univariate time series data is the substitution of the 1-dimensional distances in the recursion with 2-dimensional distances.* Therefore, the modified multidimensional DTW recursion $MDTW(i,j)$ is as follows:

$$MDTW(i,j) = distance(\overline{X_i}, \overline{Y_j}) + \min \begin{cases} MDTW(i, j-1) & \text{repeat } \overline{X_i} \\ MDTW(i-1, j) & \text{repeat } \overline{Y_j} \\ MDTW(i-1, j-1) & \text{otherwise.} \end{cases} \quad (16.3)$$

Note that the multidimensional DTW recursion is virtually identical to the univariate case, except for the term $distance(\overline{X_i}, \overline{Y_j})$ that is now a multidimensional distance between spatial coordinates. For example, one might use the Euclidean distances. The simplicity of the generalization is a result of the fact that time warping has little to do with the dimensionality of the time series. All the dimensions in the time series are warped in exactly the same way. Therefore, the 1-dimensional distance in the recursion can be substituted with multidimensional distances. It should also be pointed out that this general principle applies to most of the dynamic programming methods for computing distances between temporal series and sequences.

16.3.4.2 Similarity-Based Clustering Methods

Many conventional clustering methods, such as k-medoids and graph-based methods, are based on the similarity between data objects. Therefore, once a similarity function has been defined, these methods can be used directly for virtually any data type. It should be pointed out that these methods are very popular in different data domains, such as time series data or discrete sequence data. It was shown in Chaps. 14 and 15, how these methods may be used for time series and discrete sequence data, respectively. The approach used here is exactly analogos to the description in these chapters, except that multivariate time series similarity measures are used for computation in this case. The reader is referred to Chap. 6 for the basic description of the k-medoids and graph-based methods, as applied to multidimensional data. The main problem with similarity-based methods, is that they work best only when the trajectory segments are relatively short. For longer trajectories, it becomes more difficult to compute the similarity between pairs of objects because many portions of the trajectories may be noisy. Therefore, the choice of similarity function becomes more important. Some of the similarity functions discussed in Sect. 3.4.1 of Chap. 3, allow for gaps in the similarity computation. However, the effectiveness of these methods for multivariate time series and trajectories is highly data-specific. In general, these similarity functions are best used for trajectories of shorter lengths.

16.3.4.3 Trajectory Clustering as a Sequence Clustering Problem

Trajectory clustering methods can be performed with the same grid-based discretization methods that are used for frequent pattern mining in trajectories. A two-step approach is used. The first step is to use grid-based discretization to convert the trajectory into a 1-dimensional discrete sequence. Once this transformation has been performed, any of the sequence clustering methods discussed in Chap. 15 may be used. The overall clustering approach may be described as follows:

1. Use grid-based discretization, as discussed in Sect. 16.3.3.1, to convert the N trajectories to N discrete sequences.

2. Apply any of the sequence clustering methods of Sect. 15.3 in Chap. 15 to create clusters from the sequences.

3. Map the sequence clusters back to trajectory clusters.

As discussed in Sect. 16.3.3.1, the grid-based sequences constructed in the first step can be based on either the grid identifiers (spatial tile transformation) only, or on a combination of grid-identifiers and time-interval identifiers (spatiotemporal tile transformation). In the first case, the resulting clusters correspond to trajectories that are close together in space, but not necessarily in time. In the second case, the trajectories in a cluster will to be close together in space *and* occur at the same time. In other words, such clusters represent objects that are moving together in time.

One advantage of the sequence clustering approach, over similarity-based methods, is that many of the sequence clustering methods can ignore the irrelevant parts of the sequences in the clustering process. This is because many sequence clustering methods, such as subsequence-based clustering (Sect. 15.3.3 of Chap. 15), naturally allow for noisy gaps in the trajectories during the clustering process. This is important because longer trajectories often share significant segments in common, but may have gaps or regions where they are not similar. The ability to account for such nonmatching regions is not quite as effective with similarity function methods that compute distances between trajectories as a whole.

16.3.5 Trajectory Outlier Detection

In this problem, it is assumed that N different trajectories are available, and it is desirable to determine outlier trajectories, as those that differ significantly from the trends in the underlying data. As with all data types, the problem of trajectory outlier detection is closely related to that of trajectory clustering. In particular, both problems utilize the notion of similarity between data objects. As in the case of data clustering, one can use either a similarity-based approach, or a transformational approach to outlier detection.

16.3.5.1 Distance-Based Methods

The ability to design a distance function between trajectories provides a way to define outliers with the use of distance-based methods. In particular, the k-nearest neighbor method, or any distance-based method can easily be generalized to trajectories, once the distance function has been defined. For example, one may use the multidimensional time warping distance function to compute the distance of a trajectory to the $N-1$ other trajectories. The kth nearest neighbor distance is reported as the outlier score. Other distance-based methods such as *LOF* can also be extended to trajectory data because these methods are based only on distance values, and are agnostic to the underlying data type. As in the case of clustering, the major drawback of these methods is that it can be used effectively for shorter trajectories, but not quite as effectively in the case of longer trajectories. This is because longer trajectories will often have many noisy segments that are not truly indicative of anomalous behavior, but are disruptive to the underlying distance function.

16.3.5.2 Sequence-Based Methods

The spatial and spatiotemporal tile transformation discussed at the beginning of Sect. 16.3.3.1 can be used to transform trajectory outlier detection into sequence outlier detection. The advantage of this approach is that many methods are available for sequence outlier detection. As in the case of the other problems such as trajectory pattern mining and clustering, the approach consists of two steps:

1. Convert each of the N trajectories to sequences using either spatial tile transformation or spatiotemporal tile transformation, discussed at the beginning of Sect. 16.3.3.1.

2. Use any of the sequence outlier detection methods discussed in Sect. 15.4 of Chap. 15, to determine the outlier sequences.

3. Map the sequence outliers onto trajectory outliers.

This approach is particularly rich in terms of the types of the outliers it can find, by varying on the specific subroutines used in each of the aforementioned steps. Some examples of such variations are as follows:

- In the first step of sequence transformation, either spatial or spatiotemporal tiles may be used. When spatial tiles are used, the discovered outliers are based only on the shape of the trajectory, and they are not based on the objects moving together. From an application-centric perspective, consider the case where trajectories of taxis are tracked by GPS, and it is desirable to determine taxis that take anomalous routes relative to other taxis at any period of time. Such an application can be addressed well with spatial tiles. Spatiotemporal tiles track *online* trends. For example, for a flock of GPS-tagged animals, if a particular animal deviates from its flock, it is reported as an outlier.

- The formulations for sequence outlier detection are particularly rich. For example, sequence outlier detection allows the reporting of either *position* outliers or *combination* outliers. This is discussed in detail in Sect. 15.4 of Chap. 15. Position outliers in the transformed sequences, map onto anomalous positions in the trajectories. For example, a taxi cab making an unusual turn at a junction will be detected. On the other hand, combination outliers will map onto unusual segments of trajectories.

Thus, the sequence-based transformation is particularly useful in being able to detect a rich diversity of different types of outliers. It can determine outliers based on patterns of movements *either* over all periods, or at a particular period. It can also discover small segments of outliers in any portion of the trajectory.

16.3.6 Trajectory Classification

In this problem, it is assumed that a training data set of N labeled trajectories is provided. These are then used to construct a training model for the trajectories. The unknown class label of a test trajectory is determined with the use of this training model. Since classification is a supervised version of the clustering problem, methods for trajectory classification use similar methods to trajectory clustering. As in the case of clustering methods, either distance-based methods, or sequence-based methods may be used.

16.3.6.1 Distance-Based Methods

Several classification methods, such as nearest neighbor methods and graph-based collective classification methods, are dependent only on the notion of distances between data objects. After the distances between data objects have been defined, these classification methods are agnostic to the underlying data type.

The k-nearest neighbor method works as follows. The top-k nearest neighbors to a given test instance are determined. The dominant class label is reported as the relevant one for the test instance. Any of the multivariate extensions of time series distance functions, such as multidimensional DTW, may used for the computation process.

In graph-based methods, a k-nearest neighbor graph is constructed on the data objects. This is a semi-supervised method because the graph is constructed on a mixture of labeled and unlabeled objects. The basic discussion of graph-based methods may be found in Sect. 11.6.3 of Chap. 11. Each node corresponds to a trajectory. An undirected edge is added from node i to node j if either j is among the k nearest neighbors of i or vice versa. This results in a graph in which only a subset of the objects is labeled. The goal is to use the labeled nodes to infer the labels of the unlabeled nodes in the network. This is the collective classification problem that is discussed in detail in Sect. 19.4 of Chap. 19. When the labels on the unlabeled nodes have been determined using collective classification methods, they are mapped back to the original data objects. This approach is most effective when many test instances are simultaneously available with the training instances.

16.3.6.2 Sequence-Based Methods

In sequence-based methods, the first step is to transform the trajectories into sequences with the use of spatial or spatiotemporal tile-based methods. Once this transformation has been performed, any of the sequence classification methods discussed in Chap. 15 may be used. Therefore, the overall approach may be described as follows:

1. Convert each of the N trajectories to sequences using either the spatial tile transformation, or spatiotemporal tile transformation, discussed at the beginning of Sect. 16.3.3.1.

2. Use any of the sequence classification methods discussed in Sect. 15.6 of Chap. 15 to determine the class labels of sequences.

3. Map the sequence class labels to trajectory class labels.

The spatial tile transformation and spatiotemporal tile transformation methods provide different abilities in terms of incorporating different spatial and temporal features into the classification process. When spatial tile transformations are used, the resulting classification is not time sensitive, and trajectories from different periods can be modeled together on the basis of their *shape*. On the other hand, when the spatiotemporal tile transformation is used, the classification can only be performed on trajectories from the same approximate time period. In other words, the training and test trajectories must be drawn from the same period of time. In this case, the classification model is sensitive not only to the shape of the trajectory but also to the precise times in which their motion may have occurred. In this case, even if all the trajectories have exactly the same shape, the labels may be different because of temporal differences in speed at various times. The precise choice of the model depends on application-specific criteria.

16.4 Summary

Spatial data is common in a wide variety of applications, such as meteorological data, trajectory analysis, and disease outbreak data. This data is almost always a contextual data type, in which the data attributes are partitioned into behavioral attributes and contextual attributes. The spatial attributes may either be contextual or behavioral. These different types of data require different types of processing methods.

Contextual spatial attributes arise in the case of meteorological data where different types of spatial attributes such as temperature or pressure are measured at different spatial locations. Another example is the case of image data where the pixel values at different spatial locations are used to infer the properties of an image. An important transformation for shape-based spatial data is the centroid-sweep method that can transform a shape into time series. Another important transformation is the spatial wavelet approach that can transform spatial data into a multidimensional representation. These transformations are useful for virtually all data mining problems, such as clustering, outlier detection, or classification.

In trajectory data, the spatial attributes are behavioral, and the only contextual attribute is time. Trajectory data can be viewed as multivariate time series data. Therefore, time series distance functions can be generalized to trajectory data. This is useful in the development of a variety of data mining methods that are dependent only on the design of the distance function. Trajectory data can be transformed into sequence data with the use of tile-based transformations. Tile-based transformations are very useful because they allow the use of a wide variety of sequence mining methods for applications such as pattern mining, clustering, outlier detection, and classification.

16.5 Bibliographic Notes

The problem of spatial data mining has been studied extensively in the context of geographic data mining and knowledge discovery [388]. A detailed discussion of spatial databases may be found in [461]. The problem of search and indexing, was one of the earliest applications in the context of spatial data [443]. The centroid-sweep method for data mining of shapes is discussed in [547]. A discussion of spatial colocation pattern discovery with nonspatial behavioral attributes is found in [463]. This method has been used successfully for many data mining problems, such as clustering, classification, and outlier detection.

The problem of outlier detection from spatial data is discussed in detail in [5]. This book contains a dedicated chapter on outlier detection from spatial data. Numerous methods have been designed in the literature for spatial and spatiotemporal outlier detection [145, 146, 147, 254, 287, 326, 369, 459, 460, 462]. The algorithm for unusual shape detection was proposed in [510].

The tile-based simplification for pattern mining from trajectories was proposed in [375]. Pattern mining in trajectory data is closely related to clustering. The problem of mining periodic behaviors from trajectories is addressed in [352]. Moving object clusters have been studied as *Swarms* [351], *Flocks* [86] and *Convoys* [290]. Among these, *Swarms* provide the most relaxed definition, in which noisy gaps are allowed. In these noisy gap periods, objects from the same cluster may not move together. An algorithm for maintaining real-time communities from trajectories in social sensing applications was proposed in [429]. A method for partitioning longer trajectories into smaller segments for shape-based clustering was proposed in [338]. Anomaly monitoring from trajectories of moving object streams was studied in [117]. The *Top-Eye* method, an algorithm for monitoring the top-k anomalies in moving object trajectories, was proposed in [226]. The *TRAOD* algorithm, which discovers shape-based trajectory outliers, was proposed in [337]. A method that uses region-based and trajectory-based clustering for classification was proposed in [339].

16.6 Exercises

1. Discuss how to generalize the spatial wavelets to the case where there are n contextual attributes.

2. Implement the algorithm to construct a multidimensional representation from spatial data, with the use of wavelets.

3. Describe a method for converting shapes to a multidimensional representation.

4. Implement the algorithm for converting shapes to time series data.

5. Suppose that you had N different snapshots of sea surface temperature over successive instants in time over a spatial grid. You want to identify contiguous regions over which significant change has occurred between successive time instants. Describe an approach to identify such regions and time instants with the use of spatial wavelets.

6. Suppose the snapshots of Exercise 5 were not from successive instants in time. How would you identify spatial snapshots that were very different from the other snapshots with the use of spatial wavelets? How would you identify specific regions that are very different from the remaining data?

7. Suppose that you used the tile-based approach for finding frequent trajectory patterns. Discuss how the different constraint-based variants of sequential pattern mining map onto different constraint-based variants of sequential trajectory patterns.

8. Propose a snapshot-based clustering approach for converting trajectories to symbolic sequences. Discuss the advantages and disadvantages with respect to the tile-based approach.

9. Implement the different variations for converting trajectories to symbolic sequences with the use of the tile-based technique for frequent trajectory pattern mining.

10. Discuss how to use wavelets to perform different data mining tasks on trajectories.

Chapter 17

Mining Graph Data

"Structure is more important than content in the transmission of information."—Abbie Hoffman

17.1 Introduction

Graphs are ubiquitous in a wide variety of application domains such as bioinformatics, chemical, semi-structured, and biological data. Many important properties of graphs can be related to their structure in these domains. Graph mining algorithms can, therefore, be leveraged for analyzing various domain-specific properties of graphs. Most graphs, encountered in real applications, are one of the two types:

1. In applications such as chemical and biological data, a database of *many small graphs* is available. Each node is associated with a label that may or may not be unique to the node, depending on the application-specific scenario.

2. In applications such as the Web and social networks, a *single large graph* is available. For example, the Web can be viewed as a very large graph, in which nodes correspond to Web pages (labeled by their URLs) and edges correspond to hyperlinks between nodes.

The nature of the applications for these two types of data are quite different. Web and social network applications will be addressed in Chaps. 18 and 19, respectively. This chapter will therefore focus on the first scenario, in which many small graphs are available. A graph database may be formally defined as follows.

Definition 17.1.1 (Graph Database) *A graph database $\mathcal{D}$ is a collection of n different undirected graphs, $G_1 = (N_1, A_1) \ldots G_n = (N_n, A_n)$, such that the set of nodes in the ith graph is denoted by N_i, and the set of edges in the ith graph is denoted by A_i. Each node $p \in N_i$ is associated with a label denoted by $l(p)$.*

The labels associated with the nodes may be repeated within a single graph. For example, when each graph G_i corresponds to a chemical compound, the label of the node is the

C. C. Aggarwal, *Data Mining: The Textbook*, DOI 10.1007/978-3-319-14142-8_17

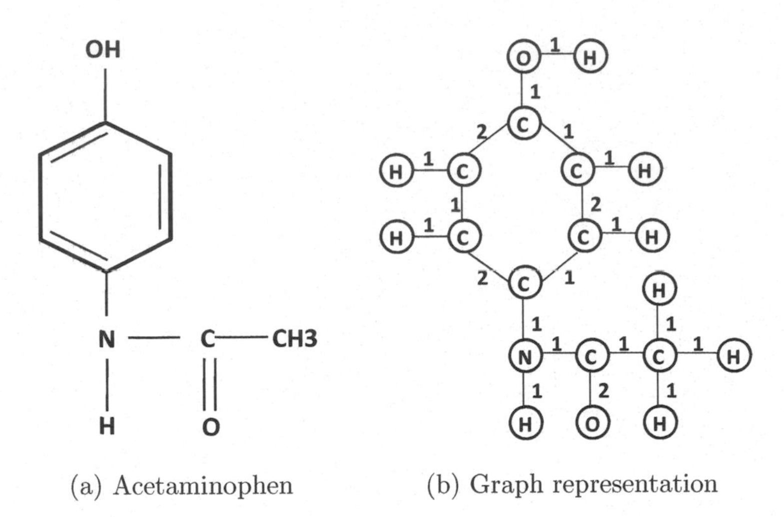

Figure 17.1: A chemical compound (*Acetaminophen*) and its associated graph representation

symbol denoting a chemical element. Because of the presence of multiple atoms of the same element, such a graph will contain label repetitions. The repetition of labels within a single graph leads to numerous challenges in graph matching and distance computation.

Graph data are encountered in many real applications. Some examples of key applications of graph data mining are as follows:

- Chemical and biological data can be expressed as graphs in which each node corresponds to an atom and a bond between a pair of atoms is represented by an edge. The edges may be weighted to reflect bond strength. An example of a chemical compound and its corresponding graph are illustrated in Fig. 17.1. Figure 17.1a shows an illustration of the chemical *acetaminophen*, a well-known analgesic. The corresponding graph representation is illustrated in Fig. 17.1b along with node labels and edge weights. In many graph mining applications, unit edge weights are assumed as a simplification.

- XML data can be expressed as attributed graphs. The relationships between different attributes of a structured record can be expressed as edges.

- Virtually any data type can be expressed as an entity-relationship graph. This provides a different way of mining conventional database records when they are expressed in the form of entity-relationship graphs.

Graph data are very powerful because of their ability to model arbitrary relationships between objects. The flexibility in graph representation comes at the price of greater computational complexity:

1. Graphs lack the "flat" structure of multidimensional or even contextual (e.g., time series) data. The latter is much easier to analyze with conventional models.

2. The repetition of labels among nodes leads to problems of *isomorphism* in computing similarity between graphs. This problem is NP-hard. This leads to computational challenges in similarity computation and graph matching.

The second issue is of considerable importance, because both matching and distance computation are fundamental subproblems in graph mining applications. For example, in a frequent subgraph mining application, an important subproblem is that of subgraph matching.

This chapter is organized as follows. Section 17.2 addresses the problem of matching and distance computation in graphs. Graph transformation methods for distance computation are discussed in Sect. 17.3. An important part of this section is the preprocessing methodologies, such as topological descriptors and kernel methods, that are often used for distance computation. Section 17.4 addresses the problem of pattern mining in graphs. The problem of clustering graphs is addressed in Sect. 17.5. Graph classification is addressed in Sect. 17.6. A summary is provided in Sect. 17.7.

17.2 Matching and Distance Computation in Graphs

The problems of matching and distance computation are closely related in the graph domain. Two graphs are said to match when a one-to-one correspondence can be established between the nodes of the two graphs, such that their labels match, and the edge presence between corresponding nodes match. The distance between such a pair of graphs is zero. Therefore, the problem of distance computation between a pair of graphs is at least as hard as that of graph matching. Matching graphs are also said to be *isomorphic*.

It should be pointed out that the term "matching" is used in two distinct contexts for graph mining, which can sometimes be confusing. For example, pairing up nodes in a single graph with the use of edges is also referred to as matching. Throughout this chapter, unless otherwise specified, our focus is not on the node matching problem, but the pairwise *graph* matching problem. This problem is also referred to as that of graph *isomorphism*.

Definition 17.2.1 (Graph Matching and Isomorphism) *Two graphs $G_1 = (N_1, A_1)$ and $G_2 = (N_2, A_2)$ are isomorphic if and only if a one-to-one correspondence can be found between the nodes of N_1 and N_2 satisfying the following properties:*

1. *For each pair of corresponding nodes $i \in N_1$ and $j \in N_2$, their labels are the same.*

$$l(i) = l(j)$$

2. *Let $[i_1, i_2]$ be a node-pair in G_1 and $[j_1, j_2]$ be the corresponding node-pair in G_2. Then the edge (i_1, i_2) exists in G_1 if and only if the edge (j_2, j_2) exists in G_2.*

The computational challenges in graph matching arise because of the repetition in node labels. For example, consider two methane molecules, illustrated in Fig. 17.2. While the unique carbon atom in the two molecules can be matched in exactly one way, the hydrogen atoms can be matched up in $4! = 24$ different ways. Two possible matchings are illustrated in Figs. 17.2a and b, respectively. In general, greater the level of label repetition in each graph is, larger the number of possible matchings will be. The number of possible matchings between a pair of graphs increases exponentially with the size of the matched graphs. For a pair of graphs containing n nodes each, the number of possible matchings can be as large as $n!$. This makes the problem of matching a pair of graphs computationally very expensive.

Lemma 17.2.1 *The problem of determining whether a matching exists between a pair of graphs, is NP-hard.*

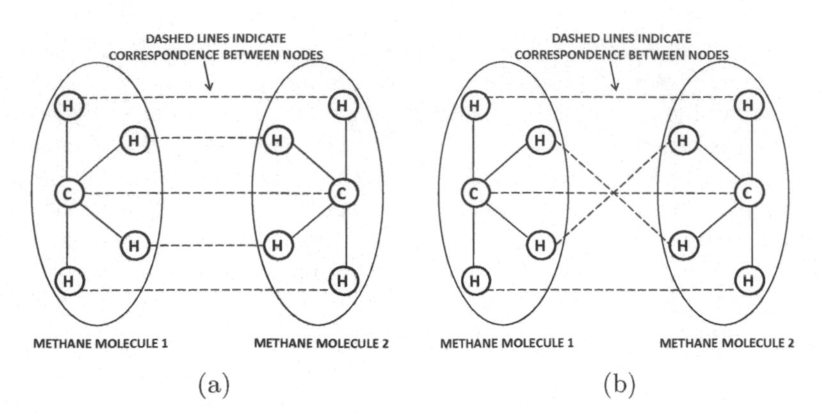

Figure 17.2: Two possible matchings between a pair of graphs representing methane molecules

The bibliographic notes contain pointers to the proof of NP-hardness. When the graphs are very large, exact matches often do not exist. However, approximate matches may exist. The level of approximation is quantified with the use of a distance function. Therefore, distance function computation between graphs is a more general problem than that of graph matching, and it is at least as difficult. This issue will be discussed in detail in the next section.

Another related problem is that of subgraph matching. Unlike the problem of exact graph matching, the query graph needs to be explicitly distinguished from the data graph in this case.

Definition 17.2.2 (Node-Induced Subgraph) *A node-induced subgraph of a graph $G = (N, A)$ is a graph $G_s = (N_s, A_s)$ satisfying the following properties:*

1. $N_s \subseteq N$
2. $A_s = A \cap (N_s \times N_s)$

In other words, all the edges in the original graph G between nodes in the subset $N_s \subseteq N$ are included in the subgraph G_s.

A subgraph isomorphism can be defined in terms of the node-induced subgraphs. A query graph G_q is a subgraph isomorphism of a data graph G, when it is an exact isomorphism of a node-induced subgraph of G.

Definition 17.2.3 (Subgraph Matching and Isomorphism) *A query graph $G_q = (N_q, A_q)$ is a subgraph isomorphism of the data graph $G = (N, A)$ if and only if the following conditions are satisfied:*

1. *Each node in N_q should be matched to a unique node with the same label in N, but each node in N may not necessarily be matched. For each node $i \in N_q$, there must exist a unique matching node $j \in N$, such that their labels are the same.*

$$l(i) = l(j)$$

2. *Let $[i_1, i_2]$ be a node-pair in G_q, and let $[j_1, j_2]$ be the corresponding node-pair in G, based on the matching discussed above. Then, the edge (i_1, i_2) exists in G_q if and only if the edge (j_1, j_2) exists in G.*

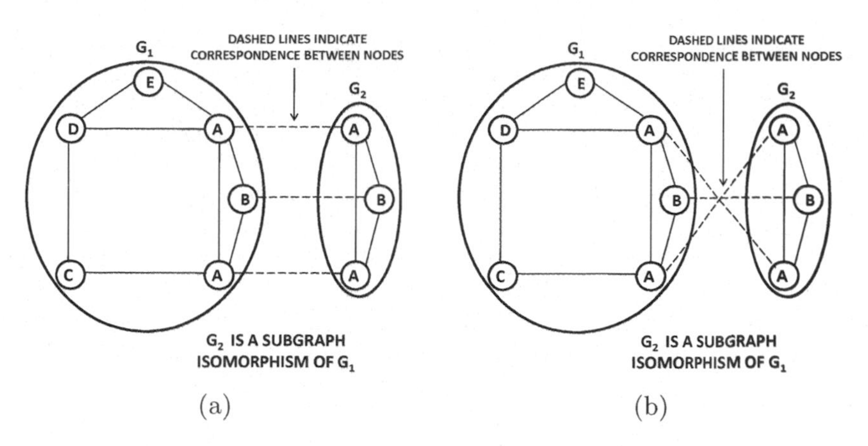

Figure 17.3: Two possible subgraph isomorphisms between a pair of graphs

The definition of subgraph isomorphism in this section assumes that *all* edges of the node-induced subgraph of the data graph are present in the query graph. In some applications, such as frequent subgraph mining, a more general definition is used, in which any subset of edges of the node-induced subgraph is also considered a subgraph isomorphism. The more general case can be handled with minor changes to the algorithm in this section. Note that the aforementioned definition allows the subgraph G_q (or G) to be disconnected. However, for practical applications, one is usually interested only in *connected* subgraph isomorphisms. Examples of two possible subgraph matchings between a pair of nodes are illustrated in Fig. 17.3. The figure also illustrates that there are two different ways for one graph to be a subgraph of the other. The problem of exact matching is a special case of subgraph matching. Therefore, the problem of subgraph matching is NP-hard as well.

Lemma 17.2.2 *The problem of subgraph matching is NP-hard.*

Subgraph matching is often used as a subroutine in applications such as frequent pattern mining. While the subgraph matching problem is a generalization of exact matching, the problem can be generalized even further to that of finding the maximum common subgraph (MCG) between a pair of graphs. This is because the MCG between two graphs is at most equal to the smaller of the two graphs when it is a subgraph of the larger one. The *MCG* or *maximum common isomorphism* between a pair of graphs is defined as follows.

Definition 17.2.4 (Maximum Common Subgraph) *A MCG between a pair of graphs* $G_1 = (N_1, A_1)$ *and* $G_2 = (N_2, A_2)$ *is a graph* $G_0 = (N_0, A_0)$ *that is a subgraph isomorphism of both* G_1 *and* G_2*, and for which the size of the node set* N_0 *is as large as possible.*

Because the MCG problem is a generalization of the graph isomorphism problem, it is NP-hard as well. In this section, algorithms for discovering subgraph isomorphisms and maximum common subgraphs will be presented. Subsequently, the relationship of these algorithms to that of distance computation between graphs will be discussed. Subgraph isomorphism algorithms can be designed to determine either *all* the subgraph isomorphisms between a query graph and a data graph, or a fast algorithm can be designed to determine *whether or not at least one* isomorphism exists.

17.2.1 Ullman's Algorithm for Subgraph Isomorphism

Ullman's algorithm is designed to determine *all* possible subgraph isomorphisms between a query graph and a data graph. It can also be used for the decision problem of determining whether or not a query graph is a subgraph isomorphism of a data graph by using an early termination criterion. Interestingly, the majority of the later graph matching algorithms are refinements of Ullman's algorithm. Therefore, this section will first present a very simplified version of Ullman's algorithm without any refinements. Subsequently, the different variations and refinements of this basic algorithm will be discussed in a separate subsection. Although the definition of subgraph isomorphisms allows the query (or data) graph to be disconnected, it is often practical and computationally expedient to focus on cases where the query and data graph are connected. Typically, small changes to the algorithm can accommodate both cases (see Exercise 14).

It will be assumed that the query graph is denoted by $G_q = (N_q, A_q)$, and the data graph is denoted by $G = (N, A)$. The first step in Ullman's algorithm is to match all possible pairs of nodes across the two graphs so that each node in the pair has the same label as the other. For each such matching pair, the algorithm expands it a node at a time with the use of a recursive search procedure. Each recursive call expands the matching subgraphs in G_q and G by one node. Therefore, one of the parameters of the recursive call is the current matching set $\mathcal{M}$ of node-pairs. Each element of $\mathcal{M}$ is a pair of matching nodes between G_q and G. Therefore, when a subgraph of m nodes has been matched between the two graphs, the set $\mathcal{M}$ contains m matched node-pairs as follows:

$$\mathcal{M} = \{(i_1^q, i_1), (i_2^q, i_2), \ldots (i_m^q, i_m)\}$$

Here, it is assumed that the node i_r^q belongs to the query graph G_q and that the node i_r belongs to the data graph G. The value of the matching set parameter $\mathcal{M}$ is initialized to the empty set at the top-level recursive call. The number of matched nodes in $\mathcal{M}$ is exactly equal to the depth of the recursive call. The recursion backtracks when either the subgraphs cannot be further matched or when G_q has been fully matched. In the latter case, the matching set $\mathcal{M}$ is reported, and the recursion backtracks to the next higher level to discover other matchings. In cases where it is not essential to determine *all possible* matchings between the pair of graphs, it is also possible to terminate the algorithm at this point. This particular exposition, however, assumes that all possible matchings need to be determined.

A simplified version of Ullman's algorithm is illustrated in Fig. 17.4. The algorithm is structured as a recursive approach that explores the space of all possible matchings between the two graphs. The input to the algorithm is the query graph G_q and the data graph G. An additional parameter $\mathcal{M}$ of this recursive call is a set containing the current matching node-pairs. While the set $\mathcal{M}$ is empty at the top-level call made by the analyst, this is not the case at lower levels of the recursion. The cardinality of $\mathcal{M}$ is exactly equal to the depth of the recursion. This is because one matching node-pair is added to $\mathcal{M}$ in each recursive call. Strictly speaking, the recursive call returns all the subgraph isomorphisms under the constraint that the matching corresponding to $\mathcal{M}$ must be respected.

The first step of the recursive procedure is to check whether the size of $\mathcal{M}$ is equal to the number of nodes in the query graph G_q. If this is indeed the case, then the algorithm reports $\mathcal{M}$ as a successful subgraph matching and backtracks out of the recursion to the next higher level to explore other matchings. Otherwise, the algorithm tries to determine further matching node-pairs to add to $\mathcal{M}$. This is the *candidate generation step.* In this

Algorithm *SubgraphMatch*(Query Graph: G_q, Data Graph: G,
Current Partially Matched Node Pairs: $\mathcal{M}$)
begin
if ($|\mathcal{M}| = |N_q|$) **then return** successful match $\mathcal{M}$;
(Case when $|\mathcal{M}| < |N_q|$**)**
$\mathcal{C}$ = Set of all label matching node pairs from (G_q, G) not in $\mathcal{M}$;
Prune $\mathcal{C}$ using heuristic methods; **(Optional efficiency optimization)**
for each pair $(i_q, i) \in \mathcal{C}$ **do**
if $\mathcal{M} \cup \{(i_q, i)\}$ is a valid partial matching
then *SubgraphMatch*$(G_q, G, \mathcal{M} \cup \{(i_q, i)\})$;
endfor
end

Figure 17.4: The basic template of Ullman's algorithm

step, all possible label matching node-pairs between G_q and G, which are not already in $\mathcal{M}$, are used to construct the candidate match set $\mathcal{C}$.

Because the number of candidate match extensions can be large, it is often desirable to prune them heuristically by using specific properties of the data graph and query graph. Some examples of such heuristics will be presented later. After the pruned set $\mathcal{C}$ has been generated, node-pairs $(i_q, i) \in \mathcal{C}$ are selected one by one, and it is checked whether they can be added to $\mathcal{M}$ to create a valid (partial) matching between the two graphs. For $\mathcal{M} \cup \{(i_q, i)\}$ to be a valid partial matching, if $i_q \in N_q$ is incident on any already matched node j_q in G_q, then i must also be incident on the matched counterpart j of j_q in G, and vice versa. If a valid partial matching exists, then the procedure is called recursively with the partial matching $\mathcal{M} \cup \{(i_q, i)\}$. After iterating through all such candidate extensions with corresponding recursive calls, the algorithm backtracks to the next higher level of the recursion.

It is not difficult to see that the procedure has exponential complexity in terms of its input size, and it is especially sensitive to the query graph size. This high complexity is because the depth of the recursion can be of the order of the query graph size, and the number of recursive branches at each level is equal to the number of matching node-pairs. Clearly, unless the number of candidate extensions is carefully controlled with more effective candidate generation and pruning, the approach will be extremely slow.

17.2.1.1 Algorithm Variations and Refinements

Although the basic matching algorithm was originally proposed by Ullman, this template has been used extensively by different matching algorithms. The different algorithms vary from one another in terms of how the size of the candidate matched pairs is restricted with careful pruning. The use of carefully selected candidate sets has a significant impact on the efficiency of the algorithm. Most pruning methods rely on a number of natural constraints that are always satisfied by two graphs in a subgraph isomorphism relationship. Some common pruning rules are as follows:

1. *Ullman's algorithm:* This algorithm uses a simple pruning rule. All node-pairs (i_q, i) are pruned from $\mathcal{C}$ in the pruning step if the degree of i is less than i_q. This is because the degree of every matching node in the query subgraph needs to be no larger than the degree of its matching counterpart in the data graph.

2. *VF2 algorithm:* In the *VF2* algorithm, those candidates (i_q, i) are pruned if i_q is not connected to already matched nodes in G_q (i.e., nodes of G_q included in $\mathcal{M}$). Subsequently, the pruning step also removes those node-pairs (i_q, i) in which i is not connected to the matched nodes in the data graph G. These pruning rules assume that the query and data graphs are connected. The algorithm also compares the number of neighbor nodes of each of i and i_q that are connected to nodes in $\mathcal{M}$ but are not included in $\mathcal{M}$. The number of such nodes in the data graph must be no smaller than the number of such nodes in the query graph. Finally, the number of neighbor nodes of each of i and i_q that are not directly connected to nodes in $\mathcal{M}$ are compared. The number of such nodes in the data graph must be no smaller than the number of such nodes in the query graph.

3. *Sequencing optimizations:* The effectiveness of the pruning steps is sensitive to the order in which nodes are added to the matching set $\mathcal{M}$. In general, nodes with rarer labels in the query graph should be selected first in the exploration of different candidate pairs in $\mathcal{C}$. Rarer labels can be matched in fewer ways across graphs. Early exploration of rare labels leads to exploration of more relevant partial matches $\mathcal{M}$ at the earlier levels of the recursion. This also helps the pruning effectiveness. Enhanced versions of *VF2* and *QuickSI* combine node sequencing and the aforementioned node pruning steps.

The reader is referred to the bibliographic notes for details of these algorithms. The definition of subgraph isomorphism in this section assumes that *all* edges of the node-induced subgraph of the data graph are present in the query graph. In some applications, such as frequent subgraph mining, a more general definition is used, in which any subset of edges of the node-induced subgraph is also considered a subgraph isomorphism. The more general case can be solved with minor changes to the basic algorithm in which the criteria to generate candidates and validate them are both relaxed appropriately.

17.2.2 Maximum Common Subgraph (MCG) Problem

The MCG problem is a generalization of the subgraph isomorphism problem. The MCG between two graphs is at most equal to the smaller of the two, when one is a subgraph of the other. The basic principles of subgraph isomorphism algorithms can be extended easily to the MCG isomorphism problem. The following will discuss the extension of the Ullman algorithm to the MCG problem. The main differences between these methods are in terms of the pruning criteria and the fact that the maximum common subgraph is continuously tracked over the course of the algorithm as the search space of subgraphs is explored.

The recursive exploration process of the MCG algorithm is identical to that of the subgraph isomorphism algorithm. The algorithm is illustrated in Fig. 17.5. The two input graphs are denoted by G_1 and G_2, respectively. As in the case of subgraph matching, the current matching in the recursive exploration is denoted by the set $\mathcal{M}$. For each matching node-pair $(i_1, i_2) \in \mathcal{M}$, it is assumed that i_1 is drawn from G_1, and i_2 is drawn from G_2. Another input parameter to the algorithm is the current best (largest) matching set of node-pairs $\mathcal{M}_{best}$. Both $\mathcal{M}$ and $\mathcal{M}_{best}$ are initialized to *null* in the initial call made to the recursive algorithm by the analyst. Strictly speaking, each recursive call determines the best matching under the constraint that the pairs in $\mathcal{M}$ must be matched. This is the reason that this parameter is set to *null* at the top-level recursive call. However, in lower level calls, the value of $\mathcal{M}$ is not *null*.

Algorithm *MCG*(Graphs: G_1, G_2, Current Partially Matched Pairs: $\mathcal{M}$,
Current Best Match: $\mathcal{M}_{best}$)
begin
$\mathcal{C}$ = Set of all label matching node pairs from (G_1, G_2) not in $\mathcal{M}$;
Prune $\mathcal{C}$ using heuristic methods; **(Optional efficiency optimization)**
for each pair $(i_1, i_2) \in \mathcal{C}$ **do**
if $\mathcal{M} \cup \{(i_1, i_2)\}$ is a valid matching
then $\mathcal{M}_{best}$ =$MCG(G_1, G_2, \mathcal{M} \cup \{(i_1, i_2)\}, \mathcal{M}_{best})$;
endfor
if $(|\mathcal{M}| > |\mathcal{M}_{best}|)$ **then return**($\mathcal{M}$) **else return**($\mathcal{M}_{best}$);
end

Figure 17.5: Maximum common subgraph (MCG) algorithm

As in the case of the subgraph isomorphism algorithm, the candidate matching node-pairs are explored recursively. The same steps of candidate extension and pruning are used in the MCG algorithm, as in the case of the subgraph isomorphism problem. However, some of the pruning steps used in the subgraph isomorphism algorithm, which are based on *subgraph* assumptions, can no longer be used. For example, in the MCG algorithm, a matching node-pair (i_1, i_2) in $\mathcal{M}$ no longer needs to satisfy the constraint that the degree of a node in one graph is greater or less than that of its matching node in the other. Because of the more limited pruning in the maximum common subgraph problem, it will explore a larger search space. This is intuitively reasonable, because the maximum common subgraph problem is a more general one than subgraph isomorphism. However, some optimizations such as expanding only to connected nodes, and sequencing optimizations such as processing rare labels earlier, can still be used.

The largest common subgraph found so far is tracked in $\mathcal{M}_{best}$. At the end of the procedure, the largest matching subgraph found so far is returned by the algorithm. It is also relatively easy to modify this algorithm to determine all possible MCGs. The main difference is that all the current MCGs can be dynamically tracked instead of tracking a single MCG.

17.2.3 Graph Matching Methods for Distance Computation

Graph matching methods are closely related to distance computation between graphs. This is because pairs of graphs that share large subgraphs in common are likely to be more similar. A second way to compute distances between graphs is by using the *edit distance*. The edit distance in graphs is analogous to the notion of the edit distance in strings. Both these methods will be discussed in this section.

17.2.3.1 MCG-based Distances

When two graphs share a large subgraph in common, it is indicative of similarity. There are several ways of transforming the MCG size into a distance value. Some of these distance definitions have also been demonstrated to be *metrics* because they are nonnegative, symmetric, and satisfy the triangle inequality. Let the MCG of graphs G_1 and G_2 be denoted by $MCS(G_1, G_2)$ with a size of $|MCS(G_1, G_2)|$. Let the sizes of the graphs G_1 and G_2

be denoted by $|G_1|$ and $|G_2|$, respectively. The various distance measures are defined as a function of these quantities.

1. *Unnormalized non-matching measure:* The unnormalized non-matching distance measure $U(G_1, G_2)$ between two graphs is defined as follows:

$$U(G_1, G_2) = |G_1| + |G_2| - 2 \cdot |MCS(G_1, G_2)| \tag{17.1}$$

   This is equal to the number of non-matching nodes between the two graphs because it subtracts out the number of matching nodes $|MCS(G_1, G_2)|$ from each of $|G_1|$ and $|G_2|$ and then adds them up. This measure is unnormalized because the value of the distance depends on the raw size of the underlying graphs. This is not desirable because it is more difficult to compare distances between pairs of graphs of varying size. This measure is more effective when the different graphs in the collection are of approximately similar size.

2. *Union-normalized distance:* The distance measure lies in the range of $(0, 1)$, and is also shown to be a metric. The union-normalized measure $UDist(G_1, G_2)$ is defined as follows:

$$UDist(G_1, G_2) = 1 - \frac{|MCS(G_1, G_2)|}{|G_1| + |G_2| - |MCS(G_1, G_2)|} \tag{17.2}$$

   This measure is referred to as the union-normalized distance because the denominator contains the number of nodes in the union of the two graphs. A different way of understanding this measure is that it normalizes the number of non-matching nodes $U(G_1, G_2)$ between the two graphs (unnormalized measure) with the number of nodes in the union of the two graphs.

$$UDist(G_1, G_2) = \frac{\text{Non-matching nodes between } G_1 \text{ and } G_2}{\text{Union size of } G_1 \text{ and } G_2}$$

   One advantage of this measure is that it is intuitively easier to interpret. Two perfectly matching graphs will have a distance of 0 from one another, and two perfectly non-matching graphs will have a distance of 1.

3. *Max-normalized distance:* This distance measure also lies in the range $(0, 1)$. The *max-normalized distance* $MDist(G_1, G_2)$ between two graphs G_1 and G_2 is defined as follows:

$$MDist(G_1, G_2) = 1 - \frac{|MCS(G_1, G_2)|}{\max\{|G_1|, |G_2|\}} \tag{17.3}$$

   The main difference from the union-normalized distance is that the denominator is normalized by the maximum size of the two graphs. This distance measure is a metric because it satisfies the triangle inequality. The measure is also relatively easy to interpret. Two perfectly matching graphs will have a distance of 0 from one another, and two perfectly non-matching graphs will have a distance of 1.

These distance measures can be computed effectively only for small graphs. For larger graphs, it becomes computationally too expensive to evaluate these measures because of the need to determine the maximum common subgraph between the two graphs.

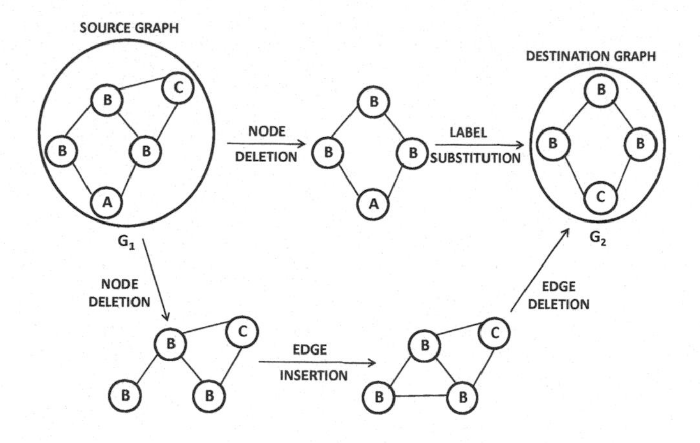

Figure 17.6: Example of two possible edits paths between graphs G_1 and G_2

17.2.3.2 Graph Edit Distance

The graph edit distance is analogous to the string edit distance, discussed in Chap. 3. The main difference is that the edit operations are specific to the graph domain. The edit distances can be applied to the nodes, the edges, or the labels. In the context of graphs, the admissible operations include (a) the insertion of nodes, (b) the deletion of nodes, (c) the label-substitution of nodes, (d) the insertion of edges, and (e) the deletion of edges. Note that the deletion of a node includes automatic deletion of all its incident edges. Each edit operation has an edit cost associated with it that is defined in an application-specific manner. In fact, the problem of learning edit costs is a challenging issue in its own right. For example, one way of learning edit costs is to use supervised distance function learning methods discussed in Chap. 3. The bibliographic notes contain pointers to some of these algorithms.

An example of two possible edit paths between graphs G_1 and G_2 is illustrated in Fig. 17.6. Note that the two paths will have different costs, depending on the costs of the constituent operations. For example, if the cost of label-substitution is very high compared to that of edge insertions and deletions, it may be more effective to use the second (lower) path in Fig. 17.6. For large and complex pairs of graphs, an exponential number of possible edit paths may exist. The edit distance $Edit(G_1, G_2)$ between two graphs is equal to the minimum cost of transforming graph G_1 to G_2 with a series of edit operations.

Definition 17.2.5 (Graph Edit Distance) *The graph edit distance $Edit(G_1, G_2)$ is the minimum cost of the edit operations to be applied to graph G_1 in order to transform it to graph G_2.*

Depending on the costs of the different operations, the edit distance is not necessarily symmetric. In other words, $Edit(G_1, G_2)$ can be different from $Edit(G_2, G_1)$. Interestingly, the edit distance is closely related to the problem of determining MCGs. In fact, for some special choices of costs, the edit distance can be shown to be equivalent to distance measures based on the maximum common subgraph. This implies that the edit-distance computation for graphs is NP-hard as well. The edit distance can be viewed as the cost of an *error-tolerant graph isomorphism*, where the "errors" are quantified in terms of the cost of edit operations. As discussed in Chap. 3, the edit-distance computation for strings and sequences can be solved polynomially using dynamic programming. The case of graphs is more difficult because it belongs to the class of NP-hard problems.

The close relationship between edit-distance computation and the MCG problem is reflected in the similar structure of their corresponding algorithms. As in the case of the maximum common subgraph problem, a recursive tree-search procedure can be used to compute the edit distance. In the following, the basic procedure for computing the edit distance will be described. The bibliographic notes contain pointers to various enhancements of this procedure.

An interesting property of the edit-distance is that it can be computed by exploring only those edit sequences in which any and all node insertion operations (together with their incident edge insertions) are performed at the end of the edit sequence. Therefore, the edit-distance algorithm maintains a series of edits $\mathcal{E}$ that are the operations to be applied to graph G_1 to transform it into a *subgraph isomorphism* G_1' of the graph G_2. By trivially adding the unmatched nodes of G_2 to G_1' and corresponding incident edges as the final step, it is possible to create G_2. Therefore, the initial part of sequence $\mathcal{E}$, without the last step, does not contain any node insertions at all. In other words, the initial part of sequence $\mathcal{E}$ may contain node deletions, node label-substitutions, edge additions, and edge deletions. An example of such an edit sequence is as follows:

$$\mathcal{E} = \text{Delete}(i_1),\ \text{Insert}(i_2, i_5),\ \text{Label-Substitute}(i_4, A \Rightarrow C),\ \text{Delete}(i_2, i_6)$$

This edit sequence illustrates the deletion of a node, followed by addition of the new edge (i_2, i_5). The label of node i_4 is substituted from A to C. Then, the edge (i_2, i_6) is deleted. The total cost of an edit sequence $\mathcal{E}$ from G_1 to a subgraph isomorphism G_1' of G_2 is equal to the sum of the edit costs of all the operations in $\mathcal{E}$, together with the cost of the node insertions and incident edge insertions that need to be performed on G_1' to create the final graph G_2.

The correctness of such an approach relies on the fact that it is always possible to arrange the optimal edit path sequence, so that the insertion of the nodes and their incident edges is performed *after* all other edge operations, node deletions and label-substitutions that transform G_1 to a subgraph isomorphism of G_2. The proof of this property follows from the fact that any *optimal* edit sequence can be reordered to push the insertion of nodes (and their incident edges) to the end, as long as an inserted node is not associated with any other edit operations (node or incident edge deletions, or label-substitutions). It is also easy to show that any edit path in which newly added nodes or edges are deleted will be suboptimal. Furthermore, an inserted node never needs to be label-substituted in an optimal path because the correct label can be set at the time of node insertion.

The overall recursive procedure is illustrated in Fig. 17.7. The inputs to the algorithm are the source and target graphs G_1 and G_2, respectively. In addition, the current edit sequence $\mathcal{E}$ being examined for further extension, and the best (lowest cost) edit sequence $\mathcal{E}_{best}$ found so far, are among the input parameters to the algorithm. These input parameters are useful for passing data between recursive calls. The value of $\mathcal{E}$ is initialized to *null* in the top-level call. While the value of $\mathcal{E}$ is *null* at the beginning of the algorithm, new edits are appended to it in each recursive call. Further recursive calls are executed with this extended sequence as the input parameter. The value of the parameter $\mathcal{E}_{best}$ at the top-level call is set to a trivial sequence of edit operations in which all nodes of G_1 are deleted and then all nodes and edges of G_2 are added.

The recursive algorithm first discovers the sequence of edits $\mathcal{E}$ that transforms the graph G_1 to a subgraph isomorphism G_1' of G_2. After this phase, the trivial sequence of node/edge insertion edits that convert G_1' to G_2 is padded at the end of $\mathcal{E}$. This step is shown in Fig. 17.7 just before the return condition in the recursive call. Because of this final padding step, the

```
Algorithm EditDistance(Graphs: G1, G2, Current Partial Edit Sequence: E,
    Best Known Edit Sequence: E_best)
begin
  if (G1 is subgraph isomorphism of G2) then begin
    Add insertion edits to E that convert G1 to G2;
    return(E);
  end;
  C = Set of all possible edits to G1 excluding node-insertions;
  Prune C using heuristic methods; (Optional efficiency optimization)
  for each edit operation e ∈ C do
  begin
    Apply e to G1 to create G1';
    Append e to E to create E';
    E_current = EditDistance(G1', G2, E', E_best);
    if (Cost(E_current) < Cost(E_best)) then E_best = E_current;
  endfor
  return(E_best);
end
```

Figure 17.7: Graph edit distance algorithm

cost of these trivial edits is always included in the cost of the edit sequence $\mathcal{E}$, which is denoted by $Cost(\mathcal{E})$.

The overall structure of the algorithm is similar to that of the MCG algorithm of Fig. 17.5. In each recursive call, it is first determined if G_1 is a subgraph isomorphism of G_2. If so, the algorithm immediately returns the current set of edits $\mathcal{E}$ after the incorporation of trivial node or edge insertions that can transform G_1 to G_2. If G_1 is not a subgraph isomorphism of G_2, then the algorithm proceeds to extend the partial edit path $\mathcal{E}$. A set of candidate edits $\mathcal{C}$ is determined, which when applied to G_1 *might* reduce the distance to G_2. In practice, these candidate edits $\mathcal{C}$ are determined heuristically because the problem of knowing the precise impact of an edit on the distance is almost as difficult as that of computing the edit distance. The simplest way of choosing the candidate edits is to consider all possible unit edits excluding node insertions. These candidate edits might be node deletions, label-substitutions and edge operations (both insertions and deletions). For a graph with n nodes, the total number of node-based candidate operations is $O(n)$, and the number of edge-based candidate operations is $O(n^2)$. It is possible to heuristically prune many of these candidate edits if it can be immediately determined that such edits can never be part of an optimal edit path. In fact, some of the pruning steps are essential to ensure finite termination of the algorithm. Some key pruning steps are as follows:

1. An edge insertion cannot be appended to the current partial edit sequence $\mathcal{E}$, if an edge deletion operation between the same pair of nodes already exists in the current partial edit path $\mathcal{E}$. Similarly, an edge which was inserted earlier cannot be deleted. An optimal edit path can never include such pairs of edits with zero net effect. This pruning step is necessary to ensure finite termination.

2. The label of a node cannot be substituted, if the label-substitution of that node exists in the current partial edit path $\mathcal{E}$. Repetitive label-substitutions of the same node are obviously suboptimal.

3. An edge can be inserted between a pair of nodes in G_1, only if at least one edge exists in G_2 between two nodes with the same labels.

4. A candidate edit should not be considered, if adding it to $\mathcal{E}$ would immediately increase the cost of $\mathcal{E}$ beyond that of $\mathcal{E}_{best}$.

5. Many other sequencing optimizations are possible for prioritizing between candidate edits. For example, all node deletions can be performed before all label-substitutions. It can be shown that the optimal edit sequence can always be arranged in this way. Similarly, label-substitutions which change the overall distribution of labels closer to the target graph may be considered first. In general, it is possible to associate a "goodness-function" with an edit, which heuristically quantifies its likelihood of finding a good edit path, when included in $\mathcal{E}$. Finding good edit paths early will ensure better pruning performance according to the aforementioned criterion (4).

The main difference among various recursive search algorithms is to use different heuristics for candidate ordering and pruning. Readers are referred to the bibliographic notes at the end of this chapter for pointers to some of these methods. After the pruned candidate edits have been determined, each of these is applied to G_1 to create G_1'. The procedure is recursively called with the pair (G_1', G_2), and an augmented edit sequence $\mathcal{E}'$. This procedure returns the best edit sequence $\mathcal{E}_{current}$ which has a prefix of $\mathcal{E}'$. If the cost of $\mathcal{E}_{current}$ is better than $\mathcal{E}_{best}$ (including trivial post-processing insertion edits for full matching), then $\mathcal{E}_{best}$ is updated to $\mathcal{E}_{current}$. At the end of the procedure, $\mathcal{E}_{best}$ is returned.

The procedure is guaranteed to terminate because repetitions in node label-substitutions and edge deletions are avoided in $\mathcal{E}$ by the pruning steps. Furthermore, the number of nodes in the edited graph is monotonically non-increasing as more edits are appended to $\mathcal{E}$. This is because $\mathcal{E}$ does not contain node insertions except at the end of the recursion. For a graph with n nodes, there are at most $\binom{n}{2}$ non-repeating edge additions and deletions and $O(n)$ node deletions and label-substitutions that can be performed. Therefore, the recursion has a finite depth of $O(n^2)$ that is also equal to the maximum length of $\mathcal{E}$. This approach t has exponential complexity in the worst case. Edit distances are generally expensive to compute, unless the underlying graphs are small.

17.3 Transformation-Based Distance Computation

The main problem with the distance measures of the previous section is that they are computationally impractical for larger graphs. A number of heuristic and kernel-based methods are used to transform the graphs into a space in which distance computations are more efficient. Interestingly, some of these methods are also *qualitatively* more effective because of their ability to focus on the relevant portions of the graphs.

17.3.1 Frequent Substructure-Based Transformation and Distance Computation

The intuition underlying this approach is that frequent graph patterns encode key properties of the graph. This is true of many applications. For example, the presence of a benzene ring (see Fig. 17.1) in a chemical compound will typically result in specific properties. Therefore, the properties of a graph can often be described by the presence of specific families of structures in it. This intuition suggests that a meaningful way of semantically describing the graph is in terms of its family of frequent substructures. Therefore, a transformation approach is used in which a text-like vector-space representation is created from each graph. The steps are as follows:

1. Apply frequent subgraph mining methods discussed in Sect. 17.4 to discover frequent subgraph patterns in the underlying graphs. This results in a "lexicon" in terms of which the graphs are represented. Unfortunately, the size of this lexicon is rather large, and many subgraphs may be redundant because of similarity to one another.

2. Select a subset of subgraphs from the subgraphs found in the first step to reduce the overlap among the frequent subgraph patterns. Different algorithms may vary in this step by using only frequent maximal subgraphs, or selecting a subset of graphs that are sufficiently nonoverlapping with one another. Create a new feature f_i for each frequent subgraph S_i that is finally selected. Let d be the total number of frequent subgraphs (features). This is the lexicon size in terms of which a text-like representation will be constructed.

3. For each graph G_i, create a vector-space representation in terms of the features $f_1 \dots f_d$. Each graph contains the features, corresponding to the subgraphs that it contains. The frequency of each feature is the number of occurrences of the corresponding subgraph in the graph G_i. It is also possible to use a binary representation by only considering presence or absence of subgraphs, rather than frequency of presence. The tf-idf normalization may be used on the vector-space representation, as discussed in Chap. 13.

After the transformation has been performed, any of the text similarity functions can be used to compute distances between graph objects. One advantage of using this approach is that it can be paired up with a conventional text index, such as the inverted index, for efficient retrieval. The bibliographic notes contain pointers to some of these methods.

This broader approach can also be used for feature transformation. Therefore, any data mining algorithm from the text domain can be applied to graphs using this approach. Later, it will be discussed how this transformation approach can be used in a more direct way by graph mining algorithms such as clustering. The main disadvantage of this approach is that subgraph isomorphism is an intermediate step in frequent substructure discovery. Therefore, the approach has exponential complexity in the worst case. Nevertheless, many fast approximations are often used to provide more efficient results without a significant loss in accuracy.

17.3.2 Topological Descriptors

Topological descriptors convert structural graphs to multidimensional data by using quantitative measures of important structural characteristics as dimensions. After the conversion has been performed, multidimensional data mining algorithms can be used on the transformed representation. This approach enables the use of a wide variety of multidimensional data mining algorithms in graph-based applications. The drawback of the approach is that the structural information is lost. Nevertheless, topological descriptors have been shown to retain important properties of graphs in the chemical domain, and are therefore used quite frequently. In general, the utility of topological descriptors in graph mining is highly domain specific. It should be pointed out that topological descriptors share a number of conceptual similarities with the frequent subgraph approach in the previous section. The

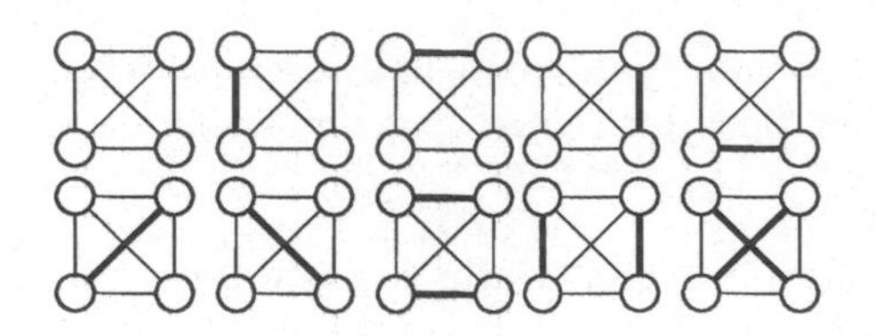

Figure 17.8: The Hosoya index for a clique of four nodes

main difference is that carefully chosen topological parameters are used to define the new feature space instead of frequent subgraphs.

Most topological descriptors are graph specific, whereas a few are node-specific. The vector of node-specific descriptors can sometimes describe the graph quite well. Node specific descriptors can also be used for enriching the labels of the nodes. Some common examples of topological descriptors are as follows:

1. *Morgan index:* This is a node-specific index that is equal to the kth order degree of a node. In other words, the descriptor is equal to the number of nodes reachable from the node within a distance of k. This is one of the few descriptors that describes nodes, rather than the complete graph. The node-specific descriptors can also be converted to a graph-specific descriptor by using the frequency histogram of the Morgan index over different nodes.

2. *Wiener index:* The Wiener index is equal to the sum of the pairwise shortest path distances between all pairs of nodes. It is therefore required to compute the all-pairs shortest path distance between different pairs of nodes.

$$W(G) = \sum_{i,j \in G} d(i,j) \tag{17.4}$$

   The Wiener index has known relationships with the chemical properties of compounds. The motivating reason for this index was the fact that it was known to be closely correlated with the boiling points of alkane molecules [511]. Later, the relationship was also shown for other properties of some families of molecules, such as their density, surface tension, viscosity, and van der Waal surface area. Subsequently, the index has also been used for applications beyond the chemical domain.

3. *Hosoya index:* The Hosoya index is equal to the number of valid pairwise node matchings in the graph. Note that the word "matching" refers to node–node matching within the same graph, rather than graph–graph matching. The matchings do not need to be maximal matchings, and even the empty matching is counted as one of the possibilities. The determination of the Hosoya index is #P-complete because an exponential number of possible matchings may exist in a graph, especially when it is dense. For example, as illustrated in Fig. 17.8, the Hosoya index for a complete graph (clique) of only four nodes is 10. The Hosoya index is also referred to as the Z-index.

4. *Estrada index:* This index is particularly useful in chemical applications for measuring the degree of protein folding. If $\lambda_1 \dots \lambda_n$ are the eigenvalues of the adjacency matrix of graph G, then the Estrada index $E(G)$ is defined as follows:

$$E(G) = \sum_{i=1}^{n} e^{\lambda_i} \tag{17.5}$$

5. *Circuit rank:* The circuit rank $C(G)$ is equal to the minimum number of edges that need to be removed from a graph in order to remove all cycles. For a graph with m edges, n nodes, and k connected components, this number is equal to $(m - n + k)$. The circuit rank is also referred to as the cyclomatic number. The cyclomatic number provides insights into the connectivity level of the graph.

6. *Randic index:* The Randic index is equal to the pairwise sum of bond contributions. If ν_i is the degree of vertex i, then the Randic index $R(G)$ is defined as follows:

$$R(G) = \sum_{i,j \in G} 1/\sqrt{\nu_i \cdot \nu_j} \tag{17.6}$$

   The Randic index is also known as the molecular connectivity index. This index is often used in the context of larger organic chemical compounds in order to evaluate their connectivity. The Randic index can be combined with the circuit rank $C(G)$ to yield the Balaban index $B(G)$:

$$B(G) = \frac{m \cdot R(G)}{C(G) + 1} \tag{17.7}$$

   Here, m is the number of edges in the network.

Most of these indices have been used quite frequently in the chemical domain because of their ability to capture different properties of chemical compounds.

17.3.3 Kernel-Based Transformations and Computation

Kernel-based methods can be used for faster similarity computation than is possible with methods such as MCG-based or edit-based measures. Furthermore, these similarity computation methods can be used directly with support vector machine (SVM) classifiers. This is one of the reasons that kernel methods are very popular in graph classification.

Several kernels are used frequently in the context of graph mining. The following contains a discussion of the more common ones. The kernel similarity $K(G_i, G_j)$ between a pair of graphs G_i and G_j is the dot product of the two graphs after hypothetically transforming them to a new space, defined by the function $\Phi(\cdot)$.

$$K(G_i, G_j) = \Phi(G_i) \cdot \Phi(G_j) \tag{17.8}$$

In practice, the value of $\Phi(\cdot)$ is not defined directly. Rather, it is defined indirectly in terms of the kernel similarity function $K(\cdot, \cdot)$. There are various ways of defining the kernel similarity.

17.3.3.1 Random Walk Kernels

In random walk kernels, the idea is to compare the label sequences induced by random walks in the two graphs. Intuitively, two graphs are similar if many sequences of labels created by random walks between pairs of nodes are similar as well. The main computational challenge is that there are an exponential number of possible random walks between pairs of nodes. Therefore, the first step is to defined a primitive kernel function $k(s_1, s_2)$ between a pair of node sequences s_1 (from G_1) and s_2 (from G_2). The simplest kernel is the identity kernel:

$$k(s_1, s_2) = I(s_1 = s_2) \tag{17.9}$$

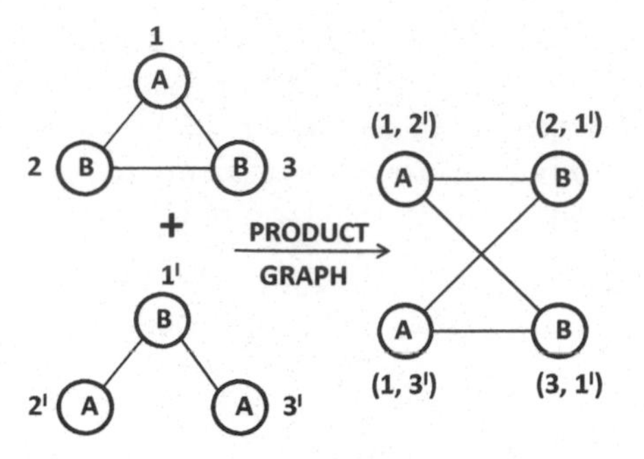

Figure 17.9: Example of the product graph

Here, $I(\cdot)$ is the indicator function that takes the value of 1 when the two sequences are the same and 0 otherwise. Then, the overall kernel similarity $K(G_1, G_2)$ is defined as the sum of the probabilities of all the primitive sequence kernels over all possible walks:

$$K(G_1, G_2) = \sum_{s_1, s_2} p(s_1|G_1) \cdot p(s_2|G_2) \cdot k(s_1, s_2) \tag{17.10}$$

Here, $p(s_i|G_i)$ is the probability of the random walk sequence s_i in the graph G_i. Note that this kernel similarity value will be higher when the same label sequences are used by the two graphs. A key challenge is to compute these probabilities because there are an exponential number of walks of a specific length, and the length of a walk may be any value in the range $(1, \infty)$.

The random walk kernel is computed using the notion of a *product graph* G_X between G_1 and G_2. The product graphs are constructed by defining a vertex $[u_1, u_2]$ between each pair of label matching vertices u_1 and u_2 in the graphs G_1 and G_2, respectively. An edge is added between a pair of vertices $[u_1, u_2]$ and $[v_1, v_2]$ in the product graph G_X if and only an edge exists between the corresponding nodes in *both* the individual graphs G_1 and G_2. In other words, the edge (u_1, v_1) must exist in G_1 and the edge (u_2, v_2) must exist in G_2. An example of a product graph is illustrated in Fig. 17.9. Note that each walk in the product graph corresponds to a pair of label-matching sequence of vertices in the two graphs G_1 and G_2. Then, if A is the binary adjacency matrix of the product graph, then the entries of A^k provide the number of walks of length k between the different pairs of vertices. Therefore, the total weighted number of walks may be computed as follows:

$$K(G_1, G_2) = \sum_{i,j} \sum_{k=1}^{\infty} \lambda^k [A^k]_{ij} = \overline{e}^T (I - \lambda A)^{-1} \overline{e} \tag{17.11}$$

Here, $\overline{e}$ is an $|G_X|$-dimensional column vector of 1s, and $\lambda \in (0, 1)$ is a discount factor. The discount factor λ should always be smaller than the inverse of the largest eigenvalue of A to ensure convergence of the infinite summation. Another variant of the random walk kernel is as follows:

$$K(G_1, G_2) = \sum_{i,j} \sum_{k=1}^{\infty} \frac{\lambda^k}{k!} [A^k]_{ij} = \overline{e}^T \exp(\lambda A) \overline{e} \tag{17.12}$$

When the graphs in a collection are widely varying in size, the kernel functions of Eqs. 17.11 and 17.12 should be further normalized by dividing with $|G_1| \cdot |G_2|$. Alternatively, in some probabilistic versions of the random walk kernel, the vectors $\overline{e}^T$ and $\overline{e}$ are replaced with starting and stopping probabilities of the random walk over various nodes in the product graph. This computation is quite expensive, and may require as much as $O(n^6)$ time.

17.3.3.2 Shortest-Path Kernels

In the shortest-path kernel, a primitive kernel $k_s(i_1, j_1, i_2, i_2)$ is defined on node-pairs $[i_1, j_1] \in G_1$ and $[i_2, j_2] \in G_2$. There are several ways of defining the kernel function $k_s(i_1, i_2, j_1, j_2)$. A simple way of defining the kernel value is to set it to 1 when the distance $d(i_1, i_2) = d(j_1, j_2)$, and 0, otherwise.

Then, the overall kernel similarity is equal to the sum of all primitive kernels over different quadruplets of nodes:

$$K(G_1, G_2) = \sum_{i_1, i_2, j_1, j_2} k_s(i_1, i_2, j_1, j_2) \tag{17.13}$$

The shortest-path kernel may be computed by applying the all-pairs shortest-path algorithm on each of the graphs. It can be shown that the complexity of the kernel computation is $O(n^4)$. Although this is still quite expensive, it may be practical for small graphs, such as chemical compounds.

17.4 Frequent Substructure Mining in Graphs

Frequent subgraph mining is a fundamental building block for graph mining algorithms. Many of the clustering, classification, and similarity search techniques use frequent substructure mining as an intermediate step. This is because frequent substructures encode important properties of graphs in many application domains. For example, consider the series of phenolic acids illustrated in Fig. 17.10. These represent a family of organic compounds with similar chemical properties. Many complex variations of this family act as signaling molecules and agents of defense in plants. The properties of phenolic acids are a direct result of the presence of two frequent substructures, corresponding to the carboxyl group and phenol group, respectively. These groups are illustrated in Fig. 17.10 as well. The relevance of such substructural properties is not restricted to the chemical domain. This is the reason that frequent substructures are often used in the intermediate stages of many graph mining applications such as clustering and classification.

The definition of a frequent subgraph is identical to the case of association pattern mining, except that a subgraph relationship is used to count the support rather than a subset relationship. Many well-known frequent substructure mining algorithms are based on the enumeration tree principle discussed in Chap. 4. The simplest of these methods is based on the *Apriori* algorithm. This algorithm is discussed in detail in Fig. 4.2 of Chap. 4. The *Apriori* algorithm uses joins to create candidate patterns of size $(k + 1)$ from frequent patterns of size k. However, because of the greater complexity of graph-structured data, the join between a pair of graphs may not result in a unique solution. For example, candidate frequent patterns can be generated by either *node extensions* or *edge extensions*. Thus, the main difference between these two variations is in terms of how frequent substructures of size k are defined and joined together to create candidate structures of size $(k + 1)$. *The "size" of a subgraph may refer to either the number of nodes in it, or the number of edges in it depending on whether node extensions or edge extensions are used.* Therefore, the following will describe the *Apriori*-based algorithm in a general way without specifically discussing

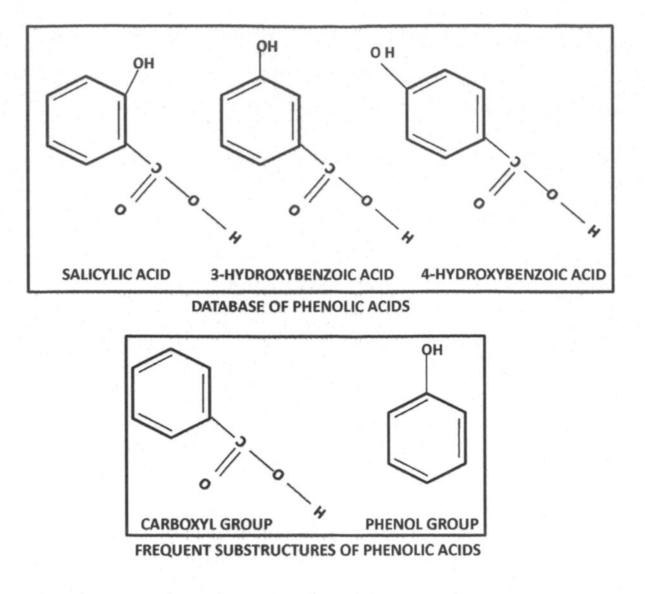

Figure 17.10: Examples of frequent substructures in a database of phenolic acids

either node extensions or edge extensions. Subsequently, the precise changes required to enable these two specific variations will be discussed.

The overall algorithm for frequent subgraph mining is illustrated in Fig. 17.11. The input to the algorithm is the graph database $\mathcal{G} = \{G_1 \dots G_n\}$ and a minimum support value *minsup*. The basic algorithm structure is similar to that of the *Apriori* algorithm, discussed in Fig. 4.2 of Chap. 4. A levelwise algorithm is used, in which candidate subgraphs $\mathcal{C}_{k+1}$ of size $(k + 1)$ are generated by using joins on graph pairs from the set of frequent subgraphs $\mathcal{F}_k$ of size k. As discussed earlier, the size of a subgraph may refer to either its nodes or edges, depending on the specific algorithm used. The two graphs need to be matching in a subgraph of size $(k - 1)$ for a join to be successfully performed. The resulting candidate subgraph will be of size $(k + 1)$. Therefore, one of the important steps of join processing, is determining whether two graphs share a subgraph of size $(k - 1)$ in common. The matching algorithms discussed in Sect. 17.2 can be used for this purpose. In some applications, where node labels are distinct and isomorphism is not an issue, this step can be performed very efficiently. On the other hand, for large graphs that have many repeating node labels, this step is slow because of isomorphism.

After the pairs of matching graphs have been identified, joins are performed on them in order to generate the candidates $\mathcal{C}_{k+1}$ of size $(k + 1)$. The different node-based and edge-based variations in the methods for performing joins will be described later. Furthermore, the *Apriori* pruning trick is used. Candidates in $\mathcal{C}_{k+1}$ that are such that any of their k-subgraphs do not exist in $\mathcal{F}_k$ are pruned. For each remaining candidate subgraph, the support is computed with respect to the graph database $\mathcal{G}$. The subgraph isomorphism algorithm discussed in Sect. 17.2 needs to be used for computing the support. All candidates in $\mathcal{C}_{k+1}$ that meet the minimum support requirement are retained in $\mathcal{F}_{k+1}$. The procedure is repeated iteratively until an empty set $\mathcal{F}_{k+1}$ is generated. At this point, the algorithm terminates, and the set of frequent subgraphs in $\cup_{i=1}^{k} \mathcal{F}_i$ is reported. Next, the two different ways of defining the size k of a graph, corresponding to node- and edge-based joins, will be described.

Algorithm *GraphApriori*(Graph Database: $\mathcal{G}$,
Minimum Support: *minsup*);
begin
$\mathcal{F}_1 = \{$ All Frequent singleton graphs $\}$;
$k = 1$;
while $\mathcal{F}_k$ is not empty **do begin**
Generate $\mathcal{C}_{k+1}$ by joining pairs of graphs in $\mathcal{F}_k$ that
share a subgraph of size $(k-1)$ in common;
Prune subgraphs from $\mathcal{C}_{k+1}$ that violate downward closure;
Determine $\mathcal{F}_{k+1}$ by support counting on $(\mathcal{C}_{k+1}, \mathcal{G})$ and retaining
subgraphs from $\mathcal{C}_{k+1}$ with support at least *minsup*;
$k = k + 1$;
end;
return$(\cup_{i=1}^{k} \mathcal{F}_i)$;
end

Figure 17.11: The basic frequent subgraph discovery algorithm is related to the *Apriori* algorithm. The reader is encouraged to compare this pseudocode with the *Apriori* algorithm described in Fig. 4.2 of Chap. 4.

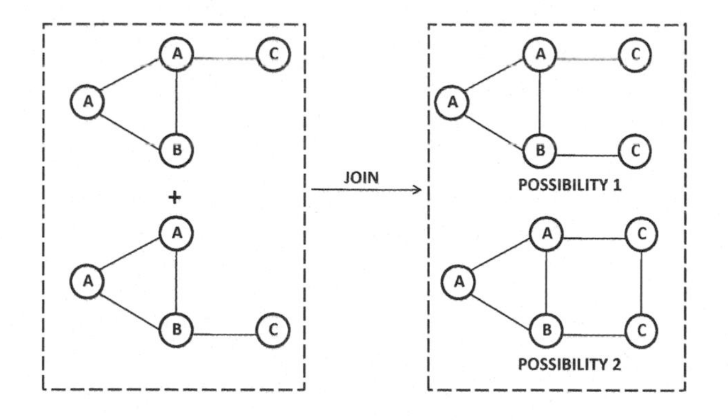

Figure 17.12: Candidates generated using node-based join of two graphs

17.4.1 Node-Based Join Growth

In the case of node-based joins, the size of a frequent subgraph in $\mathcal{F}_k$ refers to the number of *nodes* k in it. The singleton graphs in $\mathcal{F}_1$ contain a single node. These are node labels that are present in at least *minsup* graphs in the graph database $\mathcal{G}$. For two graphs from $\mathcal{F}_k$ to be joined, *a matching subgraph with* $(k-1)$ *nodes* must exist between the two graphs. This matching subgraph is also referred to as the *core*. When two k-subgraphs with $(k-1)$ common nodes are joined to create a candidate with $(k+1)$ nodes, an ambiguity exists, as to whether or not an edge exists between the two non-matching nodes. Therefore, two possible graphs are generated, depending on whether or not an edge exists between the nodes that are not common between the two. An example of the two possibilities for generating candidate subgraphs is illustrated in Fig. 17.12. While this chapter does not assume that *edge* labels are associated with graphs, the number of possible joins will be even larger when labels are associated with edges. This is because each possible edge label must be associated with the newly created edge. This will result in a larger number of candidates. Furthermore, in cases where there are isomorphic matchings of size $(k-1)$ between the two frequent subgraphs, candidates may need to be generated for each such mapping (see Exercise 8). Thus, all possible $(k-1)$ common subgraphs need to be discovered between a pair of graphs, in order to generate the candidates. Thus, the explosion in the number of candidate patterns is usually more significant in the case of frequent subgraph discovery, than in the case of frequent pattern discovery.

17.4.2 Edge-Based Join Growth

In the case of edge-based joins, the size of a frequent subgraph in $\mathcal{F}_k$ refers to the number of *edges* k in it. The singleton graphs in $\mathcal{F}_1$ contain a single edge. These correspond to edges between specific node labels that are present in at least *minsup* graphs in the database $\mathcal{G}$. In order for two graphs from $\mathcal{F}_k$ to be joined, *a matching subgraph with* $(k-1)$ *edges* needs to be present in the two graphs. The resulting candidate will contain exactly $(k+1)$ edges. Interestingly, the number of *nodes* in the candidate may not necessarily be greater than that in the individual subgraphs that are joined. In Fig. 17.13, the two possible candidates that are constructed using edge-based joins are illustrated. Note that one of the generated candidates has the same number of nodes as the original pair of graphs. As in the case of node-based joins, one needs to account for isomorphism in the process of candidate generation. Edge-based join growth tends to generate fewer candidates in total and is therefore generally more efficient. The bibliographic notes contain pointers to more details about these methods.

17.4.3 Frequent Pattern Mining to Graph Pattern Mining

The similarity between the aforementioned approach and *Apriori* is quite striking. The join-based growth strategy can also be generalized to an enumeration tree-like strategy. However, the analogous candidate tree can be generated in two different ways, corresponding to node- and edge-based extensions, respectively. Furthermore, tree growth is more complex because of isomorphism. *GraphApriori* uses a breadth-first candidate-tree generation approach as in all *Apriori*-like methods. It is also possible to use other strategies, such as depth-first methods, to grow the tree of candidates. As discussed in Chap. 4, almost all frequent pattern mining algorithms, including[1] *Apriori* and *FP-growth*, should be considered enumeration-

[1]See the discussion in Sect. 4.4.4.5 of Chap. 4.

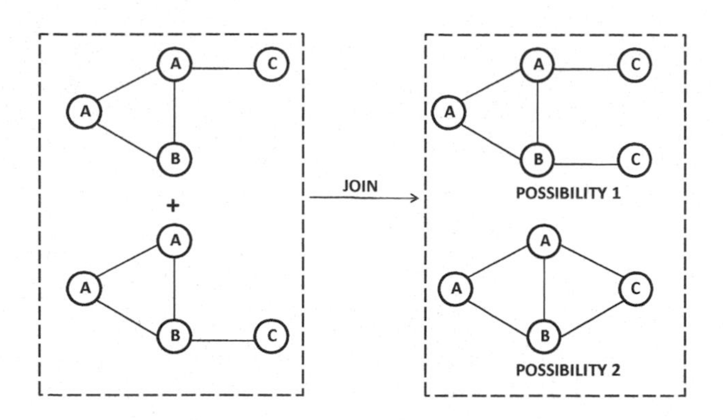

Figure 17.13: Candidates generated using edge-based join of two graphs

tree methods. Therefore, the broader principles of these algorithms can also be generalized to the growth of the candidate tree in graphs. The bibliographic notes contain pointers to these methods.

17.5 Graph Clustering

The graph clustering problem partitions a database of n graphs, denoted by $G_1 \dots G_n$, into groups. Graph clustering methods are either distance-based or frequent substructure-based. Distance-based methods are more effective for smaller graphs, in which distances can be computed robustly and efficiently. Frequent substructure-based methods are appropriate for larger graphs where distance computations become qualitatively and computationally impractical.

17.5.1 Distance-Based Methods

The design of distance functions is particularly important for virtually every complex data type because of their applicability to clustering methods, such as k-medoids and spectral methods, that are dependent only on the design of the distance function. Virtually all the complex data types discussed in Chaps. 13–16 use this general methodology for clustering. This is the reason that distance function design is usually the most fundamental problem that needs to be addressed in every data domain. Sections 17.2 and 17.3 of this chapter have discussed methods for distance computation in graphs. After a distance function has been designed, the following two methods can be used:

1. The k-medoids method introduced in Sect. 6.3.4 in Chap. 6 uses a representative-based approach, in which the distances of data objects to their closest representatives are used to perform the clustering. A set of k representatives is used, and data objects are assigned to their closest representatives by using an appropriately designed distance function. The set of k representatives is progressively optimized by using a hill-climbing approach, in which the representatives are iteratively swapped with other data objects in order to improve the clustering objective function value. The reader is referred to Chap. 6 for details of the k-medoids algorithm. A key property of this algorithm is that the computations are not dependent on the nature of the data type after the distance function has been defined.

2. A second commonly-used methodology is that of spectral methods. In this case, the individual graph objects are used to construct a single large neighborhood graph. The latter graph is a higher level similarity graph, in which each node corresponds to one of the (smaller) graph objects from the original database and the weight of the edge is equal to the similarity between the two objects. As discussed in Sect. 6.7 of Chap. 6, distances can be converted to similarity values with the use of a kernel transformation. Each node is connected to its k-nearest neighbors with an undirected edge. Thus, the problem of clustering graph *objects* is transformed to the problem of clustering *nodes* in a single large graph. This problem is discussed briefly in Sect. 6.7 of Chap. 6, and in greater detail in Sect. 19.3 of Chap. 19. Any of the network clustering or community detection algorithms can be used to cluster the nodes, although spectral methods are used quite commonly. After the node clusters have been determined, they are mapped back to clusters of graph objects.

The aforementioned methods do not work very well when the individual graph objects are large because of two reasons. It is generally computationally expensive to compute distances between large graph objects. Graph distance functions, such as matching-based methods, have a complexity that increases exponentially with graph object size. The *effectiveness* of such methods also drops sharply with increasing graph size. This is because the graphs may be similar only in some portions that repeat frequently. The rare portions of the graphs may be unique to the specific graph at hand. In fact, many small substructures may be repeated across the two graphs. Therefore, a matching-based distance function may not be able to properly compare the key features of the different graphs. One possibility is to use a substructure-based distance function, as discussed in Sect. 17.3.1. A more direct approach is to use frequent substructure-based methods.

17.5.2 Frequent Substructure-Based Methods

These methods extract frequent subgraphs from the data and use their membership in input graphs to determine clusters. The basic premise is that the frequent subgraphs are indicative of cluster membership because of their propensity to define application-specific properties. For example, in an organic chemistry application, a benzene ring (illustrated as a subgraph of Fig. 17.1a) is a frequently occurring substructure that is indicative of specific chemical properties of the compound. In an XML application, a frequent substructure corresponds to important structural relationships between entities. Therefore, the membership of such substructures in graphs is highly indicative of similarity and cluster membership. Interestingly, frequent pattern mining algorithms are also used in multidimensional clustering. An example is the *CLIQUE* algorithm (cf. Sect. 7.4.1 of Chap. 7).

In the following sections, two different methods for graph clustering will be described. The first is a generic transformational approach that can be used to apply text clustering methods to the graph domain. The second is a more direct iterative approach of relating the graph clusters to their frequent substructures.

17.5.2.1 Generic Transformational Approach

This approach transforms the graph database to a text-like domain, so that the wide variety of text clustering algorithms may be leveraged. The broad approach may be described as follows:

1. Apply frequent subgraph mining methods discussed in Sect. 17.4 in order to discover frequent subgraph patterns in the underlying graphs. Select a subset of subgraphs to

reduce overlap among the different subgraphs. Different algorithms may vary on this step by using only frequent maximal subgraphs, or selecting a subset of graphs that are sufficiently nonoverlapping with one another. Create a new feature f_i for each frequent subgraph S_i that is discovered. Let d be the total number of frequent subgraphs (features). This is the "lexicon" size in terms of which a text-like representation will be constructed.

2. For each graph G_i, create a vector-space representation in terms of the features $f_1 \ldots f_d$. Each graph contains the features corresponding to the subgraphs that it contains. The frequency of each feature is the number of occurrences of the corresponding subgraph in the graph G_i. It is also possible to use a binary representation by only considering the presence or absence of subgraphs rather than frequency of presence. Use tf-idf normalization on the vector-space representation, as discussed in Chap. 13.

3. Use any of the text-clustering algorithms discussed in Sect. 13.3 in Chap. 13, in order to discover clusters of newly created text objects. Map the text clusters to graph object clusters.

This broader approach of using text-based methods is utilized frequently with many contextual data types. For example, an almost exactly analogous approach is discussed for sequence clustering in Sect. 15.3.3 of Chap. 15. This is because a modified version of frequent pattern mining methods can be defined for most data types. It should be pointed out that, although the substructure-based transformation is discussed here, many of the kernel-based transformations and topological descriptors, discussed earlier in this chapter, may be used as well. For example, the kernel k-means algorithm can be used in conjunction with the graph kernels discussed in this chapter.

17.5.2.2 XProj: Direct Clustering with Frequent Subgraph Discovery

The *XProj* algorithm derives its name from the fact that it was originally proposed for XML graphs, and a substructure can be viewed as a PROJection of the graph. Nevertheless, the approach is not specific to XML structures, and it can be applied to any other graph domain, such as chemical compounds. The *XProj* algorithm uses the substructure discovery process as an important subroutine, and different applications may use different substructure discovery methods, depending on the data domain. Therefore, the following will provide a generic description of the *XProj* algorithm for graph clustering, although the substructure discovery process may be implemented in an application-specific way. Because the algorithm uses the frequent substructures for the clustering process, an additional input to the algorithm is the minimum support *minsup*. Another input to the algorithm is the size l of the frequent substructures mined. The size of the frequent substructures is fixed in order to ensure robust computation of similarity. These are user-defined parameters that can be tuned to obtain the most effective results.

The algorithm can be viewed as a representative approach similar to k-medoids, except that *each representative is a set of frequent substructures*. These represent the *localized substructures* of each group. The use of frequent-substructures as the representatives, instead of the original graphs, is crucial. This is because distances cannot be computed effectively between pairs of graphs, when the sizes of the graphs are larger. On the other hand, the membership of frequent substructures provides a more intuitive way of computing similarity. It should be pointed out that, unlike transformational methods, the frequent substructures

Algorithm *XProj*(Graph Database: $\mathcal{G}$, Minimum Support: *minsup*
Structural Size: l, Number of Clusters: k)
begin
Initialize clusters $\mathcal{C}_1 \dots \mathcal{C}_k$ randomly;
Compute frequent substructure sets $\mathcal{F}_1 \dots \mathcal{F}_k$ from $\mathcal{C}_1 \dots \mathcal{C}_k$;
repeat
Assign each graph $G_j \in \mathcal{G}$ to the cluster $\mathcal{C}_i$ for which the former's similarity to $\mathcal{F}_i$ is the largest $\forall i \in \{1 \dots k\}$;
Compute frequent substructure set $\mathcal{F}_i$ from $\mathcal{C}_i$ for each $i \in \{1 \dots k\}$;
until convergence;
end

Figure 17.14: The frequent subgraph-based clustering algorithm (high level description)

are local to each cluster, and are therefore better optimized. This is the main advantage of this approach over a generic transformational approach.

There are a total of k such frequent substructure sets $\mathcal{F}_1 \dots \mathcal{F}_k$, and the graph database is partitioned into k groups around these localized representatives. The algorithm is initialized with a random partition of the database $\mathcal{G}$ into k clusters. These k clusters are denoted by $\mathcal{C}_1 \dots \mathcal{C}_k$. The frequent substructures $\mathcal{F}_i$ of each of these clusters $\mathcal{C}_i$ can be determined using any frequent substructure discovery algorithm. Subsequently, each graph in $G_j \in \mathcal{G}$ is assigned to one of the representative sets $\mathcal{F}_i$ based on the similarity of G_j to each representative set $\mathcal{F}_i$. The details of the similarity computation will be discussed later. This process is repeated iteratively, so that the representative set $\mathcal{F}_i$ is generated from cluster $\mathcal{C}_i$, and the cluster $\mathcal{C}_i$ is generated from the frequent set $\mathcal{F}_i$. The process is repeated, until the change in the average similarity of each graph G_j to its assigned representative set $\mathcal{F}_i$ is no larger than a user-defined threshold. At this point, the algorithm is assumed to have converged, and it terminates. The overall algorithm is illustrated in Fig. 17.14.

It remains to be described how the similarity between a graph G_j and a representative set $\mathcal{F}_i$ is computed. The similarity between G_j and $\mathcal{F}_i$ is computed with the use of a coverage criterion. The similarity between G_j and $\mathcal{F}_i$ is equal to the fraction of frequent substructures in $\mathcal{F}_i$ that are a subgraph of G_j.

A major computational challenge is that the determination of frequent substructures in $\mathcal{F}_i$ may be too expensive. Furthermore, there may be a large number of frequent substructures in $\mathcal{F}_i$ that are highly overlapping with one another. To address these issues, the *XProj* algorithm proposes a number of optimizations. The first optimization is that the frequent substructures do not need to be determined exactly. An approximate algorithm for frequent substructure mining is designed. The second optimization is that only a subset of nonoverlapping substructures of length l are included in the sets $\mathcal{F}_i$. The details of these optimizations may be found in pointers discussed in the bibliographic notes.

17.6 Graph Classification

It is assumed that a set of n graphs $G_1 \dots G_n$ is available, but only a subset of these graphs is labeled. Among these, the first $n_t \leq n$ graphs are labeled, and the remaining $(n - n_t)$ graphs are unlabeled. The labels are drawn from $\{1 \dots k\}$. It is desired to use the labels on the training graphs to infer the labels of unlabeled graphs.

17.6.1 Distance-Based Methods

Distance-based methods are most appropriate when the sizes of the underlying graphs are small, and the distances can be computed efficiently. Nearest neighbor methods and collective classification methods are two of the distance-based methods commonly used for classification. The latter method is a transductive semi-supervised method, in which the training and test instances need to be available at the same time for the classification process. These methods are described in detail below:

1. *Nearest neighbor methods:* For each test instance, the k-nearest neighbors are determined. The dominant label from these nearest neighbors is reported as the relevant label. The nearest neighbor method for multidimensional data is described in detail in Sect. 10.8 of Chap. 10. The only modification to the method is the use of a different distance function, suited for the graph data type.

2. *Graph-based methods:* This is a semi-supervised method, discussed in Sect. 11.6.3 of Chap. 11. In graph-based methods, a higher level *neighborhood graph* is constructed from the training and test graph objects. It is important not to confuse the notion of a neighborhood graph with that of the original graph objects. The original graph objects correspond to nodes in the neighborhood graph. Each node is connected to its k nearest neighbor objects based on the distance values. This results in a graph containing both labeled and unlabeled nodes. This is the collective classification problem, for which various algorithms are described in Sect. 19.4 of Chap. 19. Collective classification algorithms can be used to derive labels of the nodes in the neighborhood graphs. These derived labels can then be mapped back to the unlabeled graph objects.

Distance-based methods are generally effective when the underlying graph objects are small. For larger graph objects, the computation of distances becomes too expensive. Furthermore, distance computations no longer remain effective from an accuracy perspective, when multiple common substructures are present in the two graphs.

17.6.2 Frequent Substructure-Based Methods

Pattern-based methods extract frequent subgraphs from the data, and use their membership in different graphs, in order to build classification models. As in the case of clustering, the main assumption is that the frequently occurring portions of graphs can related to application-specific properties of the graphs. For example, the phenolic acids of Fig. 17.10 are characterized by the two frequent substructures corresponding to the carboxyl group and the phenol group. These substructures therefore characterize important properties of a *family* or a *class* of compounds. This is generally true across many different applications beyond the chemical domain. As discussed in Sect. 10.4 of Chap. 10, frequent patterns are often used for rule-based classification, even in the "flat" multidimensional domain. As in the case of clustering, either a generic transformational approach or a more direct rule-based method can be used.

17.6.2.1 Generic Transformational Approach

This approach is generally similar to the transformational approach discussed in the previous section on clustering. However, there are a few differences that account for the impact of supervision. The broad approach may be described as follows:

1. Apply frequent subgraph mining methods discussed in Sect. 17.4 to discover frequent subgraph patterns in the underlying graphs. Select a subset of subgraphs to reduce overlap among the different subgraphs. For example, feature selection algorithms that minimize redundancy and maximize the relevance of the features may be used. Such feature selection algorithms are discussed in Sect. 10.2 of Chap. 10. Let d be the total number of frequent subgraphs (features). This is the "lexicon" size in terms of which a text-like representation will be constructed.

2. For each graph G_i, create a vector-space representation in terms of the d features found. Each graph contains the features corresponding to the subgraphs that it contains. The frequency of each feature is equal to the number of occurrences of the corresponding subgraph in graph G_i. It is also possible to use a binary representation by only considering presence or absence of subgraphs, rather than frequency of presence. Use tf-idf normalization on the vector-space representation, as discussed in Chap. 13.

3. Select any text classification algorithm discussed in Sect. 13.5 of Chap. 13 to build a classification model. Use the model to classify test instances.

This approach provides a flexible framework. After the transformation has been performed, a wide variety of algorithms may be used. It also allows the use of different types of supervised feature selection methods to ensure that the most discriminative structures are used for classification.

17.6.2.2 XRules: A Rule-Based Approach

The *XRules* method was proposed in the context of XML data, but it can be used in the context of any graph database. This is a rule-based approach that relates frequent substructures to the different classes. The training phase contains three steps:

1. In the first phase, frequent substructures with sufficient support and confidence are determined. Each rule is of the form:

   $$F_g \Rightarrow c$$

   The notation F_g denotes a frequent substructure, and c is a class label. Many other measures can be used to quantify the strength of the rule instead of the confidence. Examples include the likelihood ratio, or the cost-weighted confidence in the rare class scenario. The likelihood ratio of $F_g \Rightarrow c$ is the ratio of the fractional support of F_g in the examples containing c, to the fractional support of F_g in examples not containing c. A likelihood ratio greater than one indicates that the rule is highly likely to belong to a particular class. The generic term for these different ways of measuring the class-specific relevance is the *rule strength.*

2. In the second phase, the rules are ordered and pruned. The rules are ordered by decreasing strength. Statistical thresholds on the rule strength may be used for pruning rules with low strength. This yields a compact set $\mathcal{R}$ of ordered rules that are used for classification.

3. In the final phase, a default class is set that can be used to classify test instances not covered by any rule in $\mathcal{R}$. The default class is set to the dominant class of the set of training instances not covered by rule set $\mathcal{R}$. A graph is covered by a rule if the

left-hand side of the rule is a substructure of the graph. In the event that all training instances are covered by rule set $\mathcal{R}$, then the default class is set to the dominant class in the entire training data. In cases where classes are associated with costs, the cost-sensitive weight is used in determining the majority class.

After the training model has been constructed, it can be used for classification as follows. For a given test graph G, the rules that are fired by G are determined. If no rules are fired, then the default class is reported. Let $\mathcal{R}_c(G)$ be the set of rules fired by G. Note that these different rules may not yield the same prediction for G. Therefore, the conflicting predictions of different rules need to be combined meaningfully. The different criteria that are used to combine the predictions are as follows:

1. *Average strength:* The average strength of the rules predicting each class are determined. The class with the average highest strength is reported.

2. *Best rule:* The top rule is determined on the basis of the priority order discussed earlier. The class label of this rule is reported.

3. *Top-k average strength:* This can be considered a combination of the previous two methods. The average strength of the top-k rules of each class is used to determine the predicted label.

The *XRules* procedure uses an efficient procedure for frequent substructure discovery, and many other variations for rule quantification. Refer to the bibliographic notes.

17.6.3 Kernel SVMs

Kernel SVMs can construct classifiers with the use of kernel similarity between training and test instances. As discussed in Sect. 10.6.4 of Chap. 10, kernel SVMs do not actually need the feature representation of the data, as long as the kernel-based similarity $K(G_i, G_j)$ between any pair of graph objects is available. Therefore, the approach is agnostic to the specific data type that is used. The different kinds of graph kernels are discussed in Sect. 17.3.3 of this chapter. Any of these kernels can be used in conjunction with the *SVM* approach. Refer to Sect. 10.6.4 of Chap. 10 for details on how kernels may be used in conjunction with SVM classifiers.

17.7 Summary

This chapter studies the problem of mining graph data sets. Graph data are a challenging domain for analysis, because of the difficulty in matching two graphs when there are repetitions in the underlying labels. This is referred to as graph isomorphism. Most methods for graph matching require exponential time in the worst case. The MCG between a pair of graphs can be used to define distance measures between graphs. The edit-distance measure also uses an algorithm that is closely related to the MCG algorithm. Because of the complexity of matching algorithms in graphs, a different approach is to transform the graph database into a simpler text-like representation, in terms of which distance functions are defined. An important class of graph distance functions is graph kernels. They can be used for clustering and classification.

The frequent substructure discovery algorithm is an important building block because it can be leveraged for other graph mining problems such as clustering and classification.

The *Apriori*-like algorithms use either a node-growth strategy, or an edge-growth strategy in order to generate the candidates and corresponding frequent substructures. Most of the clustering and classification algorithms for graph data are based either on distances, or on frequent substructures. The distance-based methods include the k-medoids and spectral methods for clustering. For classification, the distance-based methods include either the k-nearest neighbor method or graph-based semi-supervised methods. Kernel-based SVM can also be considered specialized distance-based methods, in which SVMs are leveraged in conjunction with the similarity between data objects.

Frequent substructure-based methods are used frequently for graph clustering and classification. A generic approach is to transform the graphs into a new feature representation that is similar to text data. Any of the text clustering or classification algorithms can be applied to this representation. Another approach is to directly mine the frequent substructures and use them as representative sets of clusters, or antecedents of discriminative rules. The *XProj* and *XRules* algorithms are based on this principle.

17.8 Bibliographic Notes

The problem of graph matching is addressed in surveys in [26]. The Ullman algorithm for graph matching was proposed in [164]. Two other well known methods for graph-matching are *VF2* [162] and *QuickSI* [163]. Other approximate matching methods are discussed in [313, 314, 521]. The proof of NP-hardness of the graph matching problem may be found in [221, 164]. The use of the MCG for defining distance functions was studied in [120]. The relationship between the graph-edit distance and the maximum common subgraph problem is studied in detail in [119]. The graph edit-distance algorithm discussed in the chapter is a simplification of the algorithm presented in [384]. A number of fast algorithms for computing the graph edit distance are discussed in [409]. The problem of learning edit costs is studied in [408]. The survey by Bunke in [26] also discusses methods for computing the graph edit costs. A description of the use of topological descriptors in drug-design may be found in [236]. The random walk kernel is discussed in [225, 298], and the shortest-path kernel is discussed in [103]. The work in [225] also provides a generic discussion on graph kernels. The work in [42] shows that frequent substructure-based similarity computation can provide robust results in data mining applications.

The node-growth strategy for frequent subgraph mining was proposed by Inokuchi, Washio, an Motoda [282]. The edge-growth strategy was proposed by Kuramochi and Karypis [331]. The *gSpan* algorithm was proposed by Yan and Han [519] and uses a depth-first approach to build the candidate tree of graph patterns. A method that uses the vertical representation for graph pattern mining is discussed in [276]. The problem of mining frequent trees in a forest was addressed in [536]. Surveys on graph clustering and classification may be found in [26]. The *XProj* algorithm is discussed in [42], and the *XRules* algorithm is discussed in [540]. Methods for kernel SVM-based classification are discussed in the graph classification chapter by Tsuda in [26].

17.9 Exercises

1. Consider two graphs that are cliques containing an even number $2 \cdot n$ nodes. Let exactly half the nodes in each graph belong to labels A and B. What are the total number of isomorphic matchings between the two graphs?

2. Consider two graphs containing $2 \cdot n$ nodes and n distinct labels, each of which occurs twice. What is the maximum number of isomorphic matchings between the two graphs?

3. Implement the basic algorithm for subgraph isomorphism with no pruning optimizations. Test it by trying to match pairs of randomly generated graphs, containing a varying number of nodes. How does the running time vary with the size of the graph?

4. Compute the Morgan indices of order 1 and 2, for each node of the *acetaminophen* graph of Fig. 17.1. How does the Morgan index vary with the labels (corresponding to chemical elements)?

5. Write a computer program to compute each of the topological descriptors of a graph discussed in this chapter.

6. Write a computer program to execute the node-based candidate growth for frequent subgraph discovery. Refer to the bibliographic notes, if needed, for the paper describing specific details of the algorithm.

7. Write a computer program to execute the edge-based candidate growth for frequent subgraph discovery. Refer to the bibliographic notes for the paper describing specific details of the algorithm.

8. Show the different node-based joins that can be performed between the two graphs below, while accounting for isomorphism.

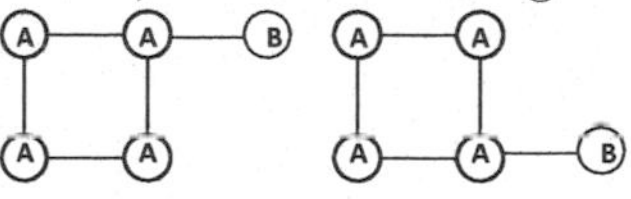

9. Show the different edge-based joins that can performed between the two graphs of Exercise 8, while accounting for isomorphism.

10. Determine the maximum number of candidates that can be generated with node-based join growth using a single pair of graphs, while accounting for isomorphism. Assume that the matching core of these graphs is a cycle of size k. What conditions in the core of the joined portion result in this scenario?

11. Discuss how the node-based growth and edge-based growth strategies translate into a candidate tree structure that is analogous to the enumeration tree in frequent pattern mining.

12. Implement a computer program to construct a text-like representation for a database of graphs, as discussed in the chapter. Use any feature selection approach of your choice of minimize redundancy. Implement a k-means clustering algorithm with this representation.

13. Repeat Exercise 12 for the classification problem. Use a naive Bayes classifier, as discussed in Chapter 10, for the final classification step and an appropriately chosen supervised feature selection method from the same chapter.

14. What changes would be require in the subgraph isomorphism algorithm for cases in which the query graph is disconnected?

Chapter 18

Mining Web Data

"Data is a precious thing, and will last longer than the systems themselves."—Tim Berners-Lee

18.1 Introduction

The Web is an unique phenomenon in many ways, in terms of its scale, the distributed and uncoordinated nature of its creation, the openness of the underlying platform, and the resulting diversity of applications it has enabled. Examples of such applications include e-commerce, user collaboration, and social network analysis. Because of the distributed and uncoordinated nature in which the Web is both created and used, it is a rich treasure trove of diverse types of data. This data can be either a source of knowledge about various subjects, or personal information about users.

Aside from the content available in the documents on the Web, the usage of the Web results in a significant amount of data in the form of user logs or Web transactions. There are two primary types of data available on the Web that are used by mining algorithms.

1. *Web content information:* This information corresponds to the Web documents and links created by users. The documents are linked to one another with hypertext links. Thus, the content information contains two components that can be mined either together, or in isolation.

   - *Document data:* The document data are extracted from the pages on the World Wide Web. Some of these extraction methods are discussed in Chap. 13.
   - *Linkage data:* The Web can be viewed as a massive graph, in which the pages correspond to nodes, and the linkages correspond to edges between nodes. This linkage information can be used in many ways, such as searching the Web or determining the similarity between nodes.

2. *Web usage data:* This data corresponds to the patterns of user activity that are enabled by Web applications. These patterns could be of various types.

C. C. Aggarwal, *Data Mining: The Textbook*, DOI 10.1007/978-3-319-14142-8_18

- *Web transactions, ratings, and user feedback:* Web users frequently buy various types of items on the Web, or express their affinity for specific products in the form of ratings. In such cases, the buying behavior and/or ratings can be leveraged to make inferences about the preferences of different users. In some cases, the user feedback is provided in the form of textual user reviews that are referred to as *opinions*.
- *Web logs:* User browsing behavior is captured in the form of Web logs that are typically maintained at most Web sites. This browsing information can be leveraged to make inferences about user activity.

These diverse data types automatically define the types of applications that are common on the Web. In coordination with the different data types, the applications are also either content- or usage-centric.

1. *Content-centric applications:* The documents and links on the Web are used in various applications such as search, clustering, and classification. Some examples of such applications are as follows:
    - *Data mining applications:* Web documents are used in conjunction with different types of data mining applications such as clustering and categorization. Such applications are used frequently by Web portals for organizing pages.
    - *Web crawling and resource discovery:* The Web is a tremendous resource of knowledge about documents on various subjects. However, this resource is widely distributed on the Internet, and it needs to be discovered and stored at a single place to make inferences.
    - *Web search:* The goal in Web search is to discover high-quality, relevant documents in response to a user-specified set of keywords. As will be evident later, the notions of quality and relevance are defined both by the linkage and content structure of the documents.
    - *Web linkage mining:* In these applications, either actual or logical representations of linkage structure on the Web are mined for useful insights. Examples of logical representations of Web structure include social and information networks. Social networks are linked networks of users, whereas information networks are linked networks of users and objects.
2. *Usage-centric applications:* The user activity on the Web is mined to make inferences. The different ways in which user activity can be mined are as follows:
    - *Recommender systems:* In these cases, preference information in the form of either ratings for product items or product buying behavior is used to make recommendations to other like-minded users.
    - *Web log analysis:* Web logs are a useful resource for Web site owners to determine relevant patterns of user browsing. These patterns can be leveraged for making inferences such as finding anomalous patterns, user interests, and optimal Web site design.

Many of the aforementioned applications overlap with other chapters in the book. For example, content-centric data mining applications have already been covered in previous chapters of this book, especially in Chap. 13 on mining text data. Some of these methods do need to be modified to account for the additional linkage data. Many linkage mining applications are discussed in Chap. 19 on social network analysis. Therefore, this chapter will focus on the applications that are not primarily covered by other chapters. Among the content-centric applications, Web crawling, search, and ranking will be discussed. Among the usage-centric applications, recommender systems and Web log mining applications will be discussed.

This chapter is organized as follows. Sect. 18.2 discusses Web crawlers and resource discovery. Search engine indexing and query-processing methods are discussed in Sect. 18.3. Ranking algorithms are presented in Sect. 18.4. Recommender systems are discussed in Sect. 18.5. Methods for mining Web logs are discussed in Sect. 18.6. The summary is presented in Sect. 18.7.

18.2 Web Crawling and Resource Discovery

Web crawlers are also referred to as *spiders* or *robots*. The primary motivation for Web crawling is that the resources on the Web are dispensed widely across globally distributed sites. While the Web browser provides a graphical user interface to access these pages in an interactive way, the full power of the available resources cannot be leveraged with the use of only a browser. In many applications, such as search and knowledge discovery, it is necessary to download all the relevant pages *at a central location*, to allow machine learning algorithms to use these resources efficiently.

Web crawlers have numerous applications. The most important and well-known application is search, in which the downloaded Web pages are indexed, to provide responses to user keyword queries. All the well-known search engines, such as Google and Bing, employ crawlers to periodically refresh the downloaded Web resources at their servers. Such crawlers are also referred to as *universal crawlers* because they are intended to crawl all pages on the Web irrespective of their subject matter or location. Web crawlers are also used for business intelligence, in which the Web sites related to a particular subject are crawled or the sites of a competitor are monitored and incrementally crawled as they change. Such crawlers are also referred to as *preferential crawlers* because they discriminate between the relevance of different pages for the application at hand.

18.2.1 A Basic Crawler Algorithm

While the design of a crawler is quite complex, with a distributed architecture and many processes or threads, the following describes a simple sequential and universal crawler that captures the essence of how crawlers are constructed.

The basic crawler algorithm, described in a very general way, uses a seed set of Universal Resource Locators (URLs) S, and a selection algorithm $\mathcal{A}$ as the input. The algorithm $\mathcal{A}$ decides which document to crawl next from a current *frontier list* of URLs. The frontier list represents URLs extracted from the Web pages. These are the candidates for pages that can eventually be fetched by the crawler. The selection algorithm $\mathcal{A}$ is important because it regulates the basic strategy used by the crawler to discover the resources. For example, if new URLs are appended to the end of the frontier list, and the algorithm $\mathcal{A}$ selects documents from the beginning of the list, then this corresponds to a breadth-first algorithm.

The basic crawler algorithm proceeds as follows. First, the seed set of URLs is added to the frontier list. In each iteration, the selection algorithm $\mathcal{A}$ picks one of the URLs from the frontier list. This URL is deleted from the frontier list and then fetched using the

Algorithm *BasicCrawler*(Seed URLs: S, Selection Algorithm: $\mathcal{A}$)
begin
  $FrontierList = S$;
  repeat
    Use algorithm $\mathcal{A}$ to select URL $X \in FrontierSet$;
    $FrontierList = FrontierList - \{X\}$;
    Fetch URL X and add to repository;
    Add all relevant URLs in fetched document X to
      end of $FrontierList$;
  until termination criterion;
end

Figure 18.1: The basic crawler algorithm

HTTP protocol. This is the same mechanism used by browsers to fetch Web pages. The main difference is that the fetching is now done by an automated program using automated selection decisions, rather than by the manual specification of a link by a user with a Web browser. The fetched page is stored in a local repository, and the URLs inside it are extracted. These URLs are then added to the frontier list, provided that they have not already been visited. Therefore, a separate data structure, in the form of a hash table, needs to be maintained to store all visited URLs. In practical implementations of crawlers, not all unvisited URLs are added to the frontier list due to Web spam, spider traps, topical preference, or simply a practical limit on the size of the frontier list. These issues will be discussed later. After the relevant URLs have been added to the frontier list, the next iteration repeats the process with the next URL on the list. The process terminates when the frontier list is empty. If the frontier list is empty, it does not necessarily imply that the entire Web has been crawled. This is because the Web is not strongly connected, and many pages are unreachable from most randomly chosen seed sets. Because most practical crawlers such as search engines are *incremental* crawlers that refresh pages over previous crawls, it is usually easy to identify unvisited seeds from previous crawls and add them to the frontier list, if needed. With large seed sets, such as a previously crawled repository of the Web, it is possible to robustly crawl most pages. The basic crawler algorithm is described in Fig. 18.1.

Thus, the crawler is a graph search algorithm that discovers the outgoing links from nodes by parsing Web pages and extracting the URLs. The choice of the selection algorithm $\mathcal{A}$ will typically result in a bias in the crawling algorithm, especially in cases where it is impossible to crawl all the relevant pages due to resource limitations. For example, a breadth-first crawler is more likely to crawl a page with many links pointing to it. Interestingly, such biases are sometimes desirable in crawlers because it is impossible for any crawler to index the entire Web. Because the indegree of a Web page is often closely related to its *PageRank*, a measure of a Web page's quality, this bias is not necessarily undesirable. Crawlers use a variety of other selection strategies defined by the algorithm $\mathcal{A}$.

1. Because most universal crawlers are incremental crawlers that are intended to refresh previous crawls, it is desirable to crawl frequently changing pages. The change frequency can be estimated from repeated previous crawls of the same page. Some resources such as news portals are updated frequently. Therefore, frequently updated pages may be selected by the algorithm $\mathcal{A}$.

2. The selection algorithm $\mathcal{A}$ may specifically choose Web pages with high *PageRank* from frontier list. The computation of *PageRank* is discussed in Sect. 18.4.1.

A practice, a combination of factors are used by the commercial crawlers employed by search engines.

18.2.2 Preferential Crawlers

In the preferential crawler, only pages satisfying a user-defined criterion need to be crawled. This criterion may be specified in the form of keyword presence in the page, a topical criterion defined by a machine learning algorithm, a geographical criterion about page location, or a combination of the different criteria. In general, an arbitrary predicate may be specified by the user, which forms the basis of the crawling. In these cases, the major change is to the approach used for updating the frontier list during crawling.

1. The Web page needs to meet the user-specified criterion in order for its extracted URLs to be added to the frontier list.

2. In some cases, the anchor text may be examined to determine the relevance of the Web page to the user-specified query.

3. In context-focused crawlers, the crawler is trained to learn the likelihood that relevant pages are within a short distance of the page, even if the Web page is itself not directly relevant to the user-specified criterion. For example, a Web page on "*data mining*" is more likely to point to a Web page on "*information retrieval*," even though the data mining page may not be relevant to the query on "*information retrieval*." URLs from such pages may be added to the frontier list. Therefore, heuristics need to be designed to learn such context-specific relevance.

Changes may also be made to the algorithm $\mathcal{A}$. For example, URLs with more relevant anchor text, or with relevant tokens in the Web address, may be selected first by algorithm $\mathcal{A}$. A URL such as `http://www.golf.com`, with the word "*golf*" in the Web address may be more relevant to the topic of "*golf*," than a URL without the word in it. The bibliographic notes contain pointers to a number of heuristics that are commonly used for preferential resource discovery.

18.2.3 Multiple Threads

When a crawler issues a request for a URL and waits for it, the system is idle, with no work being done at the crawler end. This would seem to be a waste of resources. A natural way to speed up the crawling is by leveraging concurrency. The idea is to use multiple threads of the crawler that update a shared data structure for visited URLs and the page repository. In such cases, it is important to implement concurrency control mechanisms for locking or unlocking the relevant data structures during updates. The concurrent design can significantly speed up a crawler with more efficient use of resources. In practical implementations of large search engines, the crawler is distributed geographically with each "sub-crawler" collecting pages in its geographical proximity.

18.2.4 Combatting Spider Traps

The main reason that the crawling algorithm always visits distinct Web pages is that it maintains a list of previously visited URLs for comparison purposes. However, some shopping sites create dynamic URLs in which the last page visited is appended at the end of the user sequence to enable the server to log the user action sequences within the URL for future analysis. For example, when a user clicks on the link for *page2* from `http://www.examplesite.com/page1`, the new dynamically created URL will be `http://www.examplesite.com/page1/page2`. Pages that are visited further will continue to be appended to the end of the URL, even if these pages were visited before. A natural way to combat this is to limit the maximum size of the URL. Furthermore, a maximum limit may also be placed on the number of URLs crawled from a particular site.

18.2.5 Shingling for Near Duplicate Detection

One of the major problems with the Web pages collected by a crawler is that many duplicates of the same page may be crawled. This is because the same Web page may be mirrored at multiple sites. Therefore, it is crucial to have the ability to detect near duplicates. An approach known as *shingling* is commonly used for this purpose.

A k-shingle from a document is simply a string of k consecutively occurring words in the document. A shingle can also be viewed as a k-gram. For example, consider the document comprising the following sentence:

Mary had a little lamb, its fleece was white as snow.

The set of 2-shingles extracted from this sentence is "*Mary had*", "*had a*", "*a little*", "*little lamb*", "*lamb its*", "*its fleece*", "*fleece was*", "*was white*", "*white as*", and "*as snow*". Note that the number of k-shingles extracted from a document is no longer than the length of the document, and 1-shingles are simply the set of words in the document. Let S_1 and S_2 be the k-shingles extracted from two documents D_1 and D_2. Then, the shingle-based similarity between D_1 and D_2 is simply the Jaccard coefficient between S_1 and S_2

$$J(S_1, S_2) = \frac{|S_1 \cap S_2|}{|S_1 \cup S_2|}. \tag{18.1}$$

Typically, the value of k ranges between 5 and 10 depending on the corpus size and application domain. The advantage of using k-shingles instead of the individual words (1-shingles) for Jaccard coefficient computation is that shingles are less likely than words to repeat in different documents. There are r^k distinct shingles for a lexicon of size r. For $k \geq 5$, the chances of many shingles recurring in two documents becomes very small. Therefore, if two documents have many k-shingles in common, they are very likely to be near duplicates. To save space, the individual shingles are hashed into 4-byte (32-bit) numbers that are used for comparison purposes. Such a representation also enables better efficiency.

18.3 Search Engine Indexing and Query Processing

After the documents have been crawled, they are leveraged for query processing. There are two primary stages to the search index construction:

1. *Offline stage:* This is the stage in which the search engine preprocesses the crawled documents to extract the tokens and constructs an index to enable efficient search. A quality-based ranking score is also computed for each page at this stage.

2. *Online query processing:* This preprocessed collection is utilized for online query processing. The relevant documents are accessed and then ranked using both their relevance to the query and their quality.

The preprocessing steps for Web document processing are described in Chap. 13 on mining text data. The relevant tokens are extracted and stemmed. Stop words are removed. These documents are then transformed to the vector space representation for indexing.

After the documents have been transformed to the vector space representation, an inverted index is constructed on the document collection. The construction of inverted indices is described in Sect. 5.3.1.2 of Chap. 5. The inverted list maps each word identifier to a list of document identifiers containing it. The frequency of the word is also stored with the document identifier in the inverted list. In many implementations, the position information of the word in the document is stored as well.

Aside from the inverted index that maps words to documents, an index is needed for accessing the storage location of the inverted word lists relevant to the query terms. These locations are then used to access the inverted lists. Therefore, a *vocabulary index* is required as well. In practice, many indexing methods such as hashing and tries are commonly used. Typically, a hash function is applied to each word in the query term, to yield the logical address of the corresponding inverted list.

For a given set of words, all the relevant inverted lists are accessed, and the intersection of these inverted lists is determined. This intersection is used to determine the Web document identifiers that contain all, or most of, the search terms. In cases, where one is interested only in documents containing most of the search terms, the intersection of different subsets of inverted lists is performed to determine the best match. Typically, to speed up the process, two indexes are constructed. A smaller index is constructed on only the titles of the Web page, or anchor text of pages *pointing to the page.* If enough documents are found in the smaller index, then the larger index is not referenced. Otherwise, the larger index is accessed. The logic for using the smaller index is that the title of a Web page and the anchor text of Web pages pointing to it, are usually highly representative of the content in the page.

Typically, the number of pages returned for common queries may be of the order of millions or more. Obviously, such a large number of query results will not be easy for a human user to assimilate. A typical browser interface will present only the first few (say 10) results to the human user in a single view of the search results, with the option of browsing other less relevant results. Therefore, one of the most important problems in search engine query processing is that of *ranking.* The aforementioned processing of the inverted index does provide a content-based score. This score can be leveraged for ranking. While the exact scoring methodology used by commercial engines is proprietary, a number of factors are known to influence the content-based score:

1. A word is given different weights, depending upon whether it occurs in the title, body, URL token, or the anchor text of a pointing Web page. The occurrence of the term in the title or the anchor text of a Web page pointing to that page is generally given higher weight.

2. The number of occurrences of a keyword in a document will be used in the score. Larger numbers of occurrences are obviously more desirable.

3. The prominence of a term in font size and color may be leveraged for scoring. For example, larger font sizes will be given a larger score.

4. When multiple keywords are specified, their relative positions in the documents are used as well. For example, if two keywords occur close together in a Web page, then this increases the score.

The content-based score is not sufficient, however, because it does not account for the *reputation*, or the *quality*, of the page. It is important to use such mechanisms because of the uncoordinated and open nature of Web development. After all, the Web allows anyone to publish almost anything, and therefore there is little control on the quality of the results. A user may publish incorrect material either because of poor knowledge on the subject, economic incentives, or with a deliberately malicious intent of publishing misleading information.

Another problem arises from the impact of *Web spam*, in which Web site owners intentionally serve misleading content to rank their results higher. Commercial Web site owners have significant economic incentives to ensure that their sites are ranked higher. For example, an owner of a business on golf equipment, would want to ensure that a search on the word "*golf*" ranks his or her site as high as possible. There are several strategies used by Web site owners to rank their results higher.

1. *Content-spamming:* In this case, the Web host owner fills up repeated keywords in the hosted Web page, even though these keywords are not actually visible to the user. This is achieved by controlling the color of the text and the background of the page. Thus, the idea is to maximize the content relevance of the Web page to the search engine, without a corresponding increase in the *visible* level of relevance.

2. *Cloaking:* This is a more sophisticated approach, in which the Web site serves different content to crawlers than it does to users. Thus, the Web site first determines whether the incoming request is from a crawler or from a user. If the incoming request is from a user, then the actual content (e.g., advertising content) is served. If the request is from a crawler, then the content that is most relevant to specific keywords is served. As a result, the search engine will use different content to respond to user search requests from what a Web user will actually see.

It is obvious that such spamming will significantly reduce the quality of the search results. Search engines also have significant incentives to improve the quality of their results to support their paid advertising model, in which the *explicitly marked* sponsored links appearing on the side bar of the search results are truly paid advertisements. Search engines do not want advertisements (disguised by spamming) to be served as *bona fide* results to the query, especially when such results reduce the quality of the user experience. This has led to an adversarial relationship between search engines and spammers, in which the former use reputation-based algorithms to reduce the impact of spam. At the other end of Web site owners, a *search engine optimization (SEO)* industry attempts to optimize search results by using their knowledge of the algorithms used by search engines, either through the general principles used by engines or through reverse engineering of search results.

For a given search, it is almost always the case that a small subset of the results is more informative or provides more accurate information. How can such pages be determined? Fortunately, the Web provides several natural voting mechanisms to determine the reputation of pages.

1. *Page citation mechanisms:* This is the most common mechanism used to determine the quality of Web pages. When a page is of high quality, many other Web pages point to it. A citation can be logically viewed as a vote for the Web page. While the

number of in-linking pages can be used as a rough indicator of the quality, it does not provide a complete view because it does not account for the quality of the pages pointing to it. To provide a more holistic citation-based vote, an algorithm referred to as *PageRank* is used.

2. *User feedback or behavioral analysis mechanisms:* When a user chooses a Web page from among the responses to a search result, this is clear evidence of the relevance of that page to the user. Therefore, other similar pages, or pages accessed by other similar users can be returned. Such an approach is generally hard to implement in search because of limited user-identification mechanisms. Some search engines, such as Excite, have used various forms of relevance feedback. While these mechanisms are used less often by search engines, they are nevertheless quite important for *commercial recommender systems.* In commercial recommender systems, the recommendations are made by the Web site itself during user browsing, rather than by search engines. This is because commercial sites have stronger user-identification mechanisms (e.g., user registration) to enable more powerful algorithms for inferring user interests.

Typically, the reputation score is determined using *PageRank*-like algorithms. Therefore, if $IRScore$ and $RepScore$ are the content- and reputation-based scores of the Web page, respectively, then the final ranking score is computed as a function of these scores:

$$RankScore = f(IRScore, RepScore). \tag{18.2}$$

The exact function $f(\cdot, \cdot)$ used by commercial search engines is proprietary, but it is always monotonically related to both the $IRScore$ and $RepScore$. Various other factors, such as the geographic location of the browser, also seem to play a role in the ranking.

It should be pointed out, that citation-based reputation scores are not completely immune to other types of spamming that involve coordinated creation of a large number of links to a Web page. Furthermore, the use of anchor text of *pointing* Web pages in the content portion of the rank score can sometimes lead to amusingly irrelevant search results. For example, a few years back, a search on the keyword "*miserable failure*" in the Google search engine, returned as its top result, the official biography of a previous president of the United States of America. This is because many Web pages were constructed in a coordinated way to use the anchor text "*miserable failure*" to point to this biography. This practice of influencing search results by coordinated linkage construction to a particular site is referred to as *Googlewashing.* Such practices are less often economically motivated, but are more often used for comical or satirical purposes.

Therefore, the ranking algorithms used by search engines are not perfect but have, nevertheless, improved significantly over the years. The algorithms used to compute the reputation-based ranking score will be discussed in the next section.

18.4 Ranking Algorithms

The *PageRank* algorithm uses the linkage structure of the Web for reputation-based ranking. The *PageRank* method is independent of the user query, because it only precomputes the reputation portion of the score in Eq. 18.2. The *HITS* algorithm is query-specific. It uses a number of intuitions about how authoritative sources on various topics are linked to one another in a hyperlinked environment.

18.4.1 PageRank

The *PageRank* algorithm models the importance of Web pages with the use of the citation (or linkage) structure in the Web. The basic idea is that highly reputable documents are more likely to be cited (or in-linked) by other reputable Web pages.

A random surfer model on the Web graph is used to achieve this goal. Consider a random surfer who visits random pages on the Web by selecting random links on a page. The long-term relative frequency of visits to any particular page is clearly influenced by the number of in-linking pages to it. Furthermore, the long-term frequency of visits to any page will be higher if it is linked to by other frequently visited (or *reputable*) pages. In other words, the *PageRank* algorithm models the reputation of a Web page in terms of its long-term frequency of visits by a random surfer. This long-term frequency is also referred to as the *steady-state probability*. This model is also referred to as the *random walk model*.

The basic random surfer model does not work well for all possible graph topologies. A critical issue is that some Web pages may have no outgoing links, which may result in the random surfer getting trapped at specific nodes. In fact, a probabilistic transition is not even meaningfully defined at such a node. Such nodes are referred to as *dead ends*. An example of a dead-end node is illustrated in Fig. 18.2a. Clearly, dead ends are undesirable because the transition process for *PageRank* computation cannot be defined at that node. To address this issue, two modifications are incorporated in the random surfer model. The first modification is to add links from the dead-end node (Web page) to all nodes (Web pages), including a self-loop to itself. Each such edge has a transition probability of $1/n$. This does not fully solve the problem, because the dead ends can also be defined on *groups of nodes*. In these cases, there are no outgoing links from *a group of nodes* to the remaining nodes in the graph. This is referred to as a *dead-end component*, or *absorbing component*. An example of a dead-end component is illustrated in Fig. 18.2b.

Dead-end components are common in the Web graph because the Web is not strongly connected. In such cases, the transitions at individual nodes can be meaningfully defined, but the steady-state transitions will stay trapped in these dead-end components. All the steady-state probabilities will be concentrated in dead-end components because there can be no transition out of a dead-end component after a transition occurs into it. Therefore, as long as even a minuscule probability of transition into a dead-end component[1] exists, *all* the steady-state probability becomes concentrated in such components. This situation is not desirable from the perspective of *PageRank* computation in a large Web graph, where dead-end components are not necessarily an indicator of popularity. Furthermore, in such cases, the final probability distribution of nodes in various dead-end components is not unique and it is dependent on the starting state. This is easy to verify by observing that random walks starting in different dead-end components will have their respective steady-state distributions concentrated within the corresponding components.

While the addition of edges solves the problem for dead-end nodes, an additional step is required to address the more complex issue of dead-end components. Therefore, aside from the addition of these edges, a *teleportation*, or *restart step* is used within the random surfer model. This step is defined as follows. At each transition, the random surfer may either jump to an arbitrary page with probability α, or it may follow one of the links on the page with probability $(1-\alpha)$. A typical value of α used is 0.1. Because of the use of teleportation, the

[1] A formal mathematical treatment characterizes this in terms of the *ergodicity* of the underlying Markov chains. In ergodic Markov chains, a necessary requirement is that it is possible to reach any state from any other state using a sequence of one or more transitions. This condition is referred to as *strong connectivity*. An informal description is provided here to facilitate understanding.

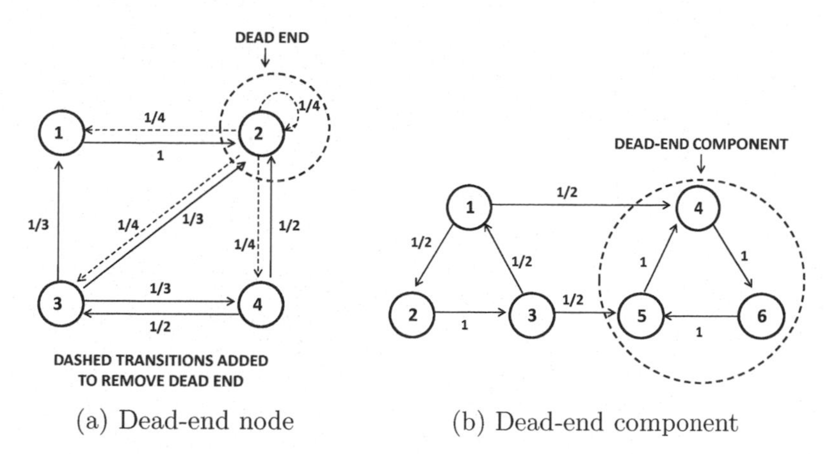

(a) Dead-end node (b) Dead-end component

Figure 18.2: Transition probabilities for *PageRank* computation with different types of dead ends

steady state probability becomes unique and independent of the starting state. The value of α may also be viewed as a *smoothing* or *damping probability*. Large values of α typically result in the steady-state probability of different pages to become more even. For example, if the value of α is chosen to be 1, then all pages will have the same steady-state probability of visits.

How are the steady-state probabilities determined? Let $G = (N, A)$ be the directed Web graph, in which nodes correspond to pages, and edges correspond to hyperlinks. The total number of nodes is denoted by n. It is assumed that A also includes the added edges from dead-end nodes to all other nodes. The set of nodes incident on i is denoted by $In(i)$, and the set of end points of the outgoing links of node i is denoted by $Out(i)$. The steady-state probability at a node i is denoted by $\pi(i)$. In general, the transitions of a Web surfer can be visualized as a *Markov chain*, in which an $n \times n$ transition matrix P is defined for a Web graph with n nodes. The *PageRank* of a node i is equal to the steady-state probability $\pi(i)$ for node i, in the Markov chain model. The probability[2] p_{ij} of transitioning from node i to node j, is defined as $1/|Out(i)|$. Examples of transition probabilities are illustrated in Fig. 18.2. These transition probabilities do not, however, account for teleportation which will be addressed[3] separately below.

Let us examine the transitions into a given node i. The steady-state probability $\pi(i)$ of node i is the sum of the probability of a teleportation into it and the probability that one of the in-linking nodes directly transitions into it. The probability of a teleportation into the node is exactly α/n because a teleportation occurs in a step with probability α, and all nodes are equally likely to be the beneficiary of the teleportation. The probability of a transition into node i is given by $(1 - \alpha) \cdot \sum_{j \in In(i)} \pi(j) \cdot p_{ji}$, as the sum of the probabilities of transitions from different in-linking nodes. Therefore, at steady-state, the probability of

[2]In some applications such as bibliographic networks, the edge (i, j) may have a weight denoted by w_{ij}. The transition probability p_{ij} is defined in such cases by $\frac{w_{ij}}{\sum_{j \in Out(i)} w_{ij}}$.

[3]An alternative way to achieve this goal is to modify G by multiplying existing edge transition probabilities by the factor $(1 - \alpha)$ and then adding α/n to the transition probability between each pair of nodes in G. As a result G will become a directed clique with bidirectional edges between each pair of nodes. Such strongly connected Markov chains have unique steady-state probabilities. The resulting graph can then be treated as a Markov chain without having to separately account for the teleportation component. This model is equivalent to that discussed in the chapter.

a transition into node i is defined by the sum of the probabilities of the teleportation and transition events are as follows:

$$\pi(i) = \alpha/n + (1-\alpha) \cdot \sum_{j \in In(i)} \pi(j) \cdot p_{ji}. \tag{18.3}$$

For example, the equation for node 2 in Fig. 18.2a can be written as follows:

$$\pi(2) = \alpha/4 + (1-\alpha) \cdot (\pi(1) + \pi(2)/4 + \pi(3)/3 + \pi(4)/2).$$

There will be one such equation for each node, and therefore it is convenient to write the entire system of equations in matrix form. Let $\overline{\pi} = (\pi(1) \ldots \pi(n))^T$ be the n-dimensional column vector representing the steady-state probabilities of all the nodes, and let $\overline{e}$ be an n-dimensional column vector of all 1 values. The system of equations can be rewritten in matrix form as follows:

$$\overline{\pi} = \alpha \overline{e}/n + (1-\alpha) P^T \overline{\pi}. \tag{18.4}$$

The first term on the right-hand side corresponds to a teleportation, and the second term corresponds to a direct transition from an incoming node. In addition, because the vector $\overline{\pi}$ represents a probability, the sum of its components $\sum_{i=1}^{n} \pi(i)$ must be equal to 1:

$$\sum_{i=1}^{n} \pi(i) = 1. \tag{18.5}$$

Note that this is a linear system of equations that can be easily solved using an iterative method. The algorithm starts off by initializing $\overline{\pi}^{(0)} = \overline{e}/n$, and it derives $\overline{\pi}^{(t+1)}$ from $\overline{\pi}^{(t)}$ by repeating the following iterative step:

$$\overline{\pi}^{(t+1)} \Leftarrow \alpha \overline{e}/n + (1-\alpha) P^T \overline{\pi}^{(t)}. \tag{18.6}$$

After each iteration, the entries of $\overline{\pi}^{(t+1)}$ are normalized by scaling them to sum to 1. These steps are repeated until the difference between $\overline{\pi}^{(t+1)}$ and $\overline{\pi}^{(t)}$ is a vector with magnitude less than a user-defined threshold. This approach is also referred to as the *power-iteration method*. It is important to understand that *PageRank* computation is expensive, and it cannot be computed on the fly for a user query during Web search. Rather, the *PageRank* values for *all* the known Web pages are precomputed and stored away. The stored *PageRank* value for a page is accessed only when the page is included in the search results for a particular query for use in the final ranking, as indicated by Eq. 18.2.

The *PageRank* values can be shown to be the n components of the largest left eigenvector[4] of the stochastic transition matrix P (see Exercise 5), for which the eigenvalue is 1. The largest eigenvalue of a stochastic transition matrix is always 1. The left eigenvectors of P are the same as the right eigenvectors of P^T. Interestingly, the largest *right* eigenvectors of the stochastic transition matrix P of an undirected graph can be used to construct *spectral embeddings* (cf. Sect. 19.3.4 of Chap. 19), which are used for network clustering.

[4]The left eigenvector $\overline{X}$ of P is a row vector satisfying $\overline{X}P = \lambda \overline{X}$. The right eigenvector $\overline{Y}$ is a column vector satisfying $P\overline{Y} = \lambda \overline{Y}$. For asymmetric matrices, the left and right eigenvectors are not the same. However, the eigenvalues are always the same. The unqualified term "eigenvector" refers to the right eigenvector by default.

18.4.1.1 Topic-Sensitive PageRank

Topic-sensitive PageRank is designed for cases in which it is desired to provide greater importance to some topics than others in the ranking process. While personalization is less common in large-scale commercial search engines, it is more common in smaller scale site-specific search applications. Typically, users may be more interested in certain combinations of topics than others. The knowledge of such interests may be available to a personalized search engine because of user registration. For example, a particular user may be more interested in the topic of automobiles. Therefore, it is desirable to rank pages related to automobiles higher when responding to queries by this user. This can also be viewed as the *personalization* of ranking values. How can this be achieved?

The first step is to fix a list of base topics, and determine a high-quality sample of pages from each of these topics. This can be achieved with the use of a resource such as the *Open Directory Project (ODP)*,[5] which can provide a base list of topics and sample Web pages for each topic. The *PageRank* equations are now modified, so that the teleportation is only performed on this sample set of Web documents, rather than on the entire space of Web documents. Let $\overline{e_p}$ be an n-dimensional personalization (column) vector with one entry for each page. An entry in $\overline{e_p}$ takes on the value of 1, if that page is included in the sample set, and 0 otherwise. Let the number of nonzero entries in $\overline{e_p}$ be denoted by n_p. Then, the *PageRank* Eq. 18.4 can be modified as follows:

$$\overline{\pi} = \alpha \overline{e_p}/n_p + (1-\alpha)P^T\overline{\pi}. \tag{18.7}$$

The same power-iteration method can be used to solve the personalized *PageRank* problem. The selective teleportations bias the random walk, so that pages in the structural locality of the sampled pages will be ranked higher. As long as the sample of pages is a good representative of different (structural) localities of the Web graph, in which pages of specific topics exist, such an approach will work well. Therefore, for each of the different topics, a separate *PageRank* vector can be precomputed and stored for use during query time.

In some cases, the user is interested in specific *combinations of* topics such as sports and automobiles. Clearly, the number of possible combinations of interests can be very large, and it is not reasonably possible or necessary to prestore every personalized *PageRank* vector. In such cases, only the *PageRank* vectors for the base topics are computed. The final result for a user is defined as a weighted linear combination of the topic-specific *PageRank* vectors, where the weights are defined by the user-specified interest in the different topics.

18.4.1.2 SimRank

The notion of *SimRank* was defined to compute the structural similarity between nodes. *SimRank* determines *symmetric* similarities between nodes. In other words, the similarity between nodes i and j, is the same as that between j and i. Before discussing *SimRank*, we define a related but slightly different asymmetric ranking problem:

Given a target node i_q and a subset of nodes $S \subseteq N$ from graph $G = (N, A)$, rank the nodes in S in their order of similarity to i_q.

Such a query is very useful in recommender systems in which users and items are arranged in the form of a bipartite graph of preferences, in which nodes corresponds to users and items, and edges correspond to preferences. The node i_q may correspond to an item node,

[5]http://www.dmoz.org.

and the set S may correspond to user nodes. Alternatively, the node i_q may correspond to a user node, and the set S may correspond to item nodes. Recommender systems will be discussed in Sect. 18.5. Recommender systems are closely related to search, in that they also perform ranking of target objects, but while taking user preferences into account.

This problem can be viewed as a limiting case of topic-sensitive *PageRank*, in which the teleportation is performed to the *single node* i_q. Therefore, the personalized *PageRank* Eq. 18.7 can be directly adapted by using the teleportation vector $\overline{e_p} = \overline{e_q}$, that is, a vector of all 0s, except for a single 1, corresponding to the node i_q. Furthermore, the value of n_p in this case is set to 1:

$$\overline{\pi} = \alpha\overline{e_q} + (1 - \alpha)P^T\overline{\pi}. \tag{18.8}$$

The solution to the aforementioned equation will provide high ranking values to nodes in the structural locality of i_q. This definition of similarity is *asymmetric* because the similarity value assigned to node j starting from query node i is different from the similarity value assigned to node i starting from query node j. Such an *asymmetric* similarity measure is suitable for *query-centered* applications such as search engines and recommender systems, but not necessarily for arbitrary network-based data mining applications. In some applications, symmetric pairwise similarity between nodes is required. While it is possible to average the two topic-sensitive *PageRank* values in opposite directions to create a symmetric measure, the *SimRank* method provides an elegant and intuitive solution.

The *SimRank* approach is as follows. Let $In(i)$ represent the in-linking nodes of i. The *SimRank* equation is naturally defined in a recursive way, as follows:

$$SimRank(i,j) = \frac{C}{|In(i)| \cdot |In(j)|} \sum_{p \in In(i)} \sum_{q \in In(j)} SimRank(p,q). \tag{18.9}$$

Here C is a constant in $(0,1)$ that can be viewed as a kind of decay rate of the recursion. As the boundary condition, the value of $SimRank(i,j)$ is set to 1 when $i = j$. When either i or j do not have in-linking nodes, the value of $SimRank(i,j)$ is set to 0. To compute *SimRank*, an iterative approach is used. The value of $SimRank(i,j)$ is initialized to 1 if $i = j$, and 0 otherwise. The algorithm subsequently updates the *SimRank* values between all node pairs iteratively using Eq. 18.9 until convergence is reached.

The notion of *SimRank* has an interesting intuitive interpretation in terms of random walks. Consider two random surfers walking *in lockstep* backward from node i and node j till they meet. Then the number of steps taken by each of them is a random variable $L(i,j)$. Then, $SimRank(i,j)$ can be shown to be equal to the expected value of $C^{L(i,j)}$. The decay constant C is used to map random walks of length l to a similarity value of C^l. Note that because $C < 1$, smaller distances will lead to higher similarity and vice versa.

Random walk-based methods are generally more robust than the shortest path distance to measure similarity between nodes. This is because random walks measures implicitly account for the *number* of paths between nodes, whereas shortest paths do not. A detailed discussion of this issue can be found in Sect. 3.5.1.2 of Chap. 3.

18.4.2 HITS

The *Hypertext Induced Topic Search (HITS)* algorithm is a *query-dependent* algorithm for ranking pages. The intuition behind the approach lies in an understanding of the typical structure of the Web that is organized into hubs and authorities.

An *authority* is a page with many in-links. Typically, it contains authoritative content on a particular subject, and, therefore, many Web users may trust that page as a resource of

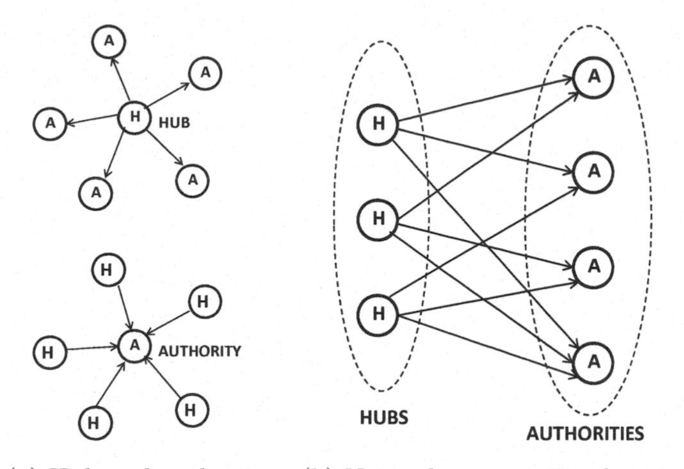

(a) Hub and authority examples (b) Network organization between hubs and authorities

Figure 18.3: Illustrating hubs and authorities

knowledge on that subject. This will result in many pages linking to the authority page. A *hub* is a page with many out-links to authorities. These represent a compilation of the links on a particular topic. Thus, a hub page provides guidance to Web users about where they can find the resources on a particular topic. Examples of the typical node-centric topology of hubs and authorities in the Web graph are illustrated in Fig. 18.3a.

The main insight used by the *HITS* algorithm is that good hubs point to many good authorities. Conversely, good authority pages are pointed to by many hubs. An example of the typical organization of hubs and authorities is illustrated in Fig. 18.3b. This mutually reinforcing relationship is leveraged by the *HITS* algorithm. For any query issued by the user, the *HITS* algorithm starts with the list of relevant pages and expands them with a *hub ranking* and an *authority ranking*.

The *HITS* algorithm starts by collecting the top-r most relevant results to the search query at hand. A typical value of r is 200. This defines the *root set* R. Typically, a query to a commercial search engine or content-based evaluation is used to determine the root set. For each node in R, the algorithm determines all nodes immediately connected (either in-linking or out-linking) to R. This provides a larger *base set* S. Because the base set S can be rather large, the maximum number of in-linking nodes to any node in R that are added to S is restricted to k. A typical value of k used is around 50. Note that this still results in a rather large base set because *each* of the possibly 200 root nodes might bring 50 in-linking nodes, along with out-linking nodes.

Let $G = (S, A)$ be the subgraph of the Web graph defined on the (expanded) base set S, where A is the set of edges between nodes in the root set S. The entire analysis of the *HITS* algorithm is restricted to this subgraph. Each page (node) $i \in S$ is assigned both a *hub score* $h(i)$ and *authority score* $a(i)$. It is assumed that the hub and authority scores are normalized, so that the sum of the squares of the hub scores and the sum of the squares of the authority scores are each equal to 1. Higher values of the score indicate better quality. The hub and authority scores are related to one another in the following way:

$$h(i) = \sum_{j:(i,j)\in A} a(j) \quad \forall i \in S \tag{18.10}$$

$$a(i) = \sum_{j:(j,i)\in A} h(j) \quad \forall i \in S. \tag{18.11}$$

The basic idea is to reward hubs for pointing to good authorities and reward authorities for being pointed to by good hubs. It is easy to see that the aforementioned system of equations reinforces this mutually enhancing relationship. This is a linear system of equations that can be solved using an iterative method. The algorithm starts by initializing $h^0(i) = a^0(i) = 1/\sqrt{|S|}$. Let $h^t(i)$ and $a^t(i)$ denote the hub and authority scores of the ith node, respectively, at the end of the tth iteration. For each $t \geq 0$, the algorithm executes the following iterative steps in the $(t+1)$th iteration:

for each $i \in S$ set $a^{t+1}(i) \Leftarrow \sum_{j:(j,i)\in A} h^t(j)$;
for each $i \in S$ set $h^{t+1}(i) \Leftarrow \sum_{j:(i,j)\in A} a^{t+1}(j)$;
Normalize L_2-norm of each of hub and authority vectors to 1;

For hub-vector $\overline{h} = [h(1) \ldots h(n)]^T$ and authority-vector $\overline{a} = [a(1) \ldots a(n)]^T$, the updates can be expressed as $\overline{a} = A^T\overline{h}$ and $\overline{h} = A\overline{a}$, respectively, when the edge set A is treated as an $|S| \times |S|$ adjacency matrix. The iteration is repeated to convergence. It can be shown that the hub vector $\overline{h}$ and the authority vector $\overline{a}$ converge in directions proportional to the dominant eigenvectors of AA^T and A^TA (see Exercise 6), respectively. This is because the relevant pair of updates can be shown to be equivalent to power-iteration updates of AA^T and A^TA, respectively.

18.5 Recommender Systems

Ever since the popularization of web-based transactions, it has become increasingly easy to collect data about user buying behaviors. This data includes information about user profiles, interests, browsing behavior, buying behavior, and ratings about various items. It is natural to leverage such data to make recommendations to customers about possible buying interests.

In the recommendation problem, the user–item pairs have *utility values* associated with them. Thus, for n users and d items, this results in an $n \times d$ matrix D of utility values. This is also referred to as the *utility-matrix*. The utility value for a user-item pair could correspond to either the buying behavior or the ratings of the user for the item. Typically, a small subset of the utility values are specified in the form of either customer buying behavior or ratings. It is desirable to use these specified values to make recommendations. The nature of the utility matrix has a significant influence on the choice of recommendation algorithm:

1. *Positive preferences only:* In this case, the specified utility matrix only contains positive preferences. For example, a specification of a "like" option on a social networking site, the browsing of an item at an online site, or the buying of a specified quantity of an item, corresponds to a positive preference. Thus, the utility matrix is sparse, with a prespecified set of positive preferences. For example, the utility matrix may contain the raw quantities of the item bought by each user, a normalized mathematical function of the quantities, or a weighted function of buying and browsing behavior. These

functions are typically specified heuristically by the analyst in an application-specific way. Entries that correspond to items not bought or browsed by the user may remain unspecified.

2. *Positive and negative preferences (ratings):* In this case, the user specifies the ratings that represent their like or dislike for the item. The incorporation of user dislike in the analysis is significant because it makes the problem more complex and often requires some changes to the underlying algorithms.

An example of a ratings-based utility matrix is illustrated in Fig. 18.4a, and an example of a positive-preference utility matrix is illustrated in Fig. 18.4b. In this case, there are six users, labeled $U_1 \ldots U_6$, and six movies with specified titles. Higher ratings indicate more positive feedback in Fig. 18.4a. The missing entries correspond to unspecified preferences in both cases. This difference significantly changes the algorithms used in the two cases. In particular, the two matrices in Fig. 18.4 have the same specified entries, but they provide very different insights. For example, the users U_1 and U_3 are very different in Fig. 18.4a because they have very different ratings for their commonly specified entries. On the other hand, these users would be considered very similar in Fig. 18.4b because these users have expressed a positive preference for the same items. The ratings-based utility provides a way for users to express negative preferences for items. For example, user U_1 does not like the movie *Gladiator* in Fig. 18.4a. There is no mechanism to specify this in the positive-preference utility matrix of Fig. 18.4b beyond a relatively ambiguous missing entry. In other words, the matrix in Fig. 18.4b is less expressive. While Fig. 18.4b provides an example of a binary matrix, it is possible for the nonzero entries to be arbitrary positive values. For example, they could correspond to the quantities of items bought by the different users.

This difference has an impact on the types of algorithms that are used in the two cases. Allowing for positive and negative preferences generally makes the problem harder. From a data collection point of view, it is also harder to infer negative preferences when they are inferred from customer behavior rather than ratings. Recommendations can also be enhanced with the use of content in the user and item representations.

1. *Content-based recommendations:* In this case, the users and items are both associated with feature-based descriptions. For example, item profiles can be determined by using the text of the item description. A user might also have explicitly specified their interests in a profile. Alternatively, their profile can be inferred from their buying or browsing behavior.

2. *Collaborative filtering:* Collaborative filtering, as the name implies, is the leveraging of the user preferences in the form of ratings or buying behavior in a "collaborative" way, for the benefit of all users. Specifically, the utility matrix is used to determine either relevant users for specific items, or relevant items for specific users in the recommendation process. A key intermediate step in this approach is the determination of similar groups of items and users. The patterns in these peer groups provide the collaborative knowledge needed in the recommendation process.

The two models are not exclusive. It is often possible to combine content-based methods with collaborative filtering methods to create a combined preference score. Collaborative filtering methods are generally among the more commonly used models and will therefore be discussed in greater detail in this section.

It is important to understand that the utility matrices used in collaborative filtering algorithms are extremely large and sparse. It is not uncommon for the values of n and d in

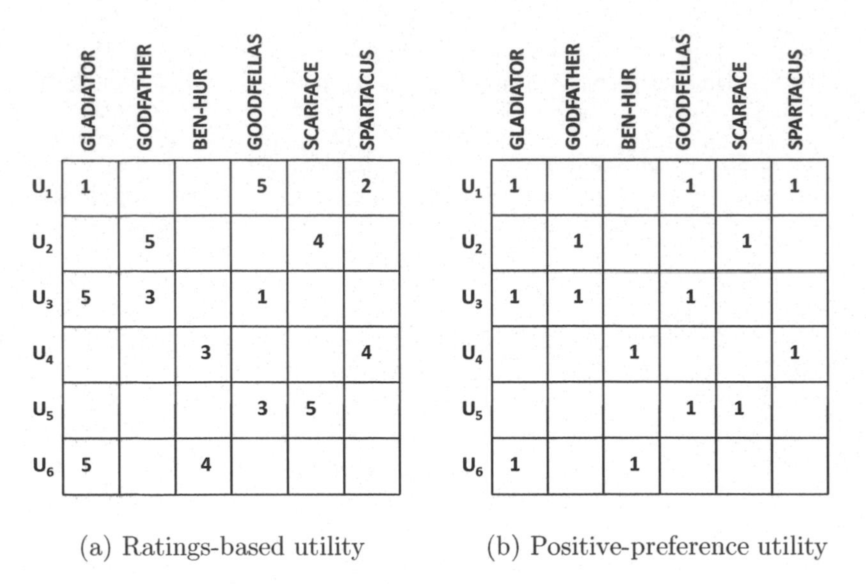

(a) Ratings-based utility (b) Positive-preference utility

Figure 18.4: Examples of utility matrices.

the $n \times d$ utility matrix to exceed 10^5. The matrix is also extremely *sparse*. For example, in a movie data set, a typical user may have specified no more than 10 ratings, out of a universe of more than 10^5 movies.

At a basic level, collaborative filtering can be viewed as a missing-value estimation or matrix completion problem, in which an incomplete $n \times d$ utility matrix is specified, and it is desired to estimate the missing values. As discussed in the bibliographic notes, many methods exist in the traditional statistics literature on missing-value estimation. However, collaborative filtering problems present a particularly challenging special case in terms of data size and sparsity.

18.5.1 Content-Based Recommendations

In content-based recommendations, the user is associated with a set of documents that describe his or her interests. Multiple documents may be associated with a user corresponding to his or her specified demographic profile, specified interests at registration time, the product description of the items bought, and so on. These documents can then be aggregated into a single textual content-based profile of the user in a vector space representation.

The items are also associated with textual descriptions. When the textual descriptions of the items match the user profile, this can be viewed as an indicator of similarity. When no utility matrix is available, the content-based recommendation method uses a simple k-nearest neighbor approach. The top-k items are found that are closest to the user textual profile. The cosine similarity with tf-idf can be used, as discussed in Chap. 13.

On the other hand, when a utility matrix is available, the problem of finding the most relevant items for a particular user can be viewed as a traditional classification problem. For each user, we have a set of *training* documents representing the descriptions of the items for which that user has specified utilities. The labels represent the utility values. The descriptions of the remaining items for that user can be viewed as the test documents for classification. When the utility matrix contains numeric ratings, the class variables are numeric. The regression methods discussed in Sect. 11.5 of Chap. 11 may be used in this case. Logistic and ordered probit regression are particularly popular. In cases where only positive preferences (rather than ratings) are available in the utility matrix, all the specified utility entries correspond to positive examples for the item. The classification is then performed only on the remaining test documents. One challenge is that only a small number of positive training examples are specified, and the remaining examples are unlabeled. In such cases, specialized classification methods using only positive and unlabeled methods may be used. Refer to the bibliographic notes of Chap. 11. Content-based methods have the advantage that they do not even require a utility matrix and leverage domain-specific content information. On the other hand, content information biases the recommendation towards items described by similar keywords to what the user has seen in the past. Collaborative filtering methods work directly with the utility matrix, and can therefore avoid such biases.

18.5.2 Neighborhood-Based Methods for Collaborative Filtering

The basic idea in neighborhood-based methods is to use either user–user similarity, or item–item similarity to make recommendations from a ratings matrix.

18.5.2.1 User-Based Similarity with Ratings

In this case, the top-k similar users to each user are determined with the use of a similarity function. Thus, for the target user i, its similarity to all the other users is computed. Therefore, a similarity function needs to be defined between users. In the case of a ratings-based matrix, the similarity computation is tricky because different users may have different scales of ratings. One user may be biased towards liking most items, and another user may be biased toward not liking most of the items. Furthermore, different users may have rated different items. One measure that captures the similarity between the rating vectors of two users is the Pearson correlation coefficient. Let $X = (x_1 \dots x_s)$ and $Y = (y_1 \dots y_s)$ be the common (specified) ratings between a pair of users, with means $\hat{x} = \sum_{i=1}^{s} x_i/s$ and $\hat{y} = \sum_{i=1}^{s} y_i/s$, respectively. Alternatively, the mean rating of a user is computed by averaging over all her specified ratings rather than using only co-rated items by the pair of users at hand. This alternative way of computing the mean is more common, and it can significantly affect the pairwise Pearson computation. Then, the Pearson correlation coefficient between the two users is defined as follows:

$$\text{Pearson}(\overline{X}, \overline{Y}) = \frac{\sum_{i=1}^{s} (x_i - \hat{x}) \cdot (y_i - \hat{y})}{\sqrt{\sum_{i=1}^{s} (x_i - \hat{x})^2} \cdot \sqrt{\sum_{i=1}^{s} (y_i - \hat{y})^2}}. \tag{18.12}$$

The Pearson coefficient is computed between the target user and all the other users. The peer group of the target user is defined as the top-k users with the highest Pearson coefficient of correlation with her. Users with very low or negative correlations are also removed from the peer group. The average ratings of each of the (specified) items of this peer group are returned as the recommended ratings. To achieve greater robustness, it is also possible to weight each rating with the Pearson correlation coefficient of its owner while computing the average. This weighted average rating can provide a prediction for the target user. The items with the highest predicted ratings are recommended to the user.

The main problem with this approach is that different users may provide ratings on different scales. One user may rate all items highly, whereas another user may rate all items negatively. The raw ratings, therefore, need to be normalized before determining the (weighted) average rating of the peer group. The normalized rating of a user is defined by subtracting her mean rating from each of her ratings. As before, the weighted average of the normalized rating of an item in the peer group is determined as a *normalized* prediction. The mean rating of the target user is then added back to the normalized rating prediction to provide a *raw* rating prediction.

18.5.2.2 Item-Based Similarity with Ratings

The main conceptual difference from the user-based approach is that peer groups are constructed in terms of *items* rather than *users*. Therefore, similarities need to be computed between items (or columns in the ratings matrix). Before computing the similarities between the columns, the ratings matrix is normalized. As in the case of user-based ratings, the average of each row in the ratings matrix is subtracted from that row. Then, the cosine similarity between the normalized ratings $\overline{U} = (u_1 \ldots u_s)$ and $\overline{V} = (v_1 \ldots v_s)$ of a pair of items (columns) defines the similarity between them:

$$\text{Cosine}(\overline{U}, \overline{V}) = \frac{\sum_{i=1}^{s} u_i \cdot v_i}{\sqrt{\sum_{i=1}^{s} u_i^2} \cdot \sqrt{\sum_{i=1}^{s} v_i^2}}. \tag{18.13}$$

This similarity is referred to as the *adjusted* cosine similarity, because the ratings are normalized before computing the similarity value.

Consider the case in which the rating of item j for user i needs to be determined. The first step is to determine the top-k most similar *items* to *item* j based on the aforementioned adjusted cosine similarity. Among the top-k matching items to item j, the ones for which user i has specified ratings are determined. The *weighted* average value of these (raw) ratings is reported as the predicted value. The weight of item r in this average is equal to the adjusted cosine similarity between item r and the target item j.

The basic idea is to leverage the user's *own* ratings in the final step of making the prediction. For example, in a movie recommendation system, the item peer group will typically be movies of a similar genre. The previous ratings history of the *same* user on such movies is a very reliable predictor of the interests of that user.

18.5.3 Graph-Based Methods

It is possible to use a random walk on the user-item graph, rather than the Pearson correlation coefficient, for defining neighborhoods. Such an approach is sometimes more effective for sparse ratings matrices. A bipartite *user-item* graph $G = (N_u \cup N_i, A)$ is constructed, where N_u is the set of nodes representing users, and N_i is the set of nodes representing items. An undirected edge exists in A between a user and an item for each nonzero entry in the utility matrix. For example, the user-item graph for both utility matrices of Fig. 18.4 is illustrated in Fig. 18.5. One can use either the personalized *PageRank* or the *SimRank* method to determine the k most similar users to a given user for user-based collaborative filtering. Similarly, one can use this method to determine the k most similar items to a given item for item-based collaborative filtering. The other steps of user-based collaborative filtering and item-based collaborative filtering remain the same.

A more general approach is to view the problem as a positive and negative *link prediction problem* on the user-item graph. In such cases, the user-item graph is augmented with positive or negative weights on edges. The normalized rating of a user for an item, after subtracting the user-mean, can be viewed as either a positive or negative weight on the edge. For example, consider the graph constructed from the ratings matrix of Fig. 18.4(a). The edge between user U_1 and the item *Gladiator* would become a negative edge because

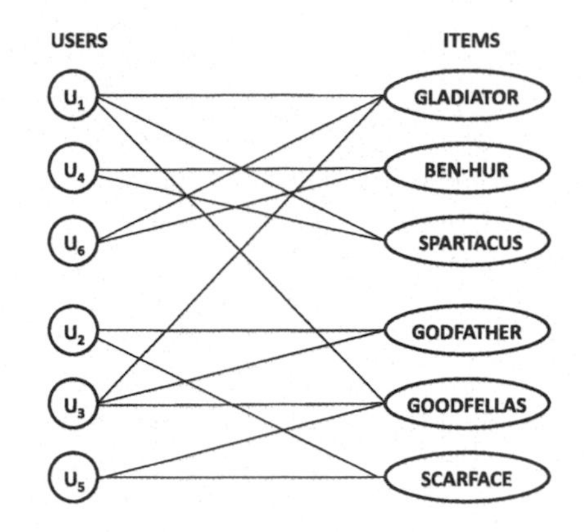

Figure 18.5: Preference graph for utility matrices of Fig. 18.4

U_1 clearly dislikes the movie *Gladiator*. The corresponding network would become a *signed* network. Therefore, the recommendation problem is that of predicting high positive weight edges between users and items in a signed network. A simpler version of the link-prediction problem with only positive links is discussed in Sect. 19.5 of Chap. 19. Refer to the bibliographic notes for link prediction methods with positive and negative links. The merit of the link prediction approach is that it can also leverage the available links between different users in a setting where they are connected by social network links. In such cases, the user-item graph no longer remains bipartite.

When users specify only positive preference values for items, the problem becomes simplified because most link prediction methods are designed for positive links. One can also use the random walks on the user-item graph to perform recommendations, rather than using it only to define neighborhoods. For example, in the case of Fig. 18.4b, the same user-item graph of Fig. 18.5 can be used in conjunction with a random-walk approach. This preference graph can be used to provide different types of recommendations:

1. The top ranking items for the user i can be determined by returning the item nodes with the largest *PageRank* in a random walk with restart at node i.

2. The top ranking users for the item j can be determined by returning the user nodes with the largest *PageRank* in a random walk with restart at node j.

The choice of the restart probability regulates the trade-off between the global popularity of the recommended item/user and the specificity of the recommendation to a particular user/item. For example, consider the case when items need to be recommended to user i. A low teleportation probability will favor the recommendation of popular items which are favored by many users. Increasing the teleportation probability will make the recommendation more specific to user i.

18.5.4 Clustering Methods

One weakness of neighborhood-based methods is the scale of the computation that needs to be performed. For *each user*, one typically has to perform computations that are proportional to *at least* the number of nonzero entries in the matrix. Furthermore, these computations need to be performed over *all* users to provide recommendations to different users.

This can be extremely slow. Therefore, a question arises, as to whether one can use clustering methods to speed up the computations. Clustering also helps address the issue of data sparsity to some extent.

Clustering methods are *exactly analogous* to neighborhood-based methods, except that the clustering is performed as a preprocessing step to define the peer groups. These peer groups are then used for making recommendations. The clusters can be defined either on users, or on items. Thus, they can be used to make either user-user similarity recommendations, or item-item similarity recommendations. For brevity, only the user-user recommendation approach is described here, although the item-item recommendation approach is exactly analogous. The clustering approach works as follows:

1. Cluster all the users into n_g groups of users using any clustering algorithm.
2. For any user i, compute the average (normalized) rating of the specified items in its cluster. Report these ratings for user i; after transforming back to the raw value.

The item–item recommendation approach is similar, except that the clustering is applied to the columns rather than the rows. The clusters define the groups of similar items (or implicitly pseudo-genres). The final step of computing the rating for a user-item combination is similar to the case of neighborhood-based methods. After the clustering has been performed, it is generally very efficient to determine all the ratings. It remains to be explained how the clustering is performed.

18.5.4.1 Adapting k-Means Clustering

To cluster the ratings matrix, it is possible to adapt many of the clustering methods discussed in Chap. 6. However, it is important to adapt these methods to sparsely specified incomplete data sets. Methods such as k-means and Expectation Maximization may be used on the normalized ratings matrix. In the case of the k-means method, there are two major differences from the description of Chap. 6:

1. In an iteration of k-means, centroids are computed by averaging each dimension over the number of specified values in the cluster members. Furthermore, the centroid itself may not be fully specified.
2. The distance between a data point and a centroid is computed only over the specified dimensions in both. Furthermore, the distance is divided by the number of such dimensions in order to fairly compare different data points.

The ratings matrix should be normalized before applying the clustering method.

18.5.4.2 Adapting Co-Clustering

The co-clustering approach is described in Sect. 13.3.3.1 of Chap. 13. Co-clustering is well suited to discovery of neighborhood sets of users and items in sparse matrices. The specified entries are treated as 1s and the unspecified entries are treated as 0s for co-clustering. An example of the co-clustering approach, as applied to the utility matrix of Fig. 18.4b, is illustrated in Fig. 18.6a. In this case, only a 2-way co-clustering is shown for simplicity. The co-clustering approach cleanly partitions the users and items into groups with a clear correspondence to each other. Therefore, user-neighborhoods and item-neighborhoods are discovered simultaneously. After the neighborhoods have been defined, the aforementioned

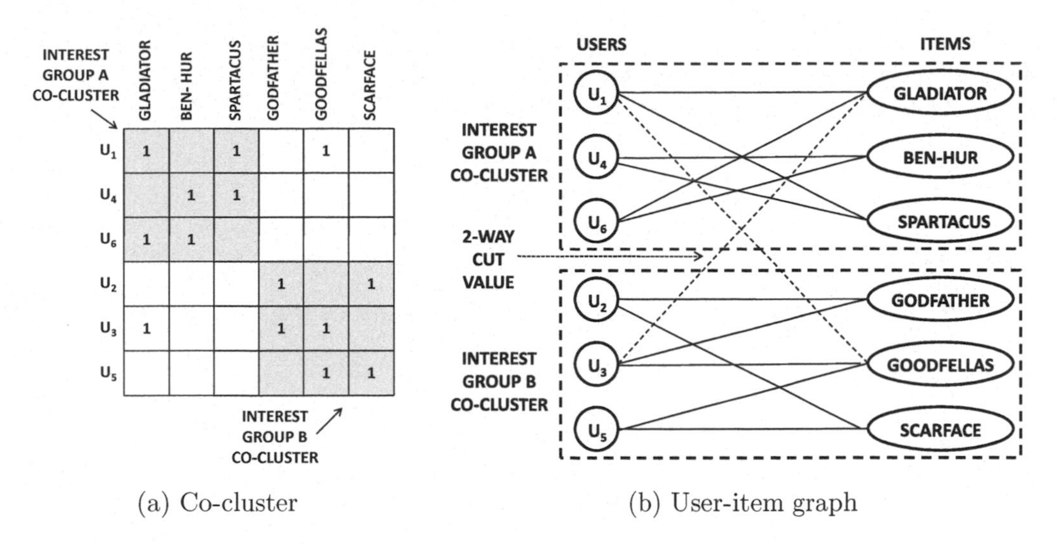

(a) Co-cluster (b) User-item graph

Figure 18.6: Co-clustering of user-item graph

user-based methods and item-based methods can be used to make predictions for the missing entries.

The co-clustering approach also has a nice interpretation in terms of the user-item graph. Let $G = (N_u \cup N_i, A)$ denote the preference graph, where N_u is the set of nodes representing users, and N_i is the set of nodes representing items. An undirected edge exists in A for each nonzero entry of the utility matrix. Then the co-cluster is a clustering of this graph structure. The corresponding 2-way graph partition is illustrated in Fig. 18.6b. Because of this interpretation in terms of user-item graphs, the approach is able to exploit item-item and user-user similarity simultaneously. Co-clustering methods are also closely related to latent factor models such as nonnegative matrix factorization that simultaneously cluster rows and columns with the use of latent factors.

18.5.5 Latent Factor Models

The clustering methods discussed in the previous section use the aggregate properties of the data to make robust predictions. This can be achieved in a more robust way with latent factor models. This approach can be used either for ratings matrices or for positive preference utility matrices. Latent factor models have increasingly become more popular in recent years. The key idea behind latent factor models is that many dimensionality reduction and matrix factorization methods summarize the correlations across rows and columns in the form of lower dimensional vectors, or *latent factors*. Furthermore, collaborative filtering is essentially a missing data imputation problem, in which these correlations are used to make predictions. Therefore, these latent factors become hidden variables that encode the correlations in the data matrix in a concise way and can be used to make predictions. A robust estimation of the k-dimensional *dominant* latent factors is often possible even from incompletely specified data, when the value of k is much less than d. This is because the more concisely defined latent factors can be estimated accurately with the sparsely specified data matrix, as long as the number of specified entries is large enough.

The n users are represented in terms of n corresponding k-dimensional factors, denoted by the vectors $\overline{U_1} \dots \overline{U_n}$. The d items are represented by d corresponding k-dimensional factors, denoted by the vectors $\overline{I_1} \ldots \overline{I_d}$. The value of k represents the reduced dimensionality of the latent representation. Then, the rating r_{ij} for user i and item j is estimated by the vector dot product of the corresponding latent factors:

$$r_{ij} \approx \overline{U_i} \cdot \overline{I_j}. \tag{18.14}$$

If this relationship is true for every entry of the ratings matrix, then it implies that the entire ratings matrix $D = [r_{ij}]_{n \times d}$ can be factorized into two matrices as follows:

$$D \approx F_{user} F_{item}^T. \tag{18.15}$$

Here F_{user} is an $n \times k$ matrix, in which the ith row represent the latent factor $\overline{U_i}$ for user i. Similarly, F_{item} is an $d \times k$ matrix, in which the jth row represents the latent factor $\overline{I_j}$ for item j. How can these factors be determined? The two key methods to use for computing these factors are singular value decomposition, and matrix factorization, which will be discussed in the sections below.

18.5.5.1 Singular Value Decomposition

Singular Value Decomposition (SVD) is discussed in detail in Sect. 2.4.3.2 of Chap. 2. The reader is advised to revisit that section before proceeding further. Equation 2.12 of Chap. 2 approximately factorizes the data matrix D into three matrices, and is replicated here:

$$D \approx Q_k \Sigma_k P_k^T. \tag{18.16}$$

Here, Q_k is an $n \times k$ matrix, Σ_k is a $k \times k$ diagonal matrix, and P_k is a $d \times k$ matrix. The main difference from the 2-way factorization format is the diagonal matrix Σ_k. However, this matrix can be included within the user factors. Therefore, one obtains the following factor matrices:

$$F_{user} = Q_k \Sigma_k \tag{18.17}$$
$$F_{item} = P_k. \tag{18.18}$$

The discussion in Chap. 2 shows that the matrix $Q_k \Sigma_k$ defines the reduced and transformed coordinates of data points in SVD. Thus, each user has a new set of a k-dimensional coordinates in a new k-dimensional basis system P_k defined by linear combinations of items. Strictly speaking, SVD is undefined for incomplete matrices, although heuristic approximations are possible. The bibliographic notes provide pointers to methods that are designed to address this issue. Another disadvantage of SVD is its high computational complexity. For nonnegative ratings matrices, $PLSA$ may be used, because it provides a probabilistic factorization similar to SVD.

18.5.5.2 Matrix Factorization

SVD is a form of matrix factorization. Because there are many different forms of matrix factorization, it is natural to explore whether they can be used for recommendations. The reader is advised to read Sect. 6.8 of Chap. 6 for a review of matrix factorization. Equation 6.30 of that section is replicated here:

$$D \approx U \cdot V^T. \tag{18.19}$$

This factorization is already directly in the form we want. Therefore, the user and item factor matrices are defined as follows:

$$F_{user} = U \quad (18.20)$$

$$F_{item} = V. \quad (18.21)$$

The main difference from the analysis of Sect. 6.8 is in how the optimization objective function is set up for incomplete matrices. Recall that the matrices U and V are determined by optimizing the following objective function:

$$J = ||D - U \cdot V^T||^2. \quad (18.22)$$

Here, $|| \cdot ||$ represents the Frobenius norm. In this case, because the ratings matrix D is only partially specified, the optimization is performed only over the *specified entries*, rather than all the entries. Therefore, the basic form of the optimization problem remains very similar, and it is easy to use any off-the-shelf optimization solver to determine U and V. The bibliographic notes contain pointers to relevant stochastic gradient descent methods. A regularization term $\lambda(||U||^2 + ||V||^2)$ containing the squared Frobenius norms of U and V may be added to J to reduce overfitting. The regularization term is particularly important when the number of specified entries is small. The value of the parameter λ is determined using cross-validation.

This method is more convenient than SVD for determining the factorized matrices because the optimization objective can be set up in a seamless way for an incompletely specified matrix no matter how sparse it might be. When the ratings are nonnegative, it is also possible to use nonnegative forms of matrix factorization. As discussed in Sect. 6.8, the nonnegative version of matrix factorization provides a number of interpretability advantages. Other forms of factorization, such as probabilistic matrix factorization and maximum margin matrix factorization, are also used. Most of these variants are different in terms of minor variations in the objective function (e.g., Frobenius norm minimization, or maximum likelihood maximization) and the constraints (e.g., nonnegativity) of the underlying optimization problem. These differences often translate to variants of the same stochastic gradient descent approach.

18.6 Web Usage Mining

The usage of the Web leads to a significant amount of *log* data. There are two primary types of logs that are commonly collected:

1. *Web server logs:* These correspond to the user activity on Web servers. Typically logs are stored in standardized format, known as the *NCSA common log format*, to facilitate ease of use and analysis by different programs. A few variants of this format, such as the *NCSA combined log format*, and *extended log format*, store a few extra fields. Nevertheless, the number of variants of the basic format is relatively small. An example of a Web log entry is as follows:

```
98.206.207.157 - - [31/Jul/2013:18:09:38 -0700] "GET /productA.pdf
HTTP/1.1" 200 328177 "-" "Mozilla/5.0 (Mac OS X) AppleWebKit/536.26
(KHTML, like Gecko) Version/6.0 Mobile/10B329 Safari/8536.25"
"retailer.net"
```

2. *Query logs:* These correspond to the queries posed by a user in a search engine. Aside from the commercial search engine providers, such logs may also be available to Web site owners if the site contains search features.

These types of logs can be used with a wide variety of applications. For example, the browsing behavior of users can be extracted to make recommendations. The area of Web usage mining is too large to be covered by a section of a single chapter. Therefore, the goal of this section is to provide an overview of how the various techniques discussed in this book can be mapped to Web usage mining. The bibliographic notes contain pointers to more detailed Web mining books on this topic. One major issue with Web log applications is that logs contain data that is not cleanly separated between different users and is therefore difficult to directly use in arbitrary application settings. In other words, significant preprocessing is required.

18.6.1 Data Preprocessing

A log file is often available as a continuous sequence of entries that corresponds to the user accesses. The entries for different users are typically interleaved with one another randomly, and it is also difficult to distinguish different sessions of the same user.

Typically, client-side cookies are used to distinguish between different user sessions. However, client-side cookies are often disabled due to privacy concerns at the client end. In such cases, only the IP address is available. It is hard to distinguish between different users on the basis of IP addresses only. Other fields, such as user agents and referrers, are often used to further distinguish. In many cases, at least a subset of the users can be identified to a reasonable level of granularity. Therefore, only the subset of the logs, where the users can be identified, is used. This is often sufficient for application-specific scenarios. The bibliographic notes contain pointers to preprocessing methods for Web logs.

The preprocessing leads to a set of *sequences* in the form of page views, which are also referred to as *click streams.* In some cases, the graph of traversal patterns, as it relates to the link structure of the pages at the site, is also constructed. For query logs, similar sequences are obtained in the form of search tokens, rather than page views. Therefore, in spite of the difference in the application scenario, there is some similarity in the nature of the data that is collected. In the following, some key applications of Web log mining will be visited briefly.

18.6.2 Applications

Click-stream data lead to a number of applications of sequence data mining. In the following, a brief overview of the various applications will be provided, along with the pointers to the relevant chapters. The bibliographic notes also contain more specific pointers.

Recommendations

Users can be recommended Web pages on the basis of their browsing patterns. In this case, it is not even necessary to use the sequence information; rather, a user-pageview matrix can be constructed from the previous browsing behavior. This can be leveraged to infer the user interest in the different pages. The corresponding matrix is typically a positive preference utility matrix. Any of the recommendation algorithms in this chapter can be used to infer the pages, in which the user is most likely to be interested.

Frequent Traversal Patterns

The frequent traversal patterns at a site provide an overview of the most likely patterns of user traversals at a site. The frequent sequence mining algorithms of Chap. 15 as well as the frequent graph pattern mining algorithms of Chap. 17 may be used to determine the paths that are most popular. The Web site owner can use these results for Web site reorganization. For example, paths that are very popular should stay as continuous paths in the Web site graph. Rarely used paths and links may be reorganized, if needed. Links may be added between pairs of pages if a sequential pattern is frequently observed between that pair.

Forecasting and Anomaly Detection

The Markovian models in Chap. 15 may be used to forecast future clicks of the user. Significant deviation of these clicks from expected values may correspond to anomalies. A second kind of anomaly occurs when an entire pattern of accesses is unusual. These types of scenarios are different from the case, where a particular page view in the sequence is considered anomalous. Hidden Markov models may be used to discover such anomalous sequences. The reader is referred to Chap. 15 for a discussion of these methods.

Classification

In some cases, the sequences from a Web log may be labeled on the basis of desirable or undesirable activity. An example of a desirable activity is when a user buys a certain product after browsing a certain sequence of pages at a site. An undesirable sequence may be indicative of an intrusion attack. When labels are available, it may be possible to perform early classification of Web log sequences. The results can be used to make *online* inferences about the future behavior of Web users.

18.7 Summary

Web data is of two types. The first type of data corresponds to the documents and links available on the Web. The second type of data corresponds to patterns of user behavior such as buying behavior, ratings, and Web logs. Each of these types of data can be leveraged for different insights.

Collecting document data from the Web is often a daunting task that is typically achieved with the use of *crawlers*, or *spiders*. Crawlers may be either universal crawlers that are used by commercial search engines, or they may be preferential crawlers, in which only topics of a particular subject are collected. After the documents are collected, they are stored and indexed in search engines. Search engines use a combination of textual similarity and reputation-based ranking to create a final score. The two most common algorithms used for ranking in search engines are the *PageRank* and *HITS* algorithms. Topic-sensitive *PageRank* is often used to compute similarity between nodes.

A significant amount of data is collected on the Web, corresponding to user-item preferences. This data can be used for making recommendations. Recommendation methods can be either content-based or user preference-based. Preference-based methods include neighborhood-based techniques, clustering techniques, graph-based techniques, and latent factor-based techniques.

Web logs are another important source of data on the Web. Web logs typically result in either sequence data or graphs of traversal patterns. If the sequential portion of the data is ignored, then the logs can also be used for making recommendations. Typical applications of Web log analysis include determining frequent traversal patterns and anomalies, and identifying interesting events.

18.8 Bibliographic Notes

Two excellent resources for Web mining are the books in [127, 357]. An early description of Web search engines, starting from the crawling to the searching phase, is provided by the founders of the Google search engine [114]. The general principles of crawling may be found in [127]. There is significant work on preferential crawlers as well [127, 357]. Numerous aspects of search engine indexing and querying are described in [377].

The *PageRank* algorithm is described in [114, 412]. The *HITS* algorithm was described in [317]. A detailed description of different variations of the *PageRank* and *HITS* algorithms may be found in [127, 343, 357, 377]. The topic-sensitive *PageRank* algorithm is described in [258], and the *SimRank* algorithm is described in [289].

Recommender systems are described well in Web and data mining books [343, 357]. In addition, general background on the topic is available in journal survey articles and special issues [2, 325]. The problem of collaborative filtering can be considered a version of the missing data imputation problem. A vast literature exists on missing data analysis [364]. Item-based collaborative filtering algorithms are discussed in [170, 445]. Graph-based methods for recommendations are discussed in [210, 277, 528]. Methods for link-prediction in signed networks are discussed in [341]. The origin of latent factor models is generally credited to a number of successful entries in the Netflix prize contest [558]. However, the use of latent factor models for estimating missing entries precedes the work in the field of recommendation analysis and the Netflix prize contest by several years [23]. This work [23] shows how *SVD* may be used for approximating missing data entries by combining it with the EM algorithm. Furthermore, the works in [272, 288, 548], which were performed earlier than the Netflix prize contest, show how different forms of matrix factorization may be used for recommendations. After the *popularization* of this approach by the Netflix prize contest, other factorization-based methods were also proposed for collaborative filtering [321, 322, 323]. Related matrix factorization models may be found in [288, 440, 456]. Latent semantic models can be viewed as probabilistic versions of latent factor models, and are discussed in [272].

Web usage mining has been described well in [357]. Both Web log mining and usage mining are described in this work. A description of methods for Web log preparation may be found in [161, 477]. Methods for anomaly detection with Web logs are discussed in [5]. Surveys on Web usage mining appear in [65, 390, 425].

18.9 Exercises

1. Implement a universal crawler with the use of a breadth-first algorithm.

2. Consider the string *ababcdef*. List all 2-shingles and 3-shingles, using each alphabet as a token.

3. Discuss why it is good to add anchor text to the Web page it points to for mining purposes, but it is often misleading for the page in which it appears.

4. Perform a Google search on "*mining text data*" and "*text data mining.*" Do you get the same top-10 search results? What does this tell you about the content component of the ranking heuristic used by search engines?

5. Show that the *PageRank* computation with teleportation is an eigenvector computation on an appropriately constructed probability transition matrix.

6. Show that the hub and authority scores in *HITS* can be computed by dominant eigenvector computations on AA^T and A^TA respectively. Here, A is the adjacency matrix of the graph $G = (S, A)$, as defined in the chapter.

7. Show that the largest eigenvalue of a stochastic transition matrix is always 1.

8. Suppose that you are told that a particular transition matrix P can be diagonalized as $P = V\Lambda V^{-1}$, where Λ is diagonal. How can you use this result to efficiently determine the k-hop transition matrix which defines the probability of a transition between each pair of nodes in k hops? What would you do for the special case when $k = \infty$? Does the result hold if we allow the entries of P and V to be complex numbers?

9. Apply the *PageRank* algorithm to the graph of Fig. 18.2b, using teleportation probabilities of 0.1, 0.2, and 0.4, respectively. What is the impact on the dead-end component (probabilities) of increasing the teleportation probabilities?

10. Repeat the previous exercise, except that the restart is performed from node 1. How are steady-state probabilities affected by increasing the teleportation probability?

11. Show that the transition matrix of the graph of Fig. 18.4.1b will have more than one eigenvector with an eigenvalue of 1. Why is the eigenvector with unit eigenvalue not unique in this case?

12. Implement the neighborhood-based approach for collaborative filtering on a ratings matrix.

13. Implement the personalized *PageRank* approach for collaborative filtering on a positive-preference utility matrix.

14. Apply the *PageRank* algorithm to the example of Fig. 18.5 by setting restart probabilities to 0.1, 0.2, and 0.4, respectively.

15. Apply the personalized *PageRank* algorithm to the example of Fig. 18.5 by restarting at node *Gladiator*, and with restart probabilities of 0.1, 0.2, and 0.4, respectively. What does this tell you about the most relevant users for the movie *Gladiator* What does this tell you about the most relevant user for the movie "*Gladiator*," who has not already watched this movie? Is it possible for the most relevant user to change with teleportation probability? What is the intuitive significance of the teleportation probability from an application-specific perspective?

16. Construct the optimization formulation for the matrix factorization problem for incomplete matrices.

17. In the bipartite graph of Fig. 18.5, what is the *SimRank* value between a user node and an item node? In this light, explain the weakness of the *SimRank* model.

Chapter 19

Social Network Analysis

"*I hope we will use the Net to cross barriers and connect cultures.*"—Tim Berners-Lee

19.1 Introduction

The tendency of humans to connect with one another is a deep-rooted social need that precedes the advent of the Web and Internet technologies. In the past, social interactions were achieved through face-to-face contact, postal mail, and telecommunication technologies. The last of these is also relatively recent when compared with the history of mankind. However, the popularization of the Web and Internet technologies has opened up entirely new avenues for enabling the seamless interaction of geographically distributed participants. This extraordinary potential of the Web was observed during its infancy by its visionary founders. However, it required a decade before the true social potential of the Web could be realized. Even today, Web-based social applications continue to evolve and create an ever-increasing amount of data. This data is a treasure trove of information about user preferences, their connections, and their influences on others. Therefore, it is natural to leverage this data for analytical insights.

Although social networks are popularly understood in the context of large online networks such as Twitter, LinkedIn, and Facebook, such networks represent only a small minority of the interaction mechanisms enabled by the Web. In fact, the traditional study of social network analysis in the field of sociology precedes the popularization of technologically enabled mechanisms. Much of the discussion in this chapter applies to social networks that extend beyond the popular notions of online social networks. Some examples are as follows:

- Social networks have been studied extensively in the field of sociology for more than a century but not from an online perspective. Data collection was rather difficult in these scenarios because of the lack of adequate technological mechanisms. Therefore, these studies were often conducted with painstaking and laborious methods for manual data collection. An example of such an effort is Stanley Milgram's famous six degrees of separation experiment in the sixties, which used postal mail between participants

C. C. Aggarwal, *Data Mining: The Textbook*, DOI 10.1007/978-3-319-14142-8_19

to test whether two arbitrary humans on the planet could be connected by a chain of six relationships. Because of the difficulty in verifying local forwards of mail, such experiments were often hard to conduct in a trustworthy way. Nevertheless, in spite of the obvious flaws in the experimental setting, these results have recently been shown to be applicable to online social networks, where the relationships between individuals are more easily quantifiable.

- A number of technological enablers, such as telecommunications, email, and electronic chat messengers, can be considered indirect forms of social networks. Such enablers result in communications between different individuals, and therefore they have a natural social aspect.

- Sites that are used for sharing online media content, such as Flickr, YouTube, or Delicious, can also be considered indirect forms of social networks, because they allow an extensive level of user interaction. In addition, social media outlets provide a number of unique ways for users to interact with one another. Examples include posting blogs or tagging each other's images. In these cases, the interaction is centered around a specific service such as content-sharing; yet many fundamental principles of social networking apply. Such social networks are extremely rich from the perspective of mining applications. They contain a tremendous amount of content such as text, images, audio, or video.

- A number of social networks can be constructed from specific kinds of interactions in professional communities. Scientific communities are organized into bibliographic and citation networks. These networks are also content rich because they are organized around publications.

It is evident that these different kinds of networks illustrate different facets of social network analysis. Many of the fundamental problems discussed in this chapter apply to these different scenarios but in different settings. Most of the traditional problems in data mining, such as clustering and classification, can also be extended to social network analysis. Furthermore, a number of more complex problem definitions are possible, such as link prediction and social influence analysis, because of the greater complexity of networks as compared to other kinds of data.

This chapter is organized as follows. Section 19.2 discusses a number of fundamental properties of social network analysis. The problem of community detection is explained in Sect. 19.3. The collective classification problem is discussed in Sect. 19.4. Section 19.5 discusses the link prediction problem. The social influence analysis problem is addressed in Sect. 19.6. The chapter summary is presented in Sect. 19.7.

19.2 Social Networks: Preliminaries and Properties

It is assumed that the social network can be structured as a graph $G = (N, A)$, where N is the set of nodes and A is the set of edges. Each individual in the social network is represented by a node in N, and is also referred to as an *actor*. The edges represent the connections between the different actors. In a social network such as Facebook, these edges correspond to friendship links. Typically, these links are undirected, although it is also possible for some "follower-based" social networks, such as Twitter, to have directed links. By default, it will be assumed that the network $G = (N, A)$ is undirected, unless otherwise specified. In some cases, the nodes in N may have content associated with them. This content may correspond to comments or other documents posted by social network users. It is assumed that the social network contains n nodes and m edges. In the following, some key properties of social networks will be discussed.

19.2.1 Homophily

Homophily is a fundamental property of social networks that is used in many applications, such as node classification. The basic idea in homophily is that nodes that are connected to one another are more likely to have similar properties. For example, a person's friendship links in Facebook may be drawn from previous acquaintances in school and work. Aside from common backgrounds, the friendship links may often imply common interests between the two parties. Thus, individuals who are linked may often share common beliefs, backgrounds, education, hobbies, or interests. This is best stated in terms of the old proverb:

Birds of a feather flock together

This property is leveraged in many network-centric applications.

19.2.2 Triadic Closure and Clustering Coefficient

Intuitively, triadic closure may be thought of as an inherent tendency of real-world networks to cluster. The principle of triadic closure is as follows:

If two individuals in a social network have a friend in common, then it is more likely that they are either connected or will eventually become connected in the future.

The principle of triadic closure implies an inherent correlation in the edge structure of the network. This is a natural consequence of the fact that two individuals connected to the same person are more likely to have similar backgrounds and also greater opportunities to interact with one another. The concept of triadic closure is related to homophily. Just as the similarity in backgrounds of connected individuals makes their properties similar, it also makes it more likely for them to be connected to the same set of actors. While homphily is typically exhibited in terms of content properties of node attributes, triadic closure can be viewed as the structural version of homophily. The concept of triadic closure is directly related to the *clustering coefficient* of the network.

The *clustering coefficient* can be viewed as a measure of the inherent tendency of a network to cluster. This is similar to the Hopkins statistic for multidimensional data (cf. Sect. 6.2.1.4 of Chap. 6). Let $S_i \subseteq N$ be the set of nodes connected to node $i \in N$ in the undirected network $G = (N, A)$. Let the cardinality of S_i be n_i. There are $\binom{n_i}{2}$ possible edges between nodes in S_i. The local clustering coefficient $\eta(i)$ of node i is the fraction of these pairs that have an edge between them.

$$\eta(i) = \frac{|\{(j,k) \in A : j \in S_i, k \in S_i\}|}{\binom{n_i}{2}} \tag{19.1}$$

The Watts–Strogatz *network average clustering coefficient* is the average value of $\eta(i)$ over all nodes in the network. It is not difficult to see that the triadic closure property increases the clustering coefficient of real-world networks.

19.2.3 Dynamics of Network Formation

Many real properties of networks are affected by how they are formed. Networks such as the World Wide Web and social networks are continuously growing over time with new nodes and edges being added constantly. Interestingly, networks from multiple domains share a number of common characteristics in the dynamic processes by which they grow. The manner in which new edges and nodes are added to the network has a direct impact on the eventual structure of the network and choice of effective mining techniques. Therefore, the following will discuss some common properties of real-world networks:

1. *Preferential attachment:* In a growing network, the likelihood of a node receiving new edges increases with its degree. This is a natural consequence of the fact that highly connected individuals will typically find it easier to make new connections. If $\pi(i)$ is the probability that a newly added node attaches itself to an existing node i in the network, then a model for the probability $\pi(i)$ in terms of the degree of node i is as follows:
$$\pi(i) \propto \text{Degree}(i)^{\alpha} \tag{19.2}$$
The value of the parameter α is dependent on the domain from which the network is drawn, such as a biological network or social network. In many Web-centric domains, a *scale-free assumption* is used. This assumption states that $\alpha \approx 1$, and therefore the proportionality is linear. Such networks are referred to as *scale-free* networks. This model is also referred to as the *Barabasi–Albert model.* Many networks, such as the World Wide Web, social networks, and biological networks, are conjectured to be scale free, although the assumption is obviously intended to be an approximation. In fact, many properties of real networks are not completely consistent with the scale-free assumption.

2. *Small world property:* Most real networks are assumed to be "small world." This means that the average path length between any pair of nodes is quite small. In fact, Milgram's experiment in the sixties conjectured that the distance between any pair of nodes is about six. Typically, for a network containing $n(t)$ nodes at time t, many models postulate that the average path lengths grow as $\log(n(t))$. This is a small number, even for very large networks. Recent experiments have confirmed that the average path lengths of large-scale networks such as Internet chat networks are quite small. As discussed below, the dynamically varying diameters have been experimentally shown to be even more constricted than the (modeled) $\log(n(t))$ growth rate would suggest.

3. *Densification:* Almost all real-world networks such as the Web and social networks add more nodes and edges over time than are deleted. The impact of adding new edges generally dominates the impact of adding new nodes. This implies that the graphs gradually densify over time, with the number of edges growing superlinearly with the number of nodes. If $n(t)$ is the number of nodes in the network at time t, and $e(t)$ is the number of edges, then the network exhibits the following *densification power law*:
$$e(t) \propto n(t)^{\beta} \tag{19.3}$$
The exponent β is a value between 1 and 2. The value of $\beta = 1$ corresponds to a network where the average degree of the nodes is not affected by the growth of the network. A value of $\beta = 2$ corresponds to a network in which the total number of

edges $e(t)$ remains a constant fraction of the complete graph of $n(t)$ nodes as $n(t)$ increases.

4. *Shrinking diameters:* In most real-world networks, as the network densifies, the average distances between the nodes shrink over time. This experimental observation is in contrast to conventional models that suggest that the diameters should increase as $\log(n(t))$. This unexpected behavior is a consequence of the fact that the addition of new edges dominates the addition of new nodes. Note that if the impact of adding new nodes were to dominate, then the average distances between nodes would increase over time.

5. *Giant connected component:* As the network densifies over time, a giant connected component emerges. The emergence of a giant connected component is consistent with the principle of preferential attachment, in which newly incoming edges are more likely to attach themselves to the densely connected and high-degree nodes in the network. This property also has a confounding impact on network clustering algorithms, because it typically leads to unbalanced clusters, unless the algorithms are carefully designed.

Preferential attachment also has a significant impact on the typical structure of online networks. It results in a small number of very high-degree nodes that are also referred to as *hubs*. The hub nodes are usually connected to many different regions of the network and, therefore, have a confounding impact on many network clustering algorithms. The notion of hubs, as discussed here, is subtly different from the notion of hubs, as discussed in the *HITS* algorithm, because it is not specific to a query or topic. Nevertheless, the intuitive notion of nodes being central points of connectivity in a network, is retained in both cases.

19.2.4 Power-Law Degree Distributions

A consequence of preferential attachment is that a small minority of high-degree nodes continue to attract most of the newly added nodes. It can be shown that the number of nodes $P(k)$ with degree k, is regulated by the following *power-law degree distribution*:

$$P(k) \propto k^{-\gamma} \tag{19.4}$$

The value of the parameter γ ranges between 2 and 3. It is noteworthy that larger values of γ lead to more small degree nodes. For example, when the value of γ is 3, the vast majority of the nodes in the network will have a degree of 1. On the other hand, when the value of γ is small, the degree distribution is less skewed.

19.2.5 Measures of Centrality and Prestige

Nodes that are central to the network have a significant impact on the properties of the network, such as its density, pairwise shortest path distances, connectivity, and clustering behavior. Many of these nodes are hub nodes, with high degrees that are a natural result of the dynamical processes of large network generation. Such actors are often more prominent because they have ties to many actors and are in a position of better influence. Their impact on network mining algorithms is also very significant. A related notion of centrality is *prestige*, which is relevant for directed networks. For example, on Twitter, an actor with a larger number of followers has greater prestige. On the other hand, following a large number of individuals does not bring any prestige but is indicative of the *gregariousness* of an actor.

The notion of *PageRank*, discussed in the previous chapter, is often used as a measure of prestige.

Measures of centrality are naturally defined for undirected networks, whereas measures of prestige are designed for directed networks. However, it is possible to generalize centrality measures to directed networks. In the following, centrality measures will be defined for undirected networks, whereas prestige measures will be defined for directed networks.

19.2.5.1 Degree Centrality and Prestige

The degree centrality $C_D(i)$ of a node i of an undirected network is equal to the degree of the node, divided by the maximum *possible* degree of the nodes. The maximum possible degree of a node in the network is one less than the number of nodes in the network. Therefore, if Degree(i) is the degree of node i, then the degree centrality $C_D(i)$ of node i is defined as follows:

$$C_D(i) = \frac{\text{Degree}(i)}{n-1} \tag{19.5}$$

Because nodes with higher degree are often hub nodes, they tend to be more central to the network and bring distant parts of the network closer together. The major problem with degree centrality is that it is rather myopic in that it does not consider nodes beyond the immediate neighborhood of a given node i. Therefore, the overall structure of the network is ignored to some extent. For example, in Fig. 19.1a, node 1 has the highest degree centrality, but it cannot be viewed as central to the network itself. In fact, node 1 is closer to the periphery of the network.

Degree *prestige* is defined for directed networks only, and uses the indegree of the node, rather than its degree. The idea is that only a high indegree contributes to the prestige because the indegree of a node can be viewed as a vote for the popularity of the node, similar to *PageRank*. Therefore, the degree prestige $P_D(i)$ of node i is defined as follows:

$$P_D(i) = \frac{\text{Indegree}(i)}{n-1} \tag{19.6}$$

For example, node 1 has the highest degree prestige in Fig. 19.1b. It is possible to generalize this notion recursively by taking into account the prestige of nodes pointing to a node, rather than simply the number of nodes. This corresponds to the rank prestige, which will be discussed later in this section.

The notion of centrality can also be extended to the node *outdegree*. This is defined as the *gregariousness* of a node. Therefore, the gregariousness $G_D(i)$ of a node i is defined as follows:

$$G_D(i) = \frac{\text{Outdegree}(i)}{n-1} \tag{19.7}$$

The gregariousness of a node defines a different qualitative notion than prestige because it quantifies the propensity of an individual to seek out new connections (such as following many other actors in Twitter), rather than his or her popularity with respect to other actors.

19.2.5.2 Closeness Centrality and Proximity Prestige

The example of Fig. 19.1a shows that the degree centrality criterion is susceptible to picking nodes on the periphery of the network with no regard to their *indirect* relationships to other nodes. In this context, closeness centrality is more effective.

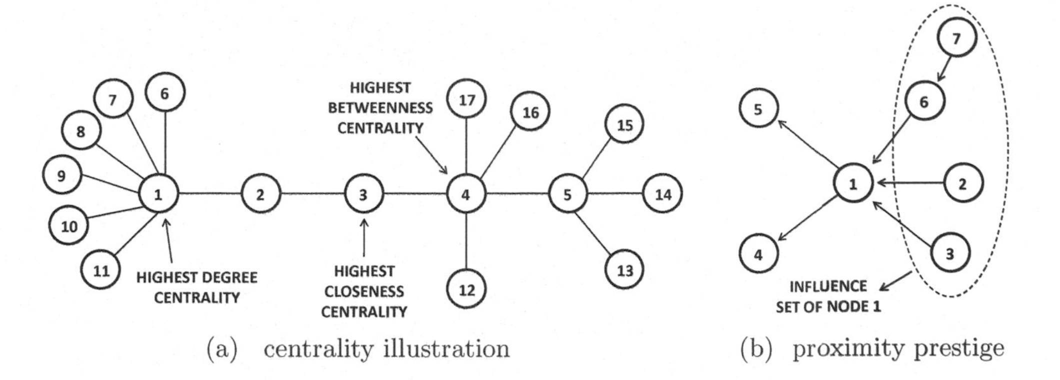

(a) centrality illustration (b) proximity prestige

Figure 19.1: Illustration of centrality and prestige

The notion of closeness centrality is meaningfully defined with respect to *undirected* and *connected* networks. The average shortest path distance, starting from node i, is denoted by AvDist(i) and is defined in terms of the pairwise shortest path distances Dist(i, j), between nodes i and j as follows:

$$\text{AvDist}(i) = \frac{\sum_{j=1}^{n} \text{Dist}(i, j)}{n-1} \tag{19.8}$$

The closeness centrality is simply the inverse of the average distance of other nodes to node i.

$$C_C(i) = 1/\text{AvDist}(i) \tag{19.9}$$

Because the value of AvDist(i) is at least 1, this measure ranges between 0 and 1. In the case of Fig. 19.1a, node 3 has the highest closeness centrality because it has the lowest average distance to other nodes.

A measure known as *proximity prestige* can be used to measure prestige in directed networks. To compute the proximity prestige of node i, the shortest path distance to node i from all other nodes is computed. Unlike undirected networks, a confounding factor in the computation is that *directed* paths may not exist from other nodes to node i. For example, no path exists to node 7 in Fig. 19.1b. Therefore, the first step is to determine the set of nodes Influence(i) that can *reach* node i with a directed path. For example, in the case of the Twitter network, Influence(i) corresponds to all recursively defined followers of node i. An example of an influence set of node 1 is illustrated in Fig. 19.1b. The value of AvDist(i) can now be computed only with respect to the influence set Influence(i).

$$\text{AvDist}(i) = \frac{\sum_{j \in \text{Influence}(i)} \text{Dist}(j, i)}{|\text{Influence}(i)|} \tag{19.10}$$

Note that distances are computed from node j to i, and not vice versa, because we are computing a prestige measure, rather than a gregariousness measure.

Both the size of the influence set and average distance to the influence set play a role in defining the proximity prestige. While it is tempting to use the inverse of the average distance, as in the previous case, this would not be fair. Nodes that have less influence should be penalized. For example, in Fig. 19.1b, node 6 has the lowest possible distance value of 1 from node 7, which is also the only node it influences. While its low average distance to its influence set suggests high prestige, its small influence set suggests that it cannot be considered a node with high prestige. To account for this, a multiplicative penalty factor is included in the measure that corresponds to the fractional size of the influence set of node i.

$$\text{InfluenceFraction}(i) = \frac{|\text{Influence}(i)|}{n-1} \tag{19.11}$$

Then, the proximity prestige $P_P(i)$ is defined as follows:

$$P_P(i) = \frac{\text{InfluenceFraction}(i)}{\text{AvDist}(i)} \tag{19.12}$$

This value also lies between 0 and 1. Higher values indicate greater prestige. The highest possible proximity prestige value of 1 is realized at the central node of a perfectly star-structured network, with a single central actor and all other actors as its (in-linking) spokes.

In the case of Fig. 19.1b, node 1 has an influence fraction of $4/6$, and an average distance of $5/4$ from the four nodes that reach it. Therefore, its proximity prestige is $4*4/(5*6) = 16/30$. On the other hand, node 6 has a better average distance of 1 to the only node that reaches it. However, because its influence fraction is only $1/6$, its proximity prestige is $1/6$ as well. This suggests that node 1 has better proximity prestige than node 6. This matches our earlier stated intuition that node 6 is not a very influential node.

19.2.5.3 Betweenness Centrality

While closeness centrality is based on notions of distances, it does not account for the *criticality* of the node in terms of the number of shortest paths that pass through it. Such notions of criticality are crucial in determining actors that have the greatest control of the flow of information between other actors in a social network. For example, while node 3 has the highest closeness centrality, it is not as critical to shortest paths between different pairs of nodes as node 4 in Fig. 19.1a. Node 4 can be shown to be more critical because it also participates in shortest paths between the pairs of nodes directly incident on it, whereas node 3 does not participate in these pairs. The other pairs are approximately the same in the two cases. Therefore, node 4 controls the flow of information between nodes 12 and 17 that node 3 does not control.

Let q_{jk} denote the number of shortest paths between nodes j and k. For graphs that are not trees, there will often be more than one shortest path between pairs of nodes. Let $q_{jk}(i)$ be the number of these pairs that pass through node i. Then, the fraction of pairs $f_{jk}(i)$ that pass through node i is given by $f_{jk}(i) = q_{jk}(i)/q_{jk}$. Intuitively, $f_{jk}(i)$ is a fraction that indicates the level of control that node i has over nodes j and k in terms of regulating the flow of information between them. Then, the betweenness centrality $C_B(i)$ is the average value of this fraction over all $\binom{n}{2}$ pairs of nodes.

$$C_B(i) = \frac{\sum_{j<k} f_{jk}(i)}{\binom{n}{2}} \tag{19.13}$$

The betweenness centrality also lies between 0 and 1, with higher values indicating better betweenness. Unlike closeness centrality, betweenness centrality can be defined for disconnected networks as well.

While the aforementioned notion of betweenness centrality is designed for nodes, it can be generalized to edges by using the number of shortest paths passing through an edge (rather than a node). For example, the edges connected to the hub nodes in Fig. 19.2 have high betweenness. Edges that have high betweenness tend to connect nodes from different clusters in the graph. Therefore, these betweenness concepts are used in many community detection algorithms, such as the Girvan–Newman algorithm. In fact, the computation of node- and edge-betweenness values is described in Sect. 19.4 on the Girvan–Newman algorithm.

19.2.5.4 Rank Centrality and Prestige

The concepts of rank centrality and prestige are defined by random surfer models. The *PageRank* score can be considered a rank centrality score in undirected networks and a rank prestige score in directed networks. Note that the *PageRank* scores are components of the largest left eigenvector of the random walk transition matrix of the social network. If the adjacency matrix is directly used instead of the transition matrix to compute the largest eigenvector, the resulting scores are referred to as *eigenvector centrality* scores. Eigenvector centrality scores are generally less desirable than *PageRank* scores because of the disproportionately large influence of high-degree nodes on the centrality scores of their neighbors.

Because the computation of these scores was already discussed in detail in Chap. 18, it will not be revisited here. The idea here is that a citation of a node by another node (such as a follower in Twitter) is indicative of prestige. Although this is also captured by degree prestige, the latter does not capture the prestige of the nodes incident on it. The *PageRank* computation can be considered a refined version of degree prestige, where the *quality* of the nodes incident on a particular node i are used in the computation of its prestige.

19.3 Community Detection

The term "community detection" is an approximate synonym for "clustering" in the context of social network analysis. The clustering of networks and graphs is also sometimes referred to as "graph partitioning" in the traditional work on network analysis. Therefore, the literature in this area is rich and includes work from many different fields. Much of the work on graph partitioning precedes the formal study of social network analysis. Nevertheless, it continues to be relevant to the domain of social networks. Community detection is one of the most fundamental problems in social network analysis. The summarization of closely related social groups is, after all, one of the most succinct and easily understandable ways of characterizing social structures.

In the social network domain, network clustering algorithms often have difficulty in cleanly separating out different clusters because of some natural properties of typical social networks.

- Multidimensional clustering methods, such as the distance-based k-means algorithm, cannot be easily generalized to networks. In small-world networks, the distances between different pairs of nodes is a small number that cannot provide a sufficiently fine-grained indicator of similarity. Rather, it is more important to use triadic closure properties of real networks, explicitly or implicitly, in the clustering process.

- While social networks usually have distinct community structures, the high-degree hub nodes connect different communities, thereby bringing them together. Examples of such hub nodes connecting up different communities are illustrated in Fig. 19.2. In this case, the nodes A, B, and C are hubs that connect up different communities. In real social networks, the structure may be even more complicated, with some of the high-degree nodes belonging to particular sets of overlapping communities.

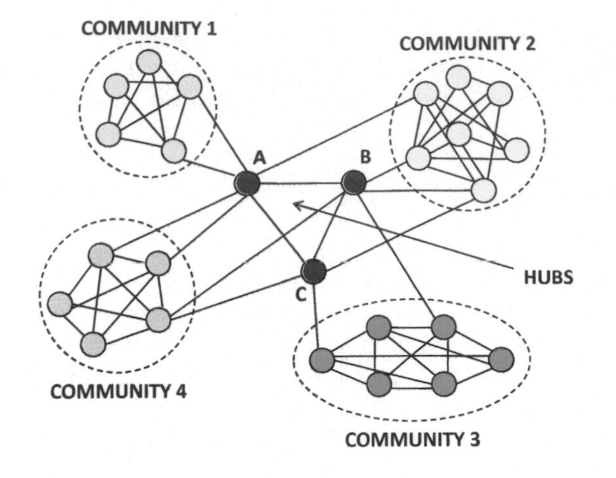

Figure 19.2: Impact of hubs on communities

- Different parts of the social network have different edge densities. In other words, the local clustering coefficients in distinct parts of the social network are typically quite different. As a result, when specific choices of parameters are used to quantify the clusters globally, it leads to unbalanced clusters because a single global parameter choice is not relevant in many network localities.

- Real social networks often have a giant component that is densely connected. This contributes further to the tendency of community detection algorithms to create imbalanced clusters, where a single cluster is the beneficiary of most of the nodes in the network.

Many network clustering algorithms have built-in mechanisms to address such issues. In the following, a discussion of some of the most well-known network clustering algorithms will be provided.

Assume that the undirected network is denoted by $G = (N, A)$. The weight of the edge (i, j) between nodes i and j, is denoted by $w_{ij} = w_{ji}$. In some cases, the inverse concept of edge costs (or lengths) is specified instead of weights. In such cases, we assume that the edge cost is denoted by c_{ij}. These values can be converted to one another by using $w_{ij} = 1/c_{ij}$, or a suitably chosen kernel function.

The problem of network clustering, or community detection, is that of partitioning the network into k sets of nodes, such that the sum of the weights of the edges with end points in different partitions is minimized. Many variations of this basic objective function are used in practice and are able to achieve different application-specific goals, such as *partition balancing* in which different clusters have similar numbers of nodes.

In the special case, where $w_{ij} = 1$, and there are no balancing constraints on partitions, the 2-way cut problem is polynomially solvable. The reader is advised to refer to the bibliographic notes for pointers to Karger's randomized minimum cut algorithm. This algorithm can determine the minimum cut in $O(n^2 \log^r(n))$ time for a network containing n nodes, where r is a constant regulating the desired level of probabilistic accuracy. However, the resulting cut is usually not balanced. Incorporating arbitrary edge weights or balancing constraints makes the problem NP-hard. Many network clustering algorithms focus on balanced 2-way partitioning of graphs. A 2-way partitioning can be recursively used to generate a k-way partitioning.

19.3.1 Kernighan–Lin Algorithm

The Kernighan–Lin algorithm is a classical method for balanced 2-way graph partitioning. The basic idea is to start with an initial partitioning of the graph into two equal[1] subsets of nodes. The algorithm then iteratively improves this partitioning, until it converges to an optimal solution. This solution is not guaranteed to be the global optimum, but it is usually a good heuristic approximation. This iterative improvement is performed by determining sequences of exchanges of nodes between partitions that improve the clustering objective function as much as possible. To evaluate the improvement in the clustering objective function by performing an exchange between a pair of nodes, some carefully chosen measures need to be continuously tracked maintained at each node. These will be discussed below.

The *internal cost* I_i of node i is the sum of the weights of edges incident on i, whose other end is present in the same partition as node i. The *external cost* E_i of node i, is the sum of the weights of the edges incident on i, whose other end is in a different partition than node i. Moving a node from one partition to the other changes its external cost to its internal cost and vice versa. Therefore, the gain D_i by moving a node i from one partition to the other is given by the difference between the external and the internal cost.

$$D_i = E_i - I_i \tag{19.14}$$

Of course, we are not interested in simply moving a node from one partition to the other, but in exchanging a pair of nodes i and j between two partitions. Then, the gain J_{ij} of exchanging nodes i and j is given by the following:

$$J_{ij} = D_i + D_j - 2 \cdot w_{ij} \tag{19.15}$$

This is simply a sum of the gains from moving nodes i and j to different partitions, with a special adjustment for the impact on the edge (i, j) that continues to be a part of the external cost of both nodes because of the exchange. The value of J_{ij} therefore quantifies the gain that one can obtain by exchanging nodes i and j. Positive values of J_{ij} result in an improvement of the objective function.

The overall algorithm uses repeated sequences of at most $(n/2)$ heuristic exchanges between the two partitions, which are designed to optimize the total gain from the exchanges. Each such sequence of at most $(n/2)$ exchanges will be referred to as an *epoch*. Each epoch proceeds as follows. A pair of nodes is found, such that the exchange leads to the maximum improvement in the objective function value. This pair of nodes is marked, although the exchange is not actually performed. The values of D_i for different nodes are recomputed, however, as if the exchange were already performed. Then, the next pair of unmarked nodes is determined with these recomputed values of D_i, for which the exchange leads to the maximum improvement in the objective function value. It should be pointed out that the gains will not always decrease, as further potential exchanges are determined. Furthermore, some intermediate potential exchanges might even have negative gain, whereas later potential exchanges might have positive gain. The process of determining potential exchange pairs is repeated until all n nodes have been paired. Any sequence of $k \leq n/2$ contiguous potential pairs, starting with the first pair, and in the same order as they were determined, is considered a valid potential k-exchange between the two partitions. Among these different possibilities, the potential k-exchange maximizing the total gain is found. If the gain is positive, then the potential k-exchange is executed. This entire process of a

[1]Without loss of generality, it can be assumed that the graph contains an even number of nodes, by adding a single dummy node.

```
Algorithm KernighanLin(Graph: G = (N, A), Weights:[w_ij])
begin
 Create random initial partition of N into N_1 and N_2;
 repeat
   Recompute D_i values for each node i ∈ N;
   Unmark all nodes in N;
   for i = 1 to n/2 do
   begin
     Select x_i ∈ N_1 and y_i ∈ N_2 to be the unmarked node pair with
                 the highest exchange-gain g(i) = J_{x_i y_i};
     Mark x_i and y_i;
     Recompute D_j for each node j, under the assumption that
           x_i and y_i will be eventually exchanged;
   end
   Determine k that maximizes G_k = Σ_{i=1}^{k} g(i);
   if (G_k > 0) then exchange {x_1 ... x_k} and
         {y_1 ... y_k} between N_1 and N_2;
 until (G_k ≤ 0);
 return(N_1, N_2);
end
```

Figure 19.3: The Kernighan–Lin algorithm

k-exchange is referred to as an *epoch*. The algorithm repeatedly executes such epochs of k-exchanges. If no such k-exchange with positive gain can be found, then the algorithm terminates. The overall algorithm is illustrated in Fig. 19.3.

The Kernighan–Lin algorithm converges rapidly to a local optimum. In fact, a very small number of epochs (fewer than five) may be required for the algorithm to terminate. Of course, there is no guarantee on the required number of epochs, considering that the problem is NP-hard. The running time of each epoch can be amortized to $O(m \cdot \log(n))$ time, where m is the number of edges and n is the number of nodes. Variants of the algorithm have been proposed to speed up the method significantly.

19.3.1.1 Speeding Up Kernighan–Lin

A fast variant of Kernighan–Lin is based on the modifications by Fiduccia and Mattheyses. This version can also handle weights associated with *both* nodes and edges. Furthermore, the approach allows the specification of the level of balance between the two partitions as a ratio. Instead of pairing nodes in an epoch to swap them, one can simply move a single node i from one partition to the other so that the gain D_i of Eq. 19.14 is as large as possible. Only nodes that can move without violating[2] the balancing constraint are considered eligible for a move at each step. After moving node i, it is marked so that it will not be considered again in the current epoch. The values of D_j on the other vertices $j \in N$ are updated to reflect this change. This process is repeated until either all nodes have been considered for a move in an epoch or the balancing criterion prevents further moves. The latter is possible

[2]Moving a node from one partition to the other will frequently cause violations unless some flexibility is allowed in the balancing ratio. In practice, a slight relaxation (or small range) of required balancing ratios may be used to ensure feasible solutions.

when the desired partition ratios are unbalanced, or the nodes do not have unit weights. Note that many potential moves in an epoch might have negative gain. Therefore, as in the original Kernighan–Lin algorithm, only the best partition created during an epoch is made final and the remaining moves are undone. A special data structure was also introduced by Fiduccia and Mattheyses to implement each epoch in $O(m)$ time, where m is the number of edges. In practice, a small number of epochs is usually required for convergence in most real-world networks, although there is no guarantee on the required number of epochs.

While the original improvement of Fiduccia and Mattheyses moves as many vertices as possible in an epoch, it was observed by Karypis and Kumar that it is not necessary to do so. Rather, one can terminate an epoch, if the partitioning objective function does not improve in a predefined number n_p of moves. These n_p moves are then undone, and the epoch terminates. The typical value of n_p chosen is 50. Furthermore, it is not always necessary to move the vertex with the largest possible gain, as long as the gain is positive. Dropping the restriction of finding a vertex with the *largest* gain improves the per-move cost significantly. The improvements from these simple modifications are significant in many scenarios.

19.3.2 Girvan–Newman Algorithm

This algorithm uses edge lengths c_{ij}, rather than the edge weights w_{ij}. The edge lengths may be viewed as in the inverse of the edge weights. In cases, where edge weights are specified, one may heuristically transform them to edge lengths by using $c_{ij} = 1/w_{ij}$, or a suitable application-specific function.

The Girvan–Newman algorithm is based on the intuition that edges with high betweenness have a tendency to connect different clusters. For example, the edges that are incident on the hub nodes in Fig. 19.2 have a high betweenness. Their high betweenness is a result of the large number of pairwise shortest paths between nodes of different communities passing through these edges. Therefore, the disconnection of these edges will result in a set of connected components that corresponds to the natural clusters in the original graph. This disconnection approach forms the basis of the Girvan–Newman algorithm.

The Girvan–Newman algorithm is a top-down hierarchical clustering algorithm that creates clusters by successively removing edges with the highest betweenness until the graph is disconnected into the required number of connected components. Because each edge removal impacts the betweenness values of some of the other edges, the betweenness values of these edges need to be recomputed after each removal. The Girvan–Newman algorithm is illustrated in Fig. 19.4.

The main challenge in the Girvan–Newman algorithm is the computation of the edge betweenness values. The computation of node betweenness values is an intermediary step in the edge-betweenness computation. Recall that all node and edge-betweenness centrality values are defined as a function of the exhaustive set of shortest paths between all source–sink pairs. These betweenness centrality values can, therefore, be decomposed into several additive components, where each component is defined by the subset of the shortest paths originating from a source node s. To compute these betweenness components, a two-step approach is used for each possible source node s:

1. The number of shortest paths from the source node s to every other node is computed.

2. The computations in the first step are used to compute the component $B_s(i)$ of the node betweenness centrality of node i, and the component $b_s(i,j)$ of the edge betweenness centrality of edge (i,j), *that correspond to the subset of shortest paths originating from a particular source node* s.

Algorithm *GirvanNewman*(Graph: $G = (N, A)$, Number of Clusters: k,
Edge lengths: $[c_{ij}]$)
begin
Compute betweenness value of all edges in graph G;
repeat
Remove edge (i, j) from G with highest betweenness;
Recompute betweenness of edges affected by removal of (i, j);
until G has k components remaining;
return connected components of G;
end

Figure 19.4: The Girvan–Newman Algorithm

These source node-specific betweenness centrality components can then be added over all possible source nodes to compute the overall betweenness centrality values.

The first step in the betweenness centrality computation is to create a graph of edges that lie on *at least* one shortest path from node s to some other node. Such edges are referred to as *tight* edges for source node s. The betweenness value component of an edge for a particular source node s can be nonzero only if that edge is tight for that source node. The Dijkstra algorithm, described in Sect. 3.5.1.1 of Chap. 3, is used to determine the shortest path distances $SP(j)$ from the source node s to node j. In order for an edge (i, j) to be tight, the following condition has to hold:

$$SP(j) = SP(i) + c_{ij} \tag{19.16}$$

Therefore, the *directed* subgraph $G^s = (N, A^s)$ of tight edges is determined, where $A^s \subseteq A$. The direction of the edge (i, j) is such that $SP(j) > SP(i)$. Therefore, the subgraph of tight edges is a *directed acyclic graph.* An example of a base graph, together with its subgraph of tight edges, is illustrated in Fig. 19.5. The edges are annotated with their lengths. In this case, node 0 is assumed to be the source node. The subgraph of tight edges will obviously vary with the choice of the source node. The shortest-path distances $SP(i)$ of node i from source node 0 are illustrated by the first component of the pair of numbers annotating the nodes in Fig. 19.5b.

The number of shortest paths $N_s(j)$ from the source node s to a given node j is relatively easy to determine from the subgraph of tight edges. This is because the number of paths to a given node is equal to the sum of the number of paths to the nodes incident on it.

$$N_s(j) = \sum_{i:(i,j)\in A^s} N_s(i) \tag{19.17}$$

The algorithm starts by setting $N_s(s) = 1$ for the source node s. Subsequently, the algorithm performs a breadth first search of the subgraph of tight edges, starting with the source node. The number of paths to each node is computed as the sum of the paths to its ancestors in the directed acyclic graph of tight edges, according to Eq. 19.17. The number of shortest paths to each node, from source node 0, is illustrated in Fig. 19.5b by the second component of the pair of numbers annotating each node.

The next step is to compute the component of the betweenness centrality for both nodes and edges starting at the source node s. Let $f_{sk}(i)$ be the fraction of shortest paths between nodes s and k, that pass through node i. Let $F_{sk}(i, j)$ be the fraction of shortest paths

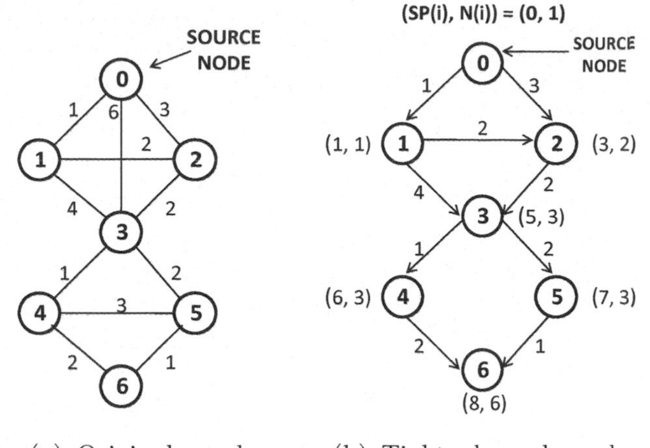

(a) Original graph (b) Tight-edge subgraph

Figure 19.5: Original graph and subgraph of tight edges

between nodes s and k, that pass through edge (i, j). The corresponding components of node betweenness centrality and edge betweenness centrality, specific to node s, are denoted by $B_s(i)$ and $b_s(i, j)$, and they are defined as follows:

$$B_s(i) = \sum_{k \neq s} f_{sk}(i) \tag{19.18}$$

$$b_s(i, j) = \sum_{k \neq s} F_{sk}(i, j) \tag{19.19}$$

It is easy to see that the *unnormalized* values[3] of the node betweenness centrality of i and the edge betweenness centrality of (i, j) may be obtained by respectively summing up each of $B_s(i)$ and $b_s(i, j)$ over the different source nodes s.

The graph G_s of tight edges is used to compute these values. The key is to set up recursive relationships between $B_s(i)$ and $b_s(i, j)$ as follows:

$$B_s(j) = \sum_{i:(i,j) \in A^s} b_s(i, j) \tag{19.20}$$

$$B_s(i) = 1 + \sum_{j:(i,j) \in A^s} b_s(i, j) \tag{19.21}$$

These relationships follow from the fact that shortest paths through a particular node always pass through exactly one of its incoming and outgoing edges, unless they end at that node. The second equation has an additional credit of 1 to account for the paths ending at node i, for which the full fractional credit $f_{si}(i) = 1$ is given to $B_s(i)$.

The source node s is always assigned a betweenness score of $B_s(s) = 0$. The nodes and edges of the directed acyclic tight graph G^s are processed "bottom up," starting at the nodes without any outgoing edges. The score $B_s(i)$ of a node i is finalized, only after the

[3]The normalized values, such as those in Eq. 19.13, may be obtained by dividing the unnormalized values by $n \cdot (n-1)$ for a network with n nodes. The constant of proportionality is irrelevant because the Girvan–Newman algorithm requires only the identification of the edge with the largest betweenness.

scores on all its outgoing edges have been finalized. Similarly, the score $b_s(i,j)$ of an edge (i,j) is finalized only after the score $B_s(j)$ of node j has been finalized. The algorithm starts by setting all nodes j without any outgoing edges to have a score of $B_s(j) = f_{sj}(j) = 1$. This is because such a node j, without outgoing edges, is (trivially) a intermediary between s and j, but it cannot be an intermediary between s and any other node. Then, the algorithm iteratively updates scores of nodes and edges in the bottom-up traversal as follows:

- *Edge Betweenness Update*: Each edge (i,j) is assigned a score $b_s(i,j)$ that is based on partitioning the score $B_s(j)$ into all the incoming edges (i,j) based on Eq. 19.20. The value of $b_s(i,j)$ is proportional to $N_s(i)$ that was computed earlier. Therefore, $b_s(i,j)$ is computed as follows.
$$b_s(i,j) = \frac{N_s(i) \cdot B_s(j)}{\sum_{k:(k,j)\in A^s} N_s(k)} \tag{19.22}$$

- *Node Betweenness Update*: The value of $B_s(i)$ is computed by summing up the values of $b_s(i,j)$ of all its outgoing edges and then adding 1, according to Eq. 19.21.

This entire procedure is repeated over all source nodes, and the values are added up. Note that this provides *unscaled values* of the node and edge betweenness, which may range from 0 to $n \cdot (n-1)$. The (aggregated) value of $B_s(i)$ over all source nodes s can be converted to $C_B(i)$ of Eq. 19.13 by dividing it with $n \cdot (n-1)$.

The betweenness values can be computed more efficiently incrementally after edge removals in the Girvan–Newman algorithm. This is because the graphs of tight edges can be computed more efficiently by the use of the incremental shortest path algorithm. The bibliographic notes contain pointers to these methods. Because most of the betweenness computations are incremental, they do not need to be performed from scratch, which makes the algorithm more efficient. However, the algorithm is still quite expensive in practice.

19.3.3 Multilevel Graph Partitioning: METIS

Most of the aforementioned algorithms are quite slow in practice. Even the spectral algorithm, discussed later in this section, is quite slow. The *METIS* algorithm was designed to provide a fast alternative for obtaining high-quality solutions. The *METIS* algorithm allows the specification of weights on *both* the nodes and edges in the clustering process. Therefore, it will be assumed that the weight on each edge (i,j) of the graph $G = (N, A)$ is denoted by w_{ij}, and the weight on node i is denoted by v_i.

The *METIS* algorithm can be used to perform either k-way partitioning or 2-way partitioning. The k-way multilevel graph-partitioning method is based on top-down 2-way recursive bisection of the graph to create k-way partitionings, although variants to perform direct k-way partitioning also exist. Therefore, the following discussion will focus on the 2-way bisection of the graph.

The *METIS* algorithm uses the principle that the partitioning of a *coarsened representation* of a graph can be used to efficiently derive an approximate partition of the original graph. The *coarsened representation* of a graph is obtained by contracting some of the adjacent nodes into a single node. The contraction may result in self-loops that are removed. Such self-loops are also referred to as *collapsed edges.* The weights of the contracted nodes are equal to the sum of the weights of the constituent nodes in the original graph. Similarly, the parallel edges across contracted nodes are consolidated into a single edge with the weights of the constituent edges added together. An example of a coarsened representation of a graph, in which some pairs of adjacent nodes are contracted, is illustrated in Fig. 19.6.

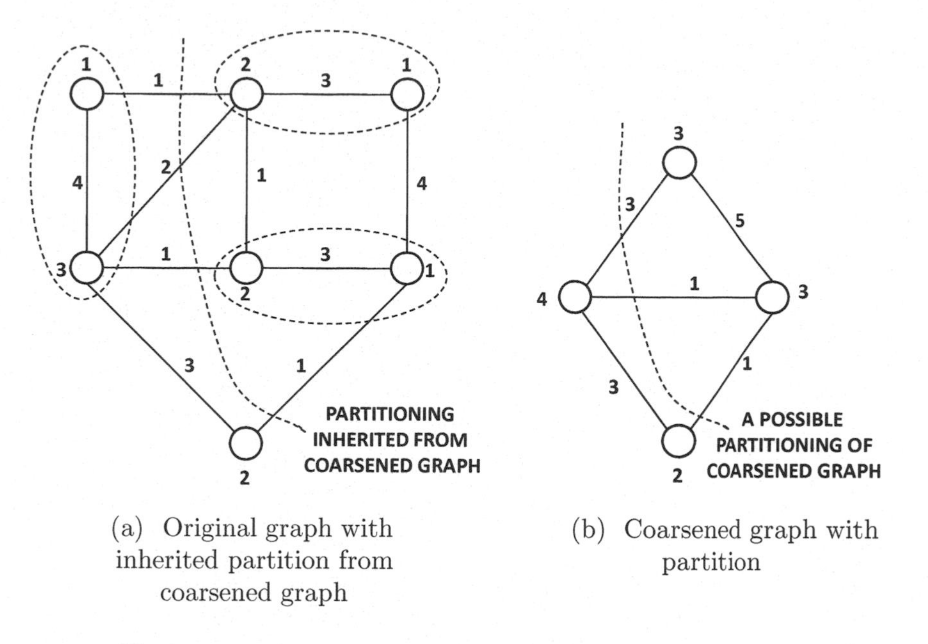

(a) Original graph with inherited partition from coarsened graph

(b) Coarsened graph with partition

Figure 19.6: Illustration of coarsening and partitioning inheritance in uncoarsening

The corresponding node weights and edge weights are also illustrated in the same figure. A good partitioning of this smaller coarsened graph maps to an approximate partitioning of the original graph. Therefore, one possible approach is to compress the original graph into a small one by using a *coarsening heuristic*, then partition this smaller graph more efficiently with any off-the-shelf algorithm, and finally map this partition onto the original graph. An example of mapping a partition on the coarsened graph to the original graph is also illustrated in Fig. 19.6. The resulting partition can be refined with an algorithm, such as the Kernighan–Lin algorithm. The multilevel scheme enhances this basic approach with *multiple levels* of coarsening and refinement to obtain a good trade-off between quality and efficiency. The multilevel partitioning scheme uses three phases:

1. *Coarsening phase:* Carefully chosen sets of nodes in the original graph $G = G_0$ are contracted to create a sequence of successively smaller graphs, $G_0, G_1, G_2 \ldots G_r$. To perform a single step of coarsening from G_{m-1} to G_m, small sets of nonoverlapping and tightly interconnected nodes are identified. Each set of tightly interconnected nodes is contracted into a single node. The heuristics for identifying these node sets will be discussed in detail later. The final graph G_r is typically smaller than a 100 nodes. The small size of this final graph is important in the context of the second partitioning phase. The different levels of coarsening created by this phase create important reference points for a later uncoarsening phase.

2. *Partitioning phase:* Any off-the-shelf algorithm can be used to create a high-quality balanced partitioning from graph G_r. Examples include the spectral approach of Sect. 19.3.4 and the Kernighan–Lin algorithm. It is much easier to obtain a high-quality partitioning with a small graph. This high-quality partitioning provides a good starting point for refinement during the uncoarsening phase. Even relatively poor partitionings of this coarsest graph often map to good partitionings on the uncontracted

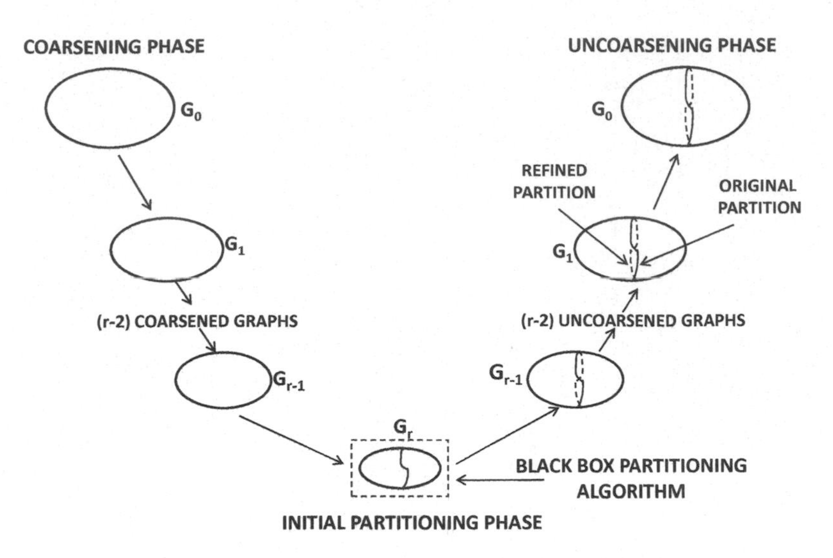

Figure 19.7: The multilevel framework [301] of METIS

graph, because the collapsed edges during coarsening are not eligible to be cut during this phase.

3. *Uncoarsening phase (refinement):* In this phase, the graphs are expanded back to their successively larger versions $G_r, G_{r-1} \ldots G_0$. Whenever the graph G_m is expanded to G_{m-1}, the latter inherits the partitioning from G_m. This inheritance is illustrated in Fig. 19.6. The fast variant of the Kernighan–Lin scheme, discussed in Sect. 19.3.1.1, is applied to this partitioning of G_{m-1} to refine it further before expanding it further to G_{m-2}. Therefore, graph G_{m-2} inherits the refined partition from G_{m-1}. Usually, the refinement phase is extremely fast because the KL-algorithm starts with a very high quality approximate partition of G_{m-1}.

A pictorial representation of the multilevel scheme, based on an illustration in [301], is provided in Fig. 19.7. Note that the second and third phases use off-the-shelf schemes that are discussed in other parts of this chapter. Therefore, the following discussion will focus only on the first phase of coarsening.

A number of techniques are used for coarsening with varying levels of complexity. In the following, a few simple schemes are described that coarsen only by matching *pairs* of nodes in a given phase. In order for a pair of nodes to be matched, they must always be connected with an edge. The coarsened graph will be at least half the size of the original graph in terms of the number of nodes. In spite of the simplicity of these coarsening methods, these schemes turn out to be surprisingly effective in the context of the overall clustering algorithm.

1. *Random edge matching:* A node i is selected at random and matched to an adjacently connected unmatched node that is also selected randomly. If no such unmatched node exists, then the vertex remains unmatched. The matching is performed, until no (adjacent) unmatched pair remains in the graph.

2. *Heavy edge matching:* As in random edge matching, a node i is selected at random and matched to an adjacently connected unmatched node. However, the difference is that the largest weight incident edge (i, j) is used to select the unmatched node j.

The intuition is that it is better to contract heavy edges because they are less likely to be part of an optimal partitioning.

3. *Heavy clique matching:* The contraction of densely connected sets of nodes in the graph will maximize the number of collapsed edges. This method tracks the weight v_i of node i, which corresponds to the number of contracted nodes it represents. Furthermore, the notation s_i denotes the sum of the weights of the collapsed edges at node i (or its precursors) in previous contraction phases. Note that if the contracted node i represents a clique in the original graph, then s_i will approach $v_i \cdot (v_i - 1)/2$. Because it is desirable to contract dense components, one must try to ensure that the value of s_i resulting from the contraction approaches its upper limit. This is achieved by computing the edge density $\mu_{ij} \in (0, 1)$ of edge (i, j):

$$\mu_{ij} = \frac{2 \cdot (s_i + s_j + w_{ij})}{(v_i + v_j) \cdot (v_i + v_j - 1)} \tag{19.23}$$

When nodes across high-density edges are contracted, they typically correspond to cliques in the *original* graph $G = G_0$, if it was unweighted. Even for weighted graphs, the use of high-edge density is generally quite effective. The nodes of the graph are visited in random order. For each node, its highest density unmatched neighbor is selected for matching. Unlike heavy edge matching, the heavy clique matching approach is not myopic to the contractions that have occurred in previous phases of the algorithm.

The multilevel scheme is effective because of its hierarchical approach, where the early clustering of coarsened graphs ensures a good initial global structure to the bisection. In other words, key components of the graph are assigned to the appropriate partitions early on, in the form of coarsened nodes. This partition is then successively improved in refinement phases. Such an approach avoids local optima more effectively because of its "big picture" approach to clustering.

19.3.4 Spectral Clustering

It is assumed that the nodes are unweighted, though the edge (i, j) is associated with the weight w_{ij}. The $n \times n$ matrix of weights is denoted by W. The spectral method uses a graph embedding approach, so that the local clustering structure of the network is preserved by the embedding of the nodes into multidimensional space. The idea is to create a multidimensional representation of the graph so that a standard k-means algorithm can be used on the transformed representation.

The simpler problem of mapping the nodes onto a 1-dimensional space will be discussed first. The generalization to the k-dimensional case is relatively straightforward. We would like to map the nodes in N into a set of 1-dimensional real values $y_1 \dots y_n$ on a line, so that the distances between these points reflect the connectivity among the nodes. Therefore, it is undesirable for nodes that are connected with high-weight edges to be mapped onto distant points on this line. This can be achieved by determining values of y_i, for which the following objective function O is minimized:

$$O = \sum_{i=1}^{n} \sum_{j=1}^{n} w_{ij} \cdot (y_i - y_j)^2 \tag{19.24}$$

This objective function penalizes the distances between y_i and y_j with weight proportional to w_{ij}. Therefore, when w_{ij} is very large, the data points y_i and y_j will be more likely to be closer to one another in the embedded space. The objective function O can be rewritten in terms of the *Laplacian matrix* L of weight matrix W. The Laplacian matrix L is defined as $\Lambda - W$, where Λ is a diagonal matrix satisfying $\Lambda_{ii} = \sum_{j=1}^{n} w_{ij}$. Let the n-dimensional column vector of embedded values be denoted by $\overline{y} = (y_1 \dots y_n)^T$. It can be shown after some algebraic rearrangement of Eq. 19.24, that the objective function O can be rewritten in terms of the Laplacian matrix:

$$O = 2\overline{y}^T L \overline{y} \tag{19.25}$$

The matrix L is positive semidefinite with nonnegative eigenvalues because the sum-of-squares objective function O is always nonnegative. We need to incorporate a scaling constraint to ensure that the trivial value of $y_i = 0$ for all i is not selected by the optimization solution. A possible scaling constraint is as follows:

$$\overline{y}^T \Lambda \overline{y} = 1 \tag{19.26}$$

The matrix Λ is incorporated in the constraint of Eq. 19.26 to achieve normalization, so that the resulting clusters are more balanced across partitions. If Λ is not used in the constraint, the result is referred to as *unnormalized* spectral clustering. In practice, the effect of this normalization is that low-degree nodes tend to clearly "pick sides" with either large positive or large negative values of y_i, whereas very high-degree nodes, which might also be hub nodes, will be embedded closer to central regions near the origin (see Exercise 7). Note that each diagonal entry Λ_{ii}, which is the sum of the weights of the edges incident on node i, can be viewed as the local density of the network at node i. It can also be shown that incorporating Λ in the constraint approximates an unnormalized embedding in which edge weights $w_{ij} = w_{ji}$ have been divided by the geometric average $\sqrt{\Lambda_{ii} \cdot \Lambda_{jj}}$ of the local densities at their end points (see Exercise 8). As discussed in Chap. 3, normalizing distance or similarity values with local densities is often helpful in obtaining high-quality results that are more accurate in their *local* context.

This constrained optimization formulation can be solved by setting the gradient of its *Lagrangian relaxation* $\overline{y}^T L \overline{y} - \lambda(\overline{y}^T \Lambda \overline{y} - 1)$ to 0. It can be shown that the resulting optimization condition is $\Lambda^{-1} L \overline{y} = \lambda \overline{y}$ where λ is the Lagrangian parameter. In other words, $\overline{y}$ is an eigenvector of $\Lambda^{-1} L$ and λ is an eigenvalue. Furthermore, this optimization condition can be used to easily show that the objective function $O = 2\overline{y}^T L \overline{y}$ evaluates to twice the eigenvalue λ for an eigenvector $\overline{y}$ satisfying this condition. Therefore, among the alternative eigenvector solutions for $\overline{y}$, the optimal solution is the smallest *nontrivial* eigenvector of the *normalized* Laplacian $\Lambda^{-1} L$. The smallest eigenvalue of $\Lambda^{-1} L$ is always 0, and it corresponds to the trivial solution where the node embedding $\overline{y}$ is proportional to $(1, 1, \dots 1)^T$. Such a trivial 1-dimensional embedding corresponds to mapping every node to the same point. This trivial eigenvector is noninformative. Therefore, it can be discarded, and it is not used in the analysis. The second smallest eigenvector then provides an optimal solution that is more informative.

This model can be generalized to a k-dimensional embedding by setting up an analogous optimization formulation with its decision variables as an $n \times k$ matrix Y with k column vectors $Y = [\overline{y_1} \dots \overline{y_k}]$ representing each dimension of the embedding. This optimization formulation minimizes the trace of the $k \times k$ matrix $Y^T L Y$ subject to the normalization constraints $Y^T \Lambda Y = I$. Because of the presence of Λ in the constraint, the columns of Y will not necessarily be orthogonal. The optimal solutions for these k column vectors can be shown to be proportional to the successive directions corresponding to the (not necessarily orthogonal) right eigenvectors of the asymmetric matrix $\Lambda^{-1} L$ with increasing eigenvalues. After discarding the first trivial eigenvector $\overline{e_1}$ with eigenvalue $\lambda_1 = 0$, this results in a set of k eigenvectors $\overline{e_2}, \overline{e_3} \ldots \overline{e_{k+1}}$, with corresponding eigenvalues $\lambda_2 \leq \lambda_3 \ldots \leq \lambda_{k+1}$. Because k eigenvectors were selected, this approach creates an $n \times k$ matrix $D_k = Y$, corresponding to a k-dimensional embedding of each of the n nodes. Note that even though the normalization constraint $Y^T \Lambda Y = I$ will not result in columns of D_k having an L_2-norm of 1, each column (eigenvector) of D_k is scaled to an L_2-norm of 1 as a post-processing[4] step. Because of this column scaling, the $n \times k$ matrix D_k does not exactly reflect the original optimization formulation in terms of Y. The resulting k-dimensional embedding preserves the clustering structure of the nodes because the optimization formulation of Y tries to minimize distances between highly connected nodes. Therefore, any multidimensional clustering algorithm discussed in Chap. 6, such as k-means, can be applied to this embedded representation to generate the clusters on the nodes. This formulation is also sometimes referred to as the *random walk version* of spectral clustering because of an interpretation in terms of random walks. It is noteworthy that the small eigenvectors of the normalized Laplacian $\Lambda^{-1}L$ are the same as the large eigenvectors of the stochastic transition matrix $\Lambda^{-1}W$ (see Exercise 15).

An *equivalent* way of setting up the spectral clustering model is to use the related vector of decision variables $\overline{z} = \sqrt{\Lambda}\overline{y}$ in the optimization formulation of Eqs. 19.25 and 19.26. This related version is referred to as the *symmetric version* of the spectral clustering model, although it is different from the random walk version only in terms of the scaling of decision variables. By setting $z = \sqrt{\Lambda}y$, it can be shown that the earlier formulation is equivalent to optimizing $\overline{z}^T \Lambda^{-1/2} L \Lambda^{-1/2} \overline{z}$ subject to $\overline{z}^T \overline{z} = 1$. We determine the smallest k (orthogonal) eigenvectors of the *symmetric normalized* Laplacian $\Lambda^{-1/2} L \Lambda^{-1/2}$, excluding the first. Each eigenvector of this matrix can also be (proportionally) obtained by pre-multiplying the aforementioned solution Y of the random walk formulation with the diagonal matrix $\sqrt{\Lambda}$. This relationship also reflects the relationship between $\overline{z}$ and $\overline{y}$. The eigenvalues are the same in both cases. For example, the first eigenvector with eigenvalue 0 will no longer be a vector of 1s, but the various entries will be proportional to $(\sqrt{\Lambda_{11}} \ldots \sqrt{\Lambda_{nn}})^T$. Because of this differential scaling of various nodes, high-degree nodes will tend to have larger (absolute) coordinate values in the symmetric version. By selecting the smallest k eigenvectors, one can generate an $n \times k$ multidimensional representation D_k of the entire set of n nodes. Just as the random walk version scales each column of D_k to unit norm in the final step, the symmetric version scales each row of D_k to unit norm. The final step of row scaling is a heuristic enhancement to adjust for the differential scaling of nodes with various degrees, and it is not a part of the optimization formulation. Interestingly, even if the rows of the random walk solution Y had been scaled to unit norm (instead of scaling the columns to unit norm), exactly the same solution would be obtained as that obtained by scaling the rows of the symmetric solution Z to unit norm (see Exercise 13).

Although the two different ways of performing spectral clustering are equivalent in terms of the optimization problem solved, there are differences in terms of the heuristic scaling adjustments. The scaling relationships are illustrated in Fig. 19.8. It is evident from Fig. 19.8 that the main *practical* difference between the two methods is regulated only by the *heuristic* scaling used in the final phase, rather than their respective optimization models. Because of the scaling variations, the clusters obtained in the two cases will not be exactly the same. The relative quality will depend on the data set at hand. These optimization problems can also be understood as linear programming relaxations of integer-programming formulations of balanced minimum cut problems. However, the minimum-cut explanation does not

[4]In practice, the *unit* eigenvectors of $\Lambda^{-1}L$ can be directly computed, and therefore an explicit post-processing step is not required.

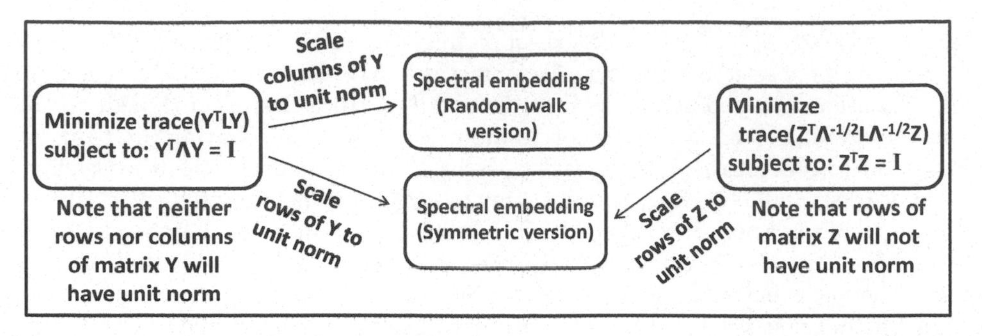

Figure 19.8: Scaling relationships between random walk and symmetric versions of spectral clustering

intuitively generalize to the relaxed version of the problem because eigenvectors have both positive and negative components.

19.3.4.1 Important Observations and Intuitions

A few observations are noteworthy about the relationships between spectral clustering, *PageRank*, and eigenvector analysis:

1. *Normalized random walk Laplacian:* The smallest *right* eigenvectors of $\Lambda^{-1}L = \Lambda^{-1}(\Lambda - W) = I - P$ are used for the random walk embedding, where P is the stochastic transition matrix of the graph. The smallest right eigenvectors of $I - P$ are the same as the largest right eigenvectors of P. The largest right eigenvector of P has eigenvalue 1. It is noteworthy that the largest *left* eigenvector of P, which also has eigenvalue 1, yields the *PageRank* of the graph. Therefore, both the left and the right eigenvectors of the stochastic transition matrix P yield different insights about the network.

2. *Normalized symmetric Laplacian:* The smallest eigenvectors of the symmetric Laplacian $\Lambda^{-1/2}(\Lambda - W)\Lambda^{-1/2}$ are the same as the largest eigenvectors of the symmetric matrix $\Lambda^{-1/2}W\Lambda^{-1/2}$. The matrix $\Lambda^{-1/2}W\Lambda^{-1/2}$ can be viewed as a normalized and sparsified similarity matrix of the graph. Most forms of nonlinear embeddings such as *SVD*, *Kernel PCA*, and *ISOMAP* are extracted as large eigenvectors of similarity matrices (cf. Table 2.3 of Chap. 2). It is the choice of the similarity matrix that regulates the varying properties of these different embeddings.

3. *Goal of normalization:* Spectral clustering is more effective when the unnormalized Laplacian L is normalized with the node-degree matrix Λ. While it is possible to explain this behavior with a cut-interpretation of spectral clustering, the intuition does not generalize easily to continuous embeddings with both positive and negative eigenvector components. A simpler way of understanding normalization is by examining the similarity matrix $\Lambda^{-1/2}W\Lambda^{-1/2}$ whose large eigenvectors yield the normalized spectral embedding. In this matrix, the edge similarities are normalized by the geometric mean of the node degrees at their end points. This can be viewed as a normalization of the edge similarities with a local measure of the network density. As discussed in Chap. 3, normalizing similarity and distance functions with local density is helpful even in the case of multidimensional data mining applications. One of the most well-known algorithms for outlier analysis in multidimensional data, referred to as *LOF*,

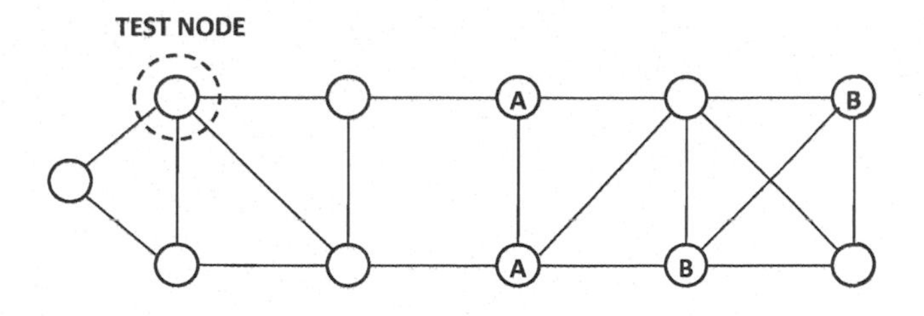

Figure 19.9: Label sparsity issues in collective classification

also uses this principle. Normalization will yield more balanced clusters in networks with widely varying density over the network.

19.4 Collective Classification

In many social networking applications, labels may be associated with nodes. For example, consider the case of a social networking application, where it is desirable to determine all individuals interested in *golf*. The labels of a small number of actors may already be available. It is desirable to use the available labels to perform the classification of nodes for which the label is not known.

The solution to this model is crucially dependent on the notion of homophily. Because nodes with similar properties are usually connected, it is reasonable to assume that this is also true of node labels. A simple solution to this problem is to examine the k labeled nodes in the proximity of a given node and report the majority label. This approach is, in fact, the network analog of a nearest neighbor classifier. However, such an approach is generally not possible in collective classification because of the sparsity of node labels. An example of a network is illustrated in Fig. 19.9, in which the two classes are labeled A and B. The remaining nodes are unlabeled. For the test node in Fig. 19.9, it is evident that it is generally closer to instances of A in the network structure, but there is no labeled node *directly* connected to the test instance.

Thus, it is evident that one must not only use the direct connections to labeled nodes, but also use the indirect connections through unlabeled nodes. Thus, collective classification in networks are always performed in a *transductive semisupervised setting*, where the test instances and training instances are classified jointly. In fact, as discussed in Sect. 11.6.3 of Chap. 11, collective classification methods can be used for semisupervised classification of any data type by transforming the data into a similarity graph. *Thus, the collective classification problem is important not only from the perspective of social network analysis, but also for semisupervised classification of any data type.*

19.4.1 Iterative Classification Algorithm

The *Iterative Classification Algorithm (ICA)* is one of the earliest classification algorithms in the literature and has been applied to a wide variety of data domains. The algorithm has the capability to use content associated with the nodes for classification. This is important because many social networks have text content associated with the nodes in the form of user posts. Furthermore, in cases where this framework is used[5] for semisupervised classification

[5]cf. Sect. 11.6.3 of Chap. 11.

Algorithm *ICA*(Graph $G = (N, A)$, Weights: $[w_{ij}]$, Node Class Labels: $\mathcal{C}$,
Base Classifier: $\mathcal{A}$, Number of Iterations: T)
begin
repeat
Extract link features at each node with current training data;
Train classifier $\mathcal{A}$ using both link and content features of
current training data and predict labels of test nodes;
Make (predicted) labels of most "certain" n_t/T
test nodes final, and add these nodes to training
data, while removing them from test data;
until T iterations;
end

Figure 19.10: The iterative classification algorithm (ICA)

of relational data with similarity graphs, the relational features continue to be available at the nodes for more effective classification.

Consider the (undirected) network $G = (N, A)$ with class labels are drawn from $\{1 \dots k\}$. Each edge $(i, j) \in A$ is associated with the weight w_{ij}. Furthermore, the content $\overline{X_i}$ is available at the node i in the form of a multidimensional feature vector. The total number of nodes is denoted by n, from which n_t nodes are unlabeled test nodes.

An important step of the *ICA* algorithm is to derive a set of *link features* in addition to the available content features in $\overline{X_i}$. The most important link features correspond to the distribution of the classes in the immediate neighborhood of the node. Therefore a feature is generated for each class, containing the fraction of its incident nodes belonging to that class. For each node i, its adjacent node j is weighted by w_{ij} for computing its credit to the relevant class. It is also possible, in principle, to derive other link features based on structural properties of the graph such as the degree of the node, *PageRank* values, number of closed triangles involving the node, or connectivity features. Such link features can be derived on the basis of an application-specific understanding of the network data set.

The basic ICA is structured as a meta-algorithm. A base classifier $\mathcal{A}$ is leveraged within an iterative framework. Many different base classifiers have been used in different implementations, such as the naive Bayes classifier, logistic regression classifier, and a neighborhood voting classifier. The main requirement is that these classifiers should be able to output a numeric score that quantifies the likelihood of a node belonging to a particular class. While the framework is independent of specific choice of classifier, the use of the naive Bayes classifier is particularly common because of the interpretation of its numeric score as a probability. Therefore, the following discussion will assume that the algorithm $\mathcal{A}$ is instantiated to the naive Bayes classifier.

The link and content features are used to train the naive Bayes classifier. For many nodes, it is difficult to robustly estimate important class-specific features, such as the fractional presence of the different classes in their neighborhood. This is a direct result of label sparsity, and it makes the class predictions of such nodes unreliable. Therefore, an iterative approach is used for augmenting the training data set. In each iteration, n_t/T (test) node labels are made "certain" by the approach, where T is a user-defined parameter controlling the maximum number of iterations. The test nodes, for which the Bayes classifier exhibits the highest class membership probabilities, are selected to be made final. These labeled test

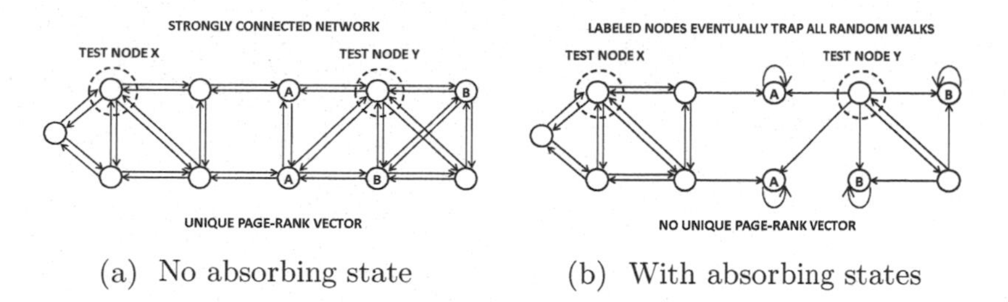

(a) No absorbing state (b) With absorbing states

Figure 19.11: Creating directed transition graphs from undirected graph of Fig. 19.9

nodes can then be added to the training data, and the classifier is retrained by extracting the link features again with the augmented training data set. The approach is repeated until the labels of all nodes have been made final. Because the labels of n_t/T nodes are finalized in each iteration, the entire process terminates in exactly T iterations. The overall pseudocode is illustrated in Fig. 19.10.

One advantage of the ICA is that it can seamlessly use content and structure in the classification process. The classifier can automatically select the most relevant features using off-the-shelf feature selection algorithms discussed in Chap. 10. This approach also has the merit that it is not strongly dependent on the notion of homophily, and can, therefore, be used for domains beyond social network analysis. Consider an adversarial relationship network in which nodes connected by links might have different labels. In such cases, the *ICA* algorithm will automatically learn the correct importance of adjacent class distributions, and therefore it will yield accurate results. This property is not true of most of the other collective classification methods, which are *explicitly* dependent on the notion of homophily. On the other hand, the errors made in the earlier phases of iterative classification can propagate and multiply in later phases because of augmented training examples with incorrect labels. This can increase the cumulative error in noisy training data sets.

19.4.2 Label Propagation with Random Walks

The label propagation method directly uses random walks on the undirected network structure $G = (N, A)$. The weight of edge (i, j) is denoted by $w_{ij} = w_{ji}$. To classify an unlabeled node i, a random walk is executed starting at node i and terminated at the first labeled node encountered. The class at which the random walk has the highest probability of termination is reported as the predicted label of node i. The intuition for this approach is that the walk is more likely to terminate at labeled nodes in the proximity of node i. Therefore, when many nodes of a particular class are located in its proximity, then the node i is more likely to be labeled with that class.

An important assumption is that the graph needs to be *label connected.* In other words, every unlabeled node needs to be able to reach a labeled node in the random walk. For undirected graphs $G = (N, A)$, this means that every connected component of the graph needs to contain at least one labeled node. In the following discussion, it will be assumed that the graph $G = (N, A)$ is undirected and label-connected.

The first step is to model the random walks in such a way that they always terminate at their *first arrival* at labeled nodes. This can be achieved by removing outgoing edges from labeled nodes and replacing them with self-loops. Furthermore, to use a random walk approach, we need to convert the undirected graph $G = (N, A)$ into a directed graph $G' = (N, A')$ with an $n \times n$ transition matrix $P = [p_{ij}]$:

1. For each undirected edge $(i, j) \in A$, directed edges (i, j) and (j, i) are added to A' between the corresponding nodes. The transition probability p_{ij} of edge (i, j) is defined as follows:
$$p_{ij} = \frac{w_{ij}}{\sum_{k=1}^{n} w_{ik}} \tag{19.27}$$
The transition probability p_{ji} of edge (j, i) is defined as follows:
$$p_{ji} = \frac{w_{ji}}{\sum_{k=1}^{n} w_{jk}} \tag{19.28}$$
For example, the directed transition graph created from the undirected graph of Fig. 19.9 is illustrated in Fig. 19.11a.

2. All outgoing edges from labeled nodes are removed from the graph G' constructed in the previous step and replaced with a self-loop of transition probability 1. Such nodes are referred to as *absorbing* nodes because they trap the random walk after an incoming transition. An example of the final transition graph is illustrated in Fig. 19.11b. Therefore, for each absorbing node i, the ith row of P is replaced with the ith row of the identity matrix.

Assume that the final $n \times n$ transition matrix is denoted by $P = [p_{ij}]$. For any absorbing node i, the value of p_{ik} is 1 only when $i = k$, and 0 otherwise. The transition matrix P does *not* have a unique steady-state probability distribution (or, *PageRank* vector), because of the presence of absorbing[6] components. The steady-state probability distribution *is dependent on the starting state of the random walk.* For example, a random walk starting at test node X in Fig. 19.11b will always eventually end at label A, whereas a walk starting with node Y might end at either label A or B. It is noteworthy that the *PageRank* computation of Sect. 18.4.1 in Chap. 18 ensures unique steady-state probabilities by using teleportation to implicitly create a strongly connected transition graph. Interestingly, the modifications that create absorbing nodes have exactly the opposite effect because the steady state probability distribution depends on the starting state. Strong connectivity of the transition graph is required to ensure a unique steady-state distribution. However, if the starting state is fixed, then each node does have a steady state probability distribution.

For any given starting node i, the steady-state probability distribution has positive values only at labeled nodes. This is because a random walk will eventually reach an absorbing node in a label-connected graph, and it will never emerge from that node. Therefore, if one can estimate the steady-state probability distribution for starting node i, then the probability values of the labeled nodes in each class can be aggregated. The class with the highest probability is reported as the relevant label of the node i.

How can the steady-state probability be computed for a particular starting node i? Let $\overline{\pi}^{(t)}$ represent the n-dimensional (row) probability vector after t steps, starting with a particular initial state $\overline{\pi}^{(0)}$. When the starting state is node i, the value of $\overline{\pi}^{(0)}$ is 1 for the ith component in this vector, and 0 otherwise. Then, we have:
$$\overline{\pi}^{(t)} = \overline{\pi}^{(t-1)} P \tag{19.29}$$

[6]In other words, the underlying Markov chain is not strongly connected, and therefore not ergodic. See the description of the *PageRank* algorithm in Chap. 18.

By recursively applying the aforementioned condition t times, and then setting $t = \infty$, it is possible to show the following:

$$\overline{\pi}^{(t)} = \overline{\pi}^{(0)} P^t \tag{19.30}$$

$$\overline{\pi}^{(\infty)} = \overline{\pi}^{(0)} P^\infty \tag{19.31}$$

How can the steady-state transition matrix P^∞ be computed? A key observation is that the largest magnitude of the eigenvalue of a stochastic matrix is always 1 (see Exercise 7 of Chap. 18). Therefore, P may be expressed as follows:

$$P = V\Delta V^{-1} \tag{19.32}$$

Here, V is an $n \times n$ matrix, whose columns contain the eigenvectors, and Δ is a diagonal matrix containing the eigenvalues, all of which have magnitude no larger than 1. Note that stochastic matrices with absorbing components will have an eigenvector with unit eigenvalue for each absorbing component. Then, by multiplying P with itself $(t-1)$ times, we get:

$$P^t = V\Delta^t V^{-1} \tag{19.33}$$

In the limit where t approaches infinity, Δ^t will contain diagonal values of only 0 or 1. Any eigenvalue in the original matrix Δ with magnitude less than 1 will approach 0 in Δ^∞. In other words, Δ^∞ can be computed easily from Δ. Therefore, if V has been computed, then P^∞ can be computed easily as well. A further optimization is that the steady-state transition matrix P^∞ can be efficiently computed by determining only the l leading eigenvectors of P, where l is the number of labeled (absorbing) nodes. Refer to the bibliographic notes for more details of this optimization.

After P^∞ has been computed, it is relatively straightforward to compute the n-dimensional node probability vector $\overline{\pi}^{(\infty)}$ that results from starting the random walk at node i. When the starting state is (unlabeled) node i, the n-dimensional vector for the starting state $\overline{\pi}^{(0)}$ contains 1 in the ith component, and 0 otherwise. According to our earlier discussion, one can compute $\overline{\pi}^{(\infty)} = \overline{\pi}^{(0)} P^\infty$. Note that $\overline{\pi}^{(\infty)}$ will contain positive probabilities only for the labeled nodes, which are also absorbing states. By summing up the probabilities in $\overline{\pi}^{(\infty)}$ of labeled nodes belonging to each class, one can obtain the probability of each class for unlabeled node i. The class with the maximum probability is reported as the relevant label.

There is, however, a simpler way of computing the class probability distributions of all unlabeled nodes in one shot, rather than having to explicitly compute P^∞, and then trying different starting vectors for $\overline{\pi}^{(0)}$. For each class $c \in \{1 \dots k\}$, let $N_c \subseteq N$ be the set of labeled nodes belonging to that class. In order for unlabeled node i to belong to class c, a walk starting at node i must end at a node in N_c. The probability of this event is given by $\sum_{j \in N_c} [P^\infty]_{ij}$. Let $\overline{Y_c}$ be a column vector with n entries such that the jth entry is 1, if node j belongs to class c, and 0, otherwise. Then, it is easy to see that the ith entry of the column vector $\overline{Z_c} = P^\infty \overline{Y_c}$ is equivalent to $\sum_{j \in N_c} [P^\infty]_{ij}$, which is the sum of the probabilities of a walk starting at unlabeled node i terminating at various nodes belonging to class c.

Therefore, we need to compute $\overline{Z_c}$ for each class $c \in \{1 \dots k\}$. Let Y be an $n \times k$ matrix for which the cth column is $\overline{Y_c}$. Similarly, let Z be an $n \times k$ matrix for which the cth column is $\overline{Z_c}$. Then Z can be obtained with simple matrix multiplication between P^∞ and Y.

$$Z = P^\infty Y \tag{19.34}$$

The class with the maximum probability in Z for unlabeled node (row) i may be reported as its class label. This approach is also referred to as the *rendezvous approach* to label propagation.

We make a few important observations. If the ith row of P is absorbing then it is the same as the ith row of the identity matrix. Therefore, premultiplying Y with P for any number of times will not change the ith row of Y. In other words, rows of Z that correspond to labeled nodes will be fixed to the corresponding rows of Y. Therefore, predictions of labeled nodes are fixed to their training labels. For unlabeled nodes, the rows of Z will always sum to 1 in label-connected networks. This is because the sum of the values in row i in Z is equal to the probability that a random walk starting at node i reaches an absorbing state. In label-connected networks, every random walk will eventually reach an absorbing state.

19.4.2.1 Iterative Label Propagation: The Spectral Interpretation

Equation 19.34 suggests a simple iterative approach for computing the label probabilities in Z, rather than computing P^{∞}. One can initialize $Z^{(0)} = Y$ and then repeatedly use the following update for increasing value of iteration index t.

$$Z^{(t+1)} = PZ^{(t)} \tag{19.35}$$

It is easy to see that $Z^{(\infty)}$ is the same as the value of Z in Eq. 19.34. For labeled (absorbing) node i, the ith row of Z will always be unaffected by the update because the ith row of P is the same as that of the identity matrix. The label-propagation update is executed to convergence. In practice, a relatively small number of iterations are required to reach convergence.

The label propagation update can be rearranged to show that the final solution Z will satisfy the following relationship at convergence:

$$(I - P)Z = 0 \tag{19.36}$$

Note that $I-P$ is simply the normalized (random walk) Laplacian of the adjacency matrix of the network G' with absorbing states. Furthermore, each column of Z is a eigenvector of this Laplacian with eigenvalue 0. In unsupervised spectral clustering, the first eigenvector with eigenvalue 0 is discarded because it is not informative. However, in collective classification, there are additional eigenvectors of $(I - P)$ with eigenvalue 0 because of the presence of absorbing states. Each class-specific column of Z contains a different eigenvector with eigenvalue 0. In fact, the label propagation solution can also be derived with an optimization formulation similar to spectral clustering on the original undirected graph G. In this case, the optimization formulation uses a similar objective function as spectral clustering *with the additional constraint* that the embedded values of all labeled nodes are fixed to 1 for a particular (say, the cth) class and they are fixed to 0 for the remaining classes. The embedded values of only the unlabeled nodes are unconstrained decision variables. The solution to each such optimization problem can be shown to be an eigenvector of $(I - P)$, with eigenvalue 0. Iterative label propagation converges to these eigenvectors.

19.4.3 Supervised Spectral Methods

Spectral methods can be used in two different ways for collective classification of graphs. The first method directly transforms the graph to multidimensional data to facilitate the use of a multidimensional classifier such as a k-nearest neighbor classifier. The embedding approach is identical to that used in spectral clustering except that the class information is incorporated within the embedding. The second method directly learns an $n \times k$ class probability matrix Z with an optimization formulation related to spectral clustering. This class probability matrix Z is similar to that derived in label propagation. Interestingly, the second method is also closely related to label propagation.

19.4.3.1 Supervised Feature Generation with Spectral Embedding

Let $G = (N, A)$ be the undirected graph with weight matrix W. The approach consists of the following steps, the first of which is to augment G with class-based supervision:

1. Add an edge with weight μ between each pair of nodes with the same label in G. If an edge already exists between a pair of such nodes, then the two edges are consolidated by adding μ to the weight of the existing edge. The resulting graph is denoted by G^+. The parameter μ controls the level of supervision from existing labels.

2. Use the spectral embedding approach of Sect. 19.3.4 to generate an r-dimensional embedding of the augmented graph G^+.

3. Apply any multidimensional classifier, such as a nearest neighbor classifier, on the embedded data.

The value of μ may be tuned with the use of cross-validation. Note that this approach does not directly learn the class probabilities. Rather, it creates a feature representation that implicitly incorporates both the homophily effects and the existing label information. This feature representation is sensitive to both network locality and label distribution. Therefore, it can be used to design an effective multidimensional classifier.

19.4.3.2 Graph Regularization Approach

The graph regularization approach learns the labels of the nodes directly with an optimization formulation related to spectral clustering. let Z be an $n \times k$ matrix of optimization variables, in which the (i, c)th entry denotes the propensity of node i to belong to label c. When the (i, c)th entry is large, it indicates that node i is more likely to belong to label c. Therefore, for the ith row of Z, the index of the largest of the k entries provides a prediction of the class label of node i. The column-vector $\overline{Z_c}$ denotes the cth column of Z for $c \in \{1 \dots k\}$. Furthermore, Y is an $n \times k$ binary matrix containing the label information. If the ith node is labeled, then exactly one entry in the ith row of Y is 1, corresponding to the relevant class label. Other entries are 0. For unlabeled nodes, all entries in the corresponding row of Y are 0. The cth column of Y is denoted by the column vector $\overline{Y_c}$.

This approach directly uses the weighted matrix W of an undirected graph $G = (N, A)$ (e.g., Fig. 19.9) rather than a directed transition graph. The variables in the matrix Z are derived with an optimization formulation related to spectral clustering. Each n-dimensional vector $\overline{Z_c}$ is viewed as a 1-dimensional embedding of the n nodes. The goal of this optimization formulation is two-fold, which is reflected in the two additive terms of the objective function:

1. *Smoothness (homophily) objective:* For each class $c \in \{1 \dots k\}$, the nodes connected with high-weight edges should be mapped to similar values in $\overline{Z_c}$. This goal is identical to the unsupervised objective function in spectral clustering. In this case, the *symmetric* Laplacian L^s is used because of its better convergence properties:

$$L^s = I - \Lambda^{-1/2} W \Lambda^{-1/2} \tag{19.37}$$

Here, Λ is a diagonal matrix in which the ith diagonal entry contains the sum of the ith row entries of the $n \times n$ weight matrix W. For brevity, we denote the normalized weight matrix by $S = \Lambda^{-1/2} W \Lambda^{-1/2}$. Therefore, the smoothness term O_s in the objective function may be written as follows:

$$O_s = \sum_{c=1}^{k} \overline{Z_c}^T L^s \overline{Z_c} = \sum_{c=1}^{k} \overline{Z_c}^T (I - S) \overline{Z_c} \tag{19.38}$$

This term is referred to as the *smoothness* term because it ensures that the *predicted* label propensities Z vary smoothly along edges, especially if the weights of the edges are large. This term can also be viewed as a *local consistency* term.

2. *Label-fitting objective:* Because the embedding Z is designed to mimic Y as closely as possible the value of $||\overline{Z_c} - \overline{Y_c}||^2$ should be as small as possible for each class c. Note that unlabeled nodes are included within $||\overline{Z_c} - \overline{Y_c}||^2$, and for those nodes, this term serves as a *regularizer*. The goal of a *regularizer* is to avoid[7] ill-conditioned solutions and overfitting in optimization models.

$$O_f = \sum_{c=1}^{k} ||\overline{Y_c} - \overline{Z_c}||^2 \tag{19.39}$$

This term can also be viewed as a *global consistency* term.

The overall objective function may be constructed as $O = O_s + \mu O_f$, where μ defines the weight of the label-fitting term. The parameter μ reflects the trade-off between the two criteria. Therefore, the overall objective function may be written as follows:

$$O = \sum_{c=1}^{k} \overline{Z_c}^T (I - S) \overline{Z_c} + \mu \sum_{c=1}^{k} ||\overline{Y_c} - \overline{Z_c}||^2 \tag{19.40}$$

To optimize this objective function, one must use the partial derivative with respect to the different decision variables in $\overline{Z_c}$ and set it to zero. This yields the following condition:

$$(I - S)\overline{Z_c} + \mu(\overline{Z_c} - \overline{Y_c}) = 0 \quad \forall c \in \{1 \dots k\} \tag{19.41}$$

Because this condition is true for each class $c \in \{1 \dots k\}$ one can write the aforementioned condition in matrix form as well:

$$(I - S)Z + \mu(Z - Y) = 0 \tag{19.42}$$

One can rearrange this optimization condition as follows:

$$Z = \frac{SZ}{1 + \mu} + \frac{\mu}{1 + \mu} Y \tag{19.43}$$

The goal is to determine a solution Z of this optimization condition. This can be achieved iteratively by initializing $Z^{(0)} = Y$ and then iteratively updating $Z^{(t+1)}$ from $Z^{(t)}$ as follows:

$$Z^{(t+1)} = \frac{SZ^{(t)}}{1 + \mu} + \frac{\mu}{1 + \mu} Y \tag{19.44}$$

[7] In this case, the regularizer ensures that no single entry in $\overline{Z_c}$ for unlabeled nodes is excessively large.

This solution is iterated to convergence. It can be shown that the approach converges to the following solution:

$$Z^{(\infty)} = \frac{\mu}{1+\mu}\left(I + \frac{S}{1+\mu} + \left(\frac{S}{1+\mu}\right)^2 + \ldots\right)Y = \frac{\mu}{1+\mu}\left(I - \frac{S}{1+\mu}\right)^{-1} Y \tag{19.45}$$

Intuitively, the matrix $\left(I - \frac{S}{1+\mu}\right)^{-1} = \left(I + \frac{S}{1+\mu} + \left(\frac{S}{1+\mu}\right)^2 + \ldots\right)$ is an $n \times n$ matrix of pairwise *weighted Katz coefficients* (cf. Definition 19.5.4) between nodes. In other words, the propensity of node i to belong to class j is predicted as a sum of its weighted Katz coefficients with respect to labeled nodes of class j. Because the Katz measure predicts links (cf. Sect. 19.5) between nodes, this approach illustrates the connections between collective classification and link prediction.

It is possible to learn the optimal value of μ with the use of cross-validation. It is noteworthy that, unlike the aforementioned label propagation algorithm with absorbing states, this approach only *biases* Z with the labels, and it does not *constrain* the rows in Z to be the same as the corresponding rows of Y for labeled nodes. In fact, the matrix Z can provide a label prediction for an already labeled node that is different from its original training label. Such cases are likely to occur when the original labeling in the training data is error-prone and noisy. The regularization approach is, therefore, more flexible and robust for networks containing noisy and error-prone training labels.

19.4.3.3 Connections with Random Walk Methods

Even though the graph regularization approach is derived using spectral methods, it is also related to random walk methods. The $n \times k$ matrix-based update Eq. 19.44 can be decomposed into k different vector-based update equations, one for each n-dimensional column $\overline{Z_c}$ of Z:

$$\overline{Z_c} = \frac{S\overline{Z_c}}{1+\mu} + \frac{\mu}{1+\mu}\overline{Y_c} \quad \forall c \in \{1 \ldots k\} \tag{19.46}$$

Each of these update equations is algebraically similar to a personalized *PageRank* equation where S replaces the transition matrix and the restart probability is $\frac{\mu}{1+\mu}$ at labeled nodes belonging to a particular class c. The vector $\overline{Y_c}$ is analogous to the personalized restart vector for class c multiplied with the number of training nodes in class c. Similarly, the vector $\overline{Z_c}$ is analogous to the personalized *PageRank* vector of class c multiplied with the number of training nodes in class c. Therefore, the class-specific Eq. 19.46 can be viewed as a personalized *PageRank* equation, scaled in proportion to the prior probability of class c. Of course, the symmetric matrix S is not truly a stochastic transition matrix because its columns do not sum to 1. Therefore, the results cannot formally be viewed as personalized *PageRank* probabilities.

Nevertheless, this algebraic similarity to personalized *PageRank* suggests the possibility of a closely related family of random walk methods, similar to label propagation. For example, instead of using a nonstochastic matrix S derived from spectral clustering, one might use a stochastic transition matrix P. In other words, Eqs. 19.27 and 19.28 are used to derive $P = \Lambda^{-1}W$. However, one difference from the transition matrix P used in label-propagation methods is that the network structure is not altered to create absorbing states. In other words, the directed transition graph of Fig. 19.11a is used, rather than that of Fig. 19.11b to derive P. Replacing S with P in Eq. 19.46 leads to a variant of the label propagation update (cf. Eq. 19.35) in which labeled nodes are no longer constrained to be predicted to their original label.

Replacing S with P^T in Eq. 19.46 leads to the (class-prior scaled) personalized *PageRank* equations. This is equivalent to executing the personalized *PageRank* algorithm k times, where the personalization vector for the cth execution restarts at labeled nodes belonging to the cth class. Each class-specific personalized *PageRank* probability is multiplied with the prior probability of that class, or, equivalently, the number of labeled training nodes in that class. For each node, the class index that yields the highest (prior-scaled) personalized *PageRank* probability is reported. The performance of these alternative methods is dependent on the data set at hand.

19.5 Link Prediction

In many social networks, it is desirable to predict future links between pairs of nodes in the network. For example, commercial social networks, such as Facebook, often recommend users as potential friends. In general, both structure and content similarity may be used to predict links between pairs of nodes. These criteria are discussed below:

- *Structural measures:* Structural measures typically use the principle of *triadic closure* to make predictions. The idea is that two nodes that share similar nodes in their neighborhoods are more likely to become connected in the future, if they are not already connected.

- *Content-based measures:* In these cases, the principle of homophily is used to make predictions. The idea is that nodes that have similar content are more likely to become linked. For example, in a bibliographic network containing scientific co-author relations, a node containing the keyword "data mining" is more likely to be connected to another node containing the keyword "machine learning."

While content-based measures have been shown to have potential in enhancing link prediction, the results are rather sensitive to the network at hand. For example, in a network such as Twitter, where the content is the form of short and noisy tweets with many nonstandard acronyms, content-based measures are not particularly effective. Furthermore, while structural connectivity usually implies content-based homophily, the reverse is not always true. Therefore, the use of content similarity has mixed results in different network domains. On the other hand, structural measures are almost always effective in different types of networks. This is because triadic closure is ubiquitous across different network domains and has more direct applicability to link prediction.

19.5.1 Neighborhood-Based Measures

Neighborhood-based measures use the number of common neighbors between a pair of nodes i and j, in different ways, to quantify the likelihood of a link between them in the future. For example, in Fig. 19.12a, Alice and Bob share four common neighbors. Therefore, it is reasonable to conjecture that a link might eventually form between them. In addition to their common neighbors, they also have their own disjoint sets of neighbors. There are different ways of normalizing neighborhood-based measures to account for the number and relative importance of different neighbors. These are discussed below.

Definition 19.5.1 (Common Neighbor Measure) *The common-neighbor measure between nodes i and j is equal to the number of common neighbors between nodes i and j. In other words, if S_i is the neighbor set of node i, and S_j is the neighbor set of node j, the common-neighbor measure is defined as follows:*

$$CommonNeighbors(i, j) = |S_i \cap S_j| \tag{19.47}$$

The major weakness of the common-neighbor measure is that it does not account for the *relative* number of common neighbors between them as compared to the number of other connections. In the example of Fig. 19.12a, Alice and Bob each have a relatively small node degree. Consider a different case in which Alice and Bob are either spammers or very popular public figures who were connected to a large number of other actors. In such a case, Alice and Bob might easily have many neighbors in common, *just by chance*. The *Jaccard measure* is designed to normalize for varying degree distributions.

Definition 19.5.2 (Jaccard Measure) *The Jaccard-based link prediction measure between nodes i and j is equal to the Jaccard coefficient between their neighbor sets S_i and S_j, respectively.*

$$JaccardPredict(i, j) = \frac{|S_i \cap S_j|}{|S_i \cup S_j|} \tag{19.48}$$

The Jaccard measure between Alice and Bob in Fig. 19.12(a) is 4/9. If the degrees of either Alice or Bob were to increase, it would result in a lower Jaccard coefficient between them. This kind of normalization is important, because of the power-law degree distributions of nodes.

The Jaccard measure adjusts much better to the variations in the degrees of the nodes *between which* the link prediction is measured. However, it does not adjust well to the degrees of their *intermediate* neighbors. For example, in Fig. 19.12a, the common neighbors of Alice and Bob are Jack, John, Jill, and Mary. However, all these common neighbors could be very popular public figures with very high degrees. Therefore, these nodes are statistically more likely to occur as common neighbors of many pairs of nodes. This makes them *less* important in the link prediction measure. The Adamic–Adar measure is designed to account for the varying importance of the different common neighbors. It can be viewed as a weighted version of the common-neighbor measure, where the weight of a common neighbor is a decreasing function of its node degree. The typical function used in the case of the Adamic–Adar measure is the inverse logarithm. In this case, the weight of the common neighbor with index k is set to $1/\log(|S_k|)$, where S_k is the neighbor set of node k.

Definition 19.5.3 (Adamic–Adar Measure) *The common-neighbor measure between nodes i and j is equal to the weighted number of common neighbors between nodes i and j. The weight of node k is defined is $1/log(|S_k|)$.*

$$AdamicAdar(i, j) = \sum_{k \in S_i \cap S_j} \frac{1}{log(|S_k|)} \tag{19.49}$$

The base of the logarithm does not matter in the previous definition, as long as it is chosen consistently for all pairs of nodes. In Fig. 19.12a, the Adamic-Adar measure between Alice and Bob is $\frac{1}{\log(4)} + \frac{1}{\log(2)} + \frac{1}{\log(2)} + \frac{1}{\log(4)} = \frac{3}{\log(2)}$.

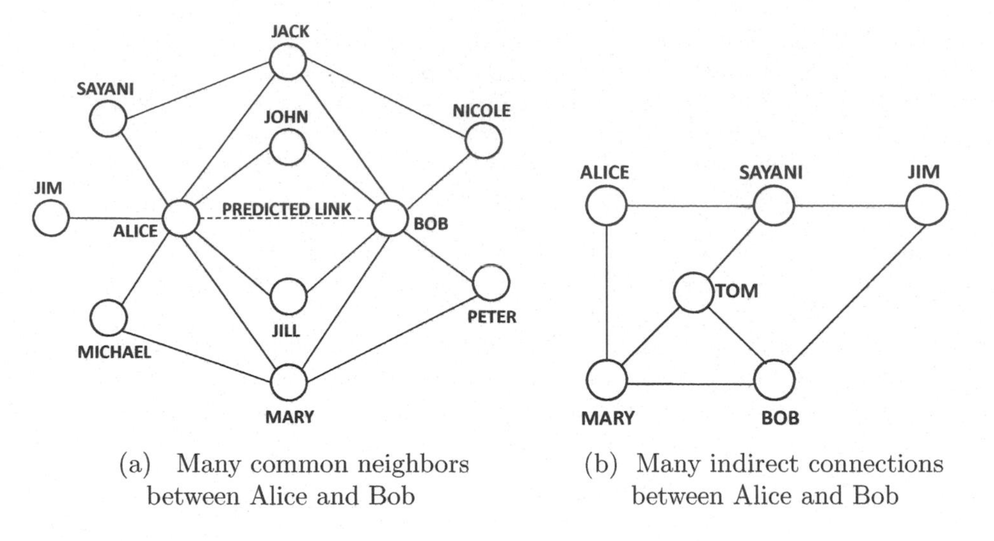

(a) Many common neighbors between Alice and Bob

(b) Many indirect connections between Alice and Bob

Figure 19.12: Examples of varying effectiveness of different link-prediction measures

19.5.2 Katz Measure

While the neighborhood-based measures provide a robust estimation of the likelihood of a link forming between a pair of nodes, they are not quite as effective when the number of shared neighbors between a pair of nodes is small. For example, in the case of Fig. 19.12b, Alice and Bob share one neighbor in common. Alice and Jim also share one neighbor in common. Therefore, neighborhood-based measures have difficulty in distinguishing between different pairwise prediction strengths in these cases. Nevertheless, there also seems to be a significant *indirect* connectivity in these cases through longer paths. In such cases, walk-based measures are more appropriate. A particular walk-based measure that is used commonly to measure the link-prediction strength is the *Katz* measure.

Definition 19.5.4 (Katz Measure) *Let $n_{ij}^{(t)}$ be the number of walks of length t between nodes i and j. Then, for a user-defined parameter $\beta < 1$, the Katz measure between nodes i and j is defined as follows:*

$$Katz(i,j) = \sum_{t=1}^{\infty} \beta^t \cdot n_{ij}^{(t)} \tag{19.50}$$

The value of β is a discount factor that de-emphasizes walks of longer length. For small enough values of β, the infinite summation of Eq. 19.50 will converge. If A is the symmetric adjacency matrix of an undirected network, then the $n \times n$ pairwise Katz coefficient matrix K can be computed as follows:

$$K = \sum_{i=1}^{\infty} (\beta A)^i = (I - \beta A)^{-1} - I \tag{19.51}$$

The eigenvalues of A^k are the kth powers of the eigenvalues of A (cf. Eq. 19.33). The value of β should always be selected to be smaller than the inverse of the largest eigenvalue of A to ensure convergence of the infinite summation. A weighted version of the measure can be computed by replacing A with the weight matrix of the graph. The Katz measure often provides prediction results of excellent quality.

It is noteworthy that the sum of the Katz coefficients of a node i with respect to other nodes is referred to as its *Katz centrality*. Other mechanisms for measuring centrality, such as closeness and *PageRank*, are also used for link prediction in a modified form. The reason for this connection between centrality and link-prediction measures is that highly central nodes have the propensity to form links with many nodes.

19.5.3 Random Walk-Based Measures

Random walk-based measures are a different way of defining connectivity between pairs of nodes. Two such measures are *PageRank* and *SimRank*. Because these methods are described in detail in Sect. 18.4.1.2 of Chap. 18, they will not be discussed in detail here.

The first way of computing the similarity between nodes i and j is with the use of the personalized *PageRank* of node j, where the restart is performed at node i. The idea is that if j is the structural proximity of i, it will have a very high personalized *PageRank* measure, when the restart is performed at node i. This is indicative of higher link prediction strength between nodes i and j. The personalized *PageRank* is an asymmetric measure between nodes i and j. Because the discussion in this section is for the case of undirected graphs, one can use the average of the values of $PersonalizedPageRank(i, j)$ and $PersonalizedPageRank(j, i)$. Another possibility is the *SimRank* measure that is already a symmetric measure. This measure computes an inverse function of the walk length required by two random surfers moving backwards to meet at the same point. The corresponding value is reported as the link prediction measure. Readers are advised to refer to Sect. 18.4.1.2 of Chap. 18 for details of the *SimRank* computation.

19.5.4 Link Prediction as a Classification Problem

The aforementioned measures are *unsupervised* heuristics. For a given network, one of these measures might be more effective, whereas another might be more effective for a different network. How can one resolve this dilemma and select the measures that are most effective for a given network?

The link prediction problem can be viewed as a classification problem by treating the presence or absence of a link between a pair of nodes as a binary class indicator. Thus, a multidimensional data record can be extracted for *each pair of nodes.* The features of this multidimensional record include all the different neighborhood-based, Katz-based, or walk-based similarities between nodes. In addition, a number of other preferential-attachment features, such as node-degrees of each node in the pair, are used. Thus, for each node pair, a multidimensional data record is constructed. The result is a positive-unlabeled classification problem, where node pairs with edges are the positive examples, and the remaining pairs are unlabeled examples. The unlabeled examples can be approximately treated as negative examples for training purposes. Because there are too many negative example pairs in large and sparse networks, only a sample of the negative examples is used. Therefore, the supervised link prediction algorithm works as follows:

1. *Training phase:* Generate a multidimensional data set containing one data record for each pair of nodes with an edge between them, and a sample of data records from pairs of nodes without edges between them. The features correspond to extracted similarity and structural features between node pairs. The class label is the presence or absence of an edge between the pair. Construct a training model on the data.

2. *Testing phase:* Convert each test node pair to a multidimensional record. Use any conventional multidimensional classifier to make label predictions.

The logistic regression method of Sect. 10.6 in Chap. 10 is a common choice for the base classifier. Cost-sensitive versions of various classifiers are commonly used because of the imbalanced nature of the underlying classification problem.

One advantage of this approach is that content features can be used in a seamless way. For example, the content similarity between a pair of nodes can be used. The classifier will automatically learn the relevance of these features in the training process. Furthermore, unlike many link prediction methods, the approach can also handle *directed networks* by extracting features in an asymmetric way. For example, instead of using node degrees, one might use indegrees and outdegrees as features. Random walk features can also be defined in an asymmetric way on directed networks, such as computing the *PageRank* of node j with restart at node i, and vice versa. In general, the supervised model is more flexible because of its ability to *learn* relationships between links and features of various types.

19.5.5 Link Prediction as a Missing-Value Estimation Problem

Section 18.5.3 of Chap. 18 discusses how link prediction can be applied to user-item graphs for recommendations. In general, both the recommendation problem and the link prediction problem may be viewed as instances of missing value estimation on matrices of different types. Recommendation algorithms are applied to user-item utility matrices, whereas link prediction algorithms are applied to incomplete adjacency matrices. All the 1s in the matrix correspond to edges. Only a small random sample of the remaining entries are set to 0, and the other entries are assumed to be unspecified. Any of the missing-value estimation methods discussed in Sect. 18.5 of Chap. 18 may be used to estimate the values of the missing entries. Among this class of methods, matrix factorization methods are among the most commonly used methods. One advantage of using these methods is that the specified matrix does not need to be symmetric. In other words, the approach can also be used for directed graphs. Refer to the bibliographic notes.

19.5.6 Discussion

The different measures have been shown to have varying levels of effectiveness over different data sets. The advantage of neighborhood-based measures is that they can be computed efficiently for very large data sets. Furthermore, they perform almost as well as the other unsupervised measures. Nevertheless, random walk-based and Katz-based measures are particularly useful for very sparse networks, in which the number of common neighbors cannot be robustly measured. Although supervision provides better accuracy, it is computationally expensive. However, supervision provides the greatest adaptability across various domains of social networks, and available side information such as content features.

In recent years, content has also been used to enhance link prediction. While content can significantly improve link prediction, it is important to point out that structural measures are far more powerful. This is because structural measures directly use the triadic properties of real networks. The triadic property of networks is true across virtually all data domains. On the other hand, content-based measures are based on "reverse homophily," where similar or link-correlated content is leveraged for predicting links. The effectiveness of this is highly network domain-specific. Therefore, content-based measures are often used in a helping role for link prediction and are rarely used in isolation for the prediction process.

19.6 Social Influence Analysis

All social interactions result in varying levels of influence between individuals. In traditional social interactions, this is sometimes referred to as "word of mouth" influence. This general principle is also true for online social networks. For example, when an actor tweets a message in Twitter, the followers of the actors are exposed to the message. The followers may often retweet the message in the network. This results in the spread of information, ideas, and opinions in the social network. Many companies view this kind of information spread as a valuable advertising channel. By tweeting a popular message to the right participants, millions of dollars worth of advertising can be generated, if the message spreads through the social network as a *cascade*. An example [532] is the famous Oreo Superbowl tweet on February 3, 2013. The power went out during the Superbowl game between the San Francisco 49ers and the Baltimore Ravens. Oreo used this opportunity to tweet the following message, along with a picture of an Oreo cookie, during the 34 min interruption: "Power out? No problem. You can still dunk in the dark." Viewers loved Oreo's message, and retweeted it thousands of times. Oreo was thus able to generate millions of dollars of advertising at zero cost, and apparently had a higher impact than *paid* television advertisements during the Superbowl.

Different actors have different abilities to influence their peers in the social network. The two most common factors that regulate the influence of an actor are as follows:

1. Their centrality within the social network structure is a crucial factor in their influence level. For example, actors with high levels of centrality are more likely to be influential. In directed networks, actors with high prestige are more likely to be influential. These measures are discussed in Sect. 19.2.

2. The edges in the network are often associated with weights that are dependent on the likelihood that the corresponding pair of actors can be influenced by each other. Depending on the diffusion model used, these weights can sometimes be directly interpreted as *influence propagation probabilities*. Several factors may determine these probabilities. For example, a well-known individual may have higher influence than lesser known individuals. Similarly, two individuals, who have been friends for a long time, are more likely to influence one another. It is often assumed that the influence propagation probabilities are already available for analytical purposes, although a few recent methods show how to estimate these probabilities in a data-driven way.

The precise impact of the aforementioned factors is quantified with the use of an influence propagation model. These are also referred to as *diffusion models*. The main goal of such models is to determine a set of seed nodes in the network, at which the dissemination of information maximizes influence. Therefore, the influence maximization problem is as follows:

Definition 19.6.1 (Influence Maximization) *Given a social network $G = (N, A)$, determine a set of k seed nodes S, influencing which will maximize the overall spread of influence in the network.*

The value of k can be viewed as a budget on the number of seed nodes that one is allowed to initially influence. This is quite consistent with real-life models, where advertisers are faced with budgets on initial advertising capacity. The goal of social influence analysis is to extend this initial advertising capacity with word-of-mouth methods.

Each model or heuristic can quantify the influence level of a node with the use of a function of S that is denoted by $f(\cdot)$. This function maps subsets of nodes to real numbers representing influence values. Therefore, after a model has been chosen for quantifying the influence $f(S)$ of a *given* set S, the optimization problem is that of determining the set S that maximizes $f(S)$. An interesting property of a very large number of influence analysis models is that the optimized function $f(S)$ is *submodular*.

What does submodularity mean? It is a mathematical way of representing the natural law of diminishing returns, as applied to sets. In other words, if $S \subseteq T$, then the *additional* influence obtained by adding an individual to set T cannot be larger than the additional influence of adding the same individual to set S. Thus, the *incremental* influence of the same individual diminishes, as larger supersets of cohorts are available as seeds. The submodularity of set S is formally defined as follows:

Definition 19.6.2 (Submodularity) *A function $f(\cdot)$ is said to be submodular, if for any pair of sets S, T satisfying $S \subseteq T$, and any set element e, the following is true:*

$$f(S \cup \{e\}) - f(S) \geq f(T \cup \{e\}) - f(T) \tag{19.52}$$

Virtually all natural models for quantifying influence turn out to be submodular. Submodularity is algorithmically convenient because a very efficient greedy optimization algorithm exists for maximizing submodular functions, as long as $f(S)$ can be evaluated for a given value of S. This algorithm starts by setting $S = \{\}$ and incrementally adds nodes to S that increase the value of $f(S)$ as much as possible. This procedure is repeated until the set S contains the required number of influencers k. The approximation level of this heuristic is based on a well-known classical result on optimization of submodular functions.

Lemma 19.6.1 *The greedy algorithm for maximizing submodular functions provides a solution with an objective function value that is at least a fraction $\left(\frac{e-1}{e}\right)$ of the optimal value. Here, e is the base of the natural logarithm.*

Thus, these results show that it is possible to optimize $f(S)$ effectively, as long as an appropriate submodular influence function $f(S)$ can be defined for a given set of nodes S.

Two common approaches for defining the influence function $f(S)$ of a set of nodes S are the *Linear Threshold Model* and the *Independent Cascade Model.* Both these diffusion models were proposed in one of the earliest works on social influence analysis. The general operational assumption in these diffusion models is that nodes are either in an active or inactive state. Intuitively, an *active* node is one which has already been influenced by the set of desired behaviors. Once a node moves to an active state, it never deactivates. Depending on the model, an active node may trigger activation of neighboring nodes either for a single time, or over longer periods. Nodes are successively activated until no more nodes are activated in a given iteration. The value of $f(S)$ is evaluated as the total number of activated nodes at termination.

19.6.1 Linear Threshold Model

In this model, the algorithm initially starts with an active set of seed nodes S and iteratively increases the number of active nodes based on the influence of neighboring active nodes. Active nodes are allowed to influence their neighbors over multiple iterations throughout the execution of the algorithm until no more nodes can be activated. The influence of neighboring nodes is quantified with the use of a linear function of the edge-specific weights b_{ij}. For each node i in the network $G = (N, A)$, the following is assumed to be true:

$$\sum_{j:(i,j)\in A} b_{ij} \leq 1 \tag{19.53}$$

Each node i is associated with a random threshold $\theta_i \sim U[0,1]$ which is fixed up front and stays constant over the course of the algorithm. The total influence $I(i)$ of the *active* neighbors of node i on it, at a given time-instant, is computed as the sum of the weights b_{ij} of all *active* neighbors of i.

$$I(i) = \sum_{j:(i,j)\in A, j \text{ is active}} b_{ij} \tag{19.54}$$

The node i becomes active in a step when $I(i) \geq \theta_i$. This process is repeated until no further nodes can be activated. The total influence $f(S)$ may be measured as the number of nodes activated by a given seed set S. The influence $f(S)$ of a given seed set S is typically computed with simulation methods.

19.6.2 Independent Cascade Model

In the aforementioned linear threshold model, once a node becomes active, it has multiple chances to influence its neighbors. The random variable θ_i was associated with a *node*, in the form of a threshold. On the other hand, in the independent cascade model, after a node becomes active, it obtains only a *single chance* to activate its neighbors, with *propagation probabilities* associated with the *edges*. The propagation probability associated with an edge is denoted by p_{ij}. In each iteration, only the *newly* active nodes are allowed to influence their neighbors, that have not already been activated. For a given node j, each of the edges (i, j) joining it to its newly active neighbors i flips a coin independently with success probability p_{ij}. If the coin toss for edge (i, j) results in a success, then the node j is activated. If node j is activated, it will get a single chance in the next iteration to influence its neighbors. In the event that no nodes are newly activated in an iteration, the algorithm terminates. The influence function value is equal to the number of active nodes at termination. Because nodes are allowed to influence their neighbors only once over the course of the algorithm, a coin is tossed for each edge at most once over the course of the algorithm.

19.6.3 Influence Function Evaluation

Both the linear threshold model and the independent cascade model are designed to compute the influence function $f(S)$ with the use of a model. The estimation of $f(S)$ is typically accomplished with simulation.

For example, consider the case of the linear threshold model. For a given seed node set S, one can use a random number generator to set the thresholds at the nodes. After the thresholds have been set, the active nodes can be labeled using any deterministic graph-search algorithm starting from the seed nodes in S and progressively activating nodes when the threshold condition is satisfied. The computation can be repeated over different sets of randomly generated thresholds, and the results may be averaged to obtain more robust estimates.

In the independent cascade model, a different simulation may be used. A coin with probability p_{ij} may be flipped for each edge. The edge is designated as *live* if the coin toss was a success. It can be shown that a node will eventually be activated by the independent cascade model, when a path of live edges exists from at least one node in S to it. This can be used to estimate the size of the (final) active set by simulation. The computation is repeated over different runs and the results are averaged.

The proof that the linear threshold model and the independent cascade model are submodular optimization problems can be found in pointers included in the bibliographic notes. However, this property is not specific to these models. Submodularity is a very natural consequence of the laws of diminishing returns, as applied to the incremental impact of individual influence in larger groups. As a result, most reasonable models for influence analysis will satisfy submodularity.

19.7 Summary

Social networks have become increasingly popular in recent years, because of their ability to connect geographically and culturally diverse participants. A significant amount of data is created because of the actions of social network participants. Much of this data are structural, in the form of relationships between different individuals.

Social network structures exhibit a number of typical properties, because of the natural dynamics of their formation. The most important similarity-based properties include triadic closure, and homophily. Typically, social networks are formed by preferential attachment, and they exhibit power-law degree distributions.

The problem of clustering social networks is challenging because of the presence of hub nodes, and the natural tendency of social networks to cluster into a single large group. Therefore, most community detection algorithms have built-in mechanisms to ensure that the underlying clusters are balanced. Clustering methods are also sometimes referred to as graph-partitioning. One of the earliest clustering methods was the Kernighan–Lin method, which uses an iterative approach for clustering. Nodes are repeatedly exchanged between partitions to iteratively improve the value of the objective function. The Girvan–Newman algorithm uses notions of betweenness centrality to generate clusters. The *METIS* algorithm generates an efficient partition by using coarsening and then creating the partitions on the coarsened representation. The spectral method uses multidimensional embeddings to generate the clusters.

In collective classification, the goal is to infer labels at the remaining vertices from the pre-existing labels at a subset of the vertices. This is a problem that has dual applicability to social network analysis and semisupervised learning. Multidimensional data sets can be transformed into similarity graphs to apply collective classification methods. The most common methods used for collective classification include iterative methods, random walk-based label propagation methods, and spectral methods.

In the link-prediction problem, the goal is to predict the links from the currently available structure and content in the network. Structural measures are generally much more effective for link-prediction than content-based measures. The structural methods use local clustering measures such as the Jaccard measure or personalized *PageRank* values for making predictions. Supervised methods are able to discriminatively determine the most relevant features for link prediction.

Social networks are often used for influencing individuals using "word-of-mouth" techniques. Typically, centrally located actors are more influential in the network. Diffusion models are used to characterize the flow of information in social networks. Two examples of such models include the linear threshold model and the independent cascade model.

19.8 Bibliographic Notes

Social network analysis has been studied extensively in the context of the field of sociology [508], though more recent work has focused on online social networks [6, 192, 532]. A detailed discussion on proximity and centrality measures may be found in [6, 192, 508, 532]. The dynamics of social network formation may be found in the excellent survey paper [69]. The derivation of the power-law with the use of the scale-free model is provided in [70]. A detailed study of the power-law in the context of the Internet topology is provided in [201]. A study of graph densification and shrinking diameters is provided in [342]. Other random graph models such as the Erdos–Renyi model and the Watts–Strogatz small-world model are discussed in [196, 509].

A detailed survey on community detection methods may be found in [212]. The minimum cut problem is polynomially solvable for some special cases. For example, the unweighted 2-way cut problem is polynomially solvable without balancing constraints [299]. The original Kernighan–Lin algorithm is presented in [312]. The enhancements to the Kernighan–Lin algorithm was discussed in [206, 301]. The Girvan–Newman algorithm discussed in this chapter is adapted from [230]. The *METIS* algorithm is presented in [301]. The normalized cut method for spectral clustering was discussed in this chapter [466]. The normalized symmetric version was proposed in [405]. More details on spectral graph theory and clustering methods may be found in [152, 371]. This chapter uses the *Laplacian eigenmap* interpretation [90] of spectral clustering, rather than the more commonly used cut interpretation, because of its comprehensive explanation of the non-integer and possibly negative eigenvector components.

ICA has been presented in the context of many different data domains, such as document data [128], and relational data [404]. Several base classifiers have been used within this framework, such as logistic regression [370] and a weighted voting classifier [373]. The discussion in this chapter is based on [404]. The iterative label propagation method was proposed in [554], and the absorbing random walk interpretation is adapted from [78]. The iterative label propagation approach [554] was originally proposed with a spectral interpretation although the random walk interpretation is also briefly discussed in the same work. Most random walk methods can also be formulated as supervised versions of spectral embeddings [530, 551, 554]. The regularization framework for collective classification is discussed in [551]. Collective classification of directed graphs is discussed in [552]. A method for incorporating content within the random walk framework is discussed in [44]. Detailed surveys on node classification methods may be found in [93, 368]. A toolkit for collective classification may be found in [427].

The link-prediction problem for social networks was proposed in [353]. The measures discussed in this chapter are based on this work. Since then, a significant amount of work has been done on incorporating content into the link prediction process. Methods that use content for link prediction may be found in [49, 64, 354, 484, 489]. The merits of supervised methods are discussed in [354], and matrix factorization methods are discussed in [383]. Recently, it has been shown how to use link prediction across multiple networks in [428]. A survey on link-prediction methods for social network analysis may be found in [63].

The problem of influence analysis in social networks was proposed in [304]. The linear threshold and independent cascade models are presented in this work. The degree-discount heuristic was proposed in [142]. A discussion of the submodularity property may be found in [403]. Other recent models for influence analysis in social networks are discussed in [45, 143, 144, 362, 488]. One of the main problems in social influence models is a difficulty in learning the influence propagation probabilities, though there has been some recent focus on this issue [235]. Recent work has also shown how influence analysis can be performed directly from the social stream [234, 482]. A survey on models and algorithms for social influence analysis may be found in [483].

19.9 Exercises

1. For the figure in Example 19.1a, compute the highest-degree centrality, closeness centrality and betweenness centrality. The nodes that take on these highest values are already marked in the figure.

2. Implement the algorithms for determining the degree centrality, closeness centrality, and betweenness centrality.

3. Implement the Kernighan–Lin algorithm.

4. Why is the balancing constraint more important in community detection algorithms, as compared to multidimensional clustering algorithms? What would the unconstrained minimum 2-way cut look like in a typical real network?

5. Consider a variation of Girvan–Newman algorithm in which edges are randomly disconnected from a network, as opposed to those with high betweenness centrality. Explain the negative impact of this change on the algorithm. Can you make minor changes to the disconnection criterion to ameliorate this impact?

6. Write an integer-programming formulation for the minimum 2-way cut problem, so that the cut is balanced in terms of the number of nodes.

7. For the random walk formulation of spectral clustering algorithm, show why the following are true:

   (a) All nontrivial eigenvectors $\overline{y}$ have both positive and negative components.

   (b) Provide an intuitive explanation why the normalization factor Λ in the constraint $\overline{y}^T \Lambda \overline{y} = 1$, increases the propensity of low-degree nodes to be embedded away from the origin, and the high-degree nodes to be embedded near the origin.

8. Suppose that all edge weights w_{ij} are discounted by the geometric mean of the weighted node-degrees at their endpoints. Write an unnormalized formulation of spectral clustering in terms of these normalized weights for discovering a 1-dimensional embedding. What effect would the weight normalization have on the embedding? Describe the algebraic similarities and differences of this formulation from the symmetric normalized formulation of spectral clustering. Discuss why the resulting eigenvectors will often be heuristically similar to those obtained with the symmetric formulation of spectral clustering.

9. Explain the relationship between the random walk label propagation and the graph regularization algorithm.

10. Discuss the connections between the link-prediction problem and network clustering.

11. Create a link prediction measure that can perform the degree normalizations performed both by the Jaccard measure and the Adamic–Adar measure.

12. Implement the linear threshold and independent cascade model for influence analysis.

13. The chapter provides a 1-dimensional formulation for the symmetric version using the column vector $\overline{z}$. Set up a generalized formulation for the symmetric version using an $n \times k$ matrix Z.

(a) Let Y be the decision variables for the random walk formulation discussed in the chapter. Show that $Z = \sqrt{\Lambda}Y$.

(b) Show that the unit-norm scaled rows of Y and Z are the same.

14. It is well known that a symmetric matrix always has real eigenvalues. Use this result to show that the stochastic transition matrix of an *undirected* graph always has real eigenvalues.

15. Show that if $(\overline{y}, \lambda)$ is an eigenvector–eigenvalue pair of the normalized Laplacian $\Lambda^{-1}(\Lambda - W)$, then $(\overline{y}, 1 - \lambda)$ is an eigenvector–eigenvalue pair of the normalized weight matrix $\Lambda^{-1}W$. Here, Λ is a diagonal matrix containing the sum of each row in the weighted adjacency matrix W.

Chapter 20

Privacy-Preserving Data Mining

"*Civilization is the progress toward a society of privacy. The savage's whole existence is public, ruled by the laws of his tribe. Civilization is the process of setting man free from men.*"—Ayn Rand

20.1 Introduction

A significant amount of application data is of a personal nature. These kind of data sets may contain sensitive information about an individual, such as his or her financial status, political beliefs, sexual orientation, and medical history. The knowledge about such personal information can compromise the privacy of individuals. Therefore, it is crucial to design data collection, dissemination, and mining techniques, so that individuals are assured of their privacy. Privacy-preservation methods can generally be executed at different steps of the data mining process:

1. *Data collection and publication:* The privacy-driven modification of a data set may be done at either the data *collection* time, or the data *publication* time. In anonymous data *collection*, a modified version of the data is collected using a software plugin within the collection platform. Therefore, the contributors of the data are assured that their data is not available even to the entity collecting the data. The implicit assumption in the collection-oriented model is that the data collector is not trusted, and therefore the privacy must be preserved at collection time. In anonymous data *publication*, the entire data set is available to a trusted entity, who has usually collected the data in the normal course of business. An example is a hospital that has collected data about its patients. Eventually, the entity may wish to release or *publish* the data to one of more third-parties for data analysis. For example, a hospital may want to use the data to study the long-term impact of various treatment alternatives. A real-world example is the Netflix prize data set [559], in which the anonymized movie ratings of users were published to advance studies on collaborative filtering algorithms. During data publication, identifying or sensitive attribute values need to either be removed or be specified approximately to preserve privacy. Generally, such publication algorithms

C. C. Aggarwal, *Data Mining: The Textbook*, DOI 10.1007/978-3-319-14142-8_20

can control the level of privacy much better than collection algorithms, because of their access to the entire data set on a trusted server.

2. *Output privacy of data mining algorithms:* Privacy can also be violated by the *output* of data mining algorithms. For example, consider a scenario where a user is allowed to determine association patterns, or otherwise query the data through a Web service, but is not provided access to the data set. In such a case, the output of the data mining and query processing algorithms provides valuable information, some of which may be private.

In some applications, organizations may wish to share their data in a private way, so that only patterns in the *shared* data may be mined, but the statistics of the *local* databases are not revealed to the participants. This problem is referred to as *distributed privacy preservation.*

In general, most forms of privacy-preserving data mining reduce the representation accuracy of the data, in order to preserve privacy. This accuracy reduction is performed in a variety of ways, such as data distortion, approximation (generalization), suppression, attribute value swapping, or microaggregation. Clearly, since the data is no longer specified exactly, this will have a detrimental impact on the quality of the data mining results. The effectiveness of the released data for mining applications is often quantified explicitly, and is referred to as its *utility.* A natural trade-off exists between privacy and utility. For example, in a case, where data values are suppressed, one might simply choose to suppress all entries. While such a solution provides perfect privacy, it offers no utility. This observation is also true for privacy-preserving publication algorithms in which noise is added to the data. When a greater amount of noise is added, a higher level of privacy is achieved, but utility is reduced. The goal of privacy-preservation methods is to maximize utility at a fixed level of privacy.

This chapter is organized as follows. Methods for privacy-preserving data collection are addressed in Sect. 20.2. Section 20.3 addresses the problem of privacy-preserving data publishing. This section includes several models such as the k-anonymity model, the ℓ-diversity model, and the t-closeness model. The problem of output privacy is addressed in Sect. 20.4. Methods for distributed and cryptographic privacy are discussed in Sect. 20.5. A summary is given in Sect. 20.6.

20.2 Privacy During Data Collection

The randomization method is designed for privacy-preservation at data *collection* time. The implicit assumption is that the data collector is not trusted, and therefore the privacy must be preserved at data collection time. The basic idea of the approach is to allow users to enter the data through a software platform that is able to add random perturbations to the data. This approach is one of the most conservative models for ensuring data privacy, because the original data records are never stored on any single server.

The random perturbations are added using a *publicly available* distribution. Examples of commonly used perturbing distributions include the uniform and the Gaussian distributions. In other words, the probability distribution used to perturb the data is specified together with the data set if and when the data collector releases the data for public use. This additional distribution information is needed to use the data effectively in the context of data mining algorithms. The basic idea is to reconstruct the distribution of the original data, by "subtracting out" the noise distribution. This aggregate distribution is then used for mining purposes. The overall approach is as follows:

1. *Privacy-preserving data collection:* In this step, random noise is added to the data while collecting data from users, with the use of a software plugin. The collected data is publicly released along with the probability distribution function (and parameters) used to add the random noise.

2. *Distribution reconstruction:* The aggregate distribution of the original data are reconstructed, by "subtracting out" the noise. Thus, at the end of this step, we will have a histogram representing the approximate probability distribution of the data values.

3. *Data mining:* Data mining methods are applied to the reconstructed distributions.

It is important to note that the last step of the process requires the design of data mining algorithms that work with probability distributions of *sets of data records*, rather than *individual* records. Thus, one disadvantage of this approach is that it requires the redesign of data mining algorithms. Nevertheless, the approach can be made to work because many data mining problems such as clustering and classification require only the probability distribution modeling of either the whole data set, or segments (e.g., different classes) of the data.

20.2.1 Reconstructing Aggregate Distributions

The reconstruction of the aggregate distribution of the original data is the key step in the randomization method. Consider the case where the original data values $x_1 \ldots x_n$ are drawn from the probability distribution $\mathbf{X}$. For each original data value x_i, a perturbation y_i is added by the software data collection tool, to yield the perturbed value z_i. The perturbation y_i is drawn from the probability distribution $\mathbf{Y}$, and is independent of $\mathbf{X}$. It is assumed that this distribution is known publicly. Furthermore, the probability distribution of the final set of perturbed values is assumed to be $\mathbf{Z}$. Therefore, the original distribution $\mathbf{X}$, the added perturbation $\mathbf{Y}$ and the final aggregate distribution $\mathbf{Z}$ are related as follows:

$$\mathbf{Z} = \mathbf{X} + \mathbf{Y}$$
$$\mathbf{X} = \mathbf{Z} - \mathbf{Y}$$

Thus, the probability distribution of $\mathbf{X}$ can be reconstructed, if the distributions of $\mathbf{Y}$ and $\mathbf{Z}$ are known explicitly. The probability distribution of $\mathbf{Y}$ is assumed to be publicly available, while discrete samples of $\mathbf{Z}$ are available in terms of $z_1 \ldots z_n$. These discrete samples are sufficient to reconstruct $\mathbf{Z}$ using a variety of methods, such as kernel density estimation. Then, the distribution for $\mathbf{X}$ can be reconstructed using the relationship shown above. The main problem with this approach emerges when the probability distribution of the perturbation $\mathbf{Y}$ has a large variance and the number n of discrete samples of $\mathbf{Z}$ is small. In such a case, the distribution of $\mathbf{Z}$ also has a large variance, and it cannot be accurately estimated with a small number of samples. Therefore, a second approach is to *directly* estimate the distribution of $\mathbf{X}$ from the discrete samples of $\mathbf{Z}$ and the known distribution of $\mathbf{Y}$.

Let $f_{\mathbf{X}}$ and $F_{\mathbf{X}}$ be the probability density and cumulative distributions functions of $\mathbf{X}$. These functions need to be *estimated* with the observed values $z_1 \ldots z_n$. Let $\hat{f}_{\mathbf{X}}$ and $\hat{F}_{\mathbf{X}}$ be the corresponding *estimated* probability density and cumulative distribution functions for $\mathbf{X}$. The key here is to use the Bayes formula with the use of observed values for $\mathbf{Z}$. Consider a simplified scenario in which only a single observed value z_1 is available. This can be used to estimate the cumulative distribution function $\hat{F}_{\mathbf{X}}(a)$ at any value of the random variable $\mathbf{X} = a$. The Bayes theorem yields the following:

$$\hat{F}_{\mathbf{X}}(a) = \frac{\int_{w=-\infty}^{w=a} f_{\mathbf{X}}(w|\mathbf{X}+\mathbf{Y}=z_1)dw}{\int_{w=-\infty}^{w=\infty} f_{\mathbf{X}}(w|\mathbf{X}+\mathbf{Y}=z_1)dw} \tag{20.1}$$

The conditional in the aforementioned equation corresponds to the fact that the sum of the data values and perturbed values is equal to z_1. Note that the expression $f_{\mathbf{X}}(w|\mathbf{X}+\mathbf{Y}=z_1)$ can be expressed in terms of the unconditional densities of $\mathbf{X}$ and $\mathbf{Y}$, as follows:

$$f_{\mathbf{X}}(w|\mathbf{X}+\mathbf{Y}=z_1) = f_{\mathbf{Y}}(z_1 - w) \cdot f_{\mathbf{X}}(w) \tag{20.2}$$

This expression uses the fact that the perturbation $\mathbf{Y}$ is independent of $\mathbf{X}$. By substituting the aforementioned expression for $f_{\mathbf{X}}(w|\mathbf{X}+\mathbf{Y}=z_1)$ in the right-hand side of Eq. 20.1, the following expression is obtained for the cumulative density of $\mathbf{X}$:

$$\hat{F}_{\mathbf{X}}(a) = \frac{\int_{w=-\infty}^{w=a} f_{\mathbf{Y}}(z_1 - w) \cdot f_{\mathbf{X}}(w)dw}{\int_{w=-\infty}^{w=\infty} f_{\mathbf{Y}}(z_1 - w) \cdot f_{\mathbf{X}}(w)dw} \tag{20.3}$$

The expression for $\hat{F}_{\mathbf{X}}(a)$ was derived using a *single* observation z_1, and needs to be generalized to the case of n different observations $z_1 \ldots z_n$. This can be achieved by averaging the previous expression over n different values:

$$\hat{F}_{\mathbf{X}}(a) = \frac{1}{n} \cdot \sum_{i=1}^{n} \frac{\int_{w=-\infty}^{w=a} f_{\mathbf{Y}}(z_i - w) \cdot f_{\mathbf{X}}(w)dw}{\int_{w=-\infty}^{w=\infty} f_{\mathbf{Y}}(z_i - w) \cdot f_{\mathbf{X}}(w)dw} \tag{20.4}$$

The corresponding density distribution can be obtained by differentiating $\hat{F}_{\mathbf{X}}(a)$. This differentiation results in the removal of the integral sign from the numerator and the corresponding instantiation of w to a. Since the denominator is a constant, it remains unaffected by the differentiation. Therefore, the following is true:

$$\hat{f}_{\mathbf{X}}(a) = \frac{1}{n} \cdot \sum_{i=1}^{n} \frac{f_{\mathbf{Y}}(z_i - a) \cdot f_{\mathbf{X}}(a)}{\int_{w=-\infty}^{w=\infty} f_{\mathbf{Y}}(z_i - w) \cdot f_{\mathbf{X}}(w)dw} \tag{20.5}$$

The aforementioned equation contains the density function $f_{\mathbf{X}}(\cdot)$ on both sides. This circularity can be resolved naturally with the use of an iterative approach. The iterative approach initializes the estimate of the distribution $f_{\mathbf{X}}(\cdot)$ to the uniform distribution. Subsequently, the estimate of this distribution is continuously updated as follows:

Set $\hat{f}_{\mathbf{X}}(\cdot)$ to be the uniform distribution;
repeat
Update $\hat{f}_{\mathbf{X}}(a) = (1/n) \cdot \sum_{i=1}^{n} \frac{f_{\mathbf{Y}}(z_i - a) \cdot \hat{f}_{\mathbf{X}}(a)}{\int_{w=-\infty}^{w=\infty} f_{\mathbf{Y}}(z_i - w) \cdot \hat{f}_{\mathbf{X}}(w)dw}$
until convergence

So far, it has been described, how to compute $f_{\mathbf{X}}(a)$ for a *particular* value of a. In order to generalize this approach, the idea is to discretize the range of the random variable $\mathbf{X}$ into k intervals, denoted by $[l_1, u_1] \ldots [l_k, u_k]$. It is assumed that the density distribution is uniform over the discretized intervals. For each such interval $[l_i, u_i]$, the density distribution is evaluated at the midpoint $a = (l_i + u_i)/2$ of the interval. Thus, in each iteration, k different values of a are used. The algorithm is terminated when the distribution does not change significantly over successive steps of the algorithm. A variety of methods can be used to compare the two distributions such as the χ^2 test. The simplest approach is to examine the average change in the density values, at the midpoints of the density distribution over successive iterations. While this algorithm is known to perform effectively in practice, it is not proven to be a optimally convergent solution. An expectation maximization (EM) method was proposed in a later work [28], which provably converges to the optimal solution.

20.2.2 Leveraging Aggregate Distributions for Data Mining

The aggregate distributions determined by the algorithm can be leveraged for a variety of data mining problems, such as clustering, classification, and collaborative filtering. This is because each of these data mining problems can be implemented with aggregate statistics of the data, rather than the original data records. In the case of the classification problem, the probability distributions of each of the classes can be reconstructed from the data. These distributions can then be used directly in the context of a naive Bayes classifier, as discussed in Chap. 10. Other classifiers, such as decision trees, can also be modified to work with aggregate distributions. The key is to use the aggregate distributions in order to design the split criterion of the decision tree. The bibliographic notes contain pointers to data mining algorithms that use the randomization method. The approach cannot be used effectively for data mining problems such as outlier detection that are dependent on individual data record values, rather than aggregate values. In general, outlier analysis is a difficult problem for most private data sets because of the tendency of outliers to reveal private information.

20.3 Privacy-Preserving Data Publishing

Privacy-preserving data *publishing* is distinct from privacy-preserving data *collection*, because it is assumed that all the records are already available to a trusted party, who might be the current owner of the data. This party then wants to release (or *publish*) this data for analysis. For example, a hospital might wish to release anonymized records about patients to study the effectiveness of various treatment alternatives.

This form of data release is quite useful, because virtually any data mining algorithm can be used on the released data. To determine sensitive information about an individual, there are two main pieces of information that an attacker (or *adversary*) must possess.

1. Who does this data record pertain to? While a straightforward way to determine the identity is to use the identifying attribute (e.g., Social Security Number), such attributes are usually stripped from the data before release. As will be discussed later, these straightforward methods of sanitization are usually not sufficient, because attackers may use other attributes, such as the age and ZIP code, to make *linkage* attacks.

2. In addition to identifying attributes, data records also contain *sensitive* attributes that most individuals do not wish to share with others. For example, when a hospital releases medical data, the records might contain sensitive disease-related attributes.

Different attributes in a data set may play different roles in either facilitating identification or facilitating sensitive information release. There are three main types of attributes:

Table 20.1: Example of a data table

SSN	Age	ZIP Code	Disease
012-345-6789	24	10598	HIV
823-627-9231	37	90210	Hepatitis C
987-654-3210	26	10547	HIV
382-827-8264	38	90345	Hepatitis C
847-872-7276	36	89119	Diabetes
422-061-0089	25	02139	HIV

1. *Explicit identifiers:* These are attributes that explicitly identify an individual. For example, the Social Security Number (SSN) of an individual can be considered an explicit identifier. Because this attribute is almost always removed in the data sanitization process, it is not relevant to the study of privacy algorithms.

2. *Pseudo-identifier or quasi-identifier (QID):* These are attributes that do not explicitly identify an individual in *isolation*, but can nevertheless be used *in combination* to identify an individual by joining them with publicly available information, such as voter registration rolls. This kind of attack is referred to as a *linkage* attack. Examples of such attributes include the Age and ZIP code. Strictly speaking, quasi-identifiers refer to the specific combination of attributes used to make a linkage attack, rather than the individual attributes.

3. *Sensitive attribute:* These are attributes that are considered private by most individuals. For example, in a medical data set, individuals would not like information about their diseases to be known publicly. In fact, many laws in the USA, such as the Health Insurance Portability and Accountability Act (HIPAA), explicitly forbid the release of such information, especially when the sensitive attributes can be linked back to specific individuals.

Most of the discussion in this chapter will be restricted to quasi-identifiers and sensitive attributes. To illustrate the significance of these attribute types, an example will be used. In Table 20.1, the medical records of a set of individuals are illustrated. The *SSN* attribute is an explicit identifier that can be utilized to identify an individual directly. Such *directly* identifying information will almost always be removed from a data set before release. However, the impact of attributes such as the age and the ZIP code on identification is quite significant. While these attributes do not directly identify an individual, they provide very useful hints, when combined with other publicly available information. For example, it is possible for a small geographic region, such as a ZIP code, to contain only one individual of a specific gender, race, and date of birth. When combined with publicly available voter registration rolls, one might be able to identify an individual from these attributes. Such a *combination* of publicly available attributes is referred to as a quasi-identifier.

To understand the power of quasi-identifiers, consider a snapshot of the voter registration rolls illustrated in Table 20.2. Even in cases, where the SSN is removed from Table 20.1 before release, it is possible to join the two tables with the use of the age and ZIP code attributes. This will provide a list of the *possible* matches for each data record. For example, Joy and Sue are the only two individuals in the voter registration rolls matching an individual with HIV in the medical release of Table 20.1. Therefore, one can tell with 50 % certainty that Joy and Sue have HIV. This is not desirable especially when an adversary

Table 20.2: Example of a snapshot of fictitious voter registration rolls

Name	Age	ZIP Code
Mary A.	38	90345
John S.	36	89119
Ann L.	31	02139
Jack M.	57	10562
Joy M.	26	10547
Victor B.	46	90345
Peter P.	25	02139
Diana X.	24	10598
William W.	37	90210
Sue G.	26	10547

has other background medical information about Joy or Sue to further narrow down the possibilities. Similarly, William is the only individual in the voter registration rolls, who matches an individual with hepatitis C in the medical release. In cases, where only one data record in the voter registration rolls matches the particular combination of age and ZIP code, sensitive medical conditions about that individual may be fully compromised. This approach is referred to as a *linkage* attack. Most anonymization algorithms focus on preventing *identity disclosure*, rather than explicitly hiding the sensitive attributes. Thus, only the attributes which can be combined to construct quasi-identifiers are changed or specified approximately in the data release, whereas sensitive attributes are released in their exact form.

Many privacy-preserving data publishing algorithms assume that the quasi-identifiers are drawn out of a set of attributes that are not sensitive, because they can only be used by an adversary by performing joins with (nonsensitive) publicly available information. This assumption may, however, not always be reasonable, when an adversary has (sensitive) background information about a target at hand. Adversaries are often familiar with their targets, and they can be assumed to have background knowledge about at least a subset of the sensitive attributes. In a medical application with multiple disease attributes, knowledge about a subset of these attributes may reveal the identity of the subject of the record. Similarly, in a movie collaborative filtering application, where anonymized ratings are released, it may be possible to obtain information about a particular user's ratings on a subset of movies, through personal interaction or other rating sources. If this combination is unique to the individual, then the other ratings of the individual are compromised as well. Thus, sensitive attributes also need to be perturbed, when background knowledge is available. Much of the work in the privacy literature assumes a rigid distinction between the role of publicly available attributes (from which the quasi-identifiers are constructed) and that of the sensitive attributes. In other words, sensitive attributes are not perturbed because it is assumed that revealing them does not incur the risk of a linkage attack with publicly available information. There are, however, a few algorithms that do not make this distinction. Such algorithms generally provide better privacy protection in the presence of background information.

In this section, several models for group-based anonymization, such as k-anonymity, ℓ-diversity, and t-closeness, will be introduced. While the recent models, such as ℓ-diversity, have certain advantages over the k-anonymity model, a good understanding of k-anonymity is crucial in any study of privacy-preserving data publishing. This is because the basic framework for most of the group-based anonymization models was first proposed in the context of the k-anonymity model. Furthermore, many algorithms for other models, such as ℓ-diversity, build upon algorithms for k-anonymization.

20.3.1 The k-Anonymity Model

The k-anonymity model is one of the oldest ones for data anonymization, and it is credited with the understanding of the concept of quasi-identifiers and their impact on data privacy. The basic idea in k-anonymization methods is to allow release of the sensitive attributes, while distorting only the attributes which are available through public sources of information. Thus, even though the sensitive attributes have been released, they cannot be linked to an individual through publicly available records. Before discussing the anonymization algorithms, some of the most common techniques for data distortion will be discussed.

1. *Suppression:* In this approach, some of the attribute values are suppressed. Depending on the algorithm used, the suppression can be done in a variety of ways. For example, one might omit *some of* the age *or* ZIP code attribute values from a few selected data records in Table 20.1. Alternatively, one might completely omit the entire record for a specific individual (row suppression) or the age attribute from all individuals (column suppression). Row suppression is often utilized to remove outlier records because such records are difficult to anonymize. Column suppression is commonly used to remove highly identifying attributes, or explicit identifiers, such as the SSN.

2. *Generalization:* In the case of generalization, the attributes are specified approximately in terms of a particular range. For example, instead of specifying Age = 26 and Location (ZIP Code) = 10547 for one of the entries of Table 20.1, one might *generalize* it to $Age \in [25, 30]$ and Location (State) = New York. By specifying the attributes approximately, it becomes more difficult for an adversary to perform linkage attacks. While numeric data can be generalized to specific ranges, the generalization of categorical data is somewhat more complicated. Typically, a generalization hierarchy of the categorical attribute values needs to be provided, for use in the anonymization process. For example, a ZIP code may be generalized to a city, which in turn may be generalized to a state, and so on. There is no unique way of specifying a domain hierarchy. Typically, it needs to be semantically meaningful, and it is specified by a domain expert as a part of the input to the anonymization process. An example of a generalization taxonomy of categorical attributes for the location attribute of Table 20.1 is provided in Fig. 20.1. This hierarchy of attribute values has a tree structure, and is referred to as a *value generalization hierarchy.* The notations $A_0 \ldots A_3$ and $Z_0 \ldots Z_4$ in Fig. 20.1 denote the domain generalizations at different levels of granularity. The corresponding *domain generalization hierarchies* are also illustrated in the Fig. 20.1 by the single path between $Z_0 \ldots Z_4$ and $A_0 \ldots A_4$.

3. *Synthetic data generation:* In this case, a synthetic data set is generated that mimics the statistical properties of the original data, at the group level. Such an approach can provide better privacy, because it is more difficult to map synthetic data records to particular groups of records. On the other hand, the data records are no longer *truthful* because they are synthetically generated.

4. *Specification as probabilistic and uncertain databases:* In this case, one might specify an individual data record as a probability distribution function. This is different from the

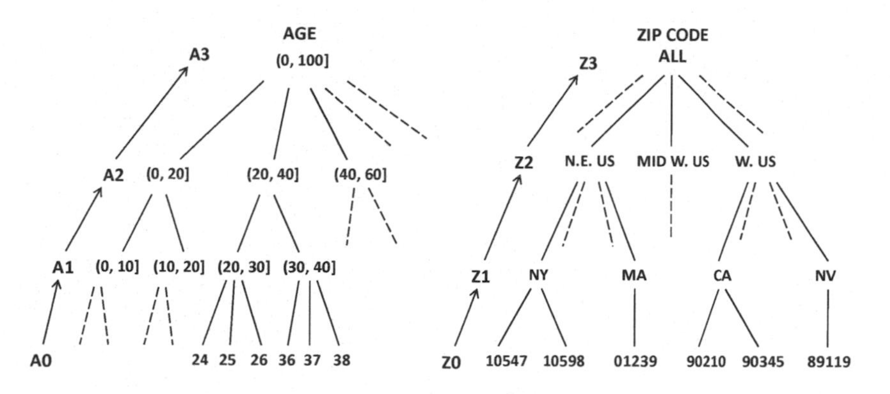

Figure 20.1: A value- and corresponding domain-generalization hierarchy for the age and ZIP code attributes

Table 20.3: Example of a 3-anonymized version of Table 20.1

Row Index	Age	ZIP Code	Disease
1	[20, 30]	Northeastern US	HIV
2	[30, 40]	Western US	Hepatitis C
3	[20, 30]	Northeastern US	HIV
4	[30, 40]	Western US	Hepatitis C
5	[30, 40]	Western US	Diabetes
6	[20, 30]	Northeastern US	HIV

aggregate distribution approach of randomization because the probability distribution is data-record specific, and is designed to ensure k-anonymity. While this approach has not been studied intensively, it has the potential to allow the use of recent advances in the field of probabilistic databases for anonymization.

Among the aforementioned methods, the generalization and suppression methods are most commonly used for anonymization. Therefore, most of the discussion in this section will be focused on these methods. First, the notion of k-anonymity will be defined.

Definition 20.3.1 (k-anonymity) *A data set is said to be k-anonymized, if the attributes of each record in the anonymized data set cannot be distinguished from at least $(k-1)$ other data records.*

This group of indistinguishable data records is also referred to as an *equivalence class*. To understand how generalization and suppression can be used for anonymization, consider the data set in Table 20.1. An example of a 3-anonymized version of this table is illustrated in Table 20.3. The SSN has been fully suppressed with column-wise suppression and replaced with an anonymized row index. Such explicit identifiers are almost always fully suppressed in anonymization. The two publicly available attributes corresponding to the age and ZIP code are now generalized and specified approximately. The subjects of the row indices 1, 3, and 6 can no longer be distinguished by using linkage attacks because their publicly available attributes are identical. Similarly, the publicly available attributes of row indices 2, 4, and 5 are identical. Thus, this table contains two *equivalence classes* containing three records each, and the data records cannot be distinguished from one another within these

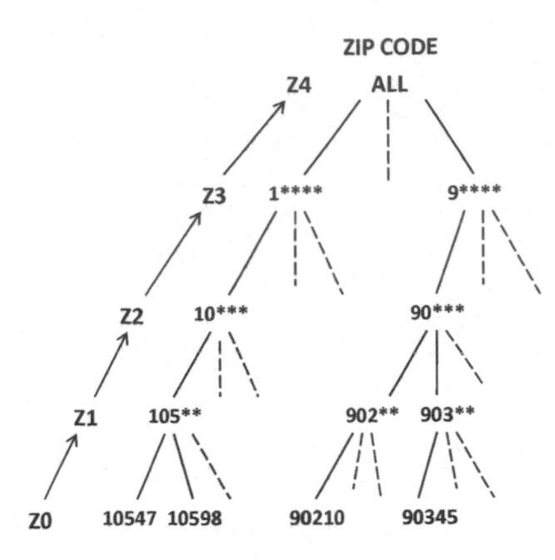

Figure 20.2: An alternate value- and corresponding domain-generalization hierarchy for the ZIP code attribute

equivalence classes. In other words, an adversary can no longer match the identification of individual data records with voter registration rolls *exactly*. If any matching is found, then it is guaranteed that at least $k = 3$ records in the data set will match any particular individual in the voter registration roll.

The ZIP code is generalized with the use of the prespecified value generalization hierarchy of Fig. 20.1. The generation of a domain generalization hierarchy for a categorical attribute can be done in several ways, and depends on the skill of the analyst responsible for the privacy modifications. An alternate example of a domain generalization hierarchy for the ZIP code attribute is illustrated in Fig. 20.2. A value generalization hierarchy on the continuous attributes does not require any special domain knowledge because it can be directly created by the analyst, using the actual distribution of the continuous values in the underlying data. This requires a simple hierarchical discretization of the continuous attributes.

The goal of the privacy-preservation algorithms is to replace the original values in the data (numeric or discrete), with one of the discrete values illustrated in the taxonomy trees of Fig. 20.1. Thus, the data is *recoded* in terms of a new set of discrete values. In most cases, the numeric attributes do retain their ordering, because the corresponding ranges are ordered. Different algorithms use different rules in the recoding process. These different ways of recoding attributes may be distinguished as follows:

- *Global versus local recoding:* In global recoding, a given attribute value is always replaced with the same discrete counterpart from the domain generalization hierarchy over all data records. Consider the aforementioned example of Fig. 20.1, in which ZIP code can be generalized either to state or region. In global recoding, the particular ZIP code value of 10547 needs to be *consistently* replaced by either *Northeastern US*, or *New York* over all the data records. However, for a different ZIP code such as 90210, a different level of hierarchy may be selected than for the 10547 value, as long as it is done consistently for a *particular* data value (e.g., 10547 or 90210) across all data records. In local recoding, different data records may use different generalizations for the same data value. For example, one data record might use *Northeastern US*, whereas another data record might use *New York* for 10547. While local recoding might *seem* to be better optimized, because of its greater flexibility, it does lose a different kind of information. In particular, because the same ZIP code might map to different values, such as *New York* and *Northeastern US*, the similarity computation between the resulting data records may be less accurate. Most of the current privacy schemes use global recoding.

- *Full-domain generalization:* Full-domain generalization is a special case of global recoding. In this approach, all values of a particular attribute are generalized to the same level of the taxonomy. For example, a ZIP code might be generalized to its state for all instances of the attribute. In other words, if the ZIP code 10547 is generalized to *New York*, then the ZIP code 90210 must be generalized to *California*. The various hierarchical alternatives for full-domain generalization of the age attribute are denoted by A_0, A_1, A_2, and A_3 in Fig. 20.1. The possible full-domain generalization levels of the ZIP code are denoted by Z_0, Z_1, Z_2, and Z_3. In this case, Z_3 represents the highest level of generalization (column suppression), and Z_0 represents the original values of the ZIP code attribute. Thus, once it is decided that the anonymization algorithm should use Z_2 for the ZIP code attribute, then *every instance of the ZIP code attribute* (Z_0) *in the data set is replaced with its generalized value in* Z_2. This is the reason that the approach is referred to as full-domain generalization, as the entire domain of data values for a particular attribute is generalized to the same level of the hierarchy. Full-domain generalization is the most common approach used in privacy-preserving data publishing.

Full-domain generalization is intuitively appealing because it ensures that the different values of an attribute have the same level of granularity throughout the data set. The earliest methods, such as Samarati's original algorithm, and *Incognito*, were all full-domain generalization algorithms.

20.3.1.1 Samarati's Algorithm

Samarati's algorithm was first proposed in the context of the definition of k-anonymity. Samarati's original AG-TS (*Attribute Generalization and Tuple Suppression*) algorithm for k-anonymity provides the basic domain generalization framework, which is the basis for group-based anonymization. It has already been discussed, how the domain generalization of a single attribute can be represented as a path. For example, the path from Z_0 to Z_3 in Fig. 20.1 represents the generalization of the ZIP code attribute. The notion of domain generalization can also be defined for *combinations* of attributes. However, in the case of attribute combinations, the relationships are no longer expressed as a path, but as a special kind of directed acyclic graph, known as a lattice. In this case, each node specifies a (full-domain) generalization level for the different attributes For example, $< A_1, Z_2 >$ denotes the domain generalization level of age to A_1 and ZIP code to Z_2. In other words, every data record is generalized to the level $< A_1, Z_2 >$. Note that $< A_1, Z_2 >$ also represents the generalization level of the (anonymized) Table 20.3 based on the domain-generalization hierarchies specified in Fig. 20.1.

Thus, each node in the lattice specifies a possible level of full-domain generalization, in terms of which the original data is represented. The edges in this graph represent the *direct* generalization relationships among these tuples of domains. A directed path in the lattice, from lower to higher levels, represents a sequence of generalizations. Conversely, a lower-level node is a specialization of a higher-level node. For example, the node $< A_1, Z_1 >$ is a direct specialization of either $< A_1, Z_2 >$, or $< A_2, Z_1 >$ because a single attribute in either can be specialized once to immediately yield $< A_1, Z_1 >$. An example of the domain generalization hierarchy for the age and ZIP code combination is illustrated in Fig. 20.3a. *The goal of the full-domain anonymization algorithm is to discover the node* $< A_i, Z_j >$ *in*

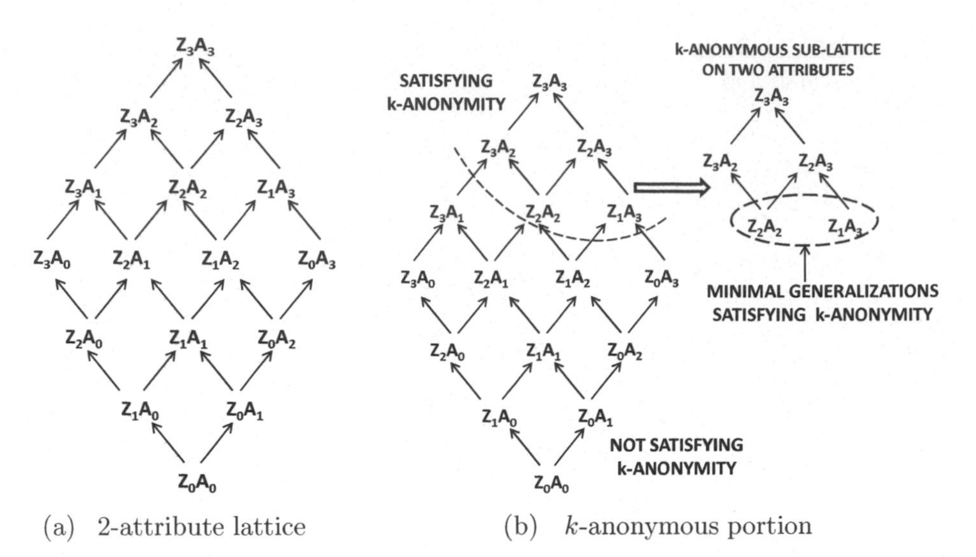

(a) 2-attribute lattice (b) k-anonymous portion

Figure 20.3: Domain generalization hierarchies over combinations of attributes

this tuple-based domain generalization hierarchy that preserves k-anonymity with the least amount of generalization. After such a node $< A_i, Z_j >$ has been discovered, the privacy algorithm generalizes all ages to the level A_i and all ZIP codes to the level Z_j.

In practice, some of the tuples may need to be suppressed in order to prevent undesirably high levels of generalization. This is because these may represent outlier tuples that cannot be incorporated in any group without significantly increasing the generalization level. For example, an individual with an age of 125 may need to be suppressed because of the outlier value of this attribute. Therefore, one of the parameters to the algorithm is a threshold $MaxSup$, which specifies the maximum number of tuples that can be suppressed. The goal is therefore to discover a node that is as low as possible in the lattice of Fig. 20.3a, such that k-anonymity is satisfied after suppressing at most $MaxSup$ tuples. The *height* of a node in the lattice is defined as its path distance in the lattice from the most specific level of representation. In the example of Fig. 20.3, the height of node $< Z_i, A_j >$ is $(i + j)$. A *minimally generalized node* may be defined as a node, for which the height is as small as possible. Therefore, in this example, one way of determining minimal generalizations, is to discover a k-anonymizable node $< Z_i, A_j >$, such that the height $(i + j)$ is as small as possible.

When there are d attributes $< Q_{i_1} \ldots Q_{i_d} >$, the sum $\sum_{k=1}^{d} i_k$ over all attributes represents the height of that particular combination of generalizations. It is easy to see that any specialization of a node $< Q_{i_1} \ldots Q_{i_d} >$ that does not satisfy k-anonymity will also not satisfy k-anonymity. Similarly, any generalization of a node satisfying k-anonymity will also satisfy k-anonymity. Therefore, the subgraph of the lattice satisfying k-anonymity and the subgraph violating k-anonymity are both connected subgraphs, and a border can be constructed between them. An example of such a border[1] is illustrated in Fig. 20.3b, and the corresponding minimal generalizations are illustrated in the same figure. Note that the minimal generalization is not unique, and that two possible minimal generalizations $< Z_2, A_2 >$ and $< Z_1, A_3 >$ are possible in this example. The reason for using minimally generalized nodes is to maximize the utility of the data for analytical algorithms. Other

[1]This border is for illustration purposes only, and does not correspond to any data set in this chapter.

more refined definitions can be used for quantifying utility that use the distribution of the attribute values more explicitly. The bibliographic notes contain pointers to some of these definitions.

Samarati's algorithm uses a simple binary search over the lattice of domain generalization tuples. Let $[0, h_{max}]$ represent the range of heights of the lattice. It is then checked whether any of the generalizations at level $h_{max}/2$ satisfies the k-anonymity constraint. If this is indeed the case, then the height $h_{max}/4$ is checked. Otherwise, the height $3 \cdot h_{max}/4$ is checked. This approach is repeated, until the lowest height at which a k-anonymous solution exists, is found. All the corresponding domain generalizations are reported, and any of these can be used for transforming the data. An important step in Samarati's algorithm is the process of using the original database to check whether a particular node in the lattice satisfies k-anonymity. However, a discussion of this step is omitted here, because similar steps are discussed below in the context of the *Incognito* algorithm.

20.3.1.2 Incognito

The lattice of Fig. 20.3 shares a number of conceptual similarities with the lattice of frequent itemset mining algorithms, as discussed in Chap. 4. Therefore, some of the anonymization algorithms for discovering full-domain generalization also have similar characteristics to those of frequent itemset mining algorithms. The *Incognito* algorithm leverages a number of principles from frequent pattern mining to efficiently discover the k-anonymous portion of the lattice.

An important observation is that the size of the lattice is exponentially related to the number of quasi-identifiers. This can lead to increasing computational complexity in many practical scenarios. While it has been shown by Meyerson and Williams [385] that optimal k-anonymization is NP-hard, it is possible to reduce the computational burden by careful exploration of the lattice. The *Incognito* algorithm is based on the observation that the k-anonymity of a subset of generalized attributes is a necessary (but not sufficient) condition for the k-anonymity of a superset of attributes with matching generalization levels of the common elements. Henceforth, this property will be referred to as ***attribute subset closure***. This property is a specific case of the *generalization property* which states that any generalization of a k-anonymous node in the lattice will always be k-anonymous.

These properties can be used to both generate candidates and prune the search process in a manner that is similar to the *Apriori* algorithm for frequent itemset mining. Therefore, nodes that are not k-anonymous with respect to a set of attributes, can be discarded, together with their specializations in the lattice hierarchy. Furthermore, generalizations of subsets of attributes that do satisfy the k-anonymity constraint, do not need to be checked because they are guaranteed to be k-anonymous.

The *Incognito* approach uses a levelwise approach, in which the following steps are repeated iteratively, until the k-anonymous sublattice containing all d attributes has been constructed. The set $\mathcal{F}_i$ denotes the set of all sublattices on i attributes that satisfies k-anonymity. The algorithm starts by initializing $\mathcal{F}_1$ to the portions of the single-attribute domain generalization hierarchies satisfying k-anonymity. This is quite simple, because single attribute hierarchies are paths. Thus, $\mathcal{F}_1$ is simply the top portion of the path, such that each generalized attribute value contains at least k tuples. Subsequently, as in frequent pattern mining, the algorithm repeatedly generates candidate sublattices in $\mathcal{C}_{i+1}$ by joining sublattices in $\mathcal{F}_i$ that have exactly $(i-1)$ attributes in common. The process of joining two sublattices will be described later. Note that $\mathcal{C}_{i+1}$ is a set of candidate sublattices on $(i+1)$ attributes. Each of these sublattices is then pruned of some of its nodes, using an

Apriori-style approach. Specifically, nodes of sublattices in $\mathcal{C}_{i+1}$ whose generalizations are not k-anonymous in $\mathcal{F}_i$ can be pruned. This step will also be described in detail later.

After candidate generation and pruning, the portion of each sublattice that satisfies k-anonymity is retained by checking the constituent nodes against the base data records. Thus, each sublattice in $\mathcal{C}_{i+1}$ reduces further in size. At this point, the set $\mathcal{C}_{i+1}$ has been transformed to the set $\mathcal{F}_{i+1}$. Thus, the following steps are repeated for increasing values of the index i:

1. Generate $\mathcal{C}_{i+1}$, the set of *candidate* sublattices on $(i + 1)$ attributes. This is achieved by joining all pairs of k-anonymous sublattices in $\mathcal{F}_i$ that share $(i-1)$ attributes. The details of a join between a pair of sublattices will be described later.

2. Prune the nodes from each sublattice in $\mathcal{C}_{i+1}$ that cannot possibly satisfy k-anonymity by using the *attribute subset closure* property with respect to the set of k-anonymous combinations in $\mathcal{F}_i$. The details of how the nodes may be pruned from a sublattice, will be described later.

3. Check each node in each (already pruned) sublattice of $\mathcal{C}_{i+1}$ against the base data, and remove those that do not satisfy k-anonymity. A node does not need to be checked, if one of its specializations already satisfies k-anonymity. This step transforms the set of candidate sublattices $\mathcal{C}_{i+1}$ to the set of k-anonymous sublattices $\mathcal{F}_{i+1}$ by removing the anonymity-violating sublattices.

If there are a total of d attributes, then the set $\mathcal{F}_d$ will contain a single sublattice of nodes satisfying k-anonymity. The nodes with the smallest height in this sublattice are reported. Note that the detailed implementation of the *Incognito* algorithm uses a slightly different approach for actually tracking the sublattices, by tracking the lattice nodes and edges in separate tables. The i-dimensional tables containing the generalization levels of lattice nodes of $\mathcal{F}_i$ are joined on their $(i-1)$ common attributes to create the $(i+1)$-dimensional tables containing the nodes of $\mathcal{C}_{i+1}$. Subsequently, the lattice edges are added between the generated nodes based on the hierarchy relationships. Nevertheless, the simpler logical description provided here matches the *Incognito* algorithm.

Next, the details of the join and pruning operations will be discussed with the use of an example. In this case, three attributes will be used for greater clarity. As discussed earlier, let A_r and Z_r represent different generalization levels of the age and ZIP code attributes, for varying values of the index r. Let P_r represent the generalization levels of an additional attribute corresponding to the profession. Higher values of the index r indicate a greater level of generalization. Consider the scenario where all three k-anonymous two-attribute sublattices on these three attributes are already available in $\mathcal{F}_2$. It is possible to use any pair of sublattices from these three possibilities, in order to perform the join. This will result in a *candidate* sublattice on all three attributes.

Consider the case, where the sublattices on (ZIP code, Age) and (ZIP code, Profession) are joined. The nodes in the new candidate sublattice will now have three attributes (ZIP code, Profession, Age) instead of two. The nodes for the new candidate sublattice are constructed by joining the nodes of the two k-anonymous sublattices. A pair of nodes $< Z_r, A_j >$ and $< Z_s, P_l >$ will be joined, if and only if $r = s$. In other words, the generalization level of the ZIP code attribute needs to be the same in both cases. This will result in the new node $< Z_r, P_l, A_j >$. In general, for pairs of nodes with k attributes, a join will be successfully executed, if and only if (a) they share $(k-1)$ attributes, and (b) the generalization levels of the $(k-1)$ common attributes are the same. An example of a join with two k-anonymous sublattices is illustrated in Fig. 20.4a.

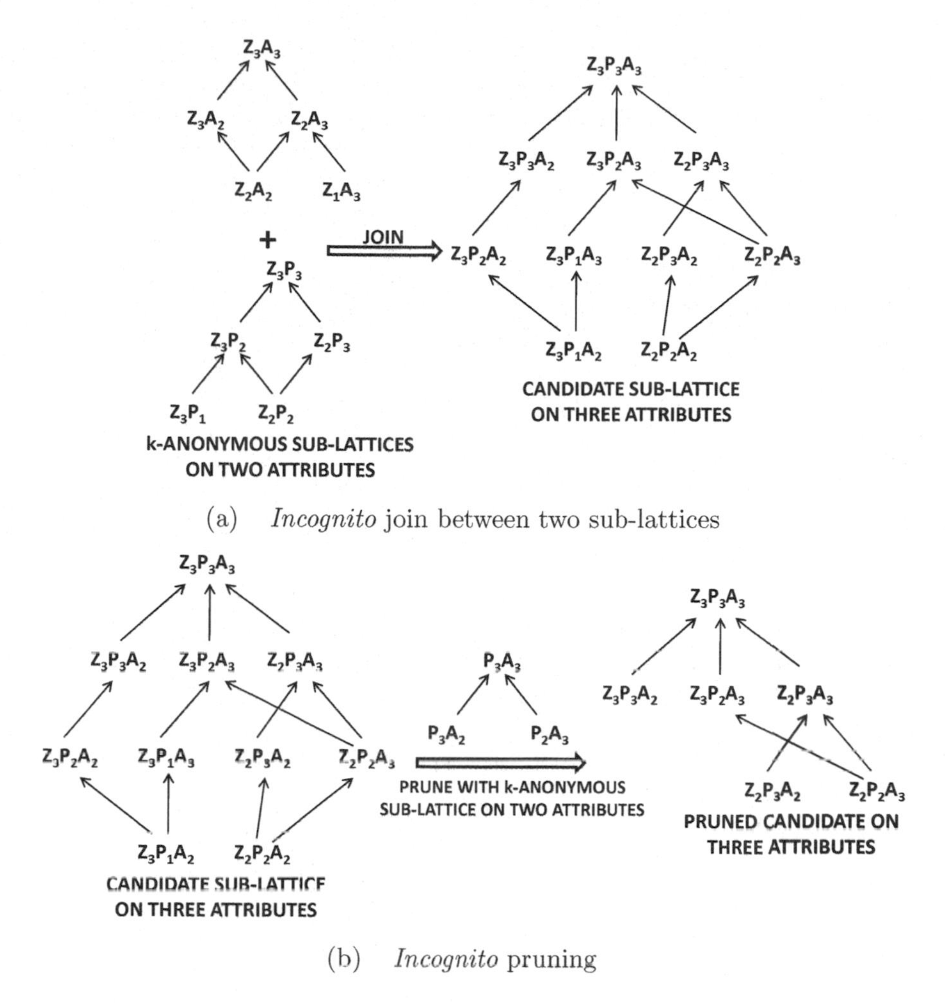

(a) *Incognito* join between two sub-lattices

(b) *Incognito* pruning

Figure 20.4: *Incognito* joins and pruning

In the previous example, the sublattice for the profession–age combination was not used for the join. However, it is still useful for pruning. This is because, if a node $< P_i, A_j >$ is not present in this sublattice, then any node of the form $< Z_m, P_i, A_j >$ will also not be k-anonymous. Therefore, such nodes can be removed from the constructed candidate sublattice together with their specializations. An example of a pruning step on the candidate sublattice is illustrated in Fig. 20.4b. This pruning is based on the attribute-subset closure property, and it is reminiscent of *Apriori* pruning in frequent itemset mining. As in the case of frequent itemset mining, all k-attribute subsets of each candidate $(k+1)$-sublattice in $\mathcal{C}_{k+1}$ need to be checked. If a node violates the closure property in any of these checks, then it is pruned.

Finally, the generated nodes in $\mathcal{C}_{k+1}$ need to checked against the original database to determine whether they satisfy k-anonymity. For example, in order to determine whether $< Z_1, A_1 >$ satisfies k-anonymity based on the value generalization in Fig. 20.1, one needs to determine the number of individuals satisfying each of the pairs of conditions such as (ZIP code $\in$ NY, $0 <$ Age ≤ 10), (ZIP code $\in$ NY, $10 <$ Age ≤ 20), (ZIP code $\in$ MA, $0 <$ Age ≤ 10), and so on. Therefore, for each node in the lattice, a vector of frequency values need to be computed. This vector is also referred to as a *frequency vector* or *frequency set.* The process of frequency vector computation can be expensive because the original database may need to be scanned to determine the number of tuples satisfying these conditions. However, several strategies can be used to reduce the burden of computation. For example, if the frequency vector of $< Z_1, A_1 >$ has already been computed, one can use *roll-up* to directly compute the frequency vectors of the generalization $< Z_2, A_1 >$ without actually scanning the database. This is because the frequency of the set (ZIP code $\in$ Northeastern US, $0 < \text{Age} \leq 10$) is the sum of the frequencies of (ZIP code $\in$ NY, $0 < \text{Age} \leq 10$), (ZIP code $\in$ NJ, $0 < \text{Age} \leq 10$), (ZIP code $\in$ MA, $0 < \text{Age} \leq 10$), and so on. The simplest approach is to use a breadth-first strategy on the lattice of each set of $(k+1)$ attributes, by determining the frequency vectors of specific (lower-level) nodes in the lattice before determining the frequency vectors of more general (higher-level) nodes. The frequency vectors of higher-level nodes can be computed efficiently from those of lower-level nodes by using the roll-up property.

Note that a separate breadth-first search needs to be performed for each subset of $(k+1)$ attributes in $\mathcal{C}_{k+1}$ to compute its frequency vectors. Furthermore, once a node has been identified by the breadth-first search to be k-anonymous, its generalizations in the lattice are guaranteed to be k-anonymous. Therefore, they are automatically marked as k-anonymous and are not explicitly checked. The original algorithm also supports a number of other optimizations, referred to as *Incognito super-roots* and *Bottom-up precomputation.* The bibliographic notes contain pointers to these methods.

20.3.1.3 Mondrian Multidimensional k-Anonymity

One of the disadvantages of the methods discussed so far is that the domain generalization hierarchies for various attributes are constructed independently as a preprocessing step. Thus, after the hierarchical discretization (domain generalization) for a numeric attribute has been fixed by the preprocessing step, it is utilized by the anonymization algorithm. This rigidity in the anonymization process creates inefficiencies in data representation, when the various data attributes are correlated in multidimensional space. For example, the salary distribution for older individuals may be different from that of younger individuals. A preprocessed domain generalization hierarchy is unable to adjust to such attribute correlations in the data set. In general, the best trade-offs between privacy and utility are achieved when the multidimensional relationships among data points are leveraged in the anonymization process. In other words, the attribute ranges for each attribute in a data point $\overline{X}$ should be generated in a dynamic way depending on the specific *multidimensional* locality of $\overline{X}$.

The *Mondrian* method generates multidimensional rectangular regions, containing at least k data points. This is achieved by recursively dividing the bounding boxes with axis-parallel cuts, until each region contains no more than k data points. This approach is not very different from the methodology used by many traditional index structures, such as kd-trees. An example of the partitioning induced by the *Mondrian* algorithm is illustrated in Fig. 20.5. In this case, a 5-anonymous partitioning is illustrated. Thus, each group contains *at least* five data points. It is easy to see that the same attribute value is represented by different ranges in different portions of the data, in order to account for the varying density of different regions. It is this flexibility that gives *Mondrian* a more compact representation of the anonymized groups than the other methods.

The *Mondrian* algorithm dynamically maintains the set $\mathcal{B}$ of multidimensional generalizations that satisfy k-anonymity and cover the data set. The *Mondrian* algorithm starts with a rectangular box B of all the data points. This represents the generalization

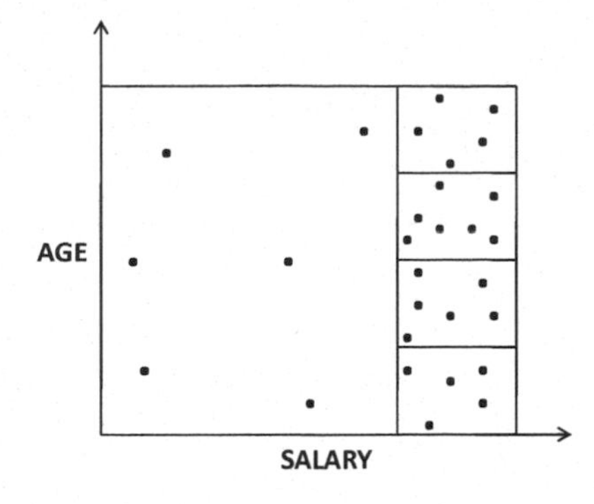

Figure 20.5: A sample 5-anonymous *Mondrian* multidimensional partitioning

of the entire data set to a single multidimensional region, and therefore trivially satisfies k-anonymity. The algorithm therefore starts by initializing $\mathcal{B} = \{B\}$. The algorithm repeatedly uses the following steps:

1. Select a rectangular region $R \in \mathcal{B}$ containing at least $2 \cdot k$ data points, such that a valid split into a pair of k-anonymous subsets exists.

2. Split the rectangular region R along any of the dimensions with an axis-parallel split, so that each of R_1 and R_2 contains at least k data points.

3. Update $\mathcal{B} \Leftarrow \mathcal{B} \cup \{R_1, R_2\} - R$

This iterative process is repeated, until the rectangular regions cannot be split any further without violating k-anonymity. There is some flexibility in the choice of the dimension for performing the split. A natural heuristic is to split the longest dimension of the selected rectangular region. After the dimension has been selected, the split should be performed so that the data points are partitioned as evenly as possible. In the absence of ties on attribute values, the data points can be divided almost equally into the two regions.

The rectangular regions in $\mathcal{B}$ define the equivalence classes that are utilized for k-anonymization. If each numeric attribute value is *unique*, it can be shown that every region will contain at most $2 \cdot k - 1$ data points. However, if there are ties among attribute values, and tied values need to be assigned to be the same partition, then an upper bound of $m + 2d \cdot (k - 1)$ can be shown on the number of data points in each partition. Here m is the number of identical copies of any data record. On the other hand, if ties on an attribute value can be flexibly assigned to any partition, then the maximum number of points in any rectangular partition, at the end of the process will be $2 \cdot k - 1$. The reader is referred to the bibliographic notes for the pointer to the proof of this bound. After the data has been divided into rectangular regions, the following approaches can be used for reporting the anonymized data points:

1. The averages along each dimension may be reported for each anonymized equivalence set.

2. The multidimensional bounding box of the data points may be reported.

The *Mondrian* algorithm has been shown to be more effective than the *Incognito* algorithm, because of the greater flexibility provided by the multidimensional approach to partitioning.

The *Mondrian* approach is naturally designed for numeric attributes with an ordering on the values. However, the approach can also be generalized to categorical attributes by designing appropriate split rules for the attributes.

20.3.1.4 Synthetic Data Generation: Condensation-Based Approach

The condensation-based approach generates synthetic data that matches the original data distribution, while maintaining k-anonymity. This means that k synthetic records are generated for each group of k records, by using the statistics of that group. The overall condensation approach may be described as follows:

1. Use any clustering approach to partition the data into groups of data records, such that each group contains at least k data records. Denote the number of created groups by m.

2. Compute the mean and covariance matrix for each group of data records. For a d-dimensional data set, the covariance matrix of a group represents the $d \times d$ covariances between pairs of attributes.

3. Compute the eigenvectors and eigenvalues of each covariance matrix. It is evident from the discussion of Principal Component Analysis (PCA) in Chap. 2, that the eigenvectors define a group-specific axis system, along which the data records are uncorrelated. The variance of the data along each eigenvector is equal to the corresponding eigenvalue. The synthetic data set to be generated, is modeled as mixture of m clusters, where the mean of each cluster is the mean of the corresponding group of original data records.

4. Generate synthetic data records for each of the m clusters. For each cluster, the number and mean of the synthetic records matches its base group. Data records are generated independently along the eigenvectors, with variance equal to the corresponding eigenvalues. The uniform distribution is typically used for synthetic data generation, because it is assumed that the data distribution does not change significantly within the small locality defined by a group. While the uniform distribution is a *local* approximation, the *global* distribution of the generated records generally matches the original data quite well.

The approach can also be generalized to data streams, by maintaining group statistics incrementally. The idea here is that group sizes are allowed to vary between k and $2 \cdot k - 1$. Whenever a group reaches the size of $2 \cdot k$, they are split into two groups. The details of group splitting will be discussed later.

To maintain the covariance statistics incrementally in the streaming scenario, an approach similar to the cluster-feature vector of *CluStream* (see Chap. 7) is used. The only difference is that the product-wise sum statistics are also maintained incrementally. For any pair of attributes i and j, the value of $\text{Sum}(i, j)$ is equal to sum of the product of attribute values i and j over the different data points. This can be easily maintained incrementally in a data stream. Then, for a set of $r \in (k, 2 \cdot k - 1)$ data points in a group, the covariance between attributes i and j may be estimated as follows:

$$\text{Covariance}(i, j) = \text{Sum}(i, j)/r - \text{Mean}(i) \cdot \text{Mean}(j) \tag{20.6}$$

$$\text{Covariance}(i, j) = \text{Sum}(i, j)/r - \text{Sum}(i) \cdot \text{Sum}(j)/r^2 \tag{20.7}$$

All the statistics in the aforementioned equation are additive, and can easily be maintained incrementally in the stream setting.

It remains to be explained how the groups are split, once the group sizes reach $2 \cdot k$. It is assumed that each group of size $2 \cdot k$ is split into two groups of size k along the longest eigenvector. The reason for choosing the longest eigenvector is to ensure the compactness of the newly created groups. The splitting of groups can be a challenge, because the original data records are not available in the streaming scenario to recalculate the statistics of each of the split groups. Therefore, an approximation (i.e., modeling assumption) is needed. The condensation approach works with the modeling assumption that the data records of a group are independently distributed along each eigenvector according to a uniform distribution. For group sizes that are much smaller than the number of points in the data set, this is not an unreasonable assumption. This is because density distributions do not change drastically over small regions of the data.

This modeling assumption of a uniform distribution is used to re-calculate the new means of each of the child groups of equal size k. This is because the range of the uniform distribution along the longest eigenvector can be approximated from its variance (eigenvalue), based on the modeling assumption. Note that the variance of a uniform distribution is one twelfth the square of its range. Therefore, if λ_{max} be the largest eigenvalue, then the range R of the uniform distribution is computed as follows:

$$R = \sqrt{12\lambda_{max}} \tag{20.8}$$

This range R is then split into two equal parts to create the two new group means. Thus, the two new group means are at a distance of $R/4$ from the old group mean in opposite directions along the longest eigenvector.

The newly created groups are assumed to have the same eigenvectors as the parent group, because the splitting is performed along an uncorrelated direction. Therefore, the directions of correlation are not assumed to change after splitting. The largest eigenvalue of the original (parent) group is replaced by an eigenvalue in each of the child groups, which is one fourth[2] the original value. Thus, if P is the $d \times d$ matrix with orthonormal columns containing the eigenvectors, and Σ is the diagonal matrix of eigenvalues (after adjustment of the largest eigenvalue), then the covariance matrix of the newly created split groups can be computed as follows:

$$C = P\Sigma P^T \tag{20.9}$$

This relationship is based on the standard PCA diagonalization discussed in Chap. 2. Note that the covariance matrices of both the split groups are the same. The covariance matrices and newly generated group means can be used to back-calculate the sum of pairwise attribute products of each group according to Eq. 20.6. Thus, as more data points arrive, these product values can continue to be updated incrementally.

The condensation-based approach is one of the few methods that can be applied to data streams with a relatively low risk of disclosure, because of its approach of using synthetic data. It is often difficult for an adversary to know which group of k synthetic records was generated from a particular base group of original records. In the case of generalization-based anonymization, it is relatively easy to identify groups of related data records, representing equivalence classes. Thus, synthetic data sets provide some additional privacy protection. Note that it is possible to generate larger data sets using this approach if needed. For example, for each group of k records, one might generate $\alpha \cdot k$ synthetic data records,

---

[2]Splitting a uniform distribution into two equal parts reduces its variance by a factor of 4.

using the statistics of that group. This scales up the size of the data with a factor of α, and further reduces the mapping between the generated data and the original data. Furthermore, additional noise can be incorporated during synthetic data generation to ensure greater protection.

These additional options do come at a price. The *truthfulness* of the published data is lost. The published data records are *synthetic* and therefore do not map onto any particular individual. In many aggregation- or modeling-based applications, this is not necessarily an issue, because the aggregate properties of the data are retained. In some medical data handling scenarios, legal restrictions may prohibit release of downgraded data, when there is a direct mapping between individuals and data records, even at a group level. The condensation approach provides a solution in some of these scenarios, because the released data records are synthetic, and are generally difficult to map onto specific groups.

The condensation approach shares a number of conceptual similarities with the *Mondrian* approach, except that it allows the use of any constrained clustering algorithm, rather than rectangular partitions constructed with single dimensional cuts. The utility of the resulting anonymization depends on the effectiveness of the clustering. Single dimensional cuts will not be able to construct high-quality clusters with increasing dimensionality. Furthermore, unlike *Mondrian*, synthetic data is generated to achieve greater anonymity.

The condensation approach does not distinguish between publicly available attributes (used in combination to construct quasi-identifiers) and sensitive attributes, and applies the approach to all the attributes. As will be evident from the subsequent discussion on the dimensionality curse in Sect. 20.3.4, the distinction between quasi-identifier and sensitive attributes is more fluid, than is often assumed in the literature on data privacy. Because it is not possible to know the level of background knowledge available to adversaries about the sensitive attributes, all attributes should be perturbed. When the sensitive attributes are released without any perturbation, they become immediately available for identification attacks, as long as background knowledge is available. For example, a number of privacy attacks on data sets such as the Netflix data set [402], have been performed using attributes that would normally not have been considered publicly available. This work [402] also makes the argument that such strong distinctions between publicly available and sensitive attributes are dangerous to make in real-world settings where the data and background knowledge available to the public continues to increase over time.

20.3.2 The ℓ-Diversity Model

While the k-anonymity model provides the basic framework for privacy-preserving data publishing, there are scenarios in which it can lead to inadvertent sensitive attribute disclosure. Consider the 3-anonymized table illustrated in Table 20.3. In this case, the row indices 1, 3, and 6 are in the same anonymized group, and cannot be distinguished from one another. However, all three individuals have the value of "HIV" on the sensitive attribute. Therefore, even though the *identity* of the specific individual from this group cannot be inferred, it can be inferred that any individual in this group has HIV. Therefore, if a voter registration roll is used to join this group to three unique individuals, then it can be inferred that all three of them have HIV. This represents a breach of sensitive *attribute* information about each of these three individuals. In other words, while the k-anonymity model prevents *identity disclosure*, it does not prevent *attribute disclosure*.

The main reason for this breach is that the sensitive information is not diverse enough within the anonymized groups. Since the goal of privacy-preserving data publishing is to prevent the revelation of sensitive information, a model that does not use the sensitive attribute values within the group formation process, cannot achieve this goal. The ℓ-diversity model is designed to ensure that the sensitive attributes within an equivalence class are sufficiently diverse.

Definition 20.3.2 (ℓ-diversity Principle) *An equivalence class is said to be ℓ diverse, if it contains ℓ "well-represented" values for the sensitive attribute. An anonymized table is said to be ℓ-diverse, if each equivalence class in it is ℓ-diverse.*

It is important to note that the notion of "well represented" can be instantiated in several different ways. Therefore, the aforementioned definition provides the basic *principle* behind this approach, but cannot be considered a hard definition. There are several ways in which the notion of "well-represented" can be instantiated. These correspond to the notions of entropy ℓ-diversity and recursive ℓ-diversity. These definitions are described below.

Definition 20.3.3 (Entropy ℓ-diversity) *Let $p_1 \ldots p_r$ be the fraction of the data records belonging to different values of the sensitive attribute in an equivalence class. The equivalence class is said to be entropy ℓ-diverse, if the entropy of its sensitive attribute value distribution is at least $log(\ell)$.*

$$-\sum_{i=1}^{r} p_i \cdot log(p_i) \geq log(\ell) \tag{20.10}$$

An anonymized table is said to satisfy entropy ℓ-diversity, if each equivalence class in it satisfies entropy ℓ-diversity.

It can be shown that the sensitive attributes in an equivalence class must have at least ℓ distinct values for the table to be ℓ-diverse (see Exercise 7). Therefore, any ℓ-diverse group has at least ℓ elements, and is ℓ-anonymous as well.

One problem with this definition of ℓ-diversity is that it may be too restrictive in many settings, especially when the distributions of the sensitive attribute values are uneven. The entropy of a table can be shown to be at least equal to the minimum entropy of the constituent equivalence classes into which it is partitioned (see Exercise 8). Therefore, to ensure ℓ-diversity of each equivalence class, the sensitive attribute distribution in the entire table must also be ℓ-diverse. This is a restrictive assumption in many settings, because most real distributions of sensitive attributes are very skewed. For example, in a medical application, the sensitive (disease) attribute is likely to have uneven frequencies between normal individuals and various diseases. Greater attribute skew reduces the (global) entropy ℓ-diversity of the sensitive-attribute distribution across the entire table. When this global ℓ-diversity is less than ℓ, it is no longer possible to create a globally ℓ-diverse partition without suppressing many data records.

Therefore, a more relaxed notion of recursive (c, ℓ)-diversity has been proposed. The basic goal of the definition is to ensure that the most frequent attribute value in an equivalence class does not dominate the less frequent sensitive values in it. An additional parameter c is used to control the relative frequency of the different values of the sensitive attribute within an equivalence class.

Definition 20.3.4 (Recursive (c, ℓ)-diversity) *Let $p_1 \ldots p_r$ be the fraction of the data records belonging to the r different values of the sensitive attribute in an equivalence class, such that $p_1 \geq p_2 \geq \ldots \geq p_r$. The equivalence class satisfies recursive (c, ℓ)-diversity, if the following is true:*

$$p_1 < c \cdot \sum_{i=\ell}^{r} p_i \tag{20.11}$$

An anonymized table is said to satisfy recursive (c, ℓ)-diversity, if each equivalence class in it satisfies entropy (c, ℓ)-diversity.

The idea is that the least frequent tail of the sensitive attribute values must contain sufficient cumulative frequency compared to the most frequent sensitive attribute value. The value of r has to be at least ℓ, for the right-hand side of the aforementioned relationship to be non-zero.

A key property of ℓ-diversity is that any generalization of an ℓ-diverse table is also ℓ-diverse. This is true for both definitions of ℓ-diversity.

Lemma 20.3.1 (Entropy ℓ-diversity monotonicity) *If a table is entropy ℓ-diverse, then any generalization of the table is entropy ℓ-diverse as well.*

Lemma 20.3.2 (Recursive (c, ℓ)-diversity monotonicity) *If a table is recursive (c, ℓ)-diverse, then any generalization of the table is recursive (c, ℓ)-diverse as well.*

The reader is advised to work out Exercises 9(a) and (b), which are related to these results. Thus, ℓ-diversity exhibits the same monotonicity property exhibited by k-anonymity algorithms. This implies that the algorithms for k-anonymity can be easily generalized to ℓ-diversity by making minor modifications. For example, both Samarati's algorithm and the *Incognito* algorithm can be adapted to the ℓ-diversity definition. The only change to any k-anonymity algorithm is as follows. Every time a table is tested for k-anonymity, it is now tested for ℓ-diversity instead. Therefore, algorithmic development of ℓ-diverse anonymization methods is typically executed by simply adapting existing k-anonymization algorithms.

20.3.3 The t-closeness Model

While the ℓ-diversity model is effective in preventing direct inference of sensitive attributes, it does not fully prevent the gain of *some* knowledge by an adversary. The primary reason for this is that ℓ-diversity does not account for the distribution of the sensitive attribute values in the original table. For example, the entropy of a set of sensitive attribute values with relative frequencies $p_1 \ldots p_r$ will take on the maximum value when $p_1 = p_2 = \ldots = p_r = 1/r$. Unfortunately, this can often represent a serious breach of privacy, when there is a significant skew in the original distribution of sensitive attribute values. Consider the example of a medical database of HIV tests, where the sensitive value takes on the two values of "HIV" or "normal," with relative proportions of $1 : 99$. In this case, a group with an equal distribution of HIV and normal patients will have the highest entropy, based on the ℓ-diversity definition.

Unfortunately, such a distribution is highly revealing when the distribution of the sensitive values in the *original data* is taken into account. Sensitive values are usually distributed in a skewed way, across most real data sets. In the medical example discussed above, it is already known that only 1 % of the patients in the *entire data set* have HIV. Thus, the equal distribution of HIV-infected and normal patients within a group, provides a significant *information gain* to the adversary. *The adversary now knows, that this small group of patients has a much higher expected chance of having HIV, than the base population.*

In this context, a notion of *Bayes optimal privacy* exists, which ensures that the additional posterior information gained after release of information is as small as possible. Unfortunately, the notion of Bayes optimal privacy is practically and computationally difficult to implement. The t-closeness model may be viewed as a practical and heuristic approach that attempts to achieve similar goals as the notion of Bayes optimal privacy. This is achieved by using the distance functions between distributions. Informally, the goal is to create an anonymization, such that the distance between the sensitive attribute distributions of each anonymized group and the base data is bounded by a user-defined threshold.

Definition 20.3.5 (t-closeness Principle) *Let $\overline{P} = (p_1 \dots p_r)$ be a vector representing the fraction of the data records belonging to the r different values of the sensitive attribute in an equivalence class. Let $\overline{Q} = (q_1 \dots q_r)$ be the corresponding fractional distributions in the full data set. Then, the equivalence class is said to satisfy t-closeness, if the following is true, for an appropriately chosen distance function $Dist(\cdot,\cdot)$:*

$$Dist(\overline{P}, \overline{Q}) \leq t \tag{20.12}$$

An anonymized table is said to satisfy t-closeness, if all equivalence classes in it satisfy t-closeness.

The previous definition does not specify any particular distance function. There are many different ways to instantiate the distance function, depending on application-specific goals. Two common instantiations of the distance function are as follows:

1. *Variational distance:* This is simply equal to half the Manhattan distance between the two distribution vectors:

$$Dist(P, Q) = \frac{\sum_{i=1}^{r} |p_i - q_i|}{2} \tag{20.13}$$

2. *Kullback-Leibler (KL) distance:* This is an information-theoretic measure that computes the difference between the cross-entropy of $(\overline{P}, \overline{Q})$, and the entropy of $\overline{P}$.

$$Dist(\overline{P}, \overline{Q}) = \sum_{i=1}^{r} (p_i \cdot \log(p_i) - p_i \cdot \log(q_i)) \tag{20.14}$$

   Note that the entropy of the first distribution is $-\sum_{i=1}^{r} p_i \cdot \log(p_i)$, whereas the cross-entropy is $-\sum_{i=1}^{r} p_i \cdot \log(q_i)$.

While these are the two most common distance measures used, other distance measures can be used in the context of different application-specific goals.

For example, one may wish to prevent scenarios in which a particular equivalence class contains semantically related sensitive attribute values. Consider the scenario, where a particular equivalence class contains diseases such as gastric ulcer, gastritis, and stomach cancer. In such cases, if a group contains only these diseases, then it provides significant information about the sensitive attribute of that group. The t-closeness method prevents this scenario by changing the distance measure, and taking the distance between different *values* of the sensitive attribute into account in the distance-computation process. In particular, the *Earth Mover Distance* can be used effectively for this scenario.

The *earth mover's distance (EMD)* is defined in terms of the "work" (or *cost*) required to transform one distribution to the other, if we allow sensitive attribute values in the original data to be flipped. Obviously, it requires less "work" to flip a sensitive value to a semantically similar value. Formally, let d_{ij} be the amount of "work" required to transform the ith sensitive value to the jth sensitive value, and let f_{ij} be the fraction of data records which are flipped from attribute value i to attribute value j. The values of d_{ij} are provided by a domain expert. Note that there are many different ways to flip the distribution $(p_1 \dots p_r)$ to the distribution $(q_1 \dots q_r)$, and it is desired to use the least cost sequence of flips to compute the distance between $\overline{P}$ and $\overline{Q}$. For example, one would rather flip "gastric ulcer" to "gastritis" rather than flipping "HIV" to "gastritis" because the former is likely to have lower cost. Therefore, f_{ij} is a variable in a linear programming optimization problem, which is constructed to minimize the overall cost of flips. For a table with r distinct sensitive attribute values, the cost of flips is given by $\sum_{i=1}^{r} \sum_{j=1}^{r} f_{ij} \cdot d_{ij}$. The earth mover's distance may be posed as an optimization problem that minimizes this objective function subject to constraints on the aggregate flips involving each sensitive attribute value. The constraints ensure that the aggregate flips do transform the distribution $\overline{P}$ to $\overline{Q}$.

$$
\begin{aligned}
Dist(\overline{P}, \overline{Q}) = \text{Minimize } & \sum_{i=1}^{r} \sum_{j=1}^{r} f_{ij} \cdot d_{ij} \\
& \text{subject to:} \\
& p_i - \sum_{j=1}^{r} f_{ij} + \sum_{j=1}^{r} f_{ji} = q_i \qquad \forall i \in \{1 \dots r\} \\
& f_{ij} \geq 0 \qquad \forall i, j \in \{1, \dots r\}
\end{aligned}
$$

The earth mover's distance has certain properties that simplify the computation of generalizations satisfying t-closeness.

Lemma 20.3.3 *Let E_1 and E_2 be two equivalence classes, and let $\overline{P_1}$, $\overline{P_2}$ be their sensitive attribute distributions. Let $\overline{P}$ be the distribution of $E_1 \cup E_2$, and $\overline{Q}$ be the global distribution of the full data set. Then, it can be shown that:*

$$
Dist(\overline{P}, \overline{Q}) \leq \frac{|E_1|}{|E_1| + |E_2|} \cdot Dist(\overline{P_1}, \overline{Q}) + \frac{|E_2|}{|E_1| + |E_2|} \cdot Dist(\overline{P_2}, \overline{Q}) \tag{20.15}
$$

This lemma is a result of the fact that the optimal objective function of a linear programming formulation is convex, and $\overline{P}$ can be expressed as a convex linear combination of $\overline{P_1}$ and $\overline{P_2}$ with coefficients $\frac{|E_1|}{|E_1|+|E_2|}$ and $\frac{|E_2|}{|E_1|+|E_2|}$, respectively. This convexity result also implies the following:

$$
Dist(\overline{P}, \overline{Q}) \leq \max\{Dist(\overline{P_1}, \overline{Q}), Dist(\overline{P_2}, \overline{Q})\}
$$

Therefore, when two equivalence classes satisfying t-closeness are merged, the merged equivalence class will also satisfy t-closeness. This implies the monotonicity property for t-closeness.

Lemma 20.3.4 (t-closeness monotonicity) *If a table satisfies t-closeness, then any generalization of the table satisfies t-closeness as well.*

The proof of this lemma follows from the fact that the generalization A of any table B, contains equivalence classes that are the union of equivalence classes in B. If each equivalence class in B already satisfies t-closeness, then the corresponding union of these equivalence classes must satisfy t-closeness. Therefore, the generalized table must also satisfy t-closeness. This monotonicity property implies that all existing algorithms for k-anonymity can be directly used for t-closeness. The k-anonymity test is replaced with a test for t-closeness.

20.3.4 The Curse of Dimensionality

As discussed at various places in this book, the curse of dimensionality causes challenges for many data mining problems. Privacy preservation is also one of the problems affected by the curse of dimensionality. There are two primary ways in which the curse of dimensionality impacts the effectiveness of anonymization algorithms:

1. *Computational challenges:* It has been shown [385], that optimal k-anonymization is NP-hard. This implies that with increasing dimensionality, it becomes more difficult to perform privacy preservation. The NP-hardness result also applies to the ℓ-diversity and t-closeness models, using a very similar argument.

2. *Qualitative challenges:* The qualitative challenges to privacy preservation are even more fundamental. Recently, it has been shown that it may be difficult to perform effective privacy preservation without losing the utility of the anonymized data records. This is an even more fundamental challenge, because it makes the privacy-preservation process less practical. The discussion of this section will be centered on this issue.

In the following, a discussion of the qualitative impact of the dimensionality curse on group-based anonymization methods will be provided. While a formal mathematical proof [10] is beyond the scope of this book, an intuitive version of the argument is presented. To understand why the curse of dimensionality increases the likelihood of breaches, one only needs to understand the well-known notion of *high dimensional data sparsity*. For ease in understanding, consider the case of numeric attributes. A generalized representation of a table can be considered a rectangular region in d-dimensional space, where d is the number of quasi-identifiers. Let $F_i \in (0, 1)$ be the fraction of the range of dimension i covered by a particular generalization. For the anonymized data set to be useful, the value of F_i should be as small as possible. However, the fractional volume of the space, covered by a generalization with fractional domain ranges of $F_1 \ldots F_d$, is given by $\prod_{i=1}^{d} F_i$. This fraction converges to 0 exponentially fast with increasing dimensionality d. As a result, the fraction of data points within the volume also reduces rapidly, especially if the correlations among the different dimensions are weak. For large enough values of d, it will be difficult to create d-dimensional regions containing at least k data points, unless the values of F_i are chosen to be close to 1. In such cases, any value of an attribute is generalized to almost the entire range of values. Such a highly generalized data set therefore loses its utility for data mining purposes. This general principle has also been shown to be true for other privacy models, such as perturbation, and ℓ-diversity. The bibliographic notes contain pointers to some of these theoretical results.

A real-world example of this scenario is the Netflix Prize data set, in which Netflix released ratings of individuals [559] for movies to facilitate the study of collaborative filtering algorithms. Many other sources of data could be found, such as the Internet Movie Database (IMDb), from which the ratings information could be matched with the Netflix prize data set. It was shown that the identity of users could be breached with a very high level of accuracy, as the number of ratings (specified dimensionality) increased [402]. Eventually, Netflix retracted the data set.

20.4 Output Privacy

The privacy-preservation process can be applied at any point in the data mining pipeline, starting with data collection, publishing, and finally, the actual application of the data mining process. The *output* of data mining algorithms can be very informative to an adversary. In particular, data mining algorithms with a large output and exhaustive data descriptions are particularly risky in the context of disclosure. For example, consider an association rule mining algorithm, in which the following rule is generated with high confidence:

$$(\text{Age} = 26, \text{ZIP Code} = 10562) \Rightarrow \text{HIV}$$

This association rule is detrimental to the privacy of an individual satisfying the condition on the left hand side of the aforementioned rule. Therefore, the discovery of this rule may result in the unforseen disclosure of private information about an individual. In general, many databases may have revealing relationships among subsets of attributes because of the constraints and strong statistical relationships between attribute values.

The problem of association rule hiding may be considered a variation of the problem of statistical disclosure control, or database inference control. In these problems, the goal is to prevent inference of sensitive values in the database from other related values. However, a crucial difference does exist between database inference control and association rule hiding. In database inference control, the focus is on hiding some of the entries, so that the privacy of other entries is preserved. In association rule hiding, the focus is on hiding the rules themselves, rather than the entries. Therefore, the privacy preservation process is applied on the output of the data mining algorithm, rather than the base data.

In association rule mining, a set of sensitive rules are specified by the system administrator. The task is to mine all association rules, such that none of the sensitive rules are discovered, but all nonsensitive rules are discovered. Association rule hiding methods are either *heuristic methods*, *border-based methods*, or *exact methods*. In the first class of methods, a subset of transactions are removed from the data. The association rules are discovered on the set of sanitized transactions. In general, if too many transactions are removed, then the remaining nonsensitive rules, which are discovered, will not reflect the true set of rules. This may lead to the discovery of rules that do not reflect the true patterns in the underlying data. In the case of *border-based methods*, the border of the frequent pattern mining algorithm is adjusted, so as to discover only nonsensitive rules. Note that when the borders of the frequent itemsets are adjusted, it will lead to the exclusion of nonsensitive rules along with the sensitive rules. The last class of problems formulates the hiding process as a constraint satisfaction problem. This formulation can be solved using integer programming. While these methods provide exact solutions, they are much slower, and their use is limited to problems of smaller size.

A related problem in output privacy is that of *query auditing*. In query auditing, the assumption is that users are allowed to issue a sequence of queries to the database. However, the response to one or more queries may sometimes lead to the compromising of sensitive information about smaller sets of individuals. Therefore, the responses to some of the queries are withheld (or *audited*) to prevent undesirable disclosure. The bibliographic notes contain specific pointers to a variety of query auditing and association rule hiding algorithms.

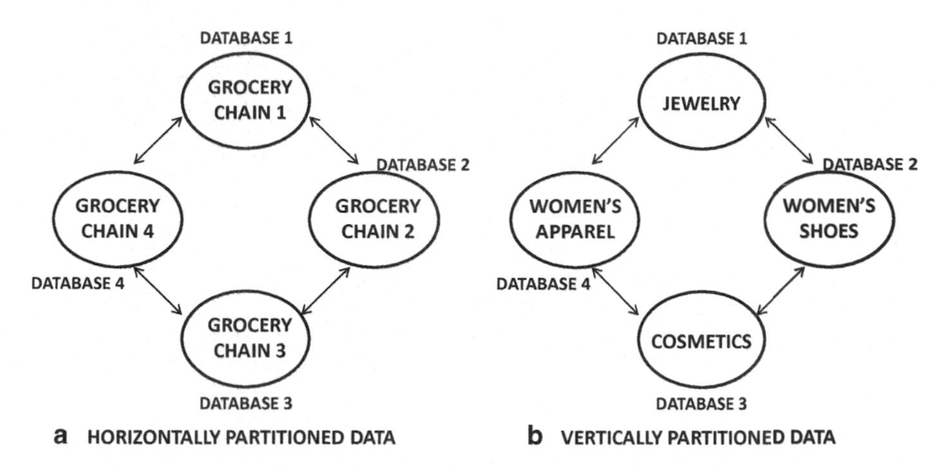

Figure 20.6: Examples of horizontally and vertically partitioned data

20.5 Distributed Privacy

In distributed privacy-preserving data mining, the goal is to mine shared insights across multiple participants owning different portions of the data, without compromising the privacy of local statistics or data records. The key is to understand that the different participants may be partially or fully adversaries/competitors, and may not wish to provide full access of their *local* data and statistics to one another. However, they might find it mutually beneficial to extract *global* insights over all the data owned by them.

The data may be partitioned either *horizontally* or *vertically* across the different participants. In horizontal partitioning, the data records owned by different adversaries have the same attributes, but different adversaries own different portions of the database. For example, a set of supermarket chains may own similar data related to customer buying behavior, but the different stores may show somewhat different patterns in their transactions because of factors specific to their particular business. In *vertical* partitioning, the different sites may contain different attributes for the same individual. For example, consider a scenario in which a database contains transactions by various customers. A particular customer may buy different kinds of items at stores containing complementary products such as jewelery, apparel, cosmetics, etc. In such cases, the aggregate association analysis across different participants can provide insights, that cannot be inferred from any particular database. Examples of horizontal and vertically partitioned data are provided in Figs. 20.6a and b, respectively.

At the most primitive level, the problem of distributed privacy-preserving data mining overlaps closely with a field in cryptography for determining secure multi-party computations. In this field, functions are computed over inputs provided by multiple recipients without actually sharing the inputs with one another. For example, in a two-party setting, Alice and Bob may have two inputs x and y, respectively, and may wish to compute the function $f(x, y)$ without revealing x or y to each other. This problem can also be generalized across k parties for computing the k argument function $h(x_1 \dots x_k)$. Many data mining algorithms may be viewed in the context of repetitive computations of primitive functions such as the scalar dot product, secure sum, secure set union, etc. For example, the scalar dot product of the binary representation of an itemset and a transaction can be used to determine whether or not that itemset is supported by that transaction. Similarly, scalar dot products can be used for similarity computations in clustering. To compute the function $f(x, y)$ or $h(x_1 \ldots, x_k)$, a *protocol* needs to be designed for exchanging information in such a way that the function is computed without compromising privacy.

A key building-block for many kinds of secure function evaluations is the 1 out of 2 oblivious-transfer protocol. This protocol involves two parties: a *sender*, and a *receiver*. The sender's input is a pair (x_0, x_1), and the receiver's input is a bit value $\sigma \in \{0, 1\}$. At the end of the process, the receiver learns x_σ only, and the sender learns nothing. In other words, the sender does not learn the value of σ.

In the oblivious transfer protocol, the sender generates two encryption keys, K_0 and K_1, but the protocol is able to ensure that the receiver knows only the decryption key for K_σ. The sender is able to generate these keys by using an encrypted input from the receiver, which encodes σ. This coded input does not reveal the value of σ to the sender, but is sufficient to generate K_0 and K_1. The sender encrypts x_0 with K_0, x_1 with K_1, and sends the encrypted data back to the receiver. At this point, the receiver can only decrypt x_σ, since this is the only input for which he or she has the decryption key. The 1 out of 2 oblivious transfer protocol has been generalized to the case of k out of N participants.

The oblivious transfer protocol is a basic building block, and can be used in order to compute several data mining primitives related to vector distances. Another important protocol that is used by frequent pattern mining algorithms is the *secure set union protocol*. This protocol allows the computation of unions of sets in a distributed way, without revealing the actual sources of the constituent elements. This is particularly useful in frequent pattern mining algorithms, because the locally large itemsets at the different sites need to be aggregated. The key in these methods is to disguise the frequent patterns at each site with enough number of fake itemsets, in order to disguise the true locally large itemsets at each site. Furthermore, it can be shown that this protocol can be generalized to compute different kinds of functions for various data mining problems on both horizontally and vertically partitioned data. The bibliographic notes contain pointers to surveys on these techniques.

20.6 Summary

Privacy-preserving data mining can be executed at different stages of the information processing pipeline, such as data collection, data publication, output publication, or distributed data sharing. The only known method for privacy protection at data collection, is the randomization method. In this method, additive noise is incorporated in the data at data collection time. The aggregate reconstructions of the data are then used for mining.

Privacy-preserving data publishing is typically performed using a group-based approach. In this approach, the sensitive attributes are treated in a different way from the attributes that are combined to construct quasi-identifiers. Only the latter types of attributes are perturbed, in order to prevent identification of the subjects of the data records. Numerous models, such as k-anonymity, ℓ-diversity, and t-closeness are used for anonymization. The eventual goal of all these methods is to prevent the release of sensitive information about individuals. When the dimensionality of the data increases, privacy preservation becomes very difficult, without a complete loss of utility.

In some cases, the output of data mining applications, such as association rule mining and query processing, may lead to release of sensitive information. Therefore, in many cases, the output of these applications may need to be restricted in to prevent the release of sensitive information. Two such well known techniques are association rule hiding, and query auditing.

In distributed privacy, the goal is to allow adversaries or semi-adversaries to collaborate in the sharing of data, for global insights. The data may be vertically partitioned across columns, or horizontally partitioned across rows. Cryptographic protocols are typically used in order to achieve this goal. The most well-known among these is the oblivious transfer protocol. Typically, these protocols are used to implement primitive data mining operations, such as the dot product. These primitive operations are then leveraged in data mining algorithms.

20.7 Bibliographic Notes

The problem of privacy-preserving data mining has been studied extensively in the statistical disclosure control and security community [1, 512]. Numerous methods, such as swapping [181], micro-aggregation [186], and suppression [179], have been proposed in the conventional statistical disclosure control literature.

The problem of privacy-preserving data mining was formally introduced in [60] to the broader data mining community. The work in [28] established models for quantification of privacy-preserving data mining algorithms. Surveys on privacy-preserving data mining may be found in [29]. The randomization method was generalized to other problems, such as association rule mining [200]. Multiplicative perturbations have also been shown to be very effective in the context of privacy-preserving data mining [140]. Nevertheless, numerous attack methods have been designed for inferring the values of the perturbed data records [11, 367].

The k-anonymity model was proposed by Samarati [442]. The binary search algorithm is also discussed in this work. This paper also set up the basic framework for group-based anonymization, which was subsequently used by all the different privacy methods. The NP-hardness of the k-anonymity problem was formally proved in [385]. A survey of k-anonymous data mining may be found in [153]. The connections between the k-anonymity problem and the frequent pattern mining problem were shown in [83]. A set enumeration method was proposed in [83] that is similar to the set enumeration methods popularly used in frequent pattern mining. The *Incognito* and *Mondrian* algorithms, discussed in this chapter, were proposed in [335] and [336]. The condensation approach to privacy-preserving data mining was proposed in [8]. Some recent methods perform a probabilistic version of k-anonymity on the data, so that the output of the anonymization is a probability distribution [9]. Thus, such an approach allows the use of probabilistic database methods on the transformed data. Many metrics have also been proposed for utility-based evaluation of private tables, rather than simply using the minimal generalization height [29, 315].

The ℓ-diversity and t-closeness models were proposed in [348] and [372], respectively with a focus on sensitive attribute disclosure. A different approach for addressing sensitive attributes is proposed in [91]. A detailed survey of many of the privacy-preserving data publishing techniques may be found in [218]. A closely related model to group-based anonymization is *differential privacy*, where the differential impact of a data record on the privacy of other data records in the database is used to perform the privacy operations [190, 191]. While differential privacy provides theoretical more robust results than many group-based models, its practical utility is yet to be realized. The curse of dimensionality in the context of anonymization problems was first observed in [10]. Subsequently, it was shown that the curse extends to other privacy models such as perturbation and ℓ-diversity [11, 12, 372].

A practical example [402] of how high-dimensional data could be used to make privacy attacks is based on the Netflix data set [559]. Interestingly, this attack uses the sensitive ratings attributes and background knowledge to make identification attacks. Recently, a few methods [514, 533] have been proposed to address the curse of dimensionality in a limited way.

The problem of output privacy is closely related to the problem of inference control and auditing in statistical databases [150]. The most common problems addressed in this domain are those of association rule hiding [497], and query auditing [399]. Distributed methods transform data mining problems into secure multi-party computation primitives [188]. Typically, these methods are dependent on the use of the oblivious transfer protocol [199, 401]. Most of these methods perform distributed privacy-preservation on either horizontally partitioned data [297] or vertically partitioned data [495]. An overview of the various privacy tools for distributed information sharing may be found in [154].

20.8 Exercises

1. Suppose that you have a 1-dimensional dataset uniformly distributed in $(0, 1)$. Uniform noise from the range $(0, 1)$ is added to the data. Derive the final shape of the perturbed distribution.

2. Suppose that your perturbed data was uniformly distributed in $(0, 1)$, and your perturbing distribution was also uniformly distributed in $(0, 1)$. Derive the original data distribution. Will this distribution be accurately reconstructed, in practice, for a finite data set?

3. Implement the Bayes distribution reconstruction algorithm for the randomization method.

4. Implement the (a) *Incognito*, and (b) *Mondrian* algorithms for k-anonymity.

5. Implement the condensation approach to k-anonymity.

6. In dynamic condensation, one of the steps is to split a group into two equal groups along the longest eigenvector. Let λ be the largest eigenvalue of the original group, $\overline{\mu}$ be the original d-dimensional mean, and $\overline{V}$ be the longest eigenvector, which is normalized to unit norm. Compute algebraic expressions for the means of the two split groups, under the uniform distribution assumption.

7. Show that the sensitive attribute in both the entropy- and recursive- ℓ-diversity models must have at least ℓ distinct values.

8 Show that the global entropy of the sensitive attribute distribution is at least equal to the minimum entropy of an equivalence class in it. [**Hint:** Use convexity of entropy]

9. Many k-anonymization algorithms such as *Incognito* depend upon the monotonicity property. Show that the monotonicity property is satisfied by (a) entropy ℓ-diversity, and (b) recursive ℓ-diversity.

10. Implement the (a) *Incognito*, and (b) *Mondrian* algorithms for entropy- and recursive ℓ-diversity, by making changes to your code in Exercise 4.

11. Show that the monotonicity property is satisfied by (a) t-closeness with variational distances, and (b) t-closeness with KL-measure.

12. Consider any group-based anonymity quantification measure $f(\overline{P})$, in which the anonymity condition is of the form $f(\overline{P}) \geq thresh$. (An example of such a measure is entropy in ℓ-diversity.) Here, $\overline{P} = (p_1 \ldots p_r)$ is the sensitive attribute distribution vector. Show that if $f(\overline{P})$ is concave, then the anonymity definition will satisfy the monotonicity property with respect to generalization. Also show that convexity ensures monotonicity in the case of anonymity conditions of the form $f(\overline{P}) \leq thresh$.

13. Implement the (a) *Incognito*, and (b) *Mondrian* algorithms for variational distance-based, and KL distance-based t-closeness, by making changes to your code for Exercise 4.

14. Suppose that you had an anonymized binary transaction database containing the items bought by different customers on a particular day. Suppose that you knew that the transactions of your family friend contained a particular subset B of items, although you did not know the other items bought by her. If every item is bought independently with probability 0.5, show that the probability that at least one of n other customers buys exactly the same pattern of items, is given by *at most* $n/2^B$. Evaluate this expression for $n = 10^4$ and $B = 20$. What does this imply in terms of the privacy of her other buying patterns?

15. Repeat Exercise 14 for movie ratings taking on one of R possible values instead of 2. Assume that each rating possibility has identical probability of $1/R$, and the ratings of different movies are independent and identically distributed. What are the corresponding probabilities of re-identification with B known ratings, and n different individuals?

16. Write a computer program to re-identify the subject of a database with B known sensitive attributes.

Bibliography

[1] N. Adam, and J. Wortman. Security-control methods for statistical databases. *ACM Computing Surveys*, 21(4), pp. 515–556, 1989.

[2] G. Adomavicius, and A. Tuzhilin. Toward the next generation of recommender systems: A survey of the state-of-the-art and possible extensions. *IEEE Transactions on Knowledge and Data Engineering*, 17(6), pp. 734–749, 2005.

[3] R. C. Agarwal, C. C. Aggarwal, and V. V. V. Prasad. A tree projection algorithm for generation of frequent item sets. *Journal of parallel and Distributed Computing*, 61(3), pp. 350–371, 2001. Also available as *IBM Research Report*, RC21341, 1999.

[4] R. C. Agarwal, C. C. Aggarwal, and V. V. V. Prasad. Depth-first generation of long patterns. *ACM KDD Conference*, pp. 108–118, 2000. Also available as "Depth first generation of large itemsets for association rules." *IBM Research Report*, RC21538, 1999.

[5] C. Aggarwal. Outlier analysis. *Springer*, 2013.

[6] C. Aggarwal. Social network data analytics. *Springer*, 2011.

[7] C. Aggarwal, and P. Yu. The igrid index: reversing the dimensionality curse for similarity indexing in high-dimensional space. *KDD Conference*, pp. 119–129, 2000.

[8] C. Aggarwal, and P. Yu. On static and dynamic methods for condensation-based privacy-preserving data mining. *ACM Transactions on Database Systems (TODS)*, 33(1), 2, 2008.

[9] C. Aggarwal. On unifying privacy and uncertain data models. *IEEE International Conference on Data Engineering*, pp. 386–395, 2008.

[10] C. Aggarwal. On k-anonymity and the curse of dimensionality, *Very Large Databases Conference*, pp. 901–909, 2005.

[11] C. Aggarwal. On randomization, public information and the curse of dimensionality. *IEEE International Conference on Data Engineering*, pp. 136–145, 2007.

[12] C. Aggarwal. Privacy and the dimensionality curse. *Privacy-Preserving Data Mining: Models and Algorithms*, Springer, pp. 433–460, 2008.

C. C. Aggarwal, *Data Mining: The Textbook*, DOI 10.1007/978-3-319-14142-8

[13] C. Aggarwal, X. Kong, Q. Gu, J. Han, and P. Yu. Active learning: a survey. *Data Classification: Algorithms and Applications*, CRC Press, 2014.

[14] C. Aggarwal. Instance-based learning: A survey. *Data Classification: Algorithms and Applications*, CRC Press, 2014.

[15] C. Aggarwal. Redesigning distance-functions and distance-based applications for high-dimensional data. *ACM SIGMOD Record*, 30(1), pp. 13–18, 2001.

[16] C. Aggarwal, and P. Yu. Mining associations with the collective strength approach. *ACM PODS Conference*, pp. 863–873, 1998.

[17] C. Aggarwal, A. Hinneburg, and D. Keim. On the surprising behavior of distance-metrics in high-dimensional space. *ICDT Conference*, pp. 420–434, 2001.

[18] C. Aggarwal. Managing and mining uncertain data. *Springer*, 2009.

[19] C. Aggarwal, C. Procopiuc, J. Wolf, P. Yu, and J. Park. Fast algorithms for projected clustering. *ACM SIGMOD Conference*, pp. 61–72, 1999.

[20] C. Aggarwal, J. Han, J. Wang, and P. Yu. On demand classification of data streams. *ACM KDD Conference*, pp. 503–508, 2004.

[21] C. Aggarwal. On change diagnosis in evolving data streams. *IEEE Transactions on Knowledge and Data Engineering*, 17(5), pp. 587–600, 2005.

[22] C. Aggarwal, and P. S. Yu. Finding generalized projected clusters in high dimensional spaces. *ACM SIGMOD Conference*, pp. 70–81, 2000.

[23] C. Aggarwal, and S. Parthasarathy. Mining massively incomplete data sets by conceptual reconstruction. *ACM KDD Conference*, pp. 227–232, 2001.

[24] C. Aggarwal. Outlier ensembles: position paper. *ACM SIGKDD Explorations*, 14(2), pp. 49–58, 2012.

[25] C. Aggarwal. On the effects of dimensionality reduction on high dimensional similarity search. *ACM PODS Conference*, pp. 256–266, 2001.

[26] C. Aggarwal, and H. Wang. Managing and mining graph data. *Springer*, 2010.

[27] C. Aggarwal, C. Procopiuc, and P. Yu. Finding localized associations in market basket data. *IEEE Transactions on Knowledge and Data Engineering*, 14(1), pp. 51–62, 2002.

[28] D. Agrawal, and C. Aggarwal. On the design and quantification of privacy-preserving data mining algorithms. *ACM PODS Conference*, pp. 247–255, 2001.

[29] C. Aggarwal, and P. Yu. Privacy-preserving data mining: models and algorithms. *Springer*, 2008.

[30] C. Aggarwal. Managing and mining sensor data. *Springer*, 2013.

[31] C. Aggarwal, and C. Zhai. Mining text data. *Springer*, 2012.

[32] C. Aggarwal, and C. Reddy. Data clustering: algorithms and applications, *CRC Press*, 2014.

[33] C. Aggarwal. Data classification: algorithms and applications. *CRC Press*, 2014.

[34] C. Aggarwal, and J. Han. Frequent pattern mining. *Springer*, 2014.

[35] C. Aggarwal. On biased reservoir sampling in the presence of stream evolution. *VLDB Conference*, pp. 607–618, 2006.

[36] C. Aggarwal. A framework for clustering massive-domain data streams. *IEEE ICDE Conference*, pp. 102–113, 2009.

[37] C. Aggarwal, and P. Yu. Online generation of association rules. *ICDE Conference*, pp. 402–411, 1998.

[38] C. Aggarwal, Z. Sun, and P. Yu. Online generation of profile association rules. *ACM KDD Conference*, pp. 129–133, 1998.

[39] C. Aggarwal, J. Han, J. Wang, and P. Yu. A framework for clustering evolving data streams, *VLDB Conference*, pp. 81–92, 2003.

[40] C. Aggarwal. Data streams: models and algorithms. *Springer*, 2007.

[41] C. Aggarwal, J. Wolf, and P. Yu. A new method for similarity indexing of market basket data. *ACM SIGMOD Conference*, pp. 407–418, 1999.

[42] C. Aggarwal, N. Ta, J. Wang, J. Feng, and M. Zaki. Xproj: A framework for projected structural clustering of XML documents. *ACM KDD Conference*, pp. 46–55, 2007.

[43] C. Aggarwal. A human-computer interactive method for projected clustering. *IEEE Transactions on Knowledge and Data Engineering*, 16(4), pp. 448–460, 2004.

[44] C. Aggarwal, and N. Li. On node classification in dynamic content-based networks. *SDM Conference*, pp. 355–366, 2011.

[45] C. Aggarwal, A. Khan, and X. Yan. On flow authority discovery in social networks. *SDM Conference*, pp. 522–533, 2011.

[46] C. Aggarwal, and P. Yu. Outlier detection for high dimensional data. *ACM SIGMOD Conference*, pp. 37–46, 2011.

[47] C. Aggarwal, and P. Yu. On classification of high-cardinality data streams. *SDM Conference*, 2010.

[48] C. Aggarwal, and P. Yu. On clustering massive text and categorical data streams. *Knowledge and information systems*, 24(2), pp. 171–196, 2010.

[49] C. Aggarwal, Y. Xie, and P. Yu. On dynamic link inference in heterogeneous networks. *SDM Conference*, pp. 415–426, 2011.

[50] C. Aggarwal, Y. Xie, and P. Yu. On dynamic data-driven selection of sensor streams. *ACM KDD Conference*, pp. 1226–1234, 2011.

[51] C. Aggarwal. On effective classification of strings with wavelets. *ACM KDD Conference*, pp. 163–172, 2002.

[52] C. Aggarwal. On abnormality detection in spuriously populated data streams. *SDM Conference*, pp. 80–91, 2005.

[53] R. Agrawal, K.-I. Lin, H. Sawhney, and K. Shim. Fast similarity search in the presence of noise, scaling, and translation in time-series databases. *VLDB Conference*, pp. 490–501, 1995.

[54] R. Agrawal, and J. Shafer. Parallel mining of association rules. *IEEE Transactions on Knowledge and Data Engineering*, 8(6), pp. 962–969, 1996. Also appears as *IBM Research Report*, RJ10004, January 1996.

[55] R. Agrawal, T. Imielinski, and A. Swami. Mining association rules between sets of items in large databases. *ACM SIGMOD Conference*, pp. 207–216, 1993.

[56] R. Agrawal, and R. Srikant. Fast algorithms for mining association rules. *VLDB Conference*, pp. 487–499, 1994.

[57] R. Agrawal, H. Mannila, R. Srikant, H. Toivonen, and A. I. Verkamo. Fast discovery of association rules. *Advances in knowledge discovery and data mining*, 12, pp. 307–328, 1996.

[58] R. Agrawal, J. Gehrke, D. Gunopulos, and P. Raghavan. Automatic subspace clustering of high dimensional data for data mining applications. *ACM SIGMOD Conference*, pp. 94–105, 1998.

[59] R. Agrawal, and R. Srikant. Mining sequential patterns. *IEEE International Conference on Data Engineering*, pp. 3–14, 1995.

[60] R. Agrawal, and R. Srikant. Privacy-preserving data mining. *ACM SIGMOD Conference*, pp. 439–450, 2000.

[61] M. Agyemang, K. Barker, and R. Alhajj. A comprehensive survey of numeric and symbolic outlier mining techniques. *Intelligent Data Analysis*, 10(6). pp. 521–538, 2006.

[62] R. Ahuja, T. Magnanti, and J. Orlin. Network flows: theory, algorithms, and applications. *Prentice Hall*, Englewood Cliffs, New Jersey, 1993.

[63] M. Al Hasan, and M. J. Zaki. A survey of link prediction in social networks. *Social network data analytics*, Springer, pp. 243–275, 2011.

[64] M. Al Hasan, V. Chaoji, S. Salem, and M. Zaki. Link prediction using supervised learning. *SDM Workshop on Link Analysis, Counter-terrorism and Security*, 2006.

[65] S. Anand, and B. Mobasher. Intelligent techniques for web personalization. *International conference on Intelligent Techniques for Web Personalization*, pp. 1–36, 2003.

[66] F. Angiulli, and C. Pizzuti. Fast Outlier detection in high dimensional spaces. *European Conference on Principles of Knowledge Discovery and Data Mining*, pp. 15–27, 2002.

[67] F. Angiulli, and F. Fassetti. Detecting distance-based outliers in streams of data. *ACM CIKM Conference*, pp. 811–820, 2007.

[68] L. Akoglu, H. Tong, J. Vreeken, and C. Faloutsos. Fast and reliable anomaly detection in categorical data. *ACM CIKM Conference*, pp. 415–424, 2012.

[69] R. Albert, and A. L. Barabasi. Statistical mechanics of complex networks. *Reviews of modern physics* 74, 1, 47, 2002.

[70] R. Albert, and A. L. Barabasi. Topology of evolving networks: local events and universality. *Physical review letters* 85, 24, pp. 5234–5237, 2000.

[71] P. Allison. Missing data. *Sage*, 2001.

[72] N. Alon, Y. Matias, and M. Szegedy. The space complexity of approximating the frequency moments. *ACM PODS Conference*, pp. 20–29, 1996.

[73] S. Altschul, T. Madden, A. Schaffer, J. Zhang, Z. Zhang, W. Miller, and D. Lipman. Gapped BLAST and PSI-BLAST: a new generation of protein database search programs. *Nucleic acids research*, 25(17), pp. 3389–3402, 1997.

[74] M. R. Anderberg. Cluster Analysis for Applications. *Academic Press*, New York, 1973.

[75] P. Andritsos, P. Tsaparas, R. J. Miller, and K. C. Sevcik. LIMBO: Scalable clustering of categorical data. *EDBT Conference*, pp. 123–146, 2004.

[76] M. Ankerst, M. M. Breunig, H.-P. Kriegel, and J. Sander. OPTICS: ordering points to identify the clustering structure. *ACM SIGMOD Conference*, pp. 49–60, 1999.

[77] A. Apostolico, and C. Guerra. The longest common subsequence problem revisited. *Algorithmica*, 2(1–4), pp. 315–336, 1987.

[78] A. Azran. The rendezvous algorithm: Multiclass semi-supervised learning with markov random walks. *International Conference on Machine Learning*, pp. 49–56, 2007.

[79] A. Banerjee, S. Merugu, I. S. Dhillon, and J. Ghosh. Clustering with Bregman divergences. *Journal of Machine Learning Research*, 6, pp. 1705–1749, 2005.

[80] S. Basu, A. Banerjee, and R. J. Mooney. Semi-supervised clustering by seeding. *ICML Conference*, pp. 27–34, 2002.

[81] S. Basu, M. Bilenko, and R. J. Mooney. A probabilistic framework for semi-supervised clustering. *ACM KDD Conference*, pp. 59–68, 2004.

[82] R. J. Bayardo Jr. Efficiently mining long patterns from databases. *ACM SIGMOD*, pp. 85–93, 1998.

[83] R. J. Bayardo, and R. Agrawal. Data privacy through optimal k-anonymization. *IEEE International Conference on Data Engineering*, pp. 217–228, 2005.

[84] R. Beckman, and R. Cook. Outliers. *Technometrics*, 25(2), pp. 119–149, 1983.

[85] A. Ben-Hur, C. S. Ong, S. Sonnenburg, B. Scholkopf, and G. Ratsch. Support vector machines and kernels for computational biology. *PLoS computational biology*, 4(10), e1000173, 2008.

[86] M. Benkert, J. Gudmundsson, F. Hubner, and T. Wolle. Reporting flock patterns. *COMGEO*, 2008

[87] D. Berndt, and J. Clifford. Using dynamic time warping to find patterns in time series. *KDD Workshop*, 10(16), pp. 359–370, 1994.

[88] K. Beyer, J. Goldstein, R. Ramakrishnan, and U. Shaft. When is "nearest neighbor" meaningful? *International Conference on Database Theory*, pp. 217–235, 1999.

[89] V. Barnett, and T. Lewis. Outliers in statistical data. *Wiley*, 1994.

[90] M. Belkin, and P. Niyogi. Laplacian eigenmaps and spectral techniques for embedding and clustering. *NIPS*, pp. 585–591, 2001.

[91] M. Bezzi, S. De Capitani di Vimercati, S. Foresti, G. Livraga, P. Samarati, and R. Sassi. Modeling and preventing inferences from sensitive value distributions in data release. *Journal of Computer Security*, 20(4), pp. 393–436, 2012.

[92] L. Bergroth, H. Hakonen, and T. Raita. A survey of longest common subsequence algorithms. *String Processing and Information Retrieval*, 2000.

[93] S. Bhagat, G. Cormode, and S. Muthukrishnan. Node classification in social networks. *Social Network Data Analytics*, Springer, pp. 115–148. 2011.

[94] M. Bilenko, S. Basu, and R. J. Mooney. Integrating constraints and metric learning in semi-supervised clustering. *ICML Conference*, 2004.

[95] C. M. Bishop. Pattern recognition and machine learning. *Springer*, 2007.

[96] C. M. Bishop. Neural networks for pattern recognition. *Oxford University Press*, 1995.

[97] C. M. Bishop. Improving the generalization properties of radial basis function neural networks. *Neural Computation*, 3(4), pp. 579–588, 1991.

[98] D. Blei, A. Ng, and M. Jordan. Latent dirichlet allocation. *Journal of Machine Learning Research*, 3: pp. 993–1022, 2003.

[99] D. Blei. Probabilistic topic models. *Communications of the ACM*, 55(4), pp. 77–84, 2012.

[100] A. Blum, and T. Mitchell. Combining labeled and unlabeled data with co-training. *Proceedings of Conference on Computational Learning Theory*, 1998.

[101] A. Blum, and S. Chawla. Combining labeled and unlabeled data with graph mincuts. *ICML Conference*, 2001.

[102] C. Bohm, K. Haegler, N. Muller, and C. Plant. Coco: coding cost for parameter free outlier detection. *ACM KDD Conference*, 2009.

[103] K. Borgwardt, and H.-P. Kriegel. Shortest-path kernels on graphs. *IEEE International Conference on Data Mining*, 2005.

[104] S. Boriah, V. Chandola, and V. Kumar. Similarity measures for categorical data: A comparative evaluation. *SIAM Conference on Data Mining*, 2008.

[105] L. Bottou, and V. Vapnik. Local learning algorithms. *Neural Computation*, 4(6), pp. 888–900, 1992.

[106] L. Bottou, C. Cortes, J. S. Denker, H. Drucker, I. Guyon, L. Jackel, Y. LeCun, U. A. Müller, E. Säckinger, P. Simard, and V. Vapnik. Comparison of classifier methods: a case study in handwriting digit recognition. *International Conference on Pattern Recognition*, pp. 77–87, 1994.

[107] J. Boulicaut, A. Bykowski, and C. Rigotti. Approximation of frequency queries by means of free-sets. *Principles of Data Mining and Knowledge Discovery*, pp. 75–85, 2000.

[108] P. Bradley, and U. Fayyad. Refining initial points for k-means clustering. *ICML Conference*, pp. 91–99, 1998.

[109] M. Breunig, H.-P. Kriegel, R. Ng, and J. Sander. LOF: Identifying density-based local outliers. *ACM SIGMOD Conference*, 2000.

[110] L. Breiman, J. Friedman, C. Stone, and R. Olshen. Classification and regression trees. *CRC press*, 1984.

[111] L. Breiman. Random forests. *Machine Learning*, 45(1), pp. 5–32, 2001.

[112] L. Breiman. Bagging predictors. *Machine Learning*, 24(2), pp. 123–140, 1996.

[113] S. Brin, R. Motwani, and C. Silverstein. Beyond market baskets: generalizing association rules to correlations. *ACM SIGMOD Conference*, pp. 265–276, 1997.

[114] S. Brin, and L. Page. The anatomy of a large-scale hypertextual web search engine. *Computer Networks*, 30(1–7), pp. 107–117, 1998.

[115] B. Bringmann, S. Nijssen, and A. Zimmermann. Pattern-based classification: A unifying perspective. *arXiv preprint, arXiv:1111.6191*, 2011.

[116] C. Brodley, and P. Utgoff. Multivariate decision trees. *Machine learning*, 19(1), pp. 45–77, 1995.

[117] Y. Bu, L. Chen, A. W.-C. Fu, and D. Liu. Efficient anomaly monitoring over moving object trajectory streams. *ACM KDD Conference*, pp. 159–168, 2009.

[118] M. Bulmer. Principles of Statistics. *Dover Publications*, 1979.

[119] H. Bunke. On a relation between graph edit distance and maximum common subgraph. *Pattern Recognition Letters*, 18(8), pp. 689–694, 1997.

[120] H. Bunke, and K. Shearer. A graph distance metric based on the maximal common subgraph.*Pattern recognition letters*, 19(3), pp. 255–259, 1998.

[121] W. Buntine. Learning Classification Trees. *Artificial intelligence frontiers in statistics*. Chapman and Hall, pp. 182–201, 1993.

[122] T. Burnaby. On a method for character weighting a similarity coefficient employing the concept of information. *Mathematical Geology*, 2(1), 25–38, 1970.

[123] D. Burdick, M. Calimlim, and J. Gehrke. MAFIA: A maximal frequent itemset algorithm for transactional databases. *IEEE International Conference on Data Engineering*, pp. 443–452, 2001.

[124] C. Burges. A tutorial on support vector machines for pattern recognition. *Data mining and knowledge discovery*, 2(2), pp. 121–167, 1998.

[125] T. Calders, and B. Goethals. Mining all non-derivable frequent itemsets. *Principles of Knowledge Discovery and Data Mining*, pp. 74–86, 2002.

[126] T. Calders, C. Rigotti, and J. F. Boulicaut. A survey on condensed representations for frequent sets. In *Constraint-based mining and inductive databases*, pp. 64–80, Springer, 2006.

[127] S. Chakrabarti. Mining the Web: Discovering knowledge from hypertext data. *Morgan Kaufmann*, 2003.

[128] S. Chakrabarti, B. Dom, and P. Indyk. Enhanced hypertext categorization using hyperlinks. *ACM SIGMOD Conference*, pp. 307–318, 1998.

[129] S. Chakrabarti, S. Sarawagi, and B. Dom. Mining surprising patterns using temporal description length. *VLDB Conference*, pp. 606–617, 1998.

[130] K. P. Chan, and A. W. C. Fu. Efficient time series matching by wavelets.*IEEE International Conference on Data Engineering*, pp. 126–133, 1999.

[131] V. Chandola, A. Banerjee, and V. Kumar. Anomaly detection: A survey. *ACM Computing Surveys*, 41(3), 2009.

[132] V. Chandola, A. Banerjee, and V. Kumar. Anomaly detection for discrete sequences: A survey. *IEEE Transactions on Knowledge and Data Engineering*, 24(5), pp. 823–839, 2012.

[133] O. Chapelle. Training a support vector machine in the primal. *Neural Computation*, 19(5), pp. 1155–1178, 2007.

[134] C. Chatfield. The analysis of time series: an introduction. *CRC Press*, 2003.

[135] A. Chaturvedi, P. Green, and J. D. Carroll. K-modes clustering, *Journal of Classification*, 18(1), pp. 35–55, 2001.

[136] N. V. Chawla, N. Japkowicz, and A. Kotcz. Editorial: Special issue on learning from imbalanced data sets. *ACM SIGKDD Explorations Newsletter*, 6(1), 1–6, 2004.

[137] N. V. Chawla, K. W. Bower, L. O. Hall, and W. P. Kegelmeyer. SMOTE: synthetic minority over-sampling technique. *Journal of Artificial Intelligence Research (JAIR)*, 16, pp. 321–356, 2002.

[138] N. Chawla, A. Lazarevic, L. Hall, and K. Bowyer. SMOTEBoost: Improving prediction of the minority class in boosting. *PKDD*, pp. 107–119, 2003.

[139] N. V. Chawla, D. A. Cieslak, L. O. Hall, and A. Joshi. Automatically countering imbalance and its empirical relationship to cost. *Data Mining and Knowledge Discovery*, 17(2), pp. 225–252, 2008.

[140] K. Chen, and L. Liu. A survey of multiplicative perturbation for privacy-preserving data mining. *Privacy-Preserving Data Mining: Models and Algorithms*, Springer, pp. 157–181, 2008.

[141] L. Chen, and R. Ng. On the marriage of L_p-norms and the edit distance. *VLDB Conference*, pp. 792–803, 2004.

[142] W. Chen, Y. Wang, and S. Yang. Efficient influence maximization in social networks. *ACM KDD Conference*, pp. 199–208, 2009.

[143] W. Chen, C. Wang, and Y. Wang. Scalable influence maximization for prevalent viral marketing in large-scale social networks. *ACM KDD Conference*, pp. 1029–1038, 2010.

[144] W. Chen, Y. Yuan, and L. Zhang. Scalable influence maximization in social networks under the linear threshold model. *IEEE International Conference on Data Mining*, pp. 88–97, 2010.

[145] D. Chen, C.-T. Lu, Y. Chen, and D. Kou. On detecting spatial outliers. *Geoinformatica*, 12: pp. 455–475, 2008.

[146] T. Cheng, and Z. Li. A hybrid approach to detect spatialtemporal outliers. *International Conference on Geoinformatics*, pp. 173–178, 2004.

[147] T. Cheng, and Z. Li. A multiscale approach for spatio-temporal outlier detection. *Transactions in GIS*, 10(2), pp. 253–263, March 2006.

[148] Y. Cheng. Mean shift, mode seeking, and clustering. *IEEE Transactions on PAMI*, 17(8), pp. 790–799, 1995.

[149] H. Cheng, X. Yan, J. Han, and C. Hsu. Discriminative frequent pattern analysis for effective classification. *ICDE Conference*, pp. 716–725, 2007.

[150] F. Y. Chin, and G. Ozsoyoglu. Auditing and inference control in statistical databases. *IEEE Transactions on Software Enginerring*, 8(6), pp. 113–139, April 1982.

[151] B. Chiu, E. Keogh, and S. Lonardi. Probabilistic discovery of time series motifs. *ACM KDD Conference*, pp. 493–498, 2003.

[152] F. Chung. Spectral Graph Theory. *Number 92 in CBMS Conference Series in Mathematics, American Mathematical Society*, 1997.

[153] V. Ciriani, S. De Capitani di Vimercati, S. Foresti, and P. Samarati. k-anonymous data mining: A survey. *Privacy-preserving data mining: models and algorithms*, Springer, pp. 105–136, 2008.

[154] C. Clifton, M. Kantarcioglu, J. Vaidya, X. Lin, and M. Y. Zhu. Tools for privacy preserving distributed data mining. *ACM SIGKDD Explorations Newsletter*, 4(2), pp. 28–34, 2002.

[155] N. Cristianini, and J. Shawe-Taylor. An introduction to support vector machines and other kernel-based learning methods. *Cambridge University Press*, 2000.

[156] W. Cochran. Sampling techniques. *John Wiley and Sons*, 2007.

[157] D. Cohn, L. Atlas, and R. Ladner. Improving generalization with active learning. *Machine Learning*, 5(2), pp. 201–221, 1994.

[158] D. Cohn, Z. Ghahramani, and M. Jordan. Active learning with statistical models. *Journal of Artificial Intelligence Research*, 4, pp. 129–145, 1996.

[159] D. Comaniciu, and P. Meer. Mean shift: A robust approach toward feature space analysis. *IEEE Transactions on PAMI*, 24(5), pp. 603–619, 2002.

[160] D. Cook, and L. Holder. Graph-based data mining. *IEEE Intelligent Systems*, 15(2), pp. 32–41, 2000.

[161] R. Cooley, B. Mobasher, and J. Srivastava. Data preparation for mining world wide web browsing patterns. *Knowledge and information systems*, 1(1), pp. 5–32, 1999.

[162] L. P. Cordella, P. Foggia, C. Sansone, and M. Vento. A (sub)graph isomorphism algorithm for matching large graphs. *IEEE Transactions on Pattern Mining and Machine Intelligence*, 26(10), pp. 1367–1372, 2004.

[163] H. Shang, Y. Zhang, X. Lin, and J. X. Yu. Taming verification hardness: an efficient algorithm for testing subgraph isomorphism. *Proceedings of the VLDB Endowment*, 1(1), pp. 364–375, 2008.

[164] J. R. Ullmann. An algorithm for subgraph isomorphism. *Journal of the ACM*, 23: pp. 31–42, January 1976.

[165] G. Cormode, and S. Muthukrishnan. An improved data stream summary: the count-min sketch and its applications. *Journal of Algorithms*, 55(1), pp. 58–75, 2005.

[166] S. Cost, and S. Salzberg. A weighted nearest neighbor algorithm for learning with symbolic features. *Machine Learning*, 10(1), pp. 57–78, 1993.

[167] T. Cover, and P. Hart. Nearest neighbor pattern classification. *IEEE Transactions on Information Theory*, 13(1), pp. 21–27, 1967.

[168] D. Cutting, D. Karger, J. Pedersen, and J. Tukey. Scatter/gather: A cluster-based approach to browsing large document collections. *ACM SIGIR Conference*, pp. 318–329, 1992.

[169] M. Dash, K. Choi, P. Scheuermann, and H. Liu. Feature selection for clustering-a filter solution. *ICDM Conference*, pp. 115–122, 2002.

[170] M. Deshpande, and G. Karypis. Item-based top-n recommendation algorithms. *ACM Transactions on Information Systems (TOIS)*, 22(1), pp. 143–177, 2004.

[171] I. Dhillon. Co-clustering documents and words using bipartite spectral graph partitioning, *ACM KDD Conference*, pp. 269–274, 2001.

[172] I. Dhillon, S. Mallela, and D. Modha. Information-theoretic co-clustering. *ACM KDD Conference*, pp. 89–98, 2003.

[173] I. Dhillon, Y. Guan, and B. Kulis. Kernel k-means: spectral clustering and normalized cuts. *ACM KDD Conference*, pp. 551–556, 2004.

[174] P. Domingos. MetaCost: A general framework for making classifiers cost-sensitive. *ACM KDD Conference*, pp. 155–164, 1999.

[175] P. Domingos. Bayesian averaging of classifiers and the overfitting problem. *ICML Conference*, pp. 223–230, 2000.

[176] P. Domingos, and G. Hulten. Mining high-speed data streams. *ACM KDD Conference*, pp. 71–80. 2000.

[177] P. Clark, and T. Niblett. The CN2 induction algorithm. *Machine Learning*, 3(4), pp. 261–283, 1989.

[178] W. W. Cohen. Fast effectve rule induction. *ICML Conference*, pp. 115–123, 1995.

[179] L. H. Cox. Suppression methodology and statistical disclosure control. *Journal of the American Statistical Association*, 75(370), pp. 377–385, 1980.

[180] E. Cohen, M. Datar, S. Fujiwara, A. Gionis, P. Indyk, R. Motwani, and C. Yang. Finding interesting associations without support pruning. *IEEE Transactions on Knowledge and Data Engineering*, 13(1), pp. 64–78, 2001.

[181] T. Dalenius, and S. Reiss. Data-swapping: A technique for disclosure control. *Journal of statistical planning and inference*, 6(1), pp. 73–85, 1982.

[182] G. Das, and H. Mannila. Context-based similarity measures for categorical databases. *PKDD Conference*, pp. 201–210, 2000.

[183] B. V. Dasarathy. Nearest neighbor (NN) norms: NN pattern classification techniques. *IEEE Computer Society Press*, 1990,

[184] S. Deerwester, S. Dumais, T. Landauer, G. Furnas, and R. Harshman. Indexing by latent semantic analysis. *JASIS*, 41(6), pp. 391–407, 1990.

[185] C. Ding, X. He, and H. Simon. On the equivalence of nonnegative matrix factorization and spectral clustering. *SDM Conference*, pp. 606–610, 2005.

[186] J. Domingo-Ferrer, and J. M. Mateo-Sanz. Practical data-oriented microaggregation for statistical disclosure control. *IEEE Transactions on Knowledge and Data Engineering*, 14(1), pp. 189–201, 2002.

[187] P. Domingos, and M. Pazzani. On the optimality of the simple bayesian classifier under zero-one loss. *Machine Learning*, 29(2–3), pp. 103–130, 1997.

[188] W. Du, and M. Atallah. Secure multi-party computation: A review and open problems. *CERIAS Tech. Report*, 2001-51, Purdue University, 2001.

[189] R. Duda, P. Hart, and D. Stork. Pattern classification. *John Wiley and Sons*, 2012.

[190] C. Dwork. Differential privacy: A survey of results. *Theory and Applications of Models of Computation*, Springer, pp. 1–19, 2008.

[191] C. Dwork. A firm foundation for private data analysis. *Communications of the ACM*, 54(1), pp. 86–95, 2011.

[192] D. Easley, and J. Kleinberg. Networks, crowds, and markets: Reasoning about a highly connected world. *Cambridge University Press*, 2010.

[193] C. Elkan. The foundations of cost-sensitive learning. *IJCAI*, pp. 973–978, 2001.

[194] R. Elmasri, and S. Navathe. *Fundamentals of Database Systems*. Addison-Wesley, 2010.

[195] L. Ertoz, M. Steinbach, and V. Kumar. A new shared nearest neighbor clustering algorithm and its applications. *Workshop on Clustering High Dimensional Data and its Applications*, pp. 105–115, 2002.

[196] P. Erdos, and A. Renyi. On random graphs. *Publicationes Mathematicae Debrecen*, 6, pp. 290–297, 1959.

[197] M. Ester, H.-P. Kriegel, J. Sander, and X. Xu. A density-based algorithm for discovering clusters in large spatial databases with noise. *ACM KDD Conference*, pp. 226–231, 1996.

[198] M. Ester, H. P. Kriegel, J. Sander, M. Wimmer, and X. Xu. Incremental clustering for mining in a data warehousing environment. *VLDB Conference*, pp. 323–333, 1998.

[199] S. Even, O. Goldreich, and A. Lempel. A randomized protocol for signing contracts. *Communications of the ACM*, 28(6), pp. 637–647, 1985.

[200] A. Evfimievski, R. Srikant, R. Agrawal, and J. Gehrke. Privacy preserving mining of association rules. *Information Systems*, 29(4), pp. 343–364, 2004.

[201] M. Faloutsos, P. Faloutsos, and C. Faloutsos. On power-law relationships of the internet topology. *ACM SIGCOMM Computer Communication Review*, pp. 251–262, 1999.

[202] C. Faloutsos, and K. I. Lin. Fastmap: A fast algorithm for indexing, data-mining and visualization of traditional and multimedia datasets. *ACM SIGMOD Conference*, pp. 163–174, 1995.

[203] W. Fan, S. Stolfo, J. Zhang, and P. Chan. AdaCost: Misclassification cost sensitive boosting. *ICML Conference*, pp. 97–105, 1999.

[204] T. Fawcett. ROC Graphs: Notes and Practical Considerations for Researchers. *Technical Report HPL-2003-4*, Palo Alto, CA, HP Laboratories, 2003.

[205] X. Fern, and C. Brodley. Random projection for high dimensional data clustering: A cluster ensemble approach. *ICML Conference*, pp. 186–193, 2003.

[206] C. Fiduccia, and R. Mattheyses. A linear-time heuristic for improving network partitions. In *IEEE Conference on Design Automation*, pp. 175–181, 1982.

[207] R. Fisher. The use of multiple measurements in taxonomic problems. *Annals of Eugenics*, 7: pp. 179–188, 1936.

[208] P. Flajolet, and G. N. Martin. Probabilistic counting algorithms for data base applications. *Journal of Computer and System Sciences*, 31(2), pp. 182–209, 1985.

[209] G. W. Flake. Square unit augmented, radially extended, multilayer perceptrons. *Neural Networks: Tricks of the Trade*, pp. 145–163, 1998.

[210] F. Fouss, A. Pirotte, J. Renders, and M. Saerens. Random-walk computation of similarities between nodes of a graph with application to collaborative recommendation. *IEEE Transactions on Knowledge and Data Engineering*, 19(3), pp. 355–369, 2007.

[211] S. Forrest, C. Warrender, and B. Pearlmutter. Detecting intrusions using system calls: alternate data models. *IEEE ISRSP*, 1999.

[212] S. Fortunato. Community Detection in Graphs. *Physics Reports*, 486(3–5), pp. 75–174, February 2010.

[213] A. Frank, and A. Asuncion. UCI Machine Learning Repository, Irvine, CA: University of California, School of Information and Computer Science, 2010. `http://archive.ics.uci.edu/ml`

[214] E. Frank, M. Hall, and B. Pfahringer. Locally weighted naive bayes. *Proceedings of the Nineteenth conference on Uncertainty in Artificial Intelligence*, pp, 249–256, 2002.

[215] Y. Freund, and R. Schapire. A decision-theoretic generalization of online learning and application to boosting. *Computational Learning Theory*, pp. 23–37, 1995.

[216] J. Friedman. Flexible nearest neighbor classification. *Technical Report, Stanford University*, 1994.

[217] J. Friedman, R. Kohavi, and Y. Yun. Lazy decision trees. *Proceedings of the National Conference on Artificial Intelligence*, pp. 717–724, 1996.

[218] B. Fung, K. Wang, R. Chen, and P. S. Yu. Privacy-preserving data publishing: A survey of recent developments. *ACM Computing Surveys (CSUR)*, 42(4), 2010.

[219] G. Gan, C. Ma, and J. Wu. Data clustering: theory, algorithms, and applications. *SIAM*, 2007.

[220] V. Ganti, J. Gehrke, and R. Ramakrishnan. CACTUS: Clustering categorical data using summaries. *ACM KDD Conference*, pp. 73–83, 1999.

[221] M. Garey, and D. S. Johnson. Computers and intractability: A guide to the theory of NP-completeness. *New York, Freeman*, 1979.

[222] H. Galhardas, D. Florescu, D. Shasha, and E. Simon. AJAX: an extensible data cleaning tool. *ACM SIGMOD Conference* 29(2), pp. 590, 2000.

[223] J. Gao, and P.-N. Tan. Converting output scores from outlier detection algorithms into probability estimates. *ICDM Conference*, pp. 212–221, 2006.

[224] M. Garofalakis, R. Rastogi, and K. Shim. SPIRIT: Sequential pattern mining with regular expression constraints. *VLDB Conference*, pp. 7–10, 1999.

[225] T. Gartner, P. Flach, and S. Wrobel. On graph kernels: Hardness results and efficient alternatives. *COLT: Kernel 2003 Workshop Proceedings*, pp. 129–143, 2003.

[226] Y. Ge, H. Xiong, Z.-H. Zhou, H. Ozdemir, J. Yu, and K. Lee. Top-Eye: Top-k evolving trajectory outlier detection. *CIKM Conference*, pp. 1733–1736, 2010.

[227] J. Gehrke, V. Ganti, R. Ramakrishnan, and W.-Y. Loh. BOAT: Optimistic decision tree construction. *ACM SIGMOD Conference*, pp. 169–180, 1999.

[228] J. Gehrke, R. Ramakrishnan, and V. Ganti. Rainforest-a framework for fast decision tree construction of large datasets. *VLDB Conference*, pp. 416–427, 1998.

[229] D. Gibson, J. Kleinberg, and P. Raghavan. Clustering categorical data: an approach based on dynamical systems. *The VLDB Journal*, 8(3), pp. 222–236, 2000.

[230] M. Girvan, and M. Newman. Community structure in social and biological networks. *Proceedings of the National Academy of Sciences*, 99(12), pp. 7821–7826.

[231] S. Goil, H. Nagesh, and A. Choudhary. MAFIA: Efficient and scalable subspace clustering for very large data sets. *ACM KDD Conference*, pp. 443–452, 1999.

[232] D. W. Goodall. A new similarity index based on probability. *Biometrics*, 22(4), pp. 882–907, 1966.

[233] K. Gouda, and M. J. Zaki. Genmax: An efficient algorithm for mining maximal frequent itemsets. *Data Mining and Knowledge Discovery*, 11(3), pp. 223–242, 2005.

[234] A. Goyal, F. Bonchi, and L. V. S. Lakshmanan. A data-based approach to social influence maximization. *VLDB Conference*, pp. 73–84, 2011.

[235] A. Goyal, F. Bonchi, and L. V. S. Lakshmanan. Learning influence probabilities in social networks. *ACM WSDM Conference*, pp. 241–250, 2011.

[236] R. Gozalbes, J. P. Doucet, and F. Derouin. Application of topological descriptors in QSAR and drug design: history and new trends. *Current Drug Targets-Infectious Disorders*, 2(1), pp. 93–102, 2002.

[237] M. Gupta, J. Gao, C. Aggarwal, and J. Han. Outlier detection for temporal data. Morgan and Claypool, 2014.

[238] S. Guha, R. Rastogi, and K. Shim. ROCK: A robust clustering algorithm for categorical attributes. *Information Systems*, 25(5), pp. 345–366, 2000.

[239] S. Guha, R. Rastogi, and K. Shim. CURE: An efficient clustering algorithm for large databases. *ACM SIGMOD Conference*, pp. 73–84, 1998.

[240] S. Guha, A. Meyerson, N. Mishra, R. Motwani, and L. O'Callaghan. Clustering data streams: Theory and practice. *IEEE Transactions on Knowledge and Data Engineering*, 15(3), pp. 515–528, 2003.

[241] D. Gunopulos, and G. Das. Time series similarity measures and time series indexing. *ACM SIGMOD Conference*, pp, 624, 2001.

[242] V. Guralnik, and G. Karypis. A scalable algorithm for clustering sequential data. *IEEE International Conference on Data Engineering*, pp. 179–186, 2001.

[243] V. Guralnik, and G. Karypis. Parallel tree-projection-based sequence mining algorithms. *Parallel Computing*, 30(4): pp. 443–472, April 2004. Also appears in *European Conference in Parallel Processing*, 2001.

[244] D. Gusfield. Algorithms on strings, trees and sequences. *Cambridge University Press*, 1997.

[245] I. Guyon (Ed.). Feature extraction: foundations and applications. *Springer*, 2006.

[246] I. Guyon, and A. Elisseeff. An introduction to variable and feature selection. *Journal of Machine Learning Research*, 3, pp. 1157–1182, 2003.

[247] M. Halkidi, Y. Batistakis, and M. Vazirgiannis. Cluster validity methods: part I. *ACM SIGMOD record*, 31(2), pp. 40–45, 2002.

[248] M. Halkidi, Y. Batistakis, and M. Vazirgiannis. Clustering validity checking methods: part II. *ACM SIGMOD Record*, 31(3), pp. 19–27, 2002.

[249] E. Han, and G. Karypis. Centroid-based document classification: analysis and experimental results. *ECML Conference*, pp. 424–431, 2000.

[250] J. Han, M. Kamber, and J. Pei. Data mining: concepts and techniques. *Morgan Kaufmann*, 2011.

[251] J. Han, G. Dong, and Y. Yin. Efficient mining of partial periodic patterns in time series database. *International Conference on Data Engineering*, pp. 106–115, 1999.

[252] J. Han, J. Pei, and Y. Yin. Mining frequent patterns without candidate generation. *ACM SIGMOD Conference*, pp. 1–12, 2000.

[253] J. Han, H. Cheng, D. Xin, and X. Yan. Frequent pattern mining: current status and future directions. *Data Mining and Knowledge Discovery*, 15(1), pp. 55–86, 2007.

[254] J. Haslett, R. Brandley, P. Craig, A. Unwin, and G. Wills. Dynamic graphics for exploring spatial data with application to locating global and local anomalies. *The American Statistician*, 45: pp. 234–242, 1991.

[255] T. Hastie, and R. Tibshirani. Discriminant adaptive nearest neighbor classification. *IEEE Transactions on Pattern Analysis and Machine Intelligence*, 18(6), pp. 607–616, 1996.

[256] T. Hastie, R. Tibshirani, and J. Friedman. The elements of statistical learning. *Springer*, 2009.

[257] V. Hautamaki, V. Karkkainen, and P. Franti. Outlier detection using k-nearest neighbor graph. *International Conference on Pattern Recognition*, pp. 430–433, 2004.

[258] T. H. Haveliwala. Topic-sensitive pagerank. *World Wide Web Conference*, pp. 517-526, 2002.

[259] D. M. Hawkins. Identification of outliers. *Chapman and Hall*, 1980.

[260] S. Haykin. Kalman filtering and neural networks. *Wiley*, 2001.

[261] S. Haykin. Neural networks and learning machines. *Prentice Hall*, 2008.

[262] X. He, D. Cai, and P. Niyogi. Laplacian score for feature selection. *Advances in Neural Information Processing Systems*, 18, 507, 2006.

[263] Z. He, X. Xu, J. Huang, and S. Deng. FP-Outlier: Frequent pattern-based outlier detection. *COMSIS*, 2(1), pp. 103–118, 2005.

[264] Z. He, X. Xu, and S. Deng. Discovering cluster-based local outliers, *Pattern Recognition Letters*, Vol 24(9–10), pp. 1641–1650, 2003.

[265] M. Henrion, D. Hand, A. Gandy, and D. Mortlock. CASOS: A subspace method for anomaly detection in high-dimensional astronomical databases. *Statistical Analysis and Data Mining*, 2012.
Online first: `http://onlinelibrary.wiley.com/enhanced/doi/10.1002/sam.11167/`

[266] A. Hinneburg, C. Aggarwal, and D. Keim. What is the nearest neighbor in high-dimensional space? *VLDB Conference*, pp. 506–516, 2000.

[267] A. Hinneburg, and D. Keim. An efficient approach to clustering in large multimedia databases with noise. *ACM KDD Conference*, pp. 58–65, 1998.

[268] A. Hinneburg, D. A. Keim, and M. Wawryniuk. HD-Eye: Visual mining of high-dimensional data. *Computer Graphics and Applications*, 19(5), pp. 22–31, 1999.

[269] A. Hinneburg, and H. Gabriel. DENCLUE 2.0: Fast clustering based on kernel-density estimation. *Intelligent Data Analysis, Springer*, pp. 70–80, 2007.

[270] D. S. Hirschberg. Algorithms for the longest common subsequence problem. *Journal of the ACM (JACM)*, 24(4), pp. 664–675, 1975.

[271] T. Hofmann. Probabilistic latent semantic indexing. *ACM SIGIR Conference*, pp. 50–57, 1999.

[272] T. Hofmann. Latent semantic models for collaborative filtering. *ACM Transactions on Information Systems (TOIS)*, 22(1), pp. 89–114, 2004.

[273] M. Holsheimer, M. Kersten, H. Mannila, and H. Toivonen. A perspective on databases and data mining, *ACM KDD Conference*, pp. 150–155, 1995.

[274] S. Hofmeyr, S. Forrest, and A. Somayaji. Intrusion detection using sequences of system calls. *Journal of Computer Security*, 6(3), pp. 151–180, 1998.

[275] D. Hosmer Jr., S. Lemeshow, and R. Sturdivant. Applied logistic regression. *Wiley*, 2013.

[276] J. Huan, W. Wang, and J. Prins. Efficient mining of frequent subgraphs in the presence of isomorphism. *IEEE ICDM Conference*, pp. 549–552, 2003.

[277] Z. Huang, X. Li, and H. Chen. Link prediction approach to collaborative filtering. *ACM/IEEE-CS joint conference on Digital libraries*, pp. 141–142, 2005.

[278] Z. Huang, and M. Ng. A fuzzy k-modes algorithm for clustering categorical data. *IEEE Transactions on Fuzzy Systems*, 7(4), pp. 446–452, 1999.

[279] G. Hulten, L. Spencer, and P. Domingos. Mining time-changing data streams. *ACM KDD Conference*, pp. 97–106, 2001.

[280] J. W. Hunt, and T. G. Szymanski. A fast algorithm for computing longest common subsequences. *Communications of the ACM*, 20(5), pp. 350–353, 1977.

[281] Y. S. Hwang, and S. Y. Bang. An efficient method to construct a radial basis function neural network classifier. *Neural Networks*, 10(8), pp. 1495–1503, 1997.

[282] A. Inokuchi, T. Washio, and H. Motoda. An apriori-based algorithm on mining frequent substructures from graph data. *Principles on Knowledge Discovery and Data Mining*, pp. 13–23, 2000.

[283] H. V. Jagadish, A. O. Mendelzon, and T. Milo. Similarity-based queries. *ACM PODS Conference*, pp. 36–45, 1995.

[284] A. K. Jain, and R. C. Dubes. Algorithms for clustering data. *Prentice-Hall, Inc.*, 1998.

[285] A. Jain, M. Murty, and P. Flynn. Data clustering: A review. *ACM Computing Surveys (CSUR)*, 31(3):264–323, 1999.

[286] A. Jain, R. Duin, and J. Mao. Statistical pattern recognition: A review. *IEEE Transactions on Pattern Analysis and Machine Intelligence,*, 22(1), pp. 4–37, 2000.

[287] V. Janeja, and V. Atluri. Random walks to identify anomalous free-form spatial scan windows. *IEEE Transactions on Knowledge and Data Engineering*, 20(10), pp. 1378–1392, 2008.

[288] J. Rennie, and N. Srebro. Fast maximum margin matrix factorization for collaborative prediction. *ICML Conference*, pp. 713–718, 2005.

[289] G. Jeh, and J. Widom. SimRank: a measure of structural-context similarity. *ACM KDD Conference*, pp. 538–543, 2003.

[290] H. Jeung, M. L. Yiu, X. Zhou, C. Jensen, and H. Shen. Discovery of convoys in trajectory databases. *VLDB Conference*, pp. 1068–1080, 2008.

[291] T. Joachims. Making Large scale SVMs practical. *Advances in Kernel Methods, Support Vector Learning*, pp. 169–184, *MIT Press*, Cambridge, 1998.

[292] T. Joachims. Training Linear SVMs in Linear Time. *ACM KDD Conference*, pp. 217–226, 2006.

[293] T. Joachims. Transductive inference for text classification using support vector machines. *International Conference on Machine Learning*, pp. 200–209, 1999.

[294] T. Joachims. Transductive learning via spectral graph partitioning. *ICML Conference*, pp. 290–297, 2003.

[295] I. Jolliffe. Principal component analysis. *John Wiley and Sons*, 2005.

[296] M. Joshi, V. Kumar, and R. Agarwal. Evaluating boosting algorithms to classify rare classes: comparison and improvements. *IEEE ICDM Conference*, pp. 257–264, 2001.

[297] M. Kantarcioglu. A survey of privacy-preserving methods across horizontally partitioned data. *Privacy-Preserving Data Mining: Models and Algorithms*, Springer, pp. 313–335, 2008.

[298] H. Kashima, K. Tsuda, and A. Inokuchi. Kernels for graphs. In *Kernel Methods in Computational Biology*, MIT Press, Cambridge, MA, 2004.

[299] D. Karger, and C. Stein. A new approach to the minimum cut problem. *Journal of the ACM (JACM)*, 43(4), pp. 601–640, 1996.

[300] G. Karypis, E. H. Han, and V. Kumar. Chameleon: Hierarchical clustering using dynamic modeling. *Computer*, 32(8), pp, 68–75, 1999.

[301] G. Karypis, and V. Kumar. A fast and high quality multilevel scheme for partitioning irregular graphs. *SIAM Journal on scientific Computing*, 20(1), pp. 359–392, 1998.

[302] G. Karypis, R. Aggarwal, V. Kumar, and S. Shekhar. Multilevel hypergraph partitioning: applications in VLSI domain. *IEEE Transactions on Very Large Scale Integration (VLSI) Systems*, 7(1), pp. 69–79, 1999.

[303] L. Kaufman, and P. J. Rousseeuw. Finding groups in data: an introduction to cluster analysis. *Wiley*, 2009.

[304] D. Kempe, J. Kleinberg, and E. Tardos. Maximizing the spread of influence through a social network. *ACM KDD Conference*, pp. 137–146, 2003.

[305] E. Keogh, S. Lonardi, and C. Ratanamahatana. Towards parameter-free data mining. *ACM KDD Conference*, pp. 206–215, 2004.

[306] E. Keogh, J. Lin, and A. Fu. HOT SAX: Finding the most unusual time series subsequence: Algorithms and applications. *IEEE ICDM Conference*, pp. 8, 2005.

[307] E. Keogh, and M. Pazzani. Scaling up dynamic time-warping for data mining applications. *ACM KDD Conference*, pp. 285–289, 2000.

[308] E. Keogh. Exact indexing of dynamic time warping. *VLDB Conference*, pp. 406–417, 2002.

[309] E. Keogh, K. Chakrabarti, M. Pazzani, and S. Mehrotra. Dimensionality reduction for fast similarity searching in large time series datanases. *Knowledge and Infomration Systems*, pp. 263–286, 2000.

[310] E. Keogh, S. Lonardi, and B. Y.-C. Chiu. Finding surprising patterns in a time series database in linear time and space. *ACM KDD Conference*, pp. 550–556, 2002.

[311] E. Keogh, S. Lonardi, and C. Ratanamahatana. Towards parameter-free data mining. *ACM KDD Conference*, pp. 206–215, 2004.

[312] B. Kernighan, and S. Lin. An efficient heuristic procedure for partitioning graphs. *Bell System Technical Journal*, 1970.

[313] A. Khan, N. Li, X. Yan, Z. Guan, S. Chakraborty, and S. Tao. Neighborhood-based fast graph search in large networks. *ACM SIGMOD Conference*, pp. 901–912, 2011.

[314] A. Khan, Y. Wu, C. Aggarwal, and X. Yan. Nema: Fast graph matching with label similarity. *Proceedings of the VLDB Endowment*, 6(3), pp. 181–192, 2013.

[315] D. Kifer, and J. Gehrke. Injecting utility into anonymized datasets. *ACM SIGMOD Conference*, pp. 217–228, 2006.

[316] L. Kissner, and D. Song. Privacy-preserving set operations. *Advances in Cryptology–CRYPTO*, pp. 241–257, 2005.

[317] J. Kleinberg. Authoritative sources in a hyperlinked environment. *Journal of the ACM (JACM)*, 46(5), pp. 604–632, 1999.

[318] S. Knerr, L. Personnaz, and G. Dreyfus. Single-layer learning revisited: a stepwise procedure for building and training a neural network. In J. Fogelman, editor, *Neurocomputing: Algorithms, Architectures and Applications.* Springer-Verlag, 1990.

[319] E. Knorr, and R. Ng. Algorithms for mining distance-based outliers in large datasets. *VLDB Conference*, pp. 392–403, 1998.

[320] E. Knorr, and R. Ng. Finding intensional knowledge of distance-based outliers. *VLDB Conference*, pp. 211–222, 1999.

[321] Y. Koren, R. Bell, and C. Volinsky. Matrix factorization techniques for recommender systems. *Computer*, 42(8), pp. 30–37, 2009.

[322] Y. Koren. Factorization meets the neighborhood: a multifaceted collaborative filtering model. *ACM KDD Conference*, pp. 426–434, 2008.

[323] Y. Koren. Collaborative filtering with temporal dynamics. *Communications of the ACM,*, 53(4), pp. 89–97, 2010.

[324] D. Kostakos, G. Trajcevski, D. Gunopulos, and C. Aggarwal. Time series data clustering. *Data Clustering: Algorithms and Applications*, CRC Press, 2013.

[325] J. Konstan. Introduction to recommender systems: algorithms and evaluation. *ACM Transactions on Information Systems*, 22(1), pp. 1–4, 2004.

[326] Y. Kou, C. T. Lu, and D. Chen. Spatial weighted outlier detection, *SIAM Conference on Data Mining*, 2006.

[327] A. Krogh, M. Brown, I. Mian, K. Sjolander, and D. Haussler. Hidden Markov models in computational biology: Applications to protein modeling. *Journal of molecular biology*, 235(5), pp. 1501–1531, 1994.

[328] J. B. Kruskal. Nonmetric multidimensional scaling: a numerical method. *Psychometrika*, 29(2), pp. 115–129, 1964.

[329] B. Kulis, S. Basu, I. Dhillon, and R. Mooney. Semi-supervised graph clustering: a kernel approach. *Machine Learning*, 74(1), pp. 1–22, 2009.

[330] S. Kulkarni, G. Lugosi, and S. Venkatesh. Learning pattern classification: a survey. *IEEE Transactions on Information Theory*, 44(6), pp. 2178–2206, 1998.

[331] M. Kuramochi, and G. Karypis. Frequent subgraph discovery. *IEEE International Conference on Data Mining*, pp. 313–320, 2001.

[332] L. V. S. Lakshmanan, R. Ng, J. Han, and A. Pang. Optimization of constrained frequent set queries with 2-variable constraints. *ACM SIGMOD Conference*, pp. 157–168, 1999.

[333] P. Langley, W. Iba, and K. Thompson. An analysis of Bayesian classifiers. *Proceedings of the National Conference on Artificial Intelligence*, pp. 223–228, 1992.

[334] A. Lazarevic, and V. Kumar. Feature bagging for outlier detection. *ACM KDD Conference*, pp. 157–166, 2005.

[335] K. LeFevre, D. J. DeWitt, and R. Ramakrishnan. Incognito: Efficient full-domain k-anonymity. *ACM SIGMOD Conference*, pp. 49–60, 2005.

[336] K. LeFevre, D. J. DeWitt, and R. Ramakrishnan. Mondrian multidimensional k-anonymity. *IEEE International Conference on Data Engineering*, pp. 25, 2006.

[337] J.-G. Lee, J. Han, and X. Li. Trajectory outlier detection: A partition-and-detect framework. *ICDE Conference*, pp. 140–149, 2008.

[338] J.-G. Lee, J. Han, and K.-Y. Whang. Trajectory clustering: a partition-and-group framework. *ACM SIGMOD Conference*, pp. 593–604, 2007.

[339] J.-G. Lee, J. Han, X. Li, and H. Gonzalez. TraClass: trajectory classification using hierarchical region-based and trajectory-based clustering. *Proceedings of the VLDB Endowment*, 1(1), pp. 1081–1094, 2008.

[340] W. Lee, and D. Xiang. Information theoretic measures for anomaly detection. *IEEE Symposium on Security and Privacy*, pp. 130–143, 2001.

[341] J. Leskovec, D. Huttenlocher, and J. Kleinberg. Predicting positive and negative links in online social networks. *World Wide Web Conference*, pp. 641–650, 2010.

[342] J. Leskovec, J. Kleinberg, and C. Faloutsos. Graphs over time: densification laws, shrinking diameters, and possible explanations. *ACM KDD Conference*, pp. 177–187, 2005.

[343] J. Leskovec, A. Rajaraman, and J. Ullman. Mining of massive datasets. *Cambridge University Press*, 2012.

[344] D. Lewis. Naive Bayes at forty: The independence assumption in information retrieval. *ECML Conference*, pp. 4–15, 1998.

[345] D. Lewis, and J. Catlett. Heterogeneous uncertainty sampling for supervised learning. *ICML Conference*, pp. 148–156, 1994.

[346] C. Li, Q. Yang, J. Wang, and M. Li. Efficient mining of gap-constrained subsequences and its various applications. *ACM Transactions on Knowledge Discovery from Data (TKDD)*, 6(1), 2, 2012.

[347] J. Li, G. Dong, K. Ramamohanarao, and L. Wong. Deeps: A new instance-based lazy discovery and classification system. *Machine Learning*, 54(2), pp. 99–124, 2004.

[348] N. Li, T. Li, and S. Venkatasubramanian. t-closeness: Privacy beyond k-anonymity and ℓ-diversity. *IEEE International Conference on Data Engineering*, pp. 106–115, 2007.

[349] W. Li, J. Han, and J. Pei. CMAR: Accurate and efficient classification based on multiple class-association rules. *IEEE ICDM Conference*, pp. 369–376, 2001.

[350] Y. Li, M. Dong, and J. Hua. Localized feature selection for clustering. *Pattern Recognition Letters*, 29(1), 10–18, 2008.

[351] Z. Li, B. Ding, J. Han, and R. Kays. Swarm: Mining relaxed temporal moving object clusters. *Proceedings of the VLDB Endowment*, 3(1–2), pp. 732–734, 2010.

[352] Z. Li, B. Ding, J. Han, R. Kays, and P. Nye. Mining periodic behaviors for moving objects. *ACM KDD Conference*, pp. 1099–1108, 2010.

[353] D. Liben-Nowell, and J. Kleinberg. The link-prediction problem for social networks. *Journal of the American Society for Information Science and Technology*, 58(7), pp. 1019–1031, 2007.

[354] R. Lichtenwalter, J. Lussier, and N. Chawla. New perspectives and methods in link prediction. *ACM KDD Conference*, pp. 243–252, 2010.

[355] J. Lin, E. Keogh, S. Lonardi, and B. Chiu. Experiencing SAX: a novel symbolic representation of time series. *Data Mining and Knowledge Discovery*, 15(2), pp. 107–144, 2003.

[356] J. Lin, E. Keogh, S. Lonardi, and P. Patel. Finding motifs in time series. *Proceedings of the 2nd Workshop on Temporal Data*, 2002.

[357] B. Liu. Web data mining: exploring hyperlinks, contents, and usage data. *Springer*, New York, 2007.

[358] B. Liu, W. Hsu, and Y. Ma. Integrating classification and association rule mining. *ACM KDD Conference*, pp. 80–86, 1998.

[359] G. Liu, H. Lu, W. Lou, and J. X. Yu. On computing, storing and querying frequent patterns. *ACM KDD Conference*, pp. 607–612, 2003.

[360] H. Liu, and H. Motoda. Feature selection for knowledge discovery and data mining. *Springer*, 1998.

[361] J. Liu, Y. Pan, K. Wang, and J. Han. Mining frequent item sets by opportunistic projection. *ACM KDD Conference*, pp. 229–238, 2002.

[362] L. Liu, J. Tang, J. Han, M. Jiang, and S. Yang. Mining topic-level influence in heterogeneous networks. *ACM CIKM Conference*, pp. 199–208, 2010.

[363] D. Lin. An Information-theoretic Definition of Similarity. *ICML Conference*, pp. 296–304, 1998.

[364] R. Little, and D. Rubin. Statistical analysis with missing data. *Wiley*, 2002.

[365] F. T. Liu, K. M. Ting, and Z.-H. Zhou. Isolation forest. *IEEE ICDM Conference*, pp. 413–422, 2008.

[366] H. Liu, and H. Motoda. Computational methods of feature selection. *Chapman and Hall/CRC*, 2007.

[367] K. Liu, C. Giannella, and H. Kargupta. A survey of attack techniques on privacy-preserving data perturbation methods. *Privacy-Preserving Data Mining: Models and Algorithms*, Springer, pp. 359–381, 2008.

[368] B. London, and L. Getoor. Collective classification of network data. *Data Classification: Algorithms and Applications*, CRC Press, pp. 399–416, 2014.

[369] C.-T. Lu, D. Chen, and Y. Kou. Algorithms for spatial outlier detection, *IEEE ICDM Conference*, pp. 597–600, 2003.

[370] Q. Lu, and L. Getoor. Link-based classification. *ICML Conference*, pp. 496–503, 2003.

[371] U. von Luxburg. A tutorial on spectral clustering. *Statistics and computing*, 17(4), pp. 395–416, 2007.

[372] A. Machanavajjhala, D. Kifer, J. Gehrke, and M. Venkitasubramaniam. ℓ-diversity: privacy beyond k-anonymity. *ACM Transactions on Knowledge Discovery from Data (TKDD)*, 1(3), 2007.

[373] S. Macskassy, and F. Provost. A simple relational classifier. *Second Workshop on Multi-Relational Data Mining (MRDM) at ACM KDD Conference*, 2003.

[374] S. C. Madeira, and A. L. Oliveira. Biclustering algorithms for biological data analysis: a survey. *IEEE/ACM Transactions on Computational Biology and Bioinformatics*. 1(1), pp. 24–45, 2004.

[375] N. Mamoulis, H. Cao, G. Kollios, M. Hadjieleftheriou, Y. Tao, and D. Cheung. Mining, indexing, and querying historical spatiotemporal data. *ACM KDD Conference*, pp. 236–245, 2004.

[376] G. Manku, and R. Motwani. Approximate frequency counts over data streams. *VLDB Conference*, pp. 346–357, 2002.

[377] C. Manning, P. Raghavan, and H. Schutze. Introduction to information retrieval. *Cambridge University Press*, Cambridge, 2008.

[378] M. Markou, and S. Singh. Novelty detection: a review, part 1: statistical approaches. *Signal Processing*, 83(12), pp. 2481–2497, 2003.

[379] G. J. McLachian. Discriminant analysis and statistical pattern recognition. *Wiler Interscience*, 2004.

[380] M. Markou, and S. Singh. Novelty detection: A review, part 2: neural network-based approaches. *Signal Processing*, 83(12), pp. 2481–2497, 2003.

[381] M. Mehta, R. Agrawal, and J. Rissanen. SLIQ: A fast scalable classifier for data mining, *EDBT Conference*, pp. 18–32, 1996.

[382] P. Melville, M. Saar-Tsechansky, F. Provost, and R. Mooney. An expected utility approach to active feature-value acquisition. *IEEE ICDM Conference*, 2005.

[383] A. K. Menon, and C. Elkan. Link prediction via matrix factorization. *Machine Learning and Knowledge Discovery in Databases*, pp. 437–452, 2011.

[384] B. Messmer, and H. Bunke. A new algorithm for error-tolerant subgraph isomprohism detection. *IEEE Transactions on Pattern Mining and Machine Intelligence*, 20(5), pp. 493–504, 1998.

[385] A. Meyerson, and R. Williams. On the complexity of optimal k-anonymization. *ACM PODS Conference*, pp. 223–228, 2004.

[386] R. Michalski, I. Mozetic, J. Hong, and N. Lavrac. The multi-purpose incremental learning system AQ15 and its testing application to three medical domains. *Proceedings of the AAAI*, pp. 1–41, 1986.

[387] C. Michael, and A. Ghosh. Two state-based approaches to program-based anomaly detection. *Computer Security Applications Conference*, pp. 21, 2000.

[388] H. Miller, and J. Han. Geographic data mining and knowledge discovery. *CRC Press*, 2009.

[389] T. M. Mitchell. Machine learning. *McGraw Hill International Edition*, 1997.

[390] B. Mobasher. Web usage mining and personalization. *Practical Handbook of Internet Computing, ed. Munindar Singh*, pp, 264–265, CRC Press, 2005.

[391] D. Montgomery, E. Peck, and G. Vining. Introduction to linear regression analysis. *John Wiley and Sons*, 2012.

[392] C. H. Mooney, and J. F. Roddick. Sequential pattern mining: approaches and algorithms. *ACM Computing Surveys (CSUR)*, 45(2), 2013.

[393] B. Moret. Decision trees and diagrams. *ACM Computing Surveys (CSUR)*, 14(4), pp. 593–623, 1982.

[394] A. Mueen, E. Keogh, Q. Zhu, S. Cash, and M. Westover. Exact discovery of time series motifs. *SDM Conference*, pp. 473–484, 2009.

[395] A. Mueen, and E. Keogh. Online discovery and maintenance of time series motifs. *ACM KDD Conference*, pp. 1089–1098, 2010.

[396] E. Muller, M. Schiffer, and T. Seidl. Statistical selection of relevant subspace projections for outlier ranking. *ICDE Conference*, pp, 434–445, 2011.

[397] E. Muller, I. Assent, P. Iglesias, Y. Mulle, and K. Bohm. Outlier analysis via subspace analysis in multiple views of the data. *IEEE ICDM Conference*, pp. 529–538, 2012.

[398] S. K. Murthy. Automatic construction of decision trees from data: A multi-disciplinary survey. *Data Mining and Knowledge Discovery*, 2(4), pp. 345–389, 1998.

[399] S. Nabar, K. Kenthapadi, N. Mishra, and R. Motwani. A survey of query auditing techniques for data privacy. *Privacy-Preserving Data Mining: Models and Algorithms*, Springer, pp. 415–431, 2008.

[400] D. Nadeau, and S. Sekine. A survey of named entity recognition and classification. *Lingvisticae Investigationes*, 30(1), 3–26, 2007.

[401] M. Naor, and B. Pinkas. Efficient oblivious transfer protocols. *SODA Conference*, pp. 448–457, 2001.

[402] A. Narayanan, and V. Shmatikov. How to break anonymity of the netflix prize dataset. *arXiv preprint cs/0610105*, 2006. http://arxiv.org/abs/cs/0610105

[403] G. Nemhauser, and L. Wolsey. Integer and combinatorial optimization. *Wiley*, New York, 1988.

[404] J. Neville, and D. Jensen. Iterative classification in relational data. *AAAI Workshop on Learning Statistical Models from Relational Data*, pp. 13–20, 2000.

[405] A. Ng, M. Jordan, and Y. Weiss. On spectral clustering analysis and an algorithm. *Advances in Neural Information Processing Systems*, pp. 849–856, 2001.

[406] R. T. Ng, L. V. S. Lakshmanan, J. Han, and A. Pang. Exploratory mining and pruning optimizations of constrained associations rules. *ACM SIGMOD Conference*, pp. 13–24, 1998.

[407] R. T. Ng, and J. Han. CLARANS: A method for clustering objects for spatial data mining. *IEEE Transactions on Knowledge and Data Engineering*, 14(5), pp. 1003–1016, 2002.

[408] M. Neuhaus, and H. Bunke. Automatic learning of cost functions for graph edit distance. *Information Sciences*, 177(1), pp. 239–247, 2007.

[409] M. Neuhaus, K. Riesen, and H. Bunke. Fast suboptimal algorithms for the computation of graph edit distance. *Structural, Syntactic, and Statistical Pattern Recognition*, pp. 163–172, 2006.

[410] K. Nigam, A. McCallum, S. Thrun, and T. Mitchell. Text classification with labeled and unlabeled data using EM. *Machine Learning*, 39(2), pp. 103–134, 2000.

[411] B. Ozden, S. Ramaswamy, and A. Silberschatz. Cyclic association rules. *International Conference on Data Engineering*, pp. 412–421, 1998.

[412] L. Page, S. Brin, R. Motwani, and T. Winograd. The PageRank citation engine: Bringing order to the web. *Technical Report*, 1999–0120, Computer Science Department, Stanford University, 1998.

[413] F. Pan, G. Cong, A. Tung, J. Yang, and M. Zaki. CARPENTER: Finding closed patterns in long biological datasets. *ACM KDD Conference*, pp. 637–642, 2003.

[414] T. Palpanas. Real-time data analytics in sensor networks. *Managing and Mining Sensor Data*, pp. 173–210, Springer, 2013.

[415] F. Pan, A. K. H. Tung, G. Cong, and X. Xu. COBBLER: Combining column and row enumeration for closed pattern discovery. *International Conference on Scientific and Statistical Database Management*, pp. 21–30, 2004.

[416] C. Papadimitriou, H. Tamaki, P. Raghavan, and S. Vempala. Latent semantic indexing: A probabilistic analysis. *ACM PODS Conference*, pp. 159–168, 1998.

[417] N. Pasquier, Y. Bastide, R. Taouil, and L. Lakhal. Discovering frequent closed itemsets for association rules. *International Conference on Database Theory*, pp. 398–416, 1999.

[418] P. Patel, E. Keogh, J. Lin, and S. Lonardi. Mining motifs in massive time series databases. *IEEE ICDM Conference*, pp. 370–377, 2002.

[419] J. Pei, J. Han, H. Lu, S. Nishio, S. Tang, and D. Yang. H-mine: Hyper-structure mining of frequent patterns in large databases. *IEEE ICDM Conference*, pp. 441–448, 2001.

[420] J. Pei, J. Han, and R. Mao. CLOSET: An efficient algorithm for mining frequent closed itemsets. *ACM SIGMOD Workshop on Research Issues in Data Mining and Knowledge Discovery*, pp, 21–30, 2000.

[421] J. Pei, J. Han, B. Mortazavi-Asl, J. Wang, H. Pinto, Q. Chen, U. Dayal, and M. C. Hsu. Mining sequential patterns by pattern-growth: The prefixspan approach. *IEEE Transactions on Knowledge and Data Engineering*, 16(11), pp. 1424–1440, 2004.

[422] J. Pei, J. Han, and L. V. S. Lakshmanan. Mining frequent patterns with convertible constraints. *ICDE Conference*, pp. 433–442, 2001.

[423] D. Pelleg, and A. W. Moore. X-means: Extending k-means with efficient estimation of the number of clusters. *ICML Conference*, pp. 727–734, 2000.

[424] M. Petrou, and C. Petrou. Image processing: the fundamentals. *Wiley*, 2010.

[425] D. Pierrakos, G. Paliouras, C. Papatheodorou, and C. Spyropoulos. Web usage mining as a tool for personalization: a survey. *User Modeling and User-Adapted Interaction*, 13(4), pp, 311–372, 2003.

[426] D. Pokrajac, A. Lazerevic, and L. Latecki. Incremental local outlier detection for data streams. *Computational Intelligence and Data Mining Conference*, pp. 504–515, 2007.

[427] S. A. Macskassy, and F. Provost. Classification in networked data: A toolkit and a univariate case study. *Joirnal of Machine Learning Research*, 8, pp. 935–983, 2007.

[428] G. Qi, C. Aggarwal, and T. Huang. Link Prediction across networks by biased cross-network sampling. *IEEE ICDE Conference*, pp. 793–804, 2013.

[429] G. Qi, C. Aggarwak, and T. Huang. Online community detection in social sensing. *ACM WSDM Conference*, pp. 617–626, 2013.

[430] J. Quinlan. C4.5: programs for machine learning. *Morgan-Kaufmann Publishers*, 1993.

[431] J. Quinlan. Induction of decision trees. *Machine Learning*, 1, pp. 81–106, 1986.

[432] D. Rafiei, and A. Mendelzon. Similarity-based queries for time series data, *ACM SIGMOD Record*, 26(2), pp. 13–25, 1997.

[433] E. Rahm, and H. Do. Data cleaning: problems and current approaches, *IEEE Data Engineering Bulletin*, 23(4), pp. 3–13, 2000.

[434] R. Ramakrishnan, and J. Gehrke. Database Management Systems. *Osborne/McGraw Hill*, 1990.

[435] V. Raman, and J. Hellerstein. Potter's wheel: An interactive data cleaning system. *VLDB Conference*, pp. 381–390, 2001.

[436] S. Ramaswamy, R. Rastogi, and K. Shim. Efficient algorithms for mining outliers from large data sets. *ACM SIGMOD Conference*, pp. 427–438, 2000.

[437] M. Rege, M. Dong, and F. Fotouhi. Co-clustering documents and words using bipartite isoperimetric graph partitioning. *IEEE ICDM Conference*, pp. 532–541, 2006.

[438] E. S. Ristad, and P. N. Yianilos. Learning string-edit distance. *IEEE Transactions on Pattern Analysis and Machine Intelligence*. 20(5), pp. 522–532, 1998.

[439] F. Rosenblatt. The perceptron: A probabilistic model for information storage and organization in the brain. *Psychological review*, 65(6), 286, 1958.

[440] R. Salakhutdinov, and A. Mnih. *Probabilistic Matrix Factorization. Advances in Neural and Information Processing Systems*, pp. 1257–1264, 2007.

[441] G. Salton, and M. J. McGill. Introduction to modern information retrieval. *McGraw Hill*, 1986.

[442] P. Samarati. Protecting respondents identities in microdata release. *IEEE Transactions on Knowledge and Data Engineering*, 13(6), pp. 1010–1027, 2001.

[443] H. Samet. The design and analysis of spatial data structures. *Addison-Wesley*, Reading, MA, 1990.

[444] J. Sander, M. Ester, H. P. Kriegel, and X. Xu. Density-based clustering in spatial databases: The algorithm gdbscan and its applications. *Data Mining and Knowledge Discovery*, 2(2), pp. 169–194, 1998.

[445] B. Sarwar, G. Karypis, J. Konstan, and J. Riedl. Item-based collaborative filtering recommendation algorithms. *World Wide Web Conference*, pp. 285–295, 2001.

[446] A. Savasere, E. Omiecinski, and S. B. Navathe. An efficient algorithm for mining association rules in large databases. *Very Large Databases Conference*, pp. 432–444, 1995.

[447] A. Savasere, E. Omiecinski, and S. Navathe. Mining for strong negative associations in a large database of customer transactions. *IEEE ICDE Conference*, pp. 494–502, 1998.

[448] C. Saunders, A. Gammerman, and V. Vovk. Ridge regression learning algorithm in dual variables. *ICML Conference*, pp. 515–521, 1998.

[449] B. Scholkopf, and A. J. Smola. Learning with kernels: support vector machines, regularization, optimization, and beyond. *Cambridge University Press*, 2001.

[450] B. Scholkopf, A. Smola, and K.-R. Muller. Nonlinear component analysis as a kernel eigenvalue problem. *Neural Computation*, 10(5), pp. 1299–1319, 1998.

[451] B. Scholkopf, and A. J. Smola. *Learning with Kernels.* MIT Press, Cambridge, MA, 2002.

[452] H. Schutze, and C. Silverstein. Projections for efficient document clustering. *ACM SIGIR Conference*, pp. 74–81, 1997.

[453] F. Sebastiani. Machine Learning in Automated Text Categorization. *ACM Computing Surveys*, 34(1), 2002.

[454] B. Settles. Active Learning. *Morgan and Claypool*, 2012.

[455] B. Settles, and M. Craven. An analysis of active learning strategies for sequence labeling tasks. *Proceedings of the Conference on Empirical Methods in Natural Language Processing (EMNLP)*, pp. 1069–1078, 2008.

[456] D. Seung, and L. Lee. Algorithms for non-negative matrix factorization. *Advances in Neural Information Processing Systems*, 13, pp. 556–562, 2001.

[457] H. Seung, M. Opper, and H. Sompolinsky. Query by committee. *Fifth annual workshop on Computational learning theory*, pp. 287–294, 1992.

[458] J. Shafer, R. Agrawal, and M. Mehta. SPRINT: A scalable parallel classifier for data mining. *VLDB Conference*, pp. 544–555, 1996.

[459] S. Shekhar, C. T. Lu, and P. Zhang. Detecting graph-based spatial outliers: algorithms and applications. *ACM KDD Conference*, pp. 371–376, 2001.

[460] S.Shekhar, C. T. Lu, and P. Zhang. A unified approach to detecting spatial outliers. *Geoinformatica*, 7(2), pp. 139–166, 2003.

[461] S. Shekhar, and S. Chawla. A tour of spatial databases. *Prentice Hall*, 2002.

[462] S. Shekhar, C. T. Lu, and P. Zhang. Detecting graph-based spatial outliers. *Intelligent Data Analysis*, 6, pp. 451–468, 2002.

[463] S. Shekhar, and Y. Huang. Discovering spatial co-location patterns: a summary of results. In *Advances in Spatial and Temporal Databases* , pp. 236–256, Springer, 2001.

[464] G. Sheikholeslami, S. Chatterjee, and A. Zhang. Wavecluster: A multi-resolution clustering approach for very large spatial databases. *VLDB Conference*, pp. 428–439, 1998.

[465] P. Shenoy, J. Haritsa, S. Sudarshan, G., Bhalotia, M. Bawa, and D. Shah. Turbo-charging vertical mining of large databases. *ACM SIGMOD Conference*, 29(2), pp. 22–35, 2000.

[466] J. Shi, and J. Malik. Normalized cuts and image segmentation. *IEEE Transactions on Pattern Analysis and Machine Intelligence*. 22(8), pp. 888–905, 2000.

[467] R. Shumway, and D. Stoffer. Time-series analysis and its applications: With R examples, *Springer*, New York, 2011.

[468] M.-L. Shyu, S.-C. Chen, K. Sarinnapakorn, and L. Chang. A novel anomaly detection scheme based on principal component classifier, *ICDM Conference*, pp. 353–365, 2003.

[469] R. Sibson. SLINK: An optimally efficient algorithm for the single-link clustering method. The Computer Journal, 16(1), pp. 30–34, 1973.

[470] A. Siebes, J. Vreeken, and M. van Leeuwen. itemsets that compress. *SDM Conference*, pp. 393–404, 2006.

[471] B. W. Silverman. Density Estimation for Statistics and Data Analysis. *Chapman and Hall*, 1986.

[472] K. Smets, and J. Vreeken. The odd one out: Identifying and characterising anomalies. *SIAM Conference on Data Mining*, pp. 804–815, 2011.

[473] E. S. Smirnov. On exact methods in systematics. *Systematic Zoology*, 17(1), pp. 1–13, 1968.

[474] P. Smyth. Clustering sequences with hidden Markov models. *Advances in Neural Information Processing Systems*, pp. 648–654, 1997.

[475] E. J. Stollnitz, and T. D. De Rose. Wavelets for computer graphics: theory and applications. *Morgan Kaufmann*, 1996.

[476] R. Srikant, and R. Agrawal. Mining quantitative association rules in large relational tables. *ACM SIGMOD Conference*, pp. 1–12, 1996.

[477] J. Srivastava, R. Cooley, M. Deshpande, and P. N. Tan. Web usage mining: Discovery and applications of usage patterns from web data. *ACM SIGKDD Explorations Newsletter*, 1(2), pp. 12–23, 2000.

[478] I. Steinwart, and A. Christmann. Support vector machines. *Springer*, 2008.

[479] A. Strehl, and J. Ghosh. Cluster ensembles—a knowledge reuse framework for combining multiple partitions. *Journal of Machine Learning Research*, 3, pp. 583–617, 2003.

[480] G. Strang. An introduction to linear algebra. *Wellesley Cambridge Press*, 2009.

[481] G. Strang, and K. Borre. Linear algebra, geodesy, and GPS. *Wellesley Cambridge Press*, 1997.

[482] K. Subbian, C. Aggarwal, and J. Srivasatava. Content-centric flow mining for influence analysis in social streams. *CIKM Conference*, pp. 841–846, 2013.

[483] J. Sun, and J. Tang. A survey of models and algorithms for social influence analysis. *Social Network Data Analytics*, Springer, pp. 177–214, 2011.

[484] Y. Sun, J. Han, C. Aggarwal, and N. Chawla. When will it happen?: relationship prediction in heterogeneous information networks. *ACM international conference on Web search and data mining*, pp. 663–672, 2012.

[485] P.-N Tan, M. Steinbach, and V. Kumar. Introduction to data mining. *Addison-Wesley*, 2005.

[486] P. N. Tan, V. Kumar, and J. Srivastava. Selecting the right interestingness measure for association patterns. *ACM KDD Conference*, pp. 32–41, 2002.

[487] J. Tang, Z. Chen, A. W.-C. Fu, and D. W. Cheung. Enhancing effectiveness of outlier detection for low density patterns. *PAKDD Conference*, pp. 535–548, 2002.

[488] J. Tang, J. Sun, C. Wang, and Z. Yang. Social influence analysis in large-scale networks. *ACM SIGKDD international conference on Knowledge discovery and data mining*, pp. 807–816, 2009.

[489] B. Taskar, M. Wong, P. Abbeel, and D. Koller. Link prediction in relational data. *Advances in Neural Information Processing Systems*, 2003.

[490] J. Tenenbaum, V. De Silva, and J. Langford. A global geometric framework for nonlinear dimensionality reduction. *Science*, 290 (5500), pp. 2319–2323, 2000.

[491] K. Ting, and I. Witten. Issues in stacked generalization. *Journal of Artificial Intelligence Research*, 10, pp. 271–289, 1999.

[492] T. Mitsa. Temporal data mining. *CRC Press*, 2010.

[493] H. Toivonen. Sampling large databases for association rules. *VLDB Conference*, pp. 134–145, 1996.

[494] V. Vapnik. The nature of statistical learning theory. *Springer*, 2000.

[495] J. Vaidya. A survey of privacy-preserving methods across vertically partitioned data. *Privacy-Preserving Data Mining: Models and Algorithms*, Springer, pp. 337–358, 2008.

[496] V. Vapnik. Statistical learning theory. *Wiley*, 1998.

[497] V. Verykios, and A. Gkoulalas-Divanis. A Survey of Association Rule Hiding Methods for Privacy. *Privacy-Preserving Data Mining: Models and Algorithms*, Springer, pp. 267–289, 2008.

[498] J. S. Vitter. Random sampling with a reservoir. *ACM Transactions on Mathematical Software (TOMS)*, 11(1), pp. 37–57, 2006.

[499] M. Vlachos, M. Hadjieleftheriou, D. Gunopulos, and E. Keogh. Indexing multidimensional time-series with support for multiple distance measures. *ACM KDD Conference*, pp. 216–225, 2003.

[500] M. Vlachos, G. Kollios, and D. Gunopulos. Discovering similar multidimensional trajectories. *IEEE International Conference on Data Engineering*, pp. 673–684, 2002.

[501] T. De Vries, S. Chawla, and M. Houle. Finding local anomalies in very high dimensional space. *IEEE ICDM Conference*, pp. 128–137, 2010.

[502] A. Waddell, and R. Oldford. Interactive visual clustering of high dimensional data by exploring low-dimensional subspaces. *INFOVIS*, 2012.

[503] H. Wang, W. Fan, P. Yu, and J. Han. Mining concept-drifting data streams using ensemble classifiers. *ACM KDD Conference*, pp. 226–235, 2003.

[504] J. Wang, J. Han, and J. Pei. Closet+: Searching for the best strategies for mining frequent closed itemsets. *ACM KDD Conference*, pp. 236–245, 2003.

[505] J. Wang, Y. Zhang, L. Zhou, G. Karypis, and C. C. Aggarwal. Discriminating subsequence discovery for sequence clustering. *SIAM Conference on Data Mining*, pp. 605–610, 2007.

[506] W. Wang, J. Yang, and R. Muntz. STING: A statistical information grid approach to spatial data mining. *VLDB Conference*, pp. 186–195, 1997.

[507] J. S. Walker. Fast fourier transforms. *CRC Press*, 1996.

[508] S. Wasserman. Social network analysis: Methods and applications. *Cambridge University Press*, 1994.

[509] D. Watts, and D. Strogatz. Collective dynamics of 'small-world' networks. *Nature*, 393 (6684), pp. 440–442, 1998.

[510] L. Wei, E. Keogh, and X. Xi. SAXually Explicit images: Finding unusual shapes. *IEEE ICDM Conference*, pp. 711–720, 2006.

[511] H. Wiener. Structural determination of paraffin boiling points. *Journal of the American Chemical Society*. 1(69). pp. 17–20, 1947.

[512] L. Willenborg, and T. De Waal. Elements of statistical disclosure control. *Springer*, 2001.

[513] D. Wolpert. Stacked generalization. *Neural Networks*, 5(2), pp. 241–259, 1992.

[514] X. Xiao, and Y. Tao. Anatomy: Simple and effective privacy preservation. *Very Large Databases Conference*, pp. 139–150, 2006.

[515] D. Xin, J. Han, X. Yan, and H. Cheng. Mining compressed frequent-pattern sets. *VLDB Conference*, pp. 709–720, 2005.

[516] Z. Xing, J. Pei, and E. Keogh. A brief survey on sequence classification. *SIGKDD Explorations Newsletter*, 12(1), pp. 40–48, 2010.

[517] H. Xiong, P. N. Tan, and V. Kumar. Mining strong affinity association patterns in data sets with skewed support distribution. *ICDM Conference*, pp. 387–394, 2003.

[518] K. Yaminshi, J. Takeuchi, and G. Williams. Online unsupervised outlier detection using finite mixtures with discounted learning algorithms, *ACM KDD Conference*,pp. 320–324, 2000.

[519] X. Yan, and J. Han. gSpan: Graph-based substructure pattern mining. *IEEE International Conference on Data Mining*, pp. 721–724, 2002.

[520] X. Yan, P. Yu, and J. Han. Substructure similarity search in graph databases. *ACM SIGMOD Conference*, pp. 766–777, 2005.

[521] X. Yan, P. Yu, and J. Han. Graph indexing: a frequent structure-based approach. *ACM SIGMOD Conference*, pp. 335–346, 2004.

[522] X. Yan, F. Zhu, J. Han, and P. S. Yu. Searching substructures with superimposed distance. *International Conference on Data Engineering*, pp. 88, 2006.

[523] J. Yang, and W. Wang. CLUSEQ: efficient and effective sequence clustering. *IEEE International Conference on Data Engineering*, pp. 101–112, 2003.

[524] D. Yankov, E. Keogh, J. Medina, B. Chiu, and V. Zordan. Detecting time series motifs under uniform scaling. *ACM KDD Conference*, pp. 844–853, 2007.

[525] N. Ye. A markov chain model of temporal behavior for anomaly detection. *IEEE Information Assurance Workshop*, pp. 169, 2004.

[526] B. K. Yi, H. V. Jagadish, and C. Faloutsos. Efficient retrieval of similar time sequences under time warping. *IEEE International Conference on Data Engineering*, pp. 201–208, 1998.

[527] B. K. Yi, N. Sidiropoulos, T. Johnson, H. V. Jagadish, C. Faloutsos, and A. Biliris. Online data mining for co-evolving time sequences. *International Conference on Data Engineering*, pp. 13–22, 2000.

[528] H. Yildirim, and M. Krishnamoorthy. A random walk method for alleviating the sparsity problem in collaborative filtering. *ACM conference on Recommender systems*, pp. 131–138, 2008.

[529] X. Yin, and J. Han. CPAR: Classification based on predictive association rules. *SIAM international conference on data mining*, pp. 331–335, 2003.

[530] S. Yu, and J. Shi. Multiclass spectral clustering. *International Conference on Computer Vision*, 2003.

[531] B. Zadrozny, J. Langford, and N. Abe. Cost-sensitive learning by cost-proportionate example weighting. *ICDM Conference*, pp. 435–442, 2003.

[532] R. Zafarani, M. A. Abbasi, and H. Liu. Social media mining: an introduction. *Cambridge University Press*, New York, 2014.

[533] H. Zakerzadeh, C. Aggarwal, and K. Barker. Towards breaking the curse of dimensionality for high-dimensional privacy. *SIAM Conference on Data Mining*, pp. 731–739, 2014.

[534] M. J. Zaki. Scalable algorithms for association mining. *IEEE Transactions on Knowledge and Data Engineering*, 12(3), pp. 372–390, 2000.

[535] M. J. Zaki. SPADE: An efficient algorithm for mining frequent sequences. *Machine learning*, 42(1–2), pp. 31–60, 2001. 31–60.

[536] M. J. Zaki, and M. Wagner Jr. Data mining and analysis: fundamental concepts and algorithms. *Cambridge University Press*, 2014.

[537] M. J. Zaki, S. Parthasarathy, M. Ogihara, and W. Li. New algorithms for fast discovery of association rules. *KDD Conference*, pp. 283–286, 1997.

[538] M. J. Zaki, and K. Gouda. Fast vertical mining using diffsets. *ACM KDD Conference*, pp. 326–335, 2003.

[539] M. J. Zaki, and C. Hsiao. CHARM: An efficient algorithm for closed itemset mining. *SIAM Conference on Data Mining*, pp. 457–473, 2002.

[540] M. J. Zaki, and C. Aggarwal. XRules: An effective algorithm for structural classification of XML data. *Machine Learning*, 62(1–2), pp. 137–170, 2006.

[541] B. Zenko. Is combining classifiers better than selecting the best one? *Machine Learning*, pp. 255–273, 2004.

[542] Y. Zhai, and B. Liu. Web data extraction based on partial tree alignment. *World Wide Web Conference*, pp. 76–85, 2005.

[543] D. Zhan, M. Li, Y. Li, and Z.-H. Zhou. Learning instance specific distances using metric propagation. *ICML Conference*, pp. 1225–1232, 2009.

[544] H. Zhang, A. Berg, M. Maire, and J. Malik. SVM-KNN: Discriminative nearest neighbor classification for visual category recognition. *Computer Vision and Pattern Recognition*, pp. 2126–2136, 2006.

[545] J. Zhang, Z. Ghahramani, and Y. Yang. A probabilistic model for online document clustering with application to novelty detection. *Advances in Neural Information Processing Systems*, pp. 1617–1624, 2004.

[546] J. Zhang, Q. Gao, and H. Wang. SPOT: A system for detecting projected outliers from high-dimensional data stream. *ICDE Conference*, 2008.

[547] D. Zhang, and G. Lu. Review of shape representation and description techniques. *Pattern Recognition*, 37(1), pp. 1–19, 2004.

[548] S. Zhang, W. Wang, J. Ford, and F. Makedon. Learning from incomplete ratings using nonnegative matrix factorization. *SIAM Conference on Data Mining*, pp. 549–553, 2006.

[549] T. Zhang, R. Ramakrishnan, and M. Livny. BIRCH: an efficient data clustering method for very large databases. *ACM SIGMOD Conference*, pp. 103–114, 1996.

[550] Z. Zhao, and H. Liu. Spectral feature selection for supervised and unsupervised learning. *ICML Conference*, pp. 1151–1157, 2007.

[551] D. Zhou, O. Bousquet, T. Lal, J. Weston, and B. Scholkopf. Learning with local and global consistency. *Advances in Neural Information Processing Systems*, 16(16), pp. 321–328, 2004.

[552] D. Zhou, J. Huang, and B. Scholkopf. Learning from labeled and unlabeled data on a directed graph. *ICML Conference*, pp. 1036–1043, 2005.

[553] F. Zhu, X. Yan, J. Han, P. S. Yu, and H. Cheng. Mining colossal frequent patterns by core pattern fusion. *ICDE Conference*, pp. 706–715, 2007.

[554] X. Zhu, Z. Ghahramani, and J. Lafferty. Semi-supervised learning using gaussian fields and harmonic functions. *ICML Conference*, pp. 912–919, 2003.

[555] X. Zhu, and A. Goldberg. Introduction to semi-supervised learning. *Morgan and Claypool*, 2009.

[556] `http://db.csail.mit.edu/labdata/labdata.html`.

[557] `http://www.itl.nist.gov/iad/mig/tests/tdt/tasks/fsd.html`.

[558] `http://sifter.org/~simon/journal/20061211.html`.

[559] `http://www.netflixprize.com/`.

Index

C. C. Aggarwal, *Data Mining: The Textbook*, DOI 10.1007/978-3-319-14142-8